Lecture Notes in Computer Science

Commenced Publication in 1973
Founding and Former Series Editors:
Gerhard Goos, Juris Hartmanis, and Jan van Leeuwen

Maciej Liśkiewicz   Rüdiger Reischuk (Eds.)

Fundamentals of Computation Theory

15th International Symposium, FCT 2005
Lübeck, Germany, August 17-20, 2005
Proceedings

 Springer

Volume Editors

Maciej Liśkiewicz
Rüdiger Reischuk
Universität zu Lübeck, Institut für Theoretische Informatik
Ratzeburger Allee 160, 23538 Lübeck, Germany
E-mail: {liskiewi,reischuk}@tcs.uni-luebeck.de

Library of Congress Control Number: 2005930271

CR Subject Classification (1998): F.1, F.2, F.4.1, I.3.5, G.2

ISSN   0302-9743
ISBN-10   3-540-28193-2 Springer Berlin Heidelberg New York
ISBN-13   978-3-540-28193-1 Springer Berlin Heidelberg New York

Springer is a part of Springer Science+Business Media

springeronline.com

© Springer-Verlag Berlin Heidelberg 2005
Printed in Germany

Typesetting: Camera-ready by author, data conversion by Scientific Publishing Services, Chennai, India
Printed on acid-free paper   SPIN: 11537311   06/3142   5 4 3 2 1 0

Preface

This volume 3623 of the Springer Lecture Notes in Computer Science is dedicated to the 15th Symposium on Fundamentals of Computation Theory FCT 2005, held in Lübeck, Germany, on August 17–20, 2005.

The FCT symposium was established in 1977 as a biennial event for researchers interested in all aspects of theoretical computer science, in particular in algorithms, complexity, and formal and logical methods. The previous FCT conferences were held at the following places: Poznań (Poland, 1977), Wendisch-Rietz (Germany, 1979), Szeged (Hungary, 1981), Borgholm (Sweden, 1983), Cottbus (Germany, 1985), Kazan (Russia, 1987), Szeged (Hungary, 1989), Gosen-Berlin (Germany, 1991), Szeged (Hungary, 1993), Dresden (Germany, 1995), Kraków (Poland, 1997), Iasi (Romania, 1999), Riga (Latvia, 2001) and Malmö (Sweden, 2003). The FCT conference series is coordinated by a steering committee. Its current members are B. Chlebus (Denver/Warsaw), Z. Esik (Szeged), M. Karpinski (Bonn), A. Lingas (Lund), M. Santha (Paris), E. Upfal (Providence) and I. Wegener (Dortmund).

The call for papers for FCT 2005 sought contributions on original research in all aspects of theoretical computer science including *design and analysis of algorithms, abstract data types, approximation algorithms, automata and formal languages, categorical and topological approaches, circuits, computational and structural complexity, circuit and proof theory, computational biology, computational geometry, computer systems theory, concurrency theory, cryptography, domain theory, distributed algorithms and computation, molecular computation, quantum computation and information, granular computation, probabilistic computation, learning theory, rewriting, semantics, logic in computer science, specification, transformation and verification, and algebraic aspects of computer science.* 105 papers have been submitted – most of them with a focus on fundamental questions in these areas of computing. Thanks to all the authors who gave the program committee the chance to select 46 top papers for presentation at the conference. An extended abstract of these results can be found in these proceedings.

Our thanks go to the 15 members of the program committee that took their duty very seriously. Each submitted paper has carefully been reviewed by at least five members – some with the help of subreferees. The Easy-Chair evaluation system turned out to be of great help in our electronic discussion and decision process – great thanks to Andrei Voronkov for developing such a useful and professional tool.

In addition to the presentation of the accepted papers, invited lectures on current research topics in theoretical computer science were given by Martin Dyer (Leeds), Martin Grohe (Berlin), and Daniel Spielman (Cambridge, MA).

We are grateful to the German Science Foundation (DFG) and the *Gesell-schaft der Freunde und Förderer der Universität zu Lübeck* for their financial support to organize FCT 2005, and all members of the Institute for Theoretical Computer Science (ITCS) of the University of Lübeck for their help to prepare and run this conference.

Rüdiger Reischuk   Maciej Liśkiewicz
Program Chair   *Conference Chair*

Lübeck, August 2005

Organization

Program Committee

Tatsuya Akutsu, *Kyoto*
Giorgio Ausiello, *Rome*
Martin Dietzfelbinger, *Ilmenau*
Tao Jiang, *Riverside*
Matthias Krause, *Mannheim*
Maciej Liśkiewicz, *Lübeck*
Damian Niwiński, *Warsaw*
Mitsunori Ogihara, *Rochester*

Jean-Eric Pin, *Paris*
Rüdiger Reischuk, *Lübeck (Chair)*
Branislav Rovan, *Bratislava*
Ludwig Staiger, *Halle*
Amnon Ta-Shma, *Tel-Aviv*
Jan Arne Telle, *Bergen*
Thomas Wilke, *Kiel*

Organizing Committee

Jan Arpe
Frank Balbach
Uwe-Jens Heinrichs
Markus Hinkelmann
Andreas Jakoby

Maciej Liśkiewicz
Claudia Mamat
Bodo Manthey
Rüdiger Reischuk
Hagen Völzer

External Referees

E. Allender
H. Alt
A. Ambainis
K. Ambos-Spies
A. Amir
G. Antoniou
F. Armknecht
J. Arpe
F. Balbach
D. Mix Barrington
R. Beigel
W. Bein
D. Berwanger
A. Beygelzimer
M. Bezem
L. Boasson

H. Bodlaender
H. L. Bodlaender
M. Bojanczyk
B. Bollig
A. de Bonis
J. Boyar
A. Brandstädt
V. Brattka
B. Brejova
T. Böhme
C. S. Calude
K. Chatterjee
T. Colcombet
A. Condon
B. Courcelle
P. Crescenzi

M. Crochemore
F. Cucker
V. Dahllöf
C. Damm
S. V. Daneshmand
P. Darondeau
J. Dassow
O. Devillers
M. Dezani
V. Diekert
W. Dosch
M. Droste
S. Dziembowski
D. Etherington
H. Fernau
E. Fischer

Table of Contents

Computational and Structural Complexity

Graphs and Complexity

Computational Game Theory

Visual Cryptography and Computational Geometry

Query Complexity

Distributed Systems

Automata and Formal Languages

Graph Algorithms

Semantics

Approximation Algorithms

Average-Case Complexity

Algorithms

Complexity II

Graph Algorithms

Automata II

Pattern Matching

The Complexity of Querying External Memory and Streaming Data

Martin Grohe[1], Christoph Koch[2], and Nicole Schweikardt[1]

[1] Institut für Informatik, Humboldt-Universität Berlin, Germany
{grohe, schweika}@informatik.hu-berlin.de
[2] Database Group, Universität des Saarlandes, Saarbrücken, Germany
koch@cs.uni-sb.de

Abstract. We review a recently introduced computation model for streaming and external memory data. An important feature of this model is that it distinguishes between *sequentially reading (streaming)* data from external memory (through main memory) and *randomly accessing* external memory data at specific memory locations; it is well-known that the latter is much more expensive in practice. We explain how a number of lower bound results are obtained in this model and how they can be applied for proving lower bounds for XML query processing.

1 Introduction

Modern computers rely on a hierarchy of storage media from tapes and disks at the bottom through what is usually called random-access memory or main memory up to various levels of (even on-CPU) memory caches at the top. The storage media at the bottom of this hierarchy are the slowest and least expensive and those at the top the fastest and dearest. The need for this *memory hierarchy* is dictated by the ever-growing amounts of data that have to be managed and processed by computers. Currently, the most pronounced performance and price gap in this hierarchy is between (random access) main memory and the next-lower level in the memory hierarchy, usually magnetic disks, which have to rely on mechanical, physically moving parts. One often refers to the upper layers above this gap by *internal memory* and the lower layers of the memory hierarchy by *external memory*. The technological reality is such that the time for accessing a given bit of information in external memory is five to six orders of magnitude greater than the time required to access a bit in internal memory.

Current external storage technology (disks and tapes) renders algorithms that can read and write their data to and from external memory in few *sequential scans* much faster than algorithms that require many random data accesses. Indeed, the time required to move a read/write head to a certain position of a disk or tape – a slow mechanical operation – is by orders of magnitude greater than actually reading a considerable amount of data stored in sequence once the read/write head has been placed at the starting position of the data in question.

Managing and processing huge amounts of data has been traditionally the domain of database research. It is generally assumed that databases have to

M. Liśkiewicz and R. Reischuk (Eds.): FCT 2005, LNCS 3623, pp. 1–16, 2005.
© Springer-Verlag Berlin Heidelberg 2005

reside in external, inexpensive storage because of their sheer size. There has been a wealth of research on query processing and optimization respecting the mentioned physical realities and distinguishing between internal and external memory (cf. e.g. [20, 10, 24, 16]). In fact, this distinction is in a sense the defining essence of database techniques.

The fundamental problems that have to be faced in processing very large datasets have generated a recent renewed interest in theoretical aspects of external memory processing. In the classical model of external memory algorithms (see, for example, [24, 16]), the cost measure is simply the number of bits read from external memory divided by the page size. This model ignores the very important distinction mentioned above between *random access* to data in particular external memory locations and *sequential scans* of the disks. More recent models focus on data processing with few sequential scans of the external memory [2, 14, 4, 12]. An important special case is *data stream processing*, in which only one sequential scan of the data is permitted. This is yet another field that has seen much activity in recent years [18, 1, 5].

In [11, 12], we introduced a formal model for external memory processing that allows to distinguish between sequential scans and random access. The two most significant cost measures in our setting are the number of random accesses to external memory and the size of the internal memory. Our model is based on standard multi-tape Turing machines. Our machines have several tapes of unbounded size, among them the input and output tapes, which represent the external memory (for example, each of these tapes may represent a disk). In addition, the machine has several tapes of restricted size which represent the internal memory.

We model the number of scans of the external memory data, respectively the number of random accesses, by the number of reversals of the Turing machine's read/write heads on the external memory tapes. Anything close to random I/O will result in a very considerable number of reversals, while a full sequential scan of an external memory tape can be effected cheaply. The reversals done by a read/write head are a clean and fundamental notion [25], but of course real external storage technology based on disks does not allow to reverse their direction of rotation. On the other hand, we can of course simulate k forward scans by $2k$ reversals in our machine model — and allowing for forward as well as backward scans makes the *lower* bound results presented in this paper even stronger.

Note that our model puts no restriction on the number of head reversals on the *internal* memory tapes, the size of the external memory tapes, or the running time of the machine.

In this paper, we give a survey of the above-mentioned machine model and strong lower bounds that were recently obtained for it [11, 12, 13]. We start in Section 2 by formally introducing the machine model and showing a number of basic properties for it. In Section 3, we consider the case of machines with only a single external memory tape. Here we can employ techniques from communication complexity to obtain lower bounds. In Section 4, we apply the results of

Section 3 to XML query processing problems. XML is a data exchange format that is currently drawing much attention in data management research. In [11], we obtained lower bounds for processing queries in the languages XQuery and XPath, the two most widely used query languages for XML data (in fact, XPath is basically a sublanguage of XQuery that is also often used in isolation and has become part of other XML-related data transformation languages such as XSLT). Section 5 goes beyond the case of machines with a single external memory tape. We will see that techniques from communication complexity fail to prove lower bounds. In [12], we introduced a new technique to establish lower bounds in this model. These are based on a new (non-uniform) machine model, so-called *list machines*, which allow us to analyze the flow of information in a Turing machine computation. The main result is a lower bound for the sorting problem.

Related Work
Most results presented in this survey are due to [11, 12, 13].

Strong lower bounds for a number of problems are known in models which permit a small number of sequential scans of the input data, but no auxiliary external memory (that is, the version of our model with no external memory tapes besides the input tape) [1, 2, 3, 4, 5, 6, 14, 17, 18]. All these lower bounds are obtained by communication complexity.

In [3], the problem of determining whether a given relational query can be evaluated scalably on a data stream or not at all is addressed. The complexity of XML query evaluation in a streaming model is also addressed in [5, 6]. The time and space complexity of XPath query evaluation in the standard (main memory) model is studied in [8, 9, 23].

Obviously, our model is also related to the *bounded reversal Turing machines*, which have been studied in classical complexity theory (see, for example, [25]). However, in bounded reversal Turing machines, the number of head reversals is limited on *all* tapes, whereas in our model there is no such restriction on the internal memory tapes. This makes our model considerably stronger. In particular, in our lower bound results we allow internal memory size that is polynomially related to the input size.

2   The Machine Model

Our model is based on standard multitape Turing machines. If not explicitly mentioned otherwise, we assume our machines to be deterministic. The machines we consider have $t + u$ tapes. The first t tapes are called *external memory tapes*; they represent external memory devices such as hard disks. The other u tapes are called *internal memory tapes*; they represent the internal memory. The first tape is always viewed as the input tape. If necessary, the machines have an additional write-only output tape. Configurations, runs, and acceptance are defined in the usual way. Figure 1 illustrates our model.

Let M be such a Turing machine and ρ a run of M. The *(internal) space* required by ρ, space(ρ), is the total number of cells on the internal memory tapes

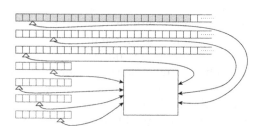

Fig. 1. Our machine model

visited during the run ρ. For $1 \leq i \leq t$, the number of *head reversals* on the i-th external memory tape, rev(ρ, i), is the number of times the read/write head on tape i changes its direction during the run ρ.

For functions $r, s : \mathbb{N} \rightarrow \mathbb{N}$ (where $\mathbb{N}$ denote the set of positive integers), we call the machine M (r, s, t)-*bounded* if it has t external memory tapes and for every run ρ of M with an input of length N and $i \in \{1, .., t\}$ we have $1 + \sum_{i=1}^{t} \text{rev}(\rho, i) \leq r(N)$ and space$(\rho) \leq s(N)$.

Definition 1. *(1) For functions $r, s : \mathbb{N} \rightarrow \mathbb{N}$ and $t \in \mathbb{N}$ we let $\text{ST}(r, s, t)$ be the class of all problems that can be decided by an (r, s, t)-bounded Turing machine.*

(2) For classes R, S of functions we let $\text{ST}(R, S, t) = \bigcup_{\substack{r \in R \\ s \in S}} \text{ST}(r, s, t)$.

Furthermore, we let $\text{ST}(R, S, O(1)) = \bigcup_{t \in \mathbb{N}} \text{ST}(R, S, t)$.

(3) We let $\text{ST}(R, S) = \text{ST}(R, S, 1)$.

While usually by "problem" we mean "decision problem", that is, language over some finite alphabet, occasionally we are more liberal and also view partial functions as problems. In particular, we may write $f \in \text{ST}(r, s, t)$ for a partial function $f : \Sigma^* \rightarrow \Sigma^*$.

Occasionally, we also consider the nondeterministic versions NST($\ldots$) of our ST($\ldots$) classes.

Note that we do not restrict the running time of an (r, s, t)-bounded machine in any way. Neither do we restrict the external space, that is, the number of cells on the external memory tapes that are visited during a run. However, it is not hard to see that implicitly the running time and hence the external space are bounded in terms of reversals and internal space:

Lemma 2 ([12]). *Let $r, s : \mathbb{N} \rightarrow \mathbb{N}$ and $t \in \mathbb{N}$, and let M be an (r, s, t)-bounded Turing machine. Then the length of every finite run of M is at most*

$$N \cdot 2^{O(r(N) \cdot (t + s(N)))}.$$

For $(r, s, 1)$-bounded Turing machines, the bound can be improved to

$$(N + r(N)) \cdot r(N) \cdot 2^{O(s(N))},$$

as an easy induction on $r(N)$ shows.

Random Access

If we think of the first t tapes of an (r, s, t)-bounded Turing machine as representing hard disks, then admitting heads to reverse their direction may not be very realistic. But as we mainly use our model to prove *lower* bounds, it does not do any harm either. Head reversals are a convenient way to simulate *random access* in our model.

Alternatively, we can explicitly include random access into our model as follows: A *random access Turing machine* is a Turing machine which has a special *address tape* on which only binary strings can be written. These binary strings are interpreted as nonnegative integers specifying external memory addresses, that is, numbers of cells on the external memory tapes. For each external memory tape $i \in \{1, \ldots, t\}$ the machine has a special state ra_i. If ra_i is entered, then in one step the head on tape i is moved to the cell that is specified by the number on the address tape, and the content of the address tape is deleted. Figure 2 illustrates the augmented model.

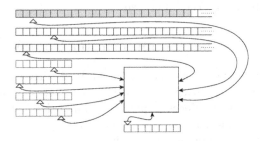

Fig. 2. Random access Turing machine

Let $q, r, s : \mathbb{N} \to \mathbb{N}$ and $t \in \mathbb{N}$. A random access Turing machine T is (q, r, s, t)-*bounded* if it is (r, s, t)-bounded (in the sense of a standard Turing machine) and, in addition, every run ρ of T on an input of length N involves at most $q(N)$ random accesses.

The address tape is considered as part of the internal memory; thus in a (q, r, s, t)-bounded random access Turing machine the length of the address tape is bounded by $s(N)$, where N is the length of the input. This implies that we can only address the first $2^{s(N)}$ cells of the external memory tapes. If during a computation, these tapes get longer, we only have random access to initial segments of length $2^{s(N)}$.

The following lemma follows from the simple observation that each random access can be simulated by moving the head to the desired position. This simulation is possible with at most two head reversals (if we want the head to be headed in the same direction before and after the simulation of the random access) and a slight space overhead.

Lemma 3 ([12]). *Let* $q, r, s : \mathbb{N} \to \mathbb{N}$ *and* $t \in \mathbb{N}$. *Then if a problem can be solved by a* (q, r, s, t)-*bounded random access Turing machine, it can also be solved by an* $(r + 2q, O(s), t)$-*bounded Turing machine.*

From now on, we will focus on standard Turing machines without address tapes.

Let us summarize the classes we have defined. For functions r, s on the natural numbers:

– $ST(O(r), O(s), O(1))$ is the class of all problems that can be solved on a machine with internal memory size $O(s(N))$ and an arbitrary (albeit constant) number of arbitrarily large external memory devices, which can be read sequentially, in addition allowing at most $O(r(N))$ head reversals and random accesses. The first external memory device contains the input data.

– $ST(O(r), O(s))$ is the class of all problems that can be solved on a machine that only has one external memory device and is otherwise restricted as above. Essentially, $ST(O(r), O(s))$ is the class of all problems that can be solved on an $O(s(N))$-space bounded machine allowing at most $O(r(N))$ sequential scans of the input.[1]

– $ST(1, O(s))$ is the class of all problems that can be solved on an $O(s(N))$-space bounded machine if the input is a data stream.

We always use N to denote the input size.

3   Lower Bounds via Communication Complexity

Lower bounds for the $ST(r, s)$-model can be obtained fairly easily by employing known results from communication complexity. The idea is to divide the input tape into two parts, which induces a split of the input data into two parts. Now we ask how much information must be communicated between the two parts to answer a specific query. Suppose we can prove a lower bound of $c(N)$ for the number of bits that need to be communicated, where N denotes the size of the input. Then we also have a lower bound $r(N) \cdot s(N) \geq \Omega(c(N))$ for functions r, s such that the query can be answered in $ST(r, s)$, because each time we cross the dividing line between the two parts of the input, we can only "transport" $O(s(N))$ bits of information.

In the basic model of communication complexity [26], two players, Alice and Bob, jointly want to evaluate $F(x, y)$, where $F : A \times B \to C$ is a function defined on finite sets A, B. Alice is given the first argument $x \in A$ and Bob the second argument $y \in B$. The two players exchange messages according to some fixed protocol until one of them has gathered enough information to compute $F(x, y)$. The *cost* of the protocol is the maximum number of bits communicated, where the maximum is taken over all argument pairs $(x, y) \in A \times B$. The *communication complexity of F* is the minimum of the costs of all protocols computing F.

A function that almost obviously has a high communication complexity is the *disjointness function* DISJ$_\ell$, defined on subsets $x, y \subseteq \{0, \dots, \ell - 1\}$ by

$$\text{DISJ}_\ell(x, y) = \begin{cases} 1 & \text{if } x \cap y = \emptyset, \\ 0 & \text{otherwise.} \end{cases}$$

---

[1] $ST(O(r), O(s))$ is slightly more powerful because the input can be overwritten and the external memory device can be used for storing auxiliary data.

Indeed, it is easy to see that the communication complexity of DISJ$_\ell$ is ℓ. It will be convenient for us to work with a slight modification of the disjointness function. For $k \leq \ell$, let DISJ$_{\ell,k}$ be the restriction of DISJ$_\ell$ to pairs of k-element subsets of $\{0, \ldots, \ell - 1\}$.

Theorem 4 ([21], cf. Example 2.12 in [15]). *The communication complexity of* DISJ$_{\ell,k}$ *is* $\Omega\left(\log\binom{\ell}{k}\right)$.

Let us now consider the following decision problem:

DISJOINT-SETS
Instance: Strings $x_1, \ldots, x_m, y_1, \ldots, y_m \in \{0,1\}^n$.
Question: Is $\{x_1, \ldots, x_m\} \cap \{y_1, \ldots, y_m\} = \emptyset$?

Formally, the input may either be specified as the string

$$x_1 \# x_2 \# \ldots \# x_m \# y_1 \# \ldots \# y_m$$

over $\{0, 1, \#\}$ or as an XML-document as described in Section 4 (see Figure 3). In both cases, the input size N is $\Theta(m \cdot n)$.

The lower bound of the following corollary follows easily from Theorem 4. To see this, we consider instances with $n = 2\log m$ and note that for such instances we have $N = \Theta(m \cdot \log m)$ and, with $\ell = 2^n$ and $k = m$,

$$\log\binom{\ell}{k} = \log\binom{m^2}{m} \geq \log\left(m^m\right) = m \cdot \log m.$$

The upper bound of the corollary is trivial.

Corollary 5. *(1)* DISJOINT-SETS $\in$ ST$(1, N)$.
(2) *For all functions* $r, s : \mathbb{N} \to \mathbb{N}$ *with* $r(N) \cdot s(N) \in o(N)$,

$$\text{DISJOINT-SETS} \notin \text{ST}(r, s).$$

In the next section, we will apply this simple result to prove lower bounds for querying XML-documents.

Next, we separate the deterministic from the nondeterministic ST-classes. We observe that the complement of DISJOINT-SETS can be decided by a $(1, n, 1)$-bounded *nondeterministic* Turing machine by guessing an input string x_i, storing it in the internal memory, and comparing with all strings y_j. Thus the restriction of the complement of DISJOINT-SETS to inputs with $n = 2 \cdot \log m$ is in NST$(1, O(\log N))$. The proof of Corollary 5 shows that it is not in ST(r, s) for all functions r, s with $r(N) \cdot s(N) \in o(N)$. Thus:

Corollary 6. *For all functions* r, s *with* $r(N) \cdot s(N) \in o(N)$

$$\text{NST}(1, O(\log N)) \not\subseteq \text{ST}(r, s).$$

The following hierarchy theorem is based on a more sophisticated result from communication complexity that deals with the number (and not only size) of messages Alice has to send Bob in a communication protocol. For functions $s : \mathbb{N} \to \mathbb{N}$ and $k \in \mathbb{N}$, let $\mathrm{ST}(k, s)$ denote the class $\mathrm{ST}(r, s)$ with $r(N) = k$ for all $N \in \mathbb{N}$.

Theorem 7 ([11], based on [7]). *For every fixed $k \in \mathbb{N}$ and all classes S of functions from $\mathbb{N}$ to $\mathbb{N}$ such that $O(\log n) \subseteq S \subseteq o\left(\frac{\sqrt{N}}{(\lg n)^3}\right)$ we have*

$$\mathrm{ST}(k, S) \subsetneq \mathrm{ST}(k{+}1, S).$$

4 Applications to XML Query Processing

In this section, we show how the results and techniques of the previous section can be applied to prove lower bounds for the complexity of XML-query processing. We assume that the reader is vaguely familiar with XML-syntax. We will only use very basic XML consisting of opening tags <t> and the corresponding closing tags </t> and plain text (no attributes, no DTDs, et cetera).

As an example, let us encode instances of the DISJOINT-SETS problem in XML. An instance $x_1, \ldots, x_m, y_1, \ldots, y_m \in \{0, 1\}^n$ is represented by the XML-document displayed in Figure 3.

```
<instance>
  <set1>
    <string> x₁ </string> ... <string> xₘ </string>
  </set1>
  <set2>
    <string> y₁ </string> ... <string> yₘ </string>
  </set2>
</instance>
```

Fig. 3. XML-representation of the two m-element sets $\{x_1, \ldots, x_m\}$ and $\{y_1, \ldots, y_m\}$

We will talk about the XML query languages XQuery and XPath, but the reader is not expected to know these languages. To avoid the awkward term "XQuery query", we refer to queries in the language XQuery as *xqueries*. An xquery Q transforms a given document D into a document $Q(D)$.

As an example, consider the xquery displayed in Figure 4. The syntax is, to a certain extent, self-explanatory. The query transforms the XML-document in Figure 3 to the document in Figure 5, where $1 \leq i_1 < i_2 < \cdots < i_\ell \leq m$ such that $\{x_1, \ldots, x_m\} \cap \{y_1, \ldots, y_m\} = \{x_{i_1}, \ldots, x_{i_\ell}\}$. Thus the query computes the intersection of the two sets. By the results of the previous section, the query cannot be evaluated by an $(r, s, 1)$-bounded Turing machine for any functions r, s with $r(N) \cdot s(N) \in o(N)$.

```
<result>
   for $x in /instance/set1/string
   where some $y in /instance/set2/string satisfies $x = $y
   return $x
</result>
```

Fig. 4. An xquery computing the intersection of two sets

```
<result>
   <string> x_{i_1} </string> ... <string> x_{i_ℓ} </string>
</result>
```

Fig. 5. Result of applying the query in Figure 4 to the document in Figure 3

We can associate the following decision problem with the evaluation problem for a query Q:

Q-Filtering
Instance: XML-document D.
Question: Is $Q(D)$ nonempty ?

Here we call an XML-document *empty* if it just consists of an opening and closing tag, as for example `<result> </result>`. We ignore whitespaces.

The Q-Filtering problem for the query Q of Figure 4 is the complement of Disjoint-Sets. Thus we get:

Corollary 8. *There is an xquery Q such that for all functions r, s with $r(N) \cdot s(N) \in o(N)$,*

$$Q\text{-Filtering} \notin \mathrm{ST}(r, s).$$

XPath is a *node selecting language*; the result of applying an XPath query to an XML-document is a set of nodes of this document. A *node* of a document is pair of corresponding opening- and closing tags. The nodes are arranged in a tree-like fashion in the obvious way. Thus we can view XML-documents as labeled rooted trees. Inner nodes are labeled by tags and leaves by the text parts of the document. Node selection in XPath works by regular expressions over the tags which specify paths in a document (tree). There is no need for the reader to know any details about the language here. XPath is mainly used as a tool in more complicated XML-related formalisms such as XQuery or XML-Schema. As far as expressive power is concerned, XPath is strictly a fragment of XQuery. For our complexity lower bounds, we only consider a small fragment of XPath called Core-XPath [8]. It is a clean logical fragment that captures the core functionality of the much larger and messier language XPath. We define Q-Filtering for XPath queries Q as for xqueries. Here, $Q(D)$ is a set of nodes of D and emptiness has the natural meaning.

To measure the complexity of Q-FILTERING for XPath queries Q, the appropriate parameter is not the size of the input document but its *height* if viewed as a tree. We use the notation $ST(r, s)$ for functions r, s depending on the height and not the size of the input document with the obvious meaning.

Theorem 9 ([11]).

(1) For every Core-XPath query Q,

$$Q\text{-}\textsc{Filtering} \in ST(1, O(h)).$$

(2) There is a Core-XPath query Q such that for all functions r, s with $r(h) \cdot s(h) \in o(h)$,

$$Q\text{-}\textsc{Filtering} \notin ST(r(h), s(h)).$$

Here h denote the height of the input document.

The upper bound is proved by standard automata theoretic techniques (implicitly, the result can be found in [19, 22]). For the lower bound, we again use the DISJOINT-SETS problem, but encoded as an XML-document in a different way than before.

5   Lower Bounds via List Machines

In this section, we turn to the classes $ST(r, s, t)$ for $t > 1$. To prove lower bounds for these classes, communication complexity based arguments as used in Section 3 utterly fail. The reason is that we can easily communicate arbitrarily many bits from one part of the input to any other part just by copying the first part to a second external memory tape and then reading it in parallel with the second part. This requires no internal memory and just two head reversals.

This idea can be used to prove that the $ST(r, s, 2)$ classes are much more powerful than the $ST(r, s) = ST(r, s, 1)$ classes. Observe that Theorem 4 not only yields a lower bound for the DISJOINT-SETS problem, but also for its restriction to input sets which are ordered in any specific way, because the communication complexity lower bound of Theorem 4 is independent of the way the two arguments are given to the function $\text{DISJ}_{k,\ell}$. Moreover, we obtained the lower bound of Corollary 5 for instances with $n = 2 \cdot \log m$. Thus for all functions $r, s : \mathbb{N} \to \mathbb{N}$ with $r(N) \cdot s(N) \in o(N)$ the following problem is not in $ST(r, s)$.

> *Instance:* Strings $x_1, \ldots, x_m, y_1, \ldots, y_m \in \{0, 1\}^{2 \cdot \log m}$ such that
> $\qquad x_m \leq x_{m-1} \leq \cdots \leq x_1$ and $y_1 \leq y_2 \leq \ldots \leq y_m$.
> *Question:* Is $\{x_1, \ldots, x_m\} \cap \{y_1, \ldots, y_m\} = \emptyset$?

Here $\leq$ denotes the lexicographical order of strings over $\{0, 1\}$.

By copying all x_i in the order they are given to the second tape and then comparing them in reverse order with the y_j, it is easy to see that the problem is in $ST(2, O(\log N), 2)$, where $N = \Theta(m \cdot \log m)$ is the size of the input. Thus we get:

Proposition 10. *For all functions* $r, s : \mathbb{N} \to \mathbb{N}$ *with* $r(N) \cdot s(N) \in o(N)$,

$$\mathrm{ST}(2, O(\log N), 2) \nsubseteq \mathrm{ST}(r, s) = \mathrm{ST}(r, s, 1).$$

Note that several external memory tapes do not help if no head reversals are permitted, that is, $\mathrm{ST}(1, s, t) = \mathrm{ST}(1, s)$ for all $s : \mathbb{N} \to \mathbb{N}$ and $t \geq 1$.

While several external memory tapes enable us to copy large segments of the input tape from one place to another, the segments themselves remain unchanged. In particular, there seems no easy way to "significantly" re-order the input or large parts of it. This leads to the idea that *sorting* should be hard even in the model with several external memory tapes.

SORT
  Input: $x_1, \ldots, x_m \in \{0, 1\}^n$.
  Output: $x_1, \ldots, x_m$ sorted in ascending lexicographical order.

It is not hard to see that the standard *merge-sort* algorithm achieves the following bound:

Proposition 11. SORT *can be solved by an* $\big(O(\log m), O(n), 3\big)$*-bounded Turing machine.*

The main lower bound result is the following:

Theorem 12 ([12]).

$$\mathrm{SORT} \notin \mathrm{ST}\left(o(\log N), \, O\left(\frac{\sqrt[5]{N}}{\log N}\right), \, O(1)\right).$$

The main ideas of the proof will be outlined in Subsection 5.1 below.

Unfortunately there is still a considerable gap between the lower bound of Theorem 12 and the upper bound of Proposition 11. However, for the important special case of sorting strings of length $O(\log m)$, or equivalently integers in the range $\{0, \ldots, m^{O(1)}\}$, with a little more effort we obtain *tight* bounds. Let us denote the restriction of SORT to instances with $n = 6 \cdot \log m$ by SHORT-SORT. The factor 6 is needed for technical reasons, any constant above 6 would work as well. Proposition 11 shows that SHORT-SORT $\in \mathrm{ST}(O(\log m), O(\log m), 3)$, which yields the upper bound of the following theorem. The lower bound is obtained by reducing a restricted version of the sorting problem for "long" strings, which is used in the proof of Theorem 12, to SHORT-SORT.

Theorem 13 ([12]). SHORT-SORT *is in* $\mathrm{ST}(O(\log N), O(\log N), 3)$, *but not in*

$$\mathrm{ST}\left(o(\log N), O\left(\sqrt[6]{N}\right), O(1)\right).$$

By further refining the techniques underlying the proof of Theorem 12, we also obtain lower bounds for the following two related decision problems:

SET-EQUALITY
Instance: $x_1, \ldots, x_m, y_1, \ldots, y_m \in \{0, 1\}^n$.
Question: Is $\{x_1, \ldots, x_m\} = \{y_1, \ldots, y_m\}$?

CHECKSORT
Instance: $x_1, \ldots, x_m, y_1, \ldots, y_m \in \{0, 1\}^n$.
Question: Is $(y_1, \ldots, y_m)$ the sorted version of $(x_1, \ldots, x_m)$, that
is, $\{x_1, \ldots, x_m\} = \{y_1, \ldots, y_m\}$ and $y_1 \leq \ldots \leq y_m$ with
respect to the lexicographical order ?

Theorem 14 ([13]).

$$\text{SET-EQUALITY}, \text{CHECKSORT} \notin \text{ST}\left(o(\log N), O\left(\frac{\sqrt[5]{N}}{\log N}\right), O(1)\right).$$

Since both SET-EQUALITY and CHECKSORT are easily reducible to SORT, similarly as for SHORT-SORT, we obtain matching upper bounds for the restrictions of the problems to input strings of length $O(\log m)$.

The complement of the restriction of SET-EQUALITY to instances with $n = 6 \cdot \log m$ can be decided by an $(1, O(\log m), 1)$-bounded nondeterministic Turing machine, which just guesses an x_i which is different from all y_j, stores it on an internal memory tape, and compares it to all y_j. Thus we get:

Corollary 15 ([13]).

$$\text{NST}(1, O(\log N), 1) \not\subseteq \text{ST}\left(o(\log N), O\left(\frac{\sqrt[5]{N}}{\log N}\right), O(1)\right).$$

On the other hand, the SET-EQUALITY problem can be expressed by the xquery of Figure 6.

```
<result>
  if ( every $x in /instance/set1/string satisfies
        some $y in /instance/set2/string satisfies $x = $y )
    and
     ( every $y in /instance/set2/string satisfies
        some $x in /instance/set1/string satisfies $x = $y )
  then <true/>
  else ()
</result>
```

Fig. 6. An xquery that checks whether two sets are equal

We therefore obtain:

Corollary 16. *There is an xquery Q such that*

$$Q\text{-FILTERING} \notin \mathrm{ST}\left(o(\log N),\ O\left(\frac{\sqrt[5]{N}}{\log N}\right),\ O(1)\right).$$

5.1 List Machines and the Proof of Theorem 12

As pointed out at the beginning of Section 5, arguments that are solely based on communication complexity do not lead to lower bound proofs for the classes $\mathrm{ST}(r, s, t)$ where $t \geq 2$ external memory tapes are available, because the second external memory tape can be used to transfer large parts of the input tape from one place to another.

On the other hand, even with additional external memory tapes, there seems no easy way of significantly *re-order* large parts of the input. In fact, it is well-known that the sorting problem cannot be solved by a *comparison exchange algorithm* that performs significantly less than $m \cdot \log m$ comparisons. Consequently, for sufficiently small $r(N)$ and $s(N)$, even with $t > 1$ external memory tapes, sorting by solely *comparing and moving around the input strings* is impossible. However, this does not lead to a proof of Theorem 12, because the *Turing machines*, on which the $\mathrm{ST}(\ldots)$ classes are based, can perform much more complicated operations than just "compare and move input strings". Indeed, many algorithms that solve certain data stream problems in a surprisingly efficient way are based on much more intricate operations (cf. [18]).

For proving the lower bound of Theorem 12 we introduce a new machine model, so-called *list machines*, which enable us to analyze the flow of information in a Turing machine computation. On the one hand, list machines can only compare and move around input strings as a whole and in this sense are "weaker" than Turing machines. We exploit this weakness in our proof that (appropriately restricted) list machines cannot sort. On the other hand, list machines are non-uniform and have a large number of tape symbols and states. In this sense, they are much stronger than Turing machines.

Here, we only describe list machines informally. For a formal definition and a precise statement of the following Simulation Lemma, we refer the reader to [12].

List machines are similar to Turing machines, with the following important differences:

- They are *non-uniform*; the input consists of m bit strings each of which has length n, for *fixed* m, n.
- They work on *lists* instead of tapes. In particular, this means that a new cell can be inserted between two existing cells.
- Instead of single symbols, list cells contain *strings* over the alphabet

$$A = I \cup \text{states} \cup \{\langle, \rangle\},$$

where $I = \{0, 1\}^n$ is the set of potential input strings.

- The transition function only determines the list machine's new state and the head movements and *not what is written into the list cells*.
- If (at least) one head moves, the information w on the current state and the content of *all* list cells that are seen by the machine's read/write heads directly before the transition, is stored on every single list as follows: On those lists whose heads are about to move a step to the left or the right, the information w overwrites the current cell entry. On each of the other lists (i.e., those whose heads do not move in the current step), a *new list cell*, containing the information w, is inserted behind the current head position.

The operation of a list machine is illustrated by Figure 7.

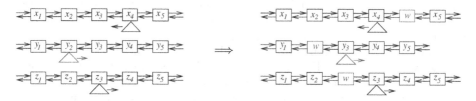

Fig. 7. A transition of a list machine. The example transition is of the form $(q, x_4, y_2, z_3) \rightarrow (q', stay, right, stay)$. The new string w that is written into the tape cells consists of the current state q and the content of the list cells read before the transition. Formally, we let $w = q\langle x_4 \rangle \langle y_2 \rangle \langle z_3 \rangle$.

The crucial fact is that Turing machines can be simulated by list machines. Informally, the *Simulation Lemma* states that every (r, s, t)-bounded Turing machine can be simulated by a family of list machines with

- $r(N)$ head reversals;
- t lists;
- $2^{O(s(N)+\log N)}$ states.

As usual, N denotes the input size.

The proof of the Simulation Lemma is the technically most difficult part of the proof of Theorem 12.

The next step in the proof of Theorem 12 is to show a lower bound for sorting on list machines. Very roughly, the idea is to analyze the *skeleton* of a list machine computation. The skeleton of a configuration or run of a list machine is obtained by replacing all input strings (of size n) by their indices (of size $\log m$). Intuitively, the skeleton determines the flow of information during a run, but not the outcome of the comparisons. Here we say that two input strings are *compared* if they appear together in a list cell in the run. Now counting arguments, based on the fact that there are not too many skeletons, show that there is are inputs $\bar{v} := (v_1, \ldots, v_m)$ and $\bar{v}' := (v_1', \ldots, v_m')$ in I^m and an index $i \in \{1, \ldots, m\}$ such that in the computation of the list machine:

- $\bar{v}$ and $\bar{v}'$ generate the same skeleton.
- $v_i \neq v_i'$, but $v_j = v_j'$ for all $j \neq i$.

– When v_i printed as part of the output, then no head reads a list cell that contains v_i (this information depends only on the skeleton), and thus the machine cannot know if is supposed to print v_i or v'_i. Thus for one of the two inputs, it will print the wrong string.

This shows that the machine cannot sort. Combined with the Simulation Lemma, it yields a proof of Theorem 12.

References

[1] G. Aggarwal, M. Datar, S. Rajagopalan, and M. Ruhl. On the streaming model augmented with a sorting primitive. In *Proceedings of the 44th Annual IEEE Symposium on Foundations of Computer Science*, pages 540–549, 2004.

[2] N. Alon, Y. Matias, and M. Szegedy. The space complexity of approximationg the frequency moments. *Journal of Computer and System Sciences*, 58:137–147, 1999.

[3] A. Arasu, B. Babcock, T. Green, A. Gupta, and J. Widom. Characterizing memory requirements for queries over continuous data streams. In *Proceedings of the 21st ACM Symposium on Principles of Database Systems*, pages 221–232, 2002.

[4] B. Babcock, S. Babu, M. Datar, R. Motwani, and J. Widom. Models and issues in data stream systems. In *Proceedings of the 21st ACM Symposium on Principles of Database Systems*, pages 1–16, 2002.

[5] Z. Bar-Yossef, M. Fontoura, and V. Josifovski. On the memory requirements of XPath evaluation over XML streams. In *Proceedings of the 23rd ACM Symposium on Principles of Database Systems*, pages 177–188, 2004.

[6] Z. Bar-Yossef, M. Fontoura, and V. Josifovski. Buffering in query evaluation over XML streams. In *Proceedings of the 24th ACM Symposium on Principles of Database Systems*, 2005. To appear.

[7] P. Duris, Z. Galil, and G. Schnitger. Lower bounds on communication complexity. *Information and Computation*, 73:1–22, 1987.

[8] G. Gottlob, C. Koch, and R. Pichler. Efficient algorithms for processing XPath queries. In *Proceedings of the 28th International Conference on Very Large Data Bases*, 2002.

[9] G. Gottlob, C. Koch, and R. Pichler. The complexity of XPath query evaluation. In *Proceedings of the 22nd ACM Symposium on Principles of Database Systems*, pages 179–190, 2003.

[10] G. Graefe. Query evaluation techniques for large databases. *ACM Computing Surveys*, 25(2):73–170, 1993.

[11] M. Grohe, C. Koch, and N. Schweikardt. Tight lower bounds for query processing on streaming and external memory data. In *Proceedings of the 32nd International Colloquium on Automata, Languages and Programming*, 2005. To appear.

[12] M. Grohe and N. Schweikardt. Lower bounds for sorting with few random accesses to external memory. In *Proceedings of the 24th ACM Symposium on Principles of Database Systems*, 2005. To appear.

[13] M. Grohe and N. Schweikardt. Unpublished manuscript, available from the authors, 2005.

[14] M. Henzinger, P. Raghavan, and S. Rajagopalan. Computing on data streams. In *External memory algorithms*, volume 50, pages 107–118. DIMACS Series In Discrete Mathematics And Theoretical Computer S cience, 1999.

[15] E. Kushilevitz and N. Nisan. *Communication Complexity*. Cambridge University Press, 1997.

[16] U. Meyer, P. Sanders, and J.F. Sibeyn, editors. *Algorithms for Memory Hierarchies*, volume 2832 of *Lecture Notes in Computer Science*. Springer-Verlag, 2003.

[17] J.J. Munro and M.S. Paterson. Selection and sorting with limited storage. *Theoretical Computer Science*, 12:315–323, 1980.

[18] S. Muthukrishnan. Data streams: algorithms and applications. In *Proceedings of the 14th Annual ACM-SIAM Symposium on Discrete Algorithms*, pages 413–413, 2003.

[19] A. Neumann and H. Seidl. Locating matches of tree patterns in forests. In V. Chandru and V. Vinay, editors, *Proceedings of the 18th Conference on Foundations of Software Technology and Theoretical Computer Science*, volume 1530 of *Lecture Notes in Computer Science*, pages 134–145, 1998.

[20] Raghu Ramakrishnan and Johannes Gehrke. *Database Management Systems*. McGraw-Hill, 2002.

[21] A.A. Razborov. Applications of matrix methods to the theory of lower bounds in computational complexity. *Combinatorica*, 10:81–93, 1990.

[22] L. Segoufin and V. Vianu. Validating streaming XML documents. In *Proceedings of the 21st ACM Symposium on Principles of Database Systems*, pages 53–64, 2002.

[23] Luc Segoufin. Typing and querying XML documents: Some complexity bounds. In *Proceedings of the 22nd ACM Symposium on Principles of Database Systems*, pages 167–178, 2003.

[24] J.F. Vitter. External memory algorithms and data structures: Dealing with massive data. *ACM Computing Surveys*, 33:209–271, 2001.

[25] K. Wagner and G. Wechsung. *Computational Complexity*. VEB Deutscher Verlag der Wissenschaften, 1986.

[26] A. Yao. Some complexity questions related to distributive computing. In *Proceedings of the 11th ACM Symposium on Theory of Computing*, pages 209–213, 1979.

The Smoothed Analysis of Algorithms

Daniel A. Spielman

Department of Mathematics, Massachusetts Institute of Technology

Abstract. We survey the progress that has been made in the smoothed analysis of algorithms.

1 Introduction

Theorists have long been challenged by the existence of remarkable algorithms that are known by scientists and engineers to work well in practice, but whose theoretical analyses have been negative or unconvincing. The root of the problem is that algorithms are usually analyzed in one of two ways: by worst-case or average-case analysis. The former can improperly suggest that an algorithm will perform poorly, while the latter can be unconvincing because the random inputs it considers may fail to resemble those encountered in practice.

Teng and I [1] introduced smoothed analysis to help explain the success of some of these algorithms and heuristics. Smoothed analysis is a hybrid of worst-case and average-case analyses that inherits advantages of both. The smoothed complexity of an algorithm is the maximum over its inputs of the expected running time of the algorithm under slight random perturbations of that input. The smoothed complexity is then measured as a function of both the input length and the magnitude of the perturbations. If an algorithm has low smoothed complexity, then it should perform well on most inputs in every neighborhood of inputs. Smoothed analysis makes sense for algorithms whose inputs are subject to slight amounts of noise in their low-order digits, which is typically the case if they are derived from measurements of real-world phenomena.

In this talk, I discuss how smoothed analysis can help explain the excellent observed behavior of the simplex method for linear programming, and provide an overview of other algorithms to which it has been applied. The attached bibliography is a list of papers on smothed analysis of which I am aware.

References

1. Spielman, D.A., Teng, S.H.: Smoothed analysis of algorithms: Why the simplex algorithm usually takes polynomial time. Journal of the ACM **51** (2004) 385–463
2. Blum, A., Dunagan, J.: Smoothed analysis of the perceptron algorithm for linear programming. In: SODA '02. (2002) 905–914
3. Dunagan, J., Spielman, D.A., Teng, S.H.: Smoothed analysis of interior point algorithms: Condition number. Available at http://arxiv.org/abs/cs.DS/0302011 (2003)

M. Liśkiewicz and R. Reischuk (Eds.): FCT 2005, LNCS 3623, pp. 17–18, 2005.

4. Sankar, A., Spielman, D.A., Teng, S.H.: Smoothed analysis of interior point algorithms: Condition number. available at `http://arxiv.org/abs/cs.NA/0310022` (2005)
5. Spielman, Teng: Smoothed analysis: Motivation and discrete models. In: WADS: 8th Workshop on Algorithms and Data Structures. (2003)
6. Damerow, V., Sohler, C.: Smoothed number of extreme points under uniform noise. In: Proceedings of the 20th European Workshop on Computational Geometry. (2004)
7. Spielman, D.A., Teng, S.H.: Smoothed analysis of termination of linear programming algorithms. Mathematical Programming **97** (2003) 375–404
8. Wschebor: Smoothed analysis of kappa(A). COMPLEXITY: Journal of Complexity **20** (2004)
9. Damerow, auf der Heide, M., Racke, Scheideler, Sohler: Smoothed motion complexity. In: ESA: Annual European Symposium on Algorithms. (2003)
10. Becchetti, Leonardi, Marchetti-Spaccamela, Schafer, Vredeveld: Average case and smoothed competitive analysis of the multi-level feedback algorithm. In: FOCS: IEEE Symposium on Foundations of Computer Science (FOCS). (2003)
11. Blum, A., Dunagan, J.: Smoothed analysis of the perceptron algorithm for linear programming. In: Proceedings of the 13th Annual ACM-SIAM Symposium On Discrete Mathematics (SODA-02), New York, ACM Press (2002) 905–914
12. Banderier, C., Mehlhorn, K., Beier, R.: Smoothed analysis of three combinatorial problems. In: Proc. of the 28th International Symposium on Mathematical Foundations of Computer Science (MFCS-03), Bratislava. (2003)
13. Flaxman, Frieze: The diameter of randomly perturbed digraphs and some applications. In: RANDOM: International Workshop on Randomization and Approximation Techniques in Computer Science, LNCS (2004)
14. Beier, R., Vöcking, B.: Typical properties of winners and losers in discrete optimization. In: Proceedings of the thirty-sixth annual ACM Symposium on Theory of Computing (STOC-04), New York, ACM Press (2004) 343–352
15. Beier, Vocking: Random knapsack in expected polynomial time. JCSS: Journal of Computer and System Sciences **69** (2004)
16. Sankar, A.: Smoothed Analysis of Gaussian Elimination. PhD thesis, M.I.T. (2004)
17. Rglin, H., Vcking, B.: Smoothed analysis of integer programming. In: Proc. 11th IPCO (Berlin). (2005)
18. Michael Krivelevich, B.S., Tetali, P.: On smoothed analysis of dense graphs and formulas. Proceedings of Random Structuctures and Algorithms (to appear)

Path Coupling Using Stopping Times

Magnus Bordewich[1], Martin Dyer[1], and Marek Karpinski[2]

[1] School of Computing, University of Leeds, Leeds LS2 9JT, UK
{dyer, magnusb}@comp.leeds.ac.uk
[2] Dept. of Computer Science, University of Bonn, 53117 Bonn, Germany
marek@cs.uni-bonn.de

Abstract. We analyse the mixing time of Markov chains using path coupling with stopping times. We apply this approach to two hypergraph problems. We show that the Glauber dynamics for independent sets in a hypergraph mixes rapidly as long as the maximum degree Δ of a vertex and the minimum size m of an edge satisfy $m \geq 2\Delta + 1$. We also state results that the Glauber dynamics for proper q-colourings of a hypergraph mixes rapidly if $m \geq 4$ and $q > \Delta$, and if $m = 3$ and $q \geq 1.65\Delta$. We give related results on the hardness of exact and approximate counting for both problems.

1 Introduction

In this paper, we develop a new approach to using path coupling with stopping times to bound the convergence of time of Markov chains. Our interest is in applying these results to randomised approximate counting. For an introduction, see [16]. We illustrate our methods by considering the approximation of the numbers of independent sets and q-colourings in hypergraphs with upper-bounded degree Δ, and lower-bounded edge size m. These problems in hypergraphs are of interest in their own right but, while approximate optimisation in this setting has received considerable attention [5,6,13,17], there has been surprisingly little work on approximate counting other than in the graph case $m = 2$. The tools we develop here may also allow study of approximate counting for several related hypergraph problems. Note, for example, that independent sets in hypergraphs correspond to *edge covers* under hypergraph duality.

Our results are achieved by considering, in the context of path coupling [4], the stopping time at which the distance first changes between two coupled chains. The first application of these stopping times to path coupling was given by Dyer, Goldberg, Greenhill, Jerrum and Mitzenmacher [9]. Their analysis was later improved by Hayes and Vigoda [12], using results closely related to those which we develop in this paper. Mitzenmacher and Niklova [19] had earlier stated a similar theorem, but could not provide a conclusive proof. Their approach had many similarities to our Theorem 2.1, but conditioning problems necessitate a proof along somewhat different lines.

Our main technical result, Theorem 2.1, shows that the chain mixes rapidly if the expected distance between the two chains has decreased at this stopping time.

M. Liśkiewicz and R. Reischuk (Eds.): FCT 2005, LNCS 3623, pp. 19–31, 2005.

We note that a similar conclusion follows from [12, Corollary 4]. However, we give a simpler and more transparent proof of this result, without initially assuming bounded stopping times as is done in the approach of [12]. As a consequence, our Theorem 2.1 will usually give a moderate improvement in the mixing time bound in comparison with [12, Corollary 4]. See Remark 2.3 below.

The problem of approximately counting independent sets in graphs has been widely studied, see for example [8,10,18,20,22], but the only previous work on the approximate counting of independent sets in *hypergraphs* seems to that of Dyer and Greenhill [10]. They showed rapid mixing to the uniform distribution of a simple Markov chain on independent sets in a hypergraph with maximum degree 3 and maximum edge size 3. However, this was the only interesting case resolved. Their results imply rapid mixing only for $m \leq \Delta/(\Delta - 2)$, which gives $m \leq 3$ when $\Delta = 3$ and $m \leq 2$ when $\Delta \geq 4$. In Theorem 3.1 we prove rapid mixing of a simple Markov chain, the *Glauber dynamics*, for any hypergraph such that $m \geq 2\Delta+1$, where m is the smallest edge size and Δ is the maximum degree. This is a marked improvement for large m. More generally, we consider the *hardcore distribution* on independent sets with *fugacity* λ. (See, for example, [10,18,22].) In [10], it is proved that rapid mixing occurs if $\lambda \leq m/((m - 1)\Delta - m)$. Here we improve this considerably for larger values of m, to $\lambda \leq (m - 1)/2\Delta$. We also give two hardness results: that computing the number of independent sets in hypergraphs is #P-complete except in trivial cases, and that there can be no approximation for the number of independent sets in a hypergraphs if the minimum edge size is at most logarithmic in Δ. It may be noted that our upper and lower bounds are exponentially different. We have no strong belief that either is close to the threshold at which approximate counting becomes possible, if such a threshold exists.

Counting q-colourings of hypergraphs was considered by Bubley [3], who showed that the Glauber dynamics was rapidly mixing if $q \geq 2\Delta$, generalising a result for graphs of Jerrum [15] and Salas and Sokal [21]. Much work has been done on improving this result for graph colourings, see [7] and its references, but little attention appears to have been given to the hypergraph case. Here we prove rapid mixing of Glauber dynamics for proper colourings of hypergraphs if $m \geq 4$, $q > \Delta$, and if $m = 3$, $q \geq 1.65\Delta$. For a precise statement of our result see Theorem 5.2. We give hardness results showing that computing the number of colourings in hypergraphs is #P-complete except in trivial cases, and that there can be no approximation for the number of colourings of hypergraphs if $q \leq (1 - 1/m)\Delta^{1/(m-1)}$. Again, there is a considerable discrepancy between the upper and lower bounds for large m.

The paper is organised as follows. Section 1.1 gives an intuitive motivation for the stopping time approach of the paper. Section 2 contains the full description and proof of Theorem 2.1 for path coupling with stopping times. We apply this to hypergraph independent sets in Section 3. Section 4 contains the related hardness results. Section 5 contains our results on the Glauber dynamics for hypergraph colouring. Finally, Section 6 gives the hardness results for counting colourings in hypergraphs. For reasons of space, most of the proofs are omitted in Sections 4–6. These may be found in [1].

1.1 Intuition

Let $\mathcal{H} = (\mathcal{V}, \mathcal{E})$ be a hypergraph of maximum degree Δ and minimum edge size m. A subset $S \subseteq \mathcal{V}$ of the vertices is *independent* if no edge is a subset of S. Let $\Omega(\mathcal{H})$ be the set of all independent sets of $\mathcal{H}$. Let λ be the *fugacity*, which weights independent sets. (See [10].) The most important case is $\lambda = 1$, which weights all independent sets equally and gives rise to the uniform distribution on all independent sets. We define the Markov chain $\mathcal{M}(\mathcal{H})$ with state space $\Omega(\mathcal{H})$ by the following transition process (*Glauber dynamics*). If the state of $\mathcal{M}$ at time t is X_t, the state at $t + 1$ is determined by the following procedure.

(i) Select a vertex $v \in \mathcal{V}$ uniformly at random,
(ii) (a) if $v \in X_t$ let $X_{t+1} = X_t \backslash \{v\}$ with probability $1/(1 + \lambda)$,
    (b) if $v \notin X_t$ and $X_t \cup \{v\}$ is independent, let $X_{t+1} = X_t \cup \{v\}$ with probability $\lambda/(1 + \lambda)$,
    (c) otherwise let $X_{t+1} = X_t$.

This chain is easily shown to be ergodic with stationary probability proportional to $\lambda^{|I|}$ for each independent set $I \subseteq \mathcal{V}$. In particular, $\lambda = 1$ gives the uniform distribution. The natural coupling for this chain is the "identity" coupling, the same transition is attempted in both copies of the chain. If we try to apply standard path coupling to this chain, we immediately run into difficulties. Consider two chains X_t and Y_t such that $Y_t = X_t \cup \{w\}$, where $w \notin X_t$ (the *change vertex*) is of degree Δ. An edge $e \in \mathcal{E}$ is *critical* in Y_t if it has only one vertex $z \in \mathcal{V}$ which is not in Y_t, and we call z *critical for* e. If each of the edges through w is critical for Y_t, then there are Δ choices of v in the transition which can be added in X_t but not in Y_t. Thus, if $\lambda = 1$, the change in the expected Hamming distance between X_t and Y_t after one step could be as high as $\frac{\Delta}{2n} - \frac{1}{n}$. Thus we obtain rapid mixing only in the case $\Delta = 2$. This case has some intrinsic interest, since the complement of an independent set corresponds, under hypergraph duality, to an *edge cover* [11] in a graph. Thus we may uniformly generate edge covers, but the scope for unmodified path coupling is obviously severely limited.

The insight on which this paper is based is as follows. Although in one step it could be more likely that a *bad vertex* (increasing Hamming distance) is chosen than a *good vertex* (decreasing Hamming distance), it is even more likely that one of the other vertices in an edge containing w is chosen and removed from the independent set. Once the edge has two unoccupied vertices other than w, then any vertex in that edge can be added in both chains. This observation enables us to show that, if T is defined to be the stopping time at which the distance between X_t and Y_t first changes, the expected distance between X_T and Y_T will be less than 1. Theorem 2.1 below shows that under these circumstances path coupling can easily be adapted to prove rapid mixing.

Having established this general result, we use it to prove that $\mathcal{M}(\mathcal{H})$ is rapidly mixing for hypergraphs with $m \geq 2\lambda\Delta + 1$. Note that, though all the results in this paper will be proved for uniform hypergraphs of edge size m, they carry through trivially for hypergraphs of minimum edge size m.

2   Path Coupling Using a Stopping Time

First we prove the main result discussed above.

Theorem 2.1. *Let $\mathcal{M}$ be a Markov chain on state space Ω. Let d be an integer valued metric on $\Omega \times \Omega$, and let (X_t, Y_t) be a path coupling for $\mathcal{M}$, where S is the set of pairs of states (X, Y) such that $\mathrm{d}(X, Y) = 1$. For any initial states $(X_0, Y_0) \in S$ let T be the stopping time given by the minimum t such that $\mathrm{d}(X_t, Y_t) \neq 1$. Suppose, for some $p > 0$, that*

(i) $\Pr(T = t \,|\, T \geq t) \geq p$, independently for each t,
(ii) $\mathbf{E}[\mathrm{d}(X_T, Y_T)] \leq \alpha < 1$.

Then $\mathcal{M}$ mixes rapidly. In particular the mixing time $\tau(\varepsilon)$ of $\mathcal{M}$ satisfies

$$\tau(\varepsilon) \;\leq\; \frac{1}{p}\frac{3}{1-\alpha}\,\ln(eD_2)\ln\left(\frac{2D_1}{\varepsilon(1-\alpha)}\right),$$

where $D_1 = \max\{\mathrm{d}(X, Y) : X, Y \in \Omega\}$ and $D_2 = \max\{\mathrm{d}(X_T, Y_T) : X_0, Y_0 \in \Omega, \mathrm{d}(X_0, Y_0) = 1\}$.

Proof. Consider the following game. In each round a gambler either wins £1, loses some amount £$(l - 1)$ or continues to the next round. If he loses £$(l - 1)$ in a game, he starts l separate (but possibly dependent) games simultaneously in an effort to win back his money. If he has several games going and loses one at a certain time, he starts l more games, while continuing with the others that did not conclude. We know that the probability he finishes a game in a given step is at least p, and the expected winnings in each game is at least $1 - \alpha$. The question is: does his return have positive expectation at any fixed time? We will show that it does. But first a justification for our interest in this game.

Each game represents a single step on the path between two states of the coupled Markov chain. We start with X_0 and Y_0 differing at a single vertex. The first game is won if the first time the distance between the coupled chains changes is in convergence. The game is lost if the distance increases to l. At that point we consider the distance l path X_t to Y_t, and the l games played represent the l steps in the path. Although these games are clearly dependent, they each satisfy the conditions given. The gambler's return at time t is one minus the length of the path at time t, so a positive expected return corresponds to an expected path length less than one. We will show that the expected path length is sufficiently small to ensure coupling.

For the initial game we define the *level* to be zero, for any other possible game we define the level to be one greater than the level of the game whose loss precipitated it. We define the random variables M_k, l_{jk} and $I_{jk(t)}$ as follows. M_k is the number of games at level k that are played, l_{jk}, for $j = 1 \ldots M_k$, is the number of games in level $k + 1$ which are started as a result of the outcome of game j in level k, and $I_{jk}(t)$ is an indicator function which takes the value 1 if game j in level k is active at time t, and 0 otherwise. Let $N(t)$ be the number of games active at time t. Then, by linearity of expectations,

$$\mathbf{E}[N(t)] = \sum_{k=0}^{\infty} \mathbf{E} \left[\sum_{j=1}^{M_k} I_{jk}(t) \right] . \tag{1}$$

We will bound this sum in two parts, splitting it at a point $k = K$ to be determined. For $k \le K$ we observe that $M_k \le D_2^k$. Since $\Pr(I_{jk}(t) = 1)$ is at most the probability that exactly $k - 1$ games of a sequence are complete at time t, regardless of outcome, we have

$$\mathbf{E} \left[\sum_{j=1}^{M_k} I_{jk}(t) \right] \le D_2^k \max_j \mathbf{E}[I_{jk}(t)]$$

$$\le D_2^k \Pr(\text{exactly } k - 1 \text{ games complete by time } t).$$

So that

$$\sum_{k=0}^{K} \mathbf{E} \left[\sum_{j=1}^{M_k} I_{jk}(t) \right] \le \sum_{k=0}^{K} D_2^k \Pr(\text{exactly } k - 1 \text{ games complete by time } t)$$

$$\le D_2^K \Pr(\text{at most } K \text{ games complete by } t). \tag{2}$$

On the other hand, for $k > K$ we observe that

$$\mathbf{E} \left[\sum_{j=1}^{M_k} I_{jk}(t) \right] \le \mathbf{E}[M_k] = \mathbf{E}_{M_{k-1}} \left[\mathbf{E}[M_k | M_{k-1}] \right]$$

$$= \mathbf{E}_{M_{k-1}} \left[\mathbf{E}[\sum_{j=1}^{M_{k-1}} l_{jk-1} | M_{k-1}] \right]$$

Since $\mathbf{E}[l_{jk-1}] \le \alpha$ for any starting conditions, we may apply this bound even when conditioning on M_{k-1}. So

$$\mathbf{E} \left[\sum_{j=1}^{M_k} I_{jk}(t) \right] \le \mathbf{E}[\alpha M_{k-1}] \le \alpha^k, \tag{3}$$

using linearity of expectation, induction and $\mathbf{E}[M_1] \le \alpha$. Putting (2) and (3) together we get

$$\mathbf{E}[N(t)] \le D_2^K \Pr(\text{at most } K \text{ games complete by } t) + \sum_{k=K+1}^{\infty} \alpha^k$$

$$= D_2^K \Pr(\text{at most } K \text{ games complete by } t) + \frac{\alpha^{K+1}}{1 - \alpha}. \tag{4}$$

We now set $K = \lfloor (\ln \alpha)^{-1} \ln(\frac{\varepsilon(1-\alpha)}{2D_1}) \rfloor$, hence the final term is at most $\varepsilon/2D_1$. The probability that a game completes in any given step is at least p. If we select a time $\tau \geq c/p$ for $c \geq K+1 \geq 1$, then the probability that at most K games are complete is clearly maximised by taking this probability to be exactly p in all games. Hence, by Chernoff's bound (see, for example, [14, Theorem 2.1]),

$$\mathbf{E}[N(\tau)] \leq D_2^K \sum_{k=0}^{K} \binom{\tau}{k} p^k (1-p)^{\tau-k} + \frac{\varepsilon}{2D_1}$$

$$\leq e^{K \ln D_2 - \frac{(c-K)^2}{2c}} + \frac{\varepsilon}{2D_1}$$

$$\leq e^{K \ln D_2 + K - c/2} + \frac{\varepsilon}{2D_1}.$$

Choosing $c = 2K \ln(eD_2) + 2 \ln \frac{2D_1}{\varepsilon}$, we obtain $\mathbf{E}[N(\tau)] < \frac{\varepsilon}{D_1}$, where $\tau = \lceil \frac{3 \ln(eD_2)}{p(1-\alpha)} \ln\left(\frac{2D_1}{\varepsilon(1-\alpha)}\right) \rceil$.

We conclude that the gambler's expected return at time τ is positive. More importantly, for any initial states $X_0, Y_0 \in \Omega$, the expected distance at time τ is at most ε by linearity of expectations, and so the probability that the chain has not coupled is at most ε. The mixing time claimed now follows by standard arguments. See, for example, [16]. □

Remark 2.2. The assumption that the stopping time occurs when the distance changes is not essential. The assumption that S contains only pairs of states at distance 1 can be removed at the expense of a more complicated proof. We clearly cannot dispense with assumption (ii), or we cannot bound mixing time. Assumption (i) may appear a restriction, but appears to be naturally satisfied in most applications. It seems more natural than the assumption of bounded stopping time, used in [12]. Assumption (i) can easily be replaced by something weaker, for example by allowing p to vary with time rather than remain constant. Provided $p \neq 0$ sufficiently often, a similar proof will be valid.

Remark 2.3. Let $\gamma = 1/(1-\alpha)$. It seems likely that D_2 will be small in comparison to γ in most applications, so we might suppose $D_2 < \gamma < D_1$. The mixing time bound from Theorem 2.1 can then be written $O(p^{-1}\gamma \log D_2 \log(D_1/\varepsilon))$. We may compare this with the bound which can be derived using [12, Corollary 4]. This can be written in similar form as $O(p^{-1}\gamma \log \gamma \log(D_1/\varepsilon))$. In such cases we obtain a reduction in the estimate of mixing time by a factor $\log \gamma / \log D_2$. In the applications below, for example, we have $D_2 = 2$ and $\gamma = \Omega(\Delta)$, so the improvement is $\Omega(\log \Delta)$.

Remark 2.4. The reason for our improvement on the result of [12] is that the use of an upper bound on the stopping time, as is done in [12], will usually underestimate the number of stopping times which occur in a long interval, and hence the mixing rate.

3   Hypergraph Independent Sets

We now use the approach of path coupling via stopping times to prove that the chain discussed in Section 1.1 is rapidly mixing. The metric used in path coupling analyses throughout the paper will be Hamming distance between the coupled chains. We prove the following theorem.

Theorem 3.1. Let λ, Δ be fixed, and let $\mathcal{H}$ be a hypergraph such that $m \geq 2\lambda\Delta + 1$. Then the Markov chain $\mathcal{M}(\mathcal{H})$ has mixing time $O(n \log n)$.

Before commencing the proof itself, we analyse the stopping time T for this problem.

3.1   Edge Process

Let X_t and Y_t be copies of $\mathcal{M}$ which we wish to couple, with $Y_0 = X_0 \cup \{w\}$. Let e be any edge containing w, with $m = |e|$. We consider only the times at which some vertex in e is chosen. The progress of the coupling on e can then be modelled by the following "game". We will call the number of unoccupied vertices in e (excluding w) *units*. At a typical step of the game we have k units, and we either win the game, win a unit, keep the same state or lose a unit. These events happen with the following probabilities: we win the game with probability $1/m$, win a unit with probability at least $(m - k - 1)/(1 + \lambda)m$, lose a unit with probability at most $\lambda k/(1 + \lambda)m$ and stay in the same state otherwise. If ever $k = 0$, we are bankrupt and we lose the game. Winning the game models the "good event" that the vertex v is chosen and the two chains couple. Losing the game models the "bad event" that the coupling increases the distance to 2. We wish to know the probability that the game ends in bankruptcy. We are most interested in the case where $k = 1$ initially, which models e being critical. Note that the value of k in the process on hypergraph independent sets dominates the value in our model, since we can always delete (win in the game), but we may not be able to insert (lose in the game) because the chosen vertex is critical in some other edge.

Let p_k denote the probability that a game is lost, given that we start with k units. We have the following system of simultaneous equations.

$$(m - 1 + 2\lambda)p_1 - (m - 2)p_2 = \lambda$$
$$-k\lambda p_{k-1} + (m - k + (k + 1)\lambda)p_k - (m - k - 1)p_{k+1} = 0 \quad (k = 2, 3, \ldots, m - 1)$$
$$(5)$$

Solving these yields the following, which may be confirmed by substituting into Equations (5).

$$p_k = \frac{1}{\binom{m-1}{k}}\left(\lambda^k - \frac{\sum_{i=1}^{k}\binom{m}{i}\lambda^{m+k-i}}{(1 + \lambda)^m - \lambda^m}\right) = \frac{\sum_{i=k+1}^{m}\binom{m}{i}\lambda^{m+k-i}}{((1 + \lambda)^m - \lambda^m)\binom{m-1}{k}} \quad (k = 1, 2, \ldots, m - 1).$$
$$(6)$$

In particular

$$p_1 = \frac{\lambda}{m-1}\left(1 - \frac{m\lambda^{m-1}}{(1+\lambda)^m - \lambda^m}\right). \tag{7}$$

3.2   The Expected Distance Between X_T and Y_T

The stopping time for the pair of chains X_t and Y_t will be when the distance between them changes, in other words either a good or bad event occurs. The probability that we observe the bad event on a particular edge e with $w \in e$ is at most p_k as calculated above. Let ξ_t denote the number of empty vertices in e at time t when the process is started with $\xi_0 = k$. Now ξ_t can never reach 0 without first reaching $k-1$ and, since the process is Markovian, it follows that

$$p_k = \Pr(\exists t\, \xi_t = 0|\xi_0 = k) = \Pr(\exists t\, \xi_t = 0|\xi_s = k-1)\Pr(\exists s\, \xi_s = k-1|\xi_0 = k) < p_{k-1}.$$

Since w is in at most Δ edges, the probability that we observe the bad event on any edge is at most Δp_1. The probability that the stopping time ends with the good event is therefore at least $1 - \Delta p_1$. The path coupling calculation is then

$$\mathbf{E}[d(X_T, Y_T)] \le 2\Delta p_1.$$

This is required to be less than 1 in ordered to apply Theorem 2.1. If $m \ge 2\lambda\Delta+1$, then by (7)

$$2\Delta p_1 = 1 - \frac{(2\lambda\Delta + 1)\lambda^{2\lambda\Delta}}{(1+\lambda)^{2\lambda\Delta+1} - \lambda^{2\lambda\Delta+1}}.$$

Proof of Theorem 3.1. The above work puts us in a position to apply Theorem 2.1. Let $m \ge 2\lambda\Delta + 1$. Then for $\mathcal{M}(\mathcal{H})$ we have

(i) $\Pr(d(X_t, Y_t) \ne 1|d(X_{t-1}, Y_{t-1}) = 1) \ge \frac{1}{n}$ for all t, and

(ii) $\mathbf{E}[d(X_T, Y_T)] < 1 - \dfrac{(2\lambda\Delta + 1)\lambda^{2\lambda\Delta}}{(1+\lambda)^{2\lambda\Delta+1} - \lambda^{2\lambda\Delta+1}}.$

Also for $\mathcal{M}(\mathcal{H})$ we have $D_1 = n$ and $D_2 = 2$. Hence by Theorem 2.1, $\mathcal{M}(\mathcal{H})$ mixes in time

$$\tau(\varepsilon) \le 6n\frac{(1+\lambda)^{2\lambda\Delta+1} - \lambda^{2\lambda\Delta+1}}{(2\lambda\Delta + 1)\lambda^{2\lambda\Delta}}\ln\left(n\varepsilon^{-1}\frac{(1+\lambda)^{2\lambda\Delta+1} - \lambda^{2\lambda\Delta+1}}{(2\lambda\Delta + 1)\lambda^{2\lambda\Delta}}\right).$$

This is $O(n\log n)$ for fixed λ, Δ.   □

Remark 3.2. In the most important case, $\lambda = 1$, we require $m \ge 2\Delta + 1$. This does not include the case $m = 3$, $\Delta = 3$ considered in [10]. We have attempted to improve the bound by employing the chain proposed by Dyer and Greenhill in [10, Section 4]. However, this gives only a marginal improvement. For large $\lambda\Delta$, we obtain convergence for $m \ge 2\lambda\Delta + \frac{1}{2} + o(1)$. For $\lambda = 1$, this gives a better bound on mixing time for $m = 2\Delta + 1$, with dependence on Δ similar to Remark 3.3 below, but does not even achieve mixing for $m = 2\Delta$. We omit the details in order to deal with the Glauber dynamics, and to simplify the analysis.

Remark 3.3. The terms in the running time which are exponential in λ, Δ would disappear if we instead took graphs for which $m \geq 2\lambda\Delta + 2$. In this case the running time would be

$$\tau(\varepsilon) \leq 6(2\lambda\Delta + 1)n \ln(n\varepsilon^{-1}(2\lambda\Delta + 1)) \leq 12(2\lambda\Delta + 1)n \ln(n\varepsilon^{-1}).$$

Furthermore, if we took graphs such that $m > (2 + \delta)\lambda\Delta$, for some $\delta > 0$, then the running time would no longer depend on λ, Δ at all, but would be $\tau(\varepsilon) \leq c_\delta n \ln(n\varepsilon^{-1})$ for some constant c_δ.

Remark 3.4. It seems that path coupling cannot show anything better than m linear in $\lambda\Delta$. Suppose the initial configuration has edges $\{w, v_1, \ldots, v_{m-2}, x_i\}$ for $i = 1, \ldots, \Delta$, with $w, v_1, \ldots, v_{m-2} \in X_0$, $x_1, \ldots, x_\Delta \notin X_0$ and w the change vertex. Consider the first step where any vertex changes state. Let $\mu = (1 + \lambda)(m - 1 + \Delta)$. The good event occurs with probability $(1 + \lambda)/\mu$, insertion of a critical vertex with probability $\lambda\Delta/\mu$, and deletion of a non-critical vertex with probability $(m - 1)/\mu$. We therefore need $(m - 1) + (1 + \lambda) \geq \lambda\Delta$, i.e. $m \geq \lambda(\Delta - 1)$, to show convergence by path coupling.

Remark 3.5. It seems we could improve our bound $m \geq 2\lambda\Delta + 1$ for rapid mixing of the Glauber dynamics somewhat if we could analyse the process on all edges simultaneously. Examination of the extreme cases, where all edges adjacent to w are otherwise independent, or where they are dependent except for one vertex (as in Remark 3.4), suggests that improvement to $(1 + o(1))\lambda\Delta$ may be possible, where the $o(1)$ is relative to $\lambda\Delta$. However, the analysis in the general case seems difficult, since edges can intersect arbitrarily.

4 Hardness Results for Independent Sets

We have established that the number of independent sets of a hypergraph can be approximated efficiently using the Markov Chain Monte Carlo technique for hypergraphs with edge size linear in Δ. We next state hardness results, in particular that exact counting is unlikely to be possible, and that unless NP=RP, there can be no *fpras* for the number of independent sets of all hypergraphs with edge size $\Omega(\log \Delta)$.

Theorem 4.1. *Let* $\mathcal{G}(m, \Delta)$ *be the class of uniform hypergraphs with minimum edge size* $m \geq 3$ *and maximum degree* Δ. *Computing the number of independent sets of hypergraphs in* $\mathcal{G}(m, \Delta)$ *is #P-complete if* $\Delta \geq 2$. *If* $\Delta \leq 1$, *it is in P.*

Let $G = (V, E)$, with $|V| = n$, be a graph with maximum degree Δ and N_i independent sets of size i ($i = 0, 2, \ldots, n$). For $\lambda > 0$ let $Z_G(\lambda) = \sum_{i=0}^{n} N_i \lambda^i$ define the *hard core partition function*. The following is a combination of results in Luby and Vigoda [18] and Berman and Karpinski [2].

Theorem 4.2. *If* $\lambda > 694/\Delta$, *there is no fpras for* $Z_G(\lambda)$ *unless NP=RP.*

We note that Theorem 4.2 could probably be strengthened using the approach of [8]. However, this has yet to be done.

Theorem 4.3. *Unless NP=RP, there is no* fpras *for counting independent sets in hypergraphs with maximum degree Δ and minimum edge size $m < 2\lg(1 + \Delta/694) - 1 = \Omega(\log \Delta)$.*

5   Hypergraph Colouring

In this section we present without proof another result obtained using Theorem 2.1. We now consider Glauber dynamics on the set of proper colourings of a hypergraph. Again our hypergraph $\mathcal{H}$ will have maximum degree Δ, minimum edge size m, and we will have a set of q colours. A colouring of the vertices of $\mathcal{H}$ is proper if no edge is monochromatic. Let $\Omega'(\mathcal{H})$ be the set of all proper q-colourings of $\mathcal{H}$. We define the Markov chain $\mathcal{C}(\mathcal{H})$ with state space $\Omega'(\mathcal{H})$ by the following transition process. If the state of $\mathcal{C}$ at time t is X_t, the state at $t+1$ is determined by

(i) selecting a vertex $v \in \mathcal{V}$ and a colour $k \in \{1, 2, \ldots, q\}$ uniformly at random,
(ii) let X_t' be the colouring obtained by recolouring v colour k
(iii) if X_t' is a proper colouring let $X_{t+1} = X_t'$
    otherwise let $X_{t+1} = X_t$.

This chain is easily shown to be ergodic with the uniform stationary distribution. Again we may use Theorem 2.1 to prove rapid mixing of this chain under certain conditions, however first we note that the following result may be obtained using standard path coupling techniques.

Theorem 5.1. *For $m \geq 4$, $q > \Delta$, the Markov chain $\mathcal{C}(\mathcal{H})$ mixes in time $O(n \log n)$.*

This leaves little room for improvement in the case $m \geq 4$, indeed it is not clear whether the Markov chain described is even ergodic for $q \leq \Delta$. The following simple construction does show that the chain is not in general ergodic if $q \leq \frac{\Delta}{m} + 1$. Let $q = \frac{\Delta}{m} + 1$, and take a hypergraph $\mathcal{H}$ on $q(m-1)$ vertices. We will group the vertices into q groups $\mathcal{V} = \mathcal{V}_1, \mathcal{V}_2, \ldots, \mathcal{V}_q$, each of size $m - 1$. Then the edge set of $\mathcal{H}$ is $E = \{\{v\} \cup \mathcal{V}_j : v \in \mathcal{V}, v \notin \mathcal{V}_j\}$. The degree of each vertex is $(q-1) + (q-1)(m-1) = \Delta$. If we now colour each group $\mathcal{V}_j$ a different colour, we obtain $q!$ distinct colourings, but for each of these the Markov chain is frozen (no transition is valid).

The case $m = 2$ is graph colouring and has been extensively studied. See, for example, [7]. This leaves the case $m = 3$, hypergraphs with 3 vertices in each edge. The standard path coupling argument only shows rapid mixing for $q \geq 2\Delta$, since there may be two vertices in each edge that can be selected and lead to a divergence of the two chains. This occurs if, of the two vertices in an edge which are not w, one is coloured red and the other blue. However, we can do better using Theorem 2.1. The proof is omitted, but again hinges on the fact that an edge is very unlikely to persist in such a critical state.

Theorem 5.2. *There exists Δ_0 such that, if $\mathcal{H}$ is a 3-uniform hypergraph with maximum degree $\Delta > \Delta_0$ and $q \geq 1.65\Delta$, the Markov chain $\mathcal{C}(\mathcal{H})$ mixes rapidly.*

6   Hardness Results for Colouring

Again we state a theorem that exact counting is #P-complete except in the few cases where it is clearly in P. Let $\mathcal{G}(m, \Delta)$ be as in Theorem 4.1.

Theorem 6.1. *Computing the number of q-colourings of hypergraphs in $\mathcal{G}(m, \Delta)$ is #P-complete if $\Delta, q > 1$. If $\Delta \leq 1$ or $q \leq 1$ it is in P.*

Again let $\mathcal{G}(m, \Delta)$ be as defined in Theorem 4.1. The hardness of approximation result, Corollary 6.3, follows directly from the following NP-completeness result.

Theorem 6.2. *Determining whether a hypergraph in $\mathcal{G}(m, \Delta)$ has any q-colouring is NP-complete for any $m > 1$ and $2 < q \leq (1 - 1/m)\Delta^{1/(m-1)}$.*

Corollary 6.3. *Unless NP=RP, there is no fpras for counting q-colourings of a hypergraphs with maximum degree Δ and minimum edge size m if $2 < q \leq (1 - 1/m)\Delta^{1/(m-1)}$.*

Remark 6.4. It is clearly a weakness that our lower bound for approximate counting is based entirely on an NP-completeness result. However, we note that the same situation pertains for graph colouring, which has been the subject of more intensive study.

7   Conclusions

We have presented an approach to the analysis of path coupling with stopping times which improves on the method of [12] in most applications. Our method may itself permit further development.

We apply the method to independent sets and q-colourings in hypergraphs with maximum degree Δ and minimum edge size m. In the case of independent sets, there seems scope for improving the bound $m \geq 2\Delta+1$, but anything better than $m \geq \Delta + o(\Delta)$ would seem to require new methods. For colourings, there is probably little improvement possible in our result $q > \Delta$ for $m \geq 4$, but many questions remain for $q \leq \Delta$. For example, even the ergodicity of the Glauber (or any other) dynamics is not clearly established. For the most interesting case, $m = 3$, the bound $q > 1.65\Delta$ (for large Δ) can almost certainly be reduced, but substantial improvement may prove difficult.

Our #P-completeness results seem best possible for both of the problems we consider. On the other hand, our lower bounds for hardness of approximate counting seem very weak in both cases, and are far from our upper bounds. These lower bounds can probably be improved, but we have no plausible conjecture as to what may be the truth.

Acknowledgments

We are grateful to Tom Hayes for commenting on an earlier draft of this paper, and to Mary Cryan for useful discussions at an early stage of this work.

References

1. M. Bordewich, M. Dyer and M. Karpinski, Path coupling using stopping times and counting independent sets and colourings in hypergraphs, (2005) http://arxiv.org/abs/math.PR/0501081.
2. P. Berman and M. Karpinski, Improved approximation lower bounds on small occurrence optimization, *Electronic Colloquium on Computational Complexity* **10** (2003), Technical Report TR03-008.
3. R. Bubley, *Randomized algorithms: approximation, generation and counting*, Springer-Verlag, London, 2001.
4. R. Bubley and M. Dyer, Graph orientations with no sink and an approximation for a hard case of #SAT, in *Proc. 8th Annual ACM-SIAM Symposium on Discrete Algorithms (SODA 1997)*, SIAM, 1997, pp. 248–257.
5. I. Dinur, V. Guruswami, S. Khot and O. Regev, A new multilayered PCP and the hardness of hypergraph vertex cover, in *Proc. 35th ACM Symposium on Theory of Computing (STOC 2003)*, ACM, 2003, pp. 595–601.
6. I. Dinur, O. Regev and C. Smyth, The hardness of 3-uniform hypergraph coloring, in *Proc. 43rd Symposium on Foundations of Computer Science (FOCS 2002)*, IEEE, 2002, pp. 33–42
7. M. Dyer, A. Frieze, T. Hayes and E. Vigoda, Randomly coloring constant degree graphs, in *Proc. 45th Annual IEEE Symposium on Foundations of Computer Science (FOCS 2004)*, IEEE, 2004, pp. 582–589.
8. M. Dyer, A. Frieze and M. Jerrum, On counting independent sets in sparse graphs, *SIAM Journal on Computing* **31** (2002), 1527–1541.
9. M. Dyer, L. Goldberg, C. Greenhill, M. Jerrum and M. Mitzenmacher, An extension of path coupling and its application to the Glauber dynamics for graph colorings, *SIAM Journal on Computing* **30** (2001), 1962–1975.
10. M. Dyer and C. Greenhill, On Markov chains for independent sets, *Journal of Algorithms* **35** (2000), 17–49.
11. M. Garey and D. Johnson, *Computer and intractability*, W. H. Freeman and Company, 1979.
12. T. Hayes and E. Vigoda, Variable length path coupling, in *Proc. 15th Annual ACM-SIAM Symposium on Discrete Algorithms (SODA 2004)*, SIAM, 2004, pp. 103–110.
13. T. Hofmeister and H. Lefmann, Approximating maximum independent sets in uniform hypergraphs, *Proc. 23rd International Symposium on Mathematical Foundations of Computer Science (MFCS 1998)*, Lecture Notes in Computer Science **1450**, Springer, 1998, pp. 562–570.
14. S. Janson, T. Łuczak and A. Ruciński, *Random graphs*, Wiley-Interscience, New York, 2000.
15. M. Jerrum, A very simple algorithm for estimating the number of k-colorings of a low-degree graph, *Random Structure and Algorithms* **7** (1995), 157–165.
16. M. Jerrum, *Counting, sampling and integrating: algorithms and complexity*, ETH Zürich Lectures in Mathematics, Birkhäuser, Basel, 2003.
17. M. Krivelevich, R. Nathaniel and B. Sudakov, Approximating coloring and maximum independent sets in 3-uniform hypergraphs, in *Proc. 12th Annual ACM-SIAM Symposium on Discrete Algorithms, (SODA 2001)*, SIAM, 2001, pp. 327–328.
18. M. Luby and E. Vigoda, Fast convergence of the Glauber dynamics for sampling independent sets, *Random Structures and Algorithms* **15** (1999), 229–241.
19. M. Mitzenmacher and E. Niklova, Path coupling as a branching process, unpublished manuscript, 2002.

20. M. Molloy, Very rapidly mixing Markov chains for 2Δ-coloring and for independent sets in a graph with maximum degree 4, *Random Structures and Algorithms* **18** (2001), 101–115.

21. J. Salas and A. Sokal, Absence of phase transition for anti-ferromagnetic Potts models via the Dobrushin uniqueness theorem, *Journal of Statistical Physics* **86** (1997), 551–579.

22. E. Vigoda, A note on the Glauber dynamics for sampling independent sets, *The Electronic Journal of Combinatorics* **8**, R8(1), 2001.

On the Incompressibility of Monotone DNFs

Matthias P. Krieger*

Johann Wolfgang Goethe-Universität Frankfurt, Institut für Informatik,
Lehrstuhl für Theoretische Informatik, Robert-Mayer-Straße 11-15,
D-60054 Frankfurt am Main, Germany
mkrieger@cs.uni-frankfurt.de

Abstract. We prove optimal lower bounds for multilinear circuits and
for monotone circuits with bounded depth. These lower bounds state
that, in order to compute certain functions, these circuits need exactly
as many OR gates as the respective DNFs. The proofs exploit a property
of the functions that is based solely on prime implicant structure. Due
to this feature, the lower bounds proved also hold for approximations of
the considered functions that are similar to slice functions. Known lower
bound arguments cannot handle these kinds of approximations. In order
to show limitations of our approach, we prove that cliques of size $n - 1$
can be detected in a graph with n vertices by monotone formulae with
$O(\log n)$ OR gates.

Our lower bound for multilinear circuits improves a lower bound due
to Borodin, Razborov and Smolensky for nondeterministic read-once
branching programs computing the clique function.

1 Introduction

In this paper we consider Boolean circuits consisting of AND and OR gates.
These circuits have variables and negated variables as inputs. Unless otherwise
noted, all gates have fanin 2. A circuit without any negated inputs is called
monotone. A circuit whose gates have fanout 1 is a *formula*. A *monom* is a
conjunction of variables and negated variables. In this paper we regard monoms
also as sets. An *implicant* of a Boolean function f is a monom that does not
evaluate to 1 unless f does. An implicant is a *prime implicant* (minterm) if
no new implicant can be obtained by removing variables or negated variables
from the conjunction. For a Boolean function f, we denote the set of its prime
implicants by $PI(f)$.

Until now the best known lower bounds for non-monotone circuits are linear.
However, there has been considerable success in proving superpolynomial lower
bounds for *monotone* circuits. Nowadays we have several powerful techniques
to prove lower bounds for monotone circuits: the method of approximations
(Razborov [1]); the method of probabilistic amplifications for estimating the
depth of monotone circuits (Karchmer and Wigderson [2]); the rank argument
for formulas (Razborov [3]) and span programs (Gál [4], Gál and Pudlák [5]).

---

* Partially supported by DFG grant SCHN 503/2-2.

Also, it is known that negation is almost powerless for so-called slice functions (see e.g. monographs [6,7,8]). The t-slice function of f is a function of the form $f \wedge T_t^n \vee T_{t+1}^n$, where T_t^n is the t-th threshold function in n variables. A superpolynomial lower bound for the *monotone* complexity of a slice function implies a lower bound of the same order for its non-monotone complexity. Unfortunately, the currently available arguments for proving monotone lower bounds seem to be incapable of yielding sufficient lower bounds for slice functions. Therefore it is justified to seek new methods for proving monotone lower bounds.

One property of t-slice functions which seems to make the known arguments unsuitable for them is that they accept *all* inputs with more than t ones. The available proof methods rely on adequate sets of inputs which are mapped to 0 by the function considered. That t-slice functions accept all inputs with more than t ones seems to be an obstacle to constructing adequate sets of rejected inputs. Therefore it is justified to seek lower bound arguments for functions of the form $f \vee T_{t+1}^n$ that share this problematic property with slice functions; because of this similarity, we will refer to functions of the form $f \vee T_{t+1}^n$ as *t-pseudoslice functions* in the sequel.

In this paper we make some steps in this direction. We propose proof methods for some restricted circuit models that avoid these shortcomings. In particular, the properties of functions that we exploit are based solely on the prime implicant structure and do not rely on any additional information about prime clauses or rejected inputs. In this sense our lower bound arguments are "asymmetric". Unlike the currently available arguments, they are applicable to certain pseudoslice functions as well.

Moreover, the lower bounds we prove are optimal for the circuit classes considered. They state that multilinear circuits and circuits with sufficiently small alternation depth require exactly as many OR gates as the DNFs of the considered functions. This means that by using these circuit types instead of DNFs, we cannot even save a single OR gate! In other words, the DNFs are "incompressible" when we restrict ourselves to the respective circuit classes.

2   Results

A Boolean circuit is *multilinear* if the inputs to each of its AND gates are computed from disjoint sets of variables. To be more precise, for a gate g let $var\,(g)$ be the set of variables that occur in the subcircuit rooted at g. A Boolean circuit is multilinear if $var\,(g_1) \cap var\,(g_2) = \emptyset$ for each of its AND gates g with inputs g_1 and g_2. Multilinear circuits have been studied in [9,10] ([10] uses a slightly less restrictive definition). Multilinear circuits are a generalization of nondeterministic read-once branching programs, which have received much attention (see e.g. monograph [11]). Boolean multilinear circuits are related to arithmetic multilinear circuits which are characterized by the restriction that the highest power of the polynomials computed at their gates is no larger than 1. Arithmetic multilinear circuits have been studied in [12,13,14]. The direct arithmetic counterpart to Boolean multilinear circuits are syntactic multilinear circuits, defined by Raz [14].

It is clear that every Boolean function f can be computed by a multilinear circuit with $|PI(f)| - 1$ OR gates: just take the DNF of f. Many functions commonly referred to have multilinear circuits that are much smaller than their DNFs. Consider the threshold function T_k^n as an example. The threshold function T_k^n has $\binom{n}{k}$ prime implicants, but can be computed by a multilinear circuit of size $O(nk)$ [11, chapter 4]. Thus, the gap between the size of a smallest multilinear circuit which computes a certain function and the size of the DNF of this function can be exponential. It is also known that the gap between multilinear complexity and monotone complexity is exponential [9].

We identify a class of functions whose multilinear circuits require exactly as many OR gates as their DNF, the so-called union-free functions. We call a monotone Boolean function *union-free* if the union of any two of its prime implicants does not contain a new prime implicant.

Theorem 1. *Let f be a monotone union-free function. Then any multilinear circuit for f must have at least $|PI(f)| - 1$ OR gates.*

In the proof of this theorem we establish the following property of union-free functions: among the optimal (with respect to the number of OR gates) circuits there is one which is a formula, and for each of its AND gates, at least one input to the gate computes a monom.

The clique function $CLIQUE(n, s)$ is the function on $\binom{n}{2}$ variables representing the edges of an undirected graph G whose value is 1 iff G contains an s-clique. The clique function is a prominent example of a union-free function.

Lemma 1. *The function $CLIQUE(n, s)$ is union-free.*

Proof. Suppose the union of two distinct s-cliques A and B contains all edges of some third clique C. Since all three cliques are distinct and have the same number of vertices, C must contain a vertex u which does not belong to A and a vertex v which does not belong to B. This already leads to a contradiction because either the vertex u (if $u = v$) or the edge $\{u, v\}$ (if $u \neq v$) of C would remain uncovered by the cliques A and B. □

Corollary 1. *Multilinear circuits for $CLIQUE(n, s)$ require $\binom{n}{s} - 1$ OR gates (just as many as the DNF of this function).*

Because nondeterministic read-once branching programs can be simulated by multilinear circuits in a natural way, Corollary 1 improves the lower bound of $\exp(\Omega(\min(s, n - s)))$ given in [15] for nondeterministic read-once branching programs computing $CLIQUE(n, s)$.

Our lower bound for multilinear circuits also holds for certain pseudoslices of union-free functions. We call a monotone function k-*homogeneous* if each of its prime implicants has k variables.

Theorem 2. *Let f be a monotone k-homogeneous union-free function. Then any monotone multilinear circuit which computes the t-pseudoslice of f such that $t \geq 2k$ must have at least $|PI(f)| - 1$ OR gates.*

The next result we discuss shows that the union-freeness property is not sufficient for proving good lower bounds for *general* monotone circuits. By Corollary 1, $CLIQUE(n, n-1)$ requires $n-1$ OR gates to be computed by a multilinear circuit. On the other hand, we have the following upper bound.

Theorem 3. *The function $CLIQUE(n, n-1)$ can be computed by a monotone formula with $O(\log n)$ OR gates.*

This is apparently the first non-trivial upper bound for the monotone complexity of the clique function. The only other upper bound for the clique function that we are aware of is given in [6] and is only for its non-monotone complexity.

A circuit has *alternation depth d* iff d is the highest number of blocks of OR gates and blocks of AND gates on paths from input to output gates. A Σ_d-circuit (respectively, Π_d-circuit) is a circuit with alternation depth d such that the output gate is an OR gate (AND gate, respectively). We give incompressibility results, similar to those for multilinear circuits, also for monotone Σ_4-circuits. A Boolean function is *s-disjoint* if any two of its prime implicants do not have s variables in common.

Theorem 4. *Let f be a monotone k-homogeneous s-disjoint function such that $|PI(f)| \leq (k/2s)^{k/2s}$. Then every monotone Σ_4-circuit for f must have at least $|PI(f)| - 1$ OR gates.*

The same also holds for any t-pseudoslice of f such that $t \geq k^2/2s$.

Let $POLY(q, s)$ be the polynomial function introduced by Andreev [16]. This function has $n = q^2$ variables corresponding to the points in the grid $GF(q) \times GF(q)$, where q is a prime power. The function $POLY(q, s)$ accepts a $q \times q$ 0-1 matrix $X = (x_{i,j})$ iff there is a polynomial $p(z)$ of degree at most $s-1$ over $GF(q)$ such that $x_{i,p(i)} = 1$ for all $i \in GF(q)$. If $s < q/2$, then $POLY(q, s)$ is another example of a union-free function.

The function $POLY(q, s)$ is q-homogeneous. This function is also s-disjoint because the graphs of two distinct polynomials of degree at most $s-1$ cannot share s points. This together with $|PI(POLY(q, s))| = q^s$ and Theorem 4 leads to the following corollary.

Corollary 2. *If $s \leq \sqrt{q}/2$, then any Σ_4-circuit for $POLY(q, s)$ must have at least $q^s - 1$ OR gates (just as many as the DNF of this function).*

The construction used in the proof of Theorem 3 yields a Π_3-formula. Hence, Theorem 4 suggests that it is harder to prove upper bounds for sufficiently disjoint functions because an efficient monotone circuit for them must be more complicated than a Σ_4-circuit. It is not even clear whether these polynomial functions $POLY(q, s)$ with $s \leq \sqrt{q}/2$ can be computed by general monotone circuits that are smaller than the respective DNFs.

The rest of the paper is devoted to the proof of our theorems.

3   Lower Bounds for Multilinear Circuits

In this section we prove Theorems 1 and 2. The following lemma allows us to restrict ourselves to *monotone* multilinear circuits. It is a special case of a theorem given in [17] for read-once nondeterministic machines.

Lemma 2. *If f is a monotone function, then any optimal multilinear circuit for f is monotone.*

Our next lemma describes a restriction of multilinear circuits which leads to exponential lower bounds for certain monotone Boolean functions. Given a prime implicant p, we show that certain variables of p can be substituted by some variables of another prime implicant p', yielding a "derived" implicant of the function. We say a path from a gate to the output of a circuit is *consistent* with a monom m if m is an implicant of all the functions computed at the gates along this path. We call a gate g *necessary* for an implicant m of a circuit S if m is not an implicant of the circuit $S_{g \to 0}$ we obtain from S by replacing g with the constant 0. Clearly, for every gate g which is necessary for an implicant m of S, there is a path from g to the output of S which is consistent with m. Let $PI_g(f)$ denote the set of prime implicants of f that g is necessary for. By $PI(g)$ we denote the set of prime implicants of the function computed at gate g.

Lemma 3 (Exchange Lemma). *Let g be a gate in a monotone multilinear circuit S for a function f and p, p' be prime implicants in $PI_g(f)$. Let $m \subseteq p$ and $m' \subseteq p'$ be distinct prime implicants in $PI(g)$.*

(i) If w is a path from g to the output of S that is consistent with p, then w is consistent with the derived monom $(p \setminus m) \cup m'$. This means in particular that the derived monom $(p \setminus m) \cup m'$ is also an implicant of f.

(ii) If f is union-free, then the identity $p = (p' \setminus m') \cup m$ holds.

(iii) If f is a t-pseudoslice of a monotone k-homogeneous union-free function f^* such that $t \geq 2k$ and p, p' are prime implicants of f^* as well, then the same identity $p = (p' \setminus m') \cup m$ also holds.

Proof. (i) We have to show that $(p \setminus m) \cup m'$ is an implicant of all functions computed along w ($g = g_1, \ldots, g_t$). We prove this by induction on the length of the path w. For $g_1 = g$ the claim is correct since $(p \setminus m) \cup m'$ is a superset of $m' \in PI(g_1)$. For the inductive step, assume that $q \in PI(g_i)$ such that $q \subseteq (p \setminus m) \cup m'$. If g_{i+1} is an OR gate, then q is an implicant of g_{i+1}. If g_{i+1} is an AND gate, then let h be the other gate feeding it. We know that p is an implicant of the function computed at g_{i+1}. Hence, there must be some $m_h \in PI(h)$ such that $m_h \subseteq p$. Because the circuit is multilinear, we have $var(g_i) \cap var(h) = \emptyset$. Gate g belongs to the subcircuit rooted at gate g_i. We conclude that $var(g) \subseteq var(g_i)$ and that $var(g) \cap var(h) = \emptyset$. Since a variable of a prime implicant of a gate must occur somewhere in the subcircuit rooted at that gate, we conclude from $m \in PI(g)$ and $m_h \in PI(h)$ that $m \cap m_h = \emptyset$. Now we can see that $q \cup m_h$, an implicant of the function computed at g_{i+1}, is a subset of $(p \setminus m) \cup m'$.

(ii) According to (i), the monom $(p \setminus m) \cup m'$ is an implicant of f. Clearly, $(p \setminus m) \cup m' \subseteq p \cup p'$. Since f is union-free, this implies $p \subseteq (p \setminus m) \cup m'$ or $p' \subseteq (p \setminus m) \cup m'$. Because m and m' are distinct prime implicants, we have $m \not\subseteq m'$ and $m \not\supseteq m'$. The inclusion $p \subseteq (p \setminus m) \cup m'$ is impossible because $m \not\subseteq m'$. So $p' \subseteq (p \setminus m) \cup m'$ holds, this implies $m' \supseteq p' \setminus p$.

Since its assumptions are symmetrical, claim (i) also implies that $(p' \setminus m') \cup m$ is an implicant of f. Arguing in the same way as above we conclude that $p \subseteq (p' \setminus m') \cup m$. Since $m' \supseteq p' \setminus p$, we have $(p' \setminus m') \cup m \subseteq p$. Because p is a prime implicant of f, this means $p = (p' \setminus m') \cup m$.

(iii) We observe that the assumptions allow us to reason the same way as in (ii). Again, the monom $(p \setminus m) \cup m'$ is an implicant of f. We have $|(p \setminus m) \cup m'| \leq 2k$ because $|p| = k$ and $|m'| \leq |p'| = k$. Thus, $(p \setminus m) \cup m'$ must also be an implicant of f^*. Since f^* is union-free, we can conclude the same way as in (ii) that $m' \supseteq p' \setminus p$. As in (ii), $(p' \setminus m') \cup m$ is an implicant of f and also of f^* since $|(p' \setminus m') \cup m| \leq 2k$. We may now proceed as in (ii) and conclude that $p = (p' \setminus m') \cup m$. $\qquad\square$

We call a monotone circuit *broom-like* if, for each of its AND gates with inputs g_1 and g_2, $|PI(g_1)| = 1$ or $|PI(g_2)| = 1$ (or both). Thus, broom-like circuits have a particularly simple structure, and there is a direct correspondence between their prime implicants and their OR gates.

Lemma 4. *Every monotone multilinear circuit S for a union-free function f can be transformed into a broom-like formula for f with at most as many OR gates as S.*

Proof. We first transform S into a broom-like multilinear *circuit* for f without an increase in the number of OR gates. For this we need to know the following.

Claim 1. *Let g be an AND gate with inputs g_1 and g_2. Then there exists m in $PI(g_1) \cup PI(g_2)$ such that $m \subseteq p$ for all $p \in PI_g(f)$.*

Proof. Suppose there is no suitable m in $PI(g_1)$. We show that then there must be an m in $PI(g_2)$ such that $m \subseteq p$ for all p in $PI_g(f)$. Since there is no suitable m in $PI(g_1)$, $PI_g(f)$ cannot be empty. We pick some arbitrary p' in $PI_g(f)$. Because p' is an implicant of the function computed at g, there must be some m'_2 in $PI(g_2)$ such that $m'_2 \subseteq p'$. We prove that in fact

$$m'_2 \subseteq p \text{ for all } p \in PI_g(f) \ .$$

We distinguish two cases. First note that there must be an m'_1 in $PI(g_1)$ such that $m'_1 \subseteq p'$.

Case 1: $m'_1 \not\subseteq p$. Then there is some m_1 in $PI(g_1)$ such that $m_1 \subseteq p$, since p is an implicant of the function computed at g. Lemma 3(ii) yields that $p = (p' \setminus m'_1) \cup m_1$. Hence, $m'_2 \subseteq p$ because $m'_2 \subseteq p'$ and $m'_1 \cap m'_2 = \emptyset$ due to the multilinearity of the circuit.

Case 2: $m'_1 \subseteq p$. Note that there must be some $p'' \in PI_g(f)$ such that $m'_1 \not\subseteq p''$ because, by our initial assumption, $m'_1 \in PI(g_1)$ cannot be a suitable choice of m. Case 1 applies to p'' because $m'_1 \not\subseteq p''$, and we conclude $m'_2 \subseteq p''$. There must be some m''_1 in $PI(g_1)$ with $m''_1 \subseteq p''$. We use Lemma 3 again and find that $p = (p'' \setminus m''_1) \cup m'_1$. Hence, $m'_2 \subseteq p$ because $m'_2 \subseteq p''$ and $m''_1 \cap m'_2 = \emptyset$ due to the multilinearity of the circuit. □

We describe a modification that can be applied to every AND gate g which prevents S from being broom-like. Let g_1 and g_2 be the gates that feed g. The gate g prevents S from being broom-like, so $|PI(g_1)| > 1$ and $|PI(g_2)| > 1$. Let m be the monom in $PI(g_i)$ ($i \in \{1,2\}$) given by Claim 1. We add a new gate h that computes m (along with the corresponding subcircuit for this computation). Then we disconnect g from g_i and feed g from h instead of g_i. Clearly, the resulting circuit S' rejects all the inputs that the original circuit rejected, since we are dealing with monotone circuits. Because S' accepts all inputs that $S_{g \to 0}$ accepts, g must be necessary for any prime implicant p of S that is not a prime implicant of S'. But according to Claim 1, after the modification every such p remains an implicant of the function computed at g. This way we obtain a broom-like multilinear circuit S^* for f without an increase in the number of OR gates.

We now describe a way of transforming a broom-like multilinear circuit S^* for f into a broom-like formula F for f without an increase in the number of OR gates.

Claim 2. *Let g be a gate in S^*. Then*
(i) there is some m in $PI(g)$ such that $m \subseteq p$ for all p in $PI_g(f)$, or
(ii) there is some path w from g to the output of S^* that is consistent with all $p \in PI_g(f)$.

Proof. We show that if (i) does not hold, then (ii) follows. This proof has a similar structure compared to the proof of the first claim. Since (i) does not hold, $PI_g(f)$ cannot be empty. So there is some $p' \in PI_g(f)$ and some path w' from g to the output of S^* that is consistent with p'. We prove that in fact

$$w' \text{ is consistent with } p \text{ for all } p \in PI_g(f) .$$

We distinguish two cases. First note that there is some $m' \in PI(g)$ with $m' \subseteq p'$ because p' is an implicant of the function computed at g.

Case 1: $m' \not\subseteq p$. There must be some $m \in PI(g)$ such that $m \subseteq p$ because p is an implicant of the function computed at g. Lemma 3 yields that $p = (p' \setminus m') \cup m$ and that w' is consistent with p.

Case 2: $m' \subseteq p$. Because (i) does not hold, there is some p'' in $PI_g(f)$ such that $m' \not\subseteq p''$. Case 1 applies to p'' because $m' \not\subseteq p''$, and we conclude that w' is consistent with p''. There must be some m'' in $PI(g)$ with $m'' \subseteq p''$. Lemma 3 tells us that $p = (p'' \setminus m'') \cup m'$ and that w' is consistent with p. □

We now describe a modification that we carry out for every gate g of S^* with fanout larger than 1 in order to reduce its fanout to 1. As with the modification for making the circuit broom-like, we only have to check the prime implicants for which g is necessary. We distinguish two cases according to Claim 2.

Case 1: There is some m in $PI(g)$ such that $m \subseteq p$ for all p in $PI_g(f)$. We remove g from the circuit and replace all wires from g by subcircuits that each compute m. The resulting circuit computes a function that is clearly implied by all prime implicants p in $PI_g(f)$.

Case 2: There is some path w from g to the output of S^* that is consistent with all p in $PI_g(f)$. We then cut all wires stemming from g that are not on path w, i.e. we replace inputs to other gates from g by the constant 0. All prime implicants in $PI_g(f)$ are preserved because after the modification w is still consistent with all of them. To see this, note that, due to the multilinearity of the circuit, every AND gate on w can have at most one input that depends on g (such an input must be on w itself). □

The following lemma enables us to count the prime implicants of monotone functions by counting the OR gates of their monotone broom-like formulas.

Lemma 5. *Let F be a monotone broom-like formula computing f. Then F has at least $|PI(f)| - 1$ OR gates.*

Proof. The lemma can be proved by induction on the size of the formula. The details are omitted. □

Theorem 1 follows immediately from Lemma 4 together with Lemma 5.

To verify Theorem 2, we use Lemma 3(iii) in place of Lemma 3(ii). The construction of Lemma 4 then yields a broom-like formula for a function $\tilde{f}$ such that $PI\left(\tilde{f}\right) \supseteq PI(f)$. The lower bound then follows with Lemma 5.

4   An Upper Bound for the Clique Function

We will use the following lemma.

Lemma 6. *Let G be a graph with n vertices. If its complement $\overline{G}$ does not contain a triangle and does not have two edges which are not incident with a common vertex, then G has an $(n-1)$-clique.*

Proof. Suppose G does not have an $n-1$-clique. Then $\overline{G}$ is not a star. Suppose $\overline{G}$ does not have two edges which are not incident with a common vertex. Choose arbitrary distinct edges e_1 and e_2 in $\overline{G}$. Let e_1 and e_2 be incident with the common vertex u. Since $\overline{G}$ is not a star, there is an edge e_3 which is not incident with u. Let e_2 and e_3 be incident with the common vertex $v \neq u$. e_1 and e_3 must share the common vertex w, which is distinct from u and v. Hence, u, v and w form a triangle in $\overline{G}$. □

Proof of Theorem 3. To design the desired Π_3-formula for $CLIQUE(n, n-1)$ we use a code $C \subseteq A^k$ for some k over an alphabet A with a constant number of symbols (independent of n) such that $|C| \geq n$ and the minimal distance d of C is larger than $3k/4$. The existence of such a code of length $k = O(\log n)$ is guaranteed by the Gilbert bound (see e.g. [18]).

We assign to each vertex x (and hence, to each $(n-1)$-clique $V \setminus \{x\}$) its own codeword $code\,(x) \in C$. For each $1 \leq i \leq k$ and $a \in A$, let $S_{i,a}$ be the intersection of all $(n-1)$-cliques whose codes have symbol a in the i-th position. Hence,

$$S_{i,a} = V \setminus \{x \in V \mid code\,(x) \text{ has symbol } a \text{ in position } i\}\,. \tag{1}$$

Let $m_{i,a}$ be the monom consisting of all variables which correspond to edges having both their endpoints in $S_{i,a}$ (if $|S_{i,a}| \leq 1$, we set $m_{i,a} = 1$). We give the following Π_3-formula F for $CLIQUE\,(n, n-1)$:

$$F = \bigwedge_{i=1}^{k} \bigvee_{a \in A} m_{i,a}\,.$$

Every $(n-1)$-clique $V \setminus \{x\}$ with $code\,(x) = (a_1, \ldots, a_k)$ is accepted by the monom $\bigwedge_{i=1}^{k} m_{i,a_i}$ because the clique $V \setminus \{x\}$ contains all the cliques S_{i,a_i}, $i = 1, \ldots, k$. Hence, every $(n-1)$-clique is accepted by F. It remains to show that F does not accept any graph without an $(n-1)$-clique.

Let G be a graph accepted by F. Then there is a sequence $a_1, \ldots, a_k$ of symbols in A such that G is accepted by the monom $\bigwedge_{i=1}^{k} m_{i,a_i}$. For a vertex $x \in V$, let

$$P_x = \{i \mid code\,(x) \text{ has symbol } a_i \text{ in position } i\}\,.$$

Since the code C has minimal distance $d > 3k/4$, this implies that for every two distinct vertices x and y,

$$|P_x \cap P_y| \leq k - d < k/4\,. \tag{2}$$

Let $\{x, y\}$ be an edge of the complement graph $\overline{G}$. Then the edge $\{x, y\}$ cannot belong to any of the monoms $m_{1,a_1}, \ldots, m_{k,a_k}$, implying that $x \notin S_{i,a_i}$ or $y \notin S_{i,a_i}$ for all $i = 1, \ldots, k$. According to (1) this means that for all $i = 1, \ldots, k$, $code\,(x)$ or $code\,(y)$ has symbol a_i at position i. So we have

$$P_x \cup P_y = [k] = \{1, \ldots, k\}\,. \tag{3}$$

Now we are able to show that G must contain an $(n-1)$-clique. We do so by showing that its complement $\overline{G}$ does not contain a triangle and does not contain a pair of vertex disjoint edges. The result then follows with Lemma 6.

Assume first that $\overline{G}$ contains a triangle with vertices u, v and w. By (3), we have that $P_u \cup P_w = [k]$ and $P_v \cup P_w = [k]$. Taking the intersection of these two equations yields

$$(P_u \cap P_v) \cup P_w = [k]\,.$$

But by (2), we have that $|P_u \cap P_v| < k/4$, so $|P_w| > 3k/4$. Similarly we obtain $|P_u| > 3k/4$, implying that $|P_u \cap P_w| > k/2$, a contradiction with (2).

Assume now that $\overline{G}$ contains a pair of vertex disjoint edges $\{u, v\}$ and $\{x, y\}$. By (3), we have $P_u \cup P_v = [k]$ and $P_x \cup P_y = [k]$. Assume w.l.o.g. that $|P_u| \geq |P_v|$. Then $|P_u| \geq k/2$. We know that

$$P_u = P_u \cap [k] = P_u \cap (P_x \cup P_y) = (P_u \cap P_x) \cup (P_u \cap P_y)\,.$$

Assume w.l.o.g. that $|P_u \cap P_x| \geq |P_u \cap P_y|$. Then $|P_u \cap P_x| \geq |P_u|/2 \geq k/4$, a contradiction with (2). $\square$

5   Lower Bounds for Monotone Σ_4-Circuits

The following lemma shows that the union-freeness property is a special case of the disjointness property. This lemma names the properties of sufficiently disjoint functions that we exploit when proving the lower bound of Theorem 4.

Lemma 7. *Let $p_1, ..., p_r$ be prime implicants of a monotone Boolean function f and m be an implicant of f. Let f be k-homogeneous and k/r-disjoint.*
   (i) If $\bigcup_{i=1}^{r} p_i \supseteq m$, then $m \supseteq p_i$ for some i.
   (ii) If $x_1, \ldots, x_r$ are variables such that $x_i \in p_i$ and $x_i \notin p_j$ for $i \neq j$, then $\bigcup_{i=1}^{r} (p_i \setminus \{x_i\})$ is not an implicant of f.

Proof. (i) There must be some prime implicant p of f with $m \supseteq p$. Since $\bigcup_{i=1}^{r} p_i \supseteq p$, p must share at least k/r variables with some p_i. Because f is k/r-disjoint, this implies $p = p_i$. Claim (ii) is a direct consequence of (i).   □

The following lemma deals with Π_3-circuits with gates of *unbounded* fanin.

Lemma 8. *Let f be a monotone k-homogeneous and s-disjoint function. If $r \leq k/2s$ and h is a function such that $h \leq f$ (i.e., f evaluates to 1 if h does) and $|PI(h) \cap PI(f)| \geq r$, then any monotone Π_3-circuit for h with bottom fanin at most $s - 1$ must have top fanin at least $(k/2s)^r$.*

Proof. Let S be a monotone Π_3-circuit with top fanin a and bottom fanin at most $s - 1$. Let $F_1, \ldots, F_a$ be the functions computed by the Σ_2-subcircuits of S that are inputs to the AND gate which is the output gate of S. The function F computed by S can be represented in the form

$$F = \bigwedge_{i=1}^{a} F_i .$$

Let $a < (k/2s)^r$. We now show that the circuit S must then make an error, i.e. that $F \neq h$. For the sake of contradiction, assume that $F = h$. We choose arbitrary prime implicants $p_1, \ldots, p_r \in PI(h) \cap PI(f)$. Our goal is to pick $x_1 \in p_1, \ldots, x_r \in p_r$ suitable for Lemma 7(ii), yielding $F \neq h$.
   We pick the x_is in the order indicated by their indices. During this process we consider the preliminary monoms

$$m_t = \bigcup_{i=1}^{t} (p_i \setminus \{x_i\}) , \ t = 1, \ldots, r .$$

The preliminary monom m_t is available after the t-th step of the process. Finally, m_r is the desired implicant needed for the contradiction with Lemma 7(ii). Let A_t denote the set of indices of the functions F_i which are not implied by m_t, i.e. $i \in A_t$ iff m_t is not an implicant of F_i.

Claim 3. *There is always a choice of x_t in order to make*

$$|A_t| \leq \frac{|A_{t-1}|}{k/2s} .$$

Proof. We describe a choice of x_t that makes A_t sufficiently small. For every i in A_{t-1} we choose some $m_i \in PI(F_i)$ with $p_t \supseteq m_i$. Every F_i has such a prime implicant because p_t is a prime implicant of $h = F$. As x_t, we pick a variable of p_t that does not belong to any other of the prime implicants $p_1, \ldots, p_r$. Since each of the prime implicants can share at most $s - 1$ variables with each of the other $r - 1$ prime implicants, the prime implicant p_t has at least $k - (s-1)(r-1)$ variables which do not belong to any of the other prime implicants. Of these "private" variables of p_t, at most $s - 1$ can belong to some particular monom m_i we chose. If we add all the occurrences of the private variables of p_t in the monoms m_i together, we count at most $(s-1)|A_{t-1}|$ occurrences. Using that p_t has at least $k - (s-1)(r-1)$ private variables, we find that at least one of these variables is in not more than $|A_{t-1}| / (k/2s)$ of the chosen monoms. This sufficiently "rare" variable is our choice of x_t. Since only those $i \in A_{t-1}$ remain in A_t for which x_t belongs to the chosen monom m_i, the desired bound for $|A_t|$ follows.   □

We now finish the proof of Lemma 8. We start with $|A_0| = a < (k/2s)^r$. According to the claim, we can always choose the $x_1, \ldots, x_r$ such that A_r is empty. This means the finally constructed monom m_r is in fact an implicant of F.   □

Proof of Theorem 4 (Sketch). Let S be a monotone Σ_4-circuit with gates of fanin 2 which computes a monotone k-homogeneous s-disjoint function f. We assume that S has the smallest possible number of OR gates. The function f can be represented, according to the structure of S, as a disjunction of functions f_i which are computed by the Π_3-subcircuits of S: $f = \bigvee f_i$. Let f_i be computed by the Π_3-circuit S_i. Without loss of generality we can assume that no Π_1-subcircuit of any S_i depends on more than $s - 1$ variables, i.e. every S_i has bottom fanin at most $s - 1$ when regarded as a circuit of unbounded fanin.

Every prime implicant of f must be a prime implicant of at least one of the f_i. Let R be the largest number of prime implicants of f that are prime implicants of one particular $f_i = h$. Let h be computed by the Π_3-circuit $S_i = H$. Under our assumption that S is optimal with respect to the number of OR gates used, we conclude that the case $2 \leq R < k/2s$ cannot occur. Otherwise, by Lemma 8, H would require at least $(k/2s)^R - 1 \geq R^2 - 1$ OR gates and could be replaced by a simple two-level circuit requiring only $R - 1$ OR gates.

In the case $R = 1$ the circuit S is essentially a DNF and needs $|PI(f)| - 1$ OR gates. In the remaining case $R \geq k/2s$ Lemma 8 yields that H has a top fanin of at least $(k/2s)^{k/2s}$. The inequality $(k/2s)^{k/2s} \geq |PI(f)|$ is assumed by Theorem 4, so the desired lower bound for the number of OR gates in S follows.

Acknowledgement. I am grateful to Stasys Jukna for helpful discussions.

References

1. Razborov, A.: Lower bounds for the monotone complexity of some Boolean functions. Sov. Math., Dokl. **31** (1985) 354–357
2. Karchmer, M., Wigderson, A.: Monotone circuits for connectivity require superlogarithmic depth. SIAM J. Discrete Math. **3** (1990) 255–265
3. Razborov, A.: Applications of matrix methods to the theory of lower bounds in computational complexity. Combinatorica **10** (1990) 81–93
4. Gál, A.: A characterization of span program size and improved lower bounds for monotone span programs. Comput. Complexity **10** (2001) 277–296
5. Gál, A., Pudlák, P.: A note on monotone complexity and the rank of matrices. Inf. Process. Lett. **87** (2003) 321–326
6. Wegener, I.: The complexity of Boolean functions. Wiley-Teubner Series in Computer Science. John Wiley & Sons Ltd., Chichester (1987)
7. Dunne, P.E.: The complexity of Boolean networks. Volume 29 of APIC Studies in Data Processing. Academic Press Ltd., London (1988)
8. Savage, J.E.: Models of computation: Exploring the power of computing. Addison-Wesley Publishing Company, Reading, MA (1998)
9. Sengupta, R., Venkateswaran, H.: Multilinearity can be exponentially restrictive (preliminary version). Technical Report GIT-CC-94-40, Georgia Institute of Technology. College of Computing (1994)
10. Ponnuswami, A.K., Venkateswaran, H.: Monotone multilinear boolean circuits for bipartite perfect matching require exponential size. In Lodaya, K., Mahajan, M., eds.: FSTTCS. Volume 3328 of Lecture Notes in Computer Science., Springer (2004) 460–468
11. Wegener, I.: Branching programs and binary decision diagrams. Theory and applications. SIAM Monographs on Discrete Mathematics and Applications (2000)
12. Nisan, N., Wigderson, A.: Lower bounds on arithmetic circuits via partial derivatives. Comput. Complexity **6** (1996/97) 217–234
13. Raz, R.: Multi-linear formulas for permanent and determinant are of super-polynomial size. In Babai, L., ed.: STOC, ACM (2004) 633–641
14. Raz, R.: Multilinear-$NC_1 \neq$ Multilinear-NC_2. In: FOCS, IEEE Computer Society (2004) 344–351
15. Borodin, A., Razborov, A.A., Smolensky, R.: On lower bounds for read-k-times branching programs. Comput. Complexity **3** (1993) 1–18
16. Andreev, A.: On a method for obtaining lower bounds for the complexity of individual monotone functions. Sov. Math., Dokl. **31** (1985) 530–534
17. Grigni, M., Sipser, M.: Monotone complexity. In Paterson, M.S., ed.: Boolean function complexity. Volume 169 of London Mathematical Society Lecture Note Series. Cambridge University Press, Cambridge (1992) 57–75
18. van Lint, J.H.: Introduction to coding theory. Volume 86 of Graduate Texts in Mathematics. Springer-Verlag, New York (1982)

Bounds on the Power of Constant-Depth Quantum Circuits*

Stephen Fenner[1], Frederic Green[2], Steven Homer[3], and Yong Zhang[1]

[1] University of South Carolina, Columbia, SC, USA
{fenner, zhang29}@cse.sc.edu
[2] Clark University, Worcester, MA, USA
fgreen@black.clarku.edu
[3] Boston University, Boston, MA, USA
homer@bu.edu

Abstract. We show that if a language is recognized within certain error bounds by constant-depth quantum circuits over a finite family of gates, then it is computable in (classical) polynomial time. In particular, for $0 < \epsilon \leq \delta \leq 1$, we define $\mathbf{BQNC}^0_{\epsilon,\delta}$ to be the class of languages recognized by constant depth, polynomial-size quantum circuits with acceptance probability either $< \epsilon$ (for rejection) or $\geq \delta$ (for acceptance). We show that $\mathbf{BQNC}^0_{\epsilon,\delta} \subseteq \mathbf{P}$, provided that $1 - \delta \leq 2^{-2d}(1 - \epsilon)$, where d is the circuit depth.

On the other hand, we adapt and extend ideas of Terhal & DiVincenzo [1] to show that, for any family $\mathcal{F}$ of quantum gates including Hadamard and CNOT gates, computing the acceptance probabilities of depth-five circuits over $\mathcal{F}$ is just as hard as computing these probabilities for arbitrary quantum circuits over $\mathcal{F}$. In particular, this implies that $\mathbf{NQNC}^0 = \mathbf{NQACC} = \mathbf{NQP} = \mathrm{coC}_=\mathbf{P}$, where $\mathbf{NQNC}^0$ is the constant-depth analog of the class $\mathbf{NQP}$. This essentially refutes a conjecture of Green et al. that $\mathbf{NQACC} \subseteq \mathbf{TC}^0$ [2].

1 Introduction

This paper investigates $\mathbf{QNC}^0$ circuits, that is, families of quantum circuits with polynomial size and constant depth, using quantum gates of bounded width. Informally speaking, we show that

1. decision problems computed by $\mathbf{QNC}^0$ circuits within certain error bounds can be computed classically in polynomial time (Corollary 2), yet
2. computing probability amplitudes of $\mathbf{QNC}^0$ circuits exactly is as hard as computing arbitrary $\#\mathbf{P}$ functions, even when we restrict the circuits to depth three over a fixed finite set of quantum gates (Theorem 1).

---

* This work was supported in part by the National Security Agency (NSA) and Advanced Research and Development Agency (ARDA) under Army Research Office (ARO) contract numbers DAAD 19-02-1-0058 (for S. Homer, and F. Green) and DAAD 19-02-1-0048 (for S. Fenner and Y. Zhang).

M. Liśkiewicz and R. Reischuk (Eds.): FCT 2005, LNCS 3623, pp. 44–55, 2005.

The second result extends and improves work of Terhal & DiVincenzo [1]. Combined with recent results of Aaronson [3] it shows that

$$\mathbf{postBQNC^0} = \mathbf{postBQP} = \mathbf{PP} \quad \text{(Theorem 2)},$$

where $\mathbf{postBQNC^0}$ (respectively, $\mathbf{postBQP}$) is the class of languages computable by $\mathbf{QNC^0}$ (respectively, polynomial-size quantum) circuits with bounded error and postselection.

Much can be done with $O(\log n)$-depth quantum circuits ($\mathbf{QNC^1}$ circuits). Moore & Nilsson showed that many circuits can be parallelized to log depth—in particular, those implementing stabilizer error-correcting codes [4]. Cleve & Watrous were able to approximate the Quantum Fourier Transform over modulus 2^n with $O(\log n)$-depth circuits [5]. At first glance, $\mathbf{QNC^0}$ circuits appear extremely weak; one might expect that nothing can be computed with $\mathbf{QNC^0}$ circuits that cannot already be computed classically in polynomial time (even in $\mathbf{NC^0}$), since each output qubit can only be connected to a constant number of input qubits. This is certainly the case for decision problems if we restrict ourselves to observing a *single* output qubit, but surprisingly, this is still open in the more reasonable case where we observe several outputs at once, then apply some classical Boolean acceptance criterion on the results. The reason why these circuits are probably hard to simulate classically is that, although each individual output probability amplitude is easy to compute, it may be the case that different output qubits are correlated with each other, and the correlation graph has a high rate of expansion. (Terhal & DiVincenzo show that this is *not* the case for a depth two circuit, which can thus be simulated easily classically [1].)

To get use out of $o(\log n)$-depth quantum circuits, people have augmented them with quantum gates of unbounded fan-in. There are a number of unbounded-width gate classes studied in the literature, most being defined in analogy with classical Boolean gates. The generalized Toffoli gate (see Section 2.1) is the quantum equivalent of the unbounded Boolean AND-gate. Likewise, there are quantum equivalents of Mod-gates and threshold gates. One particular quantum gate corresponds to something taken almost completely for granted in Boolean circuits—fan-out. A fan-out gate copies the (classical) value of a qubit to several other qubits at once.[1] Using these gates, one can define quantum versions of various classical circuit classes: the previously mentioned $\mathbf{QNC^k}$ (Moore & Nilsson [4]), $\mathbf{QAC^k}$ and $\mathbf{QACC^k}$ (Moore [6], Green et al. [2]), and $\mathbf{QTC^k}$ are analogous to $\mathbf{NC^k}$, $\mathbf{AC^k}$, $\mathbf{ACC}$, and $\mathbf{TC^k}$, respectively. The case of particular interest is when $k = 0$. All these classes are allowed constant-width gates drawn from a finite family. The classes differ in the additional gates allowed.

Although small-depth quantum circuit classes are defined analogously to Boolean classes, their properties have turned out to be quite different from their classical versions. A simple observation of Moore [6] shows that the n-qubit fan-out gate and the n-qubit parity (Mod_2) gate are equivalent up to constant depth, i.e., each can be simulated by a constant-depth circuit using the other. This is

---

[1] There is no violation of the No-Cloning Theorem here; only the classical value is copied.

completely different from the classical case, where parity cannot be computed even with $\mathbf{AC^0}$ circuits where fan-out is unrestricted [7,8]. Later, Green et al. showed that all quantum Mod_q-gates are *constant-depth equivalent* for $q > 1$, and are thus all equivalent to fan-out. Thus $\mathbf{QNC^0_f} = \mathbf{QACC^0}(q) = \mathbf{QACC^0}$ for any $q > 1$. (The f subscript means, "with fan-out.") The classical analogs of these classes are provably different. In particular, classical Mod_p and Mod_q gates are not constant-depth equivalent if p and q are distinct primes, and neither can be simulated by $\mathbf{AC^0}$ circuits [9,10].

Building on ideas in [4], Høyer & Špalek used $\mathbf{QNC^0}$ circuits with unbounded fan-out gates to parallelize a sequence of commuting gates applied to the same qubits, and thus greatly reduced the depth of circuits for various purposes [11]. They showed that threshold gates can be approximated in constant depth this way, and they can be implemented exactly if Toffoli gates are also allowed. Thus $\mathbf{QTC^0_f} = \mathbf{QACC^0}$ as well. Threshold gates, and hence fanout gates, are quite powerful; many important arithmetic operations can be computed in constant depth with threshold gates [12]. This implies that the quantum Fourier transform—the quantum part of Shor's factoring algorithm—can be approximated in constant depth using fanout gates.

All these results rely for their practicality on unbounded-width quantum gates being available, especially fan-out or some (any) Mod gate. Unfortunately, making such a gate in the lab remains a daunting prospect; it is hard enough just to fabricate a reliable CNOT gate. Much more likely in the short term is that only one- and two-qubit gates will be available, which brings us back to the now more interesting question of $\mathbf{QNC^0}$. How powerful is this class?

A handful of hardness results about simulating constant-depth quantum circuits with constant-width gates are given by Terhal & DiVincenzo [1]. They show that if one can classically efficiently simulate, via sampling, the acceptance probability of quantum circuits of depth at least three using one- and two-qubit gates, then $\mathbf{BQP} \subseteq \mathbf{AM}$. They also showed that the polynomial hierarchy collapses if one can efficiently compute the acceptance probability exactly for such circuits. (Actually, a much stronger result follows from their proof, namely, $\mathbf{P} = \mathbf{PP}$.) Their technique uses an idea of Gottesman & Chuang for teleporting CNOT gates [13] to transform an arbitrary quantum circuit with CNOT and single-qubit gates into a depth-three circuit whose acceptance probability is proportional to, though exponentially smaller than, the original circuit. Their results, however, only hold on the supposition that depth-three circuits with *arbitrary* single-qubit and CNOT gates are simulatable. We weaken their hypothesis by showing how to produce a depth-three circuit with essentially the same gates as the original circuit. In addition, we can get by with only simple qubit state teleportation [14]. Our results immediately show that the class $\mathbf{NQNC^0}$ (the constant-depth analog of $\mathbf{NQP}$, see below), is actually the same as $\mathbf{NQP}$, which is known to be as hard as the polynomial hierarchy [15]. We give this result in Section 3.1. It underscores yet another drastic difference between the quantum and classical case: while $\mathbf{AC^0}$ is well contained in $\mathbf{P}$, $\mathbf{QNC^0}$ circuits (even just depth-three) can have amazingly complex behavior. Our result is also tight; Terhal & DiVincenzo showed that the acceptance probabilities of depth-two circuits over one- and two-qubit gates are computable in polynomial time.

In Section 3.2, we give contrasting upper bounds for $\mathbf{QNC^0}$-related language classes. We show that various bounded-error versions of $\mathbf{QNC^0}$ (defined below) are contained in $\mathbf{P}$. Particularly, $\mathbf{EQNC^0} \subseteq \mathbf{P}$, where $\mathbf{EQNC^0}$ is the constant-depth analog of the class $\mathbf{EQP}$ (see below). Our proof uses elementary probability theory, together with the fact that single output qubit measurement probabilities can be computed directly, and the fact that output qubits are "largely" independent of each other. In hindsight, it is not too surprising that $\mathbf{EQNC^0} \subseteq \mathbf{P}$. $\mathbf{EQNC^0}$ sets a severe limitation on the behavior of the circuit: it must accept with certainty or reject with certainty. This containment is more surprising (to us) for the bounded-error $\mathbf{QNC^0}$ classes.

We give open questions and suggestions for further research in Section 4.

An unabridged version of this paper is available as http://www.cse.sc.edu/~fenner/papers/eqnc.ps.

2 Preliminaries

2.1 Gates and Circuits

We assume prior knowledge of basic concepts in computational complexity: polynomial time, $\mathbf{P}$, $\mathbf{NP}$, as well as the counting class $\#\mathbf{P}$ [16]. Information can be found, for example, in Papadimitriou [17]. The class $\mathbf{C_{\neq}P}$ ($\mathrm{co}\mathbf{C_{=}P}$) was defined by Wagner [18]. One way of defining $\mathbf{C_{\neq}P}$ is as follows: a language L is in $\mathbf{C_{\neq}P}$ iff there are two $\#\mathbf{P}$ functions f and g such that, for all x, $x \in L \iff f(x) \neq g(x)$. $\mathbf{C_{\neq}P}$ was shown to be hard for the polynomial hierarchy by Toda & Ogihara [19].

We will also assume some background in quantum computation and the quantum circuit model. See Nielsen and Chuang [20] for a good reference of basic concepts and notation.

Our notion of quantum circuits is fairly standard (again see, for example, [20]): a series of quantum gates, drawn from some specified set of unitary operators, acting on some specified number of qubits, labeled $1, \ldots, m$. The first few qubits are considered *input* qubits, which are assumed to be in some basis state initially (i.e., classical input); the rest are ancillæ, each assumed to be in the $|0\rangle$ state initially. Thus the initial state of the qubits is $|x, 00\cdots 0\rangle$, for some binary string x. Some arbitrary set of qubits are specified as *output* qubits, and these qubits are measured in the computational basis at the final state. We assume that the sets of input and output qubits are part of the description of the circuit. For the purposes of computing decision problems, we will say that the circuit *accepts* its input if all the output qubits are observed to be 0 in the final state. Otherwise the circuit rejects. This acceptance criterion is simple, and it is essentially the one given in [2]. Although we do not study it here, one may consider other acceptance criteria, for example, feeding the observed outputs into an arbitrary polynomial time classical computation. To our knowledge, the power of such a model has not been studied.

We let $\Pr[C(x)]$ denote the probability that C accepts input x.

If C is any quantum circuit, it will be convenient for us to define $|C|$, the *size* of C, to be the number of output qubits plus the sum of the arities of all the gates occurring in the circuit. C may be laid out by partitioning its gates into *layers* $1, \ldots, d$, such that (i) gates in the same layer act on pairwise disjoint sets of qubits, and (ii) all gates in layer i are applied before any gates in layer $i + 1$, for $1 \leq i < d$. The *depth* of C is then the smallest possible value of d. The *width* of C is the number of qubits in C.

The standard quantum complexity classes (of languages) can be defined in terms of quantum circuit families. A quantum circuit family is a sequence $\{C_n\}_{n \geq 0}$ of quantum circuits, where each C_n has n inputs. We say that $\{C_n\}$ is *uniform* if there is a (classical) polynomial-time algorithm that outputs a description of C_n on input 0^n. The classes **BQP**, **EQP**, and **NQP** are defined using polynomial size quantum circuits with gates drawn from some fixed finite universal set of gates (see [21,22,23]). It was shown in [15,24] that **NQP** $=$ **C**$_{\neq}$**P**, and is thus hard for the polynomial hierarchy.

2.2   Complexity Classes Using QNC Circuits

The circuit class **QNC** was first suggested by Moore and Nilsson [4] as the quantum analog of the class **NC** of bounded fan-in Boolean circuits with polylogarithmic depth and polynomial size. We define the class **QNC**^{k} in the same fashion as definitions in Green et al. [2] with some minor modifications, needed for technical reasons (see the unabridged paper for more details).

Definition 1 ([4]). **QNC**^{k} *is the class of quantum circuit families* $\{C_n\}_{n \geq 0}$ *for which there exists a polynomial p such that each C_n contains n input qubits and at most $p(n)$ many ancillæ. Each C_n has depth $O(\log^k n)$ and uses only single-qubit gates and CNOT gates. The single-qubit gates must be from a fixed finite set.*

Next we define the language classes **NQNC**^{k} and **EQNC**^{k}. These are **QNC**^{k} analogs of the classes **NQP** and **EQP**, respectively.

Definition 2 ([2]). *Let $k \geq 0$ be an integer.*

- **NQNC**^{k} *is the class of languages L such that there is a uniform $\{C_n\} \in$* **QNC**^{k} *such that, for all x, $x \in L \iff \Pr[C_{|x|}(x)] > 0$.*
- **EQNC**^{k} *is the class of all L such that there is a uniform $\{C_n\} \in$ **QNC**^{k} so that, for all x, $\Pr[C_{|x|}(x)] \in \{0,1\}$, and $x \in L \iff \Pr[C_{|x|}(x)] = 1$.*

Bounded-error **QAC**^{k} classes were mentioned in [2], and one can certainly ask about similar classes for **QNC**^{k} circuits. It is not obvious that there is one robust definition of **BQNC**^{0}—perhaps because it is not clear how to reduce error significantly by amplification in constant depth.[2] In the next definition, we

---

[2] One can always reduce error *classically* by just running the circuit several times on the same input. In this case, the best definition of **BQNC**^{0} may be that the gap between the allowed accept and reject probabilities should be at least 1/poly.

will try to be as general as possible while still maintaining our assumption that **0** is the only accepting output.

Definition 3. *Let ϵ and δ be functions mapping (descriptions of) quantum circuits into real numbers such that, for all quantum circuits C, $0 < \epsilon(C) \leq \delta(C) \leq 1$. We write ϵ_C and δ_C to denote $\epsilon(C)$ and $\delta(C)$, respectively. $\mathbf{BQNC}^k_{\epsilon,\delta}$ is the class of languages L such that there is a uniform $\{C_n\} \in \mathbf{QNC}^k$ such that for any string x of length n,*

$$x \in L \implies \Pr[C_n(x)] \geq \delta_{C_n},$$
$$x \notin L \implies \Pr[C_n(x)] < \epsilon_{C_n}.$$

An interesting special case is when $\epsilon_C = \delta_C = 1$, that is, the input is accepted iff the circuit accepts with probability 1, and there is no promise on the acceptance probability. One might expect that, by the symmetry of the definitions, this class $\mathbf{BQNC}^0_{1,1}$ is the same as $\mathbf{NQNC}^0$, but it is almost certainly not, as we will see.

3   Main Results

3.1   Simulating QNC⁰ Circuits Exactly is Hard

Theorem 1. $\mathbf{NQNC}^0 = \mathbf{NQP} = \mathbf{C}_{\neq}\mathbf{P}$.

As a corollary, we essentially solve an open problem of Green et al. [2]. They conjectured that $\mathbf{NQACC} \subseteq \mathbf{TC}^0$, the class of constant-depth Boolean circuits with threshold gates.

Corollary 1. $\mathbf{NQNC}^0 = \mathbf{NQNC}^k = \mathbf{NQAC}^k = \mathbf{NQACC} = \mathbf{C}_{\neq}\mathbf{P}$ *for any $k \geq 0$. Thus, $\mathbf{NQACC} \not\subseteq \mathbf{TC}^0$ unless $\mathbf{C}_{\neq}\mathbf{P} = \mathbf{TC}^0$.*

Let B be the two-qubit Bell gate, consisting of a Hadamard gate applied to the first qubit, followed by a CNOT gate applied to the first qubit as control and the second as target. Also let

which produces the EPR state $(|00\rangle + |11\rangle)/\sqrt{2}$. Theorem 1 follows immediately from the following lemma (see the unabridged paper).

Lemma 1. *For any quantum circuit C using gates drawn from any family $\mathcal{F}$, there is a depth-three quantum circuit C' of size linear in $|C|$ using gates drawn from $\mathcal{F} \cup \{B, B^\dagger\}$ such that for any input x of the appropriate length, $\Pr[C'(x)] = 2^{-m}\Pr[C(x)]$, for some $m \leq 2|C|$ depending only on C. The middle layer of C' contains each gate in C exactly once and no others. The third layer contains only $B^\dagger$-gates, and the first layer contains only B-gates, which are used only to create EPR states.*

Proof. Our construction is a simplified version of the main construction in Terhal & DiVincenzo [1], but ours is stronger in two respects, discussed below: it works for any family of gates allowed in the original circuit, and introduces no new gates except B and $B^\dagger$. To construct $\mathcal{C}'$, we start with $\mathcal{C}$ and simply insert, for each qubit q of $\mathcal{C}$, the nonadaptive teleportation module

between any two consecutive quantum gates of $\mathcal{C}$ acting on q. No further gates involve the qubits r_1 and r_2 to the right of the $B^\dagger$-gate. This module, which lacks the usual corrective Pauli gates, is a nonadaptive version of the standard single-qubit teleportation circuit [14]. It faithfully teleports the state if and only if the observed output of the $B^\dagger$-gate on the right is 00 [1]. After inserting each teleportation circuit, the gates acting before and after it are now acting on different qubits. Further, any entanglement the qubit state has with other qubits is easily seen to be preserved in the teleported qubit. The input qubits of $\mathcal{C}'$ are those of $\mathcal{C}$. The output qubits of $\mathcal{C}'$ are of two kinds: output qubits corresponding to outputs of $\mathcal{C}$ are the *original outputs*; the other outputs are the *check* qubits (in pairs) coming from the added $B^\dagger$-gates. We'll call the measurement of each such pair a *check measurement.*

In addition to the gates in $\mathcal{C}$, $\mathcal{C}'$ uses only B-gates to make the initial EPR pairs and $B^\dagger$-gates for the check measurements. A sample transformation is shown below.

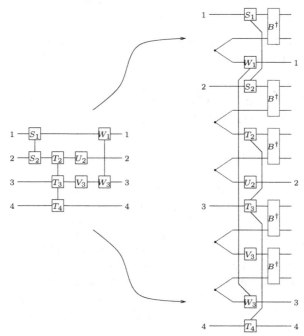

The circuit $\mathcal{C}$ on the left has five gates: S, T, U, V, and W, with subscripts added to mark which qubits each gate is applied to. The qubits in $\mathcal{C}'$ are numbered corresponding to those in $\mathcal{C}$. $\mathcal{C}'$ has depth three since it uses the first layer to make the initial EPR states and the third layer to rotate the Bell basis back to the computational basis. All the gates of $\mathcal{C}$ appear on the second layer. From the above constuction and the properties of the teleportation module, it is not hard to see that for all x of the appropriate length,

$$\Pr[\mathcal{C}(x)] = \Pr[\text{all orig. outputs of } \mathcal{C}' \text{ are } 0 \mid \text{all qubits are teleported correctly}]$$
$$= \Pr[\text{all orig. outputs of } \mathcal{C}' \text{ are } 0 \mid \text{all check meas. results are } 00]$$
$$= \Pr[\mathcal{C}'(x)] / \Pr[\text{all check meas. results are } 00],$$

since the check measurements are among the output measurements of $\mathcal{C}'$. Let k be the number of $B^\dagger$-gates on layer 3. Clearly, $k \leq |\mathcal{C}|$, and it is well-known that each check measurement will give 00 with probability $1/4$, independent of all other measurements. So the lemma follows by setting $m = 2k$.

Remarks. Using the gate teleportation apparatus of Gottesmann and Chuang [13], Terhal & DiVincenzo also construct a depth-three[3] quantum circuit $\mathcal{C}'$ out of an arbitrary circuit $\mathcal{C}$ (over CNOT and single-qubit gates) with a similar relationship of acceptance probabilities. However, they only teleport the CNOT gate, and their $\mathcal{C}'$ may contain single-qubit gates formed by compositions of arbitrary numbers of single-qubit gates from $\mathcal{C}$. (Such gates may not even be approximable in constant depth by circuits over a fixed finite family of gates.) When their construction is applied to each circuit in a uniform family, the resulting circuits are thus not generally over a finite gate set, even if the original circuits were. Our construction solves this problem by teleporting every qubit state in between all gates involving it. Besides B and $B^\dagger$, we only use the gates of the original circuit. We also are able to bypass the CNOT gate teleportation technique of [13], using instead basic single-qubit teleportation [14], which works with arbitrary gates.

Aaronson [3] has recently considered the power of postselection in quantum circuits, defining a "new" class **postBQP** and showing that, actually, **postBQP** = **PP**, thereby giving an new quantum proof of a celebrated result of Beigel, Reingold, & Spielman [25] stating that **PP** is closed under intersection. The class **postBQP** is defined using uniform families of polynomial-size quantum circuits C with two distinguished qubits: a regular output qubit q and a postselection qubit p, subject to the promise that in the final state, p is observed to be 1 with positive probability. $\Pr[C(x)]$ is then defined as the conditional probability that q is observed to be 1, given that p is observed to be 1. The acceptance criterion on $\Pr[C(x)]$ is the same as with **BQP**.

Clearly, the definition of **postBQP** remains unchanged if we swap the roles of 0 and 1 in the measurement of p, and if we allow multiple postselection qubits

---

[3] They count the depth as four, but they include the final measurement as an additional layer whereas we do not.

and condition $\Pr[C(x)]$ on their all being 0. Thus the following definition gives a reasonable constant-depth version of **postBQP**:

Definition 4. *A language L is in* **postBQNC^0** *if there is a uniform family $\{C_n\}_{n \geq 0}$ of constant-depth, polynomial-size quantum circuits (over some fixed finite universal set of gates) with output qubit q and postselection qubits $p_1, \ldots, p_k$ such that, for all inputs x of length n, $\Pr[p_1, \ldots, p_k$ of C_n are all 0$] > 0$, and the quantity $\Pr[C_n(x)] = \Pr[q$ is $1 \mid p_1, \ldots, p_k$ are all 0$]$ is in $[0, \frac{1}{3}] \cup [\frac{2}{3}, 1]$, and $x \in L$ iff $\Pr[C_n(x)] \geq 1/2$.*

The construction in the proof of Lemma 1 immediately yields the following (see the unabridged paper for a proof):

Theorem 2. **postBQNC^0** $=$ **postBQP**, *and hence* **postBQNC^0** $=$ **PP**.

3.2   Simulating QNC^0 Circuits Approximately Is Easy

In this section we prove that **BQNC$^0_{\epsilon, \delta}$** $\subseteq$ **P** for certain ϵ, δ. For convenience we will assume that all gates used in quantum circuits are either one- or two-qubit gates that have "reasonable" matrix elements—algebraic numbers, for instance. Our results apply more broadly, but they then require greater care to prove.

Definition 5. *Let C be a quantum circuit and let p and q be qubits of C. We say that q* depends *on p if there is a forward path in C starting at p before the first layer, possibly passing through gates, and ending at q after the last layer. More formally, we can define dependence by induction on the depth of C. For depth zero, q depends on p iff $q = p$. For depth $d > 0$, let C' be the same as C but missing the first layer. Then q depends on p (in C) iff there is a qubit r such that q depends on r (in C') and either $p = r$ or there is a gate on the first layer of C that involves both p and r.*

Definition 6. *For C a quantum circuit and q a qubit of C, define $D_q = \{p \mid q$ depends on $p\}$. If S is a set of qubits of C, define $D_S = \bigcup_{q \in S} D_q$. Let the dependency graph of C be the undirected graph with the output qubits of C as vertices, and with an edge between two qubits q_1 and q_2 iff $D_{q_1} \cap D_{q_2} \neq \emptyset$.*

If C has depth d, then the degree of its dependency graph is clearly less than 2^{2d}. The following lemma is straightforward.

Lemma 2. *Let C be a quantum circuit and let S and T be sets of output qubits of C. Fix an input x and bit vectors u and v with lengths equal to the sizes of S and T, respectively. Let $E_{S=u}$ (respectively $E_{T=v}$) be the event that the qubits in S (respectively T) are observed to be in the state u (respectively v) in the final state of C on input x. If $D_S \cap D_T = \emptyset$, then $E_{S=u}$ and $E_{T=v}$ are independent.*

For an algebraic number a, we let $\|a\|$ be the size of some reasonable representation of a. The results in this section follow from the next theorem.

Theorem 3. *There is a deterministic decision algorithm A which takes as input*

1. *a quantum circuit C with depth d and n input qubits,*
2. *a binary string x of length n, and*
3. *an algebraic number $t \in [0, 1]$,*

and behaves as follows: Let D be one plus the degree of the dependency graph of C. A runs in time $\text{Poly}(|C|, 2^{2^d}, \|t\|)$, and

- *if $\Pr[C(x)] \geq 1 - t$, then A accepts, and*
- *if $\Pr[C(x)] < 1 - Dt$, then A rejects.*

Note that since $D \leq 2^{2d}$, if $t < 2^{-2d}$, then A will reject when $\Pr[C(x)] < 1 - 2^{2d}t$.

Proof (of Theorem 3). On input (C, x, t) as above,

1. A computes the dependency graph $G = (V, E)$ of C and its degree, and sets D to be the degree plus one.
2. A finds a D-coloring $c : V \rightarrow \{1, \ldots, D\}$ of G via a standard greedy algorithm.
3. For each output qubit $q \in V$, A computes P_q—the probability that 0 is measured on qubit q in the final state (given input x).
4. For each color $i \in \{1, \ldots, D\}$, let $B_i = \{q \in V \mid c(q) = i\}$. A computes

$$P_{B_i} = \prod_{q \in B_i} P_q,$$

   which by Lemma 2 is the probability that all qubits colored i are observed to be 0 in the final state.
5. If $P_{B_i} \geq 1 - t$ for all i, the A accepts; otherwise, A rejects.

We show that A is correct. The proof that A runs in the given time is reasonably straightforward (see the unabridged paper for details). If $\Pr[C(x)] \geq 1 - t$, then for each $i \in \{1, \ldots, D\}$, we have $1 - t \leq \Pr[C(x)] \leq P_{B_i}$, and so A accepts. On the other hand, if $\Pr[C(x)] < 1 - Dt$, then

$$Dt < 1 - \Pr[C(x)] \leq \sum_{i=1}^{D} (1 - P_{B_i}),$$

so there must exist an i such that $t < 1 - P_{B_i}$, and thus A rejects.

Corollary 2. *Suppose ϵ and δ are polynomial-time computable, and for any quantum circuit C of depth d, $\delta_C = 1 - 2^{-2d}(1 - \epsilon_C)$. Then $\mathbf{BQNC}^0_{\epsilon,\delta} \subseteq \mathbf{P}$.*

Proof. For each C of depth d in the circuit family and each input x, apply the algorithm A of Theorem 3 with $t = 1 - \delta_C = 2^{-2d}(1 - \epsilon_C)$, noting that $D \leq 2^{2d}$.

The following two corollaries are instances of Corollary 2.

Corollary 3. *For quantum circuit C, let $\delta_C = 1 - 2^{-(2d+1)}$, where d is the depth of C. Then $\mathbf{BQNC}^0_{(1/2),\delta} \subseteq \mathbf{P}$.*

Corollary 4. $\mathbf{BQNC}^0_{1,1} \subseteq \mathbf{P}$, *and hence* $\mathbf{EQNC}^0 \subseteq \mathbf{P}$.

Corollary 4 can actually be proven directly: for each output, comput its probability of being 0. We accept iff all probabilities are 1.

We observe here that by a simple proof using our techniques, one can show that a $\mathbf{QNC}^0$ circuit cannot implement the generalized Toffoli gate, because its target depends on nonconstantly many input qubits.

4   Conclusions, Open Questions, and Further Research

Our upper bound results in Section 3.2 can be improved in certain ways. For example, the containment in $\mathbf{P}$ is easily seen to apply to $(\log\log n + O(1))$-depth circuits as well. Can we increase the depth further? We can perhaps also put $\mathbf{BQNC}^0_{\epsilon,\delta}$ into classes smaller than $\mathbf{P}$. $\mathbf{L}$ seems managable. How about $\mathbf{NC}^1$? Beyond that, we are unsure how $\mathbf{BQNC}^0_{\epsilon,\delta}$ compares with other complexity classes, and we currently know of no interesting language in this class, for any interesting values of ϵ and δ.

It would be nice to temper the admittedly bizarre conditions of Corollary 2, which are an artifact of the limitations of the algorithm in the proof of Theorem 3. Ideally, we would like to narrow the probability gap between ϵ and δ in Corollary 2 to $1/\text{poly}$, say, independent of the circuit depth, and still get containment in $\mathbf{P}$.

Finally, there are some general questions about whether we have the "right" definitions for these classes. For example, the accepting outcome is defined to be all outputs being 0. One can imagine more general accepting conditions, such as arbitrary classical polynomial-time postprocessing. If we allow this, then all our classes will obviously contain $\mathbf{P}$. If we allow arbitrary classical polynomial-time *pre*processing, then all our classes will be closed under Karp reductions.

We would like to thank David DiVincenzo, Mark Heiligman, and Scott Aaronson for helpful conversations.

References

1. Terhal, B.M., DiVincenzo, D.P.: Adaptive quantum computation, constant depth quantum circuits and arthur-merlin games. Quantum Information and Computation **4** (2004) 134–145
2. Green, F., Homer, S., Moore, C., Pollett, C.: Counting, fanout and the complexity of quantum ACC. Quantum Information and Computation **2** (2002) 35–65
3. Aaronson, S.: Quantum computing, postselection, and probabilistic polynomial-time (2004) Manuscript.
4. Moore, C., Nilsson, M.: Parallel quantum computation and quantum codes. SIAM Journal on Computing **31** (2002) 799–815

5. Cleve, R., Watrous, J.: Fast parallel circuits for the quantum Fourier transform. In: Proceedings of the 41st IEEE Symposium on Foundations of Computer Science. (2000) 526–536
6. Moore, C.: Quantum circuits: Fanout, parity, and counting (1999) Manuscript.
7. Ajtai, M.: Σ_1^1 formulæ on finite structures. Annals of Pure and Applied Logic **24** (1983) 1–48
8. Furst, M., Saxe, J.B., Sipser, M.: Parity, circuits, and the polynomial time hierarchy. Mathematical Systems Theory **17** (1984) 13–27
9. Razborov, A.A.: Lower bounds for the size of circuits of bounded depth with basis $\{\&, \oplus\}$. Math. Notes Acad. Sci. USSR **41** (1987) 333–338
10. Smolensky, R.: Algebraic methods in the theory of lower bounds for Boolean circuit complexity. In: Proceedings of the 19th ACM Symposium on the Theory of Computing. (1987) 77–82
11. Høyer, P., Špalek, R.: Quantum circuits with unbounded fan-out. In: Proceedings of the 20th Symposium on Theoretical Aspects of Computer Science. Volume 2607 of Lecture Notes in Computer Science., Springer-Verlag (2003) 234–246
12. Siu, K.Y., Bruck, J., Kailath, T., Hofmeister, T.: Depth efficient neural networks for division and related problems. IEEE Transactions on Information Theory **39** (1993) 946–956
13. Gottesman, D., Chuang, I.L.: Demonstrating the viability of universal quantum computation using teleportation and single-qubit operations. Letters to Nature **402** (1999) 390–393
14. Bennett, C.H., Brassard, G., Crépeau, C., Jozsa, R., Peres, A., Wootters, W.: Teleporting an unknown quantum state via dual classical and Einstein-Podolsky-Rosen channels. Physical Review Letters **70** (1993) 1895–1899
15. Fenner, S., Green, F., Homer, S., Pruim, R.: Determining acceptance possibility for a quantum computation is hard for the polynomial hierarchy. Proceedings of the Royal Society London A **455** (1999) 3953–3966
16. Valiant, L.: The complexity of computing the permanent. Theoretical Computer Science **8** (1979) 189–201
17. Papadimitriou, C.H.: Computational Complexity. Addison-Wesley (1994)
18. Wagner, K.: The complexity of combinatorial problems with succinct input representation. Acta Informatica **23** (1986) 325–356
19. Toda, S., Ogiwara, M.: Counting classes are at least as hard as the polynomial-time hierarchy. SIAM Journal on Computing **21** (1992) 316–328
20. Nielsen, M.A., Chuang, I.L.: Quantum Computation and Quantum Information. Cambridge University Press (2000)
21. Bernstein, E., Vazirani, U.: Quantum complexity theory. SIAM Journal on Computing **26** (1997) 1411–1473
22. Bennett, C.H., Bernstein, E., Brassard, G., Vazirani, U.: Strengths and weaknesses of quantum computation. SIAM Journal on Computing **26** (1997) 1510–1523
23. Adleman, L.M., DeMarrais, J., Huang, M.D.A.: Quantum computability. SIAM Journal on Computing **26** (1997) 1524–1540
24. Yamakami, T., Yao, A.C.C.: $NQP_C = co\text{-}C_{=}P$. Information Processing Letters **71** (1999) 63–69
25. Beigel, R., Reingold, N., Spielman, D.: PP is closed under intersection. In: Proceedings of the 23rd annual ACM Symposium on Theory of Computing. (1991) 1–9

Biautomatic Semigroups

Michael Hoffmann and Richard M. Thomas

Department of Computer Science, University of Leicester,
Leicester LE1 7RH, UK

Abstract. We consider biautomatic semigroups. There are two different definitions of a biautomatic structure for a group in the literature; whilst these definitions are not equivalent, the idea of a biautomatic group is well defined, in that a group possesses one type of biautomatic structure if and only if it possesses the other. However the two definitions give rise to different notions of biautomaticity for semigroups and we study these ideas in this paper. In particular, we settle the question as to whether automaticity and biautomaticity are equivalent for semigroups by giving examples of semigroups which are automatic but not biautomatic.

Keywords: Automata and formal languages, semigroups, automatic, biautomatic.

1 Introduction

The notion of automaticity has been widely studied in groups (see [2,6,14] for example) and some progress has been made in understanding this notion in the wider context of semigroups (see [4,5,8] for example).

Initially the definition of automaticity in semigroups took the multiplication by generators and the insertion of paddings both to be on the right (as in the conventional definition for groups). Whilst these choices make no difference for groups, it was pointed out in [9] that the different conventions do make a significant difference for semigroups.

In this paper we consider biautomatic semigroups. The notion of biautomaticity in groups has been a subject of considerable interest; however, it is still an open question as to whether an automatic group is necessarily biautomatic.

There are (at least) two different definitions of a biautomatic structure for a group in the literature; whilst these definitions are not equivalent, the idea of a biautomatic group is well defined, in that a group possesses one type of biautomatic structure if and only if it possesses the other (see Proposition 11 below). However the two definitions give rise to different notions of biautomaticity for semigroups (although they do coincide for cancellative semigroups; see Theorem 1).

We study these ideas in Section 4. In particular, we settle the question as to whether automaticity and biautomaticity are equivalent for semigroups by giving an example of a semigroup which is automatic (in all the ways that have been proposed so far) but not biautomatic in any of the ways we introduce here (see Theorem 2).

M. Liśkiewicz and R. Reischuk (Eds.): FCT 2005, LNCS 3623, pp. 56–67, 2005.

2   Preliminaries

For any finite set (or alphabet) A, let A^+ denote the set of all non-empty words over A and A^* denote the set of all words over A including the empty word ϵ. A set of words in A^* is called a *language* over A. For any word α in A^*, let $|\alpha|$ denote the length of α (where $|\epsilon|$ is taken to be 0). For any $k \in \mathbb{N}$, let A^k denote the set of all words α in A^* with $|\alpha| = k$, $A^{\leqslant k}$ the set of all words α in A^* with $|\alpha| \leqslant k$, and $A^{<k}$ the set of all words α in A^* with $|\alpha| < k$.

We now turn to semigroups; we refer the reader to [10,12] as general references. If S is a semigroup and $A \subseteq S$ is a set of generators of S, then there is a natural homomorphism $\theta : A^+ \to S$ where each word α in A^+ is mapped to the corresponding element of S; we write $S = \langle A \rangle$ if S is generated by A. Note that we will be considering semigroup generating sets even when our semigroup S is a monoid or group (i.e. sets that generate S as a semigroup). We will normally be concerned with finite sets A, so that the semigroup S is finitely generated.

Where there is no danger of confusion, we will sometimes suppress the reference to θ, simply writing α for the element of the semigroup represented by the word α. In this context, if α and β are elements of A^+, we will write $\alpha \equiv \beta$ if α and β are identical as words, and $\alpha = \beta$ if α and β represent the same element of S (i.e. if $\alpha\theta = \beta\theta$). If we wish to stress which semigroup we are working in, we will write $\alpha =_S \beta$ if α and β represent the same element of the semigroup S. We may also write $\alpha = s$ (or $\alpha =_S s$), where $\alpha \in A^+$ and $s \in S$, which says that $\alpha\theta = s$ in S.

A semigroup can be extended by adding an identity; we write S^1 for the semigroup obtained by adjoining an external identity element to S.

As in the case of automatic groups (see [6] for example), we consider automata accepting pairs (α, β) with $\alpha, \beta \in A^+$. If $\alpha \equiv a_1 a_2 \ldots a_n$ and $\beta \equiv b_1 b_2 \ldots b_m$, this is accomplished by having an automaton with input alphabet $A \times A$ and reading pairs $(a_1, b_1), (a_2, b_2)$, and so on. To deal with the case where $n \neq m$, we introduce a *padding symbol* \$. More formally, we define $\delta_A^R : A^* \times A^* \to A(2, \$)^*$, where $\$ \notin A$ and $A(2, \$) = (A \cup \{\$\}) \times (A \cup \{\$\}) - \{(\$, \$)\}$, by

$$(\alpha, \beta)\delta_A^R = \begin{cases} (a_1, b_1) \ldots (a_n, b_n) & \text{if } n = m \\ (a_1, b_1) \ldots (a_n, b_n)(\$, b_{n+1}) \ldots (\$, b_m) & \text{if } n < m \\ (a_1, b_1) \ldots (a_m, b_m)(a_{m+1}, \$) \ldots (a_n, \$) & \text{if } n > m. \end{cases}$$

We also want to have a map that inserts paddings on the left instead of on the right; so we define $\delta_A^L : A^* \times A^* \to A(2, \$)^*$ by

$$(\alpha, \beta)\delta_A^L = ((\alpha^{rev}, \beta^{rev})\delta_A^R)^{rev},$$

where α^{rev} denotes the reversal of the word α.

3   Notions of Automaticity

We now recall the definitions of the various notions of automaticity in semigroups; these will be central to the ideas explored in this paper.

If S is a semigroup generated by a finite set A, L is a regular subset of A^+ and $a \in A \cup \{\epsilon\}$, then we define:

$$L_a^\$ = \{(\alpha, \beta)\delta_A^R : \alpha, \beta \in L, \alpha a = \beta\}; \quad {}^\$L_a = \{(\alpha, \beta)\delta_A^L : \alpha, \beta \in L, \alpha a = \beta\};$$
$$_aL^\$ = \{(\alpha, \beta)\delta_A^R : \alpha, \beta \in L, a\alpha = \beta\}; \quad {}_a^\$L = \{(\alpha, \beta)\delta_A^L : \alpha, \beta \in L, a\alpha = \beta\}.$$

Given this, we make the following definition:

Definition 1. *Let S be a semigroup generated by a finite set A and suppose that L is a regular language over A that maps onto S. The pair (A, L) is said to be*
  a left-left automatic structure *for S if ${}_a^\$L$ is regular for all $a \in A \cup \{\epsilon\}$;*
  a left-right automatic structure *for S if ${}^\$L_a$ is regular for all $a \in A \cup \{\epsilon\}$;*
  a right-left automatic structure *for S if ${}_aL^\$$ is regular for all $a \in A \cup \{\epsilon\}$;*
  a right-right automatic structure *for S if $L_a^\$$ is regular for all $a \in A \cup \{\epsilon\}$.*
A finitely generated semigroup S is said to be
  left-left automatic *if it has a left-left automatic structure;*
  left-right automatic *if it has a left-right automatic structure;*
  right-left automatic *if it has a right-left automatic structure;*
  right-right automatic *if it has a right-right automatic structure.*

Note that the notion of "automatic" as defined in [4] (for example) is equivalent to the notion of "right-right automatic" here.

Remark 1. A natural question to ask is that of the relationship between the four notions given in Definition 1. In general, they are independent of each other. To be more precise, for any subset T of the four notions, there is a semigroup satisfying all the notions in T but none of the notions outside T; furthermore, there is an automatic structure which is an automatic structure for all the notions in T simultaneously [9]. □

Let $S = \langle A : R \rangle$ be a semigroup, where $R \subseteq A^+ \times A^+$, and let R^{rev} denote

$$\{(\omega_1, \omega_2) : (\omega_1^{rev}, \omega_2^{rev}) \in R\}.$$

We let S^{rev} denote the semigroup $\langle A : R^{rev} \rangle$. We recall Lemma 3.4 of [9]:

Proposition 1. *If S is a finitely generated semigroup then:*
  S is left-left automatic if and only if S^{rev} is right-right automatic.
  S is left-right automatic if and only if S^{rev} is right-left automatic.

The proof of Proposition 5.3 in [4] generalizes to all four types of automaticity and we have:

Proposition 2. *Let (A, L) be an X-Y automatic structure for a semigroup S for some X, $Y \in \{\text{left, right}\}$, K be a regular subset of L and suppose that K maps onto S; then (A, K) is an X-Y automatic structure for S.*

Remark 2. Let S be a semigroup and let (A, L) be a pair which is simultaneously an X-Y automatic structure for S for some pairs (X, Y) with X, Y $\in$ {left, right}. It follows from Proposition 2 that, if there exists a regular subset K of L such that K maps onto S, then (A, K) is also an X-Y automatic structure for S for the same set of pairs (X, Y). □

Let (A, L) be an X-Y automatic structure for S for some X, Y $\in$ {left, right}. We say that (A, L) is an *automatic structure with uniqueness* for S if L maps bijectively onto S. As in [9] one can use Proposition 2 to deduce:

Proposition 3. *If S is a semigroup with an X-Y automatic structure (A, L) for some X, Y $\in$ {left, right}, then there exists a X-Y automatic structure (A, K) with uniqueness for S with $K \subseteq L$.*

We will use the following result frequently in this paper (see Corollary 4.2 of [9]):

Proposition 4. *Let $L \subseteq A^* \times A^*$ and suppose there is a constant k such that $||\alpha| - |\beta|| \leqslant k$ for all $(\alpha, \beta) \in L$; then the set $\{(\alpha, \beta)\delta_A^L : (\alpha, \beta) \in L\}$ is regular if and only if the set $\{(\alpha, \beta)\delta_A^R : (\alpha, \beta) \in L\}$ is regular.*

The notions of automaticity are connected in groups (see Lemma 5.6 of [9]):

Proposition 5. *Let G be a group and A be a finite generating set for G which is closed under inversion; then (A, L) is a right-right automatic structure for G if and only if (A, L^{-1}) is a left-left automatic structure for G.*

Here L^{-1} denotes the language obtained from L by reversing each word and replacing each symbol by the corresponding inverse symbol. Using Proposition 5, it was shown in [9] that all four notions of automaticity coincide for groups.

We also recall Lemma 7.7 from [9]:

Proposition 6. *Let S be a cancellative semigroup with a right-right automatic structure (A, L); then there exists a constant N such that, for any $\alpha \in L$, any $a \in A \cup \{\epsilon\}$ and any $s \in S$ with $s = \alpha a$ or $sa = \alpha$, we have the following:*

1. *there exists $\beta \in L$ such that $|\beta| \leqslant |\alpha| + N$ and $s = \beta$, and*
2. *if there exists $\gamma \in L$ such that $|\gamma| > |\alpha| + N$ and $\gamma = s$, then there exist infinitely many $\eta \in L$ with $\eta = s$.*

Remark 3. As noted in [9], Proposition 6 can be generalized to the other notions of automaticity with the proviso that, when we are considering right-left or left-left automaticity, we have a constant N such that, for any $\alpha \in L$ and any $s \in S$ with $s = a\alpha$ or $as = \alpha$ for some $a \in A \cup \{\epsilon\}$, we have that the conclusions of Proposition 6 hold.   □

In cancellative semigroups, we do have some connections between these notions as follows (see Lemma 8.2 and Remark 8.3 of [9]):

Proposition 7. *If S is a cancellative semigroup with a right-right and left-left automatic structure (A, L) then (A, L) is also a right-left and a left-right automatic structure.*

Proposition 8. *For any cancellative semigroup S, we have*

$$S \text{ right-left automatic} \iff S \text{ left-left automatic,}$$
$$S \text{ left-right automatic} \iff S \text{ right-right automatic.}$$

The following result gives a new sufficient condition for right-right automaticity:

Proposition 9. *Let S be a semigroup defined by a presentation of the form $\langle \{x\} \cup A : R \rangle$ with A finite, $x \notin A$ and*

$$R = \{(\alpha_1 x, x\beta_1), (\alpha_2 x, x\beta_2), \;\; \ldots\ldots \;, (\alpha_n x, x\beta_n)\}$$

where $\alpha_i, \beta_i \in A^+$ for each i. Then S is right-right automatic if R satisfies the following two hypotheses:

(H1) $\beta_i \gamma \not\equiv \beta_j$ *for all i and j with $1 \leqslant i, j \leqslant n$ and $i \neq j$ and for all $\gamma \in A^*$;*
(H2) $\alpha_i \not\equiv \gamma_1 \beta_j \gamma_2$, *where $\mu\gamma_1 \equiv \beta_k$ and $\gamma_1 \in A^+$ and $\gamma_2, \mu \in A^*$, for all i, j and k with $1 \leqslant i, j, k \leqslant n$.*

The proof of Proposition 9 is rather lengthy and we do not include it here. The basic idea is to move all the occurrences of x in a word as far as possible to the right to get a set of normal forms that maps bijectively to the semigroup. After post-multiplying a word ω in normal form by a generator, the new word can be transformed to a word in normal form by using at most one relation for each occurrence of x in ω. This is enough to determine a set of finite state automata which only accept pairs of the form $(\omega, \omega a)$ with ω in normal form and a a generator.

Whilst the hypotheses may seem rather technical, Proposition 9 will prove to be useful when we construct semigroups that are automatic but not biautomatic (see Corollary 1 and Theorem 2 in Section 5).

4   Biautomatic Semigroups

There are (at least) two equivalent definitions of a biautomatic group in the literature. Recalling that when we talk of a "generating set" for a group G, we are referring to a set that generates G as a semigroup, we first have (as in [7] for example):

Definition 2. *Let G be a group with a finite generating set A. If L is a subset of A^+, then the pair (A, L) is said to be a biautomatic structure for G if L is regular, $L_\epsilon^\$$ is regular, and, for all $a \in A$, both $L_a^\$$ and ${}_a L^\$$ are regular. A group G is said to be biautomatic if it has a biautomatic structure.*

Remark 4. Saying that (A, L) is a biautomatic structure for G in the sense of Definition 2 is equivalent to saying that (A, L) is simultaneously both a right-right and a right-left automatic structure for G in our terminology. □

There is another definition of biautomaticity (see [6] for example); to distinguish these two concepts, we will call this one (for the moment) biautomatic[#]:

Definition 3. *Let G be a group with a finite subset A that generates G and which is closed under taking inverses. If L is a subset of A^+, then the pair (A, L) is said to be a biautomatic[#] structure for G if both (A, L) and (A, L^{-1}) are automatic structures for G. A group is said to be biautomatic[#] if it has a biautomatic[#] structure.*

Remark 5. Remember that "automatic" when applied to groups means right-right automatic in our terminology. Insisting that (A, L^{-1}) is a right-right automatic structure for G is equivalent to insisting that (A, L) is a left-left automatic structure for G by Proposition 5. So (A, L) is a biautomatic$^\sharp$ structure for G if and only if A is closed under taking inverses and (A, L) is both a right-right and a left-left automatic structure for G. Given that the notion of being a right-right or left-left automatic structure does not necessitate A being closed under taking inverses, one could drop that assumption with this formulation. □

The relationship between biautomatic and biautomatic$^\sharp$ structures is summed up in the following consequence of Proposition 7 (given Remarks 4 and 5):

Proposition 10. *If G is a group with a biautomatic$^\sharp$ structure (A, L), then (A, L) is a biautomatic structure for G.*

It was pointed out in Remark 8.4 of [9] that, if $G = \mathbb{Z} = \langle x : \rangle$, $A = \{x, X\}$ where X represents x^{-1} and $L = \{x\}^*\{xX\}^* \cup \{X\}^*\{xX\}^*$, then (A, L) is a right-right and right-left automatic structure for G but not a left-left automatic structure; so a biautomatic structure (A, L) for a group is not necessarily a biautomatic$^\sharp$ structure. Notwithstanding this, we do have (rather reassuringly):

Proposition 11. *A group G is biautomatic if and only if G is biautomatic$^\sharp$.*

Proof. If G is biautomatic$^\sharp$, then G is biautomatic by Proposition 10; so it remains to prove the converse.

If G is biautomatic then we can find a biautomatic structure (A, L) with uniqueness by Remark 2. Note that (A, L) is a right-right and a right-left automatic structure as in Remark 4; in particular, $_aL^\$$ is regular for each $a \in A$. By Remark 3, there is a constant k such that, if $(\alpha, \beta)\delta_A^R \in {}_aL^\$$, then $||\alpha| - |\beta|| \leqslant k$. Using Proposition 4, we see that $_a^\$L$ is regular for each $a \in A$, so that (A, L) is also a left-left automatic structure. So (A, L) is also a biautomatic$^\sharp$ structure by Remark 5 and then G is biautomatic$^\sharp$. □

As in [13] (for example), where the notion of a biautomatic structure was used to define biautomatic monoids, we take the two definitions of a biautomatic structure for groups to define various notions of biautomaticity in semigroups:

Definition 4. *If S is a finitely generated semigroup then (A, L) is said to be*

- *a left-biautomatic structure if (A, L) is both a left-left and a left-right automatic structure;*
- *a right-biautomatic structure if (A, L) is both a right-left and a right-right automatic structure.*

A finitely generated semigroup S is said to be

- *left-biautomatic if it has a left-biautomatic structure;*
- *right-biautomatic if it has a right-biautomatic structure.*

Note that the notion of a biautomatic structure for a group is the same as the notion of a right-biautomatic structure for a semigroup here. We are using the terms "right" and "left" to denote the sides on which we take the paddings (on the right both times or on the left both times); we do not need to refer to the side on which we take the multiplications as the point of biautomaticity is that we have a structure where we can perform multiplications by generators on both the left and the right.

Now a semigroup S is left-biautomatic if and only if S^{rev} is right-biautomatic. Since G is isomorphic to G^{rev} for a group G, the two notions coincide for groups. In fact, we can extend this to cancellative semigroups, as the next result shows:

Proposition 12. *A cancellative semigroup S is right-biautomatic if and only if S is left-biautomatic.*

Proof. Let S be right-biautomatic. By Proposition 3 and Remark 4, there exists a pair (A, L) such that (A, L) is both a right-right and a right-left automatic structure with uniqueness for S. By Proposition 6 and Remark 3, we have that there exists $k \in \mathbb{N}$ such that $| \, |\alpha| - |\beta| \, | < k$ for all α and β with $(\alpha, \beta)\delta_A^R \in L_a^\$$ and for all α and β with $(\alpha, \beta)\delta_A^R \in {}_aL^\$$ (where $a \in A$). By Proposition 4 we have that (A, L) is both a left-right and left-left automatic structure for S; hence S is left-biautomatic.

The proof of the converse is similar. $\square$

However, given Remark 1, we see that Proposition 12 does not hold for non-cancellative semigroups.

In a similar way we can define notions of biautomaticity in semigroups based on the notion of biautomatic$^\sharp$ in groups:

Definition 5. *If S is a finitely generated semigroup then (A, L) is said to be*

- *a cross-biautomatic structure if (A, L) is both a right-left and a left-right automatic structure for S;*
- *a same-biautomatic structure if (A, L) is both a left-left and a right-right automatic structure for S.*

A finitely generated semigroup S is said to be

- *cross-biautomatic if it has a cross-biautomatic structure;*
- *same-biautomatic if it has a same-biautomatic structure.*

Note that the notion of a biautomatic$^\sharp$ structure for a group is the same as the notion of a same-biautomatic structure for a semigroup. We use "cross" to denote the fact that the padding is on the opposite side to the multiplication and "same" to denote that it is on the same side as the multiplication.

Remark 6. The example given in [13] of a right-biautomatic monoid M that has no finite complete rewriting system is also left-biautomatic, same-biautomatic and cross-biautomatic. The right-right and right-left automatic structure (A, L) for M given in [13] satisfies the property that, for $\alpha, \beta \in L$ with $\alpha a = \beta$ or $a\alpha = \beta$, the difference between $|\alpha|$ and $|\beta|$ is globally bounded. Proposition 4 gives that (A, L) is also a left-right and left-left automatic structure for M. $\square$

A semigroup S is cross-biautomatic if and only if S^{rev} is cross-biautomatic and a semigroup S is same-biautomatic if and only if S^{rev} is same-biautomatic. As in the case of right-biautomatic and left-biautomatic, the notions of cross-biautomatic and same-biautomatic coincide in groups; in fact we have:

Proposition 13. *A cancellative semigroup S is cross-biautomatic if and only if S is same-biautomatic.*

Proof. Let S be same-biautomatic. By Proposition 3, there exists a pair (A, L) such that (A, L) is both a right-right and a left-left automatic structure with uniqueness for S. By Proposition 6 and Remark 3, we have that there exists $k \in \mathbb{N}$ such that $| |\alpha| - |\beta| | < k$ for all α and β such that $(\alpha, \beta)\delta_A^R \in L_a^\$$ and for all α and β such that $(\alpha, \beta)\delta_A^L \in {}_a^\$L$ (where $a \in A$). By Proposition 4 we have that (A, L) is both a left-right and a right-left automatic structure for S; hence S is cross-biautomatic.

The proof of the converse is similar. □

We now tie these together and show that all these notions of biautomaticity are equivalent for cancellative semigroups:

Theorem 1. *If S is a cancellative semigroup, then the following are equivalent:*

1. *S is right-biautomatic;*
2. *S is left-biautomatic;*
3. *S is same-biautomatic;*
4. *S is cross-biautomatic.*

Proof. By Propositions 12 and 13, we only need show that S is right-biautomatic if and only if S is same-biautomatic. The proof now follows the proofs of those two results. If S is right-biautomatic we have a pair (A, L) which is a right-right and right-left automatic structure with uniqueness, and we then use Remark 3 and Proposition 4 to show that (A, L) is also a left-left automatic structure. So S is same-biautomatic.

The proof of the converse is similar. □

Remark 7. Theorem 1 does not hold for non-cancellative semigroups. Using Remark 1 we see that we can have a semigroup that satisfies any one of our four notions of biautomaticity but none of the other three. □

5   Examples

We now consider semigroup presentations of the form

$$\wp = \langle a, b : ab^m = b^n a \rangle.$$

If we consider $\wp$ as a group presentation, we get the Baumslag-Solitar group $B(m, n)$ which is known to be automatic if and only if $m = n$ (see [6] for example). As in [11], we will refer to semigroups defined by presentations of this form as *Baumslag-Solitar semigroups*.

Proposition 14. *The Baumslag-Solitar semigroup S defined by the presentation*

$$\wp \;=\; \langle a, b : ab^m = b^n a \rangle$$

with $m > n$ is not right-left automatic.

Proof. If S is a right-left automatic semigroup, then $T = S^1$ is a right-left automatic monoid by Proposition 5.1 of [9]. As noted in [11], T embeds in the group defined by $\wp$ by the results in [1], and so T is cancellative.

In [5] it was shown that, if M is a monoid which is right-right automatic, and if A is any finite (semigroup) generating set for M, then there is a regular language L over A such that (A, L) is a right-right automatic structure for M. This was generalized to the other three notions of automaticity in [9]. So there exists an automatic structure for T over every generating set of T.

Let $A = \{a, b, e\}$ (where e represents the identity of T) and suppose that L is a regular language over A such that (A, L) is a right-left automatic structure for T. We can further assume, by Proposition 3, that (A, L) is a right-left automatic structure with uniqueness. Let N be the number of states in a finite state automaton accepting L. Let $\varphi : A^* \to L$ be defined by $\alpha\varphi \equiv \beta$ where $\beta \in L$ and $\alpha =_T \beta$. We have that

$$|b^k \varphi| \geqslant k \quad \text{for all } k. \tag{1}$$

Since L has uniqueness and T is cancellative, Remark 3 (applied k times, premultiplying by a in each case) gives that there exists a constant $c_1 > 0$ such that

$$|b^{m^p}\varphi| - |(a^k b^{m^p})\varphi| \leqslant c_1 k \quad \text{for all } k \text{ and } p. \tag{2}$$

Using the inequalities (1) and (2), we have that

$$|(a^k b^{m^p})\varphi| \geqslant |b^{m^p}\varphi| - c_1 k \geqslant m^p - c_1 k \quad \text{for all } k \text{ and } p. \tag{3}$$

Since $a^p b^{m^p}$ and $b^{n^p} a^p$ represent the same element in T, we have that

$$|(a^p b^{m^p})\varphi| = |(b^{n^p} a^p)\varphi| \quad \text{for all } p. \tag{4}$$

We can obviously obtain $b^{n^p} a^p$ from e by premultiplying e by a a total of p times, and then premultiplying the result by b a total of n^p times; so, using Remark 3 again, we have

$$|(b^{n^p} a^p)\varphi| \leqslant c_1(p + n^p) \quad \text{for all } p. \tag{5}$$

We therefore have that, for all p,

$$\begin{aligned}
c_1(p + n^p) &\geqslant |(b^{n^p} a^p)\varphi| \quad \text{by inequality (5)} \\
&= |(a^p b^{m^p})\varphi| \quad \text{by equation (4)} \\
&\geqslant m^p - c_1 p \quad \text{by inequality (3)},
\end{aligned}$$

which is a contradiction for sufficiently large p since $m > n$. Hence S is not right-left automatic. $\qquad\square$

So we have:

Corollary 1. *There exists a cancellative semigroup that is right-right automatic but not right biautomatic.*

Proof. Consider the Baumslag-Solitar semigroup $S(m, n) = \langle a, b : ab^m = b^n a \rangle$ with $m > n$. It follows from the Propositions 9 and 14 that S is not right-left automatic but that it is right-right automatic.   □

In the light of Corollary 1, one might ask whether a semigroup that satisfies more than one notion of automaticity must satisfy some notion of biautomaticity; there are obviously a great many such questions one could ask (depending on which notions one chooses) but, even if we assume that we have a cancellative monoid, the answers to all such questions are "no" as is shown by the following:

Theorem 2. *Let M be the monoid with (monoid) presentation*

$$\wp = \langle a, b, c, d, f, g, x : abcx = xdfg,\ bx = xf,\ cax = xgdfgd \rangle.$$

Then:

1. *M is a cancellative monoid.*
2. *M is right-right, right-left, left-right and left-left automatic.*
3. *M has none of the properties of being right-biautomatic, left-biautomatic, same-biautomatic or cross-biautomatic.*

Proof. We see that M embeds in the group G defined by $\wp$ by the results in [1]; in particular, M is cancellative.

Note that M is right-right automatic by Proposition 9. We can also use Proposition 9 to show that M^{rev} is right-right automatic, and so M is left-left automatic by Proposition 1. By Proposition 8, we see that M is also right-left and left-right automatic. So M satisfies all four of our notions of automaticity. However, as we will see, M is not right-biautomatic and so, by Theorem 1, M satisfies none of the four notions of biautomaticity.

Let $A = \{a, b, c, d, e, f, g, x\}$, where e represents the identity element of the monoid, and let B be any set of (semigroup) generators for M. No element of A is a product of two or more elements of M; so we must have $A \subseteq B$. Now suppose that there exists a right-biautomatic structure (B, L) for M; we can assume, by Proposition 3, that (B, L) is a right-biautomatic structure with uniqueness.

Let $\varphi : B^* \to L$ be the map defined by $\alpha\varphi \equiv \beta$ if $\beta \in L$ and $\alpha =_M \beta$. Let $\theta : B^* \to \mathbb{N}$ be the map defined by $\alpha\theta = |\alpha\varphi|$. Let N be the constant referred to in Remark 3. We then write $n \approx_i m$ (where $i, n, m \in \mathbb{N}$) if $|n - m| < iN$. For every $i \in \mathbb{N}$ we see, by Remark 3 that

$$
\begin{aligned}
((dfg)^i)\theta &\approx_1 (x(dfg)^i)\theta &&= ((abc)^i x)\theta &&\approx_1 ((abc)^i)\theta \\
&\approx_2 ((bca)^i)\theta &&\approx_1 ((bca)^i x)\theta &&= (x(fgdfgd)^i)\theta \\
&\approx_1 ((fgdfgd)^i)\theta &&\approx_2 ((dfgdfg)^i)\theta &&= ((dfg)^{2i})\theta.
\end{aligned}
$$

Hence $((dfg)^i)\theta \approx_8 ((dfg)^{2i})\theta$. As A and $B = \{b_1, \ldots, b_k\}$ are generating sets for M and $e \in A$, there exists a constant m and a map $\xi : B \to A^+$ with

$b_i\xi = \alpha_i$ where $\alpha_i =_T \beta_i$ and $|\alpha_i| = m$ for each i. Let $\gamma_i = ((dfg)^i)\varphi\xi$, so that $|\gamma_i| \approx_{8m} |\gamma_{2i}|$. There is no relation in M to rewrite a word of the form $(dfg)^i$ except for inserting or deleting occurrences of e's; hence $|\gamma_i| \geqslant 3i$. This gives that

$$3 \times 2^i \leqslant |\gamma_{2i}| \leqslant |\gamma_1| + 8miN,$$

which gives a contradiction for i big enough. Hence there is no right-biautomatic structure for M over any generating set as required.   □

There is an intriguing open problem (see [6] for example) as to whether or not an automatic group (i.e. right-right automatic group in our terminology, although all the notions coincide for groups as noted above) is necessarily biautomatic (right-biautomatic in our sense, although this is equivalent to the other notions of biautomaticity for groups by Theorem 1). Corollary 1 and Theorem 2 show that the answer to this question is "no" in the wider context of semigroups. In addition, these examples give rise to some further questions.

Let $\wp$ be an semigroup presentation, $S(\wp)$ be the semigroup defined by $\wp$ and $G(\wp)$ be the group defined by $\wp$. Assume that $S(\wp)$ embeds in $G(\wp)$; this is the case in Corollary 1 and Theorem 2 (as we mentioned above) by the results in [1]. In the first case, $G(\wp)$ is the Baumslag-Solitar group with presentation

$$\langle a, b : a^{-1}b^n a = b^m \rangle$$

with $m > n$ which is not right-right automatic. In the second case, we have the group $G = G(\wp)$ with presentation

$$\langle a, b, c, d, f, g, x : abcx = xdfg, bx = xf, cax = xgdfgd \rangle.$$

Introducing $h = dfg$, and then eliminating $d = hg^{-1}f^{-1}$ and $f = x^{-1}bx$, yields

$$\langle a, b, c, g, h, x : abcx = xh, ca = xgh^2g^{-1}x^{-1}b^{-1} \rangle.$$

We now eliminate $c = b^{-1}a^{-1}xhx^{-1}$ to get

$$\langle a, b, g, h, x : b^{-1}a^{-1}xhx^{-1}a = xgh^2g^{-1}x^{-1}b^{-1} \rangle,$$

or, equivalently,

$$\langle a, b, g, h, x : (x^{-1}abxg)^{-1}hx^{-1}abxg = h^2 \rangle.$$

If we introduce $u = x^{-1}abxg$ and then delete $g = x^{-1}b^{-1}a^{-1}xu$, we get

$$\langle a, b, h, u, x : u^{-1}hu = h^2 \rangle.$$

We see that G is a free product of the free group on the generators a, b and x with the Baumslag-Solitar group $\langle h, u : u^{-1}hu = h^2 \rangle$.

Recall (see [2] or [6] for example) that, if a group H is a free product $H_1 * H_2$ of groups H_1 and H_2, then H is (right-right) automatic if and only if both H_1 and H_2 are (right-right) automatic. In our case, since the Baumslag-Solitar group is not automatic, G is not automatic.

So these examples show that $G(\wp)$ may not be (right-right) automatic even when $S(\wp)$ satisfies all four notions of automaticity. As far as the situation the other way round is concerned, Cain [3] has constructed an example of a presentation $\wp$ such that $G(\wp)$ is (right-right) automatic but $S(\wp)$ is not.

Acknowledgements

The authors are grateful to Nik Ruškuc and Volodya Shavrukov for several helpful conversations. The authors would also like to thank Chen-Hui Chiu and Hilary Craig for all their help and encouragement.

References

1. S. I. Adjan, Defining relations and algorithmic problems for groups and semigroups, *Proceedings of the Steklov Institute of Mathematics* **85** (1966), American Mathematical Society (1967); translated from *Trudy. Mat. Inst. Steklov* **85** (1966).
2. G. Baumslag, S. M. Gersten, M. Shapiro and H. Short, Automatic groups and amalgams, *J. Pure Appl. Algebra* **76** (1991), 229–316.
3. A. J. Cain, A group-embeddable finitely generated non-automatic semigroup whose universal group is automatic, *preprint*.
4. C. M. Campbell, E. F. Robertson, N. Ruškuc and R. M. Thomas, Automatic semigroups, *Theoret. Comput. Science* **365** (2001), 365–391.
5. A. J. Duncan, E. F. Robertson and N. Ruškuc, Automatic monoids and change of generators, *Math. Proc. Cambridge Philos. Soc.* **127** (1999), 403–409.
6. D. B. A. Epstein, J. W. Cannon, D. F. Holt, S. Levy, M. S. Paterson and W. Thurston, *Word Processing in Groups* (Jones and Barlett, 1992).
7. S. M. Gersten and H. B. Short, Rational subgroups of biautomatic groups, *Annals Math.* **134** (1991), 125–158.
8. M. Hoffmann, D. Kuske, F. Otto and R. M. Thomas, Some relatives of automatic and hyperbolic groups, *in* G. M. S. Gomes, J.-E. Pin and P. V. Silva (eds.), *Semigroups, Algorithms, Automata and Languages* (World Scientific, 2002), 379–406.
9. M. Hoffmann and R. M. Thomas, Notions of automaticity in semigroups, *Semigroup Forum* **66** (2003), 337–367.
10. J. M. Howie, *Fundamentals of Semigroup Theory* (Oxford University Press, 1995).
11. D. A. Jackson, Decision and separability properties for Baumslag-Solitar semigroups, *Internat. J. Algebra Comput.* **12** (2002), 33–49.
12. G. Lallement, *Semigroups and Combinatorial Applications* (John Wiley, 1979).
13. F. Otto, A. Sattler-Klein and K. Madlener, Automatic monoids versus monoids with finite convergent presentations, *in* T. Nipkow (ed.), *Rewriting Techniques and Applications - Proceedings RTA '98* (Lecture Notes in Computer Science **1379**, Springer-Verlag, 1998), 32–46.
14. H. Short, An introduction to automatic groups, *in* J. Fountain (ed.), *Semigroups, Formal Languages and Groups* (NATO ASI Series **C466**, Kluwer 1995), 233–253.

Deterministic Automata on Unranked Trees

Julien Cristau[1], Christof Löding[2], and Wolfgang Thomas[2]

[1] LIAFA, Université Paris VII, France
[2] RWTH Aachen, Germany

Abstract. We investigate bottom-up and top-down deterministic automata on unranked trees. We show that for an appropriate definition of bottom-up deterministic automata it is possible to minimize the number of states efficiently and to obtain a unique canonical representative of the accepted tree language. For top-down deterministic automata it is well known that they are less expressive than the non-deterministic ones. By generalizing a corresponding proof from the theory of ranked tree automata we show that it is decidable whether a given regular language of unranked trees can be recognized by a top-down deterministic automaton. The standard deterministic top-down model is slightly weaker than the model we use, where at each node the automaton can scan the sequence of the labels of its successors before deciding its next move.

1 Introduction

Finite automata over finite unranked trees are a natural model in classical language theory as well as in the more recent study of XML document type definitions (cf. [Nev02]). In the theory of context-free languages, unranked trees (trees with finite but unbounded branching) arise as derivation trees of grammars in which the right-hand sides are regular expressions rather than single words ([BB02]). The feature of finite but unbounded branching appears also in the tree representation of XML documents.

The generalization of tree automata from the case of ranked label alphabets to the unranked case is simple: A transition, e.g., of a bottom-up automaton is of the form (L, a, q), allowing the automaton to assume state q at an a-labeled node with say n successors if the sequence $q_1 \ldots q_n$ of states reached at the roots of the n subtrees of these successors belongs to L. Most core results of tree automata theory (logical closure properties, decidability of non-emptiness, inclusion, and equivalence) are easily transferred to this framework of "unranked tree automata" and "regular sets of unranked trees" (cf. [BWM01, Nev02]).

For certain other results of classical tree automata theory, however, such a transfer is less obvious and does not seem to be covered by existing work. In the present paper we deal with two such questions: the problem of automaton minimization, and the definition and expressive power of top-down automata (automata working from the root to the leaves, more closely following the pattern of XML query processing than the bottom-up version). We confine ourselves to the question of tree language recognition; so we do not address models like the query automata of [NS02] or the transducers of [MSV03].

M. Liśkiewicz and R. Reischuk (Eds.): FCT 2005, LNCS 3623, pp. 68–79, 2005.

The minimization problem has be reconsidered for the unranked case because two types of automata are involved: the finite tree automaton $\mathcal{A}$ used for building up run trees (on given input trees), and the finite word automata $\mathcal{B}_L$ accepting the languages L that occur in the $\mathcal{A}$-transitions. The $\mathcal{A}$-states are the input letters to the $\mathcal{B}_L$, and the $\mathcal{B}_L$-states are needed to produce the "next $\mathcal{A}$-state" (in bottom-up mode). It is not clear a priori how and in which order to minimize these automata. Using a natural definition of $\mathcal{B}_L$-automaton (which depends on a label a and produces an $\mathcal{A}$-state as output), we show in Section 3 below that a simultaneous and efficient minimization of $\mathcal{A}$ and the $\mathcal{B}_L$ is possible, moreover resulting in a minimal tree automaton that is unique up to isomorphism.

For the question of deterministic top-down processing of input trees, we start with well known results of [Vir80, GS84] on the ranked case. The generalization to the unranked case requires introducing a finite automaton that proceeds from state q at an a-labeled node deterministically to new states $q_1 \ldots q_n$ at the n successor nodes. A natural option is to provide the numbers n and i as input in order to compute q_i. A second option, closer to the idea of XML document processing, is to provide as inputs the sequence $a_1 \ldots a_n$ of successor labels and the position i. In both cases, the simple approach to define transitions via a finite table does not suffice, instead one has to introduce appropriate transducers to implement transitions. In Section 4, we present such transducers (in the form of bimachines [Eil74, Ber79]), introduce the corresponding top-down tree automata, and show that for a regular set of unranked trees one can decide whether it is recognizable by either of these top-down tree automata. For the technical presentation we focus on the second option mentioned above. The main point is an appropriate definition of "path language", recording the possible paths of trees in a given tree language; the derived notion of "path-closed" tree language then captures those tree languages that are recognizable deterministically in top-down mode.

The paper starts (in Section 2) with some technical preliminaries, gives in Section 3 the results on minimization, in Section 4 the study of top-down automata, and closes in Section 5 with some pointers to current and future work.

We thank S. Abiteboul, F. Neven, and Th. Schwentick for comments on a preliminary version of this work.

2   Automata on Unranked Trees

In this section we define unranked trees and different models of automata running on such trees. In the following, Σ always denotes a finite alphabet, i.e., a finite set of symbols, $\mathbb{N}$ denotes the set of natural numbers, and $\mathbb{N}_{>0}$ denotes the set of positive natural numbers. For a set X we denote by X^* the set of all finite words over X. The empty word is denoted by ε.

A *tree domain* D is a non-empty, prefix-closed subset of $\mathbb{N}^*_{>0}$ satisfying the following condition: if $xi \in D$ for $x \in \mathbb{N}^*_{>0}$ and $i \in \mathbb{N}_{>0}$, then $xj \in D$ for all j with $1 \leq j \leq i$.

An *unranked tree* t over Σ (simply tree in the following) is a mapping $t :$ $\mathrm{dom}_t \to \Sigma$ with a finite tree domain dom_t. The elements of dom_t are called the *nodes* of t. For $x \in \mathrm{dom}_t$ we call nodes of the form $xi \in \mathrm{dom}_t$ with $i \in \mathbb{N}_{>0}$ the *successors* of x (where xi is the ith successor). As usual, a *leaf* of t is a node without successor. If the root of t is labeled by a, i.e., $t(\varepsilon) = a$, and if the root has k successors at which the subtrees $t_1, \ldots, t_k$ are rooted, then we denote this by $t = a(t_1 \cdots t_k)$. The set of all unranked trees over Σ is denoted by T_Σ. For $a \in \Sigma$ we denote by T_Σ^a the set of all trees from T_Σ whose root is labeled by a.

A *non-deterministic bottom-up tree automaton* ($\uparrow$NTA) $\mathcal{A} = (Q, \Sigma, \Delta, F)$ consists of a finite set Q of states, a finite input alphabet Σ, a finite set $\Delta \subseteq \mathrm{Reg}(Q) \times \Sigma \times Q$ of transitions ($\mathrm{Reg}(Q)$ denotes the set of regular languages over Q), and a set $F \subseteq Q$ of final states.

A *run* of $\mathcal{A}$ on a tree t is a function $\rho : \mathrm{dom}_t \to Q$ with the following property: for each $x \in \mathrm{dom}_t$ with n successors $x1, \ldots, xn$ there is a transition $(L, t(x), \rho(x)) \in \Delta$ such that the word $\rho(x1) \cdots \rho(xn)$ is in L. If x is a leaf, this means that there must be a transition $(L, t(x), \rho(x)) \in \Delta$ with $\varepsilon \in L$. If for some run ρ of $\mathcal{A}$ on t the root of ρ is labeled with q, then we write $t \to_{\mathcal{A}} q$. For $Q' \subseteq Q$ we write $t \to_{\mathcal{A}} Q'$ if $t \to_{\mathcal{A}} q$ for some $q \in Q'$. We call ρ *accepting* if $\rho(\varepsilon) \in F$ and say that t is accepted by $\mathcal{A}$ if there is an accepting run of $\mathcal{A}$ on t. The language $T(\mathcal{A})$ accepted by $\mathcal{A}$ is $T(\mathcal{A}) := \{t \in T_\Sigma \mid \mathcal{A} \text{ accepts } t\}$. The *regular languages of unranked trees* are those that can be accepted by $\uparrow$NTAs.

In the definition of $\uparrow$NTA we did not specify how the regular languages used in the transitions are given. First of all, note that it is not necessary to have two transitions (L_1, a, q) and (L_2, a, q) because these can be merged into a single transition $(L_1 \cup L_2, a, q)$. Usually, one then assumes that the transition function is given by a set of regular expressions or non-deterministic finite automata defining for each $q \in Q$ and $a \in \Sigma$ the language $L_{a,q}$ with $(L_{a,q}, a, q) \in \Delta$.

One can also define non-deterministic tree automata that work in a top-down fashion. For this purpose it is sufficient to view the final states as initial states. Thus, for non-deterministic automata it does not make any difference whether we consider top-down or bottom-up automata. In contrast, to obtain a deterministic model with the same expressive power as the corresponding non-deterministic model one has to consider bottom-up automata as introduced in the following. Deterministic top-down automata are treated in Section 4.

The standard definition of deterministic bottom-up tree automata ($\uparrow$DTA) is obtained by imposing a semantic restriction on the set of transitions: it is required that for each letter a and all states q_1, q_2 if there are transitions (L_1, a, q_1) and (L_2, a, q_2), then $L_1 \cap L_2 = \emptyset$. Each $\uparrow$NTA can be transformed into an equivalent $\uparrow$DTA using a standard subset construction [BWM01]. Here, we do not use this semantic approach to define determinism but require a representation of the transition function that syntactically enforces determinism. Besides the advantage of not needing any semantic restrictions, our model is obtained in a natural way when applying the subset construction to $\uparrow$NTAs. Since minimization is often applied to reduce the result of a determinization construction, the choice of this model is a natural one for our purposes.

A $\uparrow$DTA $\mathcal{A}$ is given by a tuple $\mathcal{A} = (Q, \Sigma, (\mathcal{D}_a)_{a \in \Sigma}, F)$ with Q, Σ, and F as for $\uparrow$NTA, and deterministic finite automata $\mathcal{D}_a$ with output defining the tran-

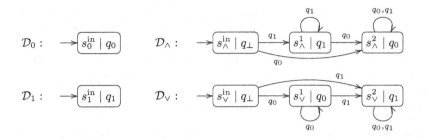

Fig. 1. A ↑DTA recognizing all ∧-∨-trees that evaluate to 1

sitions of $\mathcal{A}$. Each of the $\mathcal{D}_a$ (with $a \in \Sigma$) is of the form $\mathcal{D}_a = (S_a, Q, s_a^{\text{in}}, \delta_a, \lambda_a)$ with a finite set S_a of states, input alphabet Q, initial state s_a^{in}, transition function $\delta_a : S_a \times Q \to S_a$, and output function $\lambda_a : S_a \to Q$. As usual, we define $\delta_a^* : S_a \times Q^* \to S_a$ by $\delta_a^*(s, \varepsilon) = s$ and $\delta_a^*(s, uq) = \delta_a(\delta_a^*(s, u), q)$.

Such a ↑DTA can be transformed into the standard representation as follows. For each $q \in Q$ and $a \in \Sigma$ let $L_{a,q} = \{w \in Q^* \mid \lambda_a(\delta_a^*(s_a^{\text{in}}, w)) = q\}$. Then, the set of transitions defined by the family $(\mathcal{D}_a)_{a \in \Sigma}$ consists of all transitions $(L_{a,q}, a, q)$.

Example 1. For $\Sigma = \{\wedge, \vee, 0, 1\}$ we consider trees whose leaves are labeled by 0 or 1, and whose inner nodes are labeled by $\wedge$ or $\vee$. Such trees can be evaluated in a natural way to 0 or 1. Let T be the language of all trees that evaluate to 1. We define a ↑DTA for T with state set $Q = \{q_0, q_1, q_\perp\}$, final states $F = \{q_1\}$, and automata $\mathcal{D}_a$, $a \in \Sigma$, as depicted in Figure 1. An entry $s \mid q$ in the picture means that the output at state s is q, e.g., $\lambda_\wedge(s_\wedge^{\text{in}}) = q_\perp$. For readability we have omitted the transitions leading to rejection. All missing transitions are assumed to lead to a sink state of the respective automaton with output $q_\perp$.

3   Minimization of Deterministic Bottom-Up Automata

In this section $\mathcal{A}$ always denotes a ↑DTA $\mathcal{A} = (Q, \Sigma, (\mathcal{D}_a)_{a \in \Sigma}, F)$ with $\mathcal{D}_a = (S_a, Q, s_a^{\text{in}}, \delta_a, \lambda_a)$ for each $a \in \Sigma$. Furthermore, we let $S = \bigcup_{a \in \Sigma} S_a$. For complexity considerations we define the size of $\mathcal{A}$ as $|\mathcal{A}| = |Q| \cdot |S|$. This is a reasonable measure since the sizes of the transition functions of the automata $\mathcal{D}_a$ are of order $|Q| \cdot |S_a|$.

For minimization it is necessary to ensure that all states (from Q and S) are reachable. Note that there is an interdependence since a state in Q is reachable if it is the output of some reachable state in S, and a state in S is reachable if it can be reached by some input consisting of reachable states in Q.

Lemma 1. *The set of reachable states of a given ↑DTA can be computed in linear time.*

Proof. The algorithm maintains a set S' of reachable states from S and a set Q' of reachable states from Q. These sets are initialized to $S' = \{s_a^{\text{in}} \mid a \in \Sigma\}$

and $Q' = \{\lambda_a(s_a^{\text{in}}) \mid a \in \Sigma\}$. Starting from S' the transition graphs of the $\mathcal{D}_a$ are traversed in a breadth-first manner using only the transition labels from Q'. Whenever we encounter a state $s \in S_a$ with $q = \lambda_a(s) \notin Q'$, then q is added to Q' and the targets of the transitions with label q departing from those states in S' that have already been processed by the breadth first search are added to S'. This algorithm traverses each transition of the automata $\mathcal{D}_a$ at most once. and hence can be implemented to run in linear time.   □

From now on we assume that all states in $\mathcal{A}$ are reachable. For each $q \in Q$ we define $T_q = \{t \in \mathcal{T}_\Sigma \mid t \to_{\mathcal{A}} q\}$, and for each $a \in \Sigma$ and $s \in S_a$ we define

$$T_s = \{a(t_1 \cdots t_k) \mid \exists q_1, \ldots, q_k : t_i \in T_{q_i} \text{ and } \delta_a^*(s_a^{\text{in}}, q_1 \cdots q_k) = s\}.$$

If all states are reachable, then these sets are non-empty, and we can fix for each $q \in Q$ some $t_q \in T_q$ and for each $s \in S$ some $t_s \in T_s$.

To prove the existence of a unique minimal ↑DTA for a regular tree language we introduce two equivalence relations in the spirit of Nerode's congruence for word languages. To this aim we first define two different kinds of concatenations for trees.

The set $\mathcal{T}_{\Sigma,X}$ of pointed trees over Σ contains all trees t from $\mathcal{T}_{\Sigma \cup \{X\}}$ (for a new symbol X) such that exactly one leaf of t is labeled by X. For $t \in \mathcal{T}_{\Sigma,X}$ and $t' \in \mathcal{T}_\Sigma \cup \mathcal{T}_{\Sigma,X}$ we denote by $t \circ t'$ the tree obtained from t by replacing the leaf labeled X by t'. For $T \subseteq \mathcal{T}_\Sigma$ the equivalence relation $\sim_T \subseteq \mathcal{T}_\Sigma \times \mathcal{T}_\Sigma$ is defined by

$$t_1 \sim_T t_2 \qquad \text{iff} \qquad \forall t \in \mathcal{T}_{\Sigma,X} : t \circ t_1 \in T \Leftrightarrow t \circ t_2 \in T.$$

This relation is called 'top-congruence' in [BWM01]. In the case of ranked trees it is the natural extension of Nerode's congruence from words to trees. However, as already noted in [BWM01], for T being regular in the unranked setting it is not sufficient that $\sim_T$ is of finite index. One also has to impose a condition that ensures the regularity of the 'horizontal languages' that are used in the transition function. For this we need another concatenation operation on trees.

For trees $t = a(t_1 \ldots t_k)$ and $t' = a(t_1' \cdots t_\ell')$ let $t \odot t' = a(t_1 \cdots t_k t_1' \cdots t_\ell')$. The equivalence relation $\overset{\rightarrow}{\sim}_T$ is defined for all $a \in \Sigma$ and $t_1, t_2 \in \mathcal{T}_\Sigma^a$ by

$$t_1 \overset{\rightarrow}{\sim}_T t_2 \qquad \text{iff} \qquad \forall t \in \mathcal{T}_\Sigma^a : t_1 \odot t \sim_T t_2 \odot t$$

To simplify notation we write $[t]$ for the $\sim_T$-class of t and $[t]_\rightarrow$ for the $\overset{\rightarrow}{\sim}_T$-class of t. If T is accepted by a ↑DTA $\mathcal{A}$, then $\mathcal{A}$ has to distinguish trees that are not equivalent. This is expressed in the following lemma.

Lemma 2. *If $\mathcal{A}$ accepts the language T, then $T_q \subseteq [t_q]$ for each $q \in Q$ and $T_s \subseteq [t_s]_\rightarrow$ for each $a \in \Sigma$ and $s \in S_a$.*

The above lemma implies that $\overset{\rightarrow}{\sim}_T$ is of finite index if T is regular. On the other hand, if $\overset{\rightarrow}{\sim}_T$ is of finite index, then this ensures the regularity of what is called 'local views' in [BWM01] and hence T is regular. In the following we show that the equivalence classes of $\sim_T$ and $\overset{\rightarrow}{\sim}_T$ can be used to define a canonical minimal

$\uparrow$DTA $\mathcal{A}_T$ for T. The equivalence classes of $\sim_T$ correspond to the states of the tree automaton, and the equivalence classes of $\overset{\rightarrow}{\sim}_T$ restricted to $\mathcal{T}_\Sigma^a$ correspond to the states of the automaton defining the transitions for label a.

For the definition of $\mathcal{A}_T$ we need the following lemma stating that $\overset{\rightarrow}{\sim}_T$ refines $\sim_T$ and that $\overset{\rightarrow}{\sim}_T$ is a right-congruence (w.r.t. $\odot$). The proof of this lemma is straightforward.

Lemma 3. (a) If $t_1 \overset{\rightarrow}{\sim}_T t_2$, then $t_1 \sim_T t_2$ for all $t_1, t_2 \in \mathcal{T}_\Sigma$.
(b) If $t_1 \overset{\rightarrow}{\sim}_T t_2$ and $t_1' \sim_T t_2'$, then $t_1 \odot a(t_1') \overset{\rightarrow}{\sim}_T t_2 \odot a(t_2')$ for all $t_1, t_2 \in \mathcal{T}_\Sigma^a$ and all $t_1', t_2' \in \mathcal{T}_\Sigma$.

The assignment of $\overset{\rightarrow}{\sim}_T$-classes to $\sim_T$-classes that is induced by (a) corresponds to the mappings λ_a that assign to each state from S_a a state from Q.

The following theorem states the existence of a unique (up to isomorphism) $\uparrow$DTA for every regular language T of unranked trees. The notion of homomorphism that we use in the statement of the theorem is the natural one: a *homomorphism* from $\mathcal{A}_1 = (Q_1, \Sigma, (\mathcal{D}_a^1)_{a \in \Sigma}, F_1)$ to $\mathcal{A}_2 = (Q_2, \Sigma, (\mathcal{D}_a^2)_{a \in \Sigma}, F_2)$ maps the states from Q_1 to states from Q_2 while respecting final and non-final states, and maps for each $a \in \Sigma$ the set S_a^1 to S_a^2 (S_a^i denotes the state set of $\mathcal{D}_a^i$) while respecting the initial state, the transition function, and the output function.

Theorem 1. *For every regular $T \subseteq \mathcal{T}_\Sigma$ there is a unique minimal $\uparrow$DTA $\mathcal{A}_T$ and for each $\uparrow$DTA $\mathcal{A}$ recognizing T there is a surjective homomorphism from $\mathcal{A}$ to $\mathcal{A}_T$.*

Proof. Define $\mathcal{A}_T = (Q_T, \Sigma, (\mathcal{D}_a^T)_{a \in \Sigma}, F_T)$ by $Q_T = \mathcal{T}_\Sigma/\!\sim_T$, $F = \{[t] \mid t \in T\}$, and $\mathcal{D}_a^T = (S_a^T, Q_T, [a]_\rightarrow, \delta_a^T, \lambda_a^T)$ with $S_a^T = \mathcal{T}_\Sigma^a/\!\overset{\rightarrow}{\sim}_T$, $\delta_a^T([t]_\rightarrow, [t']) = [t \odot a(t')]_\rightarrow$ for $t \in \mathcal{T}_\Sigma^a$ and $t' \in \mathcal{T}_\Sigma$, and $\lambda_a^T([t]_\rightarrow) = [t]$. Using Lemma 3 one can easily show that $t \rightarrow_{\mathcal{A}_T} [t]$ and hence $T(\mathcal{A}_T) = T$. Furthermore, if $\mathcal{A}$ is some $\uparrow$DTA for T, then it is not difficult to see that mapping each state $q \in Q$ to $[t_q]$ and each $s \in S$ to $[t_s]_\rightarrow$ defines a surjective homomorphism from $\mathcal{A}$ to $\mathcal{A}_T$. $\qquad\square$

We now give an algorithm that computes this minimal $\uparrow$DTA $\mathcal{A}_T$ starting from any automaton $\mathcal{A}$ for T. This minimization procedure is an extension of the classical minimization procedure for finite automata (cf. [HU79]). We define equivalence relations on the state sets Q and S that correspond to the relations $\sim_T$ and $\overset{\rightarrow}{\sim}_T$ and obtain the minimal automaton by merging equivalent states. For $q_1, q_2 \in Q$ let $q_1 \sim_{\mathcal{A}} q_2$ iff $(t \circ t_{q_1} \rightarrow_{\mathcal{A}} F \Leftrightarrow t \circ t_{q_2} \rightarrow_{\mathcal{A}} F)$ for all $t \in \mathcal{T}_{\Sigma, X}$. For $a \in \Sigma$ and $s_1, s_2 \in S_a$ let $s_1 \sim_{\mathcal{A}} s_2$ iff $\lambda_a(\delta_a^*(s_1, u)) \sim_{\mathcal{A}} \lambda_a(\delta_a^*(s_2, u))$ for all $u \in Q^*$. The following lemma states that it is indeed possible to group equivalent states into a single state.

Lemma 4. *If $q_1 \sim_{\mathcal{A}} q_2$ for $q_1, q_2 \in Q$ and $s_1 \sim_{\mathcal{A}} s_2$ for $s_1, s_2 \in S_a$, then $\delta_a(s_1, q_1) \sim_{\mathcal{A}} \delta_a(s_2, q_2)$.*

For $q \in Q$ and $s \in S$ we denote by $[q]$ and $[s]$ the $\sim_{\mathcal{A}}$-class of q and s, respectively. The reduced automaton $A_\sim$ is defined as $\mathcal{A}_\sim = (Q/\!\sim_{\mathcal{A}}, \Sigma, (\mathcal{D}_a^\sim)_{a \in \Sigma}, F/\!\sim_{\mathcal{A}})$ with $\mathcal{D}_a^\sim = (S_a/\!\sim_{\mathcal{A}}, Q/\!\sim_{\mathcal{A}}, [s_a^{in}], \delta_a^\sim, \lambda_a^\sim)$, $\delta_a^\sim([s], [q]) = [\delta_a(s, q)]$, and $\lambda_a^\sim([s]) =$

$[\lambda_a(s)]$. Lemma 4 ensures that the definitions of δ_a and λ_a do not depend on the chosen representatives of the equivalence classes.

Theorem 2. *If $\mathcal{A}$ is an automaton for the language T, then $\mathcal{A}_T$ and $\mathcal{A}_\sim$ are isomorphic.*

Proof. From the definitions of $\sim_\mathcal{A}$, $\sim_T$, and $\overrightarrow{\sim}_T$ one can easily deduce that $q_1 \sim_\mathcal{A} q_2$ iff $[t_{q_1}] = [t_{q_2}]$, and $s_1 \sim_\mathcal{A} s_2$ iff $[t_{s_1}]_\to = [t_{s_2}]_\to$. This implies that in the reduced automaton every state $[q]$ can be identified with $[t_q]$ and each state $[s]$ can be identified with $[t_s]_\to$.   □

Hence, to compute the unique minimal automaton for T it suffices to compute the relation $\sim_\mathcal{A}$. The algorithm shown in Figure 2 marks all pairs of states that are not in the relation $\sim_\mathcal{A}$.

INPUT: $\uparrow$DTA $\mathcal{A} = (Q, \Sigma, (\mathcal{D}_a)_{a \in \Sigma}, F)$ with $D_a = (S_a, Q, s_a^{\text{in}}, \delta_a, \lambda_a)$

1. Mark each pair $(q_1, q_2) \in Q^2$ with $q_1 \in F \Leftrightarrow q_2 \notin F$.
2. **repeat**
3.   For each $a \in \Sigma$ mark $(s_1, s_2) \in S_a^2$ if $(\lambda_a(s_1), \lambda_a(s_2))$ is marked.
4.   For each $a \in \Sigma$ and $q \in Q$ mark $(s_1, s_2) \in S_a^2$ if $(\delta_a(s_1, q), \delta_a(s_2, q))$ is marked.
5.   For each $a \in \Sigma$ and $s \in S_a$ mark $(q_1, q_2) \in Q^2$ if $(\delta_a(s, q_1), \delta_a(s, q_2))$ is marked.
6. **until** no new pairs are marked

OUTPUT: $R = \{(q_1, q_2) \in Q^2 \mid (q_1, q_2) \text{ not marked}\}$
$ \cup \{(s_1, s_2) \in S_a^2 \mid a \in \Sigma \text{ and } (s_1, s_2) \text{ not marked}\}$

Fig. 2. Algorithm EQUIVALENT-STATES

Theorem 3. *The algorithm EQUIVALENT-STATES from Figure 2 computes for input $\mathcal{A}$ the relation $\sim_\mathcal{A}$.*

Proof. We first show that all pairs marked by the algorithm are non-equivalent. For the pairs marked in lines 1,3, and 4 this is a direct consequence of the definition of $\sim_\mathcal{A}$. For pairs (q_1, q_2) marked in line 5 the claim follows from Lemma 4 applied to q_1, q_2, and $s = s_1 = s_2$.

To show that all pairs of non-equivalent states are marked, we look at the minimal 'size' of a witness that separates the two states. For the states from Q these witnesses are pointed trees. The size we are interested in is the depth of the leaf labeled by X, i.e., for $t \in T_{\Sigma, X}$ we define $|t|_X$ to be the depth of X in t. For $q_1, q_2 \in Q$ and $n \in \mathbb{N}$ we define $q_1 \sim_n q_2$ iff $(t \circ t_{q_1} \to_\mathcal{A} F \Leftrightarrow t \circ t_{q_2} \to_\mathcal{A} F)$ for all $t \in T_{\Sigma, X}$ with $|t|_X \leq n$. For $s_1, s_2 \in S_a$ we let $s_1 \sim_n s_2$ iff $\lambda_a(\delta_a^*(s_1, u)) \sim_n \lambda_a(\delta_a^*(s_2, u))$ for all $u \in Q^*$. Using this definition one can show that a pair of states is marked in the ith iteration of the loop iff i is the minimal number such that the states are not in the relation $\sim_i$. For this we assume that each of the lines 3–5 is executed as long as there are pairs that can be marked in the respective line.   □

For readability we have used the technique of marking pairs of non-equivalent states to compute the relation $\sim_{\mathcal{A}}$. A concrete implementation of the algorithm should rather use the technique of refining an equivalence relation represented by its equivalence classes. This technique is used to improve the complexity of minimization of finite automata on words [Hop71] and can also be applied to the minimization of automata on finite ranked trees (cf. [CDG+97]). Using this technique one can obtain an algorithm running in quadratic time. A more detailed analysis on whether this bound can be improved is still to be done.

Theorem 4. *Given a $\uparrow DTA$ $\mathcal{A}$ one can compute in quadratic time the minimal $\uparrow DTA$ that is equivalent to $\mathcal{A}$.*

4 Deterministic Top-Down Automata

As in the case of ranked trees, deterministic automata that work in a top-down fashion are not as expressive as non-deterministic ones. In this section we introduce such a model for unranked trees and show that it is decidable whether a given regular tree language can be accepted by a deterministic top-down automaton. When directly adapting the definition of top-down deterministic automata on ranked trees, one obtains a model that, depending on its current state, the current label, and the number of successors of the current node, decides which states it sends to the successors. Here, we have decided to make the model a bit more expressive by allowing for a transition to take into account not only the number of successors but also their labeling. All the results from this section can be adapted in a straightforward way to the weaker model as described above.

To define the transitions as just mentioned we use a certain kind of transducer to convert the sequence of labels of the successors of a node into a sequence of states of the tree automaton. Since this transducer should have information on the whole successor sequence before deciding which state to put at a certain successor we use the formalism of bimachines (cf. [Eil74, Ber79]).

A *bimachine* is of the form $\mathcal{B} = (\Sigma, \Gamma, \overrightarrow{\mathcal{B}}, \overleftarrow{\mathcal{B}}, f)$, where Σ is the input alphabet, Γ is the output alphabet, $\overrightarrow{\mathcal{B}} = (\overrightarrow{S}, \Sigma, s_0^{\rightarrow}, \overrightarrow{\delta})$ and $\overleftarrow{\mathcal{B}} = (\overleftarrow{S}, \Sigma, s_0^{\leftarrow}, \overleftarrow{\delta})$ are deterministic finite automata over Σ (without final states), and $f : \overrightarrow{S} \times \Sigma \times \overleftarrow{S} \rightarrow \Gamma$ is the output function.

Given a word $u \in \Sigma^*$ consisting of k letters $u = a_1 \cdots a_k$, $\mathcal{B}$ produces an output $v = b_1 \cdots b_k$ over Γ that is defined as follows. Let $s_0^{\rightarrow} s_1^{\rightarrow} \cdots s_k^{\rightarrow}$ be the run of $\overrightarrow{\mathcal{B}}$ on $a_1 \cdots a_k$, and let $s_0^{\leftarrow} s_1^{\leftarrow} \cdots s_k^{\leftarrow}$ be the run of $\overleftarrow{\mathcal{B}}$ on the reversed input $a_k \cdots a_1$. Then the ith output letter b_i is given by $b_i = f(s_i^{\rightarrow}, a_i, s_{k-i+1}^{\leftarrow})$. This definition is illustrated in Figure 3.

We denote the function computed by $\mathcal{B}$ by $f_{\mathcal{B}}$. One should note that using bimachines we remain inside the domain of regular languages in the sense that $f_{\mathcal{B}}(L)$ for a regular language L is again regular. For further results on bimachines see, e.g., [Ber79] or [Eil74].

A *deterministic top-down tree automaton* ($\downarrow$DTA) uses such bimachines for its transitions. It is of the form $\mathcal{A} = (Q, \Sigma, f_{\text{in}}, (\mathcal{B}_q)_{q \in Q}, F)$, where Q is finite set

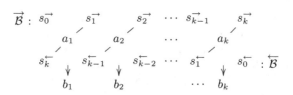

Fig. 3. Computation of a bimachine

of states, Σ is the input alphabet, $f_{\text{in}} : \Sigma \to Q$ is the initial function, $F \subseteq Q$ is a set of final states, and each $\mathcal{B}_q$ is a bimachine with input alphabet Σ and output alphabet Q.

A run of $\mathcal{A}$ on a tree t is mapping $\rho : \text{dom}_t \to Q$ such that $\rho(\varepsilon) = f_{\text{in}}(t(\varepsilon))$ and for each node x of t with $n > 0$ successors $x1, \ldots, xn$ we have $\rho(x1) \cdots \rho(xn) = f_{\mathcal{B}_{\rho(x)}}(t(x1) \cdots t(xn))$. Note that for each t there is exactly one run of $\mathcal{A}$ on t. The run ρ is accepting if each leaf is labeled with a final state. The language accepted by $\mathcal{A}$ consists of all trees t such that the run of $\mathcal{A}$ on t is accepting.

It is not difficult to see that not all regular tree languages can be recognized by $\downarrow$DTAs. Consider for example the language $T_{cd} = \{a(a(c)a(d)), a(a(d)a(c))\}$. Every $\downarrow$DTA recognizing the two trees from T_{cd} will also recognize the trees $a(a(c)a(c))$ and $a(a(d)a(d))$.

We show that it is decidable whether a given regular tree language can be recognized by a $\downarrow$DTA. The proof follows the same lines as for ranked trees using the notions of path language and path closure ([Vir80, GS84]). When a $\downarrow$DTA descends a tree, then on each path it can only see the sequence of labels of this path and on each level the sequence of labels of the siblings. For example, the information known to a $\downarrow$DTA on the leftmost path in the first tree from T_{cd} can be coded as $a \triangleright \triangledown a \triangleleft a \triangleright \triangledown \triangleleft c$. The letters outside the segments $\triangleright \cdots \triangleleft$ code the sequence of labels on the path, and the letters between the pairs $\triangleright$, $\triangleleft$ show the labels of the siblings with the position corresponding to the node of the considered path marked by $\triangledown$. This idea leads to a corresponding concept of path language. Note that this generalizes the standard concept of path language over ranked alphabets where a path code has the form $a_1 i_1 a_2 i_2 \cdots i_{\ell-1} a_\ell$ indicating that successively the successors $i_1, i_2, \ldots$ are taken. In our setting, the rank of a_j is captured by the length of the subsequent segment $\triangleright \cdots \triangleleft$ and i_j by the position of $\triangledown$ in this segment.

The alphabet we use for path languages is $\Sigma_{\text{path}} = \Sigma \cup \{\triangleright, \triangleleft, \triangledown\}$. The *path language* $\pi(t)$ of a tree t is defined inductively as $\pi(a) = a$ for each $a \in \Sigma$ and $\pi(t) = \bigcup_{i=1}^{k}\{a \triangleright a_1 \cdots a_{i-1} \triangledown a_{i+1} \cdots a_k \triangleleft w \mid w \in \pi(a_i(t_i))\}$ for $t = a(a_1(t_1) \cdots a_k(t_k))$. In this definition we allow that t_i is empty. In this case $a_i(t_i) = a_i$. The path language of $T \subseteq \mathcal{T}_\Sigma$ is $\pi(T) = \bigcup_{t \in T} \pi(t)$. The *path closure* of T is $\text{cl}(T) = \{t \in \mathcal{T}_\Sigma \mid \pi(t) \subseteq \pi(T)\}$. A language with $T = \text{cl}(T)$ is called *path closed*. In the following we show that the $\downarrow$DTA-recognizable languages are exactly the path closed regular tree languages. The proof goes through a sequence of lemmas.

Lemma 5. *If $T \subseteq \mathcal{T}_\Sigma$ is regular, then $\pi(T)$ is a regular language of words.*

Proof. Let $\mathcal{A}$ be a $\uparrow$NTA for T with one transition $(L_{a,q}, a, q)$ for each pair of state q and letter a. We assume that for each state q of $\mathcal{A}$ there is a tree t with $t \to_\mathcal{A} q$. Note that by applying a straightforward procedure for identifying reachable states, we can restrict to this case.

One can easily define a non-deterministic finite automaton $\mathcal{C}$ accepting $\pi(T)$. This automaton, on reading the first symbol of the input word w, remembers this first symbol a and guesses a final state q of $\mathcal{A}$ such that $t \to_\mathcal{A} q$ for some t with $w \in \pi(t)$. So, after this first step, $\mathcal{C}$ is in state (q, a).

The next part of the input is of the form $\rhd a_1 \cdots a_{i-1} \triangledown a_{i+1} \cdots a_k \lhd a_i$ (if it is not of this form the input is rejected). The automaton $\mathcal{C}$ guesses a sequence $q_1 \cdots q_k$ such that $q_1 \cdots q_k \in L_{a,q}$. For this purpose, on reaching the gap $\triangledown$, it guesses a_i and verifies this guess after reading $\lhd$. Furthermore, it simulates an automaton for $L_{a,q}$ to verify that the guessed sequence is indeed in $L_{a,q}$. On passing the gap $\triangledown$ it remembers the state q_i and then moves to a state (q_i, a_i) after having read $\lhd a_i$. The final states of $\mathcal{C}$ are those pairs (q, a) for which $\varepsilon \in L_{a,q}$.   □

Lemma 6. *If $T \subseteq \mathcal{T}_\Sigma$ is regular, then $cl(T)$ is recognizable by a $\downarrow$DTA.*

Proof. If T is regular, then $\pi(T)$ is a regular word language by Lemma 5. Let $\mathcal{C} = (Q, \Sigma_{\text{path}}, q_0, \delta, F)$ be a deterministic finite automaton for $\pi(T)$. We briefly sketch how to construct a $\downarrow$DTA $\mathcal{A} = (Q, \Sigma, f_{\text{in}}, (\mathcal{B}_q)_{q \in Q}, F)$ for T that simulates $\mathcal{C}$ on every path. Note that $\mathcal{A}$ has the same set of states as $\mathcal{C}$ and the same set of final states as $\mathcal{C}$. The initial function is defined by $f_{\text{in}}(a) = \delta(q_0, a)$.

For every state $q \in Q$ the behavior of $\mathcal{B}_q$ is as follows. When processing the word $a_1 \cdots a_k$ the machine $\mathcal{B}_q$ should output the sequence $q_1 \cdots q_k$ where q_i is the state reached by $\mathcal{C}$ when reading the word $\rhd a_1 \cdots a_{i-1} \triangledown a_{i+1} \cdots a_k \lhd a_i$ starting from q. To realize this idea $\mathcal{B}_q$ has for each i (1) to compute the behavior of $\mathcal{C}$ on $\rhd a_1 \cdots a_{i-1} \triangledown$ starting from q, and (2) the behavior of $\mathcal{C}$ on $a_{i+1} \cdots a_k \lhd$ starting from any state. For this purpose we define $\overrightarrow{\mathcal{B}}_q$ by $\overrightarrow{S}_q = Q \times Q$, with initial state $(\delta(q, \rhd), \delta(q, \rhd))$ and $\delta_q^\rightarrow((s_1, s_2), a) = (\delta(s_1, a), \delta(s_1, \triangledown))$. In this way at letter i we have access to the information described in (1).

To define the machine $\overleftarrow{\mathcal{B}}_q$ we denote by Q^Q the set of mappings from Q to Q and by Id_Q the identity mapping on Q. We let $\overleftarrow{S}_q = \{\text{Id}_Q\} \cup (Q^Q \times \Sigma)$ with Id_Q as initial state. Furthermore, we let $\delta_q^\leftarrow(\text{Id}_Q, a) = (h_\lhd, a)$ with $h_\lhd : Q \to Q$ defined by $h_\lhd(q) = \delta(q, \lhd)$, and for $h \in Q^Q$ and $a, a' \in \Sigma$ we let $\delta_q^\leftarrow((h, a), a') = (h', a')$ with $h' : Q \to Q$ defined by $h'(q) = h(\delta(q, a))$. This provides the information described in (2). These two informations are then combined by the output function f_q of $\mathcal{B}_q$ as follows: $f_q((q_1, q_2), a, (h, b)) = \delta(h(q_2), a)$.   □

Lemma 7. *If $T \subseteq \mathcal{T}_\Sigma$ is recognizable by a $\downarrow$DTA, then T is path closed.*

Proof. Let $\mathcal{A} = (Q, \Sigma, f_{\text{in}}, (\mathcal{B}_q)_{q \in Q}, F)$ be a $\downarrow$DTA recognizing T. Note that it is sufficient to show $cl(T) \subseteq T$ since the other inclusion always holds.

For $q \in Q$ and $a \in \Sigma$ let $\mathcal{A}_{a \to q} = (Q, \Sigma, f_{\mathrm{in}}^{a \to q}, (\mathcal{B}_q)_{q \in Q}, F)$ with $f_{\mathrm{in}}^{a \to q}(a) = q$ and $f_{\mathrm{in}}^{a \to q}(b) = f_{\mathrm{in}}(b)$ for $b \in \Sigma \setminus \{a\}$, and let $T_{a \to q}$ denote the language accepted by $\mathcal{A}_{a \to q}$. By induction on the height of t one shows that $\pi(t) \subseteq \pi(T_{a \to q})$ implies that $t \in T_{a \to q}$ for each $q \in Q$, $a \in \Sigma$, and $t \in T_\Sigma^a$. Since $T_{a \to f_{\mathrm{in}}(a)} = T$ we obtain the desired result that $t \in \mathrm{cl}(T)$ (i.e., $\pi(t) \subseteq \pi(T)$) implies that $t \in T$. The details of the induction are left to the reader. $\qquad\square$

As an immediate consequence of the previous lemmas we get the following theorem.

Theorem 5. *Let $T \subseteq T_\Sigma$ be regular. Then T is $\downarrow DTA$ recognizable if and only if T is path closed.*

To obtain the desired decidability result we simply note that the construction from Lemma 6 is effective. Furthermore, language inclusion is decidable in exponential time for finite automata over unranked trees. Most easily, this can be seen by using an encoding of unranked trees by ranked trees [Suc02] and then using algorithms for ranked tree automata (cf. [CDG+97]).

Theorem 6. *Given a regular language $T \subseteq T_\Sigma$ it is decidable whether T can be recognized by a $\downarrow DTA$. Furthermore, such a $\downarrow DTA$ can be effectively constructed.*

Let us address two restricted models of deterministic top-down tree automata that are as well natural, mutually incompatible in expressive power, and lead to completely analogous results for suitable adaptions of the notion of path language.

The first model was already indicated in the introduction. Precisely as for the case of deterministic top-down automata over ranked trees, one requires the automaton to assume states at the successor nodes of a tree node x solely on the basis of the label a at x, the state q assumed there, and the rank of a, i.e., the number n of successors, but independent of the labels of the successor nodes. This leads to a special model of bimachine where the input sequence is a word $\bullet^n$ rather than a label sequence $a_1 \ldots a_n$. Accordingly, in the definition of path language we have to use between pairs $\triangleright, \triangleleft$ just the symbol $\bullet$ instead of the label letters, besides $\triangledown$ of course.

The second model is based on a left-to-right scanning process of successor labels; so a standard sequential machine (cf. [Eil74, Ber79]) is used to produce the successor states, hence without reference to the rank (the number n of the respective successor nodes altogether). In this case, our coding of paths has to be modified by canceling the segments between a symbol $\triangledown$ and the respective next $\triangleleft$ in order to obtain results analogous Theorems 5 and 6 above.

A similar concept has been used in [MNS05] in the context of typing streaming XML documents in a single pass. Recognizability by the second model from above (using sequential machines for labeling the successors), corresponds to definability by specialized DTDs with ancestor-sibling-based types in [MNS05].

5   Conclusion

In this paper we have extended the list of properties that carry over from tree automata for ranked trees to the unranked setting. We have shown that, using appropriate definitions, it is possible to minimize bottom-up deterministic tree automata over unranked trees in quadratic time. This minimization yields unique representatives for regular languages of unranked trees that can, e.g., be used to speed up equivalence tests. We have also transferred the characterization of deterministic top-down tree languages in terms of path languages and path closure from the ranked to the unranked case. This characterization can be used to decide for a given regular language of unranked trees whether it is top-down deterministic. A refinement of the minimization algorithm and a detailed analysis of the complexity, as well as the problem of minimizing deterministic top-down automata are subject of current and future research.

References

[BB02]   J. Berstel and L. Boasson. Formal properties of XML grammars and languages. *Acta Informatica*, 38:649–671, 2002.

[Ber79]   J. Berstel. *Transductions and Context-Free Languages*. Teubner, 1979.

[BWM01]   A. Brüggemann-Klein, D. Wood, and M. Murata. Regular tree and regular hedge languages over unranked alphabets: Version 1. unfinished technical report, April 2001. http://citeseer.ist.psu.edu/451005.html.

[CDG⁺97]   H. Comon, M. Dauchet, R. Gilleron, F. Jacquemard, D. Lugiez, S. Tison, and M. Tommasi. *Tree Automata Techniques and Applications*. unpublished electronic book, 1997. http://www.grappa.univ-lille3.fr/tata.

[Eil74]   S. Eilenberg. *Automata, languages, and machines*, volume A. Academic Press, 1974.

[GS84]   F. Gécseg and M. Steinby. *Tree automata*. Akadémiai Kiadò, Budapest, 1984.

[Hop71]   J. E. Hopcroft. An *nlogn* algorithm for minimizing states in a finite automaton. *Theory of Machines and Computations*, pages 189–196, 1971.

[HU79]   J. E. Hopcroft and J. D. Ullman. *Introduction to Automata Theory, Languages, and Computation*. Addison Wesley, 1979.

[MNS05]   W. Martens, F. Neven, and Th. Schwentick. Which XML schemas admit 1-pass preorder typing? In *ICDT*, volume 3363 of *LNCS*, pages 68–82. Springer, 2005.

[MSV03]   T. Milo, D. Suciu, and V. Vianu. Typechecking for XML transformers. *Journal of Computer and System Sciences*, 66(1):66–97, 2003.

[Nev02]   F. Neven. Automata, logic, and XML. In *CSL 2002*, volume 2471 of *LNCS*, pages 2–26. Springer, 2002.

[NS02]   F. Neven and Th. Schwentick. Query automata on finite trees. *Theoretical Computer Science*, 275:633–674, 2002.

[Suc02]   D. Suciu. Typechecking for semistructured data. In *Database Programming Languages, 8th International Workshop, DBPL 2001*, volume 2397 of *LNCS*, pages 1–20. Springer, 2002.

[Vir80]   J. Virágh. Deterministic ascending tree automata I. *Acta Cybernet.*, 5:33–42, 1980.

Decidable Membership Problems for Finite Recurrent Systems over Sets of Naturals

Daniel Meister

Bayerische Julius-Maximilians-Universitaet Wuerzburg,
97074 Wuerzburg, Germany
meister@informatik.uni-wuerzburg.de

Abstract. A finite recurrent system over the power set of the natural numbers of dimension n is a pair composed of n n-ary functions over the power set of the natural numbers and an n-tuple of singleton sets of naturals. Every function is applied to the components of the tuple and computes a set of natural numbers, that might also be empty. The results are composed into another tuple, and the process is restarted. Thus, a finite recurrent system generates an infinite sequence of n-tuples of sets of natural numbers. The last component of a generated n-tuple is the output of one step, and the union of all outputs is the set defined by the system. We will consider only special finite recurrent systems: functions are built from the set operations union ($\cup$), intersection ($\cap$) and complementation ($^-$) and the arithmetic operations addition ($\oplus$) and multiplication ($\otimes$). Sum and product of two sets of natural numbers are defined elementwise. We will study two types of membership problems: given a finite recurrent system and a natural number, does the set defined by the system contain the queried number, and does the output of a specified step contain the queried number? We will determine upper and lower bounds for such problems where we restrict the allowed operations to subsets of $\{\cup, \cap, ^-, \oplus, \otimes\}$. We will show completeness results for the complexity classes NL, NP and PSPACE.

1 Introduction

Sets of natural numbers can be represented by a variety of mathematical objects. Finite sets or co-finite sets, the complements of finite sets, can be represented by words over $\{0, 1\}$, i.e., by natural numbers, with a canonical interpretation. However, large sets require large numbers in this model. If these sets possess regularities a more efficient representation would be desirable. In case of sets that are neither finite nor co-finite such a simple representation does not work at all. Stockmeyer and Meyer defined integer expressions, which are expressions built from naturals, the set operations union, intersection and complementation and an addition operation [7]. Wagner studied a hierarchical model of a similar flavour that can be understood as arithmetic circuits [9], [10]. Such concise representations however make it difficult to derive information about the set from its representation. The *membership problem* for natural numbers in general can

M. Liśkiewicz and R. Reischuk (Eds.): FCT 2005, LNCS 3623, pp. 80–91, 2005.

be understood as the problem, given a set M of natural numbers represented in a certain model and a number b, to decide whether b belongs to set M. The complexity of the membership problem heavily depends on the representation and can generally be described by the formula: the more concise the representation the more complex the membership problem.

McKenzie and Wagner recently studied a large number of membership problems [2]. Given an arithmetic circuit over sets of natural numbers involving the standard set operations union, intersection, complementation and the arithmetic operations addition and multiplication (both operations are defined on sets, and sum and product of two sets are defined elementwise) and a natural number b, does the circuit represent a set that contains b? It was shown that restricting the set of possible operations as well as restricting circuits to formulas cover a wide range of complexity classes. Here, a formula is an arithmetic circuit where every vertex has at most one successor. Their work extends past works by Stockmeyer and Meyer [7], Wagner [9] and Yang [11].

The standard approach to circuits is via functions, and circuits represent these functions efficiently. In this sense, all problems above concern such circuits but applied only to fixed inputs. Circuits of various types have been studied deeply, and they are an interesting model to obtain lower bounds complexity results. A lot of information on this subject can be found in the book by Vollmer [8]. In this paper, we combine ideas that have been sketched above to obtain set representations by special recurrent systems.

Recurrences are well-known. The sequence $1, 1, 2, 3, 5, 8, \ldots$ of numbers—the Fibonacci numbers—is generated by the simple formula $F(n+2) =_{\text{def}} F(n+1) + F(n)$ where $F(0) =_{\text{def}} F(1) =_{\text{def}} 1$. Numerical simulations of single-particle or multi-particle systems in physics use systems of recurrences instead of differential equations. Recurrences play an important role in mathematics, computer and other sciences. Though recurrences normally involve only basic arithmetic operations such as addition and multiplication over the natural or the real numbers, operations do not have to be limited to this small collection. A recurrent system over sets of natural numbers of dimension n is a pair consisting of a set of n n-ary functions $f_1, \ldots, f_n$ over sets of natural numbers and an n-tuple of naturals. Starting from singleton sets defined by the n-tuple the result of function f_i in one step is used as the i-th input in the next step (the precise definition is provided in Section 3). So, a recurrent system iteratively generates an infinite sequence of tuples of sets of natural numbers. The last component of each tuple is the output of the system in the corresponding evaluation step. Then, the union of all outputs defines a set that may be finite or infinite. The existential membership problem M_{ex} for recurrent systems asks whether there is an evaluation step such that the corresponding output contains a given number, and the exact membership problem M_{tm} asks whether a given number is contained in the result of a specified evaluation step. Functions are represented by arithmetic circuits.

We examine membership problems for recurrent systems for a restricted set of operations. Functions are built from the three known set operations and addition and multiplication. The general problems in this restricted sense are denoted by $M_{ex}(\cup, \cap, ^-, \oplus, \otimes)$ and $M_{tm}(\cup, \cap, ^-, \oplus, \otimes)$. Reducing the set of allowed operations leads to problems like $M_{ex}(\cup, \oplus)$, where functions are built only from $\cup$ and $\oplus$, or $M_{tm}(\oplus, \otimes)$. We will study the complexity of such membership problems with respect to the set of allowed operations. We will see that such problems are complete for a number of complexity classes where we will focus on NP- and PSPACE-complete problems. The general existential membership problem over $\{\cup, \cap, ^-, \oplus, \otimes\}$ is undecidable [4]; however, the exact complexity is not known in that sense that the currently best known lower bound (RE-hardness) does not meet the upper bound Σ_2. It is a most interesting question whether $M_{ex}(\cup, \cap, ^-, \oplus, \otimes)$ is coRE-hard. This would imply undecidability of the general problem considered by McKenzie and Wagner [2]. Some evidence for undecidability was given by showing that a decision algorithm would prove or disprove Goldbach's conjecture about sums of primes.

This presentation is composed as follows. In Section 3, finite recurrent systems are defined, an example is discussed and basic and supplementary results are mentioned. The following sections classify a range of membership problems for recurrent systems. Section 4 considers membership problems that are contained in P. These problems are related to number-of-paths problems in graphs, whose complexities were studied in [3]. In Section 5, NP-complete problems, such as $M_{ex}(\cap)$ and $M_{ex}(\cap, \oplus)$, are considered, Section 6 considers PSPACE-complete problems, e.g., $M_{ex}(\cup, \cap)$ and $M_{ex}(\cup, \oplus, \otimes)$, and in Section 7, problems without exact classification are delt with. The conclusions section contains a table summarising the best known upper and lower complexity bounds for all possible problems. In most cases, proofs are omitted or reduced to just the main ideas.

2   Preliminaries

We fix the alphabet $\Sigma =_{def} \{0, 1\}$. The set of all words over Σ is denoted by Σ^*. All inputs are assumed to be given as words over Σ. For definitions and notations of complexity classes we refer to the book by Papadimitriou [5]. If the computation mode is not mentioned we mean deterministic computations; nondeterminism is always indicated. The class FL contains all functions that can be computed deterministically by a Turing machine with output tape using logarithmic working space. A set A is *log-space reducible* to set B, $A \leq_m^L B$, if there is $f \in$ FL such that, for all $x \in \Sigma^*$, $x \in A \leftrightarrow f(x) \in B$. We will also say that A *reduces* to B. For complexity class $\mathcal{C}$, set A is $\leq_m^L$-*complete* for $\mathcal{C}$, if $A \in \mathcal{C}$ and $B \leq_m^L A$ for all $B \in \mathcal{C}$. We will shortly say that A is $\mathcal{C}$-complete.

Numbers. The set of the natural numbers is denoted by $\mathbb{N}$ and surely contains 0. If we talk about numbers, we always mean natural numbers. Unless otherwise stated numbers are represented in binary form. The power set of $\mathbb{N}$ is the set of all subsets of $\mathbb{N}$. For natural numbers a, b, $a \leq b$, $[a, b] =_{def} \{a, a+1, \ldots, b\}$. Two numbers are *relatively prime*, if their greatest common devisor is 1.

Theorem 1 (Chinese Remainder Theorem). *Let $b_1, \ldots, b_k$ be pairwise relatively prime numbers, and let $n_1, n_2 \in \mathbb{N}$. Let $b =_{\mathrm{def}} b_1 \cdots b_k$. Then, $n_1 \equiv n_2 \pmod{b}$ if and only if $n_1 \equiv n_2 \pmod{b_i}$ for every $i \in [1, k]$.*

For set A and two binary operations $\diamond$ and $\circ$ over A, the triple $(A, \diamond, \circ)$ is a *semiring* if $(A, \diamond)$ and $(A, \circ)$ are commutative monoids and the two distributive laws hold. For $+$ and $\cdot$ denoting addition and multiplication over $\mathbb{N}$, $(\mathbb{N}, +, \cdot)$ is a semiring. By $\mathrm{SR}(b)$ we denote the semiring $([0, b{+}1], \mathrm{sum}_b, \mathrm{prod}_b)$ where the binary operations sum_b and prod_b are defined as follows. Let $a_1, a_2 \in \mathbb{N}$.

$$\mathrm{sum}_b(a_1, a_2) =_{\mathrm{def}} \begin{cases} a_1 + a_2 & \text{, if } a_1 + a_2 \leq b \\ b{+}1 & \text{otherwise} \end{cases}$$

Similarly for prod_b. So, matrix multiplication over $\mathrm{SR}(b)$ is well-defined.

Graphs. A simple, finite, directed graph is a pair $G = (V, A)$ where V is a finite set and $A \subseteq V \times V$. For two vertices $u, v \in V$ there is a u, v-*path* in G, if there is a sequence $(x_0, \ldots, x_k)$ such that $x_0 = u$, $x_k = v$ and $(x_i, x_{i+1}) \in A$ for all $i \in [0, k{-}1]$. The *graph accessibility problem for directed graphs*, denoted by GAP, is the set of all triples (G, u, v) where G is a directed graph, u and v are vertices of G and there is a u, v-path in G. The problem GAP is NL-complete [6]. G is *acyclic*, if there is no sequence $P = (x_0, \ldots, x_n)$ for n the number of vertices of G such that P is an x_0, x_n-path in G for any pair of vertices x_0, x_n of G. The problem ACYC is the set of all directed acyclic graphs. Since GAP restricted to acyclic graphs is NL-complete, ACYC is NL-complete. For vertices u and v of G, u is a *predecessor* of v, if $(u, v) \in A$.

Circuits. Let $\mathcal{O}$ be a set of commutative operations over set M. $C = (G, g_c, \alpha)$ is an n-ary *arithmetic $\mathcal{O}$-circuit* over M for $n \geq 0$, if $G = (V, A)$ is a (simple, finite) acyclic graph, $g_C \in V$ is a specified vertex of G, the *output vertex*, and $\alpha : V \to \mathcal{O} \cup [1, n]$ such that α establishes a 1-1 correspondence between n vertices without predecessor and $[1, n]$ and all other vertices are assigned operations from $\mathcal{O}$ whose arities correspond with the numbers of predecessors of the vertices. Vertices assigned a number are called *input vertices* of C. The arithmetic $\mathcal{O}$-circuit C over M represents a function f_C over M in the following way. Let $(a_1, \ldots, a_n) \in M^n$. The value of the input vertex assigned number i is a_i, the value of vertex u where u is assigned an operation from $\mathcal{O}$ is the result of $\alpha(u)$ applied to the values of the predecessors of u. Then, $f_C(a_1, \ldots, a_n)$ is the value of the output vertex g_C. Let f_C be an n-ary function represented by circuit C, and let $f_{C_1}, \ldots, f_{C_n}$ be n'-ary functions represented by circuits $C_1, \ldots, C_n$. A circuit representation of function $f(x_1, \ldots, x_{n'}) = f_C(f_{C_1}(x_1, \ldots, x_{n'}), \ldots, f_{C_n}(x_1, \ldots, x_{n'}))$ is obtained from $C, C_1, \ldots, C_n$ by identifying the input vertices of $C_1, \ldots, C_n$ assigned the same numbers and identifying the vertices of C assigned numbers with the output vertex of the corresponding circuit C_i.

3   Finite Recurrent Systems

A recurrence is a pair composed of a function and initial values. From recurrences one can generate infinite sequences of objects by applying the function to certain of already generated objects. Usual recurrences are defined over natural, real or complex numbers and involve only basic arithmetical operations like addition and multiplication. We extend this notion to recurrent systems over sets of numbers.

Definition 1. *A* **finite recurrent system** *over sets of natural numbers of dimension $n \geq 1$ is a pair $S = (\mathcal{F}, A)$ where $\mathcal{F} =_{\text{def}} \langle f_1, \ldots, f_n \rangle$ for $f_1, \ldots, f_n$ n-ary functions over sets of natural numbers and $A \in \mathbb{N}^n$. The dimension n of S is denoted by $\dim S$.*

Let $S = (\mathcal{F}, A)$ be a finite recurrent system over sets of natural numbers where $\mathcal{F} = \langle f_1, \ldots, f_n \rangle$ and $A = (a_1, \ldots, a_n)$. We define for every $t \in \mathbb{N}$:

$$S_i(0) =_{\text{def}} S[f_i](0) =_{\text{def}} \{a_i\}, \quad i \in [1, n]$$
$$S_i(t+1) =_{\text{def}} S[f_i](t+1) =_{\text{def}} f_i(S_1(t), \ldots, S_n(t)), \quad i \in [1, n]$$
$$\mathcal{F}(t) =_{\text{def}} (S_1(t), \ldots, S_n(t))$$
$$S(t) =_{\text{def}} S_n(t).$$

So, $S_i(t)$ denotes the result of f_i in the t-th evaluation step. We can say that a finite recurrent system over sets of naturals defines or represents an infinite sequence of sets of naturals. By $[S]$ we denote the union of these sets, i.e., $[S] =_{\text{def}} \bigcup_{t \geq 0} S(t)$. We are interested in two problems that arise from our definitions. We ask whether a number b is generated in step t and whether b is generated in some step at all, i.e., contained in $[S]$.

Several authors studied membership problems of sets of natural numbers that can be built from singleton sets of natural numbers by applying the set operations union, intersection, complementation and the two arithmetic set operations addition and multiplication, denoted by $\oplus$ and $\otimes$ [7], [9], [11], [2]. Addition and multiplication on sets are defined elementwise. Let $A, B \subseteq \mathbb{N}$. Then, $A \oplus B =_{\text{def}} \{r+s : r \in A \text{ and } s \in B\}$ and $A \otimes B =_{\text{def}} \{r \cdot s : r \in A \text{ and } s \in B\}$. Let $\mathcal{O} \subseteq \{\cup, \cap, ^-, \oplus, \otimes\}$. An n-ary $\mathcal{O}$-function $f = f(x_1, \ldots, x_n)$ is a function over the variables $x_1, \ldots, x_n$ defined by using only operations from $\mathcal{O}$. An $\mathcal{O}$-function is an n-ary $\mathcal{O}$-function for some $n \geq 1$.

Definition 2. *Let $\mathcal{O} \subseteq \{\cup, \cap, ^-, \oplus, \otimes\}$. A* **finite recurrent $\mathcal{O}$-system** *$S = (\mathcal{F}, A)$ over sets of natural numbers is a finite recurrent system over sets of natural numbers where every function in $\mathcal{F}$ is an $\mathcal{O}$-function.*

Our introductory sample sequence, the sequence of Fibonacci numbers, can be generated by a finite recurrent $\{\oplus\}$-system. Let

$$\mathcal{F} =_{\text{def}} \langle f_1, f_2 \rangle \text{ where } f_1(x_1, x_2) =_{\text{def}} x_2 \text{ and } f_2(x_1, x_2) =_{\text{def}} x_1 \oplus x_2$$
$$A =_{\text{def}} (0, 1).$$

Let $S =_{\text{def}} (\mathcal{F}, A)$. Then,

$$S(0) = S_2(0) = \{1\}$$
$$S(1) = S_2(1) = f_2(S_1(0), S_2(0)) = f_2(\{0\}, \{1\}) = \{1\}$$
$$S(2) = S_2(2) = f_2(S_1(1), S_2(1)) = f_2(\{1\}, \{1\}) = \{2\},$$

and so on. We will often speak of *recurrent systems* for short, which always means finite recurrent $\{\cup, \cap, ^-, \oplus, \otimes\}$-systems over sets of naturals. Every recurrent system S defines a possibly infinite set $[S]$ of natural numbers. The *existential membership problem* M_{ex} for recurrent systems asks whether a given number is contained in the defined set, and the *exact membership problem* M_{tm} asks whether a given number is contained in the result of a specified evaluation step. We want to study the complexities of these membership problems with respect to the allowed operations. Let $\mathcal{O} \subseteq \{\cup, \cap, ^-, \oplus, \otimes\}$.

$$\mathrm{M}_{ex}(\mathcal{O}) =_{\mathrm{def}} \{(S, b) : S \text{ a recurrent } \mathcal{O}\text{-system and } b \in [S]\}$$

$$\mathrm{M}_{tm}(\mathcal{O}) =_{\mathrm{def}} \{(S, t, b) : S \text{ a recurrent } \mathcal{O}\text{-system and } b \in S(t)\}$$

Instead of writing $\mathrm{M}_{ex}(\{\cup, \cap, \oplus\})$ we will write $\mathrm{M}_{ex}(\cup, \cap, \oplus)$ for short; similarly for the other problems. The complexities of our problems strongly depend on the input representation. We assume that natural numbers are given in binary form and functions are represented by arithmetic circuits with appropriate labels. For circuits we require any (standard) encoding that permits adjacency tests of two vertices and detection of labels of vertices in logarithmic space. It can be verified in nondeterministic logarithmic space whether an input represents an $\mathcal{O}$-function for $\mathcal{O} \subseteq \{\cup, \cap, ^-, \oplus, \otimes\}$. Using our notations, McKenzie and Wagner studied the complexity of the question, for given recurrent $\mathcal{O}$-system S and number $b \geq 0$, whether $(S, 1, b) \in \mathrm{M}_{tm}(\mathcal{O})$ [2]. Their input representation additionally required a topological ordering of the vertices of the circuits, but this is only of importance for problems that are contained in NL. We will denote the problems investigated by McKenzie and Wagner by $\mathrm{MC}(\mathcal{O})$. It follows for every $\mathcal{O} \subseteq \{\cup, \cap, ^-, \oplus, \otimes\}$ that $\mathrm{M}_{tm}(\mathcal{O})$ is decidable if and only if $\mathrm{MC}(\mathcal{O})$ is decidable. The only such problems that have not yet been proved decidable are $\mathrm{MC}(\cup, \cap, ^-, \oplus, \otimes)$ and $\mathrm{MC}(^-, \oplus, \otimes)$ (see also [2]).

Proposition 1.

(i) $\mathrm{M}_{tm}(\cup, \cap, ^-, \oplus, \otimes)$ *is either decidable or not recursively enumerable.*

(ii) $\mathrm{M}_{ex}(\cup, \cap, ^-, \oplus, \otimes)$ *is recursively enumerable if and only if* $\mathrm{MC}(\cup, \cap, ^-, \oplus, \otimes)$ *is decidable.*

(iii) $\mathrm{M}_{ex}(^-, \oplus, \otimes)$ *is recursively enumerable if and only if* $\mathrm{MC}(^-, \oplus, \otimes)$ *is decidable.*

Glaßer showed that $\mathrm{MC}(\cup, \cap, ^-, \oplus, \otimes)$ is contained in $\Delta_2 = \Sigma_2 \cap \Pi_2$ [1].

Theorem 2. *[4]*

(i) $\mathrm{M}_{ex}(\cup, \cap, \oplus, \otimes)$ *is Σ_1-complete.*

(ii) $\mathrm{M}_{ex}(^-, \oplus, \otimes)$ *is Σ_1-hard.*

(iii) $\mathrm{M}_{tm}(\cup, \cap, ^-, \oplus, \otimes) \in \Delta_2$.

(iv) $\mathrm{M}_{ex}(\cup, \cap, ^-, \oplus, \otimes) \in \Sigma_2$.

4   Easiest Membership Problems

In this section we consider membership problems that are contained in P. These problems have a strong connection to graph problems concerning the numbers of paths of certain lengths between two vertices. Such problems are investigated in [3]. In the same paper connections to matrix problems are established. This matrix interpretation is also of great advantage in the study of the problem $M_{tm}(\cap, \oplus)$.

As a general model, proofs showing containment results for existential membership problems have a common structure. First, an upper bound for the complexity of deciding $M_{tm}(\mathcal{O})$ for $\mathcal{O} \subseteq \{\cup, \cap, ^-, \oplus, \otimes\}$ is given. Second, the value of t is bounded by some number r for which holds that $(S, b) \in M_{ex}(\mathcal{O})$ if and only if there is $t < r$ such that $(S, t, b) \in M_{tm}(\mathcal{O})$. Bound r normally depends on b and the dimension of S.

Lemma 1. $M_{tm}(^-)$ *is in* L.

The problem $\mathrm{NMDP}(2, \beta)$ for $\beta \geq 1$ is the set of all tuples (G, M, k, ν, u, v) where $G = (V, A)$ is a simple finite directed graph, $M \subseteq V$, $u, v \in V$, $k, \nu \in \mathbb{N}$, k is represented in binary form, ν is represented in β-ary form, and there are ν u, v-paths in G each of which containing exactly k vertices from set M. Let $\mathrm{ExNMDP}(2, \beta)$ denote the problem corresponding to $\mathrm{NMDP}(2, \beta)$ where we ask for *exactly* ν paths.

Theorem 3. *[3]*

(i) $\mathrm{NMDP}(2, 1)$ *and* $\mathrm{ExNMDP}(2, 1)$ *are* NL-*complete.*
(ii) $\mathrm{ExNMDP}(2, 2)$ *is in* P *and* $C_=L$-*hard.*

Lemma 2. $M_{tm}(\cup)$ *and* $M_{tm}(\cap)$ *are* NL-*complete.*

Proof. For showing $M_{tm}(\cup) \in$ NL and $M_{tm}(\cap) \in$ NL, both problems are reduced to $\mathrm{NMDP}(2, 1)$. Hardness of $M_{tm}(\cup)$ and $M_{tm}(\cap)$ follows by the canonical reduction from the accessibility problem for acyclic graphs.

Theorem 4. *(i)* $M_{ex}(\emptyset)$ *and* $M_{ex}(^-)$ *are in* L.
(ii) $M_{ex}(\cup)$ *is* NL-*complete.*

McKenzie and Wagner showed that $\mathrm{MC}(\otimes)$ is NL-complete and that $\mathrm{MC}(\oplus)$ is $C_=L$-complete [2].

Theorem 5. $M_{tm}(\otimes)$ *is* NL-*complete, and* $M_{tm}(\oplus)$ *is in* P *and* $C_=L$-*hard.*

Proof. Containment of both problems is shown by using $\mathrm{ExNMDP}(2, 1)$ or $\mathrm{ExNMDP}(2, 2)$ as oracle set. Hardness of both problems follows by the results of McKenzie and Wagner [2].

Let $M_{tm}^+(\cap, \otimes)$ denote the set of tuples $(S, t, b) \in M_{tm}(\cap, \otimes)$ where $b > 0$. By a construction that replaces numbers by a representation over a basis of relatively prime numbers we can show the following lemma. The same idea with a different construction was used by McKenzie and Wagner to obtain similar results [2].

Lemma 3. $M_{tm}^+(\cap, \otimes) \leq_m^P M_{tm}(\cap, \oplus)$.

A thorough analysis of recurrent $\{\cap, \oplus\}$-systems and results from linear algebra yield the following theorem. The main part of its proof shows how to decide in polynomial time whether $S(t)$ for S a recurrent $\{\cap, \oplus\}$-system is empty. This problem is not solved entirely. However, in the uncertain case the result of $S(t)$ is either empty or too large. A complete solution of the emptyness problem is of great importance for solving $M_{tm}(\cap, \otimes)$.

Theorem 6. $M_{tm}(\cap, \oplus)$ *is in* P.

Corollary 1. $M_{tm}^+(\cap, \otimes)$ *is in* P.

5   NP-Complete Membership Problems

To show hardness of the problems considered in this section, we define a new problem. This problem can be considered a generalization of the Chinese Remainder Theorem. The Chinese Remainder Theorem shows that a system of congruence equations where the moduli are pairwise relatively prime numbers has a solution that is unique in a determined interval of natural numbers. We extend this problem with respect to two aspects. Moduli are arbitrary numbers, and for each modulus we find a set of congruence equations. A solution of this *Set-system of congruence equations* fulfills one equation from each set. Formally, we define the problem SET-SCE as follows.

Solving a Set-System of Congruence Equations (SET-SCE).
INSTANCE. $((A_1, b_1), \ldots, (A_k, b_k))$ where $A_1, \ldots, A_k$ are finite sets of natural numbers, and $b_1, \ldots, b_k$ are natural numbers greater than 1 represented in unary form.
QUESTION. Are there $n \in \mathbb{N}$ and $a_1 \in A_1, \ldots, a_k \in A_k$ such that $n \equiv a_i \pmod{b_i}$ for all $i \in [1, k]$?

Note that it is not important to require binary representation of the numbers in $A_1, \ldots, A_k$. However, we assume a binary representation of them to fix a system.

Lemma 4. SET-SCE *is* NP-*hard*.

Theorem 7. $M_{ex}(\cap)$, $M_{ex}(\oplus)$, $M_{ex}(\otimes)$, $M_{ex}(\cap, \oplus)$, $M_{ex}^+(\cap, \otimes)$ *are* NP-*complete*.

Proof. We only show that $M_{ex}(\cap)$ is NP-complete by reducing SET-SCE to $M_{ex}(\cap)$. Let $\mathcal{S} =_{\text{def}} ((A_1, b_1), \ldots, (A_k, b_k))$ be an instance of SET-SCE. We assume that A_i only contains numbers that are smaller than b_i. We define a recurrent $\{\cap\}$-system $S = (\mathcal{F}, A)$ as follows. For every $i \in [1, k]$, for every $j \in [1, b_i-1]$ we define

$$f_j^{(i)}(x) =_{\text{def}} x_{j-1}^{(i)} \quad \text{and} \quad f_0^{(i)}(x) =_{\text{def}} x_{b_i-1}^{(i)}$$

where $\mathsf{x} =_{\mathrm{def}} (x_0^{(1)}, \ldots, x_{b_1-1}^{(1)}, x_0^{(2)}, \ldots, x_{b_k-1}^{(k)}, x')$. Let $A =_{\mathrm{def}} (c_0^{(1)}, \ldots, c_{b_k-1}^{(k)}, 0)$ where $c_j^{(i)} \in \{0, 1\}$ and $c_j^{(i)} = 1$ if and only if $j \in A_i$. Furthermore, let $f'(\mathsf{x}) =_{\mathrm{def}} x_0^{(1)} \cap \cdots \cap x_0^{(k)}$ and $\mathcal{F} =_{\mathrm{def}} \langle f_0^{(1)}, \ldots, f_{b_k-1}^{(k)}, f' \rangle$. It holds that $S[f_j^{(i)}](t) = 1$ if and only if $c_r^{(i)} = 1$ for $r < b_i$ and $r \equiv t - j \pmod{b_i}$. Hence, $(S, 1) \in \mathrm{M}_{ex}(\cap)$ if and only if $\mathcal{S} \in$ SET-SCE. By Lemma 4, $\mathrm{M}_{ex}(\cap)$ is NP-hard.

Corollary 2. SET-SCE *is* NP-*complete.*

It remains open not only whether $\mathrm{M}_{tm}(\cap, \otimes)$ is polynomial-time decidable but also whether $\mathrm{M}_{ex}(\cap, \otimes)$ is contained in NP. We do not know any upper bound c for t such that $0 \in [S]$ if and only if $0 \in S(t)$ for some $t < c$ where S is a recurrent $\{\cap, \otimes\}$-system.

6   PSPACE-Complete Membership Problems

This section contains three interesting results. First, we will see that the existential membership problem for finite recurrent $\{\cup, \cap\}$-systems is PSPACE-complete. Containment is a mere observation. Hardness is shown by a reduction from QBF. Astoundingly at first glance, the corresponding exact membership problem is PSPACE-complete, too. Second, we will see that $\mathrm{M}_{tm}(\cup, \oplus, \otimes)$ can be decided in polynomial space. This result is surprising when we keep in mind that $\mathrm{MC}(\cup, \oplus, \otimes)$ is PSPACE-complete [11]. Third, we will see that a recurrent $\{\cup, \oplus, \otimes\}$-system S needs at most $(b+1) \cdot 2^{n^3}$ evaluation steps to generate number b where $n =_{\mathrm{def}} \dim S$. This leads to a polynomial-space decision algorithm for $\mathrm{M}_{ex}(\cup, \oplus, \otimes)$.

Theorem 8. *[7]* QBF *is* PSPACE-*complete.*

Theorem 9. QBF $\leq_m^{\mathrm{L}} \mathrm{M}_{ex}(\cup, \cap)$.

We turn to recurrent $\{\cup, \oplus, \otimes\}$-systems.

Lemma 5. *Let* $S = (\mathcal{F}, A)$ *be a recurrent* $\{\cup, \oplus, \otimes\}$-*system,* $n =_{\mathrm{def}} \dim S$. *Let* $b \in \mathbb{N}$. *Then,* $b \in [S]$ *if and only if there is* $t < (b+1) \cdot 2^{n^3}$ *such that* $b \in S(t)$.

As we have already discussed the problems $\mathrm{M}_{tm}(\mathcal{O})$ for $\mathcal{O} \subseteq \{\cup, \cap, {}^-, \oplus, \otimes\}$ can be considered similar to the problems $\mathrm{MC}(\mathcal{O})$ with succinct input representation. For most of our problems succinctness led to an increase of complexity. With this phenomenon in mind it is surprising that we can show that $\mathrm{M}_{tm}(\cup, \oplus, \otimes)$ is solvable in polynomial space. It is known that $\mathrm{MC}(\cup, \oplus, \otimes)$ is PSPACE-complete [11].

Theorem 10. $\mathrm{M}_{tm}(\cup, \oplus, \otimes)$ *is in* PSPACE.

Theorem 11. $\mathrm{M}_{ex}(\cup, \cap)$, $\mathrm{M}_{ex}(\cup, \cap, {}^-)$, $\mathrm{M}_{ex}(\oplus, \otimes)$, $\mathrm{M}_{ex}(\cup, \oplus)$, $\mathrm{M}_{ex}(\cup, \otimes)$ *and* $\mathrm{M}_{ex}(\cup, \oplus, \otimes)$ *are* PSPACE-*complete.*

Corollary 3. $\mathrm{M}_{tm}(\cup, \cap)$, $\mathrm{M}_{tm}(\cup, \cap, {}^-)$, $\mathrm{M}_{tm}(\oplus, \otimes)$, $\mathrm{M}_{tm}(\cup, \oplus)$, $\mathrm{M}_{tm}(\cup, \otimes)$ *and* $\mathrm{M}_{tm}(\cup, \oplus, \otimes)$ *are* PSPACE-*complete.*

7   More Complicated Problems

In this final section we consider those problems that are not yet solved entirely. These are most of the problems that allow ∩- and ⊗-operations. But also $M_{tm}(\cup, \cap, ^-, \oplus)$ and $M_{ex}(\cup, \cap, ^-, \oplus)$ are still open. We will not give tight upper and lower bounds. In most cases, we obtain upper bounds by adequately restating results by McKenzie and Wagner. However, the complexity of the mentioned problem $M_{tm}(\cup, \cap, ^-, \oplus)$ can significantly be improved with respect to the corresponding result from [2]. Let us first recall some necessary results.

Theorem 12. *[2]*

  (i) $MC(\cap, \otimes)$ *is in* P.
  (ii) $MC(\cap, \oplus, \otimes)$ *is in* coNP.
  (iii) $MC(\cup, \cap, ^-, \oplus)$ *and* $MC(\cup, \cap, ^-, \otimes)$ *are in* PSPACE.
  (iv) $MC(\cup, \cap, \oplus, \otimes)$ *is* NEXP-*complete.*

Given a recurrent system S and some number t, we find a circuit representation of $S(t)$ by concatenating circuits. We obtain the following corollary.

Corollary 4. *(i)* $M_{tm}(\cap, \otimes)$ *is in* EXP.
  (ii) $M_{tm}(\cap, \oplus, \otimes)$ *is in* coNEXP.
  (iii) $M_{tm}(\cup, \cap, ^-, \oplus)$ *and* $M_{tm}(\cup, \cap, ^-, \otimes)$ *are in* EXPSPACE.
  (iv) $M_{tm}(\cup, \cap, \oplus, \otimes)$ *is in* 2−NEXP.

In case of recurrent $\{\cup, \cap, ^-, \oplus\}$-systems, we can improve the trivial exponential-space upper bound.

Lemma 6. $M_{tm}(\cup, \cap, ^-, \oplus)$ *is in* EXP.

Observe that $M_{tm}(\cup, \cap)$ reduces to $M_{tm}(\cup, \cap, \oplus)$ and $M_{tm}(^-, \oplus)$. The latter reduction is done by replacing ∩ by ⊕, $A \cup B$ by $\overline{\overline{A} \oplus \overline{B}}$, the queried number b by 0 and every other number by 1. So, $M_{tm}(\cup, \cap, \oplus)$, $M_{tm}(^-, \oplus)$ and $M_{tm}(\cup, \cap, ^-, \oplus)$ are PSPACE-hard.

Proposition 2. $M_{ex}(\cup, \cap, ^-, \oplus)$ *is in* EXPSPACE.

In a way similar to the reduction from $M_{tm}(\cup, \cap)$ to $M_{tm}(^-, \oplus)$, the former problem reduces to $M_{tm}(^-, \otimes)$. Replace every ∪ by ⊗ and every $A \cap B$ by $\overline{\overline{A} \otimes \overline{B}}$, replace the queried number b by 0 and every other number by 1. Note that no $\{^-, \otimes\}$-function on inputs only $\{0\}$ or $\{1\}$ can compute $\emptyset$, since no such function can compute a set that contains 0 and 1. This shows PSPACE-hardness of $M_{tm}(^-, \otimes)$.

8   Concluding Remarks

In this extended abstract we introduced and studied two types of membership problems for recurrent $\{\cup, \cap, ^-, \oplus, \otimes\}$-systems. Table 1 summarises our results.

Table 1. Currently best known complexity bounds for the membership problems for finite recurrent systems. The question marks stand for Δ_2 or Σ_2.

Operation set	Exact problem M_{tm}		Existential problem M_{ex}	
	Lower bound	Upper bound	Lower bound	Upper bound
∪ ∩ ⁻ ⊕ ⊗	NEXP	?	RE	?
∪ ∩ ⊕ ⊗	NEXP	2−NEXP	RE	
⁻ ⊕ ⊗	PSPACE	?	RE	?
∪ ⊕ ⊗	PSPACE			
∩ ⊕ ⊗	PSPACE	coNEXP	PSPACE	RE
⊕ ⊗	PSPACE			
∪ ∩ ⁻ ⊕	PSPACE	EXP	PSPACE	EXPSPACE
∪ ∩ ⊕	PSPACE	EXP	PSPACE	EXPSPACE
⁻ ⊕	PSPACE	EXP	PSPACE	EXPSPACE
∪ ⊕	PSPACE			
∩ ⊕	$C_=L$	P	NP	
⊕	$C_=L$	P	NP	
∪ ∩ ⁻ ⊗	PSPACE	EXPSPACE	PSPACE	RE
∪ ∩ ⊗	PSPACE	EXPSPACE	PSPACE	RE
⁻ ⊗	PSPACE	EXPSPACE	PSPACE	RE
∪ ⊗	PSPACE			
∩ ⊗	NL	EXP	NP	RE
⊗	NL		NP	
∪ ∩ ⁻	PSPACE			
∪ ∩	PSPACE			
∩	NL		NP	
∪	NL			
⁻	L			
	L			

The question marks stand for complexity classes beyond the class of recursively enumerable sets. Future work should tighten upper and lower bounds. There are especially two open problems that the author finds worth being considered: undecidability of $M_{tm}(\cup, \cap, ^-, \oplus, \otimes)$ and an interesting upper—or lower—bound for $M_{ex}(\cap, \otimes)$. McKenzie and Wagner studied the emptyness problem for some circuits as an auxiliary problem. It would be interesting to solve the emptyness problem for recurrent $\{\cap, \oplus\}$-systems.

Acknowledgements. The idea to study finite recurrent systems was the result of a discussion with Klaus Wagner when a lot of people at Würzburg were studying membership problems for arithmetic circuits. I thank Bernhard Schwarz for his help.

References

[1] CHR. GLASSER, *private communication*, 2003.

[2] P. MCKENZIE, K.W. WAGNER, *The Complexity of Membership Problems for Circuits over Sets of Natural Numbers*, Proceedings of the 20th Annual Symposium on Theoretical Aspects of Computer Science, STACS 2003, Lecture Notes in Computer Science 2607, Springer, pp. 571–582, 2003.

[3] D. MEISTER, *The complexity of problems concerning matrix powers and the numbers of paths in a graph*, manuscript.

[4] D. MEISTER, *Membership problems for recurrent systems over the power set of the natural numbers*, Technical report 336, Institut für Informatik, Bayerische Julius-Maximilians-Universität Würzburg, 2004.

[5] CH.H. PAPADIMITRIOU, *Computational Complexity*, Addison-Wesley, 1994.

[6] W.J. SAVITCH, *Relationships Between Nondeterministic and Deterministic Tape Complexities*, Journal of Computer and System Sciences 4, pp. 177–192, 1970.

[7] L.J. STOCKMEYER, A.R. MEYER, *Word Problems Requiring Exponential Time*, Proceedings of the ACM Symposium on the Theory of Computation, pp. 1–9, 1973.

[8] H. VOLLMER, *Introduction to Circuit Complexity*, Springer, 1999.

[9] K. WAGNER, *The Complexity of Problems Concerning Graphs with Regularities*, Proceedings of the 11th International Symposium on Mathematical Fondations of Computer Science, MFCS 1984, Lecture Notes in Computer Science 176, Springer, pp. 544–552, 1984.

[10] K.W. WAGNER, *The Complexity of Combinatorial Problems with Succinct Input Representation*, Acta Informatica 23, pp. 325–356, 1986.

[11] K. YANG, *Integer Circuit Evaluation Is PSPACE-Complete*, Journal of Computer and System Sciences 63, pp. 288–303, 2001.

Generic Density and Small Span Theorem

Philippe Moser*

Computer Science Department, Iowa State University,
Ames, IA 50010 USA
moser@cs.iastate.edu

Abstract. We refine the genericity concept of [1], by assigning a real number in $[0^c 1]$ to every generic set, called its generic density. We construct sets of generic density any E-computable real in $[0^c 1]$. We also introduce strong generic density, and show that it is related to packing dimension [2]. We show that all four notions are different. We show that whereas dimension notions depend on the underlying probability measure, generic density does not, which implies that every dimension result proved by generic density arguments, simultaneously holds under any (biased coin based) probability measure. We prove such a result: we improve the small span theorem of Juedes and Lutz [3], to the packing dimension [2] setting, for k-bounded-truth-table reductions, under any (biased coin) probability measure.

1 Introduction

Resource-bounded genericity [1] yields a randomness concept for the class E which interacts nicely with Lutz resource-bounded measure [4]. Informally speaking, generic sets are sets which cannot be predicted correctly infinitely often. Genericity has been used for the investigation of structural properties of NP (under appropriate assumptions) and E, see [5] for a survey; and yielded an improved version of the small span theorem of [3], to a stronger reduction notion [6], based on the relationship between measure and genericity.

Resource-bounded measure has recently been refined via effective dimension which is an effectivization of Hausdorff dimension, yielding applications in a variety of topics, including algorithmic information theory, computational complexity, prediction, and data compression [7,8,9,10,2,11]. Hausdorff dimension is a refinement of measure theory, where every measure zero class of languages is assigned a dimension, which is a real number between 0 and 1. Another widely used dimension concept in fractal geometry, known as packing dimension (or strong dimension), was effectivized in [2]. A simple characterization of strong dimension via martingales has been given in [2], where the martingales' capital is required to grow unbounded and is not allowed to decrease too much after a certain number of rounds.

In this paper we connect genericity to resource-bounded dimension by introducing a quantified version of genericity, which is a refinement of genericity, as

* This research was supported in part by Swiss National Science Foundation Grant PBGE2-104820.

M. Liśkiewicz and R. Reischuk (Eds.): FCT 2005, LNCS 3623, pp. 92–102, 2005.

resource-bounded dimension is a refinement of resource-bounded measure. The idea is that every generic set is assigned a real number between 0 and 1, called its generic density, and which corresponds to the density such a set cannot be predicted with. We construct sets of generic density any E-computable real s. Similarly to resource-bounded strong dimension [2], we also introduce strong generic density. We show that strong generic density is related to strong dimension [2], in the sense that sets with a certain amount of randomness relatively to strong dimension, keep that amount of unpredictability relatively to strong generic density.

Next we show that all these four concepts, i.e. dimension, strong dimension, generic density and strong generic density, are indeed different.

All notions exposed so far are implicitly considered within the Cantor space of all languages under the uniform probability measure. This corresponds to the random experiment in which every membership bit of a language L is chosen according to the toss of a fair coin. Probability measures other than the uniform probability measure occur naturally in applications, and the corresponding gale notion (resp. dimension notion) has been investigated in [9,2] (resp. [12]). In section 6, we highlight a main difference between generic density and resource-bounded dimension, that is whereas the latter notion is dependent on the underlying probability measure, generic density is not; a similar result for genericity vs resource-bounded measure was given in [13]. More precisely we show that if the coin in the above random experiment is biased, then for two different biases the corresponding dimension notions differ, whereas the generic density notion remains the same. This outlines a nice feature of the generic density method over martingale based dimension: proofs obtained by generic density arguments are in some sense more informative, because all dimension results proved by generic density methods (i.e. showing some class contains some s-generic set) simultaneously hold in a wide variety of probability measure spaces. Such an example is given in the last section of this paper, where a small span theorem under any biased coin based probability measure is proved.

More precisely we prove a small span theorem in the strong dimension setting, for k-bounded-truth-table reductions (k-tt-reductions are a special case of Turing reductions, where only k non-adaptive queries are allowed) under any biased coin based probability measure. The small span theorem [3] asserts that for every language L in E, either the set of languages reducible to L, called the lower span, or the set of languages to which L reduces, called the upper span, has E-measure zero. The question whether the small span theorem still holds in the resource-bounded dimension setting – i.e. can E-measure zero be replaced by E-dimension zero – was partially disproved in [14], where E-languages with both lower and upper span of E-dimension one were constructed. Nevertheless the small span theorem under polynomial many-one reductions holds for scaled dimension [15] and partially holds in the dimension setting as shown in [15], i.e. either the lower span has E-dimension zero or the upper span has E-measure zero. By adapting the proof in [6] combined with generic frequency arguments, we prove a small span theorem in the *strong* dimension setting for *k-bounded truth table* reductions, under *any* (biased coin) probability measure, i.e. we show that for any L in E, either the lower span (under k-tt-reductions) has E-β-strong dimension zero (where β denotes the sequence of biases), or the upper span has E-β-measure zero. k-bounded-truth-table reductions and n^α-tt reductions ($\alpha < 1$) were considered in [6,16], but only in the resource-bounded measure setting.

2   Preliminaries

We use standard notation for traditional complexity classes, see for instance [17,18]. Let us fix some notations for strings and languages. A *string* is an element of $\{0,1\}^n$ for some integer n. For a string x, its length is denoted by $|x|$. $s_0, s_1, s_2 \ldots$ denotes the standard enumeration of the strings in $\{0,1\}^*$ in lexicographical order, where $s_0 = \lambda$ denotes the empty string. We sometimes enumerate the strings of size n by $s_0^n, s_2^n, s_{2^n-1}^n$. Note that $|w| = 2^{O(|s_{|w|}|)}$. For a string s_i define its position by $\text{pos}(s_i) = i$. If x, y are strings, we write $x \leq y$ if $|x| < |y|$ or $|x| = |y|$ and x precedes y in alphabetical order. A *sequence* is an element of $\{0,1\}^{\mathbb{N}}$. If w is a string or a sequence and $1 \leq i \leq |w|$ then $w[i]$ and $w[s_i]$ denotes the ith bit of w. Similarly $w[i \ldots j]$ and $w[s_i \ldots s_j]$ denote the ith through jth bits.

For two string x, y, the concatenation of x and y is denoted xy. If x is a string and y is a string or a sequence extending x i.e. $y = xu$, where u is a string or a sequence, we write $x \sqsubseteq y$. We write $x \sqsubset y$ if $x \sqsubseteq y$ and $x \neq y$.

A *language* is a set of strings. A *class* is a set of languages. The cardinal of a language L is denoted $|L|$. Let n be any integer. The set of strings of size n of language L is denoted $L^{=n}$. Similarly $L^{\leq n}$ denotes the set of strings in L of size at most n. Denote by $L_k^{=n} = L \cap \{s_0^n, \cdots s_{k-1}^n\}$. We identify language L with its characteristic function χ_L, where χ_L is the sequence such that $\chi_L[i] = 1$ iff $s_i \in L$. Thus a language can be seen as a sequence in $\{0,1\}^{\mathbb{N}}$. We denote by C the Cantor space of all infinite binary sequences. $L|s_n$ denotes the initial segment of L up to s_{n-1} given by $L[s_0 \cdots s_{n-1}]$, whereas $L \upharpoonright s_n$ denotes $L[s_0 \cdots s_n]$.

We consider bounded truth-table reductions, here is a definition. Let $k \in \mathbb{N}_+$. We say language A is k-truth-table reducible to language B, denoted $A \leq_{k-tt}^p B$ if there exists a family of polynomial computable function $f : \{0,1\}^* \times \{0,1\}^k \to \{0,1\}$ (the evaluator) and $g_i : \{0,1\}^* \to \{0,1\}^*$ ($1 \leq i \leq k$, the queries), such that for every string x: $A(x) = f(x, B(g_1(x)), \cdots, B(g_k(x)))$. Such a reduction is denoted $f(g_1, \cdots, g_k)$. A is bounded truth-table reducible to B if it is k-truth-table reducible to B for some k.

For a reducibility notion r, the lower span (resp. upper span) of a language A, denoted $A^{\geq r}$ (resp. $A^{\leq r}$) is the set of languages B such that $B \leq_r A$ (resp. $A \leq_r B$).

2.1   Lutz Resource-Bounded Measure

Lutz measure on E [4] is obtained by imposing appropriate resource-bounds on a game theoretical characterization of classical Lebesgue measure, via martingales.

A martingale is a function $d : \{0,1\}^* \to \mathbb{R}_+$ such that, for every $w \in \{0,1\}^*$, $d(w) = (d(w0) + d(w1))/2$. This definition can be motivated by the following betting game in which a gambler puts bets on the successive membership bits of a hidden language A. The game proceeds in infinitely many rounds where at the end of round n, it is revealed to the gambler whether $s_n \in A$ or not. The game starts with capital 1. Then, in round n, depending on the first $n-1$ outcomes $w = \chi_A[0 \ldots n-1]$, the gambler bets a certain fraction $\epsilon_w d(w)$ of his current capital $d(w)$, that the nth word $s_n \in A$, and bets the remaining capital $(1 - \epsilon_w)d(w)$ on the complementary event $s_n \notin A$. The game is fair, i.e. the amount put on the correct event is doubled, the one put on the wrong guess is lost. The value of $d(w)$, where $w = \chi_A[0 \ldots n]$ equals the capital of the gambler after round n on language A. The player wins on a language A if he manages to make his capital arbitrarily large during the game, i.e. $\limsup_{n \to \infty} d(\chi_A[0 \ldots n]) = \infty$ and we say that martingale d succeeds on A. The success set $S^\infty[d]$ of a martingale d is the class of all languages on which d succeeds.

Lutz's idea to define a measure notion on the class E is to consider only martingales computable in a certain time bound, i.e. martingales d such that $d(w)$ can be computed in time $2^{c|s_{|w|}|}$ for some $c > 0$. Such a martingale is called E-computable. E-computable martingales are the main tool for defining a measure notion on E, as the following definition shows.

Definition 1. *A class C has E-measure zero if there is an E-computable martingale d that succeeds on every language of C.*

This property is monotone in the following sense: If class D is contained in class C, and C has E-measure zero, then D has E-measure zero.

Definition 2. *A class C has E-measure one if its complement $E - C$ has E-measure zero.*

Lutz showed in [4] that E does not have E-measure zero, which is known as the measure conservation property. Since finite unions of measure zero sets have measure zero it's impossible for a class to have both measure zero and one.

Lutz also proved in [4] that enumerable infinite unions of measure zero sets have measure zero, more precisely.

Theorem 1 (Lutz). *Suppose $\{d_i\}_{i \geq 1}$ is a set of martingales, each covering class C_i; where $d(i, w) := d_i(w)$ is computable in time $2^{c|s_{|w|}|} + i^c$ for a some constant $c > 0$. Then $\cup_{i \geq 1} C_i$ has E-measure zero.*

The following result shows that approximable martingales can be replaced by exactly computable ones.

Lemma 1. *Exact Computation Lemma [4]*
Let $d : \{0,1\}^* \to \mathbb{R}_+$ be a martingale such that there exists a family of approximations $\{\hat{d}_k\}_k$ where $\hat{d}_k(w)$ is computable in time $2^{c|s_{|w|}|} + k^c$ for some $c > 0$, and such that $|\hat{d}_k(w) - d(w)| \leq 2^{-k}$. Then there exists an E-computable martingale $d' : \{0,1\}^* \to \mathbb{Q}_+$ such that $S^\infty[d] = S^\infty[d']$.

For a survey on resource-bounded measure see [19].

2.2 Resource-Bounded Dimension

Lutz's idea for defining a dimension notion via martingales, is to perceive taxes on the martingales' wins, so that only martingales whose capital grows quickly are considered. This motivates the following definition.

Definition 3. *For a real number $s \geq 0$, a martingale is said s-successful on a language A, if $\limsup_{m \to \infty} \frac{d(A \upharpoonright s_{m-1})}{2^{(1-s)m}} = \infty$. A martingale is s-successful on a class if it is s-successful on every language of the class.*

Remark 1. Similarly d is said strongly s-successful on A, if $\limsup$ in Definition 3 is replaced by $\liminf$.

The dimension of a class is defined as the largest tax rate which can be perceived on the martingales' benefits, without preventing them from winning.

Definition 4. *Let C be any complexity class. The E-dimension of C (resp. E-strong-dimension) is the infimum over all $s \in [0,1]$, such that there exists an E-computable martingale which s-succeeds (resp. strongly s-succeeds) on C.*

Lutz proved in [7] that the E-dimension notion satisfies all three basic measure properties, namely that E has E-dimension one, every language in E has E-dimension zero, and finally enumerable infinite unions of sets of E-dimension s have E-dimension s. More precisely,

Definition 5. *Let $X, X_0, X_1, X_2, \cdots$ be complexity classes. X is a E-union of the E-dimensioned sets $X_0, X_1, X_2, \cdots$ if $X = \bigcup_{k \geq 0} X_k$, and for each $s > \sup_{k \in \mathbb{N}} \dim_E(X_k)$, there is a function $d : \mathbb{N} \times \{0,1\}^* \to [0, \infty[$ with the following properties: d is E-computable, for each $k \in \mathbb{N}$, the function $d_k(w) := d(k, w)$ is a martingale, and for each $k \in \mathbb{N}$, d_k s-succeeds on X_k.*

The following Lemma states that the E-dimension of a E-union of sets is the supremum of the E-dimension of all sets.

Lemma 2. *[7]*
Let $X, X_0, X_1, X_2, \cdots$, be a E-union of the E-dimensioned sets $X_0, X_1, X_2, \cdots$. Then $\dim_E(X) = \sup_{k \in \mathbb{N}} \dim_E(X_k)$.

3 Generic Density

Whereas Lutz resource-bounded measure is defined via martingales, genericity is defined via strategies. Here is a definition.

Definition 6. *A function $h : \{0,1\}^* \to \{0,1\}^* \cup \{\bot\}$ is a partial one-bit extension strategy, if for every string $\tau \in \{0,1\}^*$ either $h(\tau)$ is not defined, denoted $h(\tau) = \bot$, or h extends τ by one bit i.e. $h(\tau) = \tau b$ with $b \in \{0,1\}$ (the bit b is denoted $\text{ext} h(\tau)$).*

For simplicity we use the word *strategy* for partial one-bit extension strategy. We denote $h(\tau) \downarrow$ whenever $h(\tau)$ is defined, i.e. $h(\tau) \neq \perp$. We say language A meets strategy h if $h(\tau) \sqsubseteq \chi_A$ for some string $\tau \in \{0,1\}^*$. We are interested in a genericity notion on the class E. This motivates the following definition.

Definition 7. *Let $c > 0$. A strategy $h : \{0,1\}^* \rightarrow \{0,1\}^* \cup \{\perp\}$ is 2^{cn}-computable if there is a Turing machine which on input σ computes $h(\sigma)$, in time $2^{c|s_\sigma|}$.*

A strategy is E-computable if it is 2^{cn}-computable for some $c > 0$.

As mentioned earlier, we want to quantify the genericity notion of [1]. This motivates the following definition.

Definition 8. *Strategy h is s-dense along some language A, with $s \in [0,1]$, if $\limsup_{n \to \infty} |\{x \in \{s_0, s_1, \cdots, s_n\} : h(A|x) \downarrow\}| - sn = \infty$.*

Remember that strategies are supposed to predict characteristic sequences of languages, so the higher the density of a strategy is, the more prediction it tries to make. s-strongly-dense is defined similarly with $\limsup$ replaced by $\liminf$.

Let us introduce our notion of generic density.

Definition 9. *A language G is said $(s, 2^{cn})$-generic if it meets every 2^{cn}-computable strategy which is $(1-s)$-dense along G.*

Informally s-generic sets cannot be predicted correctly by strategies, and the bigger s is, the bigger the set of defeated strategy is. For $s = 1$ all strategies halting on at least a small portion of the strings are to be met, $s = 0$ is the other extreme, where only strategies halting on a huge fraction of all strings are to be met. For the genericity notion of [1], all strategies halting on at least infinitely many strings are to be met. s-strongly-generic is defined similarly with s-dense replaced by s-strongly-dense.

Definition 10. *Let $c > 0$. The 2^{cn}-generic density of a language A, denoted gendens$_{2^{cn}}(A)$, is the supremum over all $s \in [0,1]$ such that A is $(s, 2^{cn})$-generic.*

Intuitively the bigger the generic density of a sequence is, the more unpredictability it contains.

Similarly the E-generic density of A, denoted gendens$_E(A)$, is the sup over all $s \in [0,1]$ for which A is $(s, 2^{cn})$-generic for some $c > 0$. Strong generic density Gendens$_{2^{cn}}(A)$ and Gendens$_E(A)$ are defined by replacing generic with strongly generic in Definition 10.

The following result shows that s-generic sequences do exist for any computable s, but contrary to random sequences, they can be sparse.

Theorem 2. *For every E-computable real $s \in [0,1]$ and every $c \geq 1$, there exists a sparse set G such that gendens$_{2^{cn}}(G) = s$.*

Remark 2. Similar arguments show that Theorem 2 also holds by replacing gendens with Gendens.

4   Generic Density vs Resource-Bounded Dimension

As mentioned earlier, the strong generic density of a sequence is related to its strong dimension [2], more precisely every s-strongly random set is also s-strongly-generic, i.e. every set with a certain amount of randomness relatively to strong dimension, also contains a certain amount of unpredictability in regard to strong generic density.

Whereas s-generic sets are the typical sets for generic density, the following standard notion characterizes the typical sets for strong dimension.

Definition 11. *Let $s \in [0,1]$. A language R is $(s, 2^{cn})$ strongly random if no martingale computable in 2^{cn} steps is strongly s-successful on R.*

A set R is (s, E)-strongly random if it is $(s, 2^{cn})$-strongly random for every $c > 0$.

s-strongly random sets are typical because they determine the E-strong dimension of a class that contains them, as the following standard result shows.

Lemma 3. *Let $s \in [0,1]$, $c > 0$ and let C be a class of languages such that C does not contain any $(s, 2^{cn})$-strongly random languages. Then $\mathrm{Dim}_{\mathsf{E}}(C) \leq s$.*

Corollary 1. *Lemma 3 still holds if we replace strongly random with random and Dim with dim.*

The following result shows that every s-strongly random set is s-strongly-generic, i.e. quantified randomness implies quantified unpredictability. We prove a more general result in Section 6.

Theorem 3. *Let $c > 0$. Let R be $(s, 2^{(c+2)n})$-strongly random, then R is $(s, 2^{cn})$-strongly-generic.*

Corollary 2. *Let $c > 0$. Let R be $(s, 2^{(c+2)n})$-random, then R is $(s, 2^{cn})$-generic.*

The converse of Theorem 3 is not true as the following section shows.

5   Comparing the Density Notions

As the following result shows, quantified unpredictability and quantified randomness are *different* notions.

Theorem 4. *There exists a language S such that $\mathrm{Dim}_{\mathsf{E}}(S) < \mathrm{gendens}_{\mathsf{E}}(S)$ and*

$$
\begin{array}{ccc}
\mathrm{Dim}_{\mathsf{E}}(S) & < & \mathrm{Gendens}_{\mathsf{E}}(S) \\
\vee & & \vee \\
\mathrm{dim}_{\mathsf{E}}(S) & < & \mathrm{gendens}_{\mathsf{E}}(S)
\end{array}
\tag{1}
$$

And for any sequence S, Equation 1 holds with less or equal inequalities.

As the previous result shows , there exists a set whose strong dimension is smaller than its generic density. The following result shows that the converse also holds, i.e. these two notions are incomparable.

Theorem 5. *There exists a languages S such that* $\mathrm{Dim_E}(S) = 1$ *and* $\mathrm{gendens_E}(S) = 0$.

Theorem 4 and 5 yield the following corollary.

Corollary 3. $\mathrm{Dim_E}$ *and* $\mathrm{gendens_E}$ *are incomparable.*

6   Generic Density Under Different Probability Measures

In this section we highlight a main feature of generic density over resource-bounded dimension, that is whereas the latter notion depends on the underlying probability measure, generic density does not. As we shall see, this implies that dimension results obtained by generic density methods, are somehow more informative, because they hold in a wide variety of probability measure spaces. Let us give some preliminary definitions from [2]. A probability measure on the Cantor space is a function $\nu : \{0,1\}^* \to [0,1]$ such that $\nu(\lambda) = 1$ and for all strings w, $\nu(w) = \nu(w0) + \nu(w1)$. Informally, $\nu(w)$ is the probability that $w \sqsubseteq L$, where the sequence L is chosen according to ν. A bias sequence is a sequence $\boldsymbol{\beta} = (\beta_0, \beta_1, \ldots)$ of real numbers $\beta_i \in [0,1]$. Intuitively, β_i is the probability that the ith toss of a biased coin yields 1. For a bias sequence $\boldsymbol{\beta}$, define the $\boldsymbol{\beta}$-probability measure on C by $\mu^{\boldsymbol{\beta}}(w) = \prod_{i=0}^{|w|-1} \beta_i(w)$, where $\beta_i(w) = \beta_i$ if $w_i = 1$ and $1 - \beta_i$ otherwise. $\mu^{\boldsymbol{\beta}}$ represents the probability that some language L satisfies $w \sqsubseteq L$, where the ith bit of L is determined by a coin toss with bias β_i. For simplicity $\mu^{\boldsymbol{\beta}}$ is sometimes denoted $\boldsymbol{\beta}$. The usual probability measure is called the uniform probability measure, denoted $\mu(w) = 2^{-|w|}$, and corresponds to the toss of a fair coin.

Resource-bounded dimension on spaces with probability measure ν is defined via ν-s-gales, here is a definition.

Definition 12. *[2] Let ν be a probability measure on C, let $s \in [0,1]$ and $t(n) \geq 2^{O(n)}$ be a time bound. A $t(n)$-computable ν-s-gale is a function $d : \{0,1\}^* \to [0,\infty)$ such that for all strings w, $d(w)\nu^s(w) = d(w0)\nu^s(w0) + d(w1)\nu^s(w1)$ and $d(w)$ is computable in $t(|s_{|w|}|)$ steps.*

Intuitively the s in Definition 12 represents the tax taken on the martingale's wins, whereas the factors ν adjust the wins according to the probability measure ν: if some bit appears with higher probability, then the payoff while betting on this bit ought to be smaller. An E-computable ν-s-gale is a $t(n)$-computable ν-s-gale for some $t(n) = 2^{O(n)}$.

Similarly to the usual notion, the E-ν-dimension of a language L, denoted $\dim_E^\nu(L)$, is the supremum over all s such that there is an E-computable ν-s-gale d such that $\limsup_{m \to \infty} d(L|m) = \infty$. It is easy to check that Lemma 3 also holds in spaces with any biased coin based probability measure.

For $s \in [0,1]$, denote by $\mathsf{DIM}_E^\nu(\geq s)$ (resp. $\mathsf{GENDENS_E}(\geq s)$) the set of languages with E-ν-dimension (resp. E-generic density) greater than s. Denote by SDIM the strong dimension analogue and $\mathsf{SGENDENS}$ the strong-genericity analogue.

The following result requires the weighted binary entropy function $H : (0,1)^2 \to [0,\infty)$ where $H(x,y) = x \log \frac{1}{y} + (1-x) \log \frac{1}{1-y}$ which is continuous on $(0,1)^2$.

It is clear by the work of [9,2,12,12] that resource-bounded dimension depends on the underlying probability measure, i.e. for two biases sequences α, β converging to different values, $\mathsf{DIM}_E^\alpha(\geq s) \neq \mathsf{DIM}_E^\beta(\geq s)$, i.e. sequences with high dimension in a space with underlying probability α can have smaller dimension in a space with underlying probability β.

The following result shows that this is not the case for generic density, i.e. it is independent of the underlying probability measure. More precisely we show that for a sequence of biases β converging to some number β, the sequences with β-dimension $s \log(1/\beta)/H(s,\beta)$ have generic density s. The factor $\log(1/\beta)/H(s,\beta)$ that appears when going from dimension to generic density is because the payoffs are not equal whether the bit that is bet on is zero or one, i.e. if for example $\beta < 1/2$ then the probability of the bit 0 is bigger, therefore the payoff on such bits is smaller. So if the bits of the non-generic i.e. easily predicted sequence, are always predicted to be 0, the dimension has to drop, which explains this factor. Note that when $\beta = 1/2$, the factor is equal to 1, i.e. disappears.

The following result highlights an advantage of the generic density method over the martingales based one, for dimension results that are proved by showing that a class contains an s-generic set. Such results simultaneously hold in a large range of biased-coin based probability measure spaces. Such an example is given in Section 7. Note that of course such an approach is not always possible for dimension results, see for example [2].

Theorem 6. *Let $\beta = (\beta_0, \beta_1, \ldots)$ be an E-computable bias sequence, converging to $\beta \in (0, \frac{1}{2})$. Let $s \in [0,1]$, then* $\mathsf{GENDENS}_E(\geq s) \supseteq \mathsf{DIM}_E^\beta(\geq s\frac{\log(1/\beta)}{H(s,\beta)})$.

Corollary 4. *The same holds by replacing* $\mathsf{GENDENS}_E(\geq s)$ *with* $\mathsf{SGENDENS}_E(\geq s)$ *and* $\mathsf{DIM}_E^\beta(\geq s\frac{\log(1/\beta)}{H(s,\beta)})$ *with* $\mathsf{SDIM}_E^\beta(\geq s\frac{\log(1/\beta)}{H(s,\beta)})$.

7   Small Span Theorem in Dimension

In this section we prove a small span theorem for bounded-truth-table reduction, in the strong dimension setting, in spaces with any biased coin based probability measure. The proof is adapted from [6] combined with results of the previous sections. This is an example where the generic density method is more informative than the martingale based approach, because we simultaneously prove the result for any biased coin based probability measure.

To clarify the proofs, we assume that all bounded truth-table reductions are in the following normal form, where all queries are ordered in decreasing order, and redundant ones are replaced by λ.

Definition 13. *A p-k-tt reduction $f(g_1, \ldots, g_k)$ is normal if for every $x \in \{0,1\}^*$ there exists $k' \leq k$ such that $g_i(x) > g_{i+1}(x)$ for $1 \leq i \leq k'$, and $g_i(x) = \lambda$ for $i \geq k'$.*

It is easy to check that any p-k-tt reduction $f(g_1, \ldots, g_k)$ can be transformed into an equivalent normal reduction.

Definition 14. *The collision set of a p-k-tt reduction $f(g_1, \ldots, g_k)$ denoted Coll(f) is the set of strings x, for which there exists $y < x$ such that $g_i(x) = g_i(y)$ (for $i = 1, \ldots, k$) and $f_x = f_y$, where $f_x = f(x, \cdot, \ldots, \cdot)$.*

A p-k-tt reduction $f(g_1, \ldots, g_k)$ is consistent with some language A, if for all strings x, y s.t. $g_i(x) = g_i(y)$ (for $i = 1, \ldots, k$) and $f_x = f_y$, we have $A(x) = A(y)$.

Definition 15. *Let $c > 0$ be some constant, and let $f(g_1, \cdots, g_k)$ be a p-k-tt reduction. The c-rank of $f(g_1, \cdots, g_k)$ is the largest integer $1 \leq r \leq k$ such that $\exists^\infty x : \frac{|x|-c}{k} \leq |g_r(x)|$. The c-rank is zero if no such integer exists.*

The following is the main result of this section.

Theorem 7. *(Small span theorem) Let $\beta = (\beta_0, \beta_1, \ldots)$ be an E-computable bias sequence, converging to $\beta \in (0, \frac{1}{2})$. Let A in E be any language, and $k \in \mathbb{N}$. Then either $\mathrm{Dim}_E^\beta(A^{\geq_{k-tt}^p} \cap E) = 0$ or $\mu_E^\beta(A^{\leq_{k-tt}^p}) = 0$.*

8   Conclusion

We have introduced a refined notion of genericity, in the same sense that resource-bounded dimension is a refinement of resource-bounded measure. We have exhibited a relationship between generic density and dimension, as well as a main difference regarding the underlying probability measure of the Cantor space, with the consequence that generic density based proof are in some sense more informative than martingale based ones. We give an example of such a proof by showing a small span theorem in any (biased coin) probability measure space, for stronger reductions as previously considered in the dimension setting. We expect generic density to be useful for further resource-bounded dimension investigations.

Acknowledgments. I thank J. Lutz for suggesting the possible measure probability independence of generic density, and for bringing [13] to my attention.

References

1. Ambos-Spies, K.: Resource-bounded genericity. Proceedings of the Tenth Annual Structure in Complexity Theory Conference (1995) 162–181
2. Athreya, K.B., Hitchcock, J.M., Lutz, J.H., Mayordomo, E.: Effective strong dimension in algorithmic information and computational complexity. Proceedings of the Twenty-First Symposium on Theoretical Aspects of Computer Science (2004) 632–643
3. Juedes, D., Lutz, J.: The complexity and distribution of hard problems. Proceedings of the 34th FOCS Conference (1993) 177–185

4. Lutz, J.: Almost everywhere high nonuniform complexity. Journal of Computer and System Science **44** (1992) 220–258

5. Ambos-Spies, K., Mayordomo, E.: Resource-bounded measure and randomness. Lecture Notes in Pure and Applied Mathematics (1997) 1–47

6. Ambos-Spies, K., Neis, H., Terwijn, S.: Genericity and resource bounded measure for exponential time. Theoretical Computer Science **168** (1996) 3–19

7. Lutz, J.: Dimension in complexity classes. Proceedings of the 15th Annual IEEE Conference on Computational Complexity (2000) 158–169

8. Lutz, J.H.: Effective fractal dimensions. to appear in Mathematical Logic Quarterly (2003)

9. Lutz, J.: The dimensions of individual strings and sequences. Information and Computation **187** (2003) 49–79

10. Dai, J., Lathrop, J., Lutz, J., Mayordomo, E.: Finite-state dimension. Theoretical Computer Science **310** (2004) 1–33

11. Fortnow, L., Lutz, J.: Prediction and dimension. Journal of Computer and System Sciences **70** (2005) 570–589

12. Fenner, S.A.: Gales and supergales are equivalent for defining constructive hausdorff dimension. Technical Report CC/0203017, Arxiv.org (2002)

13. Lorentz, A.K., Lutz, J.H.: Genericity and randomness over feasible probability measures. Theoretical Computer Science **207** (1998) 245–259

14. Ambos-Spies, K., Merkle, W., Reimann, J., Stephan, F.: Hausdorff dimension in exponential time. In Proceedings of the 16th IEEE Conference on Computational Complexity **172** (2001) 210–217

15. Hitchcock., J.M.: Small spans in scaled dimension. Proc. 19th ann. IEEE Conference on Computational Complexity (2004) 104–112

16. Buhrman, H., van Melkebeek, D.: Hard sets are hard to find. Journal of Computer and System Sciences **59(2)** (1999) 327–345

17. Balcazar, J.L., Diaz, J., Gabarro, J.: Structural Complexity I. EATCS Monographs on Theoretical Computer Science Volume 11, Springer Verlag (1995)

18. Balcazar, J.L., Diaz, J., Gabarro, J.: Structural Complexity II. EATCS Monographs on Theoretical Computer Science Volume 22, Springer Verlag (1990)

19. Lutz, J.: The quantitative structure of exponential time. In Hemaspaandra, L., Selman, A., eds.: Complexity Theory Retrospective II. Springer (1997) 225–260

20. Moser, P.: Baire's categories on small complexity classes. 14th Int. Symp. Fundamentals of Computation Theory (2003) 333–342

21. Impagliazzo, R., Moser, P.: A zero-one law for RP. Proceedings of the 18th Conference on Computational Complexity (2003) 48–52

22. Moser, P.: Derandomization and Quantitative Complexity. PhD thesis, University of Geneva (2004)

Logspace Optimization Problems
and Their Approximability Properties

Till Tantau*

Technische Universität Berlin,
Fakultät für Elektrotechnik und Informatik,
10623 Berlin, Germany

Abstract. This paper introduces logspace optimization problems as analogues of the well-studied polynomial-time optimization problems. Similarly to them, logspace optimization problems can have vastly different approximation properties, even though the underlying decision problems have the same computational complexity. Natural problems, including the shortest path problems for directed graphs, undirected graphs, tournaments, and forests, exhibit such a varying complexity. In order to study the approximability of logspace optimization problems in a systematic way, polynomial-time approximation classes are transferred to logarithmic space. Appropriate reductions are defined and optimization problems are presented that are complete for these classes. It is shown that under the assumption L ≠ NL some logspace optimization problems cannot be approximated with a constant ratio; some can be approximated with a constant ratio, but do not permit a logspace approximation scheme; and some have a logspace approximation scheme, but cannot be solved in logarithmic space. A new natural NL-complete problem is presented that has a logspace approximation scheme.

1 Introduction

In the introduction to Chapter 13 of his book *Computational Complexity* [9], Papadimitriou writes: *'Although all NP-complete problems share the same worst-case complexity, they have little else in common. When seen from almost any other perspective, they resume their healthy, confusing diversity. Approximability is a case in point.'* The aim of the present paper is to show that if we shift our focus from time to low space complexity, NL-complete problems turn out to be as diverse as their polynomial-time brethren.

As an example, consider the problem of telling whether the distance of two vertices in a directed graph is at most d. This problem is NL-complete and this remains even true if we restrict ourselves to undirected graphs [13] or to tournaments [8] (which are directed graphs in which for any two vertices u and v there is an edge from u to v or an edge from v to u). However, when asked

---

* Supported in part by a postdoc research fellowship grant of the German Academic Exchange Service (DAAD). Work done in part at the International Computer Science Institute, Berkeley.

M. Liśkiewicz and R. Reischuk (Eds.): FCT 2005, LNCS 3623, pp. 103–114, 2005.

to *find* a path, we do not know how to do this in logarithmic space for directed graphs, but Reingold [10] has shown how to do this for undirected graphs and it is also possible for tournaments [8]. Finally, when asked to construct a path that is, say, twice as long as the shortest one, we only know how to do this in logarithmic space for tournaments [8].

lvarez and Jenner [1], see also [5], started the systematic study of logspace optimization problems by introducing the class OptL, the logspace analogue of Krentel's class OptP [7], and identifying complete problems for this class. However, the study of OptL suffers from the same problem as the study of OptP: In Krentel's approach an optimization problem is just a function mapping instances to optimal solutions. Though appealing because of its formal simplicity, this approach has the disadvantage that, similarly to decision problems, the functions hide the solutions we are actually interested in.

In the present paper, the structural research begun by lvarez and Jenner [1] and the more algorithmic results from [8] are expanded in two directions.

On the structural side, a complexity-theoretic framework for the study of logspace optimization problems is introduced that parallels the existing framework for studying polynomial-time optimization problems. In this framework, logspace optimization problems are structured entities consisting of an instance set, a solution relation, a measure function, and a type. In this framework more fine-grained results can be stated. For example, problems can be differentiated according to how well they can be approximated.

On the algorithmic side, a new NL-complete problem is presented that admits a logspace approximation scheme: the 'hot potato problem'. For this problem, a hot potato must be passed around between people for a given number of rounds (until it has cooled off). Passing the potato from one person to another causes a certain cost, which depends on the two people involved. The objective is to minimize the total cost.

This paper is organised as follows. In Section 2 the structural framework is established. Logspace optimization classes, including the classes NLO and LO and different logspace approximation classes in between, are formally defined. Following the class definitions, reductions between polynomial-time approximation problems are transfered to logarithmic space and closure properties are established. Next, it is then shown that different variants of the most valuable vertex problem (MVV) are complete for the introduced classes under approximation-preserving reductions. These completeness results also allow us to establish a class hierarchy under the assumption L $\neq$ NL. In Section 3 different versions of the NL-complete hot potato problem are presented and it is shown that they have logspace approximation schemes.

2   A Framework for Logspace Optimization Problems

In this section the framework for the study of logspace optimization problems is established. We start with a definition block, in which logspace optimization problems, logspace approximation classes, and reductions between logspace opti-

mization problems are defined formally. We then show that most of these classes
have fairly natural complete problems. Building on the completeness results we
show that the optimization classes form a proper hierarchy under the assumption
$L \neq NL$.

2.1   Polynomial-Time and Logspace Optimization Problems

Definition 1 ([2]). *An* optimization problem *is a tuple consisting of an in-
stance set $I \subseteq \Sigma^*$, a solution relation $S \subseteq I \times \Sigma^*$, a measure function $m \colon S \to$
$\mathbb{N}^+$, and a type $t \in \{\min, \max\}$.*

For example, the optimization problem MAX-CLIQUE is defined as follows: its
instance set is the set of (the codes of) all undirected graphs, the solution relation
relates graphs to cliques within these graphs, the measure function maps pairs
consisting of a graph and a clique to the size of the clique, and the type is max.
The next definition fixes basic notations and terminology.

Definition 2. *Let $P = (I, S, m, t)$ be an optimization problem. Let $x \in I$.*

1. *Let $S(x) := \{y \in \Sigma^* \mid (x, y) \in S\}$ denote the* solutions *for x.*
2. *For* minimization *problems let $m^*(x) := \min\{m(x, y) \mid y \in S(x)\}$ denote the* optimal measure *for x. For* maximization *problems let $m^*(x) :=$
   $\max\{m(x, y) \mid y \in S(x)\}$. Let $m^*(x)$ be undefined if $S(x) = \emptyset$.*
3. *Let $S^*(x) := \{y \in \Sigma^* \mid m(x, y) = m^*(x)\}$ denote the* set of optimal solutions
   for x.
4. *Let $R(x, y) := \max\{m(x, y)/m^*(x), m^*(x)/m(x, y)\}$ denote the* performance
   ratio *of the solution y.*
5. *The* existence problem *$P_{\exists \mathrm{sol}}$ is the set $\{x \mid S(x) \neq \emptyset\}$.*
6. *The* budget problems *are the sets $P_{\mathrm{opt}_<} := \{(x, z) \mid \exists y.\ m(x, y) < z\}$ and
   $P_{\mathrm{opt}_>} := \{(x, z) \mid \exists y.\ m(x, y) > z\}$.*
7. *A function $f \colon \Sigma^* \to \Sigma^*$* produces solutions *for P if for every $x \in P_{\exists \mathrm{sol}}$ we
   have $f(x) \in S(x)$. It produces* optimal solutions *if for every $x \in P_{\exists \mathrm{sol}}$ we
   have $f(x) \in S^*(x)$.*

By restricting the computational complexity of the set I, the relation S,
and the function m, different optimization classes can be defined. The two most
widely-studied ones are PO and NPO. For optimization problems (I, S, m, t) in
NPO, the set I must be decidable in polynomial time, S must also be decidable
in polynomial time and furthermore polynomially length-bounded, and m must
be computable in polynomial time. For a problem in PO there must furthermore
exist a function in FP that computes optimal solutions for it.

Let us now transfer these definitions to logarithmic space.

Definition 3. *An optimization problem (I, S, m, t) in the class NLO if*

1. *I is decidable in logarithmic space,*
2. *S is decidable in logarithmic space via a machine M_S that reads the alleged
   solution in a one-way fashion and S is polynomially length-bounded, and*

3. *m is computable in logarithmic space via a machine M_m that reads the solution and writes the output in a one-way fashion.*

A problem in NLO is furthermore in LO if there exists a function in FL that produces optimal solutions for it.

The main technical peculiarity of the definition of NLO is the one-way access to the solution tape. This restriction is necessary to ensure that a nondeterministic logspace machine can 'guess' such a solution, see the following lemma, whose straightforward proof can be found in the full version of this paper [12]. The below theorem follows directly from the definitions and the lemma.

Lemma 4. *Let $(I, S, m, t) \in \text{NLO}$. Then there exists a nondeterministic logspace machine M with at most two nondeterministic choices per state such that for every instance $x \in I$ the following holds: If $y \in S(x)$, then y, regarded as a bitstring, is a string of nondeterministic choices that makes M accept and output $m(x, y)$ on that path. If $y \notin S(x)$, then y is a string of nondeterministic choices that make M reject. All outputs $m(x, y)$ of M for an instance x on accepting paths y have the same length.*

Theorem 5. *If $P \in \text{NLO}$, then $P_{\exists \text{sol}}, P_{\text{opt}<}, P_{\text{opt}>} \in \text{NL}$. If $P \in \text{LO}$, then $P_{\exists \text{sol}}, P_{\text{opt}<}, P_{\text{opt}>} \in \text{L}$.*

Many optimization problems have the special property that the measure of any solution is polynomially bounded in the length of the instance. For instance, MAX-CLIQUE has this property since the size of the largest clique in a graph is bounded by the number of vertices of the graph. The notation C_{pb} is commonly used to denote the restriction of a class C of optimization problems to those problems (I, S, m, t) for which there exists a polynomial p with $m(x, y) \leq p(|x|)$ for all $(x, y) \in S$.

Snuggled between NLO_{pb} and NLO, there is an interesting intermediate class NLO_{do} that does not have a counter-part in the polynomial-time setting. The subscript 'do' stands for *deterministic output*.

Definition 6. *Let $P = (I, S, m, t) \in \text{NLO}$. Then $P \in \text{NLO}_{\text{do}}$ if the machine M_m that computes m has deterministic output. This means that M_m writes the first output bit only after the one-way tape containing the alleged solution has been read completely.*

As pointed out in [1], for machines that produce deterministic output, the output depends only on the last configuration reached before the first output bit is produced. Thus, such machines can only produce polynomially many different outputs, which shows $\text{NLO}_{\text{pb}} \subsetneq \text{NLO}_{\text{do}} \subsetneq \text{NLO}$. These inclusions are strict due to entirely 'syntactic reasons', no deep complexity-theoretic results are involved.

2.2 Logspace Approximation Classes

We next define logspace approximation classes as analogues of the corresponding polynomial-time approximation classes. All of the classes are subclasses of NLO by definition, just as, say, APX is a subclass of NPO by definition.

Definition 7. *Let P be an optimization problem and let $r\colon \mathbb{N} \to \mathbb{Q}$ be a function. A function $f\colon I \to \Sigma^*$ is an r-approximator for P if it produces solutions for P and $\mathrm{R}(x, f(x)) \leq r(|x|)$ for all $x \in P_{\exists\mathrm{sol}}$.*

Definition 8. *Let P be an optimization problem. A function $f\colon I \times \mathbb{N} \to \Sigma^*$ is an approximation scheme for P if for all $x \in P_{\exists\mathrm{sol}}$ and all positive integers k we have $f(x, k) \in S(x)$ and $\mathrm{R}(x, f(x, k)) \leq 1 + 1/k$.*

Definition 9. *Let $P \in \mathrm{NLO}$.*

1. *$P \in \mathrm{exp\text{-}ApxLO}$ if there exists a $2^{n^{O(1)}}$-approximator in FL for P.*
2. *$P \in \mathrm{poly\text{-}ApxLO}$ if there exists an $n^{O(1)}$-approximator in FL for P.*
3. *$P \in \mathrm{ApxLO}$ if there exists a c-approximator in FL for P for some constant c.*
4. *$P \in \mathrm{LAS}$ if there exists an approximation scheme f for P such that $f(.,k) \in$ FL for all k.*
5. *$P \in \mathrm{FLAS}$ if there exists an approximation scheme for P that can be computed in space $O(\log k \log |x|)$.*

The approximation schemes for problems in FLAS are the best schemes we can reasonably hope for; at least for problems in $\mathrm{NLO_{pb}}$ whose budget problems are NL-complete. Suppose we had an approximation scheme for such a problem that needs only space $O(\log^{1-\epsilon} k \log n)$ where $n = |x|$ is the length of the input x. Then we could decide the budget problem in space $O(\log^{2-\epsilon} n)$ since if the optimal solution is known to be an element of $\{1, \ldots, n^d\}$, then a solution of ratio less than $1 + 1/n^d$ is actually optimal. This would show $\mathrm{NL} \subseteq \mathrm{DSPACE}(\log^{2-\epsilon} n)$, which would improve Savitch's Theorem and be a major breakthrough.

Approximating means 'coming up with good solutions efficiently'. In particular, for approximable problems we are able to come up with *some* solution efficiently. Thus, for every $P \in \mathrm{exp\text{-}ApxLO}$ we have $P_{\exists\mathrm{sol}} \in \mathrm{L}$.

2.3 Reductions Between Optimization Problems

In this paper we will consider only one notion of reducibility between optimization problems: E-reductions, where the 'E' stands for *error preserving*. They were introduced in [6] for the polynomial-time case. They are slightly more restrictive than the popular AP-reductions introduced in [3]: where an E-reduction must map an optimal solution to an optimal one, this is not necessary for AP-reductions. Since E-reduction is a *stronger* reduction than AP-reduction, all completeness results established in this paper for E-reductions also hold for AP-reductions. For the definition of the logspace versions of the reductions one can simply take their standard definitions and replace 'polynomial time' by 'logarithmic space' everywhere. Unlike the definition of NLO, no special restrictions (like one-way tape access) are necessary.

Definition 10. *Let P and P' be optimization problems. We write $P \leq_{\mathrm{E}}^{\log} P'$ if there exists a triple (f, g, α), where f and g are logspace-computable functions and $\alpha \in \mathbb{Q}$, such that for all $x \in P_{\exists\mathrm{sol}}$ we have $f(x) \in P'_{\exists\mathrm{sol}}$ and for all $y \in S'(f(x))$ we have $g(x, y) \in S(x)$ and $\mathrm{R}(x, g(x, y)) - 1 \leq \alpha(\mathrm{R}'(f(x), y) - 1)$.*

All classes introduced in this paper contain both minimization and maximization problems. As is well-recognized in polynomial-time approximation theory, one must be careful when claiming that a problem is complete for classes containing both minimization and maximization problems, since reductions between them are seldomly straight-forward. This is especially true if, as in this paper, one is also interested in approximation-preserving reductions. For subclasses of NLO such reductions are possible, but it is not clear whether NLO itself has a complete problem with respect to approximation-preserving reductions.

2.4   Closure Properties of Logspace Optimization Classes

None of the introduced classes is closed under $\leq_{\mathrm{E}}^{\log}$-reductions, but the reason for this is a bit annoying: Any optimization problem (logspace or not) for which we can produce optimal solution in logarithmic space is $\leq_{\mathrm{E}}^{\log}$-reducible to all reasonable problems in $\mathrm{LO_{pb}}$. However, there exist problems outside even NPO for which we can produce optimal solutions in logspace: their solution relation is made very hard (for example NEXP-complete), but we always make the optimal solution trivial.

The following theorems show that if we restrict ourselves to problems in NLO, the introduced classes enjoy the expected closure properties. Let $\mathrm{R}_{\mathrm{E}}^{\log}(C) := \{ P \mid P \leq_{\mathrm{E}}^{\log} P' \in C \}$ denote the *reduction closure* of C.

Theorem 11. *We have* $\mathrm{R}_{\mathrm{E}}^{\log}(C) \cap \mathrm{NLO} = C$ *for* $C \in \{\mathrm{LO}, \mathrm{FLAS}, \mathrm{LAS}, \mathrm{ApxLO},$ *poly-ApxLO, exp-ApxLO*$\}$.

Proof. Let $P \leq_{\mathrm{E}}^{\log} P' \in C$ via (f, g, α) and let $P \in \mathrm{NLO}$. Let $P' \in C$ via a approximator h, an approximation scheme h, or a function h that produces optimal solutions. Then $P \in C$ via a function e with $e(x) := g\big(x, h(f(x))\big)$. The properties of E-reductions ensure that e is a sufficiently good approximator, a sufficiently good approximation scheme, or a function that produces optimal solutions. □

2.5   Problems Complete for Logspace Optimization Classes

In this section we prove that the introduced classes have $\leq_{\mathrm{E}}^{\log}$-complete problems. Many natural optimization problems like the shortest path problem for directed graphs are examples of such complete problems, but for initial completeness results it is technically easier to use a slightly more artificial problem instead: This problems asks us to find a vertex in a graph that is reachable from a start vertex and that is 'as valuable as possible' (all results on these problems are also true if we search for the *least* valuable vertex). Variants of this problem are complete for different logspace optimization and approximation classes. Due to lack of space, proofs are given in the full version of this paper [12].

Before we proceed, let us fix some graph-theoretic terminology. A *directed graph* is a pair (V, E) with $E \subseteq V \times V$. An *undirected graph* is a directed graph with a symmetric edge relation. A *walk* of length ℓ is a sequence $(v_1, \ldots, v_{\ell+1})$ of vertices such that $(v_i, v_{i+1}) \in E$ for all $i \in \{1, \ldots, \ell\}$, and for simplicity also $\ell \leq$

$|V|$. A walk is a *path* if all its vertices are distinct. The *distance function* $d_G \colon V \times V \to \mathbb{N} \cup \{\infty\}$ maps each pair (s, t) to the length of the shortest path between them (which is the same as the length of the shortest walk between them).

Problem 12. PARTIAL-MOST-VALUABLE-VERTEX (PARTIAL-MVV)

Instances. Directed graph $G = (V, E)$, a vertex $s \in V$, and a partial weight function $w \colon V \to \mathbb{N}^+$.

Solutions. Walk starting at s in G that ends at some vertex t for which $w(t)$ is defined.

Measure. Weight $w(t)$ of the vertex t at which the walk ends.

Theorem 13.

1. PARTIAL-MVV *is* $\leq_{\mathrm{E}}^{\log}$*-complete for* $\mathrm{NLO}_{\mathrm{do}}$*.*

2. PARTIAL-MVV $\in$ exp-ApxLO *iff* PARTIAL-MVV $\in$ LO *iff* L = NL.

3. PARTIAL-MVV$_{\exists\mathrm{sol}}$ *and* PARTIAL-MVV$_{\mathrm{opt}>}$ *are* $\leq_{\mathrm{m}}^{\log}$*-complete for* NL.

We can restrict the problem in several ways. First, we can allow only small weights, that is, weights taken from the set $\{1, \dots, n\}$. The resulting problem, called PARTIAL-MVV-PB, has the same properties as the unrestricted version except that it is $\leq_{\mathrm{E}}^{\log}$-complete for $\mathrm{NLO}_{\mathrm{pb}}$ instead of $\mathrm{NLO}_{\mathrm{do}}$. Second, we can require the weight function to be a total function, which yields the problem MVV, whose properties are stated in Theorem 14. Third, we can fix a rational $r > 1$ and furthermore require that the ratio of the most valuable vertex (in the whole graph, reachable or not) to the least valuable vertex is at most r. The properties of this problem, denoted MVV-RATIO-r, are stated in Theorem 15.

Theorem 14.

1. MVV *is* $\leq_{\mathrm{E}}^{\log}$*-complete for* exp-ApxLO$_{\mathrm{do}}$*.*

2. MVV $\in$ poly-ApxLO *iff* MVV $\in$ LO *iff* L = NL.

3. MVV$_{\exists\mathrm{sol}} \in \mathrm{AC}^0$ *and* MVV$_{\mathrm{opt}>}$ *is* $\leq_{\mathrm{m}}^{\log}$*-complete for* NL.

Theorem 15.

1. MVV-RATIO-r *is* $\leq_{\mathrm{E}}^{\log}$*-complete for* ApxLO$_{\mathrm{do}}$ *for all* $r > 1$.

2. MVV-RATIO-r $\in$ LAS *for any* $r > 1$, *iff* L = NL.

3. MVV-RATIO-$r_{\exists\mathrm{sol}} \in$ L *and* MVV-RATIO-$r_{\mathrm{opt}>}$ *is* $\leq_{\mathrm{m}}^{\log}$*-complete for* NL *for all* $r > 1$.

2.6   Hierarchies of Logspace Optimization Classes

We can now (nearly) establish a strict hierarchy among logspace approximation classes, assuming L $\neq$ NL. The problems studied in the previous section all had the property that they are in one of the classes named in the following theorem, but not in the one before, unless L = NL. For example, Theorem 13 tells us that PARTIAL-MVV $\in$ NLO$_{\mathrm{pb}}$, but also that it is not an element of exp-ApxLO, or of exp-ApxLO$_{\mathrm{pb}}$ for that matter, unless L = NL. The '(nearly)' refers to the fact that we are still missing a problem that separates FLAS from LO. This will be remedied in the Section 3.

Theorem 16. *Suppose* L $\neq$ NL. *Then*

LO $\subsetneq$ FLAS $\subseteq$ LAS $\subsetneq$ ApxLO $\subsetneq$ poly-ApxLO $\subsetneq$ exp-ApxLO $\subsetneq$ NLO

LO$_{do}$ $\subsetneq$ FLAS$_{do}$ $\subseteq$ LAS$_{do}$ $\subsetneq$ ApxLO$_{do}$ $\subsetneq$ poly-ApxLO$_{do}$ $\subsetneq$ exp-ApxLO$_{do}$ $\subsetneq$ NLO$_{do}$

LO$_{pb}$ $\subsetneq$ FLAS$_{pb}$ $\subseteq$ LAS$_{pb}$ $\subsetneq$ ApxLO$_{pb}$ $\subsetneq$ poly-ApxLO$_{pb}$ = exp-ApxLO$_{pb}$ $\subsetneq$ NLO$_{pb}$

We can also ask the inverse question: Do these hierarchies collapse if NL = L? In the polynomial-time setting it is well-known that P = NP iff PO = NPO. For logarithmic space, the following theorem holds, see the full version of the paper [12] for a proof.

Theorem 17. NLO$_{do}$ $\subseteq$ LO *iff* NL = L.

Note that the implication 'if NL = L, then NLO = LO' is not known to hold. For problems in NLO $-$ NLO$_{do}$ it is not clear how we can construct a solution, let alone an optimal one, in logarithmic space even with access to an NL-complete oracle, see [5] for a detailed discussion of the involved difficulties.

3   The Hot Potato Problem

In this section we study an optimization problem whose budget problem is NL-complete, but that admits a fully logspace approximation scheme. The only other know example of such a problem is the reachability problem for tournaments [8].

A hot potato (or, more realistically, a piece of news, a task, a packet, a token) must be passed around among a group of people (organizations, processors, machines) for ℓ rounds. After ℓ rounds it will have cooled off (become invalid, outdated, finished). The 'value' (or 'cost') of getting rid of the potato by passing it to another person is given by a matrix and the objective is to maximize (or minimize) the total value (cost) of the path taken by the potato. The potato is allowed to go round in cycles and may even be held by a person for any number of rounds (though, typically, this will not be attractive value-wise). Let us concentrate on the maximization version (the same results also hold for the minimization version). The hot potato problem has a fully logspace approximation scheme, but its budget problem is NL-complete.

Problem 18. MAXIMUM-HOT-POTATO-PROBLEM (MAX-HPP)
Instances. An $n \times n$ matrix A with entries drawn from $\{1, \ldots, n\}$, a number $\ell \leq n$, and a start index i_1.
Solutions. Index sequence $(i_1, \ldots, i_\ell)$.
Measure. Maximize $s_A(i_1, \ldots, i_\ell) := \sum_{p=1}^{\ell-1} A_{i_p, i_{p+1}}$.

Theorem 19.

1. *Optimal solutions for* MAX-HPP *can be computed in space* $O(\log \ell \log n)$.
2. MAX-HPP $\in$ FLAS.
3. MAX-HPP $\in$ LO *iff* L = NL.
4. MAX-HPP$_{\exists sol}$ $\in$ AC^0, MAX-HPP$_{opt>}$ *is* $\leq_m^{\log}$-*complete for* NL.

Proof. 1. We use the divide-and-conquer technique that lies at the heart of Savitch's theorem [11]. We define a two recursive procedures *reachable*(u, v, ℓ, c) and *construct*(u, v, ℓ, c). The first returns true if there exists a sequence $\sigma = (u, j_1, \ldots, j_{\ell-1}, v)$ with $s_A(\sigma) \geq c$. Note that the length of this sequence is $\ell + 1$. To check whether the sequence exists, the algorithm tries to find a $z \in V$ and a $c' \in \{1, \ldots, c-1\}$ such that both *reachable*$\left(u, z, c', \lfloor \ell/2 \rfloor\right)$ and *reachable*$\left(z, v, c - c', \ell - \lfloor \ell/2 \rfloor\right)$ hold. This check can be done recursively and, thus, in space $O(\log \ell \log n)$ if c is polynomial in n. The second algorithm returns the desired sequence, provided it exists. It, too, recursively first constructs the first part of the list and then the second part.

The algorithm that produces optimal solutions works as follows: For each $c \in \{1, \ldots, n^2\}$ it tries to find an index i such that *reachable*$(i_1, i, c, \ell - 1)$ holds. For the largest c for which this is the case, it calls *construct*$(i_1, i, c, \ell - 1)$ for the corresponding i.

2. The logspace approximation scheme for MAX-HPP works as follows: Let A, ℓ, and $r = 1 + 1/k$ be given. Let $\sigma^* = (i_1, \ldots, i_\ell)$ denote a sequence that maximizes $s_A(\sigma^*)$. Our aim is to construct a sequence σ of length ℓ such that $s_A(\sigma^*)/s_A(\sigma) \leq r = 1 + 1/k$.

For 'small' values $\ell \leq 8k^2 + 10k + 4$ we perform a brute-force search to find the sequence σ^*. By the first claim, this brute-force search takes space $O(\log k^2 \log n) = O(\log k \log n)$.

For 'large' values $\ell > 8k^2 + 10k + 4$ we find a sequence of $2k + 1$ indices $(j_1, \ldots, j_{2k+1})$ such that $c := s_A(j_1, \ldots, j_{2k+1})$ is maximal. That is, we find a short walk of maximal value in the matrix. By iterating over all possible start points, it can be found in space $O(\log k \log n)$. We output the following sequence:

$$t := (\underbrace{i_1, \ldots, i_1}_{q \text{ times}}, j_1, j_2, \ldots, j_{2k+1}, j_1, j_2, \ldots, j_{2k+1}, \ldots, j_1, \ldots, j_{2k+1}),$$

where $q = 1 + (\ell - 1) \bmod (2k + 1)$. Then $s_A(\sigma) \geq c \lfloor (\ell - 1)/(2k + 1) \rfloor$.

Now consider the sequence $\sigma^* = (i_1, \ldots, i_\ell)$. For each index $p \in \{1, \ldots, \ell - 2k\}$ we have $s_A(i_p, i_{p+1}, \ldots, i_{p+2k}) \leq c$. To see this note that if this were not the case, then we would have $s_A(i_p, i_{p+1}, \ldots, i_{p+2k}) > c$, which contradicts the optimality of the sequence $(j_1, \ldots, j_{2k+1})$. This shows $s_A(\sigma^*) \leq c \lceil \ell/2k \rceil$.

We can now bound the performance ratio of the sequence σ as follows:

$$\frac{s_A(\sigma^*)}{s_A(\sigma)} \leq \frac{c \lceil \ell/2k \rceil}{c \lfloor (\ell - 1)/(2k + 1) \rfloor} \leq \frac{\ell/2k + 1}{(\ell - 1)/(2k + 1) - 1} = \frac{(2k + 1)(\ell + 2k)}{2k(\ell - 2 - 2k)}$$

$$\leq \frac{(2k + 1)(8k^2 + 10k + 4 + 2k)}{2k(8k^2 + 10k + 4 - 2 - 2k)} = \frac{(k + 1)(4k^2 + 4k + 1)}{k(4k^2 + 4k + 1)} = 1 + \frac{1}{k}.$$

3. This follows easily from the next claim.

4. For the completeness of the budget problem, we reduce the canonically NL-complete graph accessibility problem GAP to MAX-HPP$_{\text{opt}>}$. Let (G, s, t) be an instance for GAP. Let $n := |V|$. The reduction maps this instance to the query 'Is the optimal measure for the instance $(A, n + 1, 1)$ for MAX-HPP at least $2n$?'

Here, A is a matrix (whose construction is described below) that has n^2+n rows and columns.

For the construction of the matrix A we construct a directed graph G' with $V' = \{1,\dots,n+1\} \times V$ as follows. For rows $r_1, r_2 < n+1$, we insert edges as follows: there is an edge from (r_1, v_1) to (r_2, v_2) iff $r_2 = r_1 + 1$ and $(v_1, v_2) \in E \cup \{(v,v) \mid v \in V\}$. For the edges between row n and $n+1$ there is a special rule: there is just one edge from (n,t) to $(n+1,t)$. Note that this construction ensures that any walk in G' of length n must end at $(n+1,t)$.

Let us enumerate the elements of V' via a bijection $\nu\colon V' \to \{1,\dots,|V'|\}$ with $\nu(1,s) = 1$. Construct the $|V'| \times |V'|$ matrix A as follows: let $A_{\nu(v_1'),\nu(v_2')} = 2$ if $(v_1', v_2') \in E'$ and let $A_{\nu(v_1'),\nu(v_2')} = 1$ if $(v_1', v_2') \notin E'$. Thus, A is the adjacency matrix of the graph G', only with the modification that all entries are increased by 1.

The matrix A has the following property: there is a sequence $\sigma = (j_1,\dots, j_{n+1})$ starting at $j_1 = 1$ with $s_A(\sigma) \geq 2n$ iff there is a path from s to t in G of length at most $n-1$. To see this, first assume that such a sequence exists. Since all matrix entries are at most 2 and since the sequence has length $n+1$, we have $s_A(\sigma) = 2n$. Thus, we have $A_{j_p, j_{p+1}} = 2$ for all $p \in \{1,\dots,n\}$. This means that the sequence $(v_1',\dots,v_{n+1}')$ with $\nu(v_p') = j_p$ is a walk in G'. Since the length of this walk is n, as pointed out above, it must end at $(n+1,t)$. This in turn means that there is a walk from s to t in G of length at most $n-1$. It can be obtained by removing the first components from each v_p' and omitting consecutive duplicates and cycles. Second, let us assume that there exists a path from s to t in G of length at most $n-1$. Then there exists a path $(v_1',\dots,v_{n+1}')$ from $(1,s)$ to $(n+1,t)$ in G'. Let $j_p := \nu(v_p')$. Then for each $p \in \{1,\dots,n\}$ we have $A_{j_p, j_{p+1}} = 2$. This in turn yields $s_A(j_1,\dots,j_{n+1}) = 2n$. $\qquad\square$

The above proof breaks down for the 'undirected' version of MAX-HPP, where the matrix is required to be symmetric: We cannot simply take the symmetric closure of the graph G' and then define the matrix A as above. An optimal sequence would then consist of constantly going back and forth along an edge of weight 2. However, the symmetric version turns out to be as difficult as the general problem.

Problem 20. MAXIMUM-SYMMETRIC-HOT-POTATO-PROBLEM

Instances. A symmetric $n \times n$ matrix A with entries drawn from $\{1,\dots,n\}$, a number $\ell \leq n$, and a start index i_1.

Solutions. An index sequence $(i_1,\dots,i_\ell)$.

Goal. Maximize $s_A(i_1,\dots,i_\ell) := \sum_{p=1}^{\ell-1} A_{i_p, i_{p+1}}$.

Theorem 21.

1. MAX-SYMMETRIC-HPP $\in$ FLAS.
2. MAX-SYMMETRIC-HPP $\in$ LO *iff* L = NL.
3. MAX-SYMMETRIC-HPP$_{\exists sol} \in AC^0$, MAX-SYMMETRIC-HPP$_{opt>}$ *is* $\leq_m^{\log}$-*complete for* NL.

Proof. The 'new' part of this theorem compared to Theorem 19 is the claim that the problem MAX-SYMMETRIC-HPP$_{opt>}$ is also $\leq_m^{\log}$-complete for NL. We construct the graph G' as the symmetric closure of the graph G' constructed in Theorem 19, but define the matrix A differently: For $r_1, r_2 < n + 1$, if there is an edge $v'_1 := (r_1, v_1)$ to $v'_2 := (r_2, v_2)$ in G', let $A_{\nu(v'_1),\nu(v'_2)} = n$. For the edge between $v'_1 := (n, t)$ and $v'_2 := (n+1, t)$ let $A_{\nu(v'_1),\nu(v'_2)} = n+1$ and symmetrically $A_{\nu(v'_2),\nu(v'_1)} = n + 1$. All other entries of A are 1.

This time, there is a path from s to t in G iff there is a sequence $\sigma = (j_1, \ldots, j_{n+1})$ staring at $j_1 = 1$ such that $s_A(\sigma) \geq (n-1)n + (n+1) = n^2 + 1$. To see this, first let us assume that such a sequence exists. Since it has length $n + 1$, we must have $A_{j_p, j_{p+1}} \geq n$ for all $p \in \{1, \ldots, n\}$ and we must have $A_{j_p, j_{p+1}} \geq n + 1$ at least once. This means that the sequence $v'_1, \ldots, v'_{n+1}$ with $\nu(v'_p) = j_p$ is a walk in G' ('standing still' from time to time) that visits the edge between (n, t) and $(n + 1, t)$. Thus, there is a path from s to t in G. Second, let us assume that there exists a path from s to t in G. Then for a path $(v'_1, \ldots, v'_{n+1})$ from $(1, s)$ to $(n + 1, t)$ in G' and for $j_p := \nu(v'_p)$ we have $s_A(j_1, \ldots, j_{n+1}) = (n-1)n + (n+1) = n^2 + 1$. □

In the proof of Theorem 19 the matrix has just two different entries (1 and 2), in the proof of Theorem 21 there are three (1, n, and $n+1$). We may ask whether the construction of Theorem 21 can be modified to also use only two different entries. However, this is presumably not possible: Suppose a and b with $a \leq b$ are these entries. If there exists an index i with $A_{1,i} = b$, then an optimal sequence is given by $(1, i, 1, i, \ldots)$. If there is no such index and if $A_{i,j} = b$, then an optimal sequence is given by $(1, i, j, i, j, \ldots)$. This proves the following theorem:

Theorem 22.

1. MAX-HPP-2-ENTRIES$_{\text{opt}>}$ is $\leq_{\text{m}}^{\log}$-complete for NL.
2. MAX-SYMMETRIC-HPP-3-ENTRIES$_{\text{opt}>}$ is $\leq_{\text{m}}^{\log}$-complete for NL.
3. MAX-SYMMETRIC-HPP-2-ENTRIES $\in$ LO.

4   Conclusion

Different logspace optimization problems can have different approximation properties, even though their underlying budget or existence problems have the same complexity. Research on logspace problems has been preoccupied with existence problems. The results of this paper suggest that we should broaden our perspective.

We have seen that natural logspace optimization problems can have different approximation properties. Most logspace optimization problems are complete for an appropriate logspace approximation class or for NLO. Under the assumption $L \neq NL$, we have ApxLO $\subsetneq$ NLO as demonstrated by the problem PARTIAL-MVV; we have LAS $\subsetneq$ ApxLO as demonstrated by MVV-RATIO-2; and we have LO $\subsetneq$ FLAS as demonstrated by MAX-HPP.

The framework of this paper can be applied to many problems that have already been studied in the literature. For example, Reingold [10] gives a logspace algorithm for the reachability problem in undirected graphs and the algorithm can be used to construct a path between any two vertices. We may now ask whether anything can be said about *how long* this path will be, relative to the shortest path. This path will not always be the shortest one unless NL = L, but it might well be possible that the shortest path can be approximated in logarithmic space. Another example is the reachability algorithm of Jakoby, Liśkiewicz, and Reischuk [4] for series-parallel graphs. Their algorithm can be used to show that the reachability problem for series-parallel graphs is in poly-ApxLO, but it is not clear whether their algorithm finds a *shortest* path nor is it clear how difficult finding a shortest path is. As a final example, consider the problem of telling whether an undirected graph has a cycle. This problem can be solved in logarithmic space, but how difficult is it to compute the size of the *smallest* cycle in the graph (also known as the *girth*)?

References

1. Carme Àlvarez and Birgit Jenner. A very hard log-space counting class. *Theoretical Computer Science*, 107:3–30, 1993.
2. G. Ausiello, P. Crescenzi, G. Gambosi, V. Kann, A. Marchetti-Spaccamela, and M. Protasi. *Complexity and Approximation: Combinatorial Optimization Problems and Their Approximability Properties*. Springer-Verlag, 1999.
3. Pierluigi Crescenzi, Viggo Kann, Riccardo Silvestri, and Luca Trevisan. Structure in approximation classes. *SIAM J. Computing*, 28(5):1750–1782, 1999.
4. Anreas Jakoby, Maciej Liśkiewicz, and Rüdiger Reischuk. Space efficient algorithms for series-parallel graphs. In Afonso Ferreira and Horst Reichel, editors, *Proc. 18th Annual Symp. on Theoretical Aspects of Comput. Sci.*, volume 2010 of *Lecture Notes on Comp. Sci.*, pages 339–352. Springer-Verlag, 2001.
5. Birgit Jenner and Jacobo Torán. *Complexity Theory Retrospective II*, chapter 'The Complexity of Obtaining Solutions for Problems in NP and NL', pages 155–178. Springer-Verlag, 1997.
6. S. Khanna, R. Motwani, M. Sudan, and U. Vazirani. On syntactic versus computational views of approximability. *SIAM J. Computing*, 28(1):164–191, 1998.
7. Mark W. Krentel. The complexity of optimization problems. *Journal of Computer and System Sciences*, 36(3):490–509, June 1988.
8. Arfst Nickelsen and Till Tantau. The complexity of finding paths in graphs with bounded independence number. *SIAM Journal on Computing*, 2005. To appear.
9. Christos H. Papadimitriou. *Computational Complexity*. Addison-Wesley, 1994.
10. Omer Reingold. Undirected st-connectivity in log-space. Technical Report TR04-094, Electronic Colloquium on Computational Complexity, 2004.
11. Walter J. Savitch. Relationships between nondeterministic and deterministic tape complexities. *Journal of Computer and System Sciences*, 4(2):177–192, 1970.
12. Till Tantau. Logspace optimisation problems and their approximation properties. Technical Report TR03-077, Electronic Colloquium on Computational Complexity, www.eccc.uni-trier.de/eccc, 2003.
13. Seinosuke Toda. Counting problems computationally equivalent to computing the determinant. Technical Report CSIM 91-07, Dept. Comput. Sci. and Inf. Math., Univ. Electro-Communications, Chofu-shi, Tokyo 182, Japan, May 1991.

A Faster and Simpler 2-Approximation Algorithm for Block Sorting

Wolfgang W. Bein[1], Lawrence L. Larmore[1],
Linda Morales[2], and I. Hal Sudborough[3]

[1] School of Computer Science, University of Nevada,
Las Vegas, Nevada 89154, USA**
bein@cs.unlv.edu larmore@cs.unlv.edu
[2] Computer Science Department, Texas A&M University-Commerce,
Commerce, TX 75429, USA
Linda_Morales@tamu-commerce.edu
[3] Department of Computer Science, University of Texas at Dallas,
Richardson, TX 75083, USA
hal@utdallas.edu

Abstract. Block sorting is used in connection with optical character recognition (OCR). Recent work has focused on finding good strategies which perform well in practice. Block sorting is $\mathcal{NP}$-hard and all of the previously known heuristics lack proof of any approximation ratio. We present here an approximation algorithm for the block sorting problem with approximation ratio of 2 and run time $O(n^2)$. The approximation algorithm is based on finding an optimal sequence of absolute block deletions.

Keywords: Design and analysis of algorithms; approximation algorithms; block sorting; transposition sorting; optical character recognition.

1 Introduction

Define a *permutation* of length n to be a list $x = (x_1, \ldots, x_n)$ consisting of the integers $\{1, \ldots, n\}$ where each number occurs exactly once. For any $1 \leq i \leq j \leq n$ we denote the sublist of x that starts at position i and ends with position j by $x_{i \ldots j}$. If $j < i$, let $x_{i \ldots j}$ be the empty list. We call a nonempty sublist of x consisting of consecutive integers, such as $(3, 4, 5)$ in $(6, 3, 4, 5, 2, 1)$, a *block* of x.

The *block sorting problem* is to find a minimal length sequence of *block move steps*, or simply *block moves* for short, which sorts an initial permutation x, where a block move consists of moving one block of a list to a different position in the list. For example, the list $(4, 2, 5, 1, 3)$ can be block sorted in two moves: first move (2) to obtain $(4, 5, 1, 2, 3)$, then move $(4, 5)$ to obtain $(1, 2, 3, 4, 5)$.

Sorting problems under various operations have been studied extensively. We mention work on sorting with prefix reversals [4,5,9], transpositions [1,2], and

---

** Research of these authors supported by NSF grant CCR-0312093.

M. Liśkiewicz and R. Reischuk (Eds.): FCT 2005, LNCS 3623, pp. 115–124, 2005.
© Springer-Verlag Berlin Heidelberg 2005

Table 1. How to block sort "How ? they did it do"

A F C B E D		How ? they did it do		
A C B E F D move F	How they did	it ?	do	
A C B E D and combine	How they did	–E–	do	
A B C E D move C	How did they	–E–	do	
A E D and combine	——A——	–E–	do	
A D E move D	——A——	do	–E–	
A and combine	————A————			

block moves [7,8,12,12]. In particular, block sorting is not the same as transposition sorting, and thus the 1.5-approximation to optimality obtained by Bafna and Pevzner [1,2] does not imply a 1.5-approximation to optimal block sorting. Furthermore, optimal block sorting is known to be $\mathcal{NP}$-hard [3], while the question of $\mathcal{NP}$-hardness for optimal transposition sorting is currently open. The two problems are related in the sense that every sequence of block sorting moves defines a sequence of transpositions (but not vice-versa). Thus, the study of block sorting might give further insights into transposition sorting.

Block sorting is motivated by applications in optical character recognition; see [6,10,11]. Text regions, referred to as *zones* are selected by drawing rectangles (or piecewise rectangles, polygons) around them. Here the order of zones is significant, but in practice the output generated by a zoning procedure may be different from the correct text. A situation prone to such misidentification might arise from multi-column documents, for example, as part of "de-columnizing" such a multi-column document. Zones which are not in the correct order must be further processed (sometimes by a human editor). In the OCR community the number of editing operations, such as different kinds of deletions, insertions and moves is used to define a *zoning metric*. We refer the reader to the work in [6,10,11] for details, but mention here that the block sorting problem plays a significant part in defining the zoning metric: moving the pieces to the correct order corresponds to a block sorting problem. To evaluate the performance of a given zoning procedure it is of interest to find the minimum number of steps needed to obtain the correct string from the zones generated by the zoning procedure. An example motivating such an application is given in Table 1.

The research history of the block sorting is as follows: initially, Latifi *et al.* [6] have performed various experiments to test a number of strategies that seem to perform well in practice; however, no ratio better than three has been proved for any of these schemes. (The approximation ratio of three arises trivially: if the list is not sorted, simply pick an arbitrary maximal block and move it to a position where it can be combined into a larger block. The number of maximal blocks can be decreased by at most three for any block sorting step, while the trivial algorithm decreases that number by at least one at each step, and hence has approximation ratio three.) As a next step, [3] Bein *et al.* have shown that the block sorting problem is $\mathcal{NP}$-hard [3]. In the same paper they show how to implement the strategies of [6] in linear time, which is significant in practice. However, no approximation ratio (other than the trivial) is given.

The first approximation algorithm with a non-trivial approximation ratio for the block sorting problem is given in [12,13] by Mahajan *et al.* They give an $O(n^3)$-time block sorting algorithm with approximation ratio 2. Their algorithm first solves a related problem, the *block merging problem*, by constructing a crossing graph in $O(n^3)$ and then deriving a block merging sequence. (Even though not explicitly stated in their paper, it appears that the derivation of the actual sequence does not increase the time complexity of the overall procedure beyond $O(n^3)$.) The solution to the block merging problem is then used to get a 2-approximation for block sorting.

In this paper, we improve that result by giving a quadratic time block sorting algorithm with approximation ratio 2. Central to our method is a problem, closely related to the block sorting problem, but which is in the polynomial class, the *abs-block deletion problem*. We note that the abs-block deletion problem is not equivalent to the block merging problem; see our final remarks in Section 4. We call the *deletion* of a block an *absolute block deletion*, or *abs-block deletion* for short. The *abs-block deletion problem* is the problem of finding the minimum length sequence of abs-block deletions to transform a list of distinct integers into a monotone increasing list. The *complete abs-block deletion problem* is the same, except that the final list must be empty.

As we will show in the next section, the abs-block deletion problem can be solved in $O(n^2)$ time. In Section 3 we show, given a permutation x, that (a) if there is an abs-block deletion sequence of length m for x, there is a block sorting sequence of length m for x and that (b) if there is a block sorting sequence of length m for x, then there is an abs-block deletion sequence of length at most $2m - 1$ for x. From this we derive the 2-approximation algorithm.

2 Finding Optimal Absolute Block Deletion Sequences in Quadratic Time

2.1 Preliminaries

We note the difference between absolute block deletions and relative block deletions. We make both concepts precise here.

Given a list x of distinct integers and a sublist y of x, we say y is a *relative block* (*rel-block* for short) of x if the following conditions hold:

- y is monotone increasing.
- If $r < u < s$ are integers such that r and s are in y, then either u is in y or u is not in x.

Given x, we define a *relative block deletion* (*rel-block deletion*, for short) to be the deletion of a relative block from x.

We define a *relative block deletion sequence* for x to be a sequence of rel-block deletions, starting with x and ending with a monotone sequence. From [3] we have the following result.

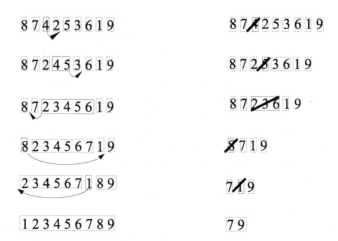

Fig. 1. Block Sorting Moves (left) and corresponding Relative Block Deletions (right)

Theorem 2.1. *Let x be a permutation. Then the minimum number of block sorting moves needed to sort x is the same as the minimum length of a rel-block deletion sequence for x.*

(Figure 1 shows a sequence of block moves with a corresponding sequence of relative block deletions.)

We say that a list y is a *subsequence* of x if y is obtained from x by deleting any number of elements. For example, $(2, 3)$ is a subsequence of $(2, 4, 3, 1)$, but not a sublist. A subsequence of x is uniquely characterized by its set of items. By a slight abuse of notation, we shall sometimes identify a subsequence with the set of its items.

Define the *closure* of a subsequence of x to be the smallest sublist of x which contains it. For example, the closure of the subsequence $(2, 3)$ of $(2, 4, 3, 1)$ is the sublist $(2, 4, 3)$. If A and A' are subsequences of a list x, we say that A and A' are *separated* if the closures of A and A' are disjoint.

An abs-block deletion sequence for a subsequence y of x consists of a sequence $A_1, \ldots, A_m$ of disjoint non-empty subsequences of y such that $y - \bigcup_{u=1}^{m} A_u$ is monotone, and each A_v is a block of the subsequence $y - \bigcup_{u<v} A_u$. For example, the minimum length abs-block deletion sequence for the list $(1, 4, 2, 5, 3)$ consists of two steps. First delete the abs-block (2), obtaining $(1, 4, 5, 3)$, then delete the abs-block $(4, 5)$, obtaining the sorted list $(1, 3)$. The left part of Figure 3 shows another example of an abs-block deletion sequence.

A complete abs-block deletion sequence for a subsequence y of x consists of an abs-block deletion sequence $A_1, \ldots, A_m$ of y such that $y - \bigcup_{u=1}^{m} A_u$ is the empty list.

2.2   A Dynamic Program for Absolute Block Deletion

We first consider the complete abs-block deletion problem for all sublists of x, which we solve in quadratic time. Once the $O(n^2)$ answers to this problem are obtained, the original abs-block deletion problem can be solved in quadratic time.

For all $1 \leq i, j \leq n$, let $t_{i,j}$ to be the minimum length of any complete abs-block deletion sequence for the sublist $x_{i...j}$. Our algorithm first computes all $t_{i,j}$ by dynamic programming. Trivially, $t_{i,i} = 1$, and $t_{i,j} = 0$ for $j < i$.

Lemma 2.2. *If $A_1, \ldots, A_m$ is an abs-block deletion sequence for a sublist of y of x, and $1 \leq u < v \leq m$, then either A_u and A_v are separated, or A_u is a subset of the closure of A_v.*

Proof. The closure of A_u cannot contain any item of A_v, since otherwise A_u could not be deleted before A_v. Thus, every item of A_v is entirely before A_u or entirely after A_u in x. If all items of A_v are before A_u or all items of A_v are after A_u, then A_u and A_v are separated. If some items of A_v are before A_u and some items are after A_u, then A_u is a subset of the closure of A_v.

Lemma 2.3. *If $A_1, \ldots, A_t, A_{t+1}, \ldots, A_m$ is any abs-block deletion sequence for a sublist y of x, and if A_t and A_{t+1} are separated, then A_t and A_{t+1} may be transposed, i.e., $A_1, \ldots, A_{t+1}, A_t, \ldots, A_m$ is an abs-block deletion sequence for y.*

Proof. For any u, let $y_u = x - \bigcup_{v<u} A_v$. By definition, A_t is a block of y_t, and A_{t+1} is a block of $y_{t+1} = y_t - A_t$. Since A_t and A_{t+1} are separated, A_{t+1} is also a block of y_t. Thus, A_{t+1} can be deleted before A_t.

Lemma 2.4. *For any $1 \leq i \leq j \leq n$, if there is a complete abs-block deletion sequence for $x_{i...j}$ of length m, then there is a complete abs-block deletion sequence for $x_{i...j}$ of length m such that x_i is deleted in the last step.*

Proof. Let $A_1, \ldots, A_m$ be a complete abs-block deletion sequence of $x_{i...j}$. Suppose that $x_i \in A_t$ for some $t < m$. For any $v > t$, x_i cannot be an item of the closure of A_v, hence, by Lemma 2.2, A_t and A_v must be separated. By Lemma 2.3, we can transpose A_t with A_v for each $v > t$ in turn, moving A_t to the end of the deletion sequence.

Let z be the inverse permutation of x, i.e., $x_i = k$ if and only if $z_k = i$. Note that z can be computed in $O(n)$ preprocessing time.

Theorem 2.5. *Let $1 \leq i \leq j \leq n$. Then*

$$t_{i,j} = \begin{cases} \min \left\{ \begin{array}{l} 1 + t_{i+1,j} \\ t_{i+1,\ell-1} + t_{\ell,j} \end{array} \right\} & \text{if } i < n \text{ and } i < \ell \leq j, \text{ where } \ell = z_{x_i+1} \\ 1 + t_{i+1,j} & \text{otherwise} \end{cases} \quad (1)$$

Proof. If $i = j$, the recurrence is trivial, so assume $i < j$. Let $a = x_i$. We first prove that the left side of (1) is less than or equal to the right side. If $A_1, \ldots, A_u$ is a complete abs-block deletion of $x_{i+1...j}$, then $A_1, \ldots, A_u, (i)$ is a complete abs-block deletion of $x_{i...j}$, hence $t_{i,j} \leq 1 + t_{i+1,j}$.

If $x_\ell = a + 1$ for some $i < \ell \leq j$, $A_1, \ldots, A_u$ is a complete abs-block deletion of $x_{i,\ell-1}$, and $A_{u+1}, \ldots, A_m$ is a complete abs-block deletion of $x_{\ell,j}$, then

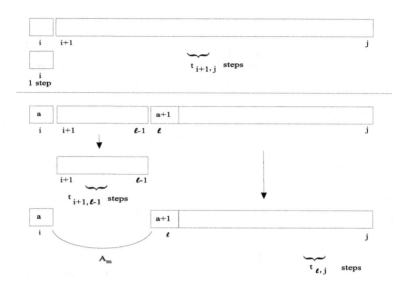

Fig. 2. The Recurrence for the $t_{i,j}$ Values

$A_1, \ldots, A_u, A_{u+1}, \ldots, A_{m-1}, A_m + \{a\}$ is a complete abs-block deletion of $x_{i,j}$. Thus, the left side of (1) is less than or equal to the right side.

We now show that the left side of (1) is greater than or equal to the right side. Let $m = t_{i,j}$, and let $A_1, \ldots, A_m$ be a minimum length complete abs-block deletion sequence of $x_{i\ldots j}$. By Lemma 2.4, we can insist that $x_i = a$ is an item of A_m. If $A_m = (a)$, then $A_1, \ldots, A_{m-1}$ is a complete abs-block deletion sequence of $x_{i+1\ldots j}$, which must be optimal by the optimality principle. Thus, $t_{i+1,j} = m-1$. Otherwise, A_m contains $a+1 = x_\ell$, where $i < \ell \leq j$. If $1 \leq t < m$, then A_t is either completely before or completely after $a+1$, since $a+1$ is deleted after A_t. By Lemma 2.3, we can permute the indices so that, for some $u < m$, $A_t \subseteq x_{i+1\ldots\ell-1}$ if $t \leq u$ and $A_t \subseteq x_{\ell\ldots j}$ for $u < t < m$. By the optimality principle, $A_1, \ldots, A_u$ is a minimum length complete abs-block deletion sequence for $x_{i+1\ldots\ell-1}$, while $A_{u+1}, \ldots, A_{m-1}, A_m - \{a\}$ is a minimum length complete abs-block deletion sequence for $x_{\ell\ldots m}$ Thus, $u = t_{i+1,\ell}$ and $m - u = t_{\ell,m}$. In any case, the left side of (1) is greater than or equal to the right side.

We now turn to the analysis of the run time of the corresponding dynamic program. We note the following crucial fact:

Lemma 2.6. *If $t_{u,v}$ are already known for all $i < u \leq v \leq j$, then $t_{i,j}$ can be computed in $O(1)$ time.*

Proof. If $i \geq j$, the result is trivial. Henceforth, assume that $i < j$. Let $m = t_{i,j}$, and let A_t be the subsequence of $x_{i\ldots j}$ that is deleted at step t of the complete abs-block deletion sequence of $x_{i\ldots j}$ of length m. By Lemma 2.4, we can assume that $x_i \in A_m$. If $|A_m| > 1$, $\ell = z_{x_i+1}$ can be found in $O(1)$ time, since we have already spent $O(n)$ preprocessing time to compute the array z. The recurrence thus takes $O(1)$ time to execute for each i, j.

Corollary 2.7. *The values of $t_{i,j}$ for all i,j can be computed in $O(n^2)$ time.*

We finally note that the optimal complete absolute block deletion sequence can be recovered by keeping appropriate pointers as the $t_{i,j}$ are computed.

We now show how to obtain a solution to the abs-block deletion problem for x in $O(n^2)$-time. Define a weighted acyclic directed graph G with one node for each $i \in \{0, \ldots n+1\}$. There is an edge of G from i to j if and only if $i < j$ and $x_i < x_j$, and the weight of that edge is $t_{i+1,j-1}$. If there is an abs-block deletion sequence of x of length m, there must be a path from 0 to $n+1$ in G of weight m, and vice-versa. Using standard dynamic programming, a minimum weight path from 0 to $n+1$ can be found in $O(n^2)$ time. Let $0 = i_0 < i_1 < \cdots < i_\ell = n+1$ be such a minimum weight path, and let $m = \sum_{u=1}^{\ell} t_{i_{u-1}+1, i_u-1}$ be the weight of that minimum path. Since every deletion is a block deletion, the entire list can be deleted to a monotone list in m block deletions. Thus we have:

Theorem 2.8. *The abs-block deletion problem can be solved in time $O(n^2)$.*

3   Absolute Block Deletion and Block Sorting

We now derive an approximation algorithm "BLOCK2" for the block sorting problem. We say that an algorithm $\mathcal{A}$ has *approximation ratio* C if, for any permutation x of $\{1, \ldots, n\}$, $\mathcal{A}$ finds a block sorting of x of length at most $C \cdot p + O(1)$, where p is the minimum number of block sorting moves needed to sort x. Given a permutation x, BLOCK2 performs the following steps:

1. Compute a minimum length abs-block deletion sequence of x. Let $A_1, \ldots, A_m$ be the blocks that are deleted in this sequence. For each $1 \le t \le m$, let a_t be the first item of A_t.
2. Let x^0 be the augmentation of x, i.e., $x_0^0 = 0$ and $x_{n+1}^0 = n+1$, and $x_i^0 = x_i$ for $1 \le i \le n$.
3. Let $M = \{0, 1, \ldots, n, n+1\} - \bigcup_{t=1}^{m} A_t$, the monotone increasing subsequence consisting of the items of x^0 that remain after those deletions. Note that $0, n+1 \in M$.
4. **Loop:** For each t from 1 to m, do the following:
   (a) Let B_t be the maximal block of x^{t-1} which contains a_t.
   (b) If a_t is not the first item of B_t, let $x^t = x^{t-1}$. Otherwise, let x^t be obtained by deleting B_t from x^{t-1} and reinserting it just after $a_t - 1$.

Figure 3 gives an example which illustrates how block sorting moves are obtained from an abs-block deletion sequence. Elements in M are underlined in the figure.

Clearly, BLOCK2 uses at most m block moves. Our result then follows from the lemma below.

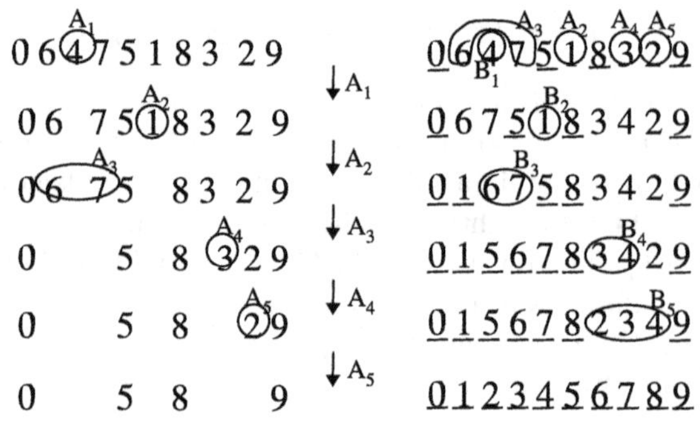

$$
\begin{array}{ll}
06\,4\,7\,5\,1\,8\,3\,\;2\,9 & 0\,6\,4\,7\,5\,1\,8\,3\,2\,9 \\
\qquad\qquad\qquad\downarrow A_1 & \\
06\;\;75\,1\,8\,3\;2\;9 & 0\,6\,7\,5\,1\,8\,3\,4\,2\,9 \\
\qquad\qquad\qquad\downarrow A_2 & \\
0\,6\;\;7\,5\;\;8\,3\,2\,9 & 0\,1\,6\,7\,5\,8\,3\,4\,2\,9 \\
\qquad\qquad\qquad\downarrow A_3 & \\
0\qquad 5\;\;8\;\;3\,2\,9 & 0\,1\,5\,6\,7\,8\,3\,4\,2\,9 \\
\qquad\qquad\qquad\downarrow A_4 & \\
0\qquad 5\;\;8\qquad 2\,9 & 0\,1\,5\,6\,7\,8\,2\,3\,4\,9 \\
\qquad\qquad\qquad\downarrow A_5 & \\
0\qquad 5\;\;8\qquad\;\; 9 & 0\,1\,2\,3\,4\,5\,6\,7\,8\,9
\end{array}
$$

Fig. 3. Obtaining Block Sorting Moves from Abs-Block Deletions

Lemma 3.1. *The list x^m is sorted.*

Proof. Assume that x^m is not sorted. Define a *jump* of x^t to be an item $x_i^t = a$ such that $x_{i-1}^t \neq a - 1$. We say that a jump $x_i^t = a$ is an *inversion* of x^t if $x_{i-1}^t > a$. Note that if a is a jump of x^t, it is a jump of x^u for all $u < t$, since no iteration of the loop of BLOCK2 creates a new jump.

We first claim that, if a is any jump of x^m, then $a \in M$. If $a \in A_t$, then a is not a jump of x^t, hence not a jump of x^m. This proves the claim.

If x^m is not sorted, then x^m must have an inversion $x_i^m = a$. By the claim, $a \in M$. Let b be the smallest item in the maximal block B of x^m that contains x_{i-1}^m. We know that $b > a$, since $x_{i-1}^m > a$, B does not contain a, and B contains all integers between b and x_{i-1}^m. By the claim, $b \in M$. But b is before a in x^m, and hence in M, since the items of M are never moved, and M is increasing: contradiction. Thus x^m is sorted.

We now show that BLOCK2 gives a 2-approximation for the block sorting problem.

Lemma 3.2. *If there is a block sorting for a permutation x which has m steps, then there is an abs-block deletion sequence for x of length at most $2m - 1$.*

Proof. Suppose we are given a block sorting sequence of x of length m. Let B_t be the block that is moved at the tth step. Let M be the monotone increasing subsequence of x consisting of all items of x which are never moved. In the construction below, we think of each B_t as a set of integers. Let $\mathcal{B}$ be the forest whose nodes are $B_1, \ldots, B_m$, where B_t is in the subtree rooted at B_u if and only if $B_t \subseteq B_u$.

We now place one credit on each root of $\mathcal{B}$ and two credits on each B_t which is not a root of $\mathcal{B}$. Thus, we place at most $2m - 1$ credits. Each B_t which is not a root then passes one credit up to its parent.

We claim that each B_t has enough credits to pay for its deletion in an abs-block deletion sequence for x, where each abs-block deletion takes one credit.

Suppose we have deleted $B_1, \ldots, B_{t-1}$. Then B_t may have been partially deleted. The remaining items form a rel-block of the remaining list, but not necessarily an abs-block. However, the number of "holes" in B_t cannot exceed the number of children of B_t in the forest $\mathcal{B}$. That is, if B_t has r children, the undeleted items of B_t form the disjoint union of at most $r + 1$ intervals, and are thus the items of at most $r + 1$ blocks in x. To delete these blocks, we use the $r + 1$ credits on B_t. After all block deletions, the remaining list is M, which is sorted.

As an immediate corollary of Lemma 3.2, we have

Theorem 3.3. *Algorithm* BLOCK2 *has an approximation ratio of* 2.

Regarding the time complexity of BLOCK2 we have:

Theorem 3.4. *The time complexity of* BLOCK2 *is* $O(n^2)$.

Proof. It takes $O(n^2)$ time to find a minimal abs-block deletion sequence. The remaining parts of the algorithm, such as additional steps to keep track of intermediate results, take $O(n^2)$ time.

4   Final Remarks and Open Problems

The block merging problem of Mahajan *et al.* [12] is defined as follows: At each step, a configuration of the problem consists of a set of lists of integers, where the set of all integers in the lists is $\{1, ..., n\}$, and no integer appears in two lists, or more than once in one list. One or more of the lists may be empty. A move consists of deleting a block from one of the lists and inserting that same block into one of the other lists in such a way that the moved block merges with another block. We are given an initial configuration, and we want to find the minimum number of moves to reach the configuration where there is only one non-empty list and it is sorted.

It is entirely possible that block merging and abs-block deletion are related, but they are not identical: the set of lists $\{(3, 7, 9), (4, 8), (1, 5), (2, 6)\}$ takes 8 steps to sort by block merging, as follows:

$$\{(3, 7, 9), (4, 8), (5), (1, 2, 6)\}$$
$$\{(1, 2, 3, 7, 9), (4, 8), (5), (6)\}$$
$$\{(7, 9), (1, 2, 3, 4, 8), (5), (6)\}$$
$$\{(7, 9), (8), (1, 2, 3, 4, 5), (6)\}$$
$$\{(7, 9), (8), \epsilon, (1, 2, 3, 4, 5, 6)\}$$
$$\{(1, 2, 3, 4, 5, 6, 7, 9), (8), \epsilon, \epsilon\}$$
$$\{(9), (1, 2, 3, 4, 5, 6, 7, 8), \epsilon, \epsilon\}$$
$$\{(1, 2, 3, 4, 5, 6, 7, 8, 9), \epsilon, \epsilon, \epsilon\}$$

However, abs-block deletion takes 4 steps to get to a monotone sequence from, $(3, 7, 9, 4, 8, 5, 1, 2, 6)$: delete (2), delete (8), delete (1), and then delete $(4, 5, 6)$.

We have given a better non-trivial approximation algorithm with a provable approximation ratio for the block sorting problem. There may be, however, room for further improvement. We mention two lines of further investigation:

1. We conjecture that a polynomial time approximation algorithm with a ratio better than 2 exists.
2. It would be interesting to see how our algorithm compares with some of the heuristics given in [6]. All of those heuristics lack proof of any approximation ratio; their advantage is that they have linear run time. Indeed, it would be desirable to give a 2-approximation with run time better than $O(n^2)$.

Finally we mention that the study of block merging, block sorting and abs-block deletions might lead to insights for other sorting problems, such as sorting by transpositions.

References

1. V. Bafna and P.A. Pevzner. Sorting by transposition. *SIAM Journal on Discrete Mathematics*, 11:224–240, 1998.
2. V. Bafna and P.A. Pevzner. Genome rearrangements and sorting by reversals. *SIAM Journal on Computing*, 25:272–289, 1999.
3. W.W. Bein, L.L Larmore, S. Latifi, and I.H Sudborough. Block sorting is hard. *International Journal of Foundations of Computer Science*, 14(3):425–437, 2003.
4. A. Caprara. Sorting by reversals is difficult. In *Proceedings 1st Conference on Computational Molecular Biology*, pages 75–83. ACM, 1997.
5. W. H. Gates and C. H. Papadimitriou. Bounds for sorting by prefix reversal. *Discrete Mathematics*, 27:47–57, 1979.
6. R. Gobi, S. Latifi, and W. W. Bein. Adaptive sorting algorithms for evaluation of automatic zoning employed in OCR devices. In *Proceedings of the 2000 International Conference on Imaging Science, Systems, and Technology*, pages 253–259. CSREA Press, 2000.
7. L.S. Heath and J.P.C. Vergara. Sorting by bounded block-moves. *Discrete Applied Mathematics*, 88:181–206, 1998.
8. L.S. Heath and J.P.C. Vergara. Sorting by short block-moves. *Algorithmics*, 28 (3):323–354, 2000.
9. H. Heydari and I. H. Sudborough. On the diameter of the pancake network. *Journal of Algorithms*, 25(1):67–94, 1997.
10. J. Kanai, S. V. Rice, and T.A. Nartker. Automatic evaluation of OCR zoning. *IEEE Transactions on Pattern Analysis and Machine Intelligence*, 17 (1):86–90, 1995.
11. S. Latifi. How can permutations be used in the evaluation of automatic zoning evaluation? In *ICEE 1993*. Amir Kabir University, 1993.
12. M. Mahajan, R. Rama, V. Raman, and S. Vijayakumar. Approximate block sorting. To appear in *International Journal of Foundations of Computer Science*.
13. M. Mahajan, R. Rama, and S. Vijayakumar. Merging and sorting by strip moves. In *Proceedings of the 23rd International Foundations and Software Technology and Theoretical Computer Science Conference (FSTTCS)*, volume 2914 of *Lecture Notes in Computer Science*, pages 314–325. Springer, 2003.

On the Power of Unambiguity in Alternating Machines

Holger Spakowski[1,*] and Rahul Tripathi[2,**]

[1] Institut für Informatik, Heinrich-Heine-Universität,
40225 Düsseldorf, Germany
spakowsk@cs.uni-duesseldorf.de
[2] Dept. of Computer Science and Engineering,
University of South Florida, Tampa, Fl 33620, USA
rahul.tripathi007@gmail.com

Abstract. Recently, the property of unambiguity in alternating Turing machines has received considerable attention in the context of analyzing globally-unique games by Aida et al. [1] and in the design of efficient protocols involving globally-unique games by Crâsmaru et al. [7]. This paper investigates the power of unambiguity in alternating Turing machines in the following settings:

1. We construct a relativized world where unambiguity based hierarchies—AUPH, UPH, and $\mathcal{UPH}$—are infinite. We construct another relativized world where UAP (unambiguous alternating polynomial-time) is not contained in the polynomial hierarchy.
2. We define the bounded-level unambiguous alternating solution class UAS(k), for every $k \geq 1$, as the class of sets for which strings in the set are accepted unambiguously by some polynomial-time alternating Turing machine N with at most k alternations, while strings not in the set either are rejected or are accepted with ambiguity by N. We construct a relativized world where, for all $k \geq 1$, $\mathrm{UP}_{\leq k} \subset \mathrm{UP}_{\leq k+1}$ and $\mathrm{UAS}(k) \subset \mathrm{UAS}(k+1)$.
3. Finally, we show that robustly k-level unambiguous polynomial-time alternating Turing machines accept languages that are computable in $\mathrm{P}^{\Sigma_k^p \oplus \mathcal{A}}$, for every oracle $\mathcal{A}$. This generalizes a result of Hartmanis and Hemachandra [11].

1 Introduction

Chandra, Kozen, and Stockmeyer [6] introduced the notion of *alternation* as a generalization of nondeterminism: Alternation allows switching of existential and universal quantifiers, whereas nondeterminism allows only existential quantifiers throughout the computation. Alternation has proved to be a central notion in

* Supported in part by the DFG under grant RO 1202/9-1.
** Supported in part by NSF grant CCF-0426761. Work done in part while affiliated with the Department of Computer Science at the University of Rochester, Rochester, NY 14627, USA.

M. Liśkiewicz and R. Reischuk (Eds.): FCT 2005, LNCS 3623, pp. 125–136, 2005.

complexity theory. For instance, the polynomial hierarchy has a characterization in terms of bounded-level alternation [6,23], the complexity class PSPACE can be characterized in terms of polynomial length-bounded alternation [6], and many important classes have characterizations based on variants of alternation (see Chapter 19 of [20]).

Unambiguity in nondeterministic computation is related to issues such as worst-case cryptography and the closure properties of #P (the class of functions that count the number of accepting paths of NP machines). The complexity class UP captures the notion of unambiguity in nondeterministic polynomial-time Turing machines. It is known that one-to-one one-way functions exist if and only if P $\neq$ UP [10,14] and that UP equals probabilistic polynomial-time if and only if #P is closed under every polynomial-time computable operation [19]. Factoring, a natural problem with cryptographic applications, belongs to UP $\cap$ coUP and is not known to belong to a subclass of UP $\cap$ coUP nontrivially.

This paper studies the power of unambiguity in alternating computations. Niedermeier and Rossmanith [18] gave the following definition of unambiguity in alternating Turing machines: An alternating Turing machine is *unambiguous* if every accepting existential configuration has exactly one move to an accepting configuration and every rejecting universal configuration has exactly one move to a rejecting configuration. They introduced a natural analog UAP (unambiguous alternating polynomial-time) of UP for alternating Turing machines. Lange and Rossmanith [17] proposed three different approaches to define a hierarchy for unambiguous computations: The alternating unambiguous polynomial hierarchy AUPH, the unambiguous polynomial hierarchy UPH, and the promise unambiguous hierarchy $\mathcal{UPH}$. Though it is known that Few $\subseteq$ UAP $\subseteq$ SPP [18] and AUPH $\subseteq$ UPH $\subseteq$ $\mathcal{UPH}$ $\subseteq$ UAP [7,17], a number of questions—such as, whether UAP is contained in the polynomial hierarchy, whether the unambiguity based hierarchies intertwine, whether these hierarchies are infinite, or whether some hierarchy is contained in a fixed level of the other hierarchy—related to these hierarchies have remained open [17]. Relatedly, Hemaspaandra and Rothe [13] showed that the existence of a sparse Turing-complete set for UP has consequences on the structure of unambiguity based hierarchies.

Recently, Aida et al. [1] introduced "uniqueness" properties for two-player games of perfect information such as Checker, Chess, and Go. A two-person perfect information game has *global uniqueness* property if every winning position of player 1 has a unique move to win and every mis-step by player 1 is punishable by a unique winning reply by player 2 throughout the course of the game. Aida et al. [1] showed that the class of languages that reduce to globally-unique games, i.e., games with global uniqueness property, is the same as the class UAP. In another recent paper, Crâsmaru et al. [7] designed a protocol by which a series of globally-unique games can be combined into a single globally-unique game, even under the condition that the result of the new game is a non-monotone function of the results of the individual games that are unknown to the players. In complexity theoretic terms, they showed that the class UAP is self-low, i.e., UAP^{UAP} = UAP. They also observed that the graph isomorphism problem, whose membership in SPP was shown by Arvind and Kurur [2], in fact belongs to the subclass UAP of SPP.

In this paper, we investigate the power of unambiguity based alternating computation in three different settings. First, we construct a relativized world in which the unambiguity based hierarchies—AUPH, UPH, and $\mathcal{UPH}$—are infinite. We construct another relativized world where UAP is not contained in the polynomial hierarchy. This latter oracle result strengthens a result (relative to an oracle, UAP differs from the second level of $\mathcal{UPH}$) of Crâsmaru et al. [7]. Our results show that proving that any of the unambiguity based hierarchies is finite or that UAP is contained in the polynomial hierarchy is impossible by relativizable proof techniques. We mention that the structure of relativized hierarchies of classes has been investigated extensively in complexity theory (see, for instance [5,12,15,16,25]) and our investigation is a work in this direction.

Second, for every $k \geq 1$, we define a complexity class UAS(k) as the class of sets for which every string in the set is accepted unambiguously by some polynomial-time alternating Turing machine N with at most k alternations, while strings not in the set either are rejected or are accepted with ambiguity by N. A variant of this class (denoted by UAS in this paper), where the number of alternations is allowed to be unbounded, was studied by Wagner [24] as the class ∇P of all sets which can be accepted by polynomial-time alternating Turing machines using partially defined AND and OR functions.[1] Beigel [3] defined the class $\text{UP}_{\leq k(n)}$ as the class of sets in NP that are accepted by nondeterministic polynomial-time Turing machines with at most $k(n)$ accepting paths on each input of length n. Beigel [3] constructed an oracle $\mathcal{A}$ such that $\text{P}^{\mathcal{A}} \subset \text{UP}^{\mathcal{A}} \subset \text{UP}^{\mathcal{A}}_{\leq k(n)} \subset \text{UP}^{\mathcal{A}}_{\leq k(n)+1} \subset \text{FewP}^{\mathcal{A}} \subset \text{NP}^{\mathcal{A}}$, for every polynomial $k(n) \geq 2$. We show that there is a relativized world $\mathcal{B}$ such that, for all $k \geq 1$, $\text{UP}^{\mathcal{B}}_{\leq k} \subset \text{UP}^{\mathcal{B}}_{\leq k+1}$, UAS($k$)$^{\mathcal{B}} \subset$ UAS($k + 1$)$^{\mathcal{B}}$, and relative to $\mathcal{B}$, the second level of $\mathcal{UPH}$ is not contained in any level of AUPH.

Finally, we investigate the power of alternating Turing machines that preserve the bounded-level unambiguity property for every oracle. We show that a polynomial-time alternating Turing machine that preserves k-level alternation unambiguously in every relativized world requires only weak oracle access in every relativized world, i.e., for every oracle $\mathcal{A}$, the language of such a machine can be computed in $\text{P}^{\Sigma^p_k \oplus \mathcal{A}}$. This is a generalization of a result of Hartmanis and Hemachandra [11], which states that if a nondeterministic polynomial-time Turing machine is robustly categorical (i.e., for no oracle and for no input, the machine has more than one accepting path), then for every oracle $\mathcal{A}$, the machine accepts a language in $\text{P}^{\text{NP} \oplus \mathcal{A}}$.

The paper is organized as follows. Section 2 describes the notations and the definitions that are relevant to this paper. In Section 3, we describe our results on relativized separations of unambiguity based hierarchies and relativized non-

---

[1] The partial counterparts AND* and OR* differ from boolean functions AND and OR, respectively, as follows: AND* is undefined for input $(0^c 0)$ and OR* is undefined for input $(1^c 1)$. Thus, these partially defined boolean functions are the unambiguous counterparts of boolean AND and OR functions, respectively.

inclusion of UAP in the polynomial hierarchy. In Section 4, for every $k \geq 1$, we define a complexity class $\text{UAS}(k)$ and study its relativized complexity w.r.t. the bounded-ambiguity class $\text{UP}_{\leq k+1}$. Finally, Section 5 includes our results on the power of robustly bounded-level unambiguous polynomial-time alternating Turing machines. (Proofs omitted due to space limitations can be found in the detailed version available at http://www.cs.rochester.edu/trs/theory-trs.html.)

2   Preliminaries

Let $\mathbb{N}^+$ denote the set of positive integers. We assume that the root of a computation tree of every alternating Turing machine (or, ATM in short) is an existential node. We recursively assign levels in a computation tree T of an ATM as follows: (a) the root of T is at level 1, (b) if a node v is assigned a level i and if v is an existential node, then the first nonexistential (i.e., universal or leaf) node w reachable along some path from v to a leaf node of T is assigned level $i + 1$, (c) if a node v is assigned a level i and if v is a universal node, then the first nonuniversal (i.e., existential or leaf) node w reachable along some path from v to a leaf node of T is assigned level $i + 1$, and (d) for all other nodes of T, the concept of levels is insignificant to this work and so the levels are undefined. We term the nonleaf nodes for which levels are defined as the *salient* nodes in the computation tree of an ATM. For any $k \in \mathbb{N}^+$, a k-level ATM is one for which, on any input, the maximum level assigned to a salient node in the computation tree of the ATM is at most k.

For every polynomial $p(.)$ and for every predicate $R(x, y, z)$ of variables x, y, z, we use $(\exists^p! y)(\forall^p! z) R(x, y, z)$ to indicate that there exists a unique value y_1 for the y variable with $|y_1| \leq p(|x|)$, such that for all values z_1 for the z variable with $|z_1| \leq p(|x|)$, $R(x, y_1, z_1)$ is true, and for all values $y_2 \neq y_1$ for the y variable with $|y_2| \leq p(|x|)$, there exists a unique value $z(y_2)$ for the z variable with $|z(y_2)| \leq p(|x|)$, such that $R(x, y_2, z(y_2))$ is false. In the same way, we interpret expressions, such as $(\exists^p! y_1)(\forall^p! y_2)(\exists^p! y_3) \dots R(x, y_1, y_2, y_3, \dots)$, with an arbitrary number of unambiguous alternations.

Definition 1 (Unambiguity Based Hierarchies [17,18]).

1. *The* alternating unambiguous polynomial hierarchy $\text{AUPH} =_{df} \bigcup_{k \geq 0} \text{AU}\Sigma_k^p$, *where* $\text{AU}\Sigma_0^p =_{df} \text{P}$ *and for every* $k \geq 1$, $\text{AU}\Sigma_k^p$ *is the class of all sets* $L \subseteq \Sigma^*$ *for which there exist a polynomial* $p(.)$ *and a polynomial-time computable predicate R such that, for all $x \in \Sigma^*$,*

$$x \in L \implies (\exists^p! y_1)(\forall^p! y_2) \dots (Q^p! y_k) R(x, y_1, y_2, \dots, y_k), \text{ and}$$
$$x \notin L \implies (\forall^p! y_1)(\exists^p! y_2) \dots (\overline{Q}^p! y_k) \neg R(x, y_1, y_2, \dots, y_k),$$

   where $Q = \exists$ and $\overline{Q} = \forall$ if k is odd, and $Q = \forall$ and $\overline{Q} = \exists$ if k is even.
2. *The* unambiguous polynomial hierarchy *is* $\text{UPH} =_{df} \bigcup_{k \geq 0} \text{U}\Sigma_k^p$, *where* $\text{U}\Sigma_0^p =_{df} \text{P}$ *and for every $k \geq 1$,* $\text{U}\Sigma_k^p =_{df} \text{UP}^{\text{U}\Sigma_{k-1}^p}$.

3. *The promise unambiguous polynomial hierarchy is* $\mathcal{UPH} =_{df} \bigcup_{k \geq 0} \mathcal{U}\Sigma_k^p,$
   where $\mathcal{U}\Sigma_0^p =_{df}$ P, $\mathcal{U}\Sigma_1^p =_{df}$ UP, *and for every* $k \geq 2$, $\mathcal{U}\Sigma_k^p$ *is the class
   of all sets* $L \in \Sigma_k^p$ *such that for some oracle NPTMs* N_1, N_2, ..., N_k,
   $L = L(N_1^{L(N_2^{\cdot^{\cdot^{L(N_k)}}})})$, *and for every* $x \in \Sigma^*$ *and for every* $1 \leq i \leq k -$
   1, $N_1^{L(N_2^{\cdot^{\cdot^{L(N_k)}}})}(x)$ *has at most one accepting path and if* N_i *asks a query*
   w *to its oracle* $L(N_{i+1}^{\cdot^{\cdot^{L(N_k)}}})$ *during the computation of* $N_1^{\cdot^{\cdot^{L(N_k)}}}(x)$, *then*
   $N_{i+1}^{\cdot^{\cdot^{L(N_k)}}}(w)$ *has at most one accepting path.*

Definition 2. [18] *UAP is the class of all sets accepted by unambiguous ATMs
in polynomial time.*

Theorem 3. *1. For all* $k \geq 0$, $\mathrm{AU}\Sigma_k^p \subseteq \mathrm{U}\Sigma_k^p \subseteq \mathcal{U}\Sigma_k^p \subseteq \Sigma_k^p$ [17].
  2. For all $k \geq 1$, $\mathrm{UP}_{\leq k} \subseteq \mathrm{AU}\Sigma_k^p \subseteq \mathrm{U}\Sigma_k^p \subseteq \mathcal{U}\Sigma_k^p \subseteq$ UAP ([7] + [17]).
  3. Few $\subseteq$ UAP $\subseteq$ SPP ([17] + [18]).

3   Relativized Separations of Unambiguity Based Hierarchies

In this section, we apply random restrictions of circuits for separating the levels
of unambiguity based hierarchies. Sheu and Long [22] constructed an oracle
$\mathcal{A}$ relative to which UP contains a language that is not in any level of the
low hierarchy in NP. Formally, Sheu and Long [22] showed that $(\exists \mathcal{A})(\forall k \geq$
$1)[\Sigma_k^{p,\mathrm{UP}^{\mathcal{A}}} \not\subseteq \Sigma_k^{p,\mathcal{A}}]$. In their proof, they introduced special kinds of random
restrictions that were motivated by, but different from, the restrictions used by
Håstad [12]. Using the random restrictions of Sheu and Long [22], we construct a
relativized world $\mathcal{A}$ in which the unambiguity based hierarchies—AUPH, UPH,
and $\mathcal{UPH}$—are infinite. This extends the separation of relativized polynomial
hierarchy [25,12] to the separations of unambiguity based relativized hierarchies.
We use the same restrictions to construct an oracle $\mathcal{A}$ relative to which UAP
is not contained in the polynomial hierarchy. Our separation results imply that
proving that any of the unambiguity based hierarchies extend up to a finite level
or proving that UAP is contained in the polynomial hierarchy is beyond the
limits of relativizable proof techniques.

We now introduce certain notions that are prevalent in the theory of circuit
lower bounds. We represent the variables of a circuit by v_z, for some $z \in \Sigma^*$. The
dual of a circuit C is obtained from C by replacing OR gates with ANDs, AND
gates with ORs, variables x_i with $\overline{x_i}$, and variables $\overline{x_j}$ with x_j. A restriction ρ
of a circuit C is a mapping from the variables of C to $\{0, 1, \star\}$. We say that
a restriction ρ of a circuit C is a *full restriction* if ρ assigns 0 or 1 to all the
variables in C. Given a circuit C and a restriction ρ, $C\lceil_\rho$ denotes the circuit
obtained from C by substituting each variable x with $\rho(x)$ if $\rho(x) \neq \star$. For every
$A \subseteq \Sigma^*$, the restriction ρ_A on the variables v_z of a circuit C is $\rho_A(v_z) = 1$ if

$z \in A$, and $\rho_A(v_z) = 0$ if $z \notin A$. The composition of two restrictions ρ_1 and ρ_2, denoted by $\rho_1\rho_2$, is defined as follows: For every $x \in \Sigma^*$, $\rho_1\rho_2(x) = \rho_2(\rho_1(x))$.

We define specialized circuits, $\Sigma_k(m)$-circuits and $\Pi_k(m)$-circuits, used for constructing relativized worlds involving Σ_k and Π_k classes.

Definition 4. *For every $m \geq 1$ and $k \geq 1$, a $\Sigma_k(m)$-circuit is a depth $k + 1$ circuit with alternating OR and AND gates such that*

1. *the top gate, i.e., the gate at level 1, is an OR gate,*
2. *the number of gates at level 1 to level $k - 1$ is bounded by 2^m,*
3. *the fanin of gates at level $k + 1$ is $\leq m$.*

A $\Pi_k(m)$-circuit is the dual circuit of a $\Sigma_k(m)$-circuit.

For every $k \geq 1$, we say that σ is a $\Sigma_k^{p,(.)}$-predicate if there exist a predicate $R(A; x, y_1, \ldots, y_k)$ over a set variable A and string variables $x, y_1, y_2, \ldots, y_k$, and a polynomial q such that the following hold: (i) $R(A; x, y_1, y_2, \ldots, y_k)$ is computable in polynomial time by a deterministic oracle Turing machine that uses A as the oracle and $\langle x, y_1, \ldots, y_k \rangle$ as the input and (ii) for every set A and string x, $\sigma(A; x)$ is true if and only if $(\exists^q y_1)(\forall^q y_2)\ldots(Q_k^q y_k)R(A; x, y_1, y_2, \ldots, y_k)$ is true, where $Q_k = \exists$ if k is odd and $Q_k = \forall$ if k is even. We say that σ is a $\Pi_k^{p,(.)}$-predicate, for $k \geq 1$, if $\neg\sigma$ is a $\Sigma_k^{p,(.)}$-predicate.

The following proposition states the relationship between $\Sigma_k^{p,(.)}$-predicates ($\Pi_k^{p,(.)}$-predicates) and $\Sigma_k(m)$-circuits (respectively, $\Pi_k(m)$-circuits).

Proposition 5 (see [15,21,22]). *Let $k \geq 1$. For every $\Sigma_k^{p,(.)}$-predicate ($\Pi_k^{p,(.)}$-predicate) σ, there is a polynomial $q(.)$ such that, for all $x \in \Sigma^*$, there is a $\Sigma_k(q(|x|))$-circuit (respectively, $\Pi_k(q(|x|))$-circuit) $C_{\sigma,x}$ with the following properties:*

1. *For every $A \subseteq \Sigma^*$, $C_{\sigma,x}\lceil_{\rho_A} = 1$ if and only if $\sigma(A; x)$ is true, and*
2. *if v_z represents a variable in $C_{\sigma,x}$, then $|z| \leq q(|x|)$.*

Let $\mathcal{B} = \{B_i\}_{i=1}^r$, where B_i's are disjoint sets that cover the variables of C, and let q be a real number between 0 and 1. Sheu and Long [22] defined two probability spaces of restrictions, $\hat{R}_{q,\mathcal{B}}^+$ and $\hat{R}_{q,\mathcal{B}}^-$, and a function g' that maps a random restriction to a restriction. A random restriction $\rho \in \hat{R}_{q,\mathcal{B}}^+$ ($\rho \in \hat{R}_{q,\mathcal{B}}^-$) is defined as follows: For every $1 \leq i \leq r$ and for every variable $x \in B_i$, let $\rho(x) = \star$ with probability q and $\rho(x) = 1$ (respectively, $\rho(x) = 0$) with probability $1 - q$. We now define the function g' for $\rho \in \hat{R}_{q,\mathcal{B}}^+$. For every $1 \leq i \leq r$, let $s_i = \star$ with probability q and let $s_i = 0$ with probability $1 - q$. Let $V_i \subseteq B_i$ be the set of variables x such that $\rho(x) = \star$. $g'(\rho)$ selects the variable v with the highest index in V_i, assigns value s_i to v, and assigns value 1 to all other variables in V_i. The function $g'(\rho)$ for $\rho \in \hat{R}_{q,\mathcal{B}}^-$ is defined in an analogous way by replacing 0 with 1 and vice versa.

Lemma 6 (Switching Lemma [22]). *Let C be a circuit consisting of an AND of ORs with bottom fanin $\leq t$. Let $\mathcal{B} = \{B_i\}_{i=1}^r$ be disjoint sets that cover the variables of C, and let q be a real number between 0 and 1. Then, for a random restriction $\rho \in \hat{R}_{q,\mathcal{B}}^+$, $\mathrm{Prob}[C\lceil_{\rho g'(\rho)}$ is not equivalent to an OR of ANDs with bottom fanin $\leq s] \leq \alpha^s$, where $\alpha < 6qt$. The above probability holds even when $\hat{R}_{q,\mathcal{B}}^+$ is replaced by $\hat{R}_{q,\mathcal{B}}^-$, or when C is an OR of ANDs and is being converted to an AND of ORs.*

Sheu and Long [22] defined a kind of restriction, called *U condition*, on the assignment of variables in certain circuits. A restriction ρ is said to satisfy the U condition if the following holds: At most one variable is assigned $\star$ or 0 in each set B_i if ρ is a random restriction from $\hat{R}_{q,\mathcal{B}}^+$, and at most one variable is assigned $\star$ or 1 in each set B_i if ρ is a random restriction from $\hat{R}_{q,\mathcal{B}}^-$ [22]. Below, we define a *global uniqueness condition* (also called *GU condition*) on *full restrictions* of any circuit C.

Definition 7. *We say that a* full restriction ρ *satisfies the* GU condition *for a circuit C, if the assignment of variables by ρ leads to the following characteristics in the computation of C:*

1. *If an OR gate G_i in C outputs 1, then there is exactly one input gate to G_i that outputs 1, and*
2. *if an AND gate G_i in C outputs 0, then there is exactly one input gate to G_i that outputs 0.*

Theorem 8. $(\exists \mathcal{A})(\forall k \geq 1)[\mathrm{AU}\Sigma_k^{p,\mathcal{A}} \not\subseteq \Pi_k^{p,\mathcal{A}}]$.

Proof. Our proof is inspired from that of Theorem 4.2 (relative to some oracle $\mathcal{D}$, for all $k \geq 1$, $\Sigma_k^{p,\mathrm{UP}^{\mathcal{D}}} \not\subseteq \Sigma_k^{p,\mathcal{D}}$) by Sheu and Long [22]. For every $k \geq 1$, we define a test language $L_k(B)$ as follows: $L_k(B) \subseteq 0^*$ such that, for every $n \in \mathbb{N}^+$,

$$0^n \in L_k(B) \implies (\exists^n! y_1)(\forall^n! y_2) \ldots (Q^n! y_k) \left[0^k 1 y_1 y_2 \ldots y_k \in B \right], \text{ and}$$

$$0^n \notin L_k(B) \implies (\forall^n! y_1)(\exists^n! y_2) \ldots (\overline{Q}^n! y_k) \left[0^k 1 y_1 y_2 \ldots y_k \notin B \right],$$

where $Q = \exists$ and $\overline{Q} = \forall$ if k is odd, and $Q = \forall$ and $\overline{Q} = \exists$ if k is even. Choose $\mathcal{O} \subseteq \Sigma^*$ such that, for every $k \geq 1$, $L_k(\mathcal{O}) = 0^*$. For every $k \geq 1$, let $\sigma_{k,1}, \sigma_{k,2}, \ldots$ be an enumeration of $\Sigma_k^{p,(\cdot)}$-predicates. In stage $\langle k, i \rangle$, we diagonalize against $\sigma_{k,i}$ and change $\mathcal{O}$ at a certain length. Finally, let $\mathcal{A} := \lim_{n \to \infty} \cup_{n \in \mathbb{N}^+} \mathcal{O}^{=n}$. We now define the stages involved in the construction of the oracle.

Stage $\langle k, i \rangle$: Choose a very large integer n so that the construction in this stage does not spoil the constructions in previous stages. Also, n must be large enough to meet the requirements in the proof of Claim 1. Set $\mathcal{O} := \mathcal{O} - \Sigma^{k(n+1)+1}$. Choose a set $B \subseteq 0^k 1 \Sigma^{kn}$ such that the following requirement is satisfied:

$$0^n \in L_k(B) \iff \sigma_{k,i}(\mathcal{O} \cup B; 0^n) \text{ is true.} \tag{1}$$

In Claim 1, we show that there is always a set $B \subseteq 0^k 1 \Sigma^{kn}$ satisfying Eqn. (1). Let $\mathcal{O} := \mathcal{O} \cup B$ and move to the next stage.
End of Stage

Clearly, the existence of a set B satisfying Eqn. (1) suffices to successfully finish stage $\langle k, i \rangle$. We now prove the statement in Claim 1.

Claim 1. *In every stage $\langle k, i \rangle$, there is a set $B \subseteq 0^k 1 \Sigma^{kn}$ satisfying Eqn. (1).*

Proof. Assume to the contrary that in some stage $\langle k, i \rangle$, Eqn. (1) is not satisfied. Then, the following holds: For every $B \subseteq 0^k 1 \Sigma^{kn}$, $0^n \in L_k(B)$ if and only if $\neg \sigma_{k,i}(\mathcal{O} \cup B; 0^n)$ is true. We define a $C(n, k)$ circuit as follows: The depth of $C(n, k)$ is k, the top gate of $C(n, k)$ is an OR gate, the fanin of all the gates at level 1 to k is 2^n, and every leaf of $C(n, k)$ is a positive variable represented by v_z, where $z \in 0^k 1 \Sigma^{kn}$. The following proposition is evident.

Proposition 9. *For every $B \subseteq 0^k 1 \Sigma^{kn}$,*

$$0^n \in L_k(B) \Longleftrightarrow [\rho_B \text{ satisfies the GU condition for } C(n,k) \text{ and } C(n,k)\lceil_{\rho_B} = 1],$$

$$and$$

$$0^n \notin L_k(B) \Longleftrightarrow [\rho_B \text{ satisfies the GU condition for } C(n,k) \text{ and } C(n,k)\lceil_{\rho_B} = 0].$$

For every $h \geq 1$, we define a family of circuits $\mathcal{F}_k^h$. Ko [15] defined a C_k^h circuit to be a depth k circuit in $\mathcal{F}_k^h$ with fanin of gates at level k exactly equal to $\sqrt{h}$ and used these circuits to separate the relativized polynomial hierarchy.

Family $\mathcal{F}_k^h$ of circuits, where $h \geq 1$: A circuit C of depth ℓ, where $1 \leq \ell \leq k$, is in $\mathcal{F}_k^h$ if and only if the following holds:

1. C has alternating OR and AND gates, and the top gate, i.e., the gate at level 1, of C is an OR gate,
2. the fanin of gates at level 1 to $\ell - 1$ is h,
3. the fanin of gates at level ℓ is $\geq \sqrt{h}$,
4. every leaf of C is a unique positive variable.

Let $C_{\sigma_{k,i}}$ be the $\Pi_k(p_i(n))$-circuit corresponding to $\neg \sigma_{k,i}((.); 0^n)$, for some polynomial $p_i(.)$. From Proposition 9, we wish to find a set $B \subseteq 0^k 1 \Sigma^{kn}$ such that (i) if ρ_B satisfies the *GU condition* for $C(n, k)$ and $C(n, k)\lceil_{\rho_B} = 1$, then $C_{\sigma_{k,i}}\lceil_{\rho_{\mathcal{O} \cup B}} = 0$, and (ii) if ρ_B satisfies the *GU condition* for $C(n, k)$ and $C(n, k)\lceil_{\rho_B} = 0$, then $C_{\sigma_{k,i}}\lceil_{\rho_{\mathcal{O} \cup B}} = 1$. Clearly, the existence of a set B satisfying (i) and (ii) suffices to prove the claim. Next, we describe our approach to show the existence of such a set B.

We define a restriction $\hat{\rho}_{\mathcal{O}}$ on $C_{\sigma_{k,i}}$ as follows: For every variable v_z in $C_{\sigma_{k,i}}$, if $z \in \mathcal{O}$ then let $\hat{\rho}_{\mathcal{O}}(v_z) = 1$, if $z \notin \mathcal{O} \cup 0^k 1 \Sigma^{kn}$ then let $\hat{\rho}_{\mathcal{O}}(v_z) = 0$, and if $z \in 0^k 1 \Sigma^{kn}$ then let $\hat{\rho}_{\mathcal{O}}(v_z) = \star$. Let $C_{\sigma_{k,i}}(\mathcal{O}) =_{df} C_{\sigma_{k,i}}\lceil_{\hat{\rho}_{\mathcal{O}}}$. Thus, the only variables v_z appearing in $C_{\sigma_{k,i}}(\mathcal{O})$ are the ones for which $z \in 0^k 1 \Sigma^{kn}$. Suppose that no set $B \subseteq 0^k 1 \Sigma^{kn}$ satisfying (i) and (ii) exists. Then, the following holds: For every $B \subseteq 0^k 1 \Sigma^{kn}$,

$$\rho_B \text{ satisfies the } GU \text{ condition for } C(n,k) \text{ and } C(n,k)\lceil_{\rho_B} = C_{\sigma_{k,i}}(\mathcal{O})\lceil_{\rho_B} = 1,$$

$$or \tag{2}$$

$$\rho_B \text{ satisfies the } GU \text{ condition for } C(n,k) \text{ and } C(n,k)\lceil_{\rho_B} = C_{\sigma_{k,i}}(\mathcal{O})\lceil_{\rho_B} = 0.$$

Lemma 10. *For every $1 \leq \ell \leq k$ and for all sufficiently large h, for any circuit $C_{\mathcal{F}} \in \mathcal{F}_k^h$ of depth ℓ, and for any $\Pi_\ell(m)$-circuit C_π, if it holds that*

(for every full restriction ρ satisfying the GU condition for $C_{\mathcal{F}})[C_{\mathcal{F}}\lceil_\rho = C_\pi\lceil_\rho]$,

then $m \geq \delta \cdot h^{1/3}$, where $\delta = 1/12$.

Since $C(n,k) \in \mathcal{F}_k^{2^n}$, $C_{\sigma_{k,i}(\mathcal{O})}$ is a $\Pi_k(p_i(n))$ circuit, and $p_i(n) = o(2^{n/3})$, we get a contradiction with Eqn. (2) and Lemma 10. ∎   (Claim 1 and Theorem 8)

Corollary 11. *There is an oracle $\mathcal{A}$ relative to which the alternating unambiguous polynomial hierarchy AUPH, the unambiguous polynomial hierarchy UPH, the promise unambiguous polynomial hierarchy $\mathcal{UPH}$, and the polynomial hierarchy PH are infinite.*

Note that Theorem 8 does not imply relativized separation of UAP from PH in any obvious way. We achieve this separation, using the proof techniques of Theorem 8, in Theorem 12.

Theorem 12. $(\exists \mathcal{A})[\mathrm{UAP}^{\mathcal{A}} \not\subseteq \mathrm{PH}^{\mathcal{A}}]$.

Crâsmaru et al. [7] showed that there is an oracle relative to which $\mathrm{UAP} \neq \mathcal{U}\Sigma_2^p$. Corollary 13 shows that in some relativized world, UAP is much more powerful than the promise unambiguous polynomial hierarchy $\mathcal{UPH}$. Thus, Corollary 13 is a strengthening of their result.

Corollary 13. *There is an oracle relative to which $\mathcal{UPH} \subset \mathrm{UAP}$.*

4   Complexity of Unambiguous Alternating Solution

Wagner studied the class $\nabla \mathrm{P}$, denoted by UAS in this paper, of all sets that are accepted by polynomial-time alternating Turing machines with partially defined AND and OR functions. UAS is a natural class with complete sets and is related to UAP in the same way as US [4] is related to UP. We define a variant of UAS, denoted by $\mathrm{UAS}(k)$, where the number of alternations allowed is bounded by some constant $k \geq 1$, instead of the unbounded number of alternations in the definition of UAS. (Thus, $\mathrm{UAS}(1)$ is the same as the unique solution class US.)

Definition 14. [24] *The class UAS, denoted by $\nabla \mathrm{P}$ in [24], is the class of all sets $L \subseteq \Sigma^*$ for which there exist polynomials $p(.)$ and $q(.)$, and a polynomial-time computable predicate R such that, for all $x \in \Sigma^*$,*

$$x \in L \Longleftrightarrow (\exists^{p}! \, y_1)(\forall^{p}! \, y_2) \ldots (Q^{p}! \, y_q) R(x, y_1, y_2, \ldots, y_q)$$

where $Q = \exists$ if $q(|x|)$ is odd and $Q = \forall$ if $q(|x|)$ is even.

The class $\mathrm{UAS}(k)$, for every $k \geq 1$, consists of all sets for which strings in the set are accepted unambiguously by some polynomial-time alternating Turing machine N with at most k alternations, while strings not in the set either are rejected or are accepted with ambiguity by N. A formal definition is as follows.

Definition 15. *The class* UAS(k), *for* $k \geq 1$, *is the class of all sets* $L \subseteq \Sigma^*$ *for which there exist a polynomial* $p(.)$ *and a polynomial-time computable predicate* R *such that, for all* $x \in \Sigma^*$,

$$x \in L \Longleftrightarrow (\exists^p! y_1)(\forall^p! y_2) \ldots (Q^p! y_k) R(x, y_1, y_2, \ldots, y_k)$$

where $Q = \exists$ *if* k *is odd and* $Q = \forall$ *if* k *is even.*

Theorem 16. *1. US $\subseteq$ UAS $\subseteq$ C$_=$P and UAS $\subseteq$ $\forall \oplus$P [24].*
2. For every $k \geq 1$, UP $\subseteq$ US $\subseteq$ UAS(k) $\subseteq$ UAS($k+1$) $\subseteq$ UAS.
3. For every $k \geq 1$, AUΣ_k^p $\subseteq$ UAS(k) $\subseteq$ P$^{\Sigma_k^p}$.

Recall from Theorem 3(2) that UP$_{\leq k}$ $\subseteq$ AUΣ_k^p, for every $k \geq 1$. Thus, it follows from Theorem 16(3) that UP$_{\leq k}$ $\subseteq$ UAS(k). However, Theorem 17 shows that relative to an oracle $\mathcal{A}$, for all $k \geq 1$, UP$_{\leq k+1}$ is not contained in UAS(k). Thus relative to the same oracle, the bounded ambiguity classes UP$_{\leq k}$ and the bounded-level unambiguous alternating solution classes UAS(k), for $k \geq 1$, form infinite hierarchies. Theorem 17 also implies that there is a relativized world where for all $k \geq 1$, UP$_{\leq k+1}$ is not contained in AUΣ_k^p. In contrast, Lange and Rossmanith [17] proved that FewP $\subseteq \mathcal{U}\Sigma_2^p$ in every relativized world. It follows that relative to the oracle of Theorem 17, for all $k \geq 1$, $\mathcal{U}\Sigma_2^p \not\subseteq$ AUΣ_k^p.

Theorem 17. $(\exists \mathcal{A})(\forall k \geq 1)[\text{UP}_{\leq k+1}^{\mathcal{A}} \not\subseteq \text{UAS}(k)^{\mathcal{A}}].$

Corollary 18. *There is an oracle $\mathcal{A}$ such that, for every $k \geq 1$, UP$_{\leq k}^{\mathcal{A}}$ $\subset$ UP$_{\leq k+1}^{\mathcal{A}}$, AU$\Sigma_k^{p,\mathcal{A}}$ $\subset$ AU$\Sigma_{k+1}^{p,\mathcal{A}}$, UAS$^{\mathcal{A}}(k)$ $\subset$ UAS$^{\mathcal{A}}(k+1)$, and $\mathcal{U}\Sigma_2^{p,\mathcal{A}}$ $\not\subseteq$ AU$\Sigma_k^{p,\mathcal{A}}$.*

5   Power of Robustly Unambiguous Alternating Machines

Hartmanis and Hemachandra [11] showed that robustly categorical nondeterministic polynomial-time Turing machines (i.e., NPTMs that for no oracle and no input have more than one accepting path) accept simple languages in the sense that, for every oracle A, the languages accepted by such machines are in P$^{\text{NP} \oplus A}$. Thus, if P = NP, then NPTMs satisfying robustly categorical property cannot separate P^A from NP^A, for any oracle A. Theorem 19 generalizes this result of Hartmanis and Hemachandra [11] and shows that, for every oracle A, robustly k-level unambiguous polynomial-time alternating Turing machines accept languages that are in P$^{\Sigma_k^p \oplus A}$. Thus, similar to the case of robustly categorical NPTMs, if P = NP, then robustly k-level unambiguous polynomial-time alternating Turing machines cannot separate P^A from $\Sigma_k^{p,A}$, and consequently cannot separate P^A from NP^A.

Theorem 19. *For all $k \in \mathbb{N}^+$, the following holds:*

$(\forall \mathcal{A})[N^{\mathcal{A}}$ *is a k-level unambiguous polynomial-time ATM*$] \Longrightarrow (\forall \mathcal{A})[L(N^{\mathcal{A}})$

$\in \text{P}^{\Sigma_k^p \oplus \mathcal{A}}].$

Corollary 20. *For all $k \in \mathbb{N}^+$, if* $P = NP$ *and* $(\forall \mathcal{A})[N^{\mathcal{A}}$ *is a k-level unambiguous polynomial-time ATM], then* $(\forall \mathcal{A})[L(N^{\mathcal{A}}) \in P^{\mathcal{A}}]$.

Crescenzi and Silvestri [8] showed that languages accepted by robustly complementary and categorical oracle NPTMs are in $P^{(UP \cup coUP) \oplus \mathcal{A}}$. In fact, their proof actually shows that the languages of such machines are in $P^{(UP \cap coUP) \oplus \mathcal{A}}$. Theorem 21 is a generalization of this result of Crescenzi and Silvestri [8] for robustly bounded-level unambiguous polynomial-time alternating Turing machines.

Theorem 21. *For all $k_i, k_j \in \mathbb{N}^+$, if for all oracles $\mathcal{A}$, $N_i^{\mathcal{A}}$ and $N_j^{\mathcal{A}}$ are, respectively, k_i-level and k_j-level unambiguous polynomial-time ATMs and* $L(N_i^{\mathcal{A}}) = \overline{L(N_j^{\mathcal{A}})}$, *then for all oracles $\mathcal{A}$, $L(N_i^{\mathcal{A}}) \in P^{(UP^{\Sigma_{k-1}^p} \cap coUP^{\Sigma_{k-1}^p}) \oplus \mathcal{A}}$, where* $k = \max\{k_i, k_j\}$.

6   Open Questions

We now mention some future research directions. Theorem 8 implies that there is a relativized world where the unambiguity based hierarchies are infinite. However, a number of questions related to the relativized structure of unambiguity based hierarchies remain open. For instance, is there a relativized world where AUPH is finite, but UPH and $\mathcal{UPH}$ are infinite? Is there a relativized world where the polynomial hierarchy is infinite, but AUPH and UPH collapse?

Hemaspaandra and Rothe [13] showed that if UP has a sparse Turing-complete set, then for every $k \geq 3$, $U\Sigma_k^p \subseteq U\Sigma_{k-1}^p$. Are there other complexity-theoretic assumptions that can help in concluding about the structure of unambiguity based hierarchies?

Fortnow [9] showed that $PH \subset SPP$ relative to a random oracle. Theorem 12 shows that there is a relativized world where $UAP \not\subseteq PH$. Can we extend the oracle separation of UAP from PH to a random oracle separation?

Aida et al. [1] and Crâsmaru et al. [7] discussed whether UAP equals SPP. In fact, Crâsmaru et al. [7] pointed out their difficulty in building an oracle $\mathcal{A}$ such that $UAP^{\mathcal{A}} \neq SPP^{\mathcal{A}}$. Can the ideas involved in oracle constructions in this paper be used to attack this problem?

Finally, is it the case that similar to robustly bounded-level unambiguous polynomial-time ATMs, robustly unbounded-level unambiguous polynomial-time ATMs require weak oracle access in every relativized world?

Acknowledgment. We thank Lane Hemaspaandra for helpful advice and guidance throughout the project.

References

1. S. Aida, M. Crâsmaru, K. Regan, and O. Watanabe. Games with uniqueness properties. *Theory of Computing Systems*, 37(1):29–47, 2004.
2. V. Arvind and P. Kurur. Graph isomorphism is in SPP. In *Proceedings of the 43rd IEEE Symposium on Foundations of Computer Science*, pages 743–750, Los Alamitos, November 16–19 2002. IEEE Computer Society.

3. R. Beigel. On the relativized power of additional accepting paths. In *Proceedings of the 4th Structure in Complexity Theory Conference*, pages 216–224. IEEE Computer Society Press, June 1989.

4. A. Blass and Y. Gurevich. On the unique satisfiability problem. *Information and Control*, 55(1–3):80–88, 1982.

5. J. Cai, T. Gundermann, J. Hartmanis, L. Hemachandra, V. Sewelson, K. Wagner, and G. Wechsung. The boolean hierarchy II: Applications. *SIAM Journal on Computing*, 18(1):95–111, 1989.

6. A. Chandra, D. Kozen, and L. Stockmeyer. Alternation. *Journal of the ACM*, 26(1), 1981.

7. M. Crâsmaru, C. Glaßer, K. Regan, and S. Sengupta. A protocol for serializing unique strategies. In *Proceedings of the 29th International Symposium on Mathematical Foundations of Computer Science*. Springer-Verlag *Lecture Notes in Computer Science #3153*, August 2004.

8. P. Crescenzi and R. Silvestri. Sperner's lemma and robust machines. *Computational Complexity*, 7:163–173, 1998.

9. L. Fortnow. Relativized worlds with an infinite hierarchy. *Information Processing Letters*, 69(6):309–313, 1999.

10. J. Grollmann and A. Selman. Complexity measures for public-key cryptosystems. *SIAM Journal on Computing*, 17(2):309–335, 1988.

11. J. Hartmanis and L. Hemachandra. Robust machines accept easy sets. *Theoretical Computer Science*, 74(2):217–225, 1990.

12. J. Håstad. *Computational Limitations of Small-Depth Circuits*. MIT Press, 1987.

13. L. Hemaspaandra and J. Rothe. Unambiguous computation: Boolean hierarchies and sparse Turing-complete sets. *SIAM Journal on Computing*, 26(3):634–653, 1997.

14. K. Ko. On some natural complete operators. *Theoretical Computer Science*, 37(1):1–30, 1985.

15. K. Ko. Relativized polynomial time hierarchies having exactly k levels. *SIAM Journal on Computing*, 18(2):392–408, 1989.

16. K. Ko. Separating the low and high hierarchies by oracles. *Information and Computation*, 90(2):156–177, 1991.

17. K.-J. Lange and P. Rossmanith. Unambiguous polynomial hierarchies and exponential size. In *Proceedings of the 9th Structure in Complexity Theory Conference*, pages 106–115. IEEE Computer Society Press, June/July 1994.

18. R. Niedermeier and P. Rossmanith. Unambiguous computations and locally definable acceptance types. *Theoretical Computer Science*, 194(1–2):137–161, 1998.

19. M. Ogiwara and L. Hemachandra. A complexity theory for feasible closure properties. *Journal of Computer and System Sciences*, 46(3):295–325, 1993.

20. C. Papadimitriou. *Computational Complexity*. Addison-Wesley, 1994.

21. M. Sheu and T. Long. The extended low hierarchy is an infinite hierarchy. *SIAM Journal on Computing*, 23(3):488–509, 1994.

22. M. Sheu and T. Long. UP and the low and high hierarchies: A relativized separation. *Mathematical Systems Theory*, 29(5):423–449, 1996.

23. L. Stockmeyer. The polynomial-time hierarchy. *Theoretical Computer Science*, 3:1–22, 1976.

24. K. Wagner. Alternating machines using partially defined "AND" and "OR". Technical Report 39, Institut für Informatik, Universität Würzburg, January 1992.

25. A. Yao. Separating the polynomial-time hierarchy by oracles. In *Proceedings of the 26th IEEE Symposium on Foundations of Computer Science*, pages 1–10, 1985.

Translational Lemmas for Alternating TMs and PRAMs

Chuzo Iwamoto, Yoshiaki Nakashiba,
Kenichi Morita, and Katsunobu Imai

Hiroshima University, Graduate School of Engineering,
Higashi-Hiroshima, 739-8527 Japan
chuzo@hiroshima-u.ac.jp

Abstract. We present translational lemmas for alternating Turing machines (ATMs) and parallel random access machines (PRAMs), and apply them to obtain tight hierarchy results on ATM- and PRAM-based complexity classes. It is shown that, for any small rational constant ϵ, there is a language which can be accepted by a $c(9 + \epsilon) \log^r n$-time $d(4 + \epsilon) \log n$-space ATM with l worktapes but not by any $c \log^r n$-time $d \log n$-space ATM with the same l worktapes if the number of tape symbols is fixed. Here, $c, d > 0$ and $r > 1$ are arbitrary rational constants, and $l \geq 2$ is an arbitrary integer. It is also shown that, for any small rational constant ϵ, there is a language which can be accepted by a $c(1+\epsilon) \log^{r_1} n$-time PRAM with n^{r_2} processors but not by any $c \log^{r_1} n$-time PRAM with $n^{r_2(1+\epsilon)}$ processors, where $c > 0$, $r_1 > 1$, and $r_2 \geq 1$ are arbitrary rational constants.

1 Introduction

The most standard models for parallel computation are alternating Turing machines (ATMs) [2], uniform circuit families [9], and parallel random access machines (PRAMs) [10]. It is well known that showing a proper hierarchy of parallel complexity classes is very difficult. A typical example is a famous open question whether $NC^k \subsetneq NC^{k+1}$. This open problem seems to be very difficult, since it is not even known whether $NC^1 \subsetneq NP$. Here, NC^k is the class of languages accepted by $O(\log^k n)$-time $O(\log n)$-space ATMs. Thus, it is open whether there is a language accepted by an $O(\log^{k+1} n)$-time $O(\log n)$-space ATM but not by any $O(\log^k n)$-time $O(\log n)$-space ATM. (Throughout this paper, all logarithms are base 2, and $\log^k n$ is the logarithm of n raised to the kth power.)

On the other hand, if constant factors are taken into account, a proper hierarchy can be derived from known results. In order to investigate "precise" complexities of ATMs, we assume that the number of worktapes is fixed $l \geq 2$ and tape symbols are 0 and 1. Under this assumption, the following simulation and separation results hold. (i) There are constants c_1 and d_1 such that every $t(n)$-time $s(n)$-space ATM can be simulated by a U_E-uniform circuit family of depth $c_1 t(n)$ and size $2^{d_1 s(n)}$ [9]. (ii) There are constants c_2, d_2 and a language L such that L can be accepted by a U_E-uniform circuit family of depth $c_2 t(n)$

M. Liśkiewicz and R. Reischuk (Eds.): FCT 2005, LNCS 3623, pp. 137–148, 2005.

and size $(z(n))^{d_2}$ but not by any U_E-uniform circuit family of depth $t(n)$ and size $z(n)$ [6]. (iii) There are constants c_3 and d_3 such that every U_E-uniform circuit family of depth $t(n)$ and size $z(n)$ can be simulated by a $c_3 t(n)$-time $d_3 \log z(n)$-space ATM if $t(n) \neq O(\log z(n) \log \log z(n))$ [9]. (In this paper, we say "$f(n) \neq O(g(n))$" if $f(n)$ grows faster than $g(n)$, i.e., $\lim_{n \to \infty} g(n)/f(n)$ exists and is 0.) Therefore, there are constants $c \geq c_1 c_2 c_3$, $d \geq d_1 d_2 d_3$, and a language which can be accepted by a $ct(n)$-time $ds(n)$-space ATM but not by any $t(n)$-time $s(n)$-space ATM. (Although the model used in [9] is multi-tape multi-symbol ATMs, a careful analysis shows that the above simulation (iii) holds for two-tape two-symbol ATMs if $t(n) \neq O(\log z(n) \log \log z(n))$, where $O(\log \log z(n))$ is the overhead of simulating a multi-tape $O(\log z(n))$-time DTM by a two-tape DTM.)

In this paper, we tighten this hierarchy of ATMs by using a translational method. Suppose that the number of worktapes is fixed $l \geq 2$ and tape symbols are 0 and 1. We show that if every $3(1+\epsilon)t_1(n)$-time $2(1+\epsilon)s_1(n)$-space ATM can be simulated by a $t_2(n)$-time $s_2(n)$-space ATM, then every $t_1(2^{kn})$-time $s_1(2^{kn})$-space ATM can be simulated by a $3(1+\epsilon)t_2(2^{kn})$-time $2(1+\epsilon)s_2(2^{kn})$-space ATM, where $\epsilon > 0$ is an arbitrary constant and $k \geq 1$ is an arbitrary integer. By using this translational lemma, it is shown that, for any small rational constant ϵ, there is a language which can be accepted by a $c(9+\epsilon) \log^r n$-time $d(4+\epsilon) \log n$-space ATM but not by any $c \log^r n$-time $d \log n$-space ATM, where $c, d > 0$ and $r > 1$ are arbitrary rational constants. Therefore, constant factors $9+\epsilon$ and $4+\epsilon$ in time and space, respectively, strictly enlarge the complexity classes defined by time- and space-bounded ATMs. Translational lemmas for deterministic and nondeterministic TMs with restricted tape alphabet size were presented in [4].

We turn our attention to the second model, uniform circuit families. The class NC^k is also defined as the set of languages accepted by U_E-uniform circuit families of depth $O(\log^k n)$ and size $n^{O(1)}$ [9]. It was shown that there is a language which can be recognized by a U_E-uniform circuit family of depth $d(1 + \epsilon)(\log n)^{r_1}$ and size $n^{r_2(1+\epsilon)}$ but not by any U_E-uniform circuit family of depth $d(\log n)^{r_1}$ and size n^{r_2}, where $\epsilon > 0$, $d > 0$, $r_1 > 1$, and $r_2 \geq 1$ are arbitrary rational constants [7]. (U_E-uniform is also called DLOGTIME-uniform using the Extended connection language.) In this hierarchy, the constant factor in depth and the exponent in size are both $1 + \epsilon$. In our ATM hierarchy, the constant factors in time and space are $9+\epsilon$ and $4+\epsilon$, respectively. The essential difference between two models is due to uniformity. One of the tasks for proving a translational lemma is to compute the value of 2^{kn} from n. If we consider the uniform circuit model, such a computation is done by circuit constructors (i.e., logarithmic-time DTMs). The complexity of these TMs is not included in circuit complexity. Namely, circuit complexity considers only depth and size of pre-constructed circuits. On the other hand, the complexity of an ATM does take account of the computation of 2^{kn}. Constant factors $9 + \epsilon$ and $4 + \epsilon$ are caused by this computation. Improving these constant factors toward $1 + \epsilon$ is an interesting future work.

It remains to consider the third model, PRAMs. The class of languages accepted by $O(\log^k n)$-time PRAMs with a polynomial number of processors is between NC^k and NC^{k+1} [8]. It was shown that there are a constant c and a language L such that L can be accepted by a $ct(n)$-time PRAM with $p(n)$ processors but not by any $t(n)$-time PRAM with $p(n)$ processors [6]. Unfortunately, the value of c was not mentioned, or no attempt was made to minimize the constant c in [6]. In this paper, we show that if every $(1 + \epsilon)t_1(n)$-time PRAM with $p_1(n)$ processors can be simulated by a $t_2(n)$-time PRAM with $p_2(n)$ processors, then every $t_1(2^{kn})$-time PRAM with $p_1(2^{kn})$ processors can be simulated by a $(1 + \epsilon)t_2(2^{kn})$-time PRAM with $p_2(2^{kn})$ processors, where $\epsilon > 0$ is an arbitrary constant and $k \geq 1$ is an arbitrary integer. By using this translational lemma, it is shown that, for any small rational constant ϵ, there is a language which can be accepted by a $c(1+\epsilon)\log^{r_1} n$-time PRAM with $n^{r_2(1+\epsilon)}$ processors but not by any $c\log^{r_1} n$-time PRAM with n^{r_2} processors, where $c > 0$, $r_1 > 1$, and $r_2 \geq 1$ are arbitrary rational constants. Here, the complexity of PRAMs is measured according to the uniform cost criterion.

In Section 2, we give definitions of ATMs and PRAMs. The main results are also given in that section. The proofs are given in Sections 3 and 4.

2 Definitions and Results

We assume familiarity with nondeterministic Turing machines [3]. Our l-worktape TM has a finite control, a semi-infinite input tape, and l semi-infinite read-write worktapes. The left end of each of the input tape and worktapes is delimited by a special end-marker \$. The tape symbols are 0 and 1 only. Initially, every cell in each worktape contains blank symbol B, the head of each worktape is placed at the leftmost cell, and the input tape contains a string over $\{0, 1\}$ followed by an infinite number of blank symbols $BB \cdots$. Non-blank symbols cannot be overwritten with blank symbol B.

The definition of an alternating TM (ATM) is mostly from [9]. The states are partitioned into existential and universal states. A tree is said to be a *computation tree* of an ATM M on a string w if its nodes are labeled with configurations of M on w such that the descendants of any non-leaf labeled by a universal (existential) configuration include all (resp. one) of the successors of that configuration. A computation tree is *accepting* if it is finite and all the leaves are accepting configurations. M *accepts* w if there is an accepting computation tree whose root is labeled with the initial configuration of M on w.

Since we consider polylogarithmic-time computations, our ATM is a so-called "random access ATM" [9]. Our ATM has no input head. Instead, it has a special *read* state, and the first worktape is regarded as an *index* tape. Whenever it enters the read state $read(a)$ with integer i written on the index tape, it halts, and accepts if and only if the ith input symbol is a.

An ATM M is defined to be $t(n)$-*time bounded* ($s(n)$-*space bounded*) if, for every accepted input w of length n, there is an accepting computation tree of height at most $t(n)$ (resp. each of whose nodes is labeled by a configuration using space at most $s(n)$). In this paper, we denote by Atime,space$(t(n), s(n))$ the class of languages accepted by $t(n)$-time $s(n)$-space ATMs with l worktapes whose tape symbols are 0 and 1. Note that we are taking account of constant factors in this paper. (Strictly speaking, Atime,space$(t(n), s(n))$ should be written as Atime,space$(t(n); s(n), l, 2)$. For simplicity of notation, we omit the numbers $l, 2$ of worktapes and symbols.)

Our PRAM is essentially the same model as defined in [10]. A PRAM has a *common memory*, $M[1], M[2], \ldots$, and a sequence of *processors* (RAMs) operating synchronously in parallel (see [1] for RAM). Each processor of a PRAM has its own *local memory*, $R[1], R[2], \ldots$, and has instructions for addition, subtraction, logical OR, AND, conditional branches based on predicates $=$ and $<$, and reading from and writing into its local memory. Processors can access to the common memory, and each processor has instructions for reading from and writing into the common memory using its local memory to specify the common memory address. If more than one processor attempts to write the same location in common memory at the same time, the lowest numbered processor succeeds. All processors have the same program.

The input string of length n is given in $M[1], M[2], \ldots, M[n]$. The computation halts when all processors have halted. The PRAM operates in time $t(n)$ if it halts within $t(n)$ steps on any input of length n. When the PRAM accepts (rejects) the input string, symbol 1 (resp. 0) appears in $M[1]$ after $t(n)$ steps. The complexity of a PRAM program is measured according to the uniform cost criterion. Let Ptime,proc$(t(n), p(n))$ be the class of languages accepted by $t(n)$-time PRAMs with $p(n)$ processors.

We first present a translational lemma for ATMs.

Lemma 1. *Suppose that* $t_1(n), t_2(n) \neq O(\log n)$ *and* $s_1(n), s_2(n) = \Theta(\log n)$ *are arbitrary functions computable by* $O(\log n)$*-time ATMs with* l *worktapes if input* n *is given as a binary string of length* $\lfloor \log n \rfloor + 1$*. Then, for any rational constant* $\epsilon > 0$,

$$\text{Atime,space}(3(1 + \epsilon)t_1(n), 2(1 + \epsilon)s_1(n)) \subseteq \text{Atime,space}(t_2(n), s_2(n))$$

implies

$$\text{Atime,space}(t_1(2^{kn}), s_1(2^{kn})) \subseteq \text{Atime,space}(3(1 + \epsilon)t_2(2^{kn}), 2(1 + \epsilon)s_2(2^{kn})),$$

where $k \geq 1$ *is an arbitrary integer, and all ATMs have only 0 and 1 as their tape symbols.*

The proof of this lemma will be given in Section 3.1. By using this lemma, the following hierarchy theorem is derived.

Theorem 1. *Suppose that* $c, d > 0$ *and* $r > 1$ *are arbitrary rational constants, and* $l \geq 2$ *is an arbitrary integer. For any small rational constant* $\epsilon > 0$*, there is a language* $L \subseteq \{0, 1\}^*$ *which can be accepted by a* $c(9+\epsilon) \log^r n$*-time* $d(4+\epsilon) \log n$*-space ATM with* l *worktapes but not by any* $c \log^r n$*-time* $d \log n$*-space ATM with* l *worktapes. Here, all ATMs have only 0 and 1 as their tape symbols.*

The proof of this theorem is given in Section 3.2. Theorem 1 implies that constant factors $9 + \epsilon$ and $4 + \epsilon$ in time and space, respectively, strictly enlarge the complexity classes defined by time- and space-bounded ATMs.

Another result of this paper is a translational lemma for PRAMs.

Lemma 2. *Suppose that $t_1(n), t_2(n) \neq O(\log n)$ and $p_1(n), p_2(n) \geq n$ are arbitrary functions computable by $O(\log n)$-time PRAMs with n processors. For any rational constant $\epsilon > 0$,*

$$\text{Ptime,proc}((1 + \epsilon)t_1(n), p_1(n)) \subseteq \text{Ptime,proc}(t_2(n), p_2(n))$$

implies

$$\text{Ptime,proc}(t_1(2^{kn}), p_1(2^{kn})) \subseteq \text{Ptime,proc}((1 + \epsilon)t_2(2^{kn}), p_2(2^{kn})),$$

where $k \geq 1$ is an arbitrary integer.

Theorem 2. *Suppose that $c > 0$, $r_1 > 1$, and $r_2 \geq 1$ are arbitrary rational constants. For any small rational constant $\epsilon > 0$, there is a language which can be accepted by a $c(1 + \epsilon) \log^{r_1} n$-time PRAM with $n^{r_2(1+\epsilon)}$ processors but not by any $c \log^{r_1} n$-time PRAM with n^{r_2} processors.*

The proofs of Lemma 2 and Theorem 2 are given in Sections 4.1 and 4.2, respectively.

3   Translation and Tight Hierarchy of Alternating TMs

3.1   Translational Lemma for Alternating TMs

In this section, we will prove Lemma 1. Let M_1 be a $t_1(2^{kn})$-time $s_1(2^{kn})$-space ATM with l worktapes. Let $L \subseteq \{0,1\}^*$ be the language accepted by M_1. We will construct a $3(1+\epsilon)t_2(2^{kn})$-time $2(1+\epsilon)s_2(2^{kn})$-space ATM M_2 with l worktapes accepting the language L. Note that all ATMs in this paper have only 0 and 1 as their tape symbols.

For all $n \geq 1$, let

$$pad(x_1 x_2 \cdots x_n) = x_1 0 x_2 0 \cdots x_n 0,$$

where $x_1 x_2 \cdots x_n \in \{0,1\}^n$. We define a language L' as

$$L' = \{pad(x)\, 11\, 0^{h-2} \mid x \in L,\, 2|x| + h = 2^{k|x|}\},$$

where 0^{h-2} is a padding sequence of length $h - 2$, and 11 is a "boundary" string between $pad(x)$ and the padding sequence 0^{h-2}.

The outline of the proof is as follows: (I) We will construct an ATM M_1' which accepts L' within time $3(1 + \epsilon)t_1(N)$ and space $2(1 + \epsilon)s_1(N)$, where N is the length of M_1''s input string. (II) From the assumption of Lemma 1 and (I), there is an ATM M_2' which accepts L' within time $t_2(N)$ and space $s_2(N)$. (III) We will construct an ATM M_2 which accepts L within time $3(1 + \epsilon)t_2(2^{kn})$ and space $2(1 + \epsilon)s_2(2^{kn})$.

(I) M_1' existentially generates (guesses) the values of n and N satisfying $2^{kn} = N$ in the second worktape. Let $bin(i)$ denote the string over $\{0, 1\}$ representing value i in binary, where the leftmost bit is the least significant bit. For example, $bin(6) = 011$. The values n and N are represented by string

$$pad(bin(n)) \, 11 \, pad(bin(N)) \, 11,$$

where 11 is a "boundary" string. M_1' starts to perform the following (1) through (4) universally.

(1) M_1' verifies whether the guessed value N is equal to the length of the input string as follows. M_1' performs (1a) and (1b) universally. (1a) M_1' generates $bin(N + 1)$ in the first worktape (= index tape). Then M_1' enters the read state $read(B)$ with the first worktape having value $N + 1$ (for simplicity, we say M_1' enters the read state with "$B, N + 1$" in the rest of this section). M_1' halts with an accepting state if the $(N + 1)$st input symbol is blank symbol B (i.e., if the input has length at most N). (1b) M_1' performs the following (i) and (ii) existentially in order to verify whether the input has length at least N. (i) M_1' enters the read state with "$0, N$". (ii) M_1' enters the read state with "$1, N$".

(2) Now M_1' can assume that it has correctly generated the value N. M_1' generates the value of 2^{kn} in the first worktape. Note that $bin(2^{kn})$ is string $00 \cdots 01$ of length $kn+1$. M_1' compares the value 2^{kn} in the first worktape with the value N in the second worktape. If $2^{kn} = N$, then M_1' halts with an accepting state (i.e., M_1' enters the read state with "$B, N + 1$").

(3) Under the assumption that the values of N and n are correct, M_1' verifies the following three conditions universally: (a) Symbol 0 appears every even cell in the first $2n$ cells in the input tape, (b) 11 are in the $(2n + 1)$st and $(2n + 2)$nd cells, and (c) each of the $(2n + 3)$rd through Nth cells contains 0. For (a), M_1' universally generates $i \in \{2, 4, \ldots, 2n\}$ in the first worktape and enters the read state with "$0, i$". (b) and (c) are left to the reader.

(4) It remains to verify whether $x = x_1 x_2 \cdots x_n \in L$. M_1' first "erases" the value N, i.e., M_1' changes all non-blank symbols following the leftmost 11 in the second worktape into 0. Note that there remains the value n (i.e., string $pad(bin(n))11$) in the first $2(\lfloor \log n \rfloor + 1) + 2$ cells in the second worktape. Now, M_1' starts to simulate M_1. Since ATMs have no input head, the simulation does not depend on the input string except when the TM is in a read state. When M_1 enters the read state with "a, i", M_1' compares i with n by sweeping non-blank symbols in the first worktape and $pad(bin(n))11$ in the second worktape. If $1 \leq i \leq n$ and a is a non-blank symbol, then M_1' enters the read state with "$a, 2i - 1$". If $i = 0$ and $a = \$$, or if $i > n$ and $a = B$, then M_1' halts with an accepting state (i.e., M_1' enters the read state with "$0, 2$"). Otherwise, M_1' halts with a rejecting state (i.e., M_1 enters the read state with "$1, 2$"). Note that the 2nd symbol of $x_1 0 x_2 0 \cdots$ is 0.

The time and space complexities of paragraphs (I), (1), (2), and (3) are bounded by $O(\log N)$. The complexity of M_1' depends on (4). The worktapes of M_1 are simulated by the corresponding worktapes of M_1'. Here, it should be noted that the second worktape of M_1' contains the value n, i.e., the string $pad(bin(n))11$ is in the first $2(\lfloor \log n \rfloor + 1) + 2$ cells. Thus, the second worktape of M_1 should be simulated by using the remaining area of M_1''s second worktape. In order to recognize the boundary string 11, symbol 0 appears every other cell in this remaining area. So, the space complexity of M_1' is $2s_1(2^{kn}) + O(\log n)$, which is bounded by $2(1 + \epsilon)s_1(N)$ because $N = 2^{kn}$ and $s_1(N) = \Theta(\log N)$.

For example, suppose that M_1 has visited five cells in the second worktape, the five cells contain 11010, and the head is placed at the third cell.

$$\$1\overset{\smile}{1}010BB\cdots$$

Then, M_1''s second worktape is

$$\$\,pad(bin(n))\,11\,0\underline{1}0\underline{1}0\overset{\smile}{0}0\underline{1}0\underline{0}1100\cdots,$$

where the end-marker $\$$ and the leftmost B in M_1's second worktape are simulated by 11 in M_1''s second worktape. (i) If M_1's head moves one position to the left, then M_1''s head moves two positions to the left; at this point, if the head is scanning 1, then the head moves one position to the left and then moves one position to the right in order to verify whether M_1's head is scanning $\$$ or not (i.e., M_1''s head is scanning 11). (ii) If M_1's head moves one position to the right, then M_1''s head moves two positions to the right. At this point, if M_1''s head reaches right "end-marker" 11, then M_1' moves 11 two positions to the right (which requires additive $6s_1(2^{kn})$ steps in total). So, the time complexity of the simulation is at most $(4t_1(2^{kn}) + 2t_1(2^{kn}))/2 + 6s_1(2^{kn}) = 3t_1(2^{kn}) + 6s_1(2^{kn})$ in total. (Note that M_1''s head moves to the left (resp. right) at most $t_1(2^{kn})/2$ times (resp. at least $t_1(2^{kn})/2$ times). This is a reason why we padded the zeros to the left of the symbols in M_1''s second worktape.) When M_1 enters the read state, M_1''s head in the second worktape must go back to the leftmost cell in order to access the value n, which requires additive $2s_1(2^{kn}) + O(\log n)$ steps. Erasing N in (4) needs $O(\log N)$ steps. Therefore, the time complexity of M_1' is $3t_1(2^{kn}) + 8s_1(2^{kn}) + O(\log n) + O(\log N)$, which is bounded by $3(1 + \epsilon)t_1(N)$ because $t_1(N) \neq O(s_1(N))$, $N = 2^{kn}$, and $s_1(N) = \Theta(\log N)$.

(II) From the assumption of the lemma and (I), there is an ATM M_2' which accepts L' within time $t_2(N)$ and space $s_2(N)$.

(III) We will construct an ATM M_2 which accepts L within time $3(1 + \epsilon)t_2(2^{kn})$ and space $2(1 + \epsilon)s_2(2^{kn})$. By a method similar to (I) and (1), M_2 correctly guesses the length n of the input string (i.e., generates $pad(bin(n))11$) in the second worktape in $O(\log n)$ steps.

M_2 starts to simulate M_2'. When M_2' enters the reading state with "a, i", M_2 compares the value i in the first worktape ($=$ index tape) with value $2n$ by sweeping non-blank symbols and $pad(bin(n))11$ in M_2's first and second worktapes, respectively. If $i \leq 2n$ and i is odd, then M_2 enters the read state with "$a, (i + 1)/2$". If $1 \leq i \leq 2n$, i is even, and $a = 0$ ($a \neq 0$), then M_2 halts with an accepting (resp. rejecting) state. If $i = 0$ and $a = \$$, then M_2 halts with an accepting state. If $i \in \{2n+1, 2n+2\}$ and $a = 1$ ($a \neq 1$), then M_2 halts with an accepting (resp. rejecting) state. If $i > 2n + 2$, then M_2 compares values i with $N = 2^{kn}$ as follows: Note that if i is represented by a binary string of length at most kn, then i is less than N. Thus, M_2 can decide whether $i \leq N$ by using the "binary counter" $pad(bin(n))11$ in the second worktape and value k in the finite control. The time and space complexities for this task are $O(s_2(N))$ and $s_2(N) + O(1)$, respectively (note that the value i in M_2' index tape is stored in space $s_2(N)$, since M_2' is $s_2(N)$-space bounded). If $i \leq N$ and $a = 0$, then M_2 halts with an accepting state. If $i > N$ and $a = B$, then M_2 halts with an accepting state. Otherwise, M_2 halts with a rejecting state.

The time and space complexities of guessing n are both $O(\log n)$. By the same analysis as in the last paragraph of task (I), the time and space complexities of M_2 are $3t_2(N) + 8s_2(N) + O(\log n) + O(s_2(N))$ and $2s_2(N) + O(\log n)$, respectively. They are bounded by $3(1 + \epsilon)t_2(2^{kn})$ and $2(1 + \epsilon)s_2(2^{kn})$, since $t_2(N) \neq O(s_2(N))$, $N = 2^{kn}$, and $s_2(N) = \Theta(\log N)$. This completes the proof of Lemma 1.

3.2   Tight Hierarchy Theorem for Alternating TMs

In this section, we will prove Theorem 1, i.e.,

$$\text{Atime,space}(c \log^r n, d \log n) \subsetneq \text{Atime,space}(c(9 + \epsilon) \log^r n, d(4 + \epsilon) \log n). \quad (1)$$

Let q be a sufficiently large integer such that $9(1 + 1/q)^{r+2} \leq 9 + \epsilon$ and $4(1 + 1/q)^3 \leq 4 + \epsilon$. In order to show the separation (1), it is enough to prove

$$\text{Atime,space}(c \log^r n, d \log n)$$
$$\subsetneq \text{Atime,space}(9c(1 + 1/q)^{r+2} \log^r n, 4d(1 + 1/q)^3 \log n).$$

The following proof is based on the same idea as in [5,7], but our proof is simpler. For example, two parameters p and q were used in [5,7], while only q is used here. Assume for contradiction that the following relation holds.

$$\text{Atime,space}(c \log^r n, d \log n)$$
$$= \text{Atime,space}(9c(1 + 1/q)^{r+2} \log^r n, 4d(1 + 1/q)^3 \log n).$$

Then,

$$\text{Atime,space}(9c(1 + 1/q)^{r+2} \log^r n, 4d(1 + 1/q)^3 \log n)$$
$$\subseteq \text{Atime,space}(c \log^r n, d \log n).$$

We regard $1 + 1/q$ in this relation as $1 + \epsilon$ in Lemma 1. By applying Lemma 1 to this relation, we obtain

$$\text{Atime,space}(3c(1 + 1/q)^{r+1}(kn)^r, 2d(1 + 1/q)^2 kn)$$
$$\subseteq \text{Atime,space}(3c(1 + 1/q)(kn)^r, 2d(1 + 1/q)kn). \quad (2)$$

Let $c' = 3c(1 + 1/q)$ and $d' = 2d(1 + 1/q)$. Then the inclusion relation (2) can be written as

$$\text{Atime,space}(c'(1 + 1/q)^r (kn)^r, d'(1 + 1/q)kn) \subseteq \text{Atime,space}(c'(kn)^r, d'kn).$$

Substituting $k = (q + i)q$ into this relation yields

$$\text{Atime,space}(c'((q + 1)(q + i)n)^r, d'(q + 1)(q + i)n) \\ \subseteq \text{Atime,space}(c'((q + i)qn)^r, d'(q + i)qn), \quad (3)$$

where i is an integer. When $i \geq 1$, $(q + i)q \leq (q + 1)(q + i - 1)$ holds. Hence, for $i \geq 1$,

$$\text{Atime,space}(c'((q + 1)(q + i)n)^r, d'(q + 1)(q + i)n) \\ \subseteq \text{Atime,space}(c'((q + 1)(q + i - 1)n)^r, d'(q + 1)(q + i - 1)n). \quad (4)$$

By substituting $i = 0$ into (3), we obtain

$$\text{Atime,space}(c'((q + 1)qn)^r, d'(q + 1)qn) \subseteq \text{Atime,space}(c'(q^2n)^r, d'q^2n). \quad (5)$$

Let $\hat{c}$ be an integer (the value of $\hat{c}$ will be fixed later). By substituting $i = 1, 2, 3, \ldots, \hat{c}q - q$ into (4), we obtain

$$\left. \begin{array}{l} \text{Atime,space}(c'((q + 1)(q + 1)n)^r, d'(q + 1)(q + 1)n) \\ \quad \subseteq \text{Atime,space}(c'((q + 1)qn)^r, d'(q + 1)qn), \\ \text{Atime,space}(c'((q + 1)(q + 2)n)^r, d'(q + 1)(q + 2)n) \\ \quad \subseteq \text{Atime,space}(c'((q + 1)(q + 1)n)^r, d'(q + 1)(q + 1)n), \\ \text{Atime,space}(c'((q + 1)(q + 3)n)^r, d'(q + 1)(q + 3)n) \\ \quad \subseteq \text{Atime,space}(c'((q + 1)(q + 2)n)^r, d'(q + 1)(q + 2)n), \\ \quad \vdots \\ \text{Atime,space}(c'((q + 1)\hat{c}qn)^r, d'(q + 1)\hat{c}qn) \\ \quad \subseteq \text{Atime,space}(c'((q + 1)(\hat{c}q - 1)n)^r, d'(q + 1)(\hat{c}q - 1)n). \end{array} \right\} \quad (6)$$

Since $r > 1$,

$$\text{Atime,space}(\hat{c} \cdot c'(q^2n)^r, \hat{c} \cdot d'q^2n) \subseteq \text{Atime,space}(c'((q + 1)\hat{c}qn)^r, d'(q + 1)\hat{c}qn). \quad (7)$$

By connecting relations (5),(6), and (7), we obtain

$$\text{Atime,space}(\hat{c} \cdot c'(q^2n)^r, \hat{c} \cdot d'q^2n) \subseteq \text{Atime,space}(c'(q^2n)^r, d'q^2n). \quad (8)$$

On the other hand, it is known [6,7,9] that, for all functions $s(n) \geq \log n$ and $t(n) \neq O(s(n) \log s(n))$ which are computable by $O(s(n))$-time TMs, there are constants $\tilde{c}$ and $\tilde{d}$ such that

$$\text{Atime,space}(t(n), s(n)) \subsetneq \text{Atime,space}(\tilde{c} \cdot t(n), \tilde{d} \cdot s(n)). \quad (9)$$

If $t(n)$ and $s(n)$ are the appropriate functions from (8) and the constants fulfill (9), and if we fix $\hat{c} \geq \max\{\tilde{c}, \tilde{d}\}$, inclusion (8) contradicts separation (9). This completes the proof of Theorem 1.

4   Translation and Tight Hierarchy of PRAMs

4.1   Translational Lemma for PRAMs

In this section, we will prove Lemma 2. Let M_1 be a $t_1(2^{kn})$-time PRAM with $p_1(2^{kn})$ processors. Let L be the language accepted by M_1. We will construct a $(1+\epsilon)t_2(2^{kn})$-time PRAM M_2 with $p_2(2^{kn})$ processors accepting the language L.

We define a language L' as $L' = \{x\$^h \mid x \in L, |x| + h = 2^{k|x|}\}$, where $\$^h$ is a padding sequence of length h, and the symbol $\$$ does not appear in x. First, we will construct a $(1 + \epsilon)t_1(N)$-time PRAM M_1' with $p_1(N)$ processors accepting L', where N is the length of M_1''s input. Processors of M_1' are denoted by $P_1', P_2', \cdots$. M_1' verifies whether the position $n + 1$ of the leftmost $\$$ satisfies $N = 2^{kn}$. Computing the value n from $N = 2^{kn}$ can be done by using addition $\log N$ times. Then, every processor P_i' ($n + 1 \leq i \leq N$) verifies whether the ith input is $\$$. Note that $p_1(N) \geq N$. Now M_1' starts to simulate $p_1(2^{kn})$-processor PRAM M_1 on the input x of length n. If M_1 accepts x (i.e., $x \in L$), then M_1' accepts the input of length N. The time complexity of M_1' is $t_1(2^{kn}) + O(\log N)$, which is bounded by $(1 + \epsilon)t_1(N)$ because $N = 2^{kn}$ and $t_1(N) \neq O(\log N)$. The number of M_1''s processors is $p_1(2^{kn}) = p_1(N)$.

From the assumption of Lemma 2 and the previous paragraph, there is a $t_2(N)$-time PRAM M_2' with $p_2(N)$ processors accepting the language L'.

Finally, we will construct a $(1 + \epsilon)t_2(2^{kn})$-time PRAM M_2 with $p_2(2^{kn})$ processors accepting L. M_2 computes the value $N = 2^{kn}$ from n by using addition kn times. Processors $P_{n+1}, P_{n+2}, \ldots, P_{2^{kn}}$ of M_2 write symbol $\$$ into common memory $M[n + 1], M[n + 2], \ldots, M[2^{kn}]$, respectively. Note that $p_2(2^{kn}) \geq 2^{kn}$. Then, M_2 starts to simulate $p_2(N)$-processor PRAM M_2' on input $x\$\$ \cdots \$$ of length $N = 2^{k|x|}$. The time complexity of M_2 is $t_2(N) + O(n)$ in total, which is bounded by $(1 + \epsilon)t_2(2^{kn})$ because $N = 2^{kn}$ and $t_2(N) \neq O(\log N)$. The number of M_2's processors is $p_2(N) = p_2(2^{kn})$. This completes the proof of Lemma 2.

4.2   Tight Hierarchy Theorem for PRAMs

In this section, we will prove Theorem 2, i.e.,

$$\text{Ptime,proc}(c \log^{r_1} n, n^{r_2}) \subsetneq \text{Ptime,proc}(c(1 + \epsilon) \log^{r_1} n, n^{r_2(1+\epsilon)}). \qquad (10)$$

The proof is similar to Theorem 1. Constant factors of time and space were translated in Section 3.2; both a constant factor and an exponent will be translated simultaneously in this section.

Let q be a sufficiently large integer such that $(1 + 1/q)^{r_1+2} \leq 1 + \epsilon$. In order to show the separation (10), it is enough to prove

$$\text{Ptime,proc}(c \log^{r_1} n, n^{r_2}) \subsetneq \text{Ptime,proc}(c(1 + 1/q)^{r_1+2} \log^{r_1} n, n^{r_2(1+1/q)}).$$

Assume for contradiction that the following relation holds.

$$\text{Ptime,proc}(c \log^{r_1} n, n^{r_2}) = \text{Ptime,proc}(c(1 + 1/q)^{r_1+2} \log^{r_1} n, n^{r_2(1+1/q)}).$$

Then,

$$\text{Ptime,proc}(c(1+1/q)^{r_1+2}\log^{r_1} n, n^{r_2(1+1/q)}) \subseteq \text{Ptime,proc}(c\log^{r_1} n, n^{r_2}).$$

We regard $1 + 1/q$ in this relation as $1 + \epsilon$ in Lemma 2. By applying Lemma 2 to this relation, we obtain

$$\text{Ptime,proc}(c(1+1/q)^{r_1+1}(kn)^{r_1}, 2^{r_2(1+1/q)kn})$$
$$\subseteq \text{Ptime,proc}(c(1+1/q)(kn)^{r_1}, 2^{r_2kn}). \quad (11)$$

Let $c' = c(1 + 1/q)$. Then the inclusion relation (11) can be written as

$$\text{Ptime,proc}(c'(1+1/q)^{r_1}(kn)^{r_1}, 2^{r_2(1+1/q)kn}) \subseteq \text{Ptime,proc}(c'(kn)^{r_1}, 2^{r_2kn}).$$

Substituting $k = (q + i)q$ into this relation yields

$$\text{Ptime,proc}(c'((q+1)(q+i)n)^{r_1}, 2^{r_2(q+1)(q+i)n})$$
$$\subseteq \text{Ptime,proc}(c'((q+i)qn)^{r_1}, 2^{r_2(q+i)qn}), \quad (12)$$

where i is an integer. When $i \geq 1$, $(q+i)q \leq (q+1)(q+i-1)$ holds. Hence, for $i \geq 1$,

$$\text{Ptime,proc}(c'((q+1)(q+i)n)^{r_1}, 2^{r_2(q+1)(q+i)n})$$
$$\subseteq \text{Ptime,proc}(c'((q+1)(q+i-1)n)^{r_1}, 2^{r_2(q+1)(q+i-1)n}). \quad (13)$$

By substituting $i = 0$ into (12), we obtain

$$\text{Ptime,proc}(c'((q+1)qn)^{r_1}, 2^{r_2(q+1)qn}) \subseteq \text{Ptime,proc}(c'(q^2n)^{r_1}, 2^{r_2q^2n}). \quad (14)$$

Let $\hat{c}$ be an integer (the value of $\hat{c}$ will be fixed later). By substituting $i = 1, 2, 3, \ldots, \hat{c}q - q$ into (13), we obtain

$$\left.\begin{array}{l}\text{Ptime,proc}(c'((q+1)(q+1)n)^{r_1}, 2^{r_2(q+1)(q+1)n}) \\ \qquad \subseteq \text{Ptime,proc}(c'((q+1)qn)^{r_1}, 2^{r_2(q+1)qn}), \\ \text{Ptime,proc}(c'((q+1)(q+2)n)^{r_1}, 2^{r_2(q+1)(q+2)n}) \\ \qquad \subseteq \text{Ptime,proc}(c'((q+1)(q+1)n)^{r_1}, 2^{r_2(q+1)(q+1)n}), \\ \text{Ptime,proc}(c'((q+1)(q+3)n)^{r_1}, 2^{r_2(q+1)(q+3)n}) \\ \qquad \subseteq \text{Ptime,proc}(c'((q+1)(q+2)n)^{r_1}, 2^{r_2(q+1)(q+2)n}), \\ \qquad\qquad \vdots \\ \text{Ptime,proc}(c'((q+1)\hat{c}qn)^{r_1}, 2^{r_2(q+1)\hat{c}qn}) \\ \qquad \subseteq \text{Ptime,proc}(c'((q+1)(\hat{c}q-1)n)^{r_1}, 2^{r_2(q+1)(\hat{c}q-1)n}).\end{array}\right\} \quad (15)$$

Since $r_1 > 1$,

$$\text{Ptime,proc}(\hat{c} \cdot c'(q^2n)^{r_1}, 2^{\hat{c} \cdot r_2 q^2 n}) \subseteq \text{Ptime,proc}(c'((q+1)\hat{c}qn)^{r_1}, 2^{r_2(q+1)\hat{c}qn}). \quad (16)$$

By connecting relations (14),(15), and (16), we obtain

$$\text{Ptime,proc}(\hat{c} \cdot c'(q^2n)^{r_1}, 2^{\hat{c} \cdot r_2 q^2 n}) \subseteq \text{Ptime,proc}(c'(q^2n)^{r_1}, 2^{r_2 q^2 n}). \quad (17)$$

On the other hand, it is known [6] that, for all functions $t(n)$ and $p(n)$ which are not constant and are computable by a $t(n)$-time PRAM with $p(n)$ processors, there is a constant $\tilde{c}$ such that

$$\text{Ptime,proc}(t(n), p(n)) \subsetneq \text{Ptime,proc}(\tilde{c} \cdot t(n), p(n)). \tag{18}$$

If $t(n)$ and $p(n)$ are the appropriate functions from (17) and the constant fulfills (18), and if we fix $\hat{c} \geq \tilde{c}$, inclusion (17) contradicts separation (18). This completes the proof of Theorem 2.

5 Conclusion

In this paper, we discussed tight hierarchy theorems for three major parallel computation models: ATMs, PRAMs, and uniform circuit families. For PRAMs and uniform circuit families, complexity classes are strictly enlarged by the constant factor $1 + \epsilon$ in time (depth) and the exponent $1 + \epsilon$ in the number of processors (size). As for ATMs, on the other hand, there is no uniform-cost criterion or no uniformity conditions. From this reason, the constant factors in time and space of the ATM-hierarchy are $9 + \epsilon$ and $4 + \epsilon$, respectively. Improving these factors toward $1 + \epsilon$ is an interesting future work.

References

1. A.V. Aho, J.E. Hopcroft, and J.D. Ullman, The Design and Analysis of Computer Algorithms, Addison-Wesley, Reading, MA, 1974.
2. A. Chandra, D. Kozen, and L. Stockmeyer, Alternation, *J. Assoc. Comput. Mach.*, **28** (1981) 114–133.
3. J.E. Hopcroft and J.D. Ullman, Introduction to Automata Theory, Languages, and Computation, Addison-Wesley, Reading, MA, 1979.
4. O.H. Ibarra, A hierarchy theorem for polynomial-space recognition, *SIAM J. Comput.* **3** 3 (1974) 184–187.
5. O.H. Ibarra, S.M.Kim, and S. Moran, Sequential machine characterizations of trellis and cellular automata and applications, *SIAM J. Comput.*, **14** 2 (1985) 426–447.
6. K. Iwama and C. Iwamoto, Parallel complexity hierarchies based on PRAMs and DLOGTIME-uniform circuits, *Proc. 11th Ann. IEEE Conf. on Computational Complexity*, Philadelphia, 1996, 24–32.
7. C. Iwamoto, N. Hatayama, K. Morita, K. Imai, and D. Wakamatsu, Hierarchies of DLOGTIME-uniform circuits, in M. Margenstern (ed.): Machines, Computations and Universality (Proc. MCU 2004, Saint-Petersburg, Sep. 21–26, 2004), LNCS 3354, Springer 2005, 211–222.
8. R.M. Karp and V. Ramachandran, Parallel algorithms for shared-memory machines. in J. van Leeuwen (ed.): Handbook of Theoretical Computer Science Vol. A, MIT Press, Amsterdam, 1990, 869–941.
9. W.L. Ruzzo, On uniform circuit complexity, *J. Comput. System Sci.*, **22** (1981) 365–383.
10. L. Stockmeyer and U. Vishkin, Simulation of parallel random access machines by circuits, *SIAM J. Comput.*, **13** 2 (1984) 409–422.

Collapsing Recursive Oracles for Relativized Polynomial Hierarchies⋆
(Extended Abstract)

Tomoyuki Yamakami

Computer Science Program, Trent University
Peterborough, Ontario, Canada K9J 7B8

Abstract. Certain recursive oracles can force the polynomial hierarchy to collapse to any fixed level. All collections of such oracles associated with each collapsing level form an infinite hierarchy, called the collapsing recursive oracle polynomial (CROP) hierarchy. This CROP hierarchy is a significant part of the extended low hierarchy (note that the assumption P = NP makes the both hierarchies coincide). We prove that all levels of the CROP hierarchy are distinct by showing "strong" types of separation. First, we prove that each level of the hierarchy contains a set that is immune to its lower level. Second, we show that any two adjacent levels of the CROP hierarchy can be separated by another level of the CROBPP hierarchy—a bounded-error probabilistic analogue of the CROP hierarchy. Our proofs extend the circuit lower-bound techniques of Yao, Håstad, and Ko.

1 Collapsing Recursive Oracles

The *polynomial hierarchy* (simply called the P hierarchy) was introduced by Meyer and Stockmeyer [13] in the early 1970s as a resource-bounded variant of the arithmetical hierarchy in recursion theory. Unable to prove the separation of any level of the P hierarchy (for instance, the separation between P and NP), Baker, Gill, and Solovay [2] looked into an "oracle model" and developed the so-called *theory of relativization*, which sheds insightful light on the mechanism of oracle access. The theory has also nurtured fundamental proof techniques and tools for, e.g., circuit lower bounds. Oracle models have been since then adopted into various schemes, such as pseudorandom functions and interactive protocols. Construction of oracles (or relativized worlds) has become a routine to assert that separating (or collapsing) complexity classes in question requires nonrelativizable proof techniques. In relativized P hierarchies, Ko [9] later discovered recursive oracles that force a relativized P hierarchy to collapse to any desired level. For simplicity, we call such an oracle a *collapsing oracle*. Of particular interest are collapsing "recursive" oracles. All PSPACE-complete sets C are examples of

---

⋆ This work was supported in part by the Natural Sciences and Engineering Research Council of Canada.

M. Liśkiewicz and R. Reischuk (Eds.): FCT 2005, LNCS 3623, pp. 149–160, 2005.

collapsing recursive oracles, which force NP^C to collapse to P^C. A more explicit recursive construction of Baker, Gill, Solovay [2] gives rise to the set A satisfying

$$A = \{\langle 1^i, x, 1^t \rangle \mid N_i^A(x) \text{ outputs 1 within } t \text{ steps}\},$$

where N_i is the ith nondeterministic oracle Turing machine, which also makes NP^A collapse to P^A. Intuitively, collapsing oracles possess rich information on the behaviors of sets in a given level. We can view such a collapsing oracle as an encoding of membership information of input strings to languages. To collapse NP^A to P^A, for instance, we need to encode into A the information on "$x \in L^A$" for any string $x \in \Sigma^*$ and any set $L^{(\cdot)} \in NP^{(\cdot)}$ so that an appropriate P^A-machine can retrieve such information from A. What kind of information should be stored inside a collapsing oracle A? How much information is necessary to store in the oracle A?

In this paper, we study the collection of collapsing recursive oracles. Our goals are to study and analyze the internal structure of a class of collapsing recursive oracles that collapse the P hierarchy to a certain level. Now, we formally introduce a collection of such classes. We call it the *collapsing recursive oracle polynomial (CROP) hierarchy*. For convenience, let REC denote the class of all recursive languages. We follow the standard convention that $\Delta_0^P = \Sigma_0^P = \Theta_0^P = P$ (thus, $\Delta_1^P = \Theta_1^P = P$).

Definition 1. *Let $k \geq 1$.*

1. $CRO\Sigma_k^P = \{A \in REC \mid \Sigma_k^P(A) = \Sigma_{k+1}^P(A)\}$.
2. $CRO\Delta_k^P = \{A \in REC \mid \Delta_k^P(A) = \Delta_{k+1}^P(A)\}$.
3. $CRO\Theta_k^P = \{A \in REC \mid \Theta_k^P(A) = \Theta_{k+1}^P(A)\}$.
4. $CROH = \bigcup_{i \geq 1} CRO\Delta_i^P$.

It is not difficult to show that $CRO\Delta_k^P \subseteq CRO\Sigma_k^P \subseteq CRO\Theta_{k+1}^P \subseteq CRO\Delta_{k+1}^P$ for each index $k \geq 1$. It is useful to note the following alternative characterization of the CROP hierarchy: for any class $\mathcal{D} \in \{\Theta_k^P, \Delta_k^P, \Sigma_k^P \mid k \geq 1\}$, $CRO \cdot \mathcal{D}$ coincides with the class $\{A \in REC \mid \Sigma_k^P(A) = co\text{-}\mathcal{D}(A)\}$. This characterization is the consequence of the so-called *upward collapse property* (e.g., $\Sigma_k^P = \Pi_k^P$ implies $\Sigma_k^P = \Sigma_{k+1}^P$) of the P hierarchy.

As mentioned earlier, any PSPACE-complete set and the set A of Baker et al. are typical examples of recursive sets that belong to the complexity class $CRO\Delta_1^P$. Even all P-T-complete sets satisfying $NP(\mathcal{C}) \subseteq \mathcal{C}$ for any given class $\mathcal{C}$ belong to $CRO\Delta_1^P$.

Lemma 1. *Let $\mathcal{C}$ be any complexity class of languages such that $NP(\mathcal{C}) \subseteq \mathcal{C}$. Any P-T-complete set for $\mathcal{C}$ is in $CRO\Delta_1^P$.*

Notice that the CROP hierarchy has been implicitly discussed in the literature. For instance, Heller [8] constructed recursive oracles A, B such that $\Delta_2^P(A) \neq \Sigma_2^P(A) = \Pi_2^P(A)$ and $\Sigma_1^P(B) \neq \Delta_2^P(B) = \Sigma_2^P(B)$. Later Bruschi [5] extended Heller's separation to any level of the P hierarchy; that is, $\Delta_k^P(A) \neq \Sigma_k^P(A) = \Pi_k^P(A)$ and $\Sigma_{k-1}^P(B) \neq \Delta_k^P(B) = \Sigma_k^P(B)$ for every $k \geq 3$. Sheu and Long [16] constructed two recursive oracles A and B such that $\Theta_k^P(A) \neq \Delta_k^P(A) = \Sigma_k^P(A)$ and $\Sigma_{k-1}^P(B) \neq \Theta_k^P(B) = \Sigma_k^P(B)$, where $k \geq 2$. These results implicitly prove that the CROP hierarchy is indeed an infinite hierarchy:

Theorem 1. *For any $k \geq 1$, $\mathrm{CRO}\Theta_k^P \neq \mathrm{CRO}\Delta_k^P \neq \mathrm{CRO}\Sigma_k^P \neq \mathrm{CRO}\Theta_{k+1}^P$.*

Theorem 1 shows a clear gap among all levels of the CROP hierarchy. In other words, a certain collapsing recursive oracle in each level of the CROP hierarchy encodes a different type of information. The theorem, however, fails to address how large such a gap is and how different such coded information is. In later sections, we strengthen the theorem by showing that much stronger separations are possible.

The CROP hierarchy is a significant "part" of the so-called *extended low hierarchy*. This hierarchy was originally introduced by Balcázar, Book, and Schöning [3] for the Σ-levels and later expanded by Allender and Hemachandra [1] and by Long and Sheu [12] to the Δ- and Θ-levels. For our discussion, we limit our attention to the recursive segment of this hierarchy; that is, all elements of the hierarchy are limited to recursive sets. We use the special prefix "r" to signify our "recursive" restriction in the following definition of the extended low hierarchy.

Definition 2. *Let $k \geq 1$.*

1. $\mathrm{rEL}\Sigma_k^P = \{A \in \mathrm{REC} \mid \exists B \in \mathrm{NP}[\Sigma_k^P(A) \subseteq \Sigma_{k-1}^P(A \oplus B)]\}$.
2. $\mathrm{rEL}\Delta_{k+1}^P = \{A \in \mathrm{REC} \mid \exists B \in \mathrm{NP}[\Delta_{k+1}^P(A) \subseteq \Delta_k^P(A \oplus B)]\}$.
3. $\mathrm{rEL}\Theta_{k+1}^P = \{A \in \mathrm{REC} \mid \exists B \in \mathrm{NP}[\Theta_{k+1}^P(A) \subseteq \Theta_k^P(A \oplus B)]\}$.

The following lemma shows the close connection between the CROP hierarchy and the extended low hierarchy.

Lemma 2. *Let $k \geq 1$.*

1. $\mathrm{CRO}\Theta_k^P \subseteq \mathrm{rEL}\Theta_{k+1}^P$, $\mathrm{CRO}\Delta_k^P \subseteq \mathrm{rEL}\Delta_{k+1}^P$, and $\mathrm{CRO}\Sigma_k^P \subseteq \mathrm{rEL}\Sigma_{k+1}^P$.
2. *If* $\mathrm{P} = \mathrm{NP}$, *then* $\mathrm{CRO}\Theta_k^P = \mathrm{rEL}\Theta_{k+1}^P$, $\mathrm{CRO}\Delta_k^P = \mathrm{rEL}\Delta_{k+1}^P$, *and* $\mathrm{CRO}\Sigma_k^P = \mathrm{rEL}\Sigma_{k+1}^P$.

Although we believe that the CROP hierarchy is different from the extended low hierarchy, we may not be able to prove the separation by currently known proof techniques because such a separation implies that $\mathrm{P} \neq \mathrm{NP}$. Nonetheless, there is certain evidence that supports the difference between the CROP hierarchy and the extended low hierarchy. We show such evidence in the subsequent section.

2   Basic Properties of the CROP Hierarchy

We highlight basic properties of the CROP hierarchy. These properties signify the importance of the hierarchy in the theory of relativization. We begin with a simple lemma concerning the density of sets in the hierarchy. A set A is called *coinfinite* if its complement $\overline{A}$ ($= \Sigma^* - A$) is infinite. If $\Delta_k^P \neq \Sigma_k^P$, then $\mathrm{CRO}\Delta_k^P$ contains no finite sets.

Lemma 3. *Assume that $\Delta_k^P \neq \Sigma_k^P$. Every set in $\mathrm{CRO}\Sigma_k^P$ is both infinite and coinfinite.*

Next, we state a closure property of the CROP hierarchy. We review the notion of *strong nondeterministic Turing reduction*. For any two sets A, B, let $A \leq_T^{sn} B$ mean that $A \in \mathrm{NP}(B) \cap \mathrm{co\text{-}NP}(B)$, which is equivalent to the inclusion $\mathrm{NP}(A) \subseteq \mathrm{NP}(B)$. Write $A \equiv_T^{sn} B$ if $A \leq_T^{sn} B$ and $B \leq_T^{sn} A$.

Lemma 4. *Let A, B be any two sets.*

1. *If $A \equiv_T^p B$ and $B \in \mathrm{CRO}\Delta_1^P$, then $A \in \mathrm{CRO}\Delta_1^P$.*
2. *Let $k \geq 2$. If $A \equiv_T^{sn} B$ and $B \in \mathrm{CRO}\Delta_k^P$, then $A \in \mathrm{CRO}\Delta_k^P$.*

The following result relates to the nonexistence of complete sets in the CROP hierarchy. The proof uses a simple diagonalization argument.

Proposition 1. *For each $k \geq 1$, $\mathrm{CRO}\Delta_k^P$ has no P-m-complete set; that is, for any set $A \in \mathrm{CRO}\Delta_k^P$, there exists a set $B \in \mathrm{CRO}\Delta_k^P$ such that $A \leq_m^p B$ and $B \not\leq_m^p A$.*

Proof. Take any set A in $\mathrm{CRO}\Delta_k^P$. Our goal is to construct a set B such that (i) $\Delta_k^P(B) = \Sigma_k^P(B)$, (ii) $A \leq_m^p B$, and (iii) $B \not\leq_m^p A$. Our oracle B is of the form $A \oplus C$ for a certain set C such that $C \not\leq_m^p A$ and $\Delta_k^P(A \oplus C) = \Sigma_k^P(A \oplus C)$, where $A \oplus C = \{0x \mid x \in A\} \cup \{1x \mid x \in C\}$. This B satisfies $B \not\leq_m^p A$ because, otherwise, $A \oplus C \leq_m^p A$ via a certain reduction f and thus $C \leq_m^p A$ via $g(x) =_{def} f(1x)$ for all $x \in \Sigma^*$, a contradiction. The set C is split into two parts: $C = C_0 \cup C_1$ with $C_0 \subseteq 0\Sigma^*$ and $C_1 \subseteq 1\Sigma^*$. The construction of C_0 and C_1 is given as follows. We fix an effective enumeration $\{f_i\}_{i \in \mathbb{N}}$ of all polynomial-time total computable functions. For convenience, set $C_0(-1) = C_1(-1) = \emptyset$.

Stage $i \geq 0$: First, we define $C_0(i)$. Assume that there exists a string $x \in \Sigma^{i-1}$ such that $f_i(0x) \in A$. We then define $C_0(i) = C_0(i-1)$. Assume otherwise that $f_i(0x) \notin A$ for any $x \in \Sigma^{i-1}$. In this case, let $C_0(i) = C_0(i-1) \cup 0\Sigma^{i-1}$. In either case, for a certain $x \in \Sigma^{i-1}$, $f_i(0x) \in A$ iff $0x \notin C$. Next, we define $C_1(i)$. Let $C(i-1) = C_0(i-1) \cup C_1(i-1)$ and consider the value of $N_i^{A \oplus C(i-1)}(\cdot)$. Define $C_1(i) = C_1(i-1) \cup \{1\langle 1^j, z, 1^t \rangle \mid |\langle 1^j, z, 1^t \rangle| = i-1 \wedge N_j^{A \oplus C(i-1)}(z) = 1$ within t steps$\}$. Note that $N_i(z)$ cannot query any string of length $> i-1$. This makes $N_j^{A \oplus C(i-1)}(z) = 1$ exactly when $1\langle 1^j, z, 1^t \rangle \in C_1(i)$.

Finally, we set $C_0 = \bigcup_{i \in \mathbb{N}} C_0(i)$, $C_1 = \bigcup_{i \in \mathbb{N}} C_1(i)$, and $C = C_0 \cup C_1$. Obviously, the set $B = A \oplus C$ satisfies the desired conditions (i)-(iii). $\qquad\square$

There is a direct connection between NP and the CROP hierarchy.

Lemma 5. *Let $k \geq 1$. (1) $\mathrm{NP} \subseteq \mathrm{CRO}\Delta_k^P \iff \Delta_k^P = \Sigma_k^P$; (2) $\mathrm{NP} \subseteq \mathrm{CRO}\Sigma_k^P \iff \Sigma_k^P = \Pi_k^P$; and (3) $\mathrm{NP} \subseteq \mathrm{CRO}\Theta_k^P \iff \Theta_k^P = \Sigma_k^P$.*

Sparse sets play an intriguing role in the CROP hierarchy. A set S is called *(polynomially) sparse* if there exists a polynomial p such that $|S \cap \Sigma^n| \leq p(n)$

for all $n \in \mathbb{N}$. Let SPARSE denote the collection of all sparse sets. As discussed earlier, the CROP hierarchy is a part of the extended low hierarchy. However, there seems to be an obvious difference between them. It is well-known in [14] that rSPARSE $\subseteq$ rELΣ_3^P; however, it is unknown whether rSPARSE $\subseteq$ CROΣ_k^P even for any $k \geq 2$. Lemma 6 and a common belief $\Delta_k^P \neq \Sigma_k^P$ for every $k \geq 1$ imply that rSPARSE $\not\subseteq$ CROΣ_k^P for any $k \geq 1$.

Long and Selman [11] proved that, for every $k \geq 2$, PH $= \Sigma_k^P$ iff PH$(A) = \Sigma_k^P(A)$ for every oracle $A \in$ SPARSE. For simplicity, let rSPARSE $=$ SPARSE$\cap$ REC. In our terminology, for every $k \geq 2$, $\Delta_k^P = \Sigma_k^P$ iff rSPARSE $\subseteq$ CROΔ_k^P. This claim is also true if we replace rSPARSE by any subclass $\mathcal{C}$ of rEXTSPARSE with $\emptyset \in \mathcal{C}$, where rEXTSPARSE is the recursive class of *extended sparse sets* defined as $\{A \in \text{REC} \mid \exists S \in \text{rSPARSE}[A \equiv_T^{sn} S]\}$. An example of such a subclass $\mathcal{C}$ is (NP $\cap$ co-NP)$/poly \cap$ REC.

Lemma 6. *Let $k \geq 2$ and let $\mathcal{C}$ be any subclass of rEXTSPARSE with $\emptyset \in \mathcal{C}$. (1) $\mathcal{C} \subseteq$ CRO$\Delta_k^P \iff \Delta_k^P = \Sigma_k^P$; (2) $\mathcal{C} \subseteq$ CRO$\Sigma_k^P \iff \Sigma_k^P = \Pi_k^P$; and (3) $\mathcal{C} \subseteq$ CRO$\Theta_k^P \iff \Theta_k^P = \Sigma_k^P$.*

On the contrary, Balcázar, Book, and Schöning [4] proved that PH is finite iff PH(A) is finite for every $A \in$ SPARSE. The main argument is that if $\exists S \in$ SPARSE$[$PH$(S) = \Sigma_k^P(S)]$ then PH $= \Sigma_{k+2}^P$. They used the following *transference property*: $A \in \Sigma_k^P/poly$ and $A \in P_{ld}(A)$ imply $\Sigma_2^P(A) \subseteq \Sigma_{k+2}^P$, where the notation "$A \in P_{ld}(A)$" means that A is polynomial-time Turing-self-reducible with length-decreasing queries. In the following lemma, we show the relationship between sparse sets in the CROP hierarchy and advice complexity classes.

Lemma 7. *Let $k \geq 2$.*

1. *rSPARSE $\cap$ CRO$\Delta_k^P \neq \emptyset \iff \Sigma_k^P \subseteq \Delta_k^P/poly$.*
2. *rSPARSE $\cap$ CRO$\Sigma_k^P \neq \emptyset \iff \Pi_k^P \subseteq \Sigma_k^P/poly$.*
3. *rSPARSE $\cap$ CRO$\Theta_k^P \neq \emptyset \iff \Sigma_k^P \subseteq \Theta_k^P/poly$.*

The proof of Lemma 7 easily follows from the argument of Balcázar et al. [4]. Lemmas 6 and 7 lead to the following consequence.

Corollary 1. *For each integer $k \geq 1$, if rSPARSE $\cap$ CRO$\Delta_k^P \neq \emptyset$, then rSPARSE $\subseteq$ CROΣ_{k+1}^P.*

Corollary 1 immediately follows from the recent result of Cai, et al. [6], who proved that $A \in \Delta_k^P/poly$ and $A \in P_{ld}(A)$ imply $S_2^P(A) \subseteq S_2^P(\Sigma_{k-1}^P)$.

3   Strong Separation by Immune Sets

The main theme of this paper is to strengthen the separation of the CROP hierarchy given in Theorem 1. Of particular interest is the question of how large the gap is between any two adjacent levels of the hierarchy. We wish to show the separation in a "stronger" sense using immune sets. We first review the notions of $\mathcal{C}$-immune sets, $\mathcal{C}$-bi-immune sets, and $\mathcal{C}$-simple sets for any complexity class $\mathcal{C}$ of languages.

Definition 3. *Let C be any complexity class of languages. A set S is called C-immune if S is infinite and no infinite subset of S belongs to C. A C-bi-immune set is a set C such that both C and $\overline{C}$ are C-immune. In addition, a set S is C-simple if S is infinite and its complement $\overline{S}$ is C-immune.*

The separation of two complexity classes by immune sets is often referred to as a *strong separation*. We show such a strong separation of any two adjacent levels of the CROP hierarchy by proving the existence of bi-immune sets at each level of the hierarchy.

Theorem 2. *Let $k \geq 1$.*

1. *There is no CRO $\cdot$ C-simple set for any $C \in \{\Theta^{\mathrm{P}}_{k+1}, \Delta^{\mathrm{P}}_k, \Sigma^{\mathrm{P}}_k\}$.*
2. *There exists a CROΔ^{p}_k-bi-immune set in CROΣ^{p}_k.*
3. *There exists a CROΣ^{p}_k-bi-immune set in CRO$\Theta^{\mathrm{p}}_{k+1}$.*
4. *There exists a CROΘ^{p}_k-bi-immune set in CROΔ^{p}_k.*

In fact, we can prove the existence of much more complex immune sets (such as hyperimmune sets in [19]).

For the proof of Theorem 2, we assume the reader's familiarity with basics of constant-depth Boolean circuits and circuit lower bound techniques of Yao [20], Håstad [7], Ko [9,10], and Sheu and Long [16]. Any gate of a Boolean circuit is limited to AND, OR, or XOR with unbounded fanin and one fanout. We set our alphabet Σ to be $\{0,1\}$. Let λ denote the empty string as well as the empty "symbol." Let $k \geq 1$. For each symbol $b \in \{0,1,\lambda\}$, let $L^{(b)}_k(A)$ denote the set $\{x \in \Sigma^* \mid \exists y_1 \in \Sigma^{|x|} \forall y_2 \in \Sigma^{|x|} \cdots Q_k y_k \in \Sigma^{|x|} [bxy_1y_2 \cdots y_k \in A]\}$, where Q_k is $\forall$ if k is even, and Q_k is $\exists$ if k is odd. Obviously, $L^{(b)}_k(A) \in \Sigma^{\mathrm{P}}_k(A)$ for any oracle A. Let $K_k(A)$ be the P-T-complete set for $\Sigma^{\mathrm{P}}_k(A)$ defined as: $K_1(A) = \{\langle 1^i, x, 1^t \rangle \mid N^A_i(x) = 1 \text{ within } t \text{ steps }\}$ and $K_k(A) = K_1(K_{k-1}(A))$ for each $k \geq 2$. For convenience, we set $K_0(A)$ to be A. Note that there exists a polynomial q satisfying the following: for every $x \in \Sigma^*$, there is a $\Sigma_k(q(|x|))$-circuit C_x such that, for every set A, $C_x\lceil_{\rho_A} = 1$ iff $x \in K_k(A)$.

Proof of Theorem 2.   1) Obviously, each level of the CROP hierarchy is closed under *complementation*. For instance, we have CROΣ^{p}_k = co-CROΣ^{p}_k. Therefore, there is no CROΣ^{p}_k-simple set.

2) Letting $k \geq 1$, we want to prove the existence of a CROΔ^{p}_k-immune set in CROΣ^{p}_k. The other claims (3)-(4) are similarly proven. For simplicity, write $L_k(A)$ for $L^{(\lambda)}_k(A)$ in this proof. Let $\{M_i\}_{i \in \mathbb{N}}$ and $\{\check{M}_i\}_{i \in \mathbb{N}}$ be respectively fixed effective enumerations of all oracle P-machines and of all deterministic Turing machines (TMs) without any specific time bound.

We build the desired oracle A by stages. During these stages, we attempt to satisfy the following two requirements.

i) $\forall x [x \in K_k(A) \iff x \notin L_k(A)]$.
ii) $\forall S \in \mathrm{REC}[|S| = \infty \wedge (S \subseteq A \vee S \subseteq \overline{A})$
    $\implies \exists y (y \in L_k(S) \iff M^{K_{k-1}(S)}_i(y) = 0)]$.

The first requirement states that $K_k(A) \in \Pi_k^P(A)$, which implies that $\Sigma_k^P(A) = \Pi_k^P(A)$ since $K_k(A)$ is P-T-complete for $\Sigma_k^P(A)$. This yields the membership that $A \in \text{CRO}\Sigma_k^P$. The second requirement states that, for any infinite recursive subset S of either A or $\overline{A}$, $L_k(S)$ is not in $\Delta_k^P(S)$. From this, it follows that $\Delta_k^P(S) \neq \Sigma_k^P(S)$, and therefore $S \notin \text{CRO}\Delta_k^P$. In other words, A is $\text{CRO}\Delta_k^P$-bi-immune. Combining the above two requirements, we obtain the desired consequence that A is $\text{CRO}\Delta_k^P$-bi-immune and is indeed in $\text{CRO}\Sigma_k^P$.

The proof is done by induction on $k \geq 1$. Since the basis case $k = 1$ is similar to the induction step $k \geq 2$, we prove only the induction step. For each pair $(i, t) \in \mathbb{N}^2$, let $\rho_{i,t}$ be the *restriction* defined as $\rho_{i,t}(x) = \check{M}_i(x)$ if $\check{M}_i(x)$ halts within t steps; otherwise, $\rho_{i,t}(x) = *$. Moreover, let $S_{\rho_{i,t}}$ be the set defined as $S_{\rho_{i,t}} = \{x \in \Sigma^* \mid \rho_{i,t}(x) = 1\}$.

Now, we formally describe the construction procedure of A by stages. We define the set $A(\ell)$ at each stage $\ell \in \mathbb{N}$ and finally set $A = \bigcup_{\ell \in \mathbb{N}} A(\ell)$. We later show that A satisfies the aforementioned two requirements.

Stage $t = 0$: Let $A(0) = \varnothing$ and $R = \mathbb{N}$. Moreover, let $b(0) = 0$.

Stage $t \geq 1$: Let $m = b(t-1)+1$. Assume that $A(t-1) \subseteq \Sigma^{<m}$. We determine the membership of all the strings of length m to A. Choose the minimal index $i \in R$ such that $i \leq t$, $\forall x \in \Sigma^{\leq m}[\rho_{i,t}(x) \neq *]$, and $S_{\rho_{i,t}}^{=m} \neq \varnothing$. If there is no such i, then skip the following and go to the next stage by setting $b(t) = b(t-1)$ and $A(t) = A(t-1)$. (This does not make the procedure an infinite loop because eventually we will find such an i.) For simplicity, write ρ for $\rho_{i,t}$ and let $S = L(\check{M}_i)$. Note that S is in general an r.e. set whereas S_ρ is recursive. Hereafter, we assume that such i exists. We need to consider two cases.

Case $m \not\equiv 0 \,(\text{mod } k + 1)$: We want to satisfy the requirement (ii). If $S_\rho^{<m} \not\subseteq A(t-1)$ and $S_\rho^{<m} \not\subseteq \overline{A(t-1)}$, then we skip the rest and go to the next stage by letting $R = R - \{i\}$, $A(t) = A(t-1)$, and $b(t) = m$, because this implies that $S \not\subseteq A$ and $S \not\subseteq \overline{A}$. Next, assume otherwise that either $S_\rho^{<m} \subseteq A(t-1)$ or $S_\rho^{<m} \subseteq \overline{A(t-1)}$. This indicates that there is a chance of either $S \subseteq A$ or $S \subseteq \overline{A}$ after the construction. Let $n = \max\{j \in \mathbb{N} \mid p_i(j) \leq m \wedge (k+1)j \leq m\}$. Note that, by the choice of n, $M_i^{K_{k-1}(\rho)}(y)\downarrow$ ("well-defined") for all $y \in \Sigma^{\leq n}$. Check whether there exists a string $y \in \Sigma^{\leq n}$ such that $y \in L_k(\rho)$ iff $M_i^{K_{k-1}(\rho)}(y) = 0$. If so, then let $A(t) = A(t-1)$. Assume otherwise. There are three cases to consider. If $S_\rho^{=m} = \Sigma^m$, then let $A(t)$ be $A(t-1) \cup 0\Sigma^{m-1}$. Thus, neither $S \subseteq A$ nor $S \subseteq \overline{A}$. Assume that this is not the case. If $S_\rho^{<m} \subseteq \overline{A(t-1)}$, then let $A(t)$ be $A(t-1) \cup S_\rho^{=m}$. This implies $S \not\subseteq \overline{A}$. If $S_\rho^{<m} \subseteq A(t-1)$, then let $A(t)$ be $A(t-1) \cup \overline{S_\rho}^{=m}$, implying $S \not\subseteq A$. Finally, let $R = R - \{i\}$ and $b(t) = m$. Go to the next stage.

Case $m \equiv 0 \,(\text{mod } k + 1)$: We target the two requirements simultaneously. Let $m = (k+1)n$. Consider all strings x in Σ^n. By employing the standard diagonalization argument, we choose a minimal set $B \subseteq \Sigma^m$ such that (a) for all $x \in \Sigma^n$, $x \notin L_k(A(t-1) \cup B)$ iff $x \in K_k(A(t-1) \cup B)$ and (b) $S_\rho^{=m} \not\subseteq A(t-1) \cup B$ and $S_\rho^{=m} \not\subseteq \Sigma^m - (A(t-1) \cup B)$. Such a set B exists because the membership question "$x \in K_k(\cdot)$?" depends only on the query strings in

$\Sigma^{\leq n}$ whereas the question "$x \in L_k(\cdot)$?" does on the strings in Σ^m. Using this B, we define $A(t) = A(t-1) \cup B$. This guarantees that $L_k(A)^{=n} = \overline{K_k(A)}^{=n}$, $S^{=m} \not\subseteq A^{=m}$, and $S^{=m} \not\subseteq \overline{A}^{=m}$. Finally, let $b(\ell) = m$ and update R as $R - \{i\}$. Go to the next stage.

By the construction of A, it is not difficult to verify that A satisfies the requirements (i)-(ii). □

How much can we strengthen Theorem 2? We exhibit a simple upper bound. For any complexity class $\mathcal{C}$, we call a set L *effectively $\mathcal{C}$-random* if, for every set $A \in \mathcal{C}$ and every polynomial p, $\left| \frac{|L^{=n} \triangle A^{=n}|}{|\Sigma^n|} - \frac{1}{2} \right| \leq \frac{1}{p(n)}$ for all but finitely many $n \in \mathbb{N}$. (This is a slightly modified version of Wilber's [18] notion of randomness.) Note that such an effectively $\mathcal{C}$-random set is $\mathcal{C}$-bi-immune. It is relatively easy to prove that, for each $k \geq 1$, no effectively $CRO\Delta_k^p$-random set exists in $CRO\Sigma_k^p$.

4 The BP-Operator and the CROBPP Hierarchy

We further introduce a bounded-error analogue of the CROP hierarchy. We first formulate this new hierarchy from Schöning's bounded-error probabilistic polynomial (BPP) hierarchy $\{BP\Delta_k^p, BP\Sigma_k^p, BP\Pi_k^p \mid k \in \mathbb{N}\}$ [15], which is induced from the P hierarchy by an application of the so-called *BP-operator*[1] as follows: $BP\Delta_k^p = BP \cdot \Delta_k^p$, $BP\Sigma_k^p = BP \cdot \Sigma_k^p$, and $BP\Pi_k^p = BP \cdot \Pi_k^p$. In particular, $BPP = BP\Delta_1^p$ and $AM = BP\Sigma_1^p$. We also define $BP\Theta_k^p$ as $BP \cdot \Theta_k^p$. The BPP hierarchy seems quite different from the P hierarchy. Most significantly, it is unknown whether the upward collapse property holds for the BPP hierarchy; for instance, the following implication is not yet proven: $BP\Sigma_k^p = BP\Pi_k^p \implies BP\Sigma_k^p = BP\Sigma_{k+1}^p$. Moreover, $BP\Sigma_k^p = BP\Pi_k^p$ might not imply $\Sigma_k^p = \Pi_k^p$, where $k \geq 1$, because Ko [10] earlier constructed an oracle A for which $BP\Sigma_k^p(A) = \Sigma_k^P(A) \neq \Pi_k^P(A)$, which immediately implies both $BP\Sigma_k^p(A) = BP\Pi_k^p(A)$ and $\Sigma_k^P(A) \neq \Pi_k^P(A)$. Now, we define the CROBPP hierarchy.

Definition 4. *Let $k \geq 1$.*

1. $CROBP\Sigma_k^p = \{A \in REC \mid BP\Sigma_k^p(A) = BP\Sigma_{k+1}^p(A)\}$.
2. $CROBP\Delta_k^p = \{A \in REC \mid BP\Delta_k^p(A) = BP\Delta_{k+1}^p(A)\}$.
3. $CROBP\Theta_{k+1}^p = \{A \in REC \mid BP\Theta_{k+1}^p(A) = BP\Theta_{k+2}^p(A)\}$.
4. $CROBPH = \bigcup_{i \geq 1} CROBP\Delta_i^p$.

Since the upward collapse property fails for relativized BPP hierarchies, the CROBPP hierarchy behaves differently from the CROP hierarchy. Inspired by the alternative characterization of the CROP hierarchy, we introduce the generic class $wCROBP \cdot \mathcal{C}$ for any relativizable complexity class $\mathcal{C}$ of languages.

---

[1] Let $\mathcal{C}$ be any complexity class. A set L is in $BP \cdot \mathcal{C}$ iff there exist a polynomial p and a set $A \in \mathcal{C}$ such that (i) for every $x \in L$, $\text{Prob}_{y \in \Sigma^{p(|x|)}}[(x^c y) \in A] \geq 3/4$ and (ii) for every $x \notin L$, $\text{Prob}_{y \in \Sigma^{p(|x|)}}[(x^c y) \notin A] \geq 3/4$.

Definition 5. *Let $\mathcal{C}$ be any relativizable complexity class of languages. Assume that there exists a number $k \in \mathbb{N}$ such that $\Sigma_k^P(A) \subseteq \mathcal{C}(A) \subseteq \Sigma_{k+1}^P(A)$ for any oracle A. Define $w\mathrm{CROBP} \cdot \mathcal{C} = \{A \in \mathrm{REC} \mid \mathrm{BP}\Sigma_{k+1}^P(A) = \text{co-BP} \cdot \mathcal{C}(A)\}$.*

The prefix "w" in Definition 5 refers to "weaker collapsing criteria." In particular, we obtain the following complexity classes: $w\mathrm{CROBP}\Sigma_k^P$, $w\mathrm{CROBP}\Delta_k^P$, and $w\mathrm{CROBP}\Theta_{k+1}^P$ for each index $k \geq 1$. The following basic inclusions hold.

Lemma 8. *Let $k \geq 1$.*

1. *$\mathrm{CRO} \cdot \mathcal{C} \subseteq \mathrm{CROBP} \cdot \mathcal{C} \subseteq w\mathrm{CROBP} \cdot \mathcal{C}$ for any class $\mathcal{C} \in \{\Theta_k^P, \Delta_k^P, \Sigma_k^P\}$.*
2. *$w\mathrm{CROBP}\Theta_k^P \subseteq \mathrm{CROBP}\Delta_k^P$.*
3. *$w\mathrm{CROBP}\Sigma_k^P \subseteq \mathrm{CROBP}\Theta_{k+1}^P$.*
4. *$w\mathrm{CROBP}\Delta_k^P \subseteq w\mathrm{CROBP}\Sigma_k^P \subseteq \mathrm{CRO}\Sigma_{k+1}^P$.*

We show the following separations, which greatly strengthen Theorem 1

Theorem 3. *Let $k \geq 1$. (1) $\mathrm{CRO}\Sigma_k^P \not\subseteq w\mathrm{CROBP}\Delta_k^P$; (2) $\mathrm{CRO}\Delta_k^P \not\subseteq w\mathrm{CROBP}\Theta_k^P$; and (3) $\mathrm{CRO}\Theta_{k+1}^P \not\subseteq w\mathrm{CROBP}\Sigma_k^P$.*

As an immediate consequence of Theorem 3, we obtain the following separation of the (weak) CROBPP hierarchy.

Corollary 2. *Let $k \geq 1$.*

1. *$\mathrm{CROBP}\Theta_k^P \neq \mathrm{CROBP}\Delta_k^P \neq \mathrm{CROBP}\Sigma_k^P \neq \mathrm{CROBP}\Theta_{k+1}^P$.*
2. *$w\mathrm{CROBP}\Theta_k^P \neq w\mathrm{CROBP}\Delta_k^P \neq w\mathrm{CROBP}\Sigma_k^P \neq w\mathrm{CROBP}\Theta_{k+1}^P$.*

Now, we give the proof of Theorem 3. In this proof, we use the terminology given in Section 3.

Proof of Theorem 3. We prove only (1). The other claims are similarly proven. Now, our goal is to prove that $\mathrm{CRO}\Sigma_k^P \not\subseteq w\mathrm{CROBP}\Delta_k^P$ for each $k \geq 1$. It suffices to construct, for each $k \geq 1$, a recursive set S such that $\mathrm{BP}\Delta_k^P(S) \not\supseteq \Sigma_k^P(S) = \Pi_k^P(S)$. Before giving the proof, we observe that $\Sigma_k^P \subseteq \mathrm{BP}\Delta_k^P$ implies that $\Sigma_{k+1}^P = \mathrm{BP}\Sigma_k^P$, which further implies $\Sigma_{k+1}^P = \Pi_{k+1}^P$. Under the common belief that all levels of the P hierarchy are different, we obtain $\Sigma_k^P \not\subseteq \mathrm{BP}\Delta_k^P$.

Note that the initial case $k = 1$ is already shown by Ko [10], who constructed a recursive oracle A such that $\mathrm{P}(A) = \mathrm{BPP}(A) \neq \mathrm{NP}(A) = \text{co-NP}(A)$. This immediately implies that $\mathrm{BPP}(A) \neq \mathrm{AM}(A)$ and $\mathrm{NP}(A) = \text{co-NP}(A)$. Hence, we obtain $\mathrm{CRONP} \not\subseteq w\mathrm{CROBPP}$. Therefore, the remaining of our proof is dedicated to the general case $k \geq 2$. This is shown by induction on $k \geq 2$. First, we prove a key lemma on a circuit lower bound necessary in our construction process. Let $[m, n]_{\mathbb{Z}} = \{m, m+1, \ldots, n\}$ for any two integers m, n with $m \leq n$.

Lemma 9. *Let $k \geq 2$. Let p, q be polynomials and let $r < 2^{n/2}$. Let $\{D_y\}_{y \in \Sigma^{p(n)}}$ be any collection of $\Delta_k(r)$-circuits. For each string x with $n < |x| \leq q(n)$, let C_x be any $\Sigma_k(|x|)$-circuit computing an $f_k^{2^{|x|}}$ function (with disjoint variable sets). Then, there exists a restriction ρ such that (i) for all strings x with $n < |x| \leq q(n)$, $C_x \lceil_\rho = *$ and (ii) $|\{y \in \Sigma^{p(n)} \mid D_y \lceil_\rho \neq *\}| \geq 2^{p(n)-n}$.*

Proof.   We begin with the basis case $k = 2$. Toward a contradiction, we assume that, for every restriction ρ, if $C_x\lceil_\rho = *$ for any x's with $n < |x| \leq q(n)$ then $|\{y \in \Sigma^{p(n)} \mid D_y\lceil_\rho \neq *\}| < 2^{p(n)-n}$. Take 2^n restrictions $\{\rho_j\}_{j=1}^{2^n}$, each of which is defined as follows: let $\rho_j(v) = *$ if variable v appears in the jth subcircuit of the top OR gate of C_x for a certain x; let $\rho_j(v) = 0$ for the other v's. Note that $C_x\lceil_{\rho_j} = *$ for any $j \in [1, 2^n]_{\mathbb{Z}}$ since C_x is a $\Sigma_2(|x|)$-circuit with all literals being distinct positive variables. The number of y's in $\Sigma^{p(n)}$ satisfying $\exists j \in [1, 2^n]_{\mathbb{Z}}(D_y\lceil_{\rho_j} \neq *)$ is less than $\sum_{j=1}^{2^n} 2^{p(n)-n} = 2^{p(n)}$. Hence, there is at least one index $y \in \Sigma^{p(n)}$ such that $D_y\lceil_{\rho_j} = *$ for all $j \in [1, 2^n]_{\mathbb{Z}}$. Fix this y.

Consider the minimal number ℓ of variables to evaluate in order to force D_y to output a Boolean value. Since D_y is a $\Delta_k(r)$-circuit, it suffices to assign values to at most r^2 variables in each subcircuit of the top OR gate of D_y. Thus, $\ell \leq r^2 < 2^n$. Take a set Q of such ℓ variables. Since $|Q| < 2^n$, there is an index j such that $\rho_j(w) \neq *$ for all $w \in Q$. This implies $D_y\lceil_{\rho_j} \neq *$, a contradiction. Therefore, the lemma holds for $k = 2$.

Next, we show the induction step. Let $k > 2$. We apply Håstad's switching lemma [7] to reduce this case to the case $k - 1$. By the switching lemma, the probability that every subcircuit $D_y\lceil_\rho$ contains a $\Delta_{k-1}(r)$-circuit is $\geq 2/3$. Moreover, the probability that every $C_u\lceil_\rho$ has a subcircuit computing an $f_{k-1}^{2^{|u|}}$ is $\geq 2/3$. Hence, by the induction hypothesis, there exists a restriction ρ such that (i) for all x's, $C_x\lceil_\rho = *$ and (ii) $|\{y \mid D_y\lceil_\rho \neq *\}| \geq 2^{p(n)-n}$. $\square$

Now, we describe how to construct the desired oracle A. Let $\{M_i\}_{i \in \mathbb{N}}$ denote any fixed effective enumeration of all oracle Δ_k^P-machines and, for each $i \in \mathbb{N}$, let p_i and q_i be two increasing polynomials such that, for any x and any $y \in \Sigma^{p_i(|x|)}$, $q_i(|x|)$ bounds the running time of $M_i^S(x, y)$ with any oracle S. We say that M is p-good at x with oracle S if either $\text{Prob}_{y \in \Sigma^{p(|x|)}}[M^S(x, y) = 1] \geq 1 - 2^{-2|x|}$ or $\text{Prob}_{y \in \Sigma^{p(|x|)}}[M^S(x, y) = 0] \geq 1 - 2^{-2|x|}$. Our requirements for the set A are described as follows.

i) $\forall x[x \in K_k(S) \iff x \notin L_k^{(1)}(S)]$.
ii) For any $i \in \mathbb{N}$, if M_i is p_i-good at every string with oracle S, then
$$\exists m \geq 1(0^m \in L_k^{(0)}(S) \iff \text{Prob}_{y \in \Sigma^{p_i(m)}}[M_i^S(0^m, y) = 0] \geq 2^{-m-2}).$$

The first requirement implies that $\Sigma_k^P(S) = \Pi_k^P(S)$ and the second requirement implies that $\Sigma_k^P(S) \not\subseteq \text{BP}\Delta_k^P(S)$.

Without loss of generality, we assume that $m \leq p_n(m) \leq q_n(m)$ for all $n, m \in \mathbb{N}$. We define the marker $t(n)$ as follows: let $t(0) = 0$ and let $t(n) = \min\{r \in \mathbb{N} \mid \exists m[t(n-1) < m \wedge q_n(m) < r < 2^m \wedge m \equiv 0 \,(\text{mod}\, k+1)]\}$ for each integer $n \geq 1$. Let $A(0) = \overline{A}(0) = \emptyset$. Assume that we are in stage $n \geq 1$. This stage consists of four steps. At this stage, we want to determine the membership of all strings in $\Sigma^{\leq t(n)} - \Sigma^{\leq t(n-1)}$ to the set A. Let $m = \min\{l \in \mathbb{N} \mid t(n-1) < l \leq t(n), l \equiv 0 \,(\text{mod}\, k+1)\}$. Initially, we set $A(n) = A(n-1)$ and $\overline{A}(n) = \overline{A}(n-1)$ and then update them during the following steps.

Step 1: Let S_n be the set of all strings x satisfying that $t(n - 1) < (k + 1)|x| + 1 < m$. Consider any string x in S_n. Let G_x be any $\Sigma_k(q(|x|))$-circuit computing $K_k(\cdot)$ on x for a certain polynomial q. In addition, let $C_x^{(0)}$ be the depth-k circuit computing $f_k^{2^{|x|}}$, which computes $L_k^{(0)}(\cdot)$ on x, and let $C_x^{(1)}$ be the circuit computing $L_k^{(1)}(\cdot)$ on x. For each $b \in \{0,1\}$, let $V_x^{(b)}$ denote the set of all variables in $C_x^{(b)}$. Note that $V_x^{(b)} \cap V_y^{(b)} = \emptyset$ if $x \neq y$. Note also that $C_x^{(b)}$'s variables are in $\Sigma^{<|x|}$. Define $V_{m,n} = \bigcup_{x \in S_n}(V_x^{(1)} \cup \Sigma^{<|x|})$. Now, we replace all variables v in G_x corresponding to $A(n) \cup \overline{A}(n)$ by 0 or 1 if v is in $A(n)$ or $\overline{A}(n)$, respectively. Choose a minimal restriction ρ such that $C_x^{(1)}\lceil_\rho \neq *$, $G_x\lceil_\rho \neq *$, and $C_x^{(1)}\lceil_\rho \neq G_x\lceil_\rho$ for all $x \in S_n$. Finally, set $A(n)$ to be $A(n) \cup \{v \in V_{m,n} \mid \rho(v) = 1\}$ and set $\overline{A}(n)$ to be $\overline{A}(n) \cup \{v \in V_{m,n} \mid \rho(v) = 0\}$. This satisfies the requirement (i) for each string $x \in S_n$.

Step 2: We satisfy the requirement (ii) in this step for 0^m. Let S_n' be the set of all strings x satisfying that $m \leq (k+1)|x| + 1 \leq t(n)$. Note that $|S_n'| \leq 2^{p_n(m)+1}$. For each $y \in \Sigma^{p_n(m)}$, let D_y be any $\Delta_k(q_n(m))$-circuit computing $M_i^{(\cdot)}(0^m, y)$. Lemma 9 gives a restriction ρ such that (i) for all $x \in S_n'$, $C_x^{(0)}\lceil_\rho = C_x^{(1)}\lceil_\rho = *$ and (ii) $|\{y \in \Sigma^{p_n(m)} \mid D_y\lceil_\rho \neq *\}| \geq 2^{p_n(m)-m}$. Choose the bit $b \in \{0,1\}$ such that $|\{y \in \Sigma^{p_n(m)} \mid D_y\lceil_\rho = b\}| \geq |\{y \in \Sigma^{p_n(m)} \mid D_y\lceil_\rho = \overline{b}\}|$. Hence, $|\{y \in \Sigma^{p_n(m)} \mid D_y\lceil_\rho = b\}| \geq 2^{p_n(m)-m-1}$. Define $V_{m,n}' = \bigcup_{x \in S_n'}(V_x^{(0)} \cup V_x^{(1)} \cup \Sigma^{\leq q_n(m)})$. There exists a restriction ρ' such that $|\{y \in \Sigma^{p_n(m)} \mid C_x^{(1)}\lceil_{\rho\rho'} = * \wedge C_x^{(0)}\lceil_{\rho\rho'} \neq * \wedge D_y\lceil_{\rho\rho'} \neq * \wedge D_y\lceil_{\rho\rho'} \neq C_x^{(0)}\lceil_{\rho\rho'}\}| \geq 2^{p_n(m)-m-2}$. Update $A(n)$ to be $A(n) \cup \{v \in V_{m,n}' \mid \rho\rho'(v) = 1\}$ and update $\overline{A}(n)$ to be $\overline{A}(n) \cup \{v \in V_{m,n}' \mid \rho\rho'(v) = 0\}$.

Step 3: We replace all variables v corresponding to $A(n) \cup \overline{A}(n)$ by 0 or 1 if $v \in A(n)$ or $\overline{A}(n)$, respectively. Step 2 left the room for finding a restriction ρ such that $C_x^{(1)}\lceil_\rho \neq *$, $G_x\lceil_\rho \neq *$, and $C_x^{(1)}\lceil_\rho \neq G_x\lceil_\rho$ for every $x \in S_n'$. Update the sets $A(n)$ and $\overline{A}(n)$ as in Step 1. The requirement (i) is thus met.

Step 4: To finish Stage n, we need to determine the values of all the strings in $\Sigma^{\leq t(n)}$ that are left unassigned. This is done simply by updating $\overline{A}(n)$ as $\overline{A}(n) \cup (\Sigma^{\leq t(n)} - \Sigma^{\leq t(n-1)} \cup A(n))$.

The above construction clearly ensures that the two requirements are met at every stage. Therefore, we obtain the desired oracle separation. $\square$

5   Further Discussion

We can further generalize our definition of collapsing recursive oracles to many other complexity classes. As an example, we define a "complementary" collapsing recursive oracle class $cCRO \cdot \mathcal{C}$ as $cCRO \cdot \mathcal{C} = \{A \in REC \mid \mathcal{C}(A) = co\text{-}\mathcal{C}(A)\}$. This definition enables us to define meaningful complexity classes, such as $cCROMA$ (induced from MA). We demonstrate the separations among CRONP, $cCROMA$, and $cCROAM$ in the following proposition.

Proposition 2. CRONP $\neq cCROMA \neq cCROAM$.

To our best knowledge, the above separations are new. The proof can be obtained by modifying the proof of Theorem 3.

Final Note. All the proofs omitted from this extended abstract will appear in its forthcoming complete version.

References

1. E. Allender and L. Hemachandra. Lower bounds for the low hierarchy. *J. ACM* **39** (1992) 234–251.
2. T. Baker, J. Gill, and R. Solovay. Relativizations of the P=?NP question. *SIAM J. Comput.* **4** (1975) 431–442.
3. J. L. Balcázar, R. V. Book, and U. Schöning. Sparse sets, lowness, and highness. *SIAM J. Comput.* **15** (1986) 739–747.
4. J. L. Balcázar, R. V. Book, and U. Schöning. The polynomial-time hierarchy and sparse oracles. *J. ACM* **33** (1986) 603–617.
5. D. Bruschi. Strong separations of the polynomial hierarchy with oracles: constructive separations by immune and simple sets. *Theoret. Comput. Sci.* **102** (1992) 215–252.
6. J. Cai, V. T. Chakaravarthy, L. A. Hemaspaandra, M. Ogihara. Some Karp-Lipton-type theorems based on S_2. Technical Report, 2001.
7. J. D. Håstad. *Computational Limitations for Small-Depth Circuits.* Ph.D. dissertation, Massachusetts Institute of Technology, 1987.
8. H. Heller. Relativized polynomial hierarchies extending two levels. *Math. Systems Theory* **17** (1984) 71–84.
9. K. Ko. Relativized polynomial time hierarchies having exactly k levels. *SIAM J. Comput.* 18 (1989) 392–408.
10. K. Ko. Separating and collapsing results on the relativized probabilistic polynomial time hierarchy. *J. ACM* **37** (1990) 415–438.
11. T. Long and A. Selman. Relativizing complexity classes with sparse oracles. *J. ACM* **33** (1986) 618–627.
12. T. J. Long and M. Sheu. A refinement of the low and high hierarchies. *Math. Systems Theory* **28** (1995) 299–327.
13. A. Meyer and L. Stockmeyer. The equivalence problem for regular expression with squaring requires exponential time. In *Proc. 13th IEEE Symposium on Switching and Automata Theory*, pp.125–129, 1972.
14. U. Schöning. Complete sets and closeness to complexity classes. *Math. Systems Theory* **19** (1986) 29–41.
15. U. Schöning. Probabilistic complexity classes and lowness. *J. Comput. System Sci.* **39** (1989) 84–100.
16. M. Sheu and T. J. Long. The extended low hierarchy is an infinite hierarchy. *SIAM J. Comput.* **23** (1994) 488–509.
17. L. Stockmeyer. On approximation algorithms for #P. *SIAM J. Comput.* **14** (1985) 849–861.
18. R. E. Wilber. Randomness and the density of hard problems. In *Proc. 24th Annual Symposium on Foundations of Computer Science*, pp.335–342, 1983.
19. T. Yamakami and T. Suzuki. Resource bounded immunity and simplify. To appear in *Theoret. Comput. Sci.*. An extended abstract appeared in *Proc. 3rd IFIP Intern. Conf. Theoretical Computer Science* (Exploring New Frontiers of Theoretical Informatics), Kluwer Academic Publishers, pp.81–95, 2004.
20. A. C. Yao. Separating the polynomial-time hierarchy by oracles. *Proc. 26th IEEE Symposium on Foundations of Computer Science*, pp.1–10, 1985.

Exact Algorithms for Graph Homomorphisms*

Fedor V. Fomin[1], Pinar Heggernes[1], and Dieter Kratsch[2]

[1] Department of Informatics, University of Bergen,
N-5020 Bergen, Norway
{Fedor.Fomin, Pinar.Heggernes}@ii.uib.no
[2] Laboratoire d'Informatique Théorique et Appliquée,
Université Paul Verlaine, 57045 Metz Cedex 01, France
kratsch@sciences.univ-metz.fr

Abstract. Graph homomorphism, also called H-coloring, is a natural generalization of graph coloring: There is a homomorphism from a graph G to a complete graph on k vertices if and only if G is k-colorable. During the recent years the topic of exact (exponential-time) algorithms for NP-hard problems in general, and for graph coloring in particular, has led to extensive research. Consequently, it is natural to ask how the techniques developed for exact graph coloring algorithms can be extended to graph homomorphisms. By the celebrated result of Hell and Nešetřil, for each fixed simple graph H, deciding whether a given simple graph G has a homomorphism to H is polynomial-time solvable if H is a bipartite graph, and NP-complete otherwise. The case where H is a cycle of length 5 is the first NP-hard case different from graph coloring. We show that, for a given graph G on n vertices and an odd integer $k \geq 5$, whether G is homomorphic to a cycle of length k can be decided in time $\min\{\binom{n}{n/k}^c \cdot 2^{n/2}\} \cdot n^{\mathcal{O}(1)}$. We extend the results obtained for cycles, which are graphs of treewidth two, to graphs of bounded treewidth as follows: If H is of treewidth at most t, then whether G is homomorphic to H can be decided in time $(2t+1)^n \cdot n^{\mathcal{O}(1)}$.

1 Introduction

Given two undirected graphs G and H, a *homomorphism* from G to H is a mapping $\varphi : V(G) \longrightarrow V(H)$ that satisfies the following: $\{x, y\} \in E(G) \implies \{\varphi(x), \varphi(y)\} \in E(H)$ for every $x, y \in V(G)$. When there is a homomorphism from G to H we say that G is *homomorphic* to H. The problem of deciding whether graph G is homomorphic to graph H is called HOM(G, H). This problem can be seen as labeling, or coloring, the vertices of G by the vertices of H, and this is why it is often also called the H-coloring problem. Note that for the special case when H is a complete graph on k vertices, G is homomorphic to H if and only if the chromatic number of G is at most k. We refer to the recent book [13] for a thorough introduction to the topic.

---

* This work is supported by the AURORA mobility programme for research collaboration between France and Norway.

M. Liśkiewicz and R. Reischuk (Eds.): FCT 2005, LNCS 3623, pp. 161–171, 2005.

For graph classes $\mathcal{G}$ and $\mathcal{H}$ we denote by HOM($\mathcal{G}, \mathcal{H}$) the restriction of the graph homomorphism problem to input graphs $G \in \mathcal{G}$ and $H \in \mathcal{H}$. If $\mathcal{G}$ or $\mathcal{H}$ is the class of all graphs then they are denoted by the placeholder '_'. The computational complexity of graph homomorphism was studied from different 'sides'.

'Left Side' of Homomorphisms. For any fixed graph G, HOM(G, _) is trivially solvable in polynomial time. Several authors independently showed that HOM($\mathcal{G}$, _) is solvable in polynomial time if all graphs in $\mathcal{G}$ have bounded treewidth. In this case polynomial-time algorithms can be obtained even for counting homomorphisms [8]. Grohe, concluding from the results of Dalmau et al. [7], showed that HOM($\mathcal{G}$, _) is solvable in polynomial time if and only if the cores of all graphs in $\mathcal{G}$ have bounded treewidth (under some parameterized complexity theoretic assumptions) [11].

'Right Side' of Homomorphisms. Hell and Nešetřil showed that for any fixed simple graph H, the problem HOM(_, H) is solvable in polynomial time if H is bipartite, and NP-complete if H is not bipartite [12]. This resolves the complexity classification of the whole right side of homomorphisms, and provides a P vs. NP dichotomy. Consequently the study of the right side of homomorphisms for undirected graphs almost stopped, as research has been mainly concentrated on finding polynomial-time algorithms for special graph classes from the 'left' side.

However for the special case of graph homomorphism, graph coloring, extensive work has been done recently resulting in faster and faster exponential-time algorithms. The recent best bounds are an $\mathcal{O}(1.3289^n)$-time algorithm for 3-coloring [4], an $\mathcal{O}(1.7504^n)$-time algorithm for 4-coloring [5], an $\mathcal{O}(2.1020^n)$-time algorithm for 5-coloring [6], and an $\mathcal{O}(2.1809^n)$-time algorithm for 6-coloring [6]. For $k \geq 7$, the k-coloring problem can be solved in time $\mathcal{O}(2.4023^n)$ [5].

Despite considerable progress on exponential-time algorithms for graph coloring problems, not much is known on exponential-time algorithms for the graph homomorphism problem. By the result of Hell and Nešetřil, HOM(_, H) is polynomial-time solvable when H is bipartite. Another 'easy' case is when $\chi(H) = \omega(H)$, i.e., the chromatic number of H is equal to its maximum clique size. It is not hard to show that in this case the HOM(_, H) problem is equivalent to the k-coloring problem with $k = \chi(H)$. Consequently the HOM(_, H) problem is equivalent to the $\chi(H)$-coloring problem for all perfect graphs H.

All this motivates us to study exact exponential-time algorithms for HOM(_, H) with graphs H satisfying $\chi(H) > \omega(H)$. Thus chordless cycles of odd length are the first natural candidates to study exponential-time algorithms for graph homomorphisms. For the cycle C_3 on 3 vertices HOM(_, C_3) is equivalent to 3-coloring, but already for the cycle C_5 on 5 vertices no better deterministic algorithm than the brute-force $\mathcal{O}^*(5^n)$ time algorithm has been known. (Throughout this paper, in addition to the standard big-Oh notation $\mathcal{O}$, we sometimes use a modified big-Oh notation $\mathcal{O}^*$ that suppresses all polynomially bounded factors. For functions f and g we write $f(n) = \mathcal{O}^*(g(n))$ if $f(n) = g(n) \cdot n^{\mathcal{O}(1)}$.)

Our Results. In this paper we initiate the study of exponential time complexity of graph homomorphism problems beyond graph coloring. We show that for an input graph G on n vertices and an odd integer $k \geq 5$, $\mathrm{HOM}(G, C_k)$ is solvable in $\mathcal{O}^*(\min\{\binom{n}{n \triangleleft k}, 2^{n \triangleleft 2}\})$ time, where C_k is the cycle on k vertices. In particular, the running time of our algorithm is $\mathcal{O}(1.64939^n)$ when $k = 5$, $\mathcal{O}(1.50700^n)$ when $k = 7$, $\mathcal{O}(1.41742^n)$ when $k = 9$, and $\mathcal{O}(\alpha^n)$ with $\alpha < \sqrt{2}$ for all $k \geq 11$. It is interesting to note that, for $k \geq 13$, our algorithm for homomorphism to C_k is faster than the fastest known 3-coloring algorithm. Hence the natural conjecture that $\mathrm{HOM}(\_, C_k)$ is at least as difficult as 3-coloring for every odd $k \geq 5$ might be mistaken. Our algorithms use 2-SAT expressions to search for suitable extensions of an initial partial homomorphism: a maximal independent set of G to be mapped to a carefully chosen subset of vertices of H. To enumerate all possible preliminary choices we use known algorithms to enumerate all maximal independent sets.

Treewidth and tree decompositions are of great importance in structural graph theory and graph algorithms. Many NP-hard problems become polynomial-time or even linear-time solvable when the input is restricted to graphs of bounded treewidth. We refer to [3] for a survey on this parameter. It seems that the treewidth can be a useful tool to design exponential-time algorithms as well. We use dynamic programming techniques similar to bounded treewidth techniques to solve $\mathrm{HOM}(G, H)$ in time $\mathcal{O}^*((2 \cdot \mathbf{tw}(H) + 1)^{|V(G)|})$, assuming that an optimal tree decomposition of H is known in advance.

2 Preliminaries

We consider undirected and simple graphs, where $V(G)$ denotes the set of vertices and $E(G)$ denotes the set of edges of a graph G. For a given subset S of $V(G)$, $G[S]$ denotes the subgraph of G induced by S, and $G - S$ denotes the graph $G[V(G) \setminus S]$. S is an *independent set* if $G[S]$ is a graph with no edges, and S is a *clique* if $G[S]$ is a complete graph. The set of neighbors of a vertex v in G is denoted by $N_G(v)$, and the set of neighbors of a vertex set S is $N_G(S) = \bigcup_{v \in S} N_G(v) \setminus S$.

K_k denotes the complete graph on k vertices and C_k denotes the chordless cycle on k vertices. A *coloring* of a graph G is a function f assigning a color to each vertex of G such that adjacent vertices have different colors. A k-coloring of a graph uses at most k colors, and the smallest number of colors in a coloring of G is denoted by $\chi(G)$. The maximum size of a clique in a graph G is denoted by $\omega(G)$

Given a mapping $\varphi : V(G) \longrightarrow V(H)$ and a set $S \subseteq V(H)$, we denote by $\varphi^{-1}(S)$ the set of all those vertices of G that are mapped to a vertex of S.

The notion of treewidth was introduced by Robertson and Seymour. A *tree decomposition* of a graph $G = (V, E)$ is a pair $(\{X_i : i \in I\}, T)$, where $\{X_i : i \in I\}$ is a collection of subsets of G (these subsets are called *bags*) and $T = (I, F)$ is a tree such that the following three conditions are satisfied:

1. $\bigcup_{i\in I} X_i = V(G)$.
2. For all $\{v,w\} \in E(G)$, there is an $i \in I$ such that $v,w \in X_i$.
3. For all $i,j,k \in I$, if j is on a path from i to k in T then $X_i \cap X_k \subseteq X_j$.

The *width* of a tree decomposition $(\{X_i : i \in I\},\ T)$ is $\max_{i\in I} |X_i| - 1$. The *treewidth* of a graph G, denoted by $\mathbf{tw}(G)$, is the minimum width over all its tree decompositions. A tree decomposition of G of width $\mathbf{tw}(G)$ is called an *optimal* tree decomposition of G.

3   Homomorphisms to Odd Cycles

Recall that, for an input graph G, $\mathrm{HOM}(G,C_k)$ is solvable in polynomial time if k is even, and NP-complete if k is odd. We study the case when $k \geq 5$ is an odd integer. Throughout the remainder of this section we assume the input graph G to be non bipartite, since every bipartite graph is homomorphic to K_2, and thus also homomorphic to C_k for all $k \geq 3$.

For a given graph G and a vertex subset $S \subseteq V(G)$, we define the levels of breadth first search starting at S as follows:

- $L_0(S) = S$;
- $L_i(S) = N_G(L_{i-1}(S)) \setminus \bigcup_{j<i} L_j(S)$, for $i > 0$.

Lemma 1. *Let $k \geq 3$ be an odd integer. A non bipartite graph $G = (V,E)$ is homomorphic to C_k if and only if there is a set $S \subset V$ such that*

- $|S| \leq |V(G)|/k$,
- *the levels $L_0(S), L_1(S), L_2(S), \ldots, L_{\lfloor \frac{k}{2}\rfloor -1}(S)$ are independent sets in G,*
- *the graph $G - S$ is bipartite, and*
- *there is a coloring of vertices of $L_1(S), L_2(S), \ldots, L_{\lfloor \frac{k}{2}\rfloor}(S)$ in Red and Blue such that every two adjacent vertices from different levels have the same color, and every two adjacent vertices from $L_{\lfloor \frac{k}{2}\rfloor}(S)$ have different colors.*

Proof. Let us choose a vertex $v \in V(C_k)$ and let $R = (v, r_1, r_2, \ldots, r_{\lfloor k/2\rfloor})$ and $B = (v, b_1, b_2, \ldots, b_{\lfloor k/2\rfloor})$ be the two edge disjoint paths in C_k of length $\lfloor k/2 \rfloor$ starting at v.

Let G be homomorphic to C_k. Since G is not bipartite, every homomorphism from G to C_k is surjective. Hence there is a homomorphism τ from G to C_k such that $|\tau^{-1}(v)| \leq |V(G)|/k$. We define $S = \tau^{-1}(v)$. We then choose a homomorphism $\varphi \colon G \longrightarrow C_k$ that minimizes

$$\sum_{1\leq i\leq \lfloor k/2\rfloor} \frac{|\varphi^{-1}(r_i)| + |\varphi^{-1}(b_i)|}{i} \tag{1}$$

subject to $\varphi^{-1}(v) = S$.

By (1), every vertex of $\varphi^{-1}(r_i)$, $i \in \{1, 2, \ldots, \lfloor k/2\rfloor - 1\}$, is adjacent in G to a vertex of $\varphi^{-1}(r_{i-1})$. In fact, suppose on the contrary that there is a vertex

$x \in \varphi^{-1}(r_i)$ that is not adjacent in G to any vertex of $\varphi^{-1}(r_{i-1})$. Then there is a homomorphism ϕ from G to C_k such that $\phi(y) = \varphi(y)$ for all $y \neq x$, and $\phi(x) = r_{i+2}$ if $i \leq \lfloor k/2 \rfloor - 2$ and $\phi(x) = b_{\lfloor k \lhd 2 \rfloor}$ if $i = \lfloor k/2 \rfloor - 1$. But the existence of such a homomorphism contradicts (1). By similar arguments, every vertex of $\varphi^{-1}(b_i)$ $i \in \{1, 2, \ldots, \lfloor k/2 \rfloor - 1\}$ is adjacent to a vertex of $\varphi^{-1}(b_{i-1})$.

Thus for every $i \in \{1, 2, \ldots, \lfloor k/2 \rfloor - 1\}$, the vertices of $\varphi^{-1}(r_i) \cup \varphi^{-1}(b_i)$ form the level $L_i(S)$ of breadth first search starting at S in G. Furthermore, each of these sets is an independent set. The graph $G - S$ is bipartite because it is homomorphic to a path. For $i \in \{1, 2, \ldots, \lfloor k/2 \rfloor\}$, we color the vertices of $\varphi^{-1}(r_i)$ in Red and the vertices of $\varphi^{-1}(b_i)$ in Blue. Such a coloring satisfies the conditions of the lemma.

Now suppose that there is a vertex set $S \subseteq V(G)$ and a breadth first search starting at S satisfying the conditions of the lemma. We construct a homomorphism from G to C_k by mapping S to v. For $i \in \{1, 2, \ldots, \lfloor k/2 \rfloor - 1\}$, all Red vertices from level $L_i(S)$ are mapped to r_i and all Blue vertices from level $L_i(S)$ are mapped to b_i. For $i \geq \lfloor k/2 \rfloor$, Red vertices from level $L_i(S)$ are mapped to $r_{\lfloor k \lhd 2 \rfloor}$ and Blue vertices from level $L_i(S)$ are mapped to $b_{\lfloor k \lhd 2 \rfloor}$. $\square$

We need the following algorithmic version of the result from [14] which is due to Byskov [5].

Proposition 1 ([5]). *All maximal independent sets in a triangle-free graph on n vertices can be listed in time $\mathcal{O}^*(2^{n \lhd 2})$.*

Lemma 2. *For a given graph G on n vertices, $\mathrm{HOM}(G, C_k)$ can be solved in $\mathcal{O}^*\left(\binom{n}{n \lhd k}\right)$ time.*

Proof. By Lemma 1, a non bipartite graph G is homomorphic to C_k if and only if there is a set $S \subseteq V(G)$ satisfying the conditions of the lemma.

For a given (independent) set S, one can decide whether S satisfies the conditions of Lemma 1 as follows.

1. Find the levels $L_0(S), L_1(S), L_2(S), \ldots, L_m(S)$ of breadth first search at S. If all sets $L_0(S), L_1(S), L_2(S), \ldots, L_{\lfloor \frac{k}{2} \rfloor - 1}(S)$ are independent sets in G proceed to step 2.
2. Check if $G - S$ is bipartite. If it is bipartite proceed to step 3.
3. To decide whether there is a coloring of $L_1(S) \cup L_2(S) \cup \cdots \cup L_{\lfloor \frac{k}{2} \rfloor}(S)$ which meets the condition of Lemma 1, we reduce the problem to 2-SAT as follows. We encode every vertex x of $L_1(S) \cup L_2(S) \cup \cdots \cup L_{\lfloor \frac{k}{2} \rfloor}(S)$ by a boolean variable x such that $x = $ TRUE means that vertex x is colored Red, and variable $x = $ FALSE means that vertex x is colored Blue. Every edge $\{x, y\}$ between $L_i(S)$ and $L_{i+1}(S)$, for each $1 \leq i \leq \lfloor \frac{k}{2} \rfloor - 1$, is encoded by two clauses $(\bar{x} \vee y)$ and $(x \vee \bar{y})$. This forces vertex x and vertex y to receive the same color. Every edge $\{u, v\}$ with both endpoints in $L_{\lfloor \frac{k}{2} \rfloor}(S)$ is encoded by two clauses $(u \vee v)$ and $(\bar{u} \vee \bar{v})$. This forces vertex u and vertex v to receive opposite colors. The corresponding 2-SAT formula is satisfiable if and only if S satisfies the conditions of Lemma 1 and there is a homomorphism from G to C_k that can be derived from S.

Consequently, for each given set S, constructing a homomorphism from G to C_k using S or concluding that S cannot be used can be done by solving the corresponding 2-SAT formula, and thus requires polynomial time. There are less than $(n/k)\binom{n}{n \triangleleft k}$ different subsets S of size at most n/k. Hence the total running time is $\mathcal{O}^*(\binom{n}{n \triangleleft k})$. $\qquad\square$

The following algorithm improves upon the running time of the previous one for $k \in \{5, 7, 9\}$.

Lemma 3. *For a given graph G on n vertices, and an odd integer $k \geq 5$, $\mathrm{HOM}(G, C_k)$ can be solved in $\mathcal{O}^*(2^{n \triangleleft 2}) = \mathcal{O}(1.41422^n)$ time.*

Proof. We may assume that $G = (V, E)$ is not bipartite. Furthermore C_k is 3-colorable and triangle-free for every odd integer $k \geq 5$. Thus G is homomorphic to C_k implies that G is 3-colorable and triangle-free.

Let $v_1, v_2, v_3, \ldots, v_{k-1}, v_k$ be the vertices of C_k, with v_i adjacent to v_{i+1} (where indices are taken modulo k). We choose the following maximal independent set of C_k: $U = \{v_2, v_4, \ldots, v_{k-3}, v_{k-1}\}$. Suppose there is a homomorphism $\varphi: G \longrightarrow C_k$. Then $\varphi^{-1}(U)$ is an independent set of G. We claim that in this case there is even a homomorphism $\psi: G \longrightarrow C_k$ such that $\psi^{-1}(U)$ is a maximal independent set of G. Let $x \in V(G) \setminus \varphi^{-1}(U)$ such that $\{x\} \cup \varphi^{-1}(U)$ is an independent set of G, and let y be a neighbor of x in G. Then $\{x, y\} \in E(G)$ implies $\{\varphi(x), \varphi(y)\} = \{v_1, v_k\}$. Thus the following modification of φ is a homomorphism from G to C_k. Let $I' \subseteq V(G) \setminus \varphi^{-1}(U)$ such that $I = I' \cup \varphi^{-1}(U)$ is a maximal independent set of G. We define a homomorphism $\psi: G \longrightarrow C_k$ such that $\psi^{-1}(U) = I'$. For every vertex $v \in V(G) \setminus I'$, we let $\psi(v) = \varphi(v)$. For every vertex $v \in I'$, we let $\psi(v) = v_2$ if $\varphi(v) = v_k$, and we let $\psi(v) = v_{k-1}$ if $\varphi(v) = v_1$.

The goal of our algorithm is to test, for every maximal independent set I of G, whether there is a homomorphism $\psi: G \longrightarrow C_k$ such that $\psi^{-1}(U) = I$. By the above claim, ψ must exist if G is homomorphic to C_k. For every maximal independent set I in G the test is done as follows: First, if $G - I$ is not bipartite, then reject I since a non bipartite graph cannot be homomorphic to $H - U$ which consists of a K_2 and $(k - 1)/2$ isolated vertices. If $G - I$ is bipartite, let A be the set of isolated vertices of $G - I$, and let J be the set of vertices in connected components of $G - I$ that have at least two vertices. Clearly $V(G) = I \cup A \cup J$. Furthermore, since G is not bipartite, $J \neq \emptyset$.

Every vertex of J must be mapped to v_1 or v_k since each component of $G[J]$ has at least two vertices. Then every vertex of $N(J)$ must be mapped to v_2 or v_{k-1}. Clearly $N(J) \subseteq I$. Following Lemma 1, we map the vertices of G in a breadth first search manner starting from J, with levels $L_0(J)=J$, $L_1(J)=N(J)$, $L_2(J)$, ..., $L_{(k-1) \triangleleft 2}(J)$. At any stage we consider only the vertices that have to be mapped due to adjacencies in G to already mapped vertices. Therefore the vertices of $L_2(J)$ must be mapped to v_3 or v_{k-2}. Clearly $L_2(J) \subseteq A$. The vertices of $L_3(J)$ must be mapped to v_4 or v_{k-3}, ..., the vertices of $L_{(k-3) \triangleleft 2}(J)$ must be mapped to $v_{(k-1) \triangleleft 2}$ or $v_{(k+3) \triangleleft 2}$, and finally the vertices of $L_{(k-1) \triangleleft 2}(J)$ must be mapped to $v_{(k+1)\lhd 2}$. Now, there may be some remaining vertices of G that are not assigned to any vertex of H by the above procedure. If $(k+1)/2$ is even, then all remaining vertices should be mapped to $v_{(k-1)\lhd 2}$ or $v_{(k+3)\lhd 2}$ if they belong to A, and to $v_{(k+1)\lhd 2}$ if they belong to I. If $(k+1)/2$ is odd, then we should do the reverse: the remaining vertices should be mapped to $v_{(k+1)\lhd 2}$ if they belong to A and to $v_{(k-1)\lhd 2}$ or $v_{(k+3)\lhd 2}$ if they belong to I. Consequently, in the end, vertices of $A \cup J$ are mapped to $V(H) \setminus U$, and vertices of I are mapped to U.

To check whether our partial mapping can be transformed into a homomorphism we shall use a 2-SAT formula. For all vertices of G except those mapped to $v_{(k+1)\lhd 2}$ there is a choice between two vertices of the host graph C_k. Furthermore adjacent vertices of G must be mapped to adjacent vertices of C_k. For every vertex x of G with $\varphi(x) \in \{v_i, v_{k-i+1}\}$ we define a boolean variable x such that variable $x = \text{TRUE}$ means that vertex x is mapped to v_i with $i = 1, 2, \ldots, (k-1)/2$, and variable $x = \text{FALSE}$ means that vertex x is mapped to v_i with $i = (k+3)/2, (k+5)/2, \ldots, k$. For each edge $\{x, y\} \in E(G[J])$, either $\varphi(x) = v_1$ and $\varphi(y) = v_k$, or vice versa. Otherwise, for each edge $\{x, y\} \in E(G)$ with $\{x, y\} \not\subseteq J$, either $\varphi(x) = v_i$ and $\varphi(y) = v_j$ with $i, j \in \{1, 2, \ldots, (k-1)/2\}$, or $\varphi(x) = v_i$ and $\varphi(y) = v_j$ with $i, j \in \{(k+3)/2, \ldots, k\}$. Therefore, for each edge $\{x, y\} \in E(G[J])$, we insert the following two clauses in our 2-SAT formula: $(\bar{x} \vee y)$ and $(x \vee \bar{y})$. For all other edges $\{x, y\} \in E(G)$, i.e., at least one of x and y does not belong to J, we insert the following two clauses in our 2-SAT formula: $(\bar{x} \vee \bar{y})$ and $(x \vee y)$.

The corresponding 2-SAT formula is satisfiable if and only if there is a homomorphism φ from G to C_k such that $\varphi^{-1}(U) = I$. Consequently, for each maximal independent set I of G, constructing a homomorphism from G to C_k using I or concluding that I cannot be used can be done by solving the corresponding 2-SAT formula, and thus requires linear time (see [1]). By Proposition 1, the number of maximal independent sets in a triangle free graph on n vertices is at most $2^{n\lhd 2}$ and all maximal independent sets of a triangle free graph can be enumerated in time $\mathcal{O}^*(2^{n\lhd 2})$. Thus the overall running time of our algorithm is $\mathcal{O}^*(2^{n\lhd 2})$. $\qquad\square$

The algorithm of Lemma 2 has running time $\mathcal{O}(1.64939^n)$ when $k = 5$, $\mathcal{O}(1.50700^n)$ when $k = 7$, and $\mathcal{O}(1.41742^n)$ when $k = 9$, and its running time is $\mathcal{O}(\alpha^n)$ with $\alpha < \sqrt{2}$ for all $k \geq 11$. Hence the algorithm of Lemma 2 is faster for all $k \geq 11$, and the algorithm of Lemma 3 is faster for $k \in \{5, 7, 9\}$. Combining Lemmata 2 and 3 we obtain the following theorem.

Theorem 1. *For a given graph G on n vertices and an odd integer $k \geq 5$,* $\text{HOM}(G, C_k)$ *can be solved in* $\mathcal{O}^*(\min\{\binom{n}{n\lhd k}, 2^{n\lhd 2}\})$ *time.*

4   Homomorphisms to Graphs of Bounded Treewidth

A tree decomposition $(\{X_i : i \in I\}, T)$ of a graph G is said to be *nice* if a root of T can be chosen such that every node $i \in I$ of T has at most two children in the rooted tree T, and

1. if a node $i \in I$ has two children j_1 and j_2 then $X_i = X_{j_1} = X_{j_2}$. (i is called a **join** node.)
2. if a node $i \in I$ has one child j, then either $X_i \subset X_j$ and $|X_i| = |X_j| - 1$ (i is called a **forget** node), or $X_j \subset X_i$ and $|X_j| = |X_i| - 1$ (i is called an **introduce** node).
3. if a node $i \in I$ is a leaf of T, then $|X_i| = 1$. (i is called a **leaf** node.)

Given a nice tree decomposition $(\{X_i : i \in I\},\ T)$, we denote by T_i the subtree of T rooted at node i, for each $i \in I$. The parent of node i is denoted by $p(i)$.

It is known that every graph G with n vertices and of treewidth at most t has a nice tree decomposition $(\{X_i : i \in I\},\ T)$ of width t such that $|I| = \mathcal{O}(t \cdot n)$. Furthermore, given a tree decomposition of G of width t, a nice tree decomposition of G of width t can be computed in time $\mathcal{O}(n)$.

There is an $\mathcal{O}(1.9601^n)$ algorithm to compute the treewidth and an optimal tree decomposition of a given graph [10]. There is also a well-known linear-time algorithm to compute the treewidth and an optimal tree decomposition for graphs of bounded treewidth [2].

We now present an algorithm to decide whether for given graphs G and H there is an homomorphism from G to H. The algorithm is based on dynamic programming on a nice tree decomposition of H.

Theorem 2. *There is an $\mathcal{O}^*((2 \cdot \mathbf{tw}(H) + 1)^{|V(G)|})$ time algorithm taking as input a graph G, a graph H, and an optimal tree decomposition of H, that solves $\mathrm{HOM}(G, H)$ and produces a homomorphism $\varphi : G \longrightarrow H$ if the answer is yes.*

Proof. Let $n = |V(G)|$ and $t = \mathbf{tw}(H)$. First our algorithm transforms the given optimal tree decomposition of H into a nice tree decomposition $(\{Y_i : i \in J\},\ U)$ of width t. Then we modify this nice tree decomposition as follows. For every non leaf node $i \in J$ of tree U we add a new **nochange** node i' as the parent of i, and we let the old parent of i in tree U become the parent of i' in the new tree. We let $X_{i'} = X_i = Y_i$. In this way we obtain a new nice tree decomposition $(\{X_i : i \in I\},\ T)$ of H of width t. In the new tree T, the parent of every node of U is a nochange node, which is more convenient for our following argumentation, because there is a difference of at most one vertex between a child and the parent of a node i in T.

We define two auxiliary subsets of vertices of H for each node $i \in I$ of T: $V_i = \cup_{j \in T_i} X_j$, and $\tilde{X}_i = X_i \cap X_{p(i)}$. Notice that $\tilde{X}_i = X_i$ if $p(i)$ is an introduce, join, or nochange node, and that $\tilde{X}_i = X_i \setminus \{u\}$ if $p(i)$ is a forget node with $X_{p(i)} = X_i \setminus \{u\}$. For r, the root of T, we define $\tilde{X}_r = X_r$.

Our algorithm computes for each node $i \in I$ of T in a bottom-up fashion all characteristics of i, defined as follows.

Definition 1. *A tuple $(S; (v_1, S_1), (v_2, S_2), \ldots, (v_{l_i}, S_{l_i}); i)$ is a characteristic of node $i \in I$ of T if $S \subseteq V(G)$ and $\{S_1, S_2, \ldots, S_{l_i}\}$ is a partition of S such that there is a homomorphism $\varphi : G[S] \longrightarrow H[V_i]$ that satisfies the following two conditions.*

- $\tilde{X}_i = \{v_1, v_2, \ldots, v_{l_i}\}$
- For every $j \in \{1, 2, \ldots, l_i\}$, $\varphi^{-1}(v_j) = S_j$.

Notice that characteristics are defined in such a way that G is homomorphic to H if and only if there is at least one characteristic for the root r of T. Furthermore the number of characteristics of a node of T is at most $\sum_{i=0}^{n} \binom{n}{i} \cdot t^i = (t+1)^n$.

For each forget, introduce, nochange, and join node $i \in I$ of T, our algorithm computes by dynamic programming all characteristics $(S; (v_1, S_1), \ldots, (v_{l_i}, S_{l_i}); i)$ of i using the full set of characteristics of i's children. Thus it suffices to describe how the full set of characteristics can be computed from the characteristics of the children for the different types of nodes in T.

Leaf Node:
Let i be a leaf node, thus $X_i = \{u\}$ for some vertex u of H. For a subset S of $V(G)$, there is a homomorphism φ from $G[S]$ to $H[V_i]$ with $V_i = \{u\}$ if and only if $\varphi^{-1}(\{u\}) = S$, and hence S is an independent set. Thus $(S; (u, S); i)$ is a characteristic of the leaf node i if and only if S is an independent set of G.

Introduce Node:
Let i be an introduce node with child j. Thus $X_i = X_j \cup \{u\}$ for some vertex $u \in V(H) \setminus V_j$, and consequently $\tilde{X}_j = X_j$. Notice that the parent of i is a nochange node, and thus $X_i = X_{p(i)}$ and $\tilde{X}_i = \tilde{X}_j \cup \{u\}$.

All characteristics of node i can be obtained by extending a characteristic of j. Since $\tilde{X}_i = \tilde{X}_j \cup \{u\}$, each characteristic of i obtained from $(S; (v_1, S_1), \ldots, (v_{l_j}, S_{l_j}); j)$ is of the form $(S \cup S'; (v_1, S_1), \ldots, (v_{l_j}, S_{l_j}), (u, S'); i)$ where $S' \subseteq V(G) \setminus S$ is an independent set in G, and for all $x \in N_{G[S]}(S')$, $\varphi(x) \in N_H(u)$. These conditions can be checked in polynomial time. Finally one characteristic of j extends to at most $2^{n-|S|}$ characteristics of i, since S' must be an independent set of $G - S$. Therefore we compute at most $\sum_{i=0}^{n} \binom{n}{i} \cdot t^i \cdot 2^{n-i} = (t+2)^n$ characteristics to obtain a full set of characteristics of an introduce node.

Forget Node:
Let i be a forget node with child j. Thus $X_i = X_j \setminus \{u\}$ for some vertex $u \in X_j$, and consequently $\tilde{X}_j = X_i$. The parent of i is a nochange node, and thus $X_i = X_{p(i)}$ and $\tilde{X}_i = \tilde{X}_j$.

$\tilde{X}_i = \tilde{X}_j$ implies that each characteristic of i can be obtained directly from a characteristic of j by simply replacing j with i. Thus each characteristic of j is a characteristic of i.

Nochange Node:
Let i be nochange node with child j. Thus $X_i = X_j$. If the parent of i is a forget node then $X_{p(i)} = X_i \setminus \{u\}$ for some vertex $u \in X_i$, and thus $\tilde{X}_i = \tilde{X}_j \setminus \{u\}$. If the parent of i is an introduce or join node, then $\tilde{X}_i = \tilde{X}_j$.

If $\tilde{X}_i = \tilde{X}_j$ then each characteristic of i extends into one characteristic of j by simply replacing j by i. Otherwise, $\tilde{X}_i = \tilde{X}_j \setminus \{u\}$ implies that each characteristic of i can be obtained from a characteristic of j, say $(S; (v_1, S_1), \ldots, (v_{l_j}, S_{l_j}); j)$, by simply removing the pair (v_q, S_q) where $u = v_q$. One obtains $(S; (v_1, S_1), \ldots, (v_{q-1}, S_{q-1}), (v_{q+1}, S_{q+1}), \ldots, (v_{l_i}, S_{l_i}); i)$.

Thus again each characteristic of j extends into a characteristic of i.

Join Node:

This is the most interesting node type. Let i be a join node with children j_1 and j_2; thus $X_i = X_{j_1} = X_{j_2}$. The parent of i is a nochange node, thus $\tilde{X}_i = \tilde{X}_{j_1} = \tilde{X}_{j_2} = X_i$.

Let $(S'; (v_1, S_1), \ldots, (v_{l_{j_1}}, S_{l_{j_1}}); j_1)$ be a characteristic of j_1. It extends into a characteristic of node i if there is a characteristic $(S''; (v_1, S_1), \ldots, (v_{l_{j_2}}, S_{l_{j_2}}); j_2)$ of j_2, i.e., both characteristics have the same set of pairs (v_i, S_i) which requires that $l_i = l_{j_1} = l_{j_2}$. In this case $(S' \cup S''; (v_1, S_1), \ldots, (v_{l_i}, S_{l_i}); i)$ is a characteristic of i, if there are no edges between $S' \setminus S''$ and $S'' \setminus S'$ in G.

Thus we compute characteristics $(S' \cup S''; (v_1, S_1), \ldots, (v_{l_i}, S_{l_i}); i)$ of i, for each subset $S' \cup S''$ of $V(G)$, each partition of S into at most t subsets, and any choice of a subset S'. Therefore we compute at most $\sum_{i=0}^{n} \binom{n}{i} \cdot t^i \cdot 2^i = (2t+1)^n$ characteristics to obtain a full set of characteristics of a join node.

Finally, notice that the number of nodes in the decomposition is a polynomial in $|V(H)|$, and that suitable data structures guarantee that the characteristics of a node can be stored such that find and insert operations can be done in polynomial time. Thus the overall running time of our algorithm is $\mathcal{O}^*((2 \cdot \mathbf{tw}(H) + 1)^{|V(G)|})$. $\square$

5   Concluding Remarks and Open Questions

For given graphs G and H, within which time bound can we solve HOM(G, H)? The trivial solution brings us $\mathcal{O}(|V(G)|^{|V(H)|})$ running time. There is a randomized $\mathcal{O}((0.4518 \cdot |V(H)|)^{|V(G)|})$-time algorithm solving HOM(G, H) which is a consequence of a more general result on constraint satisfaction problems [4].

In this paper we observed that if the right side graph H is of bounded treewidth, then HOM(G, H) can be solved in time $c^{|V(G)|} \cdot |V(H)|^{\mathcal{O}(1)}$ for some constant c. Can it be that for any graphs G and H the problem HOM(G, H) is solvable with running times   1. $f(|V(H)|) \cdot |V(G)|^{\mathcal{O}(1)}$, or   2. $f(|V(G)|) \cdot |V(H)|^{\mathcal{O}(1)}$ for some computable function $f : N \rightarrow N$? (Unfortunately) the answer to each of the questions is negative up to some widely believed assumptions in complexity theory.

In fact, for question 1, an $f(|V(H)|) \cdot |V(G)|^{\mathcal{O}(1)}$ time algorithm is also a polynomial-time algorithm for the NP-complete 3-coloring problem implying that P = NP. To answer question 2, we use the widely believed assumption from parameterized complexity [9] that the p-clique problem is not fixed parameter tractable, or in other words, that there is no algorithm for finding a clique of size p in a graph on n vertices in time $f(p) \cdot n^{\mathcal{O}(1)}$ unless FPT = W[1], a collapse of a parameterized hierarchy which is considered to be very unlikely. Since K_p is homomorphic to H if only if H has a clique of size at least p, the HOM(K_p, H) problem is equivalent to finding a p-clique in H. Therefore, the existence of an

$f(|V(G)|) \cdot |V(H)|^{\mathcal{O}(1)}$ time algorithm for HOM(G, H) would imply that the p-clique problem is fixed parameter tractable, thus FPT = W[1].

Now our question is whether a running time of $\mathcal{O}((c \cdot |V(H)|)^{|V(G)|})$ for some constant c is the best that we can hope for solving HOM(G, H). Can it be solved, say by an $\mathcal{O}(c^{|V(G)|+|V(H)|} \cdot |V(G)|^{\mathcal{O}(1)} \cdot |V(H)|^{\mathcal{O}(1)})$-time algorithm?

References

1. B. Aspvall, M. Plass, and R.E. Tarjan, A linear-time algorithm for testing the truth of certain quantified Boolean formulas. *Information Processing Letters* **8** (1979), 121–123.

2. H.L. Bodlaender, A linear-time algorithm for finding tree-decompositions of small treewidth. *SIAM J. Comput.* **25** (1996), 1305–1317.

3. H. L. Bodlaender, A partial k-arboretum of graphs with bounded treewidth. *Theoretical Computer Science* **209** (1998), 1–45.

4. R. Beigel and D. Eppstein. 3-coloring in time $O(1.3289^n)$. *Journal of Algorithms* **54** (2005), 444–453.

5. J. M. Byskov, Enumerating maximal independent sets with applications to graph colouring. *Operations Research Letters* **32** (2004), 547–556.

6. J. M. Byskov and D. Eppstein, An algorithm for enumerating maximal bipartite subgraphs. Unpublished.

7. V. Dalmau, P. G. Kolaitis, and M. Y. Vardi, Constraint satisfaction, bounded treewidth, and finite-variable logics. In *Principles and Practice of Constraint Programming (CP 2002)*, Springer-Verlag, LNCS 2470, 310–326, 2002.

8. J. Diaz, M. Serna, and D.M. Thilikos, Counting H-colorings of partial k-trees. *Theoretical Computer Science* **281** (2002), 291–309.

9. R. G. Downey and M. R. Fellows, *Parameterized complexity*, Springer-Verlag, New York, 1999.

10. F. Fomin, D. Kratsch, and I. Todinca, Exact (exponential) algorithms for treewidth and min fill-in. In *Proceedings of the 31st International Colloquium on Automata, Languages and Programming (ICALP 2004)*, Springer-Verlag, LNCS 3124, 568–580, 2004.

11. M. Grohe, The complexity of homomorphism and constraint satisfaction problems seen from the other side. In *Proceedings of the 44th Annual IEEE Symposium on Foundations of Computer Science (FOCS 2003)*, 552–561, 2003.

12. P. Hell and J. Nešetřil, On the complexity of H-coloring. *Journal of Combinatorial Theory Series B* **48** (1990), 92–110.

13. P. Hell and J. Nešetřil, Graphs and Homomorphisms. New Oxford University Press, 2004.

14. M. Hujter and Z. Tuza, The number of maximal independent sets in triangle-free graphs. *SIAM Journal on Discrete Mathematics* **6** (1993), 284–288.

Improved Algorithms and Complexity Results for Power Domination in Graphs*

Jiong Guo[1], Rolf Niedermeier[1], and Daniel Raible[2]

[1] Institut für Informatik, Friedrich-Schiller-Universität Jena,
Ernst-Abbe-Platz 2, D-07743 Jena, Germany
{guo, niedermr}@minet.uni-jena.de
[2] Wilhelm-Schickard-Institut für Informatik, Universität Tübingen,
Sand 13, D-72076 Tübingen, Germany
raibk@informatik.uni-tuebingen.de

Abstract. The POWER DOMINATING SET problem is a variant of the classical domination problem in graphs: Given an undirected graph $G = (V, E)$, find a minimum $P \subseteq V$ such that all vertices in V are "observed" by vertices in P. Herein, a vertex observes itself and all its neighbors, and if an observed vertex has all but one of its neighbors observed, then the remaining neighbor becomes observed as well. We show that POWER DOMINATING SET can be solved by "bounded-treewidth dynamic programs." Moreover, we simplify and extend several NP-completeness results, particularly showing that POWER DOMINATING SET remains NP-complete for planar graphs, for circle graphs, and for split graphs. Specifically, our improved reductions imply that POWER DOMINATING SET parameterized by $|P|$ is W[2]-hard and cannot be better approximated than DOMINATING SET.

1 Introduction

Domination is a central theme in graph theory. The basic problem is: given an undirected graph $G = (V, E)$, determine a minimum vertex set $D \subseteq V$ such that each $v \in V$ is contained in D or v is a neighbor of at least one vertex in D. The corresponding decision problem DOMINATING SET (DS) is NP-complete. Unless unlikely collapses in structural complexity theory occur, the polynomial-time approximation factor is $\Theta(\log n)$ [7]. Numerous variations of DOMINATING SET exist. For instance, the CONNECTED DOMINATING SET additionally requires that the dominating set D induces a connected subgraph of G. Opposite to DOMINATING SET, CONNECTED DOMINATING SET carries some form of *non-locality*: the correctness of the dominating set cannot be decided by locally checking every direct neighborhood. In this work, we study another "non-local" variant of domination which appears in electric power networks [8]—POWER DOMINATING SET (PDS). This variant is motivated by monitoring an electric power network,

---

* Supported by the Deutsche Forschungsgemeinschaft (DFG), Emmy Noether research group PIAF (fixed-parameter algorithms), NI 369/4.

M. Liśkiewicz and R. Reischuk (Eds.): FCT 2005, LNCS 3623, pp. 172–184, 2005.

where one is asked to place a minimum number of so-called *phase measurement units* (PMU) at some locations in the system to measure the state variables (for example, the voltage magnitude etc.). Intuitively, we have a more complex degree of non-locality in PDS. Whereas CONNECTED DOMINATING SET only refers to a property of the solution set, PDS has the non-locality in its domination mechanism: A vertex may dominate vertices at *arbitrary* distance when certain conditions are fulfilled.

For DOMINATING SET, we have one "observation rule" concerning vertices in the dominating set: these vertices observe (dominate) themselves and all their neighbors (and nothing else). The goal is to get all vertices observed by a minimum number of observers. By way of contrast, we have an additional observation rule in the case of PDS. This rule says that for an already observed vertex whose all but one neighbors are already observed, the one remaining neighbor becomes observed as well.[1] Note that the second observation rule brings in non-locality. The graph-theoretical and algorithmic study of PDS has been initiated by Haynes et al. [8]. They show that PDS is NP-complete for general graphs as well as for bipartite and chordal graphs. Moreover, they present a linear-time algorithm to solve PDS in trees. We improve their results in the following ways.

1. We present simplified and "stronger" NP-completeness proofs, giving all results of Haynes et al. in a less technical way and additionally implying NP-completeness also for planar graphs, circle graphs, and split graphs. Moreover, our simple reductions preserve parameterized complexity and approximability. So we can conclude that, in case of general graphs, PDS is W[2]-hard and it is only $\Theta(\log n)$-approximable unless unlikely collapses in structural complexity theory occur.[2]
2. We present a much simpler linear-time algorithm for PDS in trees than the one presented in [8].
3. Our main result is a dynamic programming algorithm for PDS in graphs of bounded treewidth, answering an open question of Haynes et al. [8]. Independently, fixed-parameter tractability with respect to parameter treewidth was also shown by Kneis et al. [10] using descriptive complexity tools. They express PDS in monadic second-order logic[3], which implies algorithms of highly super-exponential running time. The dependence of the combinatorial explosion on the parameter treewidth is non-elementary. In contrast, we describe and analyze a direct algorithm for PDS with much improved running time.

Let us return to the issue of non-locality. Demaine and Hajiaghayi [5], answering an open question from [1], showed that CONNECTED DOMINATING SET can be solved by bounded-treewidth dynamic programs. It was not believed in [1]

---

[1] The original definition of Haynes et al. [8] is a bit more complicated and the equivalent definition presented here is due to Kneis et al. [10].

[2] Basically the same reduction was independently shown in [10].

[3] This confirms a statement made by Petr Hliněný in a discussion with Peter Rossmanith and Rolf Niedermeier at the IWPEC 2004 meeting in Bergen, Norway, September 2004.

that such non-local properties as "connectedness" could be captured in this way. In case of PDS, the "even worse" degree of non-locality appears to make things even harder. Still, a dynamic programming solution exists. Due to the lack of space, some proofs are deferred to a long version of this extended abstract.

2   Preliminaries

All graphs in this work are simple and without self-loops. For the definitions of the considered graph classes, we refer to [4]. For a vertex v in graph G, we denote by $N_G(v)$ the open neighborhood of v in G. By $N_G[v]$, we refer to the closed neighborhood of v. This naturally generalizes to $N_G(U)$ and $N_G[U]$ for U being a set of vertices. Moreover, for notion concerning approximation algorithms we refer to [2] and for parameterized complexity we refer to [6].

To define power domination in graphs, we use two (simplified) *observation rules* due to [10].

Observation Rule 1 (OR1): A vertex in the power domination set observes itself and all of its neighbors.

Observation Rule 2 (OR2): If an observed vertex v of degree $d \geq 2$ is adjacent to $d - 1$ observed vertices, then the remaining unobserved vertex becomes observed as well.

Given an undirected graph $G = (V, E)$ and an integer $k \geq 0$, POWER DOM- INATING SET (PDS) asks whether there is a set $P \subseteq V$ with $|P| \leq k$ which observes all vertices in V with respect to the two observation rules OR1 and OR2. Herein, P is called the *power dominating set* of G. Note that every con- nected graph of maximum vertex degree 2 has a power dominating set of size 1. The classical DOMINATING SET problem can be defined by simply omitting OR2.

Lemma 1. *If G is a connected graph with at least one vertex of degree three or higher, then there is always a minimum power dominating set which contains only vertices with degree at least three.*

Lemma 1 is due to Haynes et al. [8] who also show that, given an arbitrary power dominating set P for a connected graph with maximum degree at least three, one can construct in linear time a power dominating set P' with $|P'| \leq |P|$ containing only vertices with degree at least three.

The central concept of this work are tree decompositions of graphs and their use with respect to dynamic programming as, e.g., described in [3,9].

Definition 1. *Let $G = (V, E)$ be a graph. A tree decomposition of G is a pair $\langle \{X_i \mid i \in I\}, T \rangle$, where each X_i is a subset of V, called a bag, and T is a tree with the elements of I as nodes. The following three properties must hold:*

1. $\bigcup_{i \in I} X_i = V$;
2. for every edge $\{u, v\} \in E$, there is an $i \in I$ such that $\{u, v\} \subseteq X_i$;
3. for all $i, j, k \in I$, if j lies on the path between i and k in T, then $X_i \cap X_k \subseteq X_j$.

The width *of* $\langle \{X_i \mid i \in I\}, T \rangle$ *equals* $\max\{|X_i| \mid i \in I\} - 1$. *The* treewidth *of* G *is the minimum* k *such that* G *has a tree decomposition of width* k.

Definition 2. *A tree decomposition* $\langle \{X_i \mid i \in I\}, T \rangle$ *is called a* nice tree decomposition *if* T *is rooted and the following conditions are satisfied:*

1. *Every node of the tree* T *has at most 2 children.*
2. *If a node* i *has two children* j *and* k, *then* $X_i = X_j = X_k$ *(in this case* i *is called a* JOIN NODE*).*
3. *If a node* i *has one child* j, *then one of the following holds:*
   (1) $|X_i| = |X_j| + 1$ *and* $X_j \subset X_i$ *(in this case* i *is called an* INSERT NODE*),*
   (2) $|X_i| = |X_j| - 1$ *and* $X_i \subset X_j$ *(in this case* i *is called a* FORGET NODE*).*

Every tree decomposition can be transformed into a nice tree decomposition [9, Lemma 13.1.3].

Lemma 2. *Given a width-k tree decomposition with* $O(n)$ *nodes of a graph* G, *where* n *is the number of vertices of* G, *then we can find a nice tree decomposition of* G *that also has width* k *and* $O(n)$ *nodes in* $O(n)$ *time.*

3   Power Dominating Set vs. Dominating Set

Haynes et al. [8] show the NP-completeness of PDS by giving a reduction from 3-SAT. This also shows that PDS is NP-complete even when the input graph is bipartite or chordal. We show, for general graphs, that PDS is at least as hard to solve as DS by reducing DS to PDS. Moreover, this implies the NP-completeness of PDS also in case of bipartite, chordal, circle, and planar graphs, and the inapproximability and parameterized intractability results for DS also transfer to PDS in this way. The second reduction (from VERTEX COVER) in this section shows the NP-completeness of PDS for split graphs.

Theorem 1. POWER DOMINATING SET *is NP-complete for bipartite, chordal, circle, and planar graphs.*

Proof. Since it can be easily decided whether a vertex set P is a power dominating set, PDS is in NP. To show NP-hardness, we give a reduction from DS.

Given a DS-instance $G = (V, E)$, we construct a PDS-instance $G' = (V \cup V_1, E \cup E_1)$ simply by attaching newly introduced degree-1 vertices to all vertices from V.

For a dominating set D of G, we set $P := D$. By definition, all vertices from V are observed by OR1. Applying OR2 to every vertex in V, the vertices in V_1 become observed as well. Thus, P is a power dominating set of G'.

If G' has a power dominating set P with $|P| = k$, then we can assume due to Lemma 1 that each vertex of P has degree at least three. This implies that $P \cap V_1 = \emptyset$. The proof is by contradiction. Assume that P is not a dominating set of G. Then, there is a vertex $v \in V$ with $N_G[v] \cap P = \emptyset$. We also have that $N_{G'}[v] \cap P = \emptyset$. Vertex v can get observed in G' only by applying OR2

Table 1. The first column is from [11]. Partial k-trees are the same as graphs of bounded treewidth; the polynomial-time result for this row is shown in Section 4. Empty entries mean that this has not been studied yet.

Graph classes	DOMINATING SET	POWER DOMINATING SET
bipartite	NP-c	NP-c
chordal	NP-c	NP-c
circle	NP-c	NP-c
comparability	NP-c	NP-c
planar	NP-c	NP-c
split	NP-c	NP-c
AT-free	poly. time	
cocomparability	poly. time	
distance hereditary	poly. time	
dually chordal	poly. time	
interval	poly. time	
k-polygon ($k \geq 3$)	poly. time	
partial k-tree ($k \geq 1$)	poly. time	poly. time
permutation	poly. time	
strongly chordal	poly. time	

to one of its neighbors in G'. Denote this neighbor by u. Let v' denote the newly introduced degree-1 neighbor of v in G'. It is easy to see that u cannot be v'. Hence, $u \in V$. Furthermore, u has also a degree-1 neighbor $u' \in V_1$ in G'. Since $u \notin P$ and $u' \notin P$, u' can be observed only by applying OR2 to u. However, this is impossible since u has two unobserved neighbors v and u'. Thus, P cannot be a power domination set of G', yielding a contradiction.

It is easy to verify that the reduction preserves the properties "bipartite," "chordal," "circle," and "planar." Hence, the NP-completeness of PDS follows from the NP-completeness of DS for these graph classes [11]. □

The reduction in the proof of Theorem 1 was independently achieved in [10]. Observe that it is a gap-preserving as well as a parameterized reduction. This implies that all negative results with respect to the approximability and parameterized tractability with respect to the solution size valid for DS also are valid for PDS. It is hard to approximate PDS better than $\Theta(\log n)$ [7] and PDS is W[2]-hard with the size of the power dominating set as parameter [6].

Next, we show that PDS is NP-complete for split graphs. The reduction is from the NP-complete VERTEX COVER problem: Given an undirected graph $G = (V, E)$ and an integer $k \geq 1$, is there a set $C \subseteq V$ with $|C| \leq k$ such that each edge has at least one endpoint in C?

Theorem 2. POWER DOMINATING SET *is NP-complete on split graphs.*

Proof. For a VC-instance with graph $G = (V, E)$, we construct a split graph G' from G as follows. For each edge $e = \{u, v\} \in E$ we add a new vertex w_e and two edges between w_e and u and between w_e and v to G. We denote the set of the vertices w_e by V_E. Moreover, we introduce for each vertex $v \in V$ a new degree-1 vertex v' and an edge between v and v'. The set of these vertices is denoted by V_1. Finally, we complete the graph induced by V into a clique. Note that the subgraph of G' induced by the vertices in $V_e \cup V_1$ is an independent set. Thus, G' is a split graph. The proof for the claim that G has a vertex cover of size k iff G' has a power dominating set of size k uses a similar argument as the proof of Theorem 1.   □

Altogether, Table 1 compares the computational complexity of PDS and DS.

4   Dynamic Program for Graphs of Bounded Treewidth

Haynes et al. [8] give a linear-time algorithm for PDS in trees. As a warm-up, we start with a much simpler linear-time algorithm for these "treewidth-1 graphs." Without loss of generality, we assume that the input tree T is rooted at a degree-1 node r and the *depth* of a node v is defined as the length of the path between v and r. The algorithm follows a bottom-up strategy:

1. Sort the inner nodes of T in a list L according to the non-increasing order of their depths in T;
2. **while** $L \neq \{r\}$ **do**
2.1 $v \leftarrow$ the first node in L; $L \leftarrow L \setminus \{v\}$;
2.2 **if** v has at least two unobserved children **then**
2.2.1 $P \leftarrow P \cup \{v\}$;
2.2.2 Apply the two observation rules as long as possible;
3. **if** r is unobserved **then**
3.1 $P \leftarrow P \cup \{r\}$;
4. **return** P;

The linear running time of the algorithm is due to the fact that the algorithm examines for each inner node only its children. With proper data structures, the application of the observation rules can be performed in linear time.

Our linear-time algorithm for graphs of bounded treewidth uses the same strategy as the algorithms for DS [12,1], i.e., bottom-up dynamic programming from the leaves to the root. In the following, we demonstrate that there are two difficulties associated with OR2 that make PDS "harder" than DS and which cannot be solved by a simple modification of the algorithms for DS.

Firstly, with OR2, there are more possibilities for a vertex to be observed: it can be observed by OR1 or OR2. More precisely, the application of OR2 implies that there is a certain "observation dependency" between two vertices, i.e., one vertex not in the power dominating set that becomes observed can make one of its neighbors observed. This observation dependency does not exist in DS. There, the domination status of one vertex that is not in the dominating set has no effect on the domination status of other vertices. Therefore, in order to describe this observation dependency in the dynamic programming, the three states defined in the algorithm for DS for the vertices in a bag, namely, "belonging to the dominating set," "already dominated at the current bag," and "not yet dominated at the current bag," are not sufficient. Secondly, only introducing further vertex states cannot settle the problem with the observation dependency. For example, assume that one defines the following additional state for the vertices in a bag: "already observed at the current bag by applying OR2 to one of its neighbors." Then, there could emerge a "cycle of observation dependencies:" Consider a simple cycle as the input graph, one might assign the new state "observed due to OR2" to each vertex of the cycle. This is "locally correct" but globally it is false because the reasoning is done in a circular fashion without "global justification."

Our answer to these two difficulties is to define, in addition to the vertex states, three states for the edges in the subgraph induced by the vertices in a bag. In fact, these states give one of three possible *orientations* to an undirected edge $\{u, v\}$, orienting it from u to v, orienting it from v to u, or leaving it unoriented. These orientations express observation dependencies.

4.1   Valid Orientations of Undirected Graphs

Definition 3. *An* orientation *of an undirected graph* $G = (V, E)$ *is a graph* $D = (V, E_1 \cup E_2)$ *such that for each* $\{u, v\} \in E$ *there is either a directed edge* (u, v) *or* (v, u) *in* E_1 *or an undirected edge* $\{u, v\}$ *in* E_2, *and* $\{u, v\} \in E$ *for each directed edge* $(u, v) \in E_1$ *and undirected edge* $\{u, v\} \in E_2$. *The* indegree *of a vertex* v *in* D, *denoted by* $d^-(v)$, *is defined as* $|\{(u, v) \mid (u, v) \in E_1\}|$ *and the* outdegree *of* v, *denoted by* $d^+(v)$, *as* $|\{(v, u) \mid (v, u) \in E_1\}|$. *The subgraph* $D[V']$ *induced by* $V' \subseteq V$ *in* D *is called a* suborientation.

Note that in the standard graph theory literature, an orientation of an undirected graph G is a directed graph D where there is exactly one of (u, v) and (v, u) in D for each edge $\{u, v\}$ in G. Here, we abuse the term orientation to denote a graph that results from orienting only a subset of edges.

Definition 4. *A* dependency path *in an orientation* $D = (V, E_1 \cup E_2)$ *is a subgraph of* D *consisting of a sequence of vertices and edges* $v_1, e_1, v_2, e_2, \ldots, v_i$, $i \geq 3$, *satisfying:*

(1) for all $1 \leq j, k \leq i$, $v_j = v_k \Leftrightarrow j = k$,
(2) for all $1 \leq j \leq i - 1$, *either* $e_j = (v_j, v_{j+1}) \in E_1$ *or* $e_j = \{v_j, v_{j+1}\} \in E_2$,
(3) and for all $1 \leq j \leq i - 1$, *at least one of* e_j *and* e_{j+1} *is from* E_1.

The vertices v_1 *and* v_i *are called the* tail endpoint *and the* head endpoint *of the dependency path, respectively. A* dependency cycle *in an orientation is a dependency path with an edge between* v_1 *and* v_i *which can be undirected only if* $(v_1, v_2) \in E_1$ *and* $(v_{i-1}, v_i) \in E_1$; *otherwise, it is directed. A* directed path *is a dependency path with only directed edges.*

Observe that a dependency path contains at least one directed edge.

Definition 5. *A* valid orientation $D = (V, E_1 \cup E_2)$ *of an undirected graph* $G = (V, E)$ *is an orientation such that* $\forall v \in V$: $d^-(v) \leq 1$, $\forall v \in V$: $d^-(v) = 1 \Rightarrow d^+(v) \leq 1$, *and there is no dependency cycle in* D. *We call the set of vertices with* $d^-(v) = 0$ *the* origin *of* D.

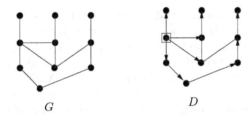

G D

Fig. 1. An example of a valid orientation D of an undirected graph G. The origin O of D contains only one vertex, marked with a rectangular box.

See Figure 1 for an example of a valid orientation. Note that one can easily decide in $O(|E_1 \cup E_2|)$ time whether an orientation D is valid and, if so, find the origin of D. The following lemma is also easy to show.

Lemma 3. *Let $D = (V, E_1 \cup E_2)$ denote a valid orientation with origin $O \subseteq V$.*
(1) For each vertex $v \in (V \setminus O)$, there is exactly one directed path from the vertices in O to v.
(2) Two directed paths from O to two vertices in $V \setminus O$ are vertex-disjoint except for their tail endpoints in O.

The following orientation problem is a reformulation of PDS: Given an undirected graph $G = (V, E)$ and an integer $k \geq 0$, VALID ORIENTATION WITH MINIMUM ORIGIN (VOMO) asks whether there is a vertex subset $O \subseteq V$ with $|O| \leq k$ such that G has a valid orientation with O as origin.

Lemma 4. *An undirected graph G has a power dominating set P iff G has a valid orientation with P as origin.*

With VOMO and Lemma 4, we have an alternative formulation of PDS. The advantage of VOMO is that, by giving each undirected edge an orientation, a dependency cycle, which corresponds to a cycle of observation dependencies, can be easily detected in the dynamic programming on the tree decomposition.

4.2   Dynamic Programming on Tree Decompositions

Given an undirected, connected graph $G = (V, E)$ with $V := \{v_1, v_2, \ldots, v_n\}$ and a nice tree decomposition $\langle \{X_i \mid i \in V(T)\}, T \rangle$ of G with treewidth k. Let T_i denote the subtree of T rooted at node i and G_i denote the subgraph of G induced by the vertices in the bags in T_i, i.e., $G_i := G[\bigcup_{j \in V(T_i)} X_j]$. Furthermore, we use Y_i to denote $(\bigcup_{j \in V(T_i)} X_j) \setminus X_i$ and set $G'_i := G[X_i]$.

Definition of States: During the bottom-up process of our dynamic programming algorithm, it computes for a bag X_i the possible valid orientations of the subgraph G_i and stores the origin sizes of the valid orientations. Here, the valid orientations of G_i are characterized by the *bag states*. A bag state s is a combination of the states for all ordered pairs of vertices in X_i, the states of vertices in X_i, and the states of edges in G'_i. In the following, we will define the states for vertex pairs, for vertices, and for edges, respectively. Herein, we use $s(uv)$, $s(v)$, and $s(e)$ to denote the state of an ordered pair of vertices $v \in X_i$ and $u \in X_i$, the state of vertex $v \in X_i$, and the state of an edge $e \in E(G'_i)$.

The decisive point in the dynamic programming is to detect the dependency cycles in the orientations D_i for G_i while reaching bag X_i. The dependency cycles inside suborientation $D_i[X_i]$ can be detected by exhaustive search in $D_i[X_i]$ which contains at most k vertices. However, the detection of dependency cycles over some vertices in Y_i depends on, for each pair of vertices u and v in X_i with $u \neq v$, the information about the dependency paths between u and v in $D_i[Y_i \cup \{u, v\}]$. We solve this problem by defining for each pair of vertices in X_i several states reflecting the information about such dependency paths. Herein, let $V(p)$ be the set of the vertices on the dependency path p. We use p_{uv} to denote a dependency path with u and v as the tail and head endpoints, respectively. The path p_{uv} is called a dependency path *from u to v*.

Definition 6. *There are four types of dependency paths from u to v, p_{uv}, in an orientation D:*

Type 1: The first edge (between u and its successor on p_{uv}) as well as the last edge (between v and its predecessor on p_{uv}) of p_{uv} are directed edges.
Type 2: The first edge of p_{uv} is a directed edge but the last edge is not.
Type 3: The last edge of p_{uv} is a directed edge but the first edge is not.
Type 4: The last edge as well as the first edge of p_{uv} are undirected edges.

In addition, the function g maps a dependency path p to one of the four types, i.e., $g(p) \in \{ Type\ 1,\ Type\ 2,\ Type\ 3,\ Type\ 4 \}$.

Observe that, if there are two dependency paths p_{uv} and p_{vu} in a valid orientation D, then $g(p_{uv}) \neq$ Type 1 and $g(p_{vu}) \neq$ Type 1, and at least one of p_{uv} and p_{vu} is Type 4 or one is Type 2 and the other is Type 3; otherwise, there is a dependency cycle in the suborientation $D[V(p_{uv}) \cup V(p_{vu})]$ and we say that there is a "path type conflict." The states for an ordered vertex pair uv with $u \neq v$ in X_i are defined according to the possible types of dependency paths from u to v in $D_i[Y_i \cup \{u, v\}]$: there are 16 states for an ordered vertex pair uv according to the 16 possible subsets of $\{$Type 1, Type 2, Type 3, Type 4$\}$. With these vertex pair states, we can simply detect a dependency cycle at a bag X_i: check, for each ordered vertex pair uv with $u \in X_i$ and $v \in X_i$, whether there is a dependency path from v to u in $D_i[X_i]$ whose type together with one type in $s(uv)$ builds a path type conflict.

Next, we define five vertex states $s(v)$ for every vertex v in a bag X_i:

- $s(v) = 1$: there is exactly one directed edge from a vertex in Y_i to v and no directed edge from v to vertices in Y_i;
- $s(v) = 2$: there is exactly one directed edge from a vertex in Y_i to v and there is exactly one directed edge from v to a vertex in Y_i;
- $s(v) = 3$: there is no directed edge between v and the vertices in Y_i;

- $s(v) = 4$: there are at least two directed edges from v to the vertices in Y_i and no directed edge from the vertices in Y_i to v;
- $s(v) = 5$: there is exactly one directed edge from v to a vertex in Y_i and no directed edge from the vertices in Y_i to v.

Furthermore, we define three edge states $s(e)$ for the edges $e = \{u, v\}$ in G_i':

- $s(e) = uv$: edge e is directed from u to v;
- $s(e) = vu$: edge e is directed from v to u;
- $s(e) = \perp$: edge e is undirected.

As a consequence, for a bag X_i with $|X_i| \leq k$, we have at most $16^{k^2} \cdot 5^k \cdot 3^{k^2}$ bag states. In the following, we say that a valid orientation D is *under the restriction of a bag state s* of the bag X_i if D satisfies the following conditions:

- the orientations in D of the edges in G_i' coincide with their states in s;
- for each ordered vertex pair uv with $u \in X_i$ and $v \in X_i$, the types of the dependency paths from u to v in $D[Y_i \cup \{u, v\}]$ coincide with $s(uv)$; and
- for each vertex $v \in X_i$, the orientations of the edges between v and vertices in Y_i coincide with $s(v)$.

In the bottom-up dynamic programming, we use a mapping A_i for each bag X_i, which stores, for each bag state, the minimum size of the origins of all possible valid orientations of G_i under the restriction of the bag state. Due to the following easy-to-prove lemma, in the computation of the values for A_i, we do not count vertices v for the size of an origin with $d^-(v) = 0$ and $d^+(v) \leq 1$.

Lemma 5. *For a connected, undirected graph $G = (V, E)$, there is always a valid orientation with origin $O \subseteq V$ such that each $v \in O$ has at least two neighbors in $V \setminus O$ which are not neighbors of other vertices in O.*

In the following, we use valid(G_i', s_i) to denote the procedure deciding whether the edge states $s_i(e)$ of the edges e in G_i' form a valid orientation of G_i'. Since G_i' contains at most k vertices, valid(G_i', s_i) needs at most $O(k!)$ time by enumerating all cycles in G_i'. Procedure valid(G_i', s_i) returns true if s_i implies a valid orientation of G_i'; otherwise, it returns false.

Initialization: For each leaf node i with bag X_i of the tree decomposition T, we initialize A_i as follows. For each bag state s_i, we set $A_i(s_i) := +\infty$ if

$$(\exists v \in X_i : s_i(v) \neq 3) \vee (\exists uv : u \in X_i \wedge v \in X_i \wedge s_i(uv) \neq \emptyset) \vee (\text{valid}(G_i', s_i) = \text{false});$$

and, otherwise,

$$A_i(s_i) := |\{v \in X_i \mid \exists e = \{v, u\} \in E(G_i') : \exists e' = \{v, w\} \in E(G_i') :$$
$$u \neq w \wedge s_i(e) = vu \wedge s_i(e') = vw\}|.$$

Updating Process: After the initialization, we visit the bags of our tree decomposition bottom-up from the leaves to the root, evaluating the corresponding mappings in each step. For each bag state s_i of a bag X_i, do the following:

- Check whether the edge states contained in s_i form a valid orientation of G_i'.
- Compute the set S^{s_i} containing the "compatible bag states" s_j (or "compatible bag state pairs" for join nodes) of the child bag X_j. The formal definition of compatible bag states (and compatible bag state pairs) will be given individually for forget, insert, and join nodes.
- For each compatible bag state s_j (or each compatible bag state pair) in S^{s_i}, check whether a valid orientation for G_j under the restriction of s_j is a valid orientation for G_i under the restriction of s_i.
- Based on the mappings A_j for all compatible bag states (or bag state pairs) in S^{s_i}, evaluate $A_i(s_i)$.

Due to the lack of space, we only treat the (most interesting) case of "Insert Nodes" of the tree decomposition here.

Insert Nodes: Suppose that node i is an insert node with child node j, $X_i := \{x_{j_1}, \ldots, x_{j_{n_j}}, x\}$ and $X_j := \{x_{j_1}, \ldots, x_{j_{n_j}}\}$.

Procedure valid(G_i', s_i) can decide whether the edge states contained in s_i form a valid orientation of G_i' in $O(k!)$ time by enumerating all cycles. For each bag state s_i of node i, we consider the set S^{s_i} containing the compatible bag states s_j of node j which satisfy

$$(\forall e \in E(G_j') : s_i(e) = s_j(e)) \wedge (\forall v : s_i(v) = s_j(v)) \wedge (\forall uv : s_i(uv) = s_j(uv)),$$

where $u, v \in X_j$.

Note that the introduction of a new vertex x does not change the number of directed edges between a vertex $v \in X_j$ and the vertices in Y_i; thus, $s_i(v) = s_j(v)$. The types of the dependency paths remain the same for each ordered vertex pair uv with $u \in X_j$ and $v \in X_j$, i.e., $s_i(uv) = s_j(uv)$. Moreover, since there is no edge between the new vertex x in X_i and the vertices in Y_i, we set $S^{s_i} := \emptyset$ for bag states s_i if $s_i(x) \neq 3$ or if there is a vertex $v \in X_j$ with $s_i(vx) \neq \emptyset$ or $s_i(xv) \neq \emptyset$.

The most difficult task here is to decide whether a valid orientation of G_j under the restriction of a bag state in S^{s_i} is still a valid orientation of G_i under the restriction of s_i. Herein, we have to take into account the states of the edges incident to x in G_i'. Observe that the conditions for a valid orientation can be violated either by a vertex in X_i having x as neighbor or by a dependency cycle over vertex x. Thus, based on the information saved in $s_j(v)$ for vertices $v \in X_j$ and in $s_j(uv)$ for the ordered vertex pairs with $u \in X_j$ and $v \in X_j$, this task can be fulfilled by firstly checking for each vertex v in X_i having x as neighbor whether v violates the vertex-degree conditions of a valid orientation, i.e., $d^-(v) \leq 1$ and $d^-(v) = 1 \Rightarrow d^+(v) \leq 1$, by taking $s_j(v)$ and $s_i(e)$ for $e = \{v, x\}$ into account. Secondly, check for every two vertices u and v in X_i whether they now have a dependency path over x in $D_i[X_i]$ built by the edge states in s_i whose type together with one path type in $s_j(uv)$ or $s_j(vu)$ forms a path type conflict. It is easy to see that these two checks can be done in $O(k!)$ time by enumerating all dependency paths in $D_i[X_i]$. Finally, if for a bag state s_i there is no s_j in S^{s_i} which passes both checks, then $S^{s_i} := \emptyset$.

Evaluate the mapping A_i of X_i as follows: for each bag state s_i, set

$$A_i(s_i) := \min_{s_j \in S^{s_i}} \{A_j(s_j) + f_{s_i}(x) + |B|\},$$

where $f_{s_i}(x) = 1$ if there exist at least two edges $e = \{u, x\}$ and $e' = \{v, x\}$ in G_i' with $s_i(e) = xu$ and $s_i(e') = xv$; otherwise, $f_{s_i}(x) = 0$. The set B contains the vertices $v \in X_j$ which either have $s_j(v) = 5$ and exactly one edge $e = \{v, x\}$ in G_i' with $s_i(e) = vx$ or have $s_j(v) = 3$ and exactly two edges $e = \{v, x\}$ and $e' = \{u, v\}$ in G_i' with $s_i(e) = vx$ and $s_i(e') = vu$. Roughly speaking, $f_{s_i}(x) = 1$ covers the case that x has to be added to the origin and the set B contains the vertices which are in the origin of the valid orientation of G_i but not in the origin of the valid orientation of G_j. Note that, if $S^{s_i} = \emptyset$ for a s_i, then $A_i(s_i) = +\infty$.

When reaching the root of T, we can determine the minimum size of the origin of a valid orientation of G in the mappings A of the root node. Together with Lemma 4, we obtain our main result.

Theorem 3. *For an n-vertex graph with a given width-k tree decomposition,* POWER DOMINATING SET *can be solved in $O(c^{k^2} \cdot n)$ time for a constant c.*

5   Conclusion

There are several avenues for future work. Table 1 contains several empty entries concerning the complexity of POWER DOMINATING SET in particular graph classes we did not address but which have been addressed for DOMINATING SET [11]. In particular, is there a "significant" difference between the complexities of DS and PDS? How do fixed-parameter tractability results for DS in planar graphs transfer to PDS? Are there nontrivial data reduction rules for PDS? So far, we are not aware of a graph class where DS is polynomial-time solvable and PDS is NP-complete or vice versa.

References

1. J. Alber, H. L. Bodlaender, H. Fernau, T. Kloks, and R. Niedermeier. Fixed parameter algorithms for Dominating Set and related problems on planar graphs. *Algorithmica*, 33(4):461–493, 2002.
2. G. Ausiello, P. Crescenzi, G. Gambosi, V. Kann, A. Marchetti-Spaccamela, and M. Protasi. *Complexity and Approximation — Combinatorial Optimization Problems and Their Approximability Properties*. Springer, 1999.
3. H. L. Bodlaender. Treewidth: Algorithmic techniques and results. In *Proc. 22nd MFCS*, volume 1295 of *LNCS*, pages 19–36. Springer, 1997.
4. A. Brandstädt, V. B. Le, and J. P. Spinrad. *Graph Classes: a Survey*. SIAM Monographs on Discrete Mathematics and Applications, 1999.
5. E. D. Demaine and M. Hajiaghayi. Bidimensionality: New connections between FPT algorithms and PTASs. In *Proc. 16th SODA*, pages 590–601, 2005.
6. R. G. Downey and M. R. Fellows. *Parameterized Complexity*. Springer, 1999.

7. U. Feige. A threshold of ln n for approximating Set Cover. *Journal of the ACM*, 45(4):634–652, 1998.
8. T. W. Haynes, S. M. Hedetniemi, S. T. Hedetniemi, and M. A. Henning. Domination in graphs: applied to electric power networks. *SIAM Journal on Discrete Mathematics*, 15(4):519–529, 2002.
9. T. Kloks. *Treewidth: Computations and Approximations*, volume 842 of *Lecture Notes in Computer Science*. Springer-Verlag, 1994.
10. J. Kneis, D. Mölle, S. Richter, and P. Rossmanith. Parameterized power domination complexity. Technical Report AIB-2004-09, Department of Computer Science, RWTH Aachen, Dec. 2004.
11. D. Kratsch. Algorithms. In T. W. Haynes, S. T. Hedetniemi, and P. J. Slater, editors, *Domination in Graphs: Advanced Topics*, pages 191–231. Marcel Dekker, 1998.
12. J. A. Telle and A. Proskurowski. Practical algorithms on partial k-trees with an application to domination-like problems. In *Proc. 3rd WADS*, volume 709 of *LNCS*, pages 610–621. Springer, 1993.

Clique-Width for Four-Vertex Forbidden Subgraphs

Andreas Brandstädt[1], Joost Engelfriet[2],
Hoàng-Oanh Le[3], and Vadim V. Lozin[4]

[1] Institut für Informatik, Universität Rostock, D-18051 Rostock, Germany
ab@informatik.uni-rostock.de
[2] LIACS, Leiden University, P.O. Box 9512, 2300 RA Leiden, The Netherlands
engelfri@liacs.nl
[3] Fachbereich Informatik, Technische Fachhochschule Berlin,
D-13353 Berlin, Germany
oanhle@tfh-berlin.de
[4] RUTCOR, Rutgers University, 640 Bartholomew Rd., Piscataway,
NJ 08854-8003, USA
lozin@rutcor.rutgers.edu

Abstract. Clique-width of graphs is a major new concept with respect to efficiency of graph algorithms. The notion of clique-width extends the one of treewidth, since bounded treewidth implies bounded clique-width. We give a complete classification of all graph classes defined by forbidden induced subgraphs of at most four vertices with respect to bounded or unbounded clique-width.

1 Introduction

Recently, in connection with graph grammars, in [17] the notion of clique-width of a graph was introduced which, by now, has attracted much attention since, in [18], Courcelle, Makowsky and Rotics have shown that every graph problem expressible in LinEMSOL($\tau_{1,L}$) (a variant of Monadic Second Order Logic) is linear-time solvable on graphs with bounded clique-width if the input graph is given together with a k-expression defining it. (The time complexity may be sublinear with respect to the size of the input graph G if the input is just a k-expression of G).

Various NP-complete problems such as Vertex Cover, Maximum Weight Stable Set (MWS), Maximum Weight Clique, Steiner Tree, Domination, k-colorability for fixed $k \geq 3$ and Maximum Induced Matching are LinEMSOL ($\tau_{1,L}$) expressible.

Restricting the input to some graph classes defined by forbidding small graphs leads to polynomial time algorithms for some problems; for example, Minty [28] gave a polynomial time algorithm for the MWS problem on claw-free graphs (see also [33]), Randerath [31] and Randerath et al. [32] discussed k-colorability of graph classes defined by small forbidden subgraphs, and Corneil et al. [15]

M. Liśkiewicz and R. Reischuk (Eds.): FCT 2005, LNCS 3623, pp. 185–196, 2005.
© Springer-Verlag Berlin Heidelberg 2005

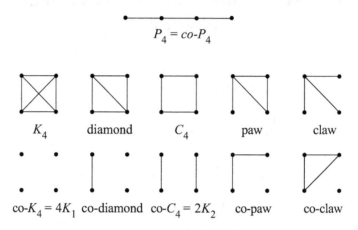

Fig. 1. All four-vertex graphs

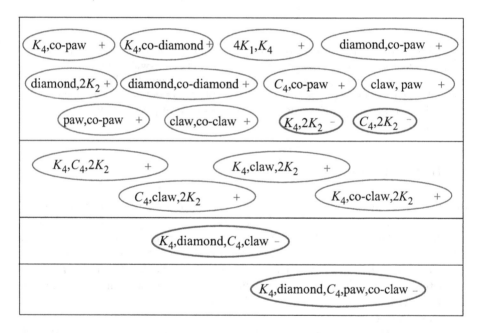

Fig. 2. Essential classes for all combinations of forbidden 4-vertex graphs; $+$ $(-)$ denotes bounded (unbounded) clique-width

described how MWS and related problems can be solved bottom-up along the cotree of a P_4-free graph (also called *cograph*).

It is known that a graph is P_4-free if and only if its clique-width is at most 2, and a 2-expression can be found in linear time along its cotree. Thus, it is a natural question to ask which other forbidden 4-vertex graphs (and which of their combinations) will lead to bounded clique-width. Figure 1 contains all 4-vertex graphs.

In [11], the clique-width (and some structure results) of $(H,\text{co-}H)$-free graphs was described for any 4-vertex graph H (see Theorem 2). Thus e.g., the (diamond, co-diamond)-free graphs and the (claw,co-claw)-free graphs have bounded clique-width (and simple structure). See also [1] for bounded clique-width of (claw,co-claw)-free graphs as well as (claw,paw)-free graphs.

We extend these results by giving a complete classification of all graph classes defined in terms of some forbidden graphs with at most four vertices with respect to bounded or unbounded clique-width. This is done by identifying 14 inclusion-maximal classes of bounded clique-width and four inclusion-minimal classes of unbounded clique-width (see Figure 2). In particular, it will turn out that for a 4-vertex graph H, the class of H-free graphs has bounded clique-width if and only if H is the P_4, and every class defined by six (or more) forbidden 4-vertex graphs has bounded clique-width. This also continues research done in [2,3,4,5,6,7,8,10,11,12,13,23] and is partially based on some of the results of these papers.

For space limitations, almost all proofs in this paper are omitted but can be found in the corresponding full version available on the first author's homepage.

2 Basic Notions

Throughout this paper, let $G = (V, E)$ with vertex set V and edge set E be a finite undirected graph without self-loops and multiple edges and let $|V| = n$, $|E| = m$. For a vertex $v \in V$, let $N(v) = \{u \mid uv \in E\}$.

Disjoint vertex sets X, Y *form a join* (*co-join, respectively* if for all pairs $x \in X$, $y \in Y$, $xy \in E$ ($xy \notin E$, respectively) holds. We will also say that X has a join to Y, that there is a join between X and Y, or that X and Y are connected by join (and similarly for co-join). Subsequently, we will consider join and co-join also as operations, i.e., the co-join operation for disjoint vertex sets X and Y is the disjoint union of the subgraphs induced by X and Y, and the join operation for X and Y consists of the co-join operation for X and Y followed by adding all edges xy, $x \in X$, $y \in Y$.

A vertex $z \in V$ *distinguishes* vertices $x, y \in V$ if $zx \in E$ and $zy \notin E$. A vertex set $M \subseteq V$ is a *module* if no vertex from $V \setminus M$ distinguishes two vertices of M, i.e., every vertex $v \in V \setminus M$ has either a join or a co-join to M. A module is *trivial* if it is either the empty set, a one-vertex set or the entire vertex set V. A graph is *prime* if it contains only trivial modules. The notion of module plays a crucial role in the *modular* (or *substitution*) *decomposition* of graphs (and other discrete structures) which is of basic importance for the design of efficient algorithms - see e.g. [29] for modular decomposition of discrete structures and its algorithmic use and [27] for a linear-time algorithm constructing the modular decomposition tree of a given graph.

For $U \subseteq V$, let $G[U]$ denote the subgraph of G induced by U. Throughout this paper, all subgraphs are understood to be induced subgraphs. Let $\mathcal{F}$ denote a set of graphs. A graph G is $\mathcal{F}$-*free* if none of its induced subgraphs is in $\mathcal{F}$.

A vertex set $U \subseteq V$ is *stable* (or *independent*) in G if the vertices in U are pairwise nonadjacent. Let co-$G = \overline{G} = (V, \overline{E})$ denote the *complement graph* of G. A vertex set $U \subseteq V$ is a *clique* in G if U is a stable set in $\overline{G}$. Let K_ℓ denote the clique with ℓ vertices, and let ℓK_1 denote the stable set with ℓ vertices.

For $k \geq 1$, let P_k denote a chordless path with k vertices and $k - 1$ edges, and for $k \geq 3$, let C_k denote a chordless cycle with k vertices and k edges. The $2K_2$ is the complement of C_4 (see Figure 1). A graph is *chordal* if it contains no induced C_k, $k \geq 4$.

A graph is a *split graph* if its vertex set is partitionable into a clique and a stable set.

Lemma 1 ([20]). *G is a split graph if and only if G is $(2K_2, C_4, C_5)$-free.*

We will also need the following classes of graphs:

- G is a *thin spider* if its vertex set is partitionable into a clique C and a stable set S with $|C| = |S|$ or $|C| = |S| + 1$ such that the edges between C and S are a matching and at most one vertex is not covered by the matching. The thin spider with 6 vertices is also called *net*. A graph is a *thick spider* if it is the complement of a thin spider. The complement of the net is called the *3-sun*.

- G is *matched co-bipartite* if its vertex set is partitionable into two cliques C_1, C_2 with $|C_1| = |C_2|$ or $|C_1| = |C_2| - 1$ such that the edges between C_1 and C_2 are a matching and at most one vertex is not covered by the matching. G is *co-matched bipartite* if G is the complement of a matched co-bipartite graph.

- A bipartite graph $B = (X, Y, E)$ is a *bipartite chain graph* [34] if there is an ordering $x_1, x_2, \ldots, x_k$ of all vertices in X such that $N(x_i) \subseteq N(x_j)$ for all $1 \leq i < j \leq k$. (Note that then also the neighborhoods of the vertices from Y are ordered by set inclusion.) If, moreover, $|X| = |Y| = k$ and $N(x_i) = \{y_1, \ldots, y_i\}$ for all $1 \leq i \leq k$, then B is prime. G is a *co-bipartite chain graph* if it is the complement of a bipartite chain graph.

- G is a *tractable graph* if G is $(4K_1, C_4, \text{claw})$-free and its vertex set can be partitioned into four (possibly empty) pairwise disjoint vertex sets Q_1, Q_2, Q_3 and Q_4 which induce cliques in G such that there are no edges between Q_i and Q_{i+2} for both $i = 1$ and $i = 2$. Note that $G[Q_i \cup Q_{i+1}]$, $i \in \{1, \ldots, 4\}$, are co-bipartite chain graphs. The P_6 and the C_7 are examples of (prime) tractable graphs.

3   Cographs, Clique-Width and Logical Expressibility of Problems

The P_4-free graphs (also called *cographs*) play a fundamental role in graph decomposition; see [16] for linear time recognition of cographs, [14,15,16] for more information on P_4-free graphs and [9] for a survey on this graph class and related ones. For a cograph G, either G or its complement is disconnected, and the *cotree* of G expresses how the graph is recursively generated from single vertices by repeatedly applying join and co-join operations. Note that the cographs are those graphs whose modular decomposition tree contains only join and co-join nodes as internal nodes.

Based on the following operations on vertex-labeled graphs, namely

(i) create a vertex u labeled by integer ℓ, denoted by $\ell(u)$,
(ii) disjoint union (i.e., co-join), denoted by $\oplus$,
(iii) join between all vertices with label i and all vertices with label j for $i \neq j$, denoted by $\eta_{i,j}$, and
(iv) relabeling all vertices of label i by label j, denoted by $\rho_{i \to j}$,

the notion of *clique-width* $cwd(G)$ of a graph G is defined in [17] as the minimum number of labels which are necessary to generate G by using the operations $(i) - (iv)$. It is easy to see that cographs are exactly the graphs whose clique-width is at most two. A *k-expression* for a graph G of clique-width k describes how G is recursively generated by repeatedly applying the operations $(i) - (iv)$ using at most k different labels. Observe that, trivially, the clique-width of a graph with n vertices is at most n. The following result by Johansson gives a slightly sharper bound.

Lemma 2 ([24]). *If G has n vertices then $cwd(G) \leq n-k$ as long as $2^k+2k \leq n$.*

Thus, the clique-width of a graph with nine vertices is at most seven.

Proposition 1 ([18,19]).

(i) *The clique-width $cwd(G)$ of a graph G is the maximum of the clique-width of its prime induced subgraphs.*
(ii) $cwd(\overline{G}) \leq 2 \cdot cwd(G)$.

In [18], it is shown that every problem expressible in a certain kind of Monadic Second Order Logic, called $LinEMSOL(\tau_{1,L})$, is linear-time solvable on any graph class with bounded clique-width for which a k-expression can be constructed in linear time. Roughly speaking, $MSOL(\tau_1)$ is Monadic Second Order Logic with quantification over subsets of vertices but not of edges; $MSOL(\tau_{1,L})$ is $MSOL(\tau_1)$ with additional vertex labels, and $LinEMSOL(\tau_{1,L})$ is the variant of $MSOL(\tau_{1,L})$ which allows to search for sets of vertices which are optimal with respect to some linear evaluation functions.

Theorem 1 ([18]). *Let $\mathcal{C}$ be a class of graphs of clique-width at most k such that there is an $\mathcal{O}(f(|E|, |V|))$ algorithm, which for each graph G in $\mathcal{C}$, constructs a k-expression defining it. Then for every $LinEMSOL(\tau_{1,L})$ problem on $\mathcal{C}$, there is an algorithm solving this problem in time $\mathcal{O}(f(|E|, |V|))$.*

The next, straightforward, proposition was already stated in [3].

Proposition 2. *The clique-width is at most 3 for chordless paths as well as for their complements, 4 for chordless cycles as well as for their complements,*

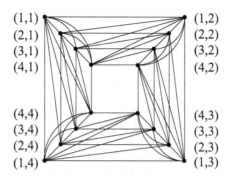

Fig. 3. The typical graph T_4

3 *for thin spiders, 4 for thick spiders, 3 for bipartite chain graphs, 3 for co-bipartite chain graphs, 4 for matched co-bipartite as well as for co-matched bipartite graphs, and corresponding k-expressions, $k \in \{3,4\}$, can be obtained in linear time.*

Finally in this section, we mention that tractable graphs have bounded clique-width which is used later for the proof that $(4K_1,C_4,\text{claw})$-free graphs have bounded clique-width. As a preparing step, we define *typical graphs* T_n (see Figure 3 for an example):

Let T_n be the graph with vertex set $\{1,2,\ldots,n\} \times \{1,2,3,4\}$ and edge set

$$\{(s,t)(x,y) \mid (y = t+1 \text{ and } x \le s) \text{ or } (x \ne s \text{ and } y = t)\}$$

(index arithmetic modulo 4). We call $\{1,2,\ldots,n\} \times \{i\}$ the *ith column* of T_n, $i \in \{1,2,3,4\}$. Note that the four columns of T_n are cliques, T_n is $4K_1$- and claw-free and there are no edges between non-consecutive columns but T_n is not tractable since it contains C_4.

Lemma 3. *The clique-width of typical graphs is at most 8.*

Proof. We give an 8-expression τ_n for the typical graph T_n: For $1 \le i \le n$, let τ_i be the expression defined inductively as follows:

$\tau_1 := \eta_{4,1}(\eta_{3,4}(\eta_{2,3}(\eta_{1,2}(1(1,1) \oplus 2(1,2) \oplus 3(1,3) \oplus 4(1,4)))));$
for $i := 2$ **to** n **do**
  begin
    $\alpha := \tau_{i-1} \oplus 5(i,1) \oplus 6(i,2) \oplus 7(i,3) \oplus 8(i,4);$
    $\beta := \eta_{1,5}(\eta_{2,6}(\eta_{3,7}(\eta_{4,8}(\alpha))));$
    $\gamma := \eta_{5,2}(\eta_{5,6}(\eta_{6,3}(\eta_{6,7}(\eta_{7,4}(\eta_{7,8}(\eta_{8,1}(\eta_{8,5}(\beta))))))));$
    $\tau_i := \rho_{5 \to 1}(\rho_{6 \to 2}(\rho_{7 \to 3}(\rho_{8 \to 4}(\gamma))))$
  end

Obviously, τ_n constructs T_n.   □

Lemma 4. *Every tractable graph with n vertices is an induced subgraph of T_n.*

Corollary 1. *The clique-width of tractable graphs is at most 8.*

4   Bounded Clique-Width

By Ramsey theory (see [22]), it is known that $(K_4, \overline{K_4})$-free graphs have at most 17 vertices. Moreover, the following is known:

Theorem 2 ([11]). *Let G be a prime graph.*

(i) *If G is (diamond,co-diamond)-free then G or $\overline{G}$ is a matched co-bipartite graph or G has at most nine vertices.*

(ii) *If G is (claw,co-claw)-free then G or $\overline{G}$ is an induced path or cycle or G has at most nine vertices.*

(iii) *If G is (paw,co-paw)-free then G is a P_4 or C_5.*

The following theorem collects the other inclusion-maximal classes of bounded clique-width defined by two forbidden 4-vertex subgraphs.

Theorem 3. *The following classes have bounded clique-width:*

(i)   *(diamond,co-paw)-free graphs;*

(ii) *(diamond,$2K_2$)-free graphs;*

(iii) *($2K_2$,paw)-free graphs;*

(iv) *(claw,paw)-free graphs;*

(v)   *(K_4,co-paw)-free graphs;*

(vi) *(K_4,co-diamond)-free graphs.*

The inclusion-maximal classes of bounded clique-width defined by three forbidden 4-vertex graphs are the following:

Theorem 4.

(i)   *Prime (K_4,C_4,$2K_2$)-free graphs have at most nine vertices;*

(ii) *Prime (C_4,claw,$2K_2$)-free graphs are thin spiders or have at most six vertices;*

(iii) *Prime (K_4,claw,$2K_2$)-free graphs have at most nine vertices.*

The most involved case is the following:

Theorem 5. *The clique-width of ($4K_1$,C_4,claw)-free graphs is bounded.*

If G is a $4K_1$-free graph then obviously, G is C_j-free for $j \geq 8$. The proof of Theorem 5 (given on 9 pages of the full version of this paper) can be done by a suitable case analysis:

Lemma 5.

(i)   *Prime ($4K_1$, claw)-free chordal graphs containing a net have at most seven vertices.*

(ii) *A prime ($4K_1$, claw, net)-free chordal graph containing a 3-sun is a 3-sun itself.*

(iii) *Prime ($4K_1$, claw)-free chordal graphs are tractable or have at most seven vertices.*

(iv) *The clique-width of prime* $(4K_1,C_4,claw)$-*free graphs containing* C_5 *is bounded.*

(v) *The clique-width of prime* $(4K_1,C_4,C_5,claw)$-*free graphs containing* C_6 *is bounded.*

(vi) *A prime* $(4K_1, C_4, C_5, C_6, claw)$-*free graph containing* C_7 *is the* C_7 *itself.*

In case (iv), we show that the graph can be partitioned into five tractable graphs in a suitable way, and in case (v), we show that the graph can be partitioned into twelve co-bipartite chain graphs in a suitable way.

5   Unbounded Clique-Width

In this section, we identify four inclusion-minimal classes of unbounded clique-width which are defined by forbidden 4-vertex graphs. Two of them are an immediate consequence of the following result by Makowsky and Rotics:

Theorem 6 ([26]). *The following graph classes have unbounded clique-width:*

(i) *split graphs;*
(ii) $H_{n,q}$ *grids.*

The $H_{n,q}$ grid is a variant of the $n \times n$ square grid. Theorem 6, Lemma 1 and the fact that $H_{n,q}$ grids contain no C_4 and no K_3, and thus no K_4, diamond, paw, and co-claw, imply:

Corollary 2. *The following classes have unbounded clique-width:*

(i) $(C_4,2K_2)$-*free graphs;*
(ii) $(K_4, diamond, C_4, paw, co$-$claw)$-*free graphs.*

For the other two cases, we show the following theorem:

Theorem 7. *The following classes have unbounded clique-width:*

(i) $(K_4,2K_2)$-*free graphs;*
(ii) $(K_4,diamond,C_4,claw)$-*free graphs.*

Theorem 7 (i) is an immediate consequence of the following stronger result:

Theorem 8. K_4-*free co-chordal graphs and, equivalently,* $4K_1$-*free chordal graphs, have unbounded clique-width.*

For the proof of Theorem 8, we analyze the following family of grids. For each integer $n \geq 1$, let G_n be the graph with vertex set $A \cup B \cup C$ and edge set $E_1 \cup E_2 \cup E_3$ defined as follows:

Consider a $(n+1) \times (n+1)$ square grid with vertex v_{ij} in row i and column j, $i,j \in \{0,\dots,n\}$. Omit the vertex v_{00}, take the vertices of the 0-th column as the A-vertices $a_i = v_{i0}$, take the vertices of the 0-th row as the B-vertices $b_i = v_{0i}$ and take the other vertices as the C-vertices $c_{ij} = v_{ij}$, $1 \leq i,j \leq n$.

The edges between A, B and C are defined as follows.

- A and B induce a complete bipartite graph; E_1 is the set of all edges $a_i b_j$, $1 \le i, j \le n$.
- Every vertex $a_i \in A$ is adjacent to every j-th row, $1 \le j \le i$; E_2 is the set of all edges $a_i v_{j\ell}$, $1 \le j \le i$, $1 \le \ell, i \le n$.
- Every vertex $b_i \in B$ is adjacent to every j-th column, $1 \le j \le i$; E_3 is the set of all edges $b_i v_{\ell j}$, $1 \le j \le i$, $1 \le \ell, i \le n$.

Thus, A, B, and C are stable sets, $|A| = |B| = n$, and $|C| = n^2$ such that $A \cup B$ induces a complete bipartite graph, and $A \cup C$ as well as $B \cup C$ induce a bipartite chain graph.

Observation 1. *For each $n \ge 1$, G_n is a K_4-free co-chordal graph.*

Lemma 6. *For each $n \ge 1$, $cwd(G_n) \ge n$.*

This may be surprising at first glance, since G_n consists of a pair of bipartite chain graphs $G[A \cup C]$, $G[B \cup C]$ with the property that A and B form a join, and is thus quite close to bipartite chain graphs whose clique-width is at most 3 but G_n has unbounded clique-width.

Observation 1 and Lemma 6 together show that K_4-free co-chordal graphs have unbounded clique-width which finishes the proof of Theorem 8.

Note that in exactly the same way, unbounded clique-width can be shown for the modified graphs G'_n, $n \ge 1$, which are defined like G_n but, instead of a join between A and B, there is a co-join between A and B. Obviously, these graphs G'_n, $n \ge 1$, are bipartite, and it is straightforward to show that they are P_8-free.

Corollary 3. *P_8-free bipartite graphs have unbounded clique-width.*

This is in contrast to the bounded clique-width of P_6-free bipartite graphs (see [21,25]). It seems to be an open question whether P_7-free bipartite graphs have unbounded clique-width.

The proof of Theorem 7 (ii) follows similar grid techniques.

6 Conclusion

The preceding results give a complete classification of all graph classes defined by forbidden induced subgraphs of four vertices with respect to bounded or unbounded clique-width.

Let $\mathcal{F}$ denote the set of the 10 graphs with four vertices (see Figure 1) different from P_4 (since the clique-width of P_4-free graphs is at most 2, we exclude P_4 from $\mathcal{F}$). For $\mathcal{F}' \subseteq \mathcal{F}$, there are 1024 classes of $\mathcal{F}'$-free graphs. For $\mathcal{F}' = \emptyset$, the class of $\mathcal{F}'$-free graphs is the class of all graphs (which has unbounded clique-width).

Recall that by Proposition 1 (ii), a class has bounded clique-width if and only if the class of its complement graphs has bounded clique-width. Thus, for each pair of classes of $(F_1, \ldots, F_k)$-free graphs and (co-F_1, $\ldots$, co-F_k)-free graphs, it suffices to mention only one of them (see Figure 2).

Note that any subclass of a class of bounded clique-width has bounded clique-width as well, whereas any superclass of a class of unbounded clique-width has unbounded clique-width as well. Thus, the classification is obtained by two types of key results:

- bounded clique-width of inclusion-maximal classes;
- unbounded clique-width of inclusion-minimal classes

such that all the other classes will be a subclass of a class of bounded clique-width or a superclass of a class of unbounded clique-width. These key results are Theorems 3, 4, and 5 for bounded clique-width and Corollary 2 and Theorems 7 and 8 for unbounded clique-width.

Then, a straightforward case analysis leads to a complete classification of all other classes defined by any subset of forbidden induced 4-vertex subgraphs. Thus, for $|\mathcal{F}'| = 1$, all classes have unbounded clique-width, for $|\mathcal{F}'| = 2$, there are 10 inclusion-maximal classes of bounded clique-width and two inclusion-minimal classes of unbounded clique-width, for $|\mathcal{F}'| = 3$, there are four inclusion-maximal classes of bounded clique-width, and for $|\mathcal{F}'| = 4$ and $|\mathcal{F}'| = 5$, there is one inclusion-minimal class of unbounded clique-width (see Figure 2).

For $6 \leq |\mathcal{F}'| \leq 10$, all classes of $\mathcal{F}'$-free graphs have bounded clique-width.

Now we consider forbidden subgraphs of at most four vertices. There are only four subgraphs H containing three vertices: the K_3, its complement $3K_1$, the P_3 and its complement. If a graph is P_3-free or $\overline{P_3}$-free then it is a cograph. Thus, the only interesting cases are the combinations of K_3 with four-vertex graphs.

- The class of (K_3,C_4)-free graphs has unbounded clique-width since $H_{n,q}$ grids are (K_3,C_4)-free (see Theorem 6).
- The class of (K_3,claw)-free graphs has bounded clique-width since these graphs are (claw,co-claw)-free (see Theorem 2).
- The class of $(K_3,\text{co-paw})$-free graphs has bounded clique-width since $(K_4,\text{co-paw})$-free graphs have bounded clique-width (see Theorem 3).
- The class of $(K_3,2K_2)$-free graphs has bounded clique-width since $(P_5,\text{diamond})$-free graphs have bounded clique-width (see [2]).
- The class of $(K_3,\text{co-diamond})$-free graphs has bounded clique-width since $(K_4,\text{co-diamond})$-free graphs have bounded clique-width (see Theorem 3).
- The class of $(K_3,4K_1)$-free graphs has bounded clique-width since, by Ramsey theory, $(K_3,4K_1)$-free graphs have at most eight vertices.

This implies that for all $H \in \{\text{claw, co-paw, } 2K_2, \text{ co-diamond, } 4K_1\}$, the class of (K_3,C_4,H)-free graphs has bounded clique-width.

As mentioned in Section 3, bounded clique-width has important algorithmic consequences. However, for the time bounds, it is essential to determine corresponding k-expressions efficiently. For $(K_4,\text{co-diamond})$-free graphs and for $(K_4,\text{co-claw},2K_2)$-free graphs, however, we do not have linear-time algorithms for determining corresponding k-expressions.

Open Problems

– What is the time complexity of determining k-expressions of (K_4,co-diamond)-free graphs and of (K_4,co-claw,$2K_2$)-free graphs?
– What is the clique-width of P_7-free bipartite graphs? (cf. Corollary 3)

Acknowledgements. The first and third author are grateful to Van Bang Le, Tilo Klembt, Suhail Mahfud and Thomas Szymczak for helpful discussions. We are also grateful to Sang-il Oum [30] for a simplified proof of Lemma 4.

References

1. R. Boliac and V.V. Lozin, On the clique-width of graphs in hereditary classes, *Proceedings of ISAAC* 2002, *Lecture Notes in Computer Science* 2518 (2002) 44-54
2. A. Brandstädt, (P_5,diamond)-Free Graphs Revisited: Structure and Linear Time Optimization, *Discrete Applied Math.* 138 (2004) 13-27
3. A. Brandstädt, F.F. Dragan, H.-O. Le and R. Mosca, New graph classes of bounded clique-width, Extended abstract in: *Conference Proceedings WG'2002, Lecture Notes in Computer Science* 2573, 57-67; full version appeared electronically in *Theory of Computing Systems*, 2004
4. A. Brandstädt, C.T. Hoàng and V.B. Le, Stability number of bull- and chair-free graphs revisited, *Discrete Applied Math.* 131 (2003) 39-50
5. A. Brandstädt and D. Kratsch, On the structure of (P_5,gem)-free graphs; *Discrete Applied Math.* 145 (2005) 155-166
6. A. Brandstädt, H.-O. Le and R. Mosca, Chordal co-gem-free graphs and (P_5,gem)-free graphs have bounded clique-width, *Discrete Applied Math.* 145 (2005) 232-241
7. A. Brandstädt, H.-O. Le and R. Mosca, Gem- and co-gem-free graphs have bounded clique-width, *Internat. J. of Foundations of Computer Science* 15 (2004) 163-185
8. A. Brandstädt, H.-O. Le and J.-M. Vanherpe, Structure and Stability Number of (Chair, Co-P, Gem)-Free Graphs, *Information Processing Letters* 86 (2003) 161-167
9. A. Brandstädt, V.B. Le and J.P. Spinrad, Graph Classes: A Survey, *SIAM Monographs on Discrete Math. Appl.*, Vol. 3, SIAM, Philadelphia (1999)
10. A. Brandstädt and V.V. Lozin, On the linear structure and clique-width of bipartite permutation graphs, *Ars Combinatoria* Vol. LXVII (2003) 273-281
11. A. Brandstädt and S. Mahfud, Maximum Weight Stable Set on graphs without claw and co-claw (and similar graph classes) can be solved in linear time, *Information Processing Letters* 84 (2002) 251-259
12. A. Brandstädt and R. Mosca, On the Structure and Stability Number of P_5- and Co-Chair-Free Graphs, *Discrete Applied Math.* 132 (2004) 47-65
13. A. Brandstädt and R. Mosca, On Variations of P_4-Sparse Graphs, *Discrete Applied Math.* 129 (2003) 521-532
14. D.G. Corneil, H. Lerchs and L.K. Stewart-Burlingham, Complement reducible graphs, *Discrete Applied Math.* 3 (1981) 163-174
15. D.G. Corneil, Y. Perl and L.K. Stewart, Cographs: recognition, applications, and algorithms, *Congressus Numer.* 43 (1984) 249-258
16. D.G. Corneil, Y. Perl and L.K. Stewart, A linear recognition algorithm for cographs, *SIAM J. Computing* 14 (1985) 926-934
17. B. Courcelle, J. Engelfriet and G. Rozenberg, Handle-rewriting hypergraph grammars, *J. Comput. Syst. Sciences*, 46 (1993) 218-270

18. B. Courcelle, J.A. Makowsky and U. Rotics, Linear time solvable optimization problems on graphs of bounded clique width, *Theory of Computing Systems* 33 (2000) 125-150

19. B. Courcelle and S. Olariu, Upper bounds to the clique-width of graphs, *Discrete Appl. Math.* 101 (2000) 77-114

20. S. Földes and P.L. Hammer, Split graphs, *Congres. Numer.* 19 (1977) 311-315

21. J.-L. Fouquet, V. Giakoumakis and J.-M. Vanherpe, Bipartite graphs totally decomposable by canonical decomposition, *Internat. J. Foundations of Computer Science* 10 (1999) 513-533

22. A.M. Gleason and R.E. Greenwood, Combinatorial relations and chromatic graphs, *Canadian J. Math.* 7 (1955) 1-7

23. M.C. Golumbic and U. Rotics, On the clique-width of some perfect graph classes, *Internat. J. of Foundations of Computer Science* 11 (2000) 423-443

24. Ö. Johansson, Clique-Decomposition, NLC-Decomposition, and Modular Decomposition - Relationships and Results for Random Graphs, *Congressus Numerantium* 132 (1998), 39-60; http://www.nada.kth.se/~ojvind/papers/CGTC98.pdf

25. V.V. Lozin, Bipartite graphs without a skew star, *Discrete Math.* 257 (2002) 83-100

26. J.A. Makowsky and U. Rotics, On the Clique-Width of Graphs with Few P_4's, *Int. J. Foundations of Computer Science* 10 (1999) 329-348

27. R.M. McConnell and J.P. Spinrad, Modular decomposition and transitive orientation, *Discrete Math.* 201 (1999) 189-241

28. G.J. Minty, On maximal independent sets of vertices in claw-free graphs, *J. Combin. Theory (B)* 28 (1980) 284-304

29. R.H. Möhring and F.J. Radermacher, Substitution decomposition for discrete structures and connections with combinatorial optimization, *Annals of Discrete Math.* 19 (1984) 257-356

30. Sang-il Oum, Personal communication, 2004

31. B. Randerath, The Vizing bound for the chromatic number based on forbidden pairs, *Dissertation Thesis* 1998, RWTH Aachen

32. B. Randerath, I. Schiermeyer and M. Tewes, Three-colourability and forbidden subgraphs. II: polynomial algorithms, *Discrete Math.* 251 (2002) 137-153

33. N. Sbihi, Algorithme de recherche d'un stable de cardinalité maximum dans un graphe sans étoile, *Discrete Math.* 29 (1980) 53-76

34. M. Yannakakis, The complexity of the partial order dimension problem, *SIAM J. Algebraic and Discrete Methods* 3 (1982) 351-358

On the Complexity of Uniformly Mixed Nash Equilibria and Related Regular Subgraph Problems

Vincenzo Bonifaci, Ugo Di Iorio, and Luigi Laura

Department of Computer and System Sciences,
University of Rome "La Sapienza",
Via Salaria, 113 - 00198 Rome, Italy
¶bonifaci, diiorio, laura⟨⟩dis.uniroma1.it

Abstract. We investigate the complexity of finding uniformly mixed Nash equilibria (that is, equilibria in which all played strategies are played with the same probability). We show that, even in very simple win/lose bimatrix games, deciding the existence of uniformly mixed equilibria in which the support of one (or both) of the players is at most or at least a given size is an NP-complete problem. Motivated by these results, we also give NP-completeness results for problems related to finding a regular induced subgraph of a certain size or regularity in a given graph, which can be of independent interest.

Classification: computational complexity, game theory, graph theory.

1 Introduction

The recent interaction between Game Theory and Theoretical Computer Science has led to a deep study of the computational issues underlying the basic game theoretic notions. A prominent object of these studies is the hardness of computing Nash equilibria in non-cooperative games [8]. Despite these attempts, however, the precise complexity of finding a Nash equilibrium in a given game is still unknown. Even in the two player case, the best algorithm known has an exponential worst-case running time [9]. Furthermore, when one requires equilibria with simple additional properties, the problem immediately becomes NP-hard [2, 5].

Motivated by these negative results, recent studies considered the problem of computing classes of simpler equilibria, such as pure equilibria [3]. Here we consider *uniformly mixed* equilibria, that is, Nash equilibria in which all the strategies played with nonzero probability by a player are played with the same probability. Uniformly mixed equilibria can be viewed, in a sense, as falling between pure and mixed Nash equilibria; playing a uniformly mixed strategy is probably the simplest way of mixing pure strategies.

Despite this apparent simplicity, we give NP-completeness results which hold even for a very constrained class of games, called *imitation simple bimatrix games*

M. Liśkiewicz and R. Reischuk (Eds.): FCT 2005, LNCS 3623, pp. 197–208, 2005.
© Springer-Verlag Berlin Heidelberg 2005

[1, 6]. An imitation simple bimatrix game is a two player game in which the payoffs of both players are in the set $\{0, 1\}$ and the payoff of one of the players (the *imitator*) is 1 if and only if he makes the same move as the opponent. Obviously, this can only strengthen our results, which continue to hold in the case of general games. Specifically, we show that it is NP-complete to decide if a given imitation simple bimatrix game has a uniformly mixed Nash equilibrium in which one (or both) of the players has a support of size at most, or at least, or precisely equal to a given size k.

Finally, motivated by the relation between these problems and the problem of deciding the existence of a regular induced subgraph with certain properties in a given graph, we give NP-completeness results for other natural variations of this last problem. In particular, we prove that it is NP-complete to decide if a graph has an induced regular subgraph of size at least k, or if it has an induced regular subgraph of regularity at least d, where k or d are given in the input.

The structure of this paper is as follows. In Section 2, we give the necessary definitions and notation. Then, in Section 3, we explain how the game-theoretic results follow from the graph-theoretic hardness results. The actual reductions are presented in Section 4, where we also present the completeness results for the other regular subgraph problems.

2   Definitions and Notation

In this section we explain the basic game-theoretical and graph-theoretical notions that we will use and we introduce our notation.

A *bimatrix game* is specified by two $n \times n$ matrices A and B, where n is the number of *pure* strategies; we will identify the set of pure strategies with the set $N = \{1, 2, \ldots, n\}$. The first player is called the *row* player and the second player is called the *column* player. If the row player plays strategy i and the column player strategy j, the payoff will be A_{ij} for the first player and B_{ij} for the second player.

A *mixed* strategy is a probability distribution over pure strategies, that is, a vector $x \in \mathbb{R}^n$ such that $\sum_i x_i = 1$ and for every $i \in N$, $x_i \geq 0$. The *support* supp(x) of a mixed strategy x is the set of pure strategies i such that $x_i > 0$. When the row player plays mixed strategy x and the column player plays mixed strategy y, their expected payoffs will be, respectively, $x^t A y$ and $x^t B y$ (x^t is the transpose of vector x). A mixed strategy x will be called *uniformly mixed*, or *uniform*, if, for every $i \in$ supp(x), $x_i = 1/|$supp$(x)|$.

A *Nash equilibrium* [7] of the game (A, B) is a pair of mixed strategies (x, y) such that for all mixed strategies $\overline{x}$ and $\overline{y}$, $x^t A y \geq \overline{x}^t A y$ and $x^t B y \geq x^t B \overline{y}$. A *Nash equilibrium strategy* for a player is a mixed strategy that is played in some Nash equilibrium by that player. A *uniformly mixed* Nash equilibrium is an equilibrium in which both players play uniformly mixed strategies.

We will consider only *imitation, simple* bimatrix games. A bimatrix game is *simple* if the entries of the matrices A and B can be only 0 or 1. A bimatrix game is an *imitation* game if one of the players, called the *imitator* has payoff

1 if he plays the same pure strategy as the opponent and payoff 0 otherwise. We will assume the row player to be the imitator. Thus, in an imitation simple bimatrix game matrix A is the identity matrix I. Actually, we will consider only imitation simple games (I, M) where M is the adjacency matrix of some simple undirected graph; that is, M is symmetric and for every i, $M_{ii} = 0$. Notice that in such games, the two players cannot win at the same time, that is, at least one player has payoff zero. However, they can both lose.

We now describe our graph-theoretical notation. Given a simple undirected graph $G = (V, E)$, we will use $G(S)$ to denote the subgraph induced by the vertices in the subset $S \subseteq V$. As a shorthand for $S \cup \{v\}$ (where $v \in V$), we will write $S + v$. We will use $d_{G(S)}(x)$ to denote the degree of node $x \in S$ in the subgraph of G induced by S.

3   Game-Theoretical Results

In this section we formulate our results on uniformly mixed equilibria and explain their connection with regular subgraph problems.

Let $G = (V, E)$ be an undirected graph. We will say that a subset S of vertices determines a *dominant-regularity induced subgraph* if there is a positive integer r such that

(i) $G(S)$ is r-regular;
(ii) for every $v \in V \setminus S$, the degree of v in $G(S + v)$ is at most r.

Our results on equilibria are based on the following Lemma.

Lemma 1. *Let (I, M) be an imitation simple bimatrix game where M is the adjacency matrix of some undirected graph G. Then the uniform Nash equilibrium strategies of the row player in (I, M) are in one-to-one correspondence with the dominant-regularity induced subgraphs of G. Moreover, for every equilibrium (x, y) such that x is uniformly mixed, there is a uniformly mixed symmetric equilibrium (x, x).*

Proof. Let S be a dominant-regularity induced subgraph with regularity r. Consider the unique uniformly mixed strategy x having support S (that is, $x_i = 1/|S|$ if $i \in S$ and $x_i = 0$ otherwise). By definition of S, we have that $|S|x^t M$ is a vector giving the degrees of every node v in the graph $G(S + v)$. But then $x^t M$ is maximal on coordinates $i \in S$; thus, if the row player plays x, the column player has no incentive to deviate from x. But if the second player plays x, the vector of incentives for the first player is $Ix = x$ and hence (x, x) is a uniformly mixed Nash equilibrium for (I, M).

For the other direction, let (x, y) be a Nash equilibrium such that x is uniformly mixed. Since the game is an imitation game, it can be easily checked that the support of x has to be included in the support of y. Let $S = \text{supp}(x)$. Since the column player has no incentive to deviate, for every $l \in N$ and for every i in the support of y, and in particular for every $i \in S$, $(x^t M)_i \geq (x^t M)_l$. Now $|S|(x^t M)_i = \sum_{j \in S} M_{ji}$ so we have

$$\sum_{j \in S} M_{ji} \geq \sum_{j \in S} M_{jl}$$

for every $i \in S$ and $l \in N$. But this last quantity is exactly the degree of l in $G(S + l)$. Thus S induces a dominant-regularity subgraph in G.

Reasoning as in the first part of the proof we can conclude that (x, x) has to be an equilibrium.   □

We can now state our main results.

Theorem 1. *It is* NP-*complete to decide, given a game and an integer k, whether the game has Nash equilibria in which the row player plays a uniformly mixed strategy having a support of size*

(i) at most k;
(ii) at least k;
(iii) equal to k.

Proof. The theorem follows from Lemma 1 and the fact that the graph-theoretical problems corresponding to (i) and (ii) (namely, finding a dominant-regularity induced subgraph of size at most or at least k) are NP-complete as shown in Section 4. Result (iii) follows[1] from a trivial reduction of (i) to (iii): since the maximum value of k is n, we can solve (i) by calling n times the procedure which decides (iii), with k taking all the possible values in $\{1, 2, \ldots, n\}$.   □

Theorem 2. *It is* NP-*complete to decide, given a game and an integer k, whether the game has Nash equilibria in which both players play the same uniformly mixed strategy having a support of size*

(i) at most k;
(ii) at least k;
(iii) equal to k.

Proof. Similar to the proof of Theorem 1, except that we also use the last part of Lemma 1.   □

Finally, we notice that the problem of deciding the existence of a uniformly mixed Nash equilibrium has always a positive answer. Indeed, such an equilibrium can be found by the following greedy algorithm for finding a dominant-regularity induced subgraph: let $S = \{v_0\}$ where v_0 is any vertex in G. Either $G(S)$ is a dominant-regularity induced subgraph (in which case we can stop), or there must be a vertex v_1 such that its degree in $G(S + v_1)$ is 1; in this case we add v_1 to S. We continue in this fashion, by adding to S vertices v_j such that the degree of v_j in $G(S + v_j)$ is j. If we reach $j = n - 1$ then the algorithm stops anyway since $G(N)$ would be a clique.

---

[1] This problem (like the others) is easily in NP since once the supports have been guessed, the equilibrium can be found by solving an appropriate linear program.

4   Hardness Results

In this section we give the NP-completeness results for the various regular sub-graph problems. We observe that, despite a superficial similarity, these results do not follow from the results of Yannakakis about induced subgraphs with an hereditary property [10], since regularity and dominant-regularity are not hereditary. All of the problems we consider are trivially in NP since a nondeterministic machine can guess the vertices of the induced subgraph and easily check the required conditions in polynomial time.

We first consider the following problem.

Problem 1. **MINIMUM DOMINANT-REGULARITY INDUCED SUBGRAPH (MIN-DIS)**

INSTANCE: a graph $G = (V, E)$ and an integer $k \leq |V|$.

QUESTION: Is there a subset $\tilde{V} \subseteq V$, with $|\tilde{V}| \leq k$, such that i) $G(\tilde{V})$ is a r-regular graph for some r and ii) $\forall v \in V - \tilde{V}$ it holds that $d_{G(\tilde{V})}(v) \leq r$?

Our reduction is from 3-Satisfiability [4], defined below.

Problem 2. **3-SATISFIABILITY (3SAT)**

INSTANCE: Collection $C = \{c_1, c_2, \ldots, c_m\}$ of clauses on a finite set $X = \{x_1, x_2, \ldots, x_n\}$ of variables such that $|c_i| = 3$ for $1 \leq i \leq m$.

QUESTION: Is there a truth assignment for U that satisfies all the clauses in C?

Theorem 3. MIN-DIS *is* NP-*complete.*

Proof. We transform 3SAT to MIN-DIS. Without loss of generality we assume that no clause includes both a literal and its opposite; we map a generic instance of 3SAT, i.e. a set of variables U and a set of clauses C to an instance of MIN-DIS in the following way:

- for each variable $x_i \in U$ we add two nodes x_i and $\overline{x}_i$ to V; we refer to this set of nodes as X;
- for each clause $c_j \in C$ we add one node c_j to V; for simplicity we refer to this set as C;
- we connect each node x_i to each other node in X except itself and its opposite $\overline{x}_i$;
- we connect each clause c_j to all the nodes in X except the three nodes that correspond to the literals that form the clause c_j itself;
- we pose $k = |U|$.

We now want to show that a) every solution of this instance of 3SAT is mapped to a solution of MIN-DIS and b) every solution of MIN-DIS is the preimage of a solution of the corresponding instance of 3SAT.

Proof of a). Consider a truth assignment that is solution of the instance of 3SAT. Let us denote by T the subset of X that corresponds to the literals that has the value **true** in this solution of 3SAT. Note that $|T| = n$, since for each variable only one between x_i and $\overline{x}_i$ has the value **true**. Let $\tilde{V} = T$, it is easy to verify that the graph $G(\tilde{V})$, induced by $\tilde{V}$ is a solution of MIN-DIS because: i) $G(\tilde{V})$ is $(n-1)$-regular, and ii) $\forall x \in X - T$ we have $d_{G(\tilde{V})}(x) = n - 1$ and $\forall c \in C$ we have $n - 3 \leq \tilde{d}(c) \leq n - 1$ because at least one (at most three) of the literals of the clause is **true**.

Proof of b). We first prove that $\tilde{V} \subset X$ (this means $\tilde{V} \cap C = \emptyset$). Let us assume that it exists $c \in C \cap \tilde{V}$; by the definition of MIN-DIS it follows that the graph $G(\tilde{V})$ is $d_{G(\tilde{V})}(c)$-regular and $d_{G(\tilde{V})}(c) \leq n$; note that in this case it holds, for at least one variable x_i, that both $x_i \notin \tilde{V}$ and $\overline{x}_i \notin \tilde{V}$ (otherwise $|\tilde{V}| \geq n + 1$ and this is not possible because $\tilde{V}$ is a solution). At least one between x_i and $\overline{x}_i$ is connected to c (because we assumed that no clause includes both a literal and its opposite), wlog assume that is x_i, but this means that $d_{G(\tilde{V})}(x_i) > d_{G(\tilde{V})}(c)$, therefore $\tilde{V}$ is not a solution because it violates condition ii) of MIN-DIS.

Now from $\tilde{V}$ we can derive a truth assignment that satisfies the formula. Note that for each couple $x_i, \overline{x}_i$ *at most* one literal is in $\tilde{V}$, otherwise, since $|\tilde{V}| = n$, there would be one couple $x_j, \overline{x}_j$ where both the nodes are outside $\tilde{V}$ and, therefore, they both violates the condition ii) of MIN-DIS because their degree is higher than the one of the nodes in $\tilde{V}$. The same argument holds to show that, for each couple $x_i, \overline{x}_i$ *at least* one node must belong to $\tilde{V}$ otherwise both nodes would violate condition ii). Therefore we have that, for each couple $x_i, \overline{x}_i$, exactly one is in $\tilde{V}$, and for each clause $c \in C$ there exists at least one $x \in \tilde{V}$ such that x and c do not share an edge, and this implies that each clause has at least one literal that satisfies it.   □

We now consider the maximization version of MIN-DIS.

Problem 3. **MAXIMUM DOMINANT-REGULARITY INDUCED SUBGRAPH (MAX-DIS)**

INSTANCE: a graph $G = (V, E)$ and an integer $k \leq |V|$.
QUESTION: Is there a subset $\tilde{V} \subseteq V$, with $|\tilde{V}| \geq k$, such that i) $G(\tilde{V})$ is a r-regular graph for some r and ii) $\forall v \in V - \tilde{V}$ it holds that $d_{G(\tilde{V})}(v) \leq r$?

The NP-hardness proof is by a reduction from Exact Cover by 3-Sets [4].

Problem 4. **EXACT COVER BY 3-SETS (X3C)**

INSTANCE: A finite set X with $|X| = 3q$ and a collection C of 3-element subsets of X.
QUESTION: Does C contain an *exact cover* for X, that is, a subcollection $C' \subseteq C$ such that every element of X occurs in exactly one member of C'?

Theorem 4. MAX-DIS *is* NP-*complete.*

Proof. We transform X3C to MAX-DIS. We build an instance of MAX-DIS from a generic instance of X3C:

- for each element $x \in X$ we add one node, that we still denote by x, to V; we denote by X the set of these nodes;
- for each collection $c \in C$ we add on node , that we still denote by c, to V; we denote by C the set of these nodes;
- (note that $V = X \cup C$)
- we fix an order over the nodes in X and we connect each node with the previous and the following one; the last node is connected to the first one, therefore all the nodes are connected in a cycle;
- we connect each node $c \in C$ to a node $x \in X$ if the subset c includes the element x;
- we pose $k = 4|X| = 4q$.

To prove that this is a transformation, we show that a) every solution of this instance of X3C is mapped to a solution of MAX-DIS and b) every solution of MAX-DIS is the preimage of a solution of the corresponding instance of X3C.

Proof of a). Let us still denote by C' the set of nodes in the instance of MAX-DIS that corresponds to the subsets solution of X3C. It holds that $\tilde{V} = X \cup C'$ is a solution of MAX-DIS because:

- $|\tilde{V}| = |X| + |C'| = 3q + q = 4q = k$;
- $G(\tilde{V})$ is 3-regular because i) each $x_i \in X$ is connected to the nodes x_{i-1}, x_{i+1} and exactly one node $c \in C'$ that corresponds to the subset including X_i in the solution of X3C; ii) each $c \in C'$, by definition, is connected to exactly three nodes in X;
- for each $c \in C - C'$ we have $d_{G(\tilde{V})}(c) = 3$, therefore condition ii) of problem MAX-DIS holds.

Proof of b). If we denote by r the regularity of graph $G(\tilde{V})$, it is easy to see that $r \leq 3$ because $|\tilde{V}| = 4q > 3q = |X|$ therefore it must exist one node $c \in \tilde{V} \cap C$ for which it holds $d_{G(\tilde{V})}(c) \leq d_{G(V)}(c) = 3$. We now prove the following points:

1. $r \neq 0$;
2. $r \neq 1$;
3. $r \neq 2$;
4. $X \subset \tilde{V}$ and there exists an exact cover of set X in the instance of X3C.

(1) To show that $r \neq 0$ we recall that at least one node $c \in C$ is also in $\tilde{V}$ and therefore it must exist one node $x \in X$, connected to c, for which $d_{G(\tilde{V})}(c) \geq 1$.

(2) Now let us assume $r = 1$, this means that $G(\tilde{V})$ is a matching. Since the nodes in C are not connected each other, at least one node of each couple must be an element x of X, therefore at least one half of $\tilde{V}$ are nodes in X, i.e. $|\tilde{V}| \leq 2|X \cap \tilde{V}|$. But from $k = 4q \leq |\tilde{V}| \leq 2|X \cap \tilde{V}|$ we derive that $|X \cap \tilde{V}| \geq 2q > 3/2q = |X|/2$; if this holds it means that more than half of X is in $\tilde{V}$, consequently at least one node x_i is in $\tilde{V}$ together with x_{i-1}, x_{i+1}, and the graph $G(\tilde{V})$ is not regular. So $r \neq 1$.

Note that, for any subsets $V', V'' \subseteq V$ such that $V = V' \cup V''$, we have

$$d_{G(V)}(v) = d_{G(V')}(v) + d_{G(V'')}(v) - d_{G(V' \cap V'')}(v). \tag{1}$$

(3) Now we assume $r = 2$. If $X \subset \tilde{V}$ there would not be any $v \in \tilde{V} \cap C$, contrarily to what we observed above. Let l be the number of nodes x that do not belong to $\tilde{V}$, i.e. $l = |X - \tilde{V}| > 0$ and let $j = |C \cap \tilde{V}| > 0$; by Equation 1 we see that

$$\sum_{x \in X \cap \tilde{V}} d_{G(\tilde{V})}(x) = \sum_{x \in X \cap \tilde{V}} d_{G(X \cap \tilde{V})}(x) + \sum_{x \in X \cap \tilde{V}} d_{G((C+x) \cap \tilde{V})}(x) \qquad (2)$$

$$\sum_{x \in X \cap \tilde{V}} d_{G(\tilde{V})}(x) = 2(3q - l) \qquad (3)$$

$$\sum_{x \in X \cap \tilde{V}} d_{G(X \cap \tilde{V})}(x) \geq 2(3q) - 4l = 6q - 4l \qquad (4)$$

the last one because every node $x_i \notin \tilde{V}$ decreases the overall degree by at most 4, and this happens only if both x_{i-1} and x_{i+1} belong to $\tilde{V}$; furthermore

$$\sum_{x \in X \cap \tilde{V}} d_{G((C+x) \cap \tilde{V})}(x) = 2j \geq 2(q + l)$$

where $j = |C \cap \tilde{V}| = |V| - (3q - l) \geq 4q - 3q + l$. This implies $2(3q - l) \geq 6q - 4l + 2(q + l) = 2(3q - l) + 2q > 2(3q - l)$, but this is not possible; therefore $r \neq 2$.

(4) From what we said before we know that $G(\tilde{V})$ is 3-regular. Again let $l = |X - \tilde{V}| > 0$, from the above considerations and Equation 2 we can write

$$\sum_{x \in X \cap \tilde{V}} d_{G(\tilde{V})}(x) = 3(3q - l)$$

$$\sum_{x \in X \cap \tilde{V}} d_{G(X \cap \tilde{V})}(x) \geq 2(3q) - 4l = 6q - 4l$$

$$\sum_{x \in X \cap \tilde{V}} d_{G((C+x) \cap \tilde{V})}(x) = 3j \geq 3(q + l)$$

analogously as the proof of previous point it follows $3(3q-l) \geq 6q-4l+3(q+l) = 3(3q - l) + 2l > 3(3q - l)$; therefore $|X - \tilde{V}| = 0$, i.e. $X \subset \tilde{V}$. Now $G(\tilde{V})$ being 3-regular implies that each $x \in X$ is connected to one and only one node $c \in \tilde{V}$: if we denote by C' the subsets corresponding to these nodes $c \in \tilde{V}$ it holds that C' is an exact cover of the set X. $\qquad \square$

The following two problems, although not directly related to equilibrium problems, are of independent interest.

Problem 5. **MAXIMUM SIZE REGULAR INDUCED SUBGRAPH (MAX-SRIS)**

   INSTANCE: A graph $G(V, E)$ and an integer $k \leq |V|$.

   QUESTION: Does there exist a set $\tilde{V} \subseteq V$ such that $|\tilde{V}| \geq k$ and $G(\tilde{V})$ is regular?

Theorem 5. MAX-SRIS *is* NP-*complete*.

Proof. Let us consider a generic instance of 3SAT consisting of a formula F with m clauses over n variables ($n \geq 1$); we assume, wlog, that $m = 2^q$ for some integer $q > 1$: note that it is always possible to build a formula F', satisfiable if and only if F is, by adding at most m copies of a clause of F. We create the corresponding instance of MAX-SRIS as follows:

- for each clause c_i we add three nodes, denoted by $c_{i,1}$, $c_{i,2}$ and $c_{i,3}$, *one for each literal in c_i*; let us denote by L the set of all these nodes.
- for each clause c_i we also add three auxiliary nodes, denoted by $c_{i,0}$, $c'_{i,0}$ and $c'_{i,1}$;
- for $1 \leq i \leq m$ we connect
  - $c_{i,0}$ with $c_{i,1}$, $c_{i,2}$ and $c_{i,3}$;
  - $c'_{i,1}$ with $c_{i,1}$, $c_{i,2}$ and $c_{i,3}$;
  - $c'_{i,0}$ with $c_{i,0}$ and $c'_{i,1}$;
- we add two binary trees T_1 and T_2, where $|T_1| = |T_2| = 2m - 1$. Note that both T_1 and T_2 have m leaves; for $1 \leq i \leq m$ we connect the i-th leaf of T_1 with $c_{i,0}$, $c_{i,1}$, $c_{i,2}$ e $c_{i,3}$; the i-th leaf of T_2 is connected with $c'_{i,0}$ e $c'_{i,1}$; finally we connect together the two roots;
- for $1 \leq i,j \leq m$ and $1 \leq t, t' \leq 3$ we connect $c_{i,t}$ with $c_{j,t'}$ if and only if they correspond to opposed literals.
- we pose $k = 8m - 2$

Note that, in the graph defined above, the following nodes have degree 3: all the internal nodes in T_1, all the nodes in T_2 and the nodes $c'_{i,0}$. Let us denote by Q the set of all the other nodes, i.e. the nodes x such that $d_{G(V)}(x) \neq 3$. We will show that a) every solution of this instance of 3SAT is mapped to a solution of MAX-SRIS and b) every solution of MAX-SRIS is the preimage of a solution of the corresponding instance of 3SAT.

Proof of a). Let s be a truth assignment function that satisfies F; consider now, for each clause c_i only one literal $c_{i,t}$ which is **true**. Let $\tilde{V}$ include all the nodes in $V - L$ together with the m nodes, in L, corresponding to the **true** valued chosen literals. It holds $|\tilde{V}| = 8m - 2 = k$; since s satisfies F there can't be an edge between any two nodes in $\tilde{V} \cap L$. Therefore $\forall v \in \tilde{V}$ we have that $d_{G(\tilde{V})}(v) = 3$, i.e. $G(\tilde{V})$ is regular.

Proof of b). Given $\tilde{V}$, solution of MAX-SRIS, such that $G(\tilde{V})$ is r-regular and $|\tilde{V}| \geq 8m - 2$, we show that this implies that F is satisfiable by proving the following points:

1. $r \geq 3$
2. $\tilde{V} \not\subseteq Q$
3. from $\tilde{V}$ it is possible to derive a truth assignment that satisfies the formula.

(1) Let us assume that $r \leq 2$; note that at most three nodes between $c_{i,0}$, $c_{i,1}$, $c_{i,2}$, $c_{i,3}$ and the i-th leaf of T_1 belong to $\tilde{V}$. Furthermore not all the internal nodes of T_1 and the nodes of T_2 (remember that the size of both T_1 and T_2 is $2m-1$) can be included in $\tilde{V}$ because there will be at least two nodes (the roots)

with degree 3. It follows that $|\tilde{V}| < 3m + (3m - 2) + 2m = 8m - 2$, where $2m$ is the total number of nodes $c'_{i,0}$ and $c'_{i,1}$.

(2) If $\tilde{V} = Q$, for every $1 \leq i \leq m$, $d_{G(\tilde{V})}(c_{i,0}) = 4$ and $d_{G(\tilde{V})}(c'_{i,1}) = 3$, it follows that $G(\tilde{V})$ is not regular. If $\tilde{V} \subset Q$ then $|\tilde{V}| < 6m \leq 8m - 2$.

(3) From the previous points it follows that $|\tilde{V} - Q| > 0$, therefore it must exists in $\tilde{V}$ one node x such that $d_{G(V)}(x) = 3$; from (1) all the nodes connected to x must belong to $\tilde{V}$, and, in particular, at least one node either in T_1 or in T_2; due to the recursive structure of the trees all the nodes of T_1 and T_2 must belong to $\tilde{V}$. Let us focus on the leaves of T_2, their degree must be 3, and this implies that $\forall i \quad c'_{i,0} \in \tilde{V}$ and $c'_{i,1} \in \tilde{V}$ where $1 \leq i \leq m$; analogously for each $c'_{i,0}$ we have $c_{i,0} \in \tilde{V}$. Hence, for each triple $c_{i,1}$, $c_{i,2}$ and $c_{i,3}$, exactly one node must belong to $\tilde{V}$ so that, for each i, $d_{G(\tilde{V})}(c'_{i,1}) = d_{G(\tilde{V})}(\text{leaves of } T_1) = 3$.

Let us denote by $\tilde{L}$ the set of nodes in L that belong to $\tilde{V}$, i.e. $\tilde{L} = \tilde{V} \cap L$. Note that i) $|\tilde{L}| = m$, and there is exactly one node for each clause, and ii) there can't be any edge between two nodes in $\tilde{L}$ otherwise their degree would be greater than 3. Therefore from $\tilde{L}$ we can derive a truth assignment by setting **true** the literals of F corresponding to nodes in $\tilde{L}$: F is satisfied because i) there is exactly one literal for each clause and ii) there is at most one between a literal and its opposite.   □

Problem 6. **MAXIMUM REGULARITY REGULAR INDUCED SUB-GRAPH (MAX-RRIS)**

INSTANCE: A graph $G(V, E)$ and an integer $r \leq |V|$.

QUESTION: Does there exist a set $\tilde{V} \subseteq V$ such that $G(\tilde{V})$ is γ-regular, with $\gamma \geq r$?

Theorem 6. MAX-RRIS *is* NP-*complete.*

Proof. We transform 3SAT to MAX-RRIS. The instance of 3SAT is a formula F with m clauses and n variables ($n \geq 1$) and, without loss of generality, let us assume that $m \geq 2$ and there is no clause c that includes both a literal and its opposite. Note that, for each formula F, it exists F', with $m' = 3(m-1)$ clauses and $n' = n+1$ variables, such that F' is satisfiable if and only if F is satisfiable: we build F' by adding $2m - 3$ identical clauses that includes only one literal, corresponding to a new variable, repeated three times. We now transform the generic instance of F' into an instance of MAX-RRIS in the following way:

- for each clause c_i we add
  - three nodes, that we denote by $c_{i,1}$, $c_{i,2}$ and $c_{i,3}$; *they represent the literals of the clause* and we refer to all of them, for $1 \leq i \leq m'$, as the set L;
  - three nodes a_i, b_i and c_i; let A, B e C denotes, respectively, the set of all the nodes a_i, b_i and c_i for $1 \leq i \leq m'$;
  - the following set of nodes: $A_{i,1}$, $A_{i,2}$, $A_{i,3}$, B_i, C_i, $C_{i,1}$, $C_{i,2}$ and $C_{i,3}$; each of these sets has cardinality equals to m'.
- we add two nodes, denoted by r_1 and r_2;
- we connect each node in the sets $A_{i,1}$, $A_{i,2}$, $A_{i,3}$, B_i, C_i, $C_{i,1}$ $C_{i,2}$ and $C_{i,3}$ with all the nodes in the same set, i.e. all the sets are m' elements cliques;

- we connect
  - each node a_i with all the nodes in $A_{i,1} \cup A_{i,2} \cup A_{i,3}$;
  - each node b_i with all the nodes in B_i;
  - each node c_i with all the nodes in C_i;
  - each node $c_{i,1}$ with all the nodes in $C_{i,1}$;
  - each node $c_{i,2}$ with all the nodes in $C_{i,2}$;
  - each node $c_{i,3}$ with all the nodes in $C_{i,3}$;
- we fix an ordering over the nodes of each set $X \in \{A_{i,1}, A_{i,2}, A_{i,3}, B_i, C_i, C_{i,1}, C_{i,2}, C_{i,3}\}$, let us denote by $X[i]$ the i-th node of X, with $1 \leq i \leq m'$; we then connect
  - each node $A_{i,t}[j]$ with the node $C_{i,t}[j]$ for all i, j, t such that $1 \leq i, j \leq m'$ and $1 \leq t \leq 3$;
  - each node $B_i[j]$ with the node $C_i[j]$ for all i, j such that $1 \leq i, j \leq m'$
- we connect each node c_i with the nodes $c_{i,1}$, $c_{i,2}$ and $c_{i,3}$;
- we connect r_1 with each node a_i and the node r_2 with each node b_i;
- we connect r_1 to r_2
- *each literal is connected to its opposite*: we connect each node $c_{i,t}$ to the ones $c_{\iota,\tau}$ that correspond to its opposite literal;
- let $r = m' + 1$.

We now show that is F' is satisfiable then it exists $\tilde{V}$ solution of the corresponding instance of MAX-RRIS. Let s be a truth assignment function satisfies F'. We build $\tilde{V}$ in the following way: for each clause we take one and only one node $c_{i,t}$ such that the corresponding literal has the value `true` in s; for each of such nodes we include i) the set $C_{i,t}$ of the connected nodes and ii) the set $A_{i,t}$ connected to $C_{i,t}$; then we include all the nodes a_i (i.e. the set A), all the nodes c_i (i.e. the set C), all the nodes b_i (i.e. the set B), all the sets B_i and C_i, and, finally, the nodes r_1 and r_2. It is easy to verify that the set $\tilde{V}$ induces a $(m' + 1)$-regular graph on G (because the nodes $c_{i,t}$ are chosen according to a truth assignment function there is no one connected to its opposite).
Now let us assume that it exists $\tilde{V} \subseteq V$ solution of MAX-RRIS; to show that this implies that F is satisfiable we prove the following points:

1. $\tilde{V} \not\subseteq L$
2. For each set $X \in \{A_{i,1}, A_{i,2}, A_{i,3}, B_i, C_i, C_{i,1}, C_{i,2}, C_{i,3}\}$, for all the value of j such that $1 \leq j \leq m'$ it holds that $X[j] \in \tilde{V}$ if and only if $X \subset \tilde{V}$. Moreover $A_{i,t} \subset \tilde{V}$ if and only if $C_{i,t} \subset \tilde{V}$ where $1 \leq i \leq m'$ and $1 \leq t \leq 3$; it also holds that $C_i \subset \tilde{V}$ if and only if $B_i \subset \tilde{V}$ where $1 \leq i \leq m'$;
3. for each $a_i \in A$ it holds that $a_i \in \tilde{V}$ if and only if it exists one and only one set $A_{i,t}$, amongst the sets $A_{i,1}, A_{i,2}$ and $A_{i,3}$ such that $A_{i,t} \subset \tilde{V}$; moreover $c_i \in \tilde{V}$ if and only if it exists $C_i \subset \tilde{V}$;
4. From $L \cap \tilde{V}$ is possible to derive a truth assignment that satisfies F.

(1) Each node $c_{i,t}$ is connected to at most $3(m-1)$ nodes of L, where $3(m-1) < 3m - 2 = m' + 1$: we built F' from F, and only the nodes that correspond to the literals in F (that has m clauses and therefore $3m$ literals) can be connected

together. If we consider only nodes in L it is not possible to reach the minimum degree $r = m' + 1$.

(2) Note that each set form a clique of size m', and every node in the set is also connected to only two external nodes. Therefore to reach the minimum degree r if a node $X[j]$ is in $\tilde{V}$ also all the ones connected to it must belong to $\tilde{V}$ and this, in particular, implies that $X \subset \tilde{V}$.

(3) Consider a node $a_i \in \tilde{V}$, from $m' + 1 = 3m - 2 > 3$ we can derive that it must exists one node $A_{i,t}[j] \in \tilde{V}$, where $1 \leq j \leq m'$ and $1 \leq t \leq 3$. From (2) it follows that $A_{i,t} \subset \tilde{V}$, and also that $C_{i,t} \subset \tilde{V}$. Assume now that there are two distinct nodes $A_{i,t}[j]$ and $A_{i,t'}[j']$, where $t \neq t'$, that belong to $\tilde{V}$; from (2) we derive $A_{i,t} \subset \tilde{V}$ and $A_{i,t'} \subset \tilde{V}$, but since a_i is connected to both the sets, it would have a degree $d_{G(\tilde{V})}(a_i) \geq 2m > m + 1$, and, therefore, the induced graph would not be a regular one. This proves the uniqueness of $A_{i,t}$. Moreover, if it holds that $c_i \in \tilde{V}$, since $m' + 1 = 3m - 2 > 3$, it must exist $C_i[j] \in \tilde{V}$, and this, from (2), implies $C_i \subset \tilde{V}$.

(4) From (1) we can derive that it must exist $x \in \tilde{V} - L$. Since $d_{G(V)}(r_1) = d_{G(V)}(r_2) = d_{G(V)}(b_i) = m' + 1$, from (2) and (3) we can derive that, for each triples $c_{i,1}, c_{i,2}, c_{i,3}$, it exists only one t such that $c_{i,t} \in \tilde{V}$ and $C_{i,t} \subset \tilde{V}$. This implies that there can't be in $\tilde{V} \cap L$ two nodes connected each other: among the corresponding literals there aren't, in the set, both a literal and its opposite. Therefore it is possible to set true all the literals whose corresponding nodes are in $\tilde{V} \cap L$; this truth assignment is feasible and satisfies F' and, hence, it satisfies F.   □

References

[1] B. Codenotti and D. Stefankovic. On the computational complexity of Nash equilibria for $(0 \cdot 1)$ bimatrix games. *Inform. Proc. Letters*, 94(3):145–150, 2005.

[2] V. Conitzer and T. Sandholm. Complexity results about Nash equilibria. In *Proc. 18th Int. Joint Conf. on Artificial Intelligence*, pages 765–771, 2003.

[3] A. Fabrikant, C. H. Papadimitriou, and K. Talwar. The Complexity of Pure Nash Equilibria. In *Proc. 36th Symp. on Theory of Computing*, pages 604–612, 2004.

[4] M. Garey and D. Johnson. *Computers and intractability. A guide to the theory of NP-completeness*. Freeman, 1978.

[5] I. Gilboa and E. Zemel. Nash and correlated equilibria: some complexity considerations. *Games and Economic Behavior*, 1(1):80–93, 1989.

[6] A. McLennan and R. Tourky. Simple complexity from imitation games. Available at http://www.econ.umn.edu/~mclennan/Papers/papers.html, 2005.

[7] J. F. Nash. Equilibrium points in n-person games. In *Proc. of the National Academy of Sciences*, volume 36, pages 48–49, 1950.

[8] C. H. Papadimitriou. On the complexity of the parity argument and other inefficient proofs of existence. *Journal of Computer and Systems Sciences*, 48(3):498–532, 1994.

[9] R. Savani and B. von Stengel. Exponentially many steps for finding a Nash equilibrium in a bimatrix game. In *Proc. 45th Symp. Foundations of Computer Science*, pages 258–267, 2004.

[10] M. Yannakakis. Node- and edge-deletion NP-complete problems. In *Proc. 10th Symp. Foundations of Computer Science*, pages 253–264, 1978.

Simple Stochastic Games and P-Matrix Generalized Linear Complementarity Problems⋆

Bernd Gärtner and Leo Rüst

Institute of Theoretical Computer Science,
ETH Zürich, CH-8092 Zürich, Switzerland
¶gaertner, ruestle⟨◊⟩@inf.ethz.ch

Abstract. We show that the problem of finding optimal strategies for both players in a simple stochastic game reduces to the generalized linear complementarity problem (GLCP) with a P-matrix, a well-studied problem whose hardness would imply NP = co-NP. This makes the rich GLCP theory and numerous existing algorithms available for simple stochastic games. As a special case, we get a reduction from binary simple stochastic games to the P-matrix linear complementarity problem (LCP).

1 Introduction

Simple stochastic games (SSG) form a subclass of general stochastic games, introduced by Shapley in 1953 [1]. SSG are two-player games on directed graphs, with certain random moves. If both players play optimally, their respective strategies assign values $v(i)$ to the vertices i, with the property that the first player wins with probability $v(i)$, given the game starts at vertex i. For a given start vertex s, the optimization problem associated with the SSG is to compute the *game value* $v(s)$; the decision problem asks whether the game value is at least $1/2$.

Previous Work. Condon was first to study the complexity-theoretic aspects of SSG [2]. She showed that the decision problem is in NP ∩ co-NP. This is considered as evidence that the problem is not NP-complete, because the existence of an NP-complete problem in NP ∩ co-NP would imply NP = co-NP. Despite this evidence and a lot of research, the question whether a polynomial time algorithm exists remains open.

SSG are significant because they allow polynomial-time reductions from other interesting classes of games. Zwick and Paterson proved a reduction from *mean payoff games* [3] which in turn admit a reduction from *parity games*, a result of Puri [4].

In her survey article from 1992, Condon reviews a number of algorithms for the optimization problem (and shows some of them to be incorrect) [5]. These algorithms compute optimal strategies for both players (we will say that they *solve* the game). For none of these algorithms, the (expected) worst-case behavior is known to be better than exponential in the number of graph vertices.

⋆ The authors acknowledge support from the Swiss Science Foundation (SNF), Project No. 200021-100316/1.

M. Liśkiewicz and R. Reischuk (Eds.): FCT 2005, LNCS 3623, pp. 209–220, 2005.

Ludwig was first to show that simple stochastic games can be solved in *subexponential* time [6], in the *binary* case where all outdegrees of the underlying graph are two. Under the known polynomial-time reduction from the general case to the binary case [3], Ludwig's algorithm becomes exponential, though. Björklund et al. [7] and independently Halman [8] established a subexponential algorithm also in the general case.

These subexponential methods had originally been developed for *linear programming* and the more general class of *LP-type problems*, independently by Kalai [9,10] as well as Matoušek, Sharir and Welzl [11]. Ludwig's contribution was to extract the combinatorial structure underlying binary SSG, and to show that this structure allows the subexponential algorithms to be applied. Halman was first to show that the problem of finding an optimal strategy for one of the players can actually be formulated as an LP-type problem [12]. Given a strategy, the other player's best response can be computed by a linear program. In a later result, Halman avoided linear programming by computing the other player's best response again by an LP-type algorithm. This resulted in *strongly* subexponential algorithms, the best known to date [8].

Independently, Björklund et al. arrived at subexponential methods by showing that SSG (as well as mean payoff and parity games) give rise to very specific LP-type problems [7]. Their contribution was to map all three classes of games to the single combinatorial problem of optimizing a *completely local-global* function over the Cartesian product of sets. Along with this, they also carried out an extensive study concerning the combinatorial properties of such functions.

Our Contribution. In this paper, we show that the problem of solving a general (not necessarily binary) simple stochastic game can be written as a *generalized linear complementarity problem* (GLCP) with a P-matrix. The GLCP, as introduced by Cottle and Dantzig [13], consists of a *vertical block* $(m \times n)$-*matrix* M where $m \geq n$ and a right-hand side m-vector q. M and q are partitioned in conformity into n horizontal blocks M^i and q^i, $i = 1, \ldots n$, where the size of block i is $m_i \times n$ in M and m_i in q. Solving a GLCP means to find a nonnegative m-vector w partitioned in conformity with M and q and a nonnegative n-vector z such that

$$w - Mz = q, \tag{1}$$

$$\prod_{j=1}^{m_i} w_j^i z_i = 0, \quad \forall i \in \{1, \ldots, n\}.$$

Here and in the following, w_j^i is the j-th element in the i-th block of w. M_j^i will denote the j-th row of the i-th block of M. A *representative submatrix* of M is an $(n \times n)$-matrix whose i-th row is M_j^i for some $j \in \{1, \ldots, m_i\}$. M is defined to be a P-matrix, if all principal minors of all its representative submatrices are positive [13].

In this paper, we will consider SSG with vertices of arbitrary outdegree and with *average* vertices determining the next vertex according to an arbitrary probability distribution. This is a natural generalization of binary SSG introduced by Condon [2]. A binary SSG will reduce to the more popular *linear complementarity problem* (LCP) where $m_i = 1$ for all n blocks.

The fact that there is a connection between games and LCP is not entirely surprising, as for example bimatrix games can be formulated as LCP [14]. Also Cottle, Pang and Stone [15, Section 1.2] list a simple game on Markov chains as an application for LCP, and certain (very easy) SSG are actually of the type considered.

LCP and methods for solving them are well-studied for general matrices M, and for specific matrix classes. The book by Cottle, Pang and Stone is the most comprehensive source for the rich theory of LCP, and for the various algorithms that have been developed to solve general and specific LCP [15]. A lot of results carry over to the GLCP. The significance of the class of P-matrices comes from the fact that M is a P-matrix if and only if the GLCP has a unique solution (w, z) for *any* right-hand side q [16]. Given this, the fact that our reduction yields a P-matrix already follows from Shapley's results. His class of games contains a superclass of SSG for which our reduction may yield *any* right-hand side q in (1). Shapley's theorem proving uniqueness of *game values* then implies that the matrix M in (1) must be a P-matrix. Our result provides an alternative proof of Shapley's theorem, specialized to SSG, and it makes the connection to matrix theory explicit.

No polynomial-time methods are known to solve P-matrix LCP, but Megiddo has shown that NP-hardness of the problem implies NP = co-NP [17], meaning that the problem has an unresolved complexity status, similar to that of SSG. Megiddo's proof easily carries over to P-matrix GLCP.

Gärtner et al. proved that the combinatorial structure of P-matrix GLCP is very similar to the structure derived by Björklund et al. for the games [18]. The latter authors also describe a reduction to what they call *controlled linear programming* [19]; controlled linear programs are easily mapped to (non-standard) LCP. Independently from our work, Björklund et al. have made this mapping explicit by deriving LCP-formulations for mean payoff games [20]. Their reduction is very similar to ours, but the authors do not prove that the resulting matrices are P-matrices, or belong to some other known class. In fact, Björklund et al. point out that the matrices they get are in general not P-matrices, and this stops them from further investigating the issue. We have a similar phenomenon here: Applying our reduction to *non-stopping* SSG (see next section), we may also obtain matrices that are not P-matrices. The fact that comes to our rescue is that the stopping assumption incurs no loss of generality. It would be interesting to see whether the matrices of Björklund et al. are actually P-matrices as well, after some transformation applied to the mean payoff game.

Matrix Classes and Algorithms. Various solution methods have been devised for GLCP, and when we specialize to the P-matrix GLCP we get from SSG, some of them are already familiar to the game community. Most notably, *principal pivot algorithms* in the GLCP world correspond to *switching* algorithms. Such an algorithm maintains a pair of strategies for both players and gradually improves them by locally switching to a different behavior. There exist examples of SSG where switching algorithms may cycle [5]. However, if switching is defined with respect to only one player (where after each switch the optimal counterstrategy of the other player is recomputed by solving a linear program), cycling is not possible. The latter is the setup of Björklund et al. [7].

In order to assess the power of the GLCP approach, we must understand the class of matrices resulting from SSG. Our result that these are P-matrices puts SSG into the realm of 'well-behaved' GLCP, but it does not give improved runtime bounds, let alone a polynomial-time algorithm. The major open question resulting from our approach is therefore the following: Can we characterize the subclass of P-matrices resulting from SSG? Is this subclass equal (or related) to some known class? In order to factor out the peculiarities of our reduction, we should require the subclass to be closed under scaling of rows and/or columns. Without having a concrete example, we believe that we obtain a proper subclass of the class of all P-matrices. This is because the GLCP restricted to (the variables coming from) any one of the two players is easy to solve by linear programming, a phenomenon that will not occur for a generic P-matrix.

The class of *hidden K-matrices* is one interesting subclass of P-matrices for which the GLCP can be solved in polynomial time. Mohan and Neogy [21] have generalized results by Pang [22] and Mangasarian [23,24] to show that a vertical block hidden K-matrix can be recognized in polynomial time through a linear program and that the solution to this linear program can in turn be used to set up another linear program for solving the GLCP itself.

The matrices we get from SSG are in general not hidden K-matrices. Still, properties of the subclass of matrices we ask for might allow their GLCP to be solved in polynomial time.

2   Simple Stochastic Games

We are given a finite directed graph G whose vertex set has the form

$$V = \{\mathbf{1}, \mathbf{0}\} \cup V_{max} \cup V_{min} \cup V_{avg},$$

where $\mathbf{1}$, the *1-sink*, and $\mathbf{0}$, the *0-sink*, are the only two vertices with no outgoing edges. For reasons that become clear later, we allow multiple edges in G (in which case G is actually a multigraph).

Vertices in V_{max} belong to the first player which we call the *max player*, while vertices in V_{min} are owned by the second player, the *min player*. Vertices in V_{avg} are *average* vertices. For $i \in V \setminus \{\mathbf{1}, \mathbf{0}\}$, we let $\mathcal{N}(i)$ be the set of neighbors of i along the outgoing edges of i. The elements of $\mathcal{N}(i)$ are $\{\eta_1(i), \ldots, \eta_{|\mathcal{N}(i)|}(i)\}$. An average vertex i is associated with a probability distribution $\mathcal{P}(i)$ that assigns to each outgoing edge (i, j) of i a probability $p_{ij} > 0$, $\sum_{j \in \mathcal{N}(i)} p_{ij} = 1$.

The SSG defined by G is played by moving a token from vertex to vertex, until it reaches either the 1-sink or the 0-sink. If the token is at vertex i, it is moved according to the following rules.

vertex type	rule
$i = \mathbf{1}$	the game is over and the max player wins
$i = \mathbf{0}$	the game is over and the min player wins
$i \in V_{max}$	the max player moves the token to a vertex in $\mathcal{N}(i)$
$i \in V_{min}$	the min player moves the token to a vertex in $\mathcal{N}(i)$
$i \in V_{avg}$	the token moves to a vertex in $\mathcal{N}(i)$ according to $\mathcal{P}(i)$

An SSG is called *stopping*, if no matter what the players do, the token eventually reaches $\mathbf{1}$ or $\mathbf{0}$ with probability 1, starting from *any* vertex. In a stopping game, there are no directed cycles involving only vertices in $V_{max} \cup V_{min}$. The following is well-known and has first been proved by Shapley [1], see also the papers by Condon [2,5]. Our reduction yields an independent proof of part (i).

Lemma 1. *Let G define a stopping SSG.*

(i) There are unique numbers $v(i), i \in G$, satisfying the equations

$$v(i) = \begin{cases} 1, & i = \mathbf{1} \\ 0, & i = \mathbf{0} \\ \max_{j \in \mathcal{N}(i)} (v(j)), & i \in V_{max} \\ \min_{j \in \mathcal{N}(i)} (v(j)), & i \in V_{min} \\ \sum_{j \in \mathcal{N}(i)} p_{ij} v(j), & i \in V_{avg} \end{cases} \quad . \tag{2}$$

(ii) The value $v(i)$ is the probability for reaching the 1-sink from vertex i, if both players play optimally.

For a discussion about what it means that 'both players play optimally', we refer to Condon's paper [5]. The important point here is that computing the numbers $v(i)$ solves the optimization version of the SSG in the sense that for every possible start vertex s, we know the *value* $v(s)$ of the game. It also solves the decision version which asks whether $v(s) \geq 1/2$. Additionally, the lemma shows that there are *pure* optimal strategies that can be read off the numbers $v(i)$: If v is a solution to (2), then an optimal pure strategy is given by moving from vertex i along one outoing edge to a vertex j with $v(j) = v(i)$.

The stopping assumption can be made without loss of generality: In a non-stopping game, replace every edge (i, j) by a new average vertex t_{ij} and new edges (i, t_{ij}) (with the same probability as (i, j) if $i \in V_{avg}$), (t_{ij}, j) with probability $1 - \epsilon$ and $(t_{ij}, \mathbf{0})$ with probability ϵ. Optimal strategies to this stopping game (which are given by the $v(i)$ values) correspond to optimal strategies in the original game if ϵ is chosen small enough [2].

3   Reduction to P-Matrix GLCP

In the following, we silently assume that G defines a stopping SSG and that every non-sink vertex of G has at least two outgoing edges (a vertex of outdegree 1 can be removed from the game without affecting the values of other vertices). In order to solve (2), we first write down an equivalent system of linear equations and inequalities, along with (nonlinear) *complementarity* conditions for certain pairs of variables. The system has one variable x_i for each vertex i and one *slack variable* y_{ij} for each edge (i, j) with $i \in V_{max} \cup V_{min}$. It has equality constraints

$$x_i = \begin{cases} 1, & i = 1 \\ 0, & i = 0 \\ y_{ij} + x_j, & i \in V_{max}, j \in \mathcal{N}(i) \\ -y_{ij} + x_j, & i \in V_{min}, j \in \mathcal{N}(i) \\ \sum_{j \in \mathcal{N}(i)} p_{ij} x_j, & i \in V_{avg}, \end{cases} \qquad (3)$$

inequality constraints

$$y_{ij} \geq 0, \quad i \in V_{max} \cup V_{min}, j \in \mathcal{N}(i), \qquad (4)$$

and complementarity constraints

$$\prod_{j \in \mathcal{N}(i)} y_{ij} = 0, \quad i \in V_{max} \cup V_{min} \qquad (5)$$

to model the max- and min-behavior in (2).

The statement of Lemma 1 (i) is equivalent to the statement that the system consisting of (3), (4) and (5) has a unique solution $x = (x_1, \ldots, x_n)$, and we will prove the latter statement. From the solution x, we can recover the game values via $v(i) = x_i$, and we also get the y_{ij}. Note that edges with $y_{ij} = 0$ in the solution correspond to *strategy edges* of the players.

It turns out that the variables x_i are redundant, and in order to obtain a proper GLCP formulation, we will remove them. For variables $x_i, i \notin V_{avg}$, this is easy.

Definition 1. *Fix $i \in V$.*

(i) The first path *of i is the unique directed path that starts from i, consists only of edges $(j, \eta_1(j))$ with $j \in V_{max} \cup V_{min}$, and ends at some vertex in $\{1, 0\} \cup V_{avg}$. The* second path *is defined analogously, with edges of the form $(j, \eta_2(j))$.*

(ii) The substitution S_i of x_i is recursively defined as the linear polynomial

$$S_i = \begin{cases} 1, & i = 1 \\ 0, & i = 0 \\ y_{i\eta_1(i)} + S_{\eta_1(i)}, & i \in V_{max} \\ -y_{i\eta_1(i)} + S_{\eta_1(i)}, & i \in V_{min} \\ x_i, & i \in V_{avg} \end{cases} \qquad (6)$$

(iii) $\bar{S}_i$ is the homogeneous polynomial obtained from S_i by removing the constant term (which is 0 or 1, as a consequence of (ii)).

Note that the first and the second path always exist, because there are no cycles involving only vertices in $V_{max} \cup V_{min}$ and each non-sink vertex has outdegree at least 2. The substitution S_i expresses x_i in terms of the y-variables associated with the first path edges of i, and in terms of the substitution of the last vertex on the first path, which is either an average vertex, or a sink.

Lemma 2. *The following system of equations is equivalent to (3).*

$$
\begin{aligned}
y_{ij} &= y_{i\eta_1(i)} + S_{\eta_1(i)} - S_j, & i \in V_{max}, j \in \mathcal{N}(i) \setminus \eta_1(i) \\
y_{ij} &= y_{i\eta_1(i)} - S_{\eta_1(i)} + S_j, & i \in V_{min}, j \in \mathcal{N}(i) \setminus \eta_1(i) \\
0 &= x_i - \sum_{j \in \mathcal{N}(i)} p_{ij} S_j, & i \in V_{avg}.
\end{aligned}
\tag{7}
$$

Proof. By induction on the length of the first path, it can be shown that in every feasible solution to (3), x_i has the same value as its substitution. The system (3) therefore implies (7). Vice versa, given any solution to (7), we can simply set

$$
x_i =
\begin{cases}
1, & i = \mathbf{1} \\
0, & i = \mathbf{0} \\
y_{i\eta_1(i)} + S_{\eta_1(i)}, & i \in V_{max} \\
-y_{i\eta_1(i)} + S_{\eta_1(i)}, & i \in V_{min},
\end{cases}
$$

to guarantee that x_i and S_i have the same value for all i. Then, the equations of (7) imply that x_i satisfies (3), for all i. □

3.1   A Non-standard GLCP

Let us assume that $V_{max} \cup V_{min} = \{1, \ldots, u\}, V_{avg} = \{u + 1, \ldots, n\}$. Moreover, for $i, j \in \{1, \ldots, u\}$ and $i < j$, we assume that there is no directed path from j to i that avoids average vertices. This is possible, because G restricted to $V \setminus V_{avg}$ is acyclic, by our stopping assumption. In other words, the order $1, \ldots, u$ topologically sorts the vertices in $V_{max} \cup V_{min}$, with respect to the subgraph induced by $V \setminus V_{avg}$. Defining vectors

$$
z = (y_{1\eta_1(1)}, \ldots, y_{u\eta_1(u)})^T, \quad w = (w^1, \ldots, w^u)^T, \quad x = (x_{u+1}, \ldots, x_n)^T, \tag{8}
$$

where $w^i = (y_{i\eta_2(i)}, \ldots, y_{i\eta_{|\mathcal{N}(i)|}(i)})$ is the vector consisting of the y_{ij} for all $j \in \mathcal{N}(i) \setminus \eta_1(i)$, conditions (4), (5) and (7)— and therefore the problem of computing the $v(i)$—can now be written as

$$
\begin{aligned}
&\text{find } w, z \\
&\text{subject to } w \geq 0, z \geq 0,
\end{aligned}
$$

$$
\prod_{j=1}^{|\mathcal{N}(i)|-1} w^i_j z_i = 0, \quad i \in V_{max} \cup V_{min} \tag{9}
$$

$$
\begin{pmatrix} w \\ 0 \end{pmatrix} - \begin{pmatrix} Q & C \\ A & B \end{pmatrix} \begin{pmatrix} z \\ x \end{pmatrix} = \begin{pmatrix} s \\ t \end{pmatrix},
$$

where

$$P = \begin{pmatrix} Q & C \\ A & B \end{pmatrix}, \quad q = \begin{pmatrix} s \\ t \end{pmatrix}$$

are a suitable matrix and a suitable vector. The vertical block matrix Q is partitioned according to w and encodes the connections between player vertices (along first paths), whereas the square matrix B encodes the connections between average vertices. A and C describe how player and average vertices interconnect.

3.2   The Structure of the Matrix P

The following three lemmas are needed to show that P is a P-matrix, i.e. all principal minors of all representative submatrices of P are positive.

Lemma 3. Q *is a* P-*matrix.*

Proof. Every representative submatrix of Q is upper-triangular, with all diagonal entries being equal to 1. The latter fact is a direct consequence of (7), and the former follows from our topological sorting: For any $k \in \mathcal{N}(j) \setminus \eta_1(j)$, the variable $y_{i\eta_1(i)}$ cannot occur in the equation of (7) for y_{jk}, $j > i$, because this would mean that i is on the first path of either $\eta_1(j) > j > i$ or $k > j > i$. Thus, every representative submatrix has determinant 1.   □

It can even be shown that Q is a hidden K-matrix [25].

Lemma 4. B *is a* P-*matrix.*

Proof. We may assume that the average vertices are 'topologically sorted' in the following sense. For $i, j \in \{u + 1, \ldots, n\}$ and $i \le j$, there is a neighbor $k \in \mathcal{N}(i)$ such that the first path of k avoids j. To construct this order, we use our stopping assumption again. Assume we have built a prefix of the order. Starting the game in one of the remaining average vertices, and with both players always moving the token along edges $(i, \eta_1(i))$, we eventually reach a sink. The last of the remaining average vertices on this path is the next vertex in our order.

Theorem 3.11.10 in the book by Cottle, Pang and Stone [15] states that a square matrix M with all off-diagonal entries nonpositive is a P-matrix if there exists a positive vector s such that $Ms > 0$. As B has all off-diagonal entries nonpositive by construction (last line of (7)), it thus remains to provide the vector s. We define s to be the monotone increasing vector

$$s_t = 1 - \epsilon^t, \quad t = 1, \ldots, |V_{avg}|,$$

where $\epsilon = \min_{i,j} p_{ij}$ is the smallest probability occuring in the average vertices' probability distributions over the outgoing edges. $s > 0$ as $0 < \epsilon \le 1/2$, and we claim that also $Bs > 0$. Consider the row of B corresponding to average vertex i, see (7). Its diagonal entry is a positive number x. We have $\epsilon \le x \le 1$ by our stopping assumption ($x < 1$ occurs if the first path of any neighbor of i comes back to i). The off-diagonal values of the row are all nonpositive and sum up to at least $-x$. Our topological sorting on the average vertices implies that the row elements to the right of x sum up to $-(x - \epsilon)$ at least. Assume that the diagonal element x is at position t in the row. Under these considerations, the value of the scalar product of the row with s is minimized if the last element of the row has value $-(x - \epsilon)$ and the element at position $t - 1$ has value $-\epsilon$. As claimed, by the following formula the value is then positive (at least for $1 < t < |V_{avg}|$, but the cases $t = 1$ and $t = |V_{avg}|$ can be checked in the same way):

$$-\epsilon(1 - \epsilon^{t-1}) + x(1 - \epsilon^t) - (x - \epsilon)(1 - \epsilon^{|V_{avg}|}) = \epsilon^t(1 - x) + \epsilon^{|V_{avg}|}(x - \epsilon) > 0.$$

□

We note that B is even a K-matrix. K-matrices form a proper subclass of hidden K-matrices [15]

Property 1. An $(n \times n)$ representative submatrix

$$P_{rep} = \begin{pmatrix} Q_{rep} & C_{rep} \\ A & B \end{pmatrix}$$

of P is given by a representative submatrix Q_{rep} of Q and C_{rep} which consists of the rows of C corresonding to the rows of Q_{rep}. P_{rep} corresponds to a subgame where edges have been deleted such that every player vertex has exactly two outgoing edges.

Such a subgame is a slightly generalized binary SSG, as average vertices can have more than two outgoing edges and arbitrary probability distributions on them.

Lemma 5. *Using elementary row operations, we can transform a representative submatrix P_{rep} of P into a matrix P'_{rep} of the form*

$$P'_{rep} = \begin{pmatrix} Q_{rep} & C_{rep} \\ 0 & B' \end{pmatrix}$$

with B' being a P-matrix.

Proof. We process the rows of the lower part (AB) of P_{rep} one by one. In the following, first and second paths (and thus also substitutions $\bar{S}_i$) are defined w.r.t. the subgame corresponding to P_{rep}.

For $k \in \{1, \dots, n\}$, let R_k be the k-th row of P_{rep} and assume that we are about to process $R_i, i \in \{u + 1, \dots, n\}$. According to (7), we have

$$R_i \begin{pmatrix} z \\ x \end{pmatrix} = x_i - \sum_{j \in \mathcal{N}(i)} p_{ij} \bar{S}_j. \tag{10}$$

We will eliminate the contribution of $\bar{S}_j$ for all $j \in \mathcal{N}(i)$, by adding suitable multiples of rows $R_k, k \in \{1, \dots, u\}$. For such a k, (7) together with (6) implies

$$R_k \begin{pmatrix} z \\ x \end{pmatrix} = \begin{cases} \bar{S}_k - \bar{S}_{\eta_1(i)}, & k \in V_{max} \\ \bar{S}_{\eta_1(i)} - \bar{S}_k, & k \in V_{min} \end{cases}. \tag{11}$$

Let $\eta_2^*(j)$ be the last vertex on the second path of j. Summing up (11) over all vertices on the second path of vertex k with suitable multiples from $\{1, -1\}$, the sum telescopes, and we get that for all $j \in \{1, \ldots, u\}$, $\bar{S}_j - \bar{S}_{\eta_2^*(j)}$ is obtainable as a linear combination of the row vectors

$$R_k \begin{pmatrix} z \\ x \end{pmatrix}, \quad k \leq u.$$

Actually, if j is an average vertex or a sink, we have $\eta_2^*(j) = j$, so that $\bar{S}_j - \bar{S}_{\eta_2^*(j)} = 0$ is also obtainable as a (trivial) linear combination in this case.

Thus, adding $(\bar{S}_j - \bar{S}_{\eta_2^*(j)})p_{ij}$ to (10) for all $j \in \mathcal{N}(i)$ transforms our current matrix into a new matrix whose i-th row has changed and yields

$$R_i' \begin{pmatrix} z \\ x \end{pmatrix} = x_i - \sum_{j \in \mathcal{N}(i)} p_{ij} \bar{S}_{\eta_2^*(j)}. \tag{12}$$

Moreover, this transformation is realized through elementary row operations. Because (12) does not contain any y-variables anymore, we get the claimed structure after all rows $R_i, i \in \{u+1, \ldots, n\}$ have been processed.

We still need to show that B' is a P-matrix, but this is easy. B encodes for each average vertex the average vertices reached along the first paths of its successors. According to (12), B' does the same thing, but replacing first paths with second paths. The two situations are obviously completely symmetric, so the fact that B is a P-matrix also yields that B' is a P-matrix. Note that in order to obtain a monotone vector s as in the proof of Lemma 4, we need to reshuffle rows and columns so that the corresponding vertices are 'topologically sorted' according to second paths. □

Lemma 6. *P is a P-matrix.*

Proof. We show that every representative submatrix P_{rep} of P is a P-matrix. By Lemma 5, $\det(P_{rep}) = \det(P_{rep}') = \det(Q_{rep})\det(B')$, so P_{rep} has positive determinant as both Q and B' are P-matrices by Lemmas 3 and 4. To see that all proper principal minors are positive, we can observe that any principal submatrix of P_{rep} is again the matrix resulting from a SSG. The subgame corresponding to a principal submatrix can be derived from the SSG by deleting vertices and redirecting edges. This may generate multiple edges, which is the reason why we allowed them in the definition of the SSG. (The easy details are omitted.)   □

3.3 A Standard GLCP

Problem (9) is a non-standard GLCP because there are variables $\{x$ with no complementarity conditions. But knowing that B is regular (as B is a P-matrix), we can express x in terms of z and obtain an equivalent standard GLCP, whose matrix is a *Schur complement* of B in P.

$$\begin{aligned}
&\text{find } w, z \\
&\text{subject to } w \geq 0, z \geq 0, \\
&\qquad\qquad w^T z = 0, \\
&\qquad\qquad w - (Q - CB^{-1}A)z = s - CB^{-1}t.
\end{aligned} \tag{13}$$

Lemma 7. $Q - CB^{-1}A$, *is a* P-*matrix.*

Proof. We have to show that every representative submatrix of $Q - CB^{-1}A$ is a P-matrix. Such submatrices are derived through $Q_{rep} - C_{rep}B^{-1}A$. It thus suffices to show that $Q_{rep} - C_{rep}B^{-1}A$ is a P-matrix, given that P_{rep} (as defined in Property 1) is a P-matrix. This is well known (see for example Tsatsomeros [26]). □

We have finally derived our main theorem:

Theorem 1. *A simple stochastic game is reducible in polynomial time to a generalized linear complementarity problem with a* P-*matrix.*

This theorem also provides a proof of Lemma 1: going through our chain of reductions again yields that the equation system in Lemma 1 (i) for the values $v(i)$ has a unique solution if and only if the GLCP (13) has a unique solution for the slack variables y_{ij}. The latter holds because the matrix of (13) is a P-matrix.

As mentioned earlier, the reduction works for a superclass of SSGin which edges are associated with a payoff. But for general stochastic games as introduced by Shapley [1], the reduction (as described in this paper) is not possible. This follows from two facts. First, optimal strategies for stochastic games are generally non-pure. Second, it is possible to get irrational solutions (vertex values) for the stochastic game even if all input data is rational. This is not possible for GLCP.

Acknowledgments

We thank Walter Morris for providing us with many insights about matrix classes and LCP algorithms. We also thank Nir Halman for inspiring discussions.

References

1. Shapley, S.: Stochastic games. Proceedings of the National Academy of Sciences, U.S.A. **39** (1953) 1095–1100
2. Condon, A.: The complexity of stochastic games. Information & Computation **96** (1992) 203–224
3. Zwick, U., Paterson, M.: The complexity of mean payoff games on graphs. Theor. Comput. Sci. **158** (1996) 343–359
4. Puri, A.: Theory of hybrid systems and discrete event systems. PhD thesis, University of California at Berkeley (1995)
5. Condon, A.: On algorithms for simple stochastic games. In Cai, J.Y., ed.: Advances in Computational Complexity Theory. Volume 13 of DIMACS Series in Discrete Mathematics and Theoretical Computer Science. American Mathematical Society (1993) 51–73
6. Ludwig, W.: A subexponential randomized algorithm for the simple stochastic game problem. Information and Computation **117** (1995) 151–155
7. Björklund, H., Sandberg, S., Vorobyov, S.: Randomized subexponential algorithms for infinite games. Technical Report 2004-09, DIMACS: Center for Discrete Mathematics and Theoretical Computer Science, Rutgers University, NJ (2004)

8. Halman, N.: Discrete and Lexicographic Helly Theorems and Their Relations to LP-type problems. PhD thesis, Tel-Aviv University (2004)
9. Kalai, G.: A subexponential randomized simplex algorithm. In: Proc. 24th annu. ACM Symp. on Theory of Computing. (1992) 475–482
10. Kalai, G.: Linear programming, the simplex algorithm and simple polytopes. Math. Programming **79** (1997) 217–233
11. Matoušek, J., Sharir, M., Welzl, E.: A subexponential bound for linear programming. Algorithmica **16** (1996) 498–516
12. Halman, N.: An EGLP formulation for the simple stochastic game problem, or a comment on the paper: *A subexponential randomized algorithm for the Simple Stochastic Game problem* by W. Ludwig. Technical Report RP-SOR-01-02, Department of Statistics and Operations Research (2001)
13. Cottle, R.W., Dantzig, G.B.: A generalization of the linear complementarity problem. Journal on Combinatorial Theory **8** (1970) 79–90
14. von Stengel, B.: Computing equilibria for two-person games. In: Handbook of Game Theory. Volume 3. Elsevier Science Publishers (North-Holland) (2002)
15. Cottle, R.W., Pang, J., Stone, R.E.: The Linear Complementarity Problem. Academic Press (1992)
16. Szanc, B.P.: The Generalized Complementarity Problem. PhD thesis, Rensselaer Polytechnic Institute, Troy, NY (1989)
17. Megiddo, N.: A note on the complexity of P-matrix LCP and computing an equilibrium. Technical report, IBM Almaden Research Center, San Jose (1988)
18. Gärtner, B., Morris, Jr., W.D., Rüst, L.: Unique sink orientations of grids. In: Proc. 11th Conference on Integer Programming and Combinatorial Optimization (IPCO). Volume 3509 of Lecture Notes in Computer Science. (2005) 210–224
19. Björklund, H., Svensson, O., Vorobyov, S.: Controlled linear programming for infinite games. Technical Report 2005-13, DIMACS: Center for Discrete Mathematics and Theoretical Computer Science, Rutgers University, NJ (2005)
20. Björklund, H., Svensson, O., Vorobyov, S.: Linear complementarity algorithms for mean payoff games. Technical Report 2005-05, DIMACS: Center for Discrete Mathematics and Theoretical Computer Science, Rutgers University, NJ (2005)
21. Mohan, S.R., Neogy, S.K.: Vertical block hidden Z-matrices and the generalized linear complementarity problem. SIAM J. Matrix Anal. Appl. **18** (1997) 181–190
22. Pang, J.: On discovering hidden Z-matrices. In Coffman, C.V., Fix, G.J., eds.: Constructive Approaches to Mathematical Models. Proceedings of a conference in honor of R. J. Duffin, New York, Academic Press (1979) 231–241
23. Mangasarian, O.L.: Linear complementarity problems solvable by a single linear program. Math. Program. **10** (1976) 263–270
24. Mangasarian, O.L.: Generalized linear complementarity problems as linear programs. Oper. Res.-Verf. **31** (1979) 393–402
25. Gärtner, B., Rüst, L.: Properties of vertical block matrices. Manuscript (2005)
26. Tsatsomeros, M.J.: Principal pivot transforms: Properties and applications. Linear Algebra and Its Applications **307** (2000) 151–165

Perfect Reconstruction of Black Pixels Revisited

Hans Ulrich Simon*

Fakultät für Mathematik, Ruhr-Universität Bochum,
D-44780 Bochum, Germany
simon@lmi.rub.de

Abstract. It has been shown recently in [5] that the visual secret sharing scheme proposed in [1] leads to the largest possible visual contrast among all schemes that perfectly reconstruct black pixels. The main purpose of this paper is to demonstrate that the largest optimal contrast (for this kind of schemes) equals the smallest possible error when we try to approximate a polynomial of degree k on $k + 1$ interpolation points by a polynomial of degree $k - 1$. Thus, the problem of finding a contrast-optimal scheme with perfect reconstruction of black pixels boils down to a well-known problem (with a well-known solution) in Approximation Theory. A second purpose of this paper is to present a tight asymptotic analysis for the contrast parameter. Furthermore, the connection between visual cryptography and approximation theory discussed in this paper (partially known before) may also find some interest in its own right.

1 Introduction

Visual cryptography and k-out-of-n secret sharing schemes are notions introduced by Naor and Shamir in [9]. A sender wishing to transmit a secret message distributes n transparencies among n recipients, where the transparencies contain seemingly random pictures. A k-out-of-n scheme achieves the following situation. If any k recipients stack their transparencies together, then a secret message is revealed visually. On the other hand, if only $k - 1$ recipients stack their transparencies, or analyze them by any other means, they are not able to obtain any information about the secret message. The reader interested in more background information about secret sharing schemes is referred to [9].

An important parameter associated with a scheme is its *contrast*. It attains values in the range from 0 to 1 that indicate the clarity with which the message becomes visible. Value 1 means that black pixels are visualized "perfectly black" and white pixels "perfectly white". Contrast 0 means that black and white pixels cannot be distinguished in the visualization. In general, the contrast parameter measures to which extent black pixels appear "more black" than white pixels in the reproduced image.

---

* This work was supported in part by the IST Programme of the European Community, under the PASCAL Network of Excellence, IST-2002-506778. This publication only reflects the authors' views.

M. Liśkiewicz and R. Reischuk (Eds.): FCT 2005, LNCS 3623, pp. 221–232, 2005.

A general construction of a k-out-of-n secret sharing scheme whose contrast comes close to optimality was provided by a series of three papers. In [3] (full paper in [4]), the authors present a linear program $LP(k, n)$ whose optimal solution represents a contrast-optimal k-out-of-n secret sharing scheme such that its value, say $C(k, n)$, equals the largest possible contrast. In [6] (full paper in [7]), it is shown that $4^{-(k-1)} \leq C(k, n) \leq 4^{-(k-1)} \frac{n^k}{n(n-1)\cdots(n-(k-1))}$. This implies that $C(k, n)$ equals $4^{-(k-1)}$ in the limit when n approaches infinity. These bounds were proven by revealing a central relation between the largest possible contrast in a secret sharing scheme and the smallest possible error in problems occurring in Approximation Theory. This work has been extended in [8], where it is shown how one can actually *construct* an almost contrast-optimal scheme (a question that had been left open in [6]).

In [12], the authors (motivated by studies on visual perception) propose a variant of secret sharing schemes where black pixels are always perfectly reconstructed in the reproduced image. This notion was further developed in [2,1]. In [5], it is shown that the scheme proposed in [1] leads to the largest possible contrast, say $C_{PB}(k, n)$, among all schemes with perfect reconstruction of black pixels. The authors show furthermore that

$$C_{PB}(k, n) = \frac{2}{1 + \sum_{i=0}^{k-1} \binom{n-k+i}{i}\binom{n}{k-1-i}} \tag{1}$$

and present the following bound in closed form:

$$\frac{2}{1 + \frac{n}{n-k+1}\binom{n-1}{k-1}2^{k-1}} < C_{PB}(k, n) < \frac{2}{2^{k-1}\binom{n-1}{k-1}} \tag{2}$$

The main purpose of this paper is to demonstrate that the deep connections between Visual Cryptography and Approximation Theory allow for a surprisingly easy computation of $C_{PB}(k, n)$. In fact, we show (by means of duality) that $C_{PB}(k, n)$ equals the smallest possible error when we try to approximate a polynomial of degree k on $k + 1$ interpolation points by a polynomial of degree $k - 1$. This, however, is a classical problem in approximation theory[1] and has a well-known solution.

Clearly, the formula for $C_{PB}(k, n)$ revealed by the approximation-theoretic considerations coincides with (1). Note that the bounds in (2) are only tight for fixed k and n approaching infinity. If $k = n$, then lower and upper bound differ by factor n. A second purpose of this paper is to tighten this asymptotic analysis. We present lower and upper bounds in closed form that differ at most by factor $1 + 1/n$ (for every choice of k and n).

The paper is structured as follows. Section 2 recalls the definition of a secret sharing scheme from [9] and the linear program $LP(k, n)$ from [3], and makes some notational conventions. Section 3 recalls some classical notions and facts from Approximation Theory. The central Section 4 is devoted to the connection

---

[1] Actually a relative of the problems corresponding to ordinary schemes without perfect reconstruction of black pixels.

between Visual Cryptography and Approximation Theory. Here, we present the approximation-theoretic interpretation of $C_{PB}(k, n)$ and show that it is actually a well-known parameter in disguise. The final section 5 presents the improved asymptotic analysis for $C_{PB}(k, n)$.

There are clearly more papers in Visual Cryptography (part of which deal with still other variants of the basic problem) than we are able to mention properly in this abstract. The reader interested in finding more pointers to the relevant literature is referred to the full paper.

2 Definitions and Notations

For the sake of completeness, we recall the definition of visual secret sharing schemes given in [9]. In the sequel, we simply refer to them under the notion *VSSS*. For a 0-1–vector v, let $H(v)$ denote the Hamming weight of v, i.e., the number of ones in v.

Definition 1 (VSSS,PBVSSS). *A k-out-of-n VSSS $C = (C_0, C_1)$ with m sub-pixels, contrast C and threshold d consists of two collections of Boolean $n \times m$ matrices $C_0 = [C_{0,1}, \ldots, C_{0,r}]$ and $C_1 = [C_{1,1}, \ldots, C_{1,s}]$, such that the following properties are valid:*

1. *For any matrix $S \in C_0$, the OR v of any k out of the n rows of S satisfies $H(v) \leq d - mC$.*
2. *For any matrix $S \in C_1$, the OR v of any k out of the n rows of S satisfies $H(v) \geq d$.*
3. *For any $q < k$ and any q-element subset $\{i_1, \ldots, i_q\} \subseteq \{1, \ldots, n\}$, the two collections of $q \times m$ matrices D_0 and D_1 obtained by restricting each $n \times m$ matrix in C_0 and C_1 to rows $i_1, \ldots, i_q$ are indistinguishable in the sense that they contain the same matrices with the same relative frequencies.*

If $d = m$ the VSSS is called visual secret sharing scheme with perfect reconstruction of black pixels *(or simply PBVSSS).*

k-out-of-n schemes are used in the following way to achieve the situation described in the introduction. The sender translates every pixel of the secret image into n sets of subpixels in the following way. If the sender wishes to transmit a white pixel, then she chooses one of the matrices from C_0 according to the uniform distribution. In the case of a black pixel, one of the matrices from C_1 is chosen. For all $1 \leq i \leq n$, recipient i obtains the i-th row of the chosen matrix as an array of subpixels, where a 1 in the row corresponds to a black subpixel and a 0 corresponds to a white subpixel. The subpixels are arranged in a fixed pattern, e.g. a rectangle. (Note that in this model, stacking transparencies corresponds to "computing" the OR of the subpixel arrays.)

The third condition in Definition 1 is often referred to as the "security property" which guarantees that any $k - 1$ of the recipients cannot obtain any information out of their transparencies. The "contrast property", represented by the first two conditions in Definition 1, guarantees that k recipients are able to recognize black pixels visually since any array of subpixels representing a black pixel contains a "significant" amount of black subpixels more than any array representing a white pixel. If the threshold parameter d equals m, any subpixel of a black pixel is black.

Definition 2 (Generator Matrices). *We say that a VSSS has generator matrices $G_0, G_1 \in \{0,1\}^{n \times m}$ if C_i consists of the $m!$ matrices that are obtained from G_i by column permutations.*

It is well-known that any VSSS can be normalized (without changing the contrast) such as to have generator matrices.

In [3], it was shown that the largest possible contrast $C(k, n)$ in a k-out-of-n VSSS coincides with the optimal value in the following linear program (with variables $\xi_0, \ldots, \xi_n$ and $\eta_0, \ldots, \eta_n$):

Linear Program LP(k,n)

max $\sum_{j=0}^{n-k} \binom{n-k}{j} \binom{n}{j}^{-1} (\xi_j - \eta_j)$ **subject to**
1. For $j = 0, \ldots, n$: $\xi_j \geq 0, \eta_j \geq 0$.
2. $\sum_{j=0}^{n} \xi_j = \sum_{j=0}^{n} \eta_j = 1$
3. For $l = 0, \ldots, k - 1$: $\sum_{j=l}^{n-k+l+1} \binom{n-k+1}{j-l} \binom{n}{j}^{-1} (\xi_j - \eta_j) = 0$.

Moreover, the variables ξ_j, η_j have the following interpretation for a VSSS that achieves the largest possible contrast with generator matrices G_0 and G_1:

- ξ_j is the fraction of columns of Hamming weight j in G_0.
- η_j is the fraction of columns of Hamming weight j in G_1.

Note that we obtain a PBVSSS iff any column in G_1 has Hamming weight at least $n - k + 1$ (such that any subpixel is colored black for at least 1-out-of-k recipients). Thus, in order to obtain a linear program whose optimal value coincides with the largest possible contrast in a k-out-of-n PBVSSS, we have only to add the following constraints to $LP(k, n)$:

4. $\eta_0 = \eta_1 = \cdots = \eta_{n-k} = 0$.

We denote this linear program as $LP_{PB}(k, n)$.

The following sections only use this linear program (and do not explicitly refer to Definition 1. We assume the reader to be familiar with the theory of linear programming (including the concept of duality). However, we do not go beyond the contents of standard books (like [10], for instance).

We make the following notational conventions. A vector with components v_i is denoted as $\boldsymbol{v}$. The all-zeros vector is denoted as $\boldsymbol{0}$. The analogous convention holds for $\boldsymbol{1}$. For sake of simplicity, we do not properly distinguish between row and column vectors when the meaning is well understood without this distinction. For instance, the inner product of two vectors $\boldsymbol{u}, \boldsymbol{v}$ is simply written as $\boldsymbol{uv}$.

We close this section with the definition of a central notion in this paper. A vector $(w_0, w_1, \ldots, w_n)$ is the *evaluation vector of function f* if $w_j = f(j)$ for $j = 0, \ldots, n$.

3   Prerequisites from Approximation Theory

In this subsection, we mention several classical results from Approximation Theory. The corresponding proofs can be looked up in any standard book (e.g., Chapter 1 in [11]).

Let P_d denote the set of polynomials (over the reals) of degree at most d. With each finite set $J \subset \mathbb{R}$, each continuous function $f : \mathbb{R} \to \mathbb{R}$, and each nonnegative integer d, we associate the following quantity:

$$E_d(f, J) = \min_{q \in P_d} \max_{x \in J} |f(x) - q(x)|. \tag{3}$$

$f - q$ is called *the error function (of polynomial q as approximation for f)*, and

$$E(f, J, q) = \max_{x \in J} |f(x) - q(x)|$$

is called *approximation error (of polynomial q as approximation for f on domain J)*. Given f and J, term $E_d(f, J)$ can therefore be interpreted as the smallest possible approximation error that can be achieved by a polynomial of degree at most d.

Assume $|J| \geq d + 1$. It is well-known that the minimum in equation (3) is attained for a unique polynomial in P_d, called the *best approximating polynomial of degree at most d for f on domain J* and denoted by $q_{d,f,J}$, or simply by q_* if d, f, J are evident from context.[2]

In this paper, we will be mainly interested in the following situation:

– f is a polynomial of degree k with leading coefficient λ_f.
– The approximating polynomials have degree at most $k - 1$.
– Domain J consists of $k + 1$ points, say $x_0 < x_1 < \cdots < x_k$.

Consider the auxiliary functions ω and I given by

$$\omega(x) := \prod_{i=0}^{k}(x - x_i) \text{ and } I(x) := \sum_{i=0}^{k} \prod_{j \neq i}(x - x_j) \cdot \frac{(-1)^i}{\omega'(x_i)} , \tag{4}$$

where the first derivation ω' of ω satisfies the equations

$$\omega'(x) = \sum_{i=0}^{k} \prod_{j \neq i}(x - x_j) , \; \omega'(x_i) = \prod_{j \neq i}(x_i - x_j) \text{ and sign}(\omega'(x_i)) = (-1)^{k-i} . \tag{5}$$

From these properties of ω, it is easy to infer the following properties of I:

1. $I(x)$ is the polynomial of degree k that takes value $(-1)^i$ on x_i for $i = 0, 1, \ldots, k$.
2. The $k - 1$ local extrema of I (zeroes of I') are found in the open intervals (x_i, x_{i+1}) for $i = 0, \ldots, k - 1$, respectively. This implies that $I(x) \leq +1$ for $x_i \leq x \leq x_{i+1}$ and even i. Symmetrically, $I(x) \geq -1$ for $x_i \leq x \leq x_{i+1}$ and odd i.

---

[2] Clearly, we have approximation error 0 if $|J| = d + 1$.

3. For the leading coefficient of I, denoted by λ_I, we get

$$\lambda_I = \sum_{i=0}^{k} \frac{(-1)^i}{\omega'(x_i)} \text{ and } \text{sign}(\lambda_I) = (-1)^k \ . \tag{6}$$

With these notations, the following holds (e.g., see [11]):

Theorem 1. *The best approximating polynomial for f on J is given by*

$$q_*(x) := f - \frac{\lambda_f}{\lambda_I} I \in P_{k-1} \ .$$

The error polynomial

$$r_*(x) := f(x) - q_*(x) = \frac{\lambda_f}{\lambda_I} I$$

satisfies

$$r_*(x_i) = (-1)^i \frac{\lambda_f}{\lambda_I}$$

such that

$$E_{k-1}(f, J) = \frac{|\lambda_f|}{|\lambda_I|} \ .$$

For ease of later reference, we say that r_* *alternates on J with* alternation type $\pm$ if $\lambda_f / \lambda_I > 0$. Symmetrically, r_* alternates on J with *alternation type* $\mp$ if $\lambda_f / \lambda_I < 0$.

For index i ranging from 0 to k, define

$$E_k^{\pm}(f, J, q) := \min\{s|\ f(x_i) - q(x_i) \leq +s \text{ if } i \text{ is even and}$$
$$f(x_i) - q(x_i) \geq -s \text{ if } i \text{ is odd}\} \ ,$$
$$E_k^{\pm}(f, J) := \min_{q \in P_{k-1}} E_k^{\pm}(f, J, q) \ ,$$
$$E_k^{\mp}(f, J, q) := \min\{s|\ f(x_i) - q(x_i) \geq -s \text{ if } i \text{ is even and}$$
$$f(x_i) - q(x_i) \leq +s \text{ if } i \text{ is odd}\} \ ,$$
$$E_k^{\mp}(f, J) := \min_{q \in P_{k-1}} E_k^{\mp}(f, J, q) \ .$$

Intuitively, $E_{k-1}^{\pm}(f, J, q)$ is a kind of one-sided approximation error, where q (as an approximating polynomial for f) is penalized on x_i only if it differs from $f(x_i)$ in the "wrong direction". By symmetry, the analogous remark is valid for $E_{k-1}^{\mp}(f, J, q)$. With these notations, the following holds (e.g., see [8]):

Theorem 2. *If $r_* = f - q_*$ alternates on J with alternation type $\alpha \in \{\pm, \mp\}$, then q_* (on top of being the minimizer for $E_{k-1}(f, J, q)$) is also the (unique) minimizer for $E_{k-1}^{\alpha}(f, J, q)$. In particular,*

$$E_{k-1}^{\alpha}(f, J) = E_{k-1}(f, J) \ .$$

We briefly note that $E_{k-1}^{\alpha}(f, J, q_*) = 0$ if $r_* = f - q_*$ alternates on J with the alternation type opposite to α.

4   Approximation Error and Contrast

4.1   Known Results

As explained in section 2, the largest possible contrast in a k-out-of-n scheme is the optimal value in linear program $LP(k, n)$. The particular structure of $LP(k, n)$ is captured in more abstract terms by the following definition. We say that a linear program LP is of *type BAP (Best Approximating Polynomial)* if there exists a matrix $A \in \mathbb{R}^{k \times (1+n)}$ and a vector $c \in \mathbb{R}^{n+1}$ such that LP (with variables $\boldsymbol{\xi} = (\xi_0, \dots, \xi_n)$ and $\boldsymbol{\eta} = (\eta_0, \dots, \eta_n)$) can be written in the following form:

The primal linear program $LP(A, c)$ of type BAP

max $c(\boldsymbol{\xi} - \boldsymbol{\eta})$ subject to
(LP1) $\boldsymbol{\xi} \geq 0, \boldsymbol{\eta} \geq 0$
(LP2) $\sum_{j=0}^{n} \xi_j = \sum_{j=0}^{n} \eta_j = 1$
(LP3) $A(\boldsymbol{\xi} - \boldsymbol{\eta}) = 0$

Furthermore, A and c must satisfy the following conditions:

(LP4) c is the evaluation vector of a polynomial of degree k.
(LP5) Matrix $A \in \mathbb{R}^{k \times (1+n)}$ has rank k, i.e., its row vectors are linearly independent.
(LP6) Each row vector in A is the evaluation vector of a polynomial of degree at most $k - 1$.

Conditions (LP5) and (LP6) imply that the vector space spanned by the rows of A consists of all evaluation vectors of polynomials from P_{k-1}.
We introduce the notation

$$n^{\underline{k}} = n(n-1) \cdots (n - (k-1))$$

for so-called "falling factorial powers" and proceed with the following result:

Lemma 1 ([6]). *The linear program $LP(k, n)$ is of type BAP. The leading coefficient of the polynomial f with evaluation vector c is $(-1)^k / n^{\underline{k}}$.*

In order to discuss subproblems of $LP(A, c)$, we introduce the following general notation. For all $J_1, J_2 \subseteq \{0, 1, \dots, n\}$, let $LP(A, c | J_1, J_2)$ be the linear program resulting from $LP(A, c)$ by adding the constraints $\xi_j = 0$ for all $j \notin J_1$ and $\eta_j = 0$ for all $j \notin J_2$. With these notations the following holds:

Theorem 3 ([8]). *Let $LP(A, c)$ be a problem of type BAP and let f denote the polynomial of degree k with evaluation vector c. Let $J_1, J_2 \subseteq \{0, 1, \dots, n\}$. Then the following holds for $LP(A, c | J_1, J_2)$ and its dual $DLP(A, c | J_1, J_2)$:*

1. *$DLP(A, c | J_1, J_2)$ is equivalent to the problem of finding a polynomial $q \in P_{k-1}$ with minimal loss. Here, the loss induced by q, is the smallest $s \in \mathbb{R}$ (strictly speaking: 2 times the smallest $s \in \mathbb{R}$) such that*

$$\forall j \in J_1 : f(j) - q(j) \leq +s \text{ and } \forall j \in J_2 : f(j) - q(j) \geq -s .$$

2. *If s_* is the optimal value for $DLP(A, c | J_1, J_2)$, then $2s^*$ is the optimal value for $LP(A, c | J_1, J_2)$.*

The loss induced by q (as defined in Theorem 3) is again a kind of one-sided error. We may therefore wonder whether Theorem 2 (relating one- and two-sided errors with each other) applies. This is however not readily possible for the following reasons:

- $J := J_1 \cup J_2$ may consist of more than $k + 1$ points.
- The ordered sequence of points in J is not necessarily distributed among J_1 and J_2 in an alternating fashion.

We will however see in the next subsection that the special problems corresponding to PBVSSS's are equivalent to problems that *are* covered by Theorem 2.

4.2   Linear Programs Induced by PBVSSS's

Recall that the problem $LP_{PB}(k, n)$ coincides with $LP(k, n)$ augmented by the constraints $\eta_0 = \cdots = \eta_{n-k} = 0$. In our general notation from the previous subsection, this is the problem $LP(A, c | \{0, \ldots, n\}, \{n - k + 1, \ldots, n\})$. In this section, we abstract again from the specific form of matrix A and vector c in $LP_{PB}(k, n)$. Instead we consider the subproblem $LP(A, c | \{0, \ldots, n\}, \{n - k + 1, \ldots, n\})$ of an arbitrary problem $LP(A, c)$ of type BAP. In section 5, we come back to the specific problem $LP_{PB}(k, n)$ and extrapolate the largest possible contrast $C_{PB}(k, n)$ from our general results.

Let $f \in P_k$ denote the polynomial with evaluation vector c. According to Theorem 3, the dual of $LP(A, c | \{0, \ldots, n\}, \{n - k + 1, \ldots, n\})$ looks as follows:

Problem 1: Find the polynomial $q \in P_{k-1}$ that minimizes s subject to

$$\forall j = 0, \ldots, n : \qquad f(j) - q(j) \leq +s$$
$$\forall j = n - k + 1, \ldots, n : f(j) - q(j) \geq -s .$$

For technical reasons, we consider also the following two problems:

Problem 2: Find the polynomial $q \in P_{k-1}$ that minimizes s subject to

$$\forall j = 0, n - k + 1, n - k + 2, \ldots, n : |f(j) - q(j)| \leq s .$$

Problem 3: Find the polynomial $q \in P_{k-1}$ that minimizes s subject to

$$\forall j = 0, n - k + 2, n - k + 4, \ldots : f(j) - q(j) \leq +s$$
$$\forall j = n - k + 1, n - k + 3, \ldots : \quad f(j) - q(j) \geq -s .$$

Note that the optimal solution q_* for problem 2 and its error polynomial $r_* = f - q_*$ are given in Theorem 1. Now, we show that the same polynomial q_* is also the optimal solution for problems 1 and 3:

Lemma 2. *Assume that $LP(A, c)$ is a problem of type BAP such that c is the evaluation vector of a polynomial $f \in P_k$. Assume furthermore that $\text{sign}(\lambda_f) = (-1)^k$ for the leading coefficient λ_f of f. Then, the optimal solution q_* for problem 2 is also an optimal solution for problems 1 and 3.*

Proof. For sake of brevity, set $x_0 = 0$, $x_i = n - k + i$ for $i = 1, \ldots, k$, and $J := \{x_0, x_1, \ldots, x_k\}$.

We first consider problems 2 and 3. Note that the optimal value for problem 2 is $E_{k-1}(f, J)$. The optimal value for problem 3 is $E_{k-1}^{\pm}(f, J)$. We know from Theorem 2 that both optimal values coincide (and are actually achieved by the same polynomial) if $r_* = f - q_*$ alternates on J with alternation type $\pm$, i.e., if $\lambda_f / \lambda_I > 0$. This is however the case since $\text{sign}(\lambda_I) = (-1)^k$ and λ_f has the same sign by assumption. Thus, problems 2 and 3 are equivalent.

Let's now bring problem 1 into play. Clearly, problem 3 (and thus problem 2) is a relaxation of problem 1. Thus, the equivalence between problems 2 and 1 would follow if the optimal solution q_* for problem 2 were a feasible solution (of the same cost) for problem 1. This can be seen as follows. Recall from Theorem 1 that the error polynomial has the form $r_*(x) = f(x) - q_*(x) = s_* I(x)$ with $s_* = |\lambda_f| / |\lambda_I| = \lambda_f / \lambda_I$ as the optimal value. According to the properties of I, we have $I(x) \leq 1$ for all $x \in [x_0, x_1] = [0, n - k + 1]$. Thus, $r_*(x) \leq s_*$ for all $x \in [0, n - k + 1]$. Thus, q_*, s_* satisfy the constraints for $j = 1, \ldots, n - k$ within problem 1. The remaining constraints within problem 1 are obviously satisfied.

Corollary 1. *All three problems have the same optimal value λ_f / λ_I, respectively. The corresponding primal problems have the optimal value $2\lambda_f / \lambda_I$, respectively.*

5   Calculation of the Optimal Contrast

We now return to our special problem $LP_{PB}(k, n)$. We will first calculate $C_{PB}(k, n)$ exactly (and clearly confirm formula (1) from [5]). Our main technical contribution in this section can be seen in the following bounds on $C_{PB}(k, n)$ (being in closed form and matching each other up to factor $1 + 1/n$):

$$\frac{2}{1 + \binom{n-1}{k-1} 2^{k-1} \frac{n+1}{n - (k-1)/2}} < C_{PB}(k, n) < \frac{2}{1 + \binom{n-1}{k-1} 2^{k-1} \frac{n}{n - (k-1)/2}} \qquad (7)$$

As mentioned in the beginning of section 4.2, $LP_{PB}(k, n)$ coincides with the problem $LP(A, c | J_1, J_2)$ for $J_1 = \{0, 1, \ldots, n\}$ and $J_2 = \{n - k + 1, n - k + 2, \ldots, n\}$. According to Lemma 1, c is the evaluation vector of a polynomial $f \in P_k$ with leading coefficient $\lambda_f = (-1)^k / n^k$.

In Corollary 1 (that we would like to apply), the optimal value of $LP(A, c | J_1, J_2)$ is given in terms of λ_f (here: $(-1)^k / n^k$) and λ_I. For this reason, we first calculate λ_I. Let $I(x)$ be the function obtained from (4) by setting $x_0 := 0$ and $x_i := n - k + i$ for $i = 1, \ldots, k$. According to (6) and (5), the leading coefficient of I is given by $\lambda_I = \sum_{i=0}^{k} \frac{(-1)^i}{\prod_{j \neq i}(x_i - x_j)}$. Plugging in the concrete values $x_0 = 0$ and $x_i = n - k + i$ for $i = 1, \ldots, k$, we obtain

$$\lambda_I = (-1)^k \left(\frac{1}{n^k} + \sum_{i=1}^{k} \frac{1}{(i-1)!(k-i)!(n-k+i)} \right) . \qquad (8)$$

We may now apply Corollary 1 and obtain $C_{PB}(k,n) = 2\lambda_f/\lambda_I$. However, just plugging in the formulas $\lambda_f = (-1)^k/n^{\underline{k}}$ and (8) leads to a result which is not easy to grasp. In the sequel, we try to find simple (exact and approximate) re-formations.

It will be technically convenient to study the expression λ_I/λ_f. Plugging in the definitions of λ_I and λ_f, we get $\lambda_I/\lambda_f = 1 + S_{k,n}$ for

$$S_{k,n} = \sum_{i=1}^{k} \frac{n^{\underline{k}}}{(i-1)!(k-i)!(n-k+i)} \; .$$

Note that $C_{PB}(k,n)$ can be written in terms of $S_{k,n}$ as follows:

$$C_{PB}(k,n) = \frac{2}{\lambda_I/\lambda_f} = \frac{2}{1 + S_{k,n}} \tag{9}$$

Clearly, the challenge is to evaluate $S_{k,n}$. To this end, we proceed as follows:

$$S_{k,n} = \sum_{i=1}^{k} \frac{n^{\underline{k}}}{(i-1)!(k-i)!(n-k+i)} \tag{10}$$

$$= \sum_{i=0}^{k-1} \binom{n}{k-i-1}\binom{n+i-k}{i} \tag{11}$$

$$= \sum_{i=0}^{k-1} \binom{n}{i}\binom{n-1-i}{k-1-i} \tag{12}$$

$$= \sum_{i=0}^{k-1} \frac{n}{n-i}\binom{n-1}{i}\binom{n-1-i}{k-1-i} \tag{13}$$

$$= \sum_{i=0}^{k-1} \frac{n}{n-i}\frac{(n-1)!}{i!(k-1-i)!(n-k)!} \tag{14}$$

$$= \frac{(n-1)!}{(n-k)!(k-1)!}\sum_{i=0}^{k-1} \frac{n}{n-i}\frac{(k-1)!}{i!(k-1-i)!} \tag{15}$$

$$= \binom{n-1}{k-1}2^{k-1}\sum_{i=0}^{k-1} \frac{n}{n-i}\binom{k-1}{i}2^{-(k-1)} \tag{16}$$

In (11), we made use of

$$\frac{n^{\underline{k}}}{(i-1)!(k-i)!(n-k+i)} = \frac{n^{\underline{k-i}}(n-k+i)(n-k+i-1)^{\underline{i-1}}}{(k-i)!(n-k+i)(i-1)!}$$

$$= \binom{n}{k-i}\binom{n-k+i-1}{i-1}$$

and shifted index i such that it now ranges from 0 to $k-1$. In (12), we substituted $k-1-i$ for i. In (13), we made use of

$$\binom{n}{i} = \frac{n}{n-i}\binom{n-1}{i}$$

and in (14), we have rewritten the product of two binomial coefficients as trinomial coefficient:

$$\binom{n-1}{i}\binom{n-1-i}{k-1-i} = \frac{(n-1)!}{i!(k-1-i)!(n-k)!}$$

Equations (15) and (16) are straightforward. Note that (9) and (11) lead to the formula (1) for $C_{PB}(k,n)$ that was proven in [5].

From (12), we get immediately

$$S_{1,n} = 1 \ , \quad S_{2,n} = 2n - 1 \ , \quad S_{n,n} = 2^n - 1 \ . \tag{17}$$

Equation (16) shows that $S_{k,n}$ can be interpreted as $\binom{n-1}{k-1}2^{k-1}$ times the expectation of $n/(n-i)$ when i is binomially distributed with respect to parameters $k-1, 1/2$, i.e., i represents the number of "heads" in $k-1$ independent tosses of a fair coin. Since the binomial distribution is centered around its average, the following seems to be a good guess for $S_{k,n}$:

$$\tilde{S}_{k,n} := \binom{n-1}{k-1}2^{k-1}\frac{n}{n-(k-1)/2} \tag{18}$$

For example,

$$\tilde{S}_{1,n} = 1 \ , \quad \tilde{S}_{2,n} = \frac{4n(n-1)}{2n-1} \ , \quad \tilde{S}_{n,n} = \frac{n}{n+1}2^n = 2^n\left(1 - \frac{1}{n+1}\right) \ . \tag{19}$$

We could clearly use Chernov's inequalities to bound in probability the absolute difference between $S_{k,n}$ and $\tilde{S}_{k,n}$. However, we can do much better:

Lemma 3. $S_{k,n}/\tilde{S}_{k,n}$ *strictly increases with* k.

The somewhat technical proof can be looked up in the full paper. We conclude from Lemma 3 and from (17) and (19) that

$$1 = \frac{S_{1,n}}{\tilde{S}_{1,n}} \le \frac{S_{k,n}}{\tilde{S}_{k,n}} \le \frac{S_{n,n}}{\tilde{S}_{n,n}} = \frac{2^n - 1}{2^n(1 - 1/(n+1))} < \frac{1}{1 - 1/(n+1)} = \frac{n+1}{n} = 1 + \frac{1}{n} \ .$$

For all $2 \le k \le n$, we can use $2/(1 + \tilde{S}_{k,n})$ as a strict upper bound and $2/\left(1 + \tilde{S}_{k,n}(n+1)/n\right)$ as a strict lower bound on $C_{PB}(k,n) = 2/(1 + S_{k,n})$. This leads to the bounds (7) on $C_{PB}(k,n)$ that we announced in the beginning of this section.

We finally briefly note without proof the following nice equality:

$$C_{PB}(k,n) = \sum_{i=0}^{k-1}\binom{n-k+i}{i}2^i$$

References

1. Carlo Blundo, Annalisa De Bonis, and Alfredo De Santis. Improved schemes for visual cryptography. *Design, Codes and Cryptography*, 24(3):255–278, 2001.
2. Carlo Blundo and Alfredo De Santis. Visual cryptography schemes with perfect reconstruction of black pixels. *Computer and Graphics*, 22(4):449–455, 1998.
3. Thomas Hofmeister, Matthias Krause, and Hans U. Simon. Contrast-optimal k out of n secret sharing schemes in visual cryptography. In *Proceedings of the 3rd International Conference on Computing and Combinatorics — COCOON '97*, pages 176–186. Springer Verlag, 1997.
4. Thomas Hofmeister, Matthias Krause, and Hans U. Simon. Contrast-optimal k out of n secret sharing schemes in visual cryptography. *Theoretical Computer Science*, 240(2):471–485, 2000.
5. Hiroki Koga and Etsuyo Ueda. The optimal $(t^c n)$-threshold visual secret sharing scheme with perfect reconstruction of black pixels. Submitted.
6. Matthias Krause and Hans U. Simon. Determining the optimal contrast for secret sharing schemes in visual cryptography. In *Proceedings of the 4th Latin American Conference on Theoretical Informatics*, pages 280–291, 2000.
7. Matthias Krause and Hans U. Simon. Determining the optimal contrast for secret sharing schemes in visual cryptography. *Combinatorics, Probability and Computing*, 12(3):285–299, 2003.
8. Christian Kuhlmann and Hans U. Simon. Construction of visual secret sharing schemes with almost optimal contrast. In *Proceedings of the Eleventh Annual ACM-SIAM Symposium on Discrete Algorithms*, pages 263–272, 2000.
9. Moni Naor and Adi Shamir. Visual cryptography. In *Proceedings of the Conference on Advances in Cryptology — EUROCRYPT '94*, pages 1–12. Springer-Verlag, LNCS 950, 1994.
10. Christos H. Papadimitriou and Kenneth Steiglitz. *Combinatorial Optimization: Algorithms and Complexity*. Prentice Hall, 1982.
11. Theodore J. Rivlin. *An Introduction to the Approximation of Functions*. Blaisdell Publishing Company, 1969.
12. Eric R. Verheul and Henk C. A. van Tilborg. Constructions and properties of k out of n visual secrete sharing schemes. *Design, Codes and Cryptography*, 11(2):179–196, 1997.

Adaptive Zooming in Point Set Labeling

Sheung-Hung Poon[1,*] and Chan-Su Shin[2,**]

[1] Department of Mathematics and Computer Science, TU Eindhoven,
5600 MB, Eindhoven, The Netherlands
spoon@win.tue.nl
[2] School of Electronics and Information, HUFS,
Mohyun-myun, Yongin-si, Gyunggi-do, Korea
cssin@hufs.ac.kr

Abstract. A set of points shown on the map usually represents special sites like cities or towns in a country. If the map in the interactive geographical information system (GIS) is browsed by users on the computer screen or on the web, the points and their labels can be viewed in a query window at different resolutions by zooming in or out according to the users' requirements. How can we make use of the information obtained from different resolutions to avoid doing the whole labeling from scratch every time the zooming factor changes? We investigate this important issue in the interactive GIS system. In this paper, we build low-height hierarchies for one and two dimensions so that optimal and approximating solutions for adaptive zooming queries can be answered efficiently. To the best of our knowledge, no previous results have been known on this issue with theoretical guarantees.

Keywords: Computational geometry, GIS, map-labeling, zooming.

1 Introduction

Point set labeling is a classical and important issue in the geographic information systems (GIS). An extensive bibliography about the map labeling can be found in [13]. The ACM Computational Geometry Impact Task Force report [5] identifies the label placement as an important research area. Nowadays, user interactivity is extremely crucial in such systems, especially for those systems available on the web. For the success of the interactivity and real-time navigation on maps in the system, the internal paradigm of the database needs to be carefully designed so that the system adjusts accordingly to satisfy the user requirements and efficiently answer the user queries.

Several aspects of the interactivity and adaption for GIS have been studied in [2,3,11,16]. In [11,9,14], it is pointed out that the zooming operation in the interactive GIS is an important issue. Petzold et al. [11] considered the problem

* This research was supported in part by the Netherlands' Organisation for Scientific Research (NWO) under project no. 612-065-307.
** This research was supported in part by the Hankuk University of Foreign Studies Research Fund of 2005.

M. Liśkiewicz and R. Reischuk (Eds.): FCT 2005, LNCS 3623, pp. 233–244, 2005.

of zooming the map by using a data structure called the reactive conflict graph. Its purpose is to minimize the dynamic query time after extensive preprocessing. At the preprocessing stage, they created a complete graph between any pair of points. Each edge of the graph stores the scaling ratio when the labels of the two points start to overlap. Firstly this process is definitely slow. Secondly at any specific zooming factor, this process cannot guarantee the size of the query output when comparing to the optimal size at that resolution. The obvious reason is that this data structure does not store any clue about the optimal solution at a specific resolution.

At any resolution, we consider the following problem. Given n distinct points $P = \{p_i : 1 \leq i \leq n\}$ in the plane, each p_i is associated with a constant number, say κ, of axis-parallel rectangular labels of unit height and of width ω_i such that p_i lies on the left boundary of its κ labels. The goal is to maximize the number of non-overlapping labels for P. We call this problem the κ-fixed-position problem. We note that the one dimensional version of this problem considers all the points of P lying on the x-axis. Even the 1-fixed problem in two dimensions is NP-complete [7] although Roy et al. [12] showed that a special variant of the one-fixed-position problem can be solved in $O(n \log n)$ time. Moreover, several variants of the above stated problem are proven to be NP-complete [7,8,10]. Agarwal, van Kreveld and Suri [1] showed that a 2-approximation of the κ-fixed-position problem can be computed in $O(n \log n)$ time, and a $(1 + \epsilon)$-approximation of the problem for any $\epsilon > 0$ can be computed in $O(n \log n + n^{\frac{2}{\epsilon} - 1})$ time. Chan [4] improved the running time for finding a $(1 + \epsilon)$-approximation to $O(n \log n + n\Delta^{\frac{1}{\epsilon} - 1})$, where $\Delta \leq n$ denotes the maximum number of labels a point lies inside. Moreover, several sliding versions of this problem were extensively studied in [15]. In this paper, we define the zooming problem properly and precisely, and the we build a low-height hierarchy for efficient adaptive queries with theoretically guaranteed output.

A *zooming* on a set of point means that while the point-to-point distances are scaled by a constant factor, the label sizes of the points remain fixed. The *zooming query within a rectangular query window* W is that given any zooming scale, we want to find the optimal solution for the κ-fixed-position solution for the labels completely inside the query window W. Instead of directly considering the zooming problem, we consider another equivalent problem. Now suppose we do not perform any zooming, meaning that we fix the point-to-point distances, and we instead scale the font-size of the label texts by a constant scaling factor. The *font-scaling query within a rectangular query window* W is that by applying any scaling factor on the font of the label texts, we want to find the optimal solution for the κ-fixed-position solution for those labels completely inside the query window W. It is clear that our original zooming problem is equivalent to the font-scaling problem. For the simplicity of notations, we will only consider the font-scaling problem for the rest of the paper. See Figure 1 for an example of labels in two different font sizes, in which optimal sets of labels are drawn with solid lines. We denote the scaling factor at a resolution γ by ρ_γ. The point set is said to be at the *coarser* resolution and has a *larger* scaling factor if it

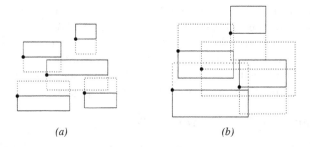

(a)   (b)

Fig. 1. An example when $\kappa = 2$. The font size/scaling factor in the right figure (b) is twice as larger as that in the left figure (a). Note that the collections of all the solid labels in both figures are optimal solutions.

has a relatively larger font size; otherwise the point set is said to be at *finer* resolution and has a *smaller* scaling factor. We will use these two terminologies interchangeably to mean the same thing. For example, in Figure 1, the point set in Figure 1 (a) lies at a resolution α finer than the resolution β at which the point set in Figure 1 (b) lies. This means that the left figure has a scaling factor ρ_α smaller than the scaling factor ρ_β of the right figure.

In this paper, in order to achieve efficient adaptive zooming querying, the main backbone structure we build is a hierarchy of $O(\log n)$ levels, where each level represents one resolution, the lowest level has the finest resolution, and the resolutions become coarser and coarser when the levels in the hierarchy increase. On each level of the hierarchy, we store some data structures so that we can efficiently find the optimal or approximating solutions for adaptive zooming queries for the resolutions between any pair of consecutive levels. In one dimension, we build an $O(\log n)$-height hierarchy in $O(n \log n)$ time and in $O(n)$ space as stated in Theorem 1, and we can answer a zooming query for optimal solution efficiently as stated in Theorem 2. In two dimensions, we build an $O(\log n)$-height hierarchy in $O(n^2)$ time and in $O(n \log n)$ space as stated in Theorem 3, and we can answer a zooming query for approximating solution efficiently as stated in Theorem 4. In Section 2, we investigate the one-dimensional zooming problem. The two-dimensional version is studied in Section 3. Finally we conclude in Section 4.

2   Adaptive Zooming in One Dimension

Consider all the points p_i in P lies on x-axis. Each point p_i can choose to take any label say σ_i from its κ fixed-position choices. Label σ_i at point p_i can be represented as an interval $[p_i, p_i + \omega_i]$. Now finding the maximum number of non-overlapping labels is equivalent to finding the maximum independent set of the intervals $[p_i, p_i + \omega_i]$ where ω_i is the width of σ_i. The optimal solution can be computed using a greedy algorithm as described below. First, sort the right endpoints $p_i + \omega_i$ of the labels. Then select the label successively whose right endpoint is leftmost and moreover which does not intersect the label just selected. This algorithm runs in $O(n \log n)$ time. For any subset S of P, let

$OPT_\gamma(S)$ denote the optimal solution at resolution γ obtained by running the above greedy algorithm on the labels at points in S. We denote $OPT_\gamma(P)$ by simply OPT_γ. Using the same greedy fashion, we describe a way to find the optimal solution at some resolution β by making use of the optimal solution at a finer resolution α in the following section. It will later serve as a main subroutine to build the hierarchy and to answer the zooming queries.

2.1   Computing $OPT_\beta(S)$ from OPT_α

Assume that the label at p_i is represented as intervals $[p_i, p'_{i,\alpha}]$ and $[p_i, p'_{i,\beta}]$ on x-axis at resolutions α and β respectively. It is clear that $p'_{i,\beta} > p'_{i,\alpha}$ as $\rho_\beta > \rho_\alpha$. Let $[q_k, q'_k]$ denote the kth label of OPT_α in the order from left to right. The algorithm to construct $OPT_\beta(S)$ in a greedy fashion by using OPT_α is presented in Algorithm $ComputeOPT$. In the algorithm, B_k denotes the subset of labels for points in S at resolution β such that the right endpoints (in sorted order) of the labels lie inside the interval $[q'_k, q'_{k+1})$ for some $1 \leq k \leq |OPT_\alpha|$. Note that B_k includes $[q_k, q'_k]$ itself. We assume $q'_{|OPT_\alpha|+1} = \infty$. Moreover, σ denotes the most recently selected label in the solution $OPT_\beta(S)$ by the algorithm.

Algorithm $ComputeOPT(S)$
Input. OPT_α and a set $S \subset P$.
Output. $OPT_\beta(S)$.
1.   **for** each label $[p_i, p'_{i,\beta}]$ for points in S,
2.   **do** Put $[p_i, p'_{i,\beta}]$ into B_k if $p'_{i,\beta} \in [q'_k, q'_{k+1})$ for some k.
3.   Initialize $OPT_\beta(S) = \emptyset$, and $\sigma = $ **nil**.
4.   **for** each B_k by incrementing k iteratively,
5.   **do** Select the label σ' from B_k such that σ' does not overlap σ and has the leftmost right endpoint among the labels of B_k.
6.   Put σ' into $OPT_\beta(S)$.
7.   Set $\sigma = \sigma'$.

The idea is simply that as all the labels in B_k intersect q'_k, there is at most one label in B_k can be selected to put into $OPT_\beta(S)$. Note that binary search is used to put $[p_i, p'_{i,\beta}]$ into B_k in the first for-loop. The algorithm runs in $O(|S| \log |OPT_\alpha|)$ time.

Lemma 1. *$ComputeOPT(S)$ computes the optimal solution $OPT_\beta(S)$ at resolution β by making use of OPT_α in $O(|S| \log |OPT_\alpha|)$ time.*

2.2   Building $O(\log n)$-Height Hierarchy

At each level of the hierarchy, we store the optimal solution at the resolution of the current level. The lowest level corresponds to the finest resolution, at which no labels can overlap, and the highest level corresponds to the coarsest resolution, at which the optimal solution has only a constant size. Between any pair of consecutive levels with resolutions α, β where $\rho_\beta > \rho_\alpha$, we have to determine a scaling factor $\rho = \rho_\beta/\rho_\alpha$ so that the size of the optimal solution $|OPT_\beta|$ drops significantly, say by a constant factor, from $|OPT_\alpha|$. If this can be done for each pair of consecutive levels, it results in an $O(\log n)$-height hierarchy. We show below how this can be done.

Building the Lowest Level. We need to decide a resolution, at which no labels overlap. Observe that $|p_{i+1} - p_i|/\omega_i$ is the minimum scaling ratio for the label at p_i to intersect the label at p_{i+1}. Let $\rho = \min_i\{|p_{i+1} - p_i|/\omega_i\}$. If we scale all labels by a factor a little smaller than ρ, no labels can overlap anymore. Thus we set $\rho - \epsilon$ (where $\epsilon > 0$ is small) to be the scaling factor for the lowest level. This step takes $O(n \log n)$ time as we need to first sort the points in P.

Building One Level Higher. As we have just discussed, to construct a level higher with resolution β from a level with finer resolution α, we need to decide a scaling factor $\rho = \rho_\beta/\rho_\alpha$ such that $|OPT_\beta|$ is a constant fraction of $|OPT_\alpha|$.

Let $\sigma_k = [q_k, q_k']$ be the kth label of OPT_α in the order from left to right. We associate A_k to σ_k, where A_k is the subset of labels of points in P at resolution α such that their right endpoints lie inside interval $[q_k', q_{k+1}')$. Note that A_k includes σ_k itself. For convenience, we set $q_{|OPT_\alpha|+1}' = \infty$. For each label $\sigma = [p, p']$ in A_k, observe that $\rho_k(\sigma) = (q_{k+1}' - p)/(p' - p)$ is the smallest scaling ratio for σ to intersect q_{k+1}'. Thus $\rho_k = \max_{\sigma \in A_k} \rho_k(\sigma)$ for A_k is the smallest scaling ratio such that all labels in A_k intersect q_{k+1}'. Note that as $\sigma_k \in A_k$, $\rho_k(\sigma) \leq \rho_k$. We call the label of A_k which constitutes ρ_k the *dominating label* of A_k. Then it is clear that the following observation holds.

Lemma 2. *Consider A_k associated with a label $\sigma_k = [q_k, q_k'] \in OPT_\alpha$. If $\Delta_k = [\delta_k, \delta_k']$ is the dominating label of A_k, then $q_k \leq \delta_k$.*

The above observation says that the dominating label of A_k has its left endpoint in the right of q_k. We set $\rho(= \rho_\beta/\rho_\alpha)$ to be the median value of all the ρ_k's. Then we claim that there are constants c_1 and c_2 such that $c_1|OPT_\alpha| \leq |OPT_\beta| \leq c_2|OPT_\alpha|$ as stated in the following lemma. Remark that we will assume all ρ_k's are different for simplicity to convey our idea. In fact, if some ρ_k's are the same, the following arguments still hold although the constants would deviate slightly.

Lemma 3. $\frac{1}{4}|OPT_\alpha| \leq |OPT_\beta| \leq \frac{3}{4}|OPT_\alpha|$.

Proof. We first prove the former part of the inequality. Let $L = \{A_k | \rho_k > \rho\}$, where we suppose the elements of L are ordered from left to right. Then $|L| = |OPT_\alpha|/2$ as by definition ρ is the median value of all the ρ_k. Note that at most one label from $A_k \in L$ can be selected for any labeling. Now we claim that the dominating labels of every other sets A_k in L do not overlap. Suppose A_i, A_j, A_k ($i < j < k$) be three consecutive sets in L. Let $\Delta_i = [\delta_i, \delta_i']$ and $\Delta_k = [\delta_k, \delta_k']$ be the dominating labels of A_i and A_k respectively. At resolution β, Let $\Delta_{i,\beta} = [\delta_i, \delta_{i,\beta}']$ and $\Delta_{k,\beta} = [\delta_k, \delta_{k,\beta}']$ be the scaled labels of Δ_i and Δ_k at resolution β respectively. We claim that $\Delta_{i,\beta}$ does not intersect $\Delta_{k,\beta}$. It suffice for us to prove $\delta_{i,\beta}' < \delta_k$. As $A_i \in L$, $\delta_{i,\beta} \leq q_{i+1}' \leq q_j'$. On the other hand, by Lemma 2, $q_k \leq \delta_k$ as Δ_k is dominating A_k. Also it is clear

$q'_j < q_{j+1} \leq q_k$. Thus we have $\delta_{i,\beta} < q_k$. This implies if we select all the dominating labels of every other sets A_k in L from left to right, they cannot overlap. Hence $|OPT_\beta| \geq |L|/2 \geq \frac{1}{4}|OPT_\alpha|$.

We then prove the latter part of the inequality. Let $S = \{A_k | \rho_k \leq \rho\}$. Then $|S| = |OPT_\alpha|/2$, and $|OPT_\alpha| = |S| + |L|$. Let us also divide the OPT_β into two subsets L' and S' where $L' = \{\sigma \in OPT_\beta \mid \sigma \in A_k \text{ for some } A_k \in L\}$, and $S' = \{\sigma \in OPT_\beta \mid \sigma \in A_k \text{ for some } A_k \in S\}$. Then $|OPT_\beta| = |L'| + |S'|$. As a label σ in $A_k \in S'$ must overlap q'_{k+1}, which means that σ overlaps all the labels in A_{k+1}. Thus if a label σ in $A_k \in S'$ lies in OPT_β, then no labels in A_{k+1} can lie in OPT_β. We call A_{k+1} is *abandoned* when σ is selected in OPT_β. Now we put all the abandoned sets A_{k+1} (due to those labels from all the A_k in S' selected into OPT_β) into sets L_a or S_a such that $L_a \subset L$ and $S_a \subset S$. Then we have that $|S'| \leq |L_a| + |S_a|$ as each $A_k \in S$ can contribute at most one label in OPT_β. Also it holds that $|L'| + |L_a| \leq |L| = |OPT_\alpha|/2$ and $|S'| + |S_a| \leq |S| = |OPT_\alpha|/2$. If $|S_a| \geq |OPT_\alpha|/4$, then $|OPT_\beta| = |L'| + |S'| \leq |L'| + (|S| - |S_a|) \leq |L| + (|S| - |S_a|) \leq |OPT_\alpha|/2 + |OPT_\alpha|/4 = \frac{3}{4}|OPT_\alpha|$. Otherwise when $|S_a| < |OPT_\alpha|/4$, $|OPT_\beta| = |L'| + |S'| \leq |L'| + (|L_a| + |S_a|) \leq |L| + |S_a| \leq \frac{3}{4}|OPT_\alpha|$. □

Building the Whole Hierarchy. Build the lowest level takes $O(n \log n)$ time. We then construct the levels one by one upwards. To construct one level higher with resolution β from the level with resolution α, we need to first determine the scaling factor $\rho = \rho_\beta/\rho_\alpha$ as described previously so that $\frac{1}{4}|OPT_\alpha| \leq |OPT_\beta| \leq \frac{3}{4}|OPT_\alpha|$. This takes $O(n \log n)$ time. Then we can construct OPT_β in time $n \log |OPT_\alpha|$ by using *ComputeOPT(P)*. We compute levels upwards until we reach a level at which the size of the optimal solution is a constant, and we stop. This gives us a $O(\log n)$-height hierarchy. It needs $O(n \log^2 n)$ time and $O(n)$ space in total. We summarize these in the following theorem.

Theorem 1. *A hierarchy of height $O(\log n)$ for the adaptive zooming query problem in one dimension can be built in time $O(n \log^2 n)$ using $O(n)$ space.*

2.3 Adaptive Querying

With the low-height hierarchy, it is possible for us to answer zooming queries efficiently at any resolution.

Theorem 2. *Given a zooming query Q with window $W = [q, q']$ at resolution γ. Let OPT_γ be the optimal set of non-overlapping labels for points in P at resolution γ by running the greedy algorithm. Then with the $O(\log n)$-height hierarchy, the optimal solution for Q can be computed in $O(|\Phi_\gamma^W| \log(|OPT_\gamma|) + \log \log n)$ time, where Φ_γ^W is the set of labels intersecting the window W at resolution γ.*

Proof. First by binary search, use ρ_γ to locate the consecutive levels of resolutions α and β (where $\rho_\alpha < \rho_\gamma < \rho_\beta$) in the hierarchy. As the height of the hierarchy is $O(\log n)$, the location is done in $O(\log \log n)$ time.

Then we search for q and q', the endpoints of W. Suppose $q'_1, q'_2, \ldots, q'_{|OPT_\alpha|}$ be the right endpoints of the greedy solution at resolution α. We need to locate

q and q' in these right endpoints. This needs $O(\log|OPT_\alpha|) = O(\log|OPT_\gamma|)$ time. Suppose q lies in $[q_i', q_{i+1}']$ and q' lies in $[q_j', q_{j+1}']$.

For each label whose right endpoints lying inside $[q_k', q_{k+1}']$ (where $i+1 \leq k \leq j$), check whether it completely lies inside W. So we can collect the set Ξ_γ^W of all the labels completely lying inside W at resolution γ. This needs $O(|\Phi_\gamma^W|)$ time.

Finally we use Ξ_γ^W to compute the optimal set of non-overlapping labels in W at resolution γ by using $ComputeOPT(\Xi_\gamma^W)$. This takes $O(|\Xi_\gamma^W|\log|OPT_\alpha|)$ $= O(|\Phi_\gamma^W|\log|OPT_\gamma|)$ time. $\qquad\square$

3   Adaptive Zooming in Two Dimensions

We then extend our idea to build the low-height hierarchy in two dimensions for efficient adaptive zooming queries. At each level of the hierarchy, we store the stabbing line structures as used in [1,4]. This helps us build the hierarchy and efficiently answer adaptive zooming queries. Let OPT_γ denote the optimal solution the labels of P at resolution γ.

Suppose all the labels have unit height at the current resolution α. We suppose a label does not include its lower boundary for convenience. We can stab all the labels by a set of horizontal lines $\ell_1, \ell_2, \ldots, \ell_k$, ordered from top to bottom, satisfying three conditions: (i) each ℓ_i must stab at least one label; (ii) a label must intersect exactly one stabbing line; and (iii) two consecutive stabbing lines are separated with distance at least one. Let A_i be the set of labels stabbed by ℓ_i, and let $OPT_\alpha(A_i)$ be optimal labeling for labels A_i by running the one dimensional greedy algorithm on $\{\sigma \cap \ell_i | \sigma \in A_i\}$. We define the *stabbing line structure* $\mathcal{L}_\alpha$ at resolution α to be the set all the stabbing lines ℓ_i at resolution α together with A_i and $OPT_\alpha(A_i)$.

Unlike in one dimension, OPT_α cannot be derived easily from the stabbing line data structure $\mathcal{L}_\alpha$ in two dimensions. However, a 2-approximation to OPT_α can be obtained easily from $\mathcal{L}_\alpha$. Let X_α^{odd} (resp., X_α^{even}) be the union of $OPT_\alpha(A_i)$ for odd i (resp., for even i). As any pair of consecutive stabbing lines is separated by a distance at least one, the labels in X_α^{odd} never overlap those in X_α^{even} and vice versa. Thus if we take the maximum-size labeling of $OPT(X_\alpha^{odd})$ and $OPT(X_\alpha^{even})$, it is a 2-approximation for OPT_α. Moreover, $\mathcal{L}_\alpha$ can help us find the $(1+\epsilon)$-approximation [1,4]. We then describe a way to find $\mathcal{L}_\beta$ at resolution β (which is coarser than α) by making use of $\mathcal{L}_\alpha$. This will serve as the main subroutine to build the hierarchy.

3.1   Computing $\mathcal{L}_\beta$ from $\mathcal{L}_\alpha$

For convenience, we assume $\rho_\alpha = 1$, $\rho_\beta = \rho$, and labels at resolution α has unit height. Let ℓ be any horizontal line at resolution β with y-coordinate $y(\ell)$. Let B_ℓ be the set of labels that intersect ℓ at resolution β. Let H be the horizontal strip bounded by the horizontal lines at $y(\ell) - \rho$ and at $y(\ell) + \rho$. Suppose that $\{\ell_i, \ldots, \ell_j\}(i < j)$ is the set of the stabbing lines at resolution α lying inside H.

Observe that the labels in B_ℓ can only be members of $A_i, \ldots, A_j$ at resolution α. Now suppose that S_ℓ is the ordered sequence of the right endpoints of the intervals in $OPT_\alpha(A_i), \ldots, OPT_\alpha(A_j)$ projected onto ℓ. Then we can obtain the optimal labeling $OPT_\beta(B_\ell)$ for the labels in B_ℓ by executing $ComputeOPT(B_\ell)$ using the points in S_ℓ as separators to partition the labels in groups. This takes $O(|B_\ell| \log |S_\ell|)$ time.

Now we describe how we draw the stabbing lines $\ell_1', \ell_2', \ldots$ from top to bottom to stab all the labels at resolution β. First, we draw the first line $\ell_1' = \ell_1$, and collect $B_{\ell_1'}$ and compute $OPT_\beta(B_{\ell_1'})$ as described in the previous paragraph. We then draw the second stabbing line ℓ_2' with y-coordinate $y(\ell_1') - \rho$ if it intersects some labels at resolution β. Otherwise we set ℓ_2' to be the stabbing line ℓ_i below and nearest to the y-coordinate $y(\ell_1') - \rho$. We continue this process until all labels at resolution β are stabbed. Note that the right endpoints of labels in $OPT_\alpha(A_i)$ at resolution α may be used twice to compute $OPT_\beta(B_\ell)$ for two consecutive stabbing lines at resolution β. In total, it takes $O(n \log |OPT_\alpha|)$ time to compute $\mathcal{L}_\beta$. We summarize the result as the following lemma.

Lemma 4. *Given $\mathcal{L}_\alpha$ at resolution α. Then $\mathcal{L}_\beta$ at a coarser resolution β can be computed from $\mathcal{L}_\alpha$ in $O(n \log |OPT_\alpha|)$ time.*

3.2   Building $O(\log n)$-Height Hierarchy

Building the Lowest Level. We build the lowest level at which the labels at distinct points do not overlap. By considering the projections of the labels onto x- and y-axes respectively, it is not hard to decide a resolution such that for each pair of points, either the x-projections or the y-projections of their labels do not overlap. This can be done in $O(n \log n)$ time.

Building One Level Higher. We have known how to construct the stabbing line structure $\mathcal{L}_\beta$ for a resolution β by making use of $\mathcal{L}_\alpha$ at a finer resolution α if the scaling factor $\rho = \rho_\beta / \rho_\alpha$ is known. In order to have a low-height hierarchy, it suffices for us to find a scaling factor ρ such that $|OPT_\beta|$ is a constant fraction of $|OPT_\alpha|$. For convenience, we assume that the height of labels at resolution α is unit.

At resolution α, a set A_ℓ of labels that intersect a stabbing line ℓ is partitioned into several groups by labels in $OPT_\alpha(A_\ell)$ (as in the one-dimensional case). Each of the groups consists of labels whose right endpoints lie between the right endpoints of two consecutive labels in $OPT_\alpha(A_\ell)$. The intersection of labels in such a group in A_ℓ is called a *kernel* (denoted by K_α), and those labels in that group are said to be *associated* with the kernel K_α. Let $\mathcal{K}_\alpha^{odd}$ and $\mathcal{K}_\alpha^{even}$ be the collections of all the kernels intersecting odd and even stabbing lines at resolution α, respectively. Let $\mathcal{K}_\alpha = \mathcal{K}_\alpha^{odd} \cup \mathcal{K}_\alpha^{even}$. Let A_{odd} and A_{even} be the set of all labels intersecting odd and even stabbing lines at resolution α, respectively. We then use the interactions of the scaled versions of the kernels in $\mathcal{K}_\alpha$ to decide the scaling factor ρ.

The labels at resolution β are obtained by scaling the labels of resolution α by factor ρ. The kernels in $\mathcal{K}_\alpha^{odd}$ and $\mathcal{K}_\alpha^{even}$ are enlarged to the kernels at resolution

β and we denote the corresponding sets of enlarged kernels at resolution β simply by $\mathcal{K}_{odd}$ and $\mathcal{K}_{even}$ respectively. Let $\mathcal{K} = \mathcal{K}_{odd} \cup \mathcal{K}_{even}$. The labels in A_{odd} and A_{even} become B_{odd} and B_{even} respectively. We also denote the scaled version of kernel K_α by K. Two kernels (resp. labels) are said to be of the *same parity* if they are contained in the same kernel (resp. label) collection $\mathcal{K}_{odd}$ or $\mathcal{K}_{even}$ (resp. A_{odd} or A_{even}). For each kernel K_α at resolution α, scale it until it intersects the left sides of first three other kernels of the same parity. We denote this scaling ratio for K_α by $\rho(K_\alpha)$. We set the scaling factor ρ to be the $(\frac{10|\mathcal{K}_\alpha|}{11})$-th smallest value of $\rho(K_\alpha)$ for all kernels $K_\alpha \in \mathcal{K}_\alpha$. With this ratio ρ, we claim that the optimal labeling at resolution β is a constant fraction of that at resolution α. Note that as the one-dimensional case, we will assume all the $\rho(K_\alpha)$ are distinct for convenience to convey our idea.

For a kernel K, let $R(K)$ be a label whose right side constitutes the right side of K. We call $R(K)$ the *right representative* of K. We denote the height and width of a kernel or label by $\tau(\cdot)$ and $\omega(\cdot)$ respectively. Also we denote by $x_l(K)$ and $x_r(K)$ the x-coordinates of the left and right sides of a kernel K respectively. First of all, the following two observations are clear.

Lemma 5. *Suppose $K \in \mathcal{K}$ at resolution β is obtained by scaling ρ times a kernel K_α at resolution α. Then its height $\tau(K_\alpha) \leq \tau(K) \leq \tau(K_\alpha) + (\rho - 1)$, and its width $\omega(K) \geq \omega(K_\alpha) + \left(\frac{\rho-1}{\rho}\right)\omega(R(K))$. Moreover if $R(K)$ is different from $R(K_\alpha)$, then $x_l(R(K)) > x_l(R(K_\alpha))$.*

Lemma 6. *Let J, K be two non-intersecting kernels on the same stabbing line ordered from left to right at resolution β, where J, K are obtained by scaling ρ times the kernels J_α, K_α at resolution α respectively. Then $x_r(R(J)) < x_r(K)$ and $x_l(J) < x_l(R(K))$.*

Let S_{odd} (resp. L_{odd}) be the subset of kernels K in $\mathcal{K}_{odd}$ with $\rho(K_\alpha)$ smaller (resp. not smaller) than ρ. Similarly, we define S_{even} and L_{even}. Let $S = S_{odd} \cup S_{even}$ and $L = L_{odd} \cup L_{even}$. The following lemma tells us that there is a large set of non-intersecting kernels in L_{odd} or L_{even}.

Lemma 7. *For any $i \in \{odd, even\}$, each kernel in L_i can intersect at most $1.5\rho + 12$ kernels in L_i.*

Proof. Let K be any kernel in L_i. For a kernel J in L_i intersecting K, we put it into I_1 if $x_l(J) \leq x_l(K)$ and into I_2 otherwise. As K intersects the left sides of all kernels in I_2, $|I_2| \leq 3$.

To determine $|I_1|$, we divide I_1 into three subsets depending on whether the kernels in I_1 intersect the supporting lines ℓ^t and ℓ^b of the top and bottom sides of K. If those kernels intersects ℓ^t (resp., ℓ^b), then put them into I_1^t (resp. I_1^b). Otherwise, i.e., if they lie between ℓ^t and ℓ^b, they are put into I_1^m. We first claim that $|I_1^t| \leq 3$. Suppose for the contradiction that $|I_1^t| \geq 4$. All the kernels in I_1^t must contain the top-left corner of K. This means that the kernel J in I_1^t with the smallest $x_l(J)$ would intersect the left sides of at least four kernels of same parity. This implies that $J \not\subseteq L_i$, which is a contradiction. Thus $|I_1^t| \leq 3$. A similar argument proves $|I_1^b| \leq 3$.

Then we bound $|I_1^m|$. As the height of K is at most ρ. There are at most $\frac{\rho}{2} + 1$ stabbing lines with the same parity between ℓ^t and ℓ^b. As on any of these stabbing line, there are at most three non-intersecting kernels in L_i stabbed before K, $|I_1^m| \leq 3(\frac{\rho}{2} + 1) = 1.5\rho + 3$.

Considering all together, $|I_1| + |I_2| \leq 1.5\rho + 12$. ☐

Let N_i be any maximal subset of non-intersecting kernels in L_i for $i \in \{odd, even\}$. By Lemma 7, we have $|N_i| \geq |L_i|/(1.5\rho + 13)$. Although no two kernels in N_i intersect each other, their right representatives may intersect. The following lemma proves that the number of those right representatives which intersect $R(K)$ for a kernel $K \in N_i$ is bounded above by $O(\rho)$. The argument is similar to Lemma 7 by packing kernels and labels around $R(K)$. This in turn implies that there are at least $\Omega(|L_i|)$ non-intersecting labels in B_i as stated in Lemma 9.

Lemma 8. *Let $K \in N_i$ be the kernel with its right representative label $R(K)$ of the shortest width among all other kernels in N_i. Then $R(K)$ can intersect the right representative labels of less than $6\rho + 4$ kernels in N_i.*

Lemma 9. *There are at least $\frac{|L_i|}{(3\rho+2)(3\rho+26)}$ non-intersecting labels in B_i for any $i \in \{odd, even\}$.*

Now we are well-equipped to show the main lemma, whose proof uses the similar idea as Lemma 3 in one dimensional case.

Lemma 10. *There exist constants $0 < c_1, c_2 < 1$ such that $c_1|OPT_\alpha| \leq |OPT_\beta| \leq c_2|OPT_\alpha|$.*

Note that the details of the proofs of Lemma 8, 9 and 10 are omitted in this preliminary version. We then describe the algorithm to compute the scaling factor ρ, and analyze its running time.

For each kernel K_α, $\rho(K_\alpha, K'_\alpha)$ can be determined in $|K_\alpha| \cdot |K'_\alpha|$ time, where $|K_\alpha|$ is the number of the associated labels of K_α. For a fixed K_α, to compute all $\rho(K_\alpha, K'_\alpha)$ for different K'_α, it takes $n|K_\alpha|$ time. To determine the third smallest value $\rho(K_\alpha)$ out of all $\rho(K_\alpha, K'_\alpha)$, it requires at most $3|K_\alpha|$ time. In total, to determine $\rho(K_\alpha)$, it takes at most $n(|K_\alpha| + 3)$ time.

To determine all $\rho(K_\alpha)$ for all K_α, it takes time $n(|K_\alpha|+3)$ summing over all kernels $K_\alpha \in \mathcal{K}_\alpha$. This takes time $O(n^2)$ to determine all $\rho(K_\alpha)$. Furthermore, the $\frac{10}{11}m$-th value ρ of all $\rho(K_\alpha)$ can be determined in $O(n \log n)$ time. In all, ρ can be computed in $O(n^2)$ time.

Auxiliary Structures for Efficient Querying. In order to efficiently locate all the labels intersecting a specific window, we associate a range tree R_α with each hierarchy level say at resolution α. We collect all the intersection points S of the boundaries of all kernels with all the stabbing lines. We then build a 2-dimensional range tree R_α on the point set S. This takes $O(|S| \log(|S|)) =$

$O(|OPT_\alpha| \log |OPT_\alpha|)$ time and space [6]. The query time to report all the points from S inside a rectangular window W is $O(\log(|S|) + k) = O(\log(|OPT_\alpha|) + k)$ where k is the number of points inside W. Remark that we suppose the technique of fractional cascading is applied to the range tree; otherwise, the query time can go up to $O(\log^2(|S|) + k)$.

Building the whole hierarchy. Combining the above statements, we have the following theorem.

Theorem 3. *A hierarchy of height $O(\log n)$ for the adaptive zooming query problem in two dimensions can be built in $O(n^2 \log n)$ time using $O(n \log n)$ space.*

3.3   Adaptive Querying

Theorem 4. *Given any zooming query Q with window $W = [x, x'] \times [y, y']$ at resolution γ. Let OPT_γ be the optimal set of non-overlapping labels for points in P at resolution γ. Let Φ_γ^W be the set of labels intersecting the window W at resolution γ. Suppose the $O(\log n)$-height hierarchy is given. Then*

(i) *The 2-approximation for Q can be computed in $O(|\Phi_\gamma^W| \log(|OPT_\gamma|) + \log \log n)$ time.*

(ii) *The $(1+\epsilon)$-approximation for Q can be computed in $O(|\Phi_\gamma^W|^{\frac{1}{\epsilon}} + \log(|OPT_\gamma|) + \log \log n)$ time.*

Proof. First by binary search, use ρ_γ to locate the consecutive levels of resolutions α and β where $\rho_\alpha < \rho_\gamma < \rho_\beta$ in the hierarchy. As the height of the hierarchy is $O(\log n)$, the location is done in $O(\log \log n)$ time.

Search the auxiliary range tree R_α at resolution α to find all points lying inside W. This takes $O(\log(|OPT_\alpha|) + k)$ time, where k is the number of points of R_α inside W. Note that $\log(|OPT_\alpha|) = O(\log(|OPT_\gamma|))$. As the labels intersecting W at resolution α will continue intersecting W at resolution γ, we have $k = O(|\Phi_\gamma^W|)$. For groups of labels corresponding to these k kernels, we check one by one whether they are inside W or not. So we can collect all the labels Ξ_γ^W completely lying inside W at resolution γ in $O(|\Phi_\gamma^W|)$ time.

(i) We use Ξ_γ^W to compute the 2-approximation solution for Q by applying the one-dimensional greedy algorithm on related stabbing lines. This takes $O(|\Xi_\gamma^W| \log |OPT_\alpha|) = O(|\Phi_\gamma^W| \log |OPT_\gamma|)$ time.

(ii) We use Ξ_γ^W to compute the $(1+\epsilon)$-approximation for Q by applying the algorithm by Chan [4]. This takes $O(|\Xi_\gamma^W|^{\frac{1}{\epsilon}}) = O(|\Phi_\gamma^W|^{\frac{1}{\epsilon}})$ time.   □

4   Conclusion and Discussion

In this paper, we build low-height hierarchies for one and two dimensions for answering adaptive zooming queries efficiently. In the model we have considered, the labels at any point are restricted to several fixed positions lying on the right of the point. One interesting research direction is to extend our results to point labeling for more general models, for example sliding models. Can some notion of point importance be added into the data structure? Can we build a hierarchy for a subdivision map to help us query the map area in a window at any resolution?

References

1. P.K. Agarwal, M. van Kreveld, and S. Suri. Label placement by maximum independent set in rectangles. *Computational Geometry: Theory and Applications*, vol. 11, pp. 209-218, 1998.
2. M. Arikawa, Y. Kambayashi, and H. Kai. Adaptive geographic information media using display agents for name placement. *Trans. Journal of the Geographic Information Systems Association*, 1997. in Japanese.
3. M. Arikawa, H. Kawakita, and Y. Kambayashi. An environment for generating interactive maps with compromises between users' requirements and limitations of display screens. *Trans. Journal of the Geographic Information Systems Association*, 2, March 1993. in Japanese.
4. T. Chan. A Note on Maximum Independent Sets in Rectangle Intersection Graphs. *Information Processing Letters*, 89(1), 19–23, 2004.
5. B. Chazelle and 36 co-authors. The computational geometry impact task force report. In B. Chazelle, J. E. Goodman, and R. Pollack, editors, *Advances in Discrete and Computational Geometry*, vol. 223, pages 407-463. American Mathematical Society, Providence, 1999.
6. M. de Berg, M. van Kreveld, M. Overmars, and O. Schwarzkopf. *Computational Geometry: Algorithms and Applications*. Springer-Verlag, Heidelberg, 1997.
7. M. Formann and F. Wagner. A packing problem with applications to lettering of maps. In *Proc. 7th Annu. ACM Sympos. Comput. Geom.*, pages 281–288, 1991.
8. D. E. Knuth and A. Raghunathan. The problem of compatible representatives. *SIAM J. Discrete Math.*, 5(3), 422–427, 1992.
9. MapFactory. http://home.coqui.net/rjvaasjo/index.htm
10. J. Marks and S. Shieber. The computational complexity of cartographic label placement. Technical report, Harvard CS, 1991.
11. I. Petzold, L. Plumer, and M. Heber. Label placement for dynamically generated screen maps. In *Proceedings of the Ottawa ICA*, pp. 893-903, 1999.
12. S. Roy, P.P. Goswami, S. Das, and S.C. Nandy. Optimal Algorithm for a Special Point-Labeling Problem In *Proc. 8th Scandinavian Workshop on Algorithm Theory (SWAT'02)*, pp. 110-120, 2002.
13. T. Strijk and A. Wolff. www.math-inf.uni-greifswald.de/map-labeling/bibliography/
14. M. van Kreveld. http://www.cs.uu.nl/centers/give/geometry/autocarto/index.html
15. M. van Kreveld, T. Strijk, and A. Wolff. Point labeling with sliding labels. *Computational Geometry: Theory and Applications*, vol. 13, pp. 21-47, 1999.
16. M. van Kreveld, R. van Oostrum, and J. Snoeyink. Efficient settlement selection for interactive display. In *Proc. Auto-Carto 13: ACSM/ASPRS Annual Convention Technical Papers*, pp. 287-296, 1997.

On the Black-Box Complexity
of Sperner's Lemma

Katalin Friedl[1], Gábor Ivanyos[2], Miklos Santha[3], and Yves F. Verhoeven[3,4]

[1] BUTE, H-1521 Budapest, P.O.Box 91, Hungary
friedl@cs.bme.hu
[2] MTA SZTAKI, H-1518 Budapest, P.O. Box 63, Hungary
Gabor.Ivanyos@sztaki.hu
[3] CNRS–LRI, UMR 8623, Université Paris XI, 91405 Orsay, France
{santha, verhoeve}@lri.fr
[4] ENST, 46 rue Barrault, 75013 Paris, France
verhoeve@enst.fr

Abstract. We present several results on the complexity of various forms of Sperner's Lemma in the black-box model of computing. We give a deterministic algorithm for Sperner problems over pseudo-manifolds of arbitrary dimension. The query complexity of our algorithm is linear in the separation number of the skeleton graph of the manifold and the size of its boundary. As a corollary we get an $O(\sqrt{n})$ deterministic query algorithm for the black-box version of the problem **2D-SPERNER**, a well studied member of Papadimitriou's complexity class PPAD. This upper bound matches the $\Omega(\sqrt{n})$ deterministic lower bound of Crescenzi and Silvestri. The tightness of this bound was not known before. In another result we prove for the same problem an $\Omega(\sqrt[4]{n})$ lower bound for its probabilistic, and an $\Omega(\sqrt[8]{n})$ lower bound for its quantum query complexity, showing that all these measures are polynomially related.

Classification: computational and structural complexity, quantum computation and information.

1 Introduction

Papadimitriou defined in [18,19] the complexity classes PPA, PPAD, and PSK in order to classify total search problems which have always a solution. The class PSK was renamed PPADS in [5]. These classes can be characterized by some underlying combinatorial principles. The class Polynomial Parity Argument (PPA) is the class of NP search problems, where the existence of the solution is guaranteed by the fact that in every finite graph the number of vertices with odd degree is even. The class PPAD is the directed version of PPA, and its basic search problem is the following: in a directed graph, where the in-degree and the out-degree of every vertex is at most one, given a source, find another source or a sink. In the class PPADS the basic search problem is more restricted than in PPAD: given a source, find a sink.

M. Liśkiewicz and R. Reischuk (Eds.): FCT 2005, LNCS 3623, pp. 245–257, 2005.

These classes are in fact subfamilies of TFNP, the family of all total NP-search problems, introduced by Megiddo and Papadimitriou [17]. Other important subclasses of TFNP are Polynomial Pigeonhole Principle (PPP) and Polynomial Local Search (PLS). The elements of PPP are problems which by their combinatorial nature obey the pigeonhole principle and therefore have a solution. In a PLS problem, one is looking for a local optimum for a particular objective function, in some neighborhood structure. All these classes are interesting because they contain search problems not known to be solvable in polynomial time, but which are also somewhat easy in the sense that they can not be NP-hard unless NP = co-NP.

Another point that makes the parity argument classes interesting is that there are several natural problems from different branches of mathematics that belong to them. For example, in a graph with odd degrees, when a Hamiltonian path is given, a theorem of Smith [26] ensures that there is another Hamiltonian path. It turns out that finding this second path belongs to the class PPA [19]. A search problem coming from a modulo 2 version of Chevalley's theorem [19] from number theory is also in PPA. Complete problems in PPAD are the search versions of Brouwer's fixed point theorem, Kakutani's fixed point theorem, Borsuk-Ulam theorem, and Nash equilibrium (see [19]).

The classical Sperner's Lemma [23] states that in a triangle with a regular triangulation whose vertices are labeled with three colors, there is always a trichromatic triangle. This lemma is of special interest since some customary proofs for the above topological fixed point theorems rely on its combinatorial content. However, it is unknown whether the corresponding search problem, that Papadimitriou [19] calls **2D-SPERNER**, is complete in PPAD. Variants of Sperner's Lemma also give rise to other problems in the parity argument classes. Papadimitriou [19] has proven that a 3-dimensional analogue of **2D-SPERNER** is in fact complete in PPAD.

The study of query complexities of the black-box versions of several problems in TFNP is an active field of research. Several recent results point into the direction that quantum algorithms can give only a limited speedup over deterministic ones in this framework. The collision lower bound of Aaronson [1] and Shi [21] about PPP, and the recent result of Santha and Szegedy [20] on PLS imply that the respective deterministic and quantum complexities are polynomially related. As a consequence, if an efficient quantum algorithm exists for a problem in these classes, it must exploit its specific structure. In a related issue, Buresh-Oppenheim and Morioka [8] have obtained relative separation results among PLS and the polynomial parity argument classes.

2   Results

A *black-box problem* is a relation $R \subseteq S \times T$ where T is a finite set and $S \subseteq \Sigma^n$ for some finite set Σ. The oracle input is a function $x \in S$, hidden by a black-box, such that x_i, for $i \in \{1, \ldots, n\}$ can be accessed via a query parameterized by i. The output of the problem is some $y \in T$ such that $(x, y) \in R$. A special case is the *functional oracle problem* when the relation is given by a function $A : S \to T$, the (unique) output is then $A(x)$. We say that A is *total* if $S = \Sigma^n$.

In the query model of computation each query adds one to the complexity of the algorithm, but all other computations are free. The state of the computation is represented by three registers, the query register $i \in \{1, \ldots, n\}$, the answer register $a \in \Sigma$, and the work register z. The computation takes place in the vector space spanned by all basis states $|i\rangle|a\rangle|z\rangle$. In the *quantum query model* introduced by Beals et al. [4] the state of the computation is a complex combination of all basis states which has unit length in the norm l_2. In the randomized model it is a non-negative real combination of unit length in the norm l_1, and in the deterministic model it is always one of the basis states.

The query operation O_x maps the basis state $|i\rangle|a\rangle|z\rangle$ into the state $|i\rangle|(a + x_i) \bmod |\Sigma|\rangle|z\rangle$ (here we identify Σ with the residue classes $\bmod|\Sigma|$). Non-query operations are independent of x. A k-*query algorithm* is a sequence of $(k + 1)$ operations $(U_0, U_1, \ldots, U_k)$ where U_i is unitary in the quantum and stochastic in the randomized model, and it is a permutation in the deterministic case. Initially the state of the computation is set to some fixed value $|0\rangle|0\rangle|0\rangle$, and then the sequence of operations $U_0, O_x, U_1, O_x, \ldots, U_{k-1}, O_x, U_k$ is applied. A quantum or randomized algorithm computes (with two-sided error) R if the observation of the appropriate last bits of the work register yield some $y \in T$ such that $(x, y) \in R$ with probability at least $2/3$. Then $\mathsf{QQC}(R)$ (resp. $\mathsf{RQC}(R)$) is the smallest k for which there exists a k-query quantum (resp. randomized) algorithm which computes R. In the case of deterministic algorithms of course exact computation is required, and the deterministic query complexity $\mathsf{DQC}(R)$ is defined then analogously. We have $\mathsf{DQC}(R) \geq \mathsf{RQC}(R) \geq \mathsf{QQC}(R)$.

Beals et al. [4] have shown that in the case of total functional oracle problems the deterministic and quantum complexities are polynomially related. For several partial functional problems exponential quantum speedups are known [10,22].

In this paper we will give several results about Sperner problems in the black-box framework. In Section 5, we will prove that the deterministic query complexity of **REGULAR 2-SPM**, the black-box version of **2D-SPERNER** is $O(\sqrt{n})$. This matches the deterministic $\Omega(\sqrt{n})$ lower bound of Crescenzi and Silvestri [9]. The tightness of this bound was not known before. In fact, this result is the corollary of a general algorithm that solves the Sperner problems over pseudo-manifolds of arbitrary dimension. The complexity analysis of the algorithm will be expressed in **Theorem 4** in two combinatorial parameters of the pseudo-manifold: the size of its boundary and the separation number of its skeleton graph. In Section 6, we show that quantum, probabilistic, and deterministic query complexities of **REGULAR 2-SPM** are polynomially related. More precisely, in **Theorem 8** we will prove that its randomized complexity is $\Omega(\sqrt[4]{n})$ and that its quantum complexity is $\Omega(\sqrt[8]{n})$. This result is analogous to the polynomial relations obtained for the respective query complexities of PPP and PLS. Because of lack of space, most proofs are absent from this extended abstract, but can be found in the full paper [12].

3   Mathematical Background on Simplicial Complexes

For an undirected graph $G = (V, E)$, and for a subset $V' \subseteq V$ of the vertices, we denote by $G[V']$ the induced subgraph of G by V'. A graph $G'' = (V'', E'')$ is a subgraph of G, in notation $G'' \subseteq G$, if $V'' \subseteq V$ and $E'' \subseteq E$. The ring $\mathbb{Z}/(2)$ denotes the ring with 2 elements.

Definition 1 (Simplicial Complex). *A simplicial complex K is a non-empty collection of subsets of a finite set U, such that whenever $S \in K$ then $S' \in K$ for every $S' \subseteq S$. An element S of K of cardinality $d + 1$ is called a d-simplex. A d'-simplex $S' \subseteq S$ is called a d'-face of S. We denote by K_d the set of d-simplices of K. An* elementary d-complex *is a simplicial complex that contains exactly one d-simplex and its subsets. The* dimension *of K, denoted by $\dim(K)$, is the largest d such that K contains a d-simplex. The elements of K_0 are called the* vertices *of K, and the elements of K_1 are called the* edges *of K. The skeleton graph $G_K = (V_K, E_K)$ is the graph whose vertices are the vertices of K, and the edges are the edges of K.*

Without loss of generality, we suppose that U consists of integers, and we identify $\{u\}$ with u, for $u \in U$.

Fact 1. *Let d be a positive integer. If S is an elementary d-complex, then G_S is the complete graph.*

Definition 2 (Oriented Simplex). *For every positive integer n, we define an equivalence relation $\equiv_n$ over $\mathbb{Z}^n$, by $a \equiv_n b$ if there exists an even permutation σ such that $\sigma \cdot a = b$. For every $a \in \mathbb{Z}^n$ we denote by $[a]_{\equiv_n}$ the equivalence class of a for $\equiv_n$. The two equivalence classes of the orderings of the 0-faces of a simplex are called its* orientations. *An* oriented simplex *is a pair formed of a simplex and one of its orientations.*

For an oriented d-simplex $(S, [\tau]_{\equiv_{d+1}})$, where τ is an ordering of the 0-faces of S, and a permutation σ over $\{1, \ldots, d + 1\}$, we denote by $\sigma \cdot (S, [\tau]_{\equiv_{d+1}})$ the oriented d-simplex $(S, [\sigma \cdot \tau]_{\equiv_{d+1}})$. For every integer d, and every simplicial complex K whose simplices have been oriented, we denote by K_d the set of oriented d-simplices of K. From now on, S may denote an oriented or a non-oriented simplex. When S is an oriented simplex, $\bar{S}$ will denote the same simplex with the opposite orientation. We also define $S^{(i)}$ to be S if i is even, and to be $\bar{S}$ if i is odd. We will often specify an oriented simplex by an ordering of its 0-faces.

Definition 3. *Let $S = (v_0, \ldots, v_d)$ be an oriented d-simplex. For every $0 \leq i \leq d$, for every $(d-1)$-face $\{v_0, \ldots, v_{i-1}, v_{i+1}, \ldots, v_d\}$ of S, the* induced orientation *is the oriented $(d-1)$-simplex $(v_0, \ldots, v_{i-1}, v_{i+1}, \ldots, v_d)^{(i)}$.*

Definition 4. *Let K be a simplicial complex whose simplices have been oriented, and let R be a ring. We define $C_d(K; R)$ as the submodule of the free R-module over the d-simplices of K with both possible orientations, whose elements are of* the form $\sum_{S\in K_d}(c_S \cdot S + c_{\bar{S}} \cdot \bar{S})$, with $c_S \in R$, satisfying the relation $c_S = -c_{\bar{S}}$. The elements of $C_d(K;R)$ are called d-chains. For every oriented simplex S of K, we denote by $\langle S \rangle$ the element $S - \bar{S}$ of $C_d(K;R)$.

Let S be an oriented d-simplex $(v_0, v_1, \ldots v_d)$ of K. The algebraic boundary of $\langle S \rangle$, denoted by $\partial_d \langle S \rangle$, is the $(d-1)$-chain of $C_{d-1}(K;R)$ defined as $\partial_d \langle S \rangle = \sum_{i=0}^{d}(-1)^i \langle (v_0, \ldots, v_{i-1}, v_{i+1}, \ldots, v_d) \rangle$.

Since $\partial_d \langle S \rangle = -\partial_d \langle \bar{S} \rangle$, the operator ∂_d has been correctly defined on a basis of $C_d(K;R)$ and can therefore be uniquely extended into a homomorphism $\partial_d : C_d(K;R) \to C_{d-1}(K;R)$. The proof of the next Lemma is straightforward.

Lemma 1. *Let S be an oriented d-simplex of a simplicial complex K. Denote by F_S the set of $(d-1)$-faces of S, and for every $S' \in F_S$ by $\tau_{S'}^S$ the induced orientation on S'. Then $\partial_d \langle S \rangle = \sum_{S' \in F_S} \langle (S', \tau_{S'}^S) \rangle$.*

Following an early version of a paper of Bloch [7], in the next definition we generalize the notion of pseudo-manifold, without the usual requirements of connectivity and pure dimensionality.

Definition 5. *A simplicial complex $\mathcal{M}$ is a pseudo d-manifold, for a positive integer d, if (i) $\mathcal{M}$ is a union of elementary d-complexes, and (ii) every $(d-1)$-simplex in $\mathcal{M}$ is a $(d-1)$-face of at most two d-simplices of $\mathcal{M}$. The boundary of $\mathcal{M}$ is the set of elementary $(d-1)$-complexes in $\mathcal{M}$ that belong exactly to one d-simplex of $\mathcal{M}$. We denote it by $\partial \mathcal{M}$. A pseudo d-manifold $\mathcal{M}$ is said to be orientable if it is possible to assign an orientation to each d-simplex of $\mathcal{M}$, such that for all $(d-1)$-simplex of $\mathcal{M}$ that is not on its boundary the orientations induced by the two d-simplices to which it belongs are opposite. Such a choice of orientations for all the d-simplices of $\mathcal{M}$ makes $\mathcal{M}$ oriented.*

If the d-simplices of $\mathcal{M}$ are oriented, then there is a natural orientation of the $(d-1)$-simplices of $\partial \mathcal{M}$, where each $(d-1)$-simplex has the orientation induced by the oriented d-simplex of which it is a $(d-1)$-face. Notice that if $\mathcal{M}$ is a pseudo d-manifold, then $\partial \mathcal{M}$ need not be a pseudo $(d-1)$-manifold. From now, all the simplicial complexes will be pseudo-manifolds. Observe that if $R = \mathbb{Z}/(2)$, then for any oriented d-simplex S, we have $\langle S \rangle = \langle \bar{S} \rangle$.

Definition 6. *Given a simplicial complex K of dimension d, the standard d-chain $\widehat{K}$ of K will be defined depending on whether K is oriented as follows:*
– if K is non-oriented, then $\widehat{K} = \sum_{S\in K_d} \langle (S, \tau_S) \rangle \in C_d(K, \mathbb{Z}/(2))$, for an arbitrary choice of orientations τ_S of the d-simplices S in K,
– if K is oriented, then $\widehat{K} = \sum_{S\in K_d} \langle (S, \tau_S) \rangle \in C_d(K, \mathbb{Z})$ where τ_S is the orientation of S in K.

Fact 2. *Let d be an integer, and let $\mathcal{M}$ be a pseudo d-manifold. Then, if $\mathcal{M}$ is not oriented the equality $\widehat{\partial \mathcal{M}} = \partial_d \widehat{\mathcal{M}}$ holds in $C_{d-1}(\partial \mathcal{M}, \mathbb{Z}/(2))$, and if $\mathcal{M}$ is oriented the equality $\widehat{\partial \mathcal{M}} = \partial_d \widehat{\mathcal{M}}$ holds in $C_{d-1}(\partial \mathcal{M}, \mathbb{Z})$.*

4 Sperner Problems

Definition 7. *Let K be a simplicial complex. A* labeling *of K is a mapping ℓ of the vertices of K into the set $\{0, \ldots, \dim(K)\}$. If a simplex S of K is labeled with all possible labels, then we say that S is* fully labeled.

A labeling ℓ naturally maps every oriented d-simplex $S = (v_0, \ldots, v_d)$ to the equivalence class $\ell(S) = [(\ell(v_0), \ldots, \ell(v_d))]_{\equiv_{d+1}}$.

Definition 8. *Given a labeling ℓ of a simplicial complex K, and an integer $0 \le d \le \dim(K)$, we define the d-dimensional flow $N_d[\langle S \rangle]$ by $N_d[\langle S \rangle] = 1$ if $\ell(S) = [(0, 1, 2 \ldots, d)]_{\equiv_{d+1}}$, $N_d[\langle S \rangle] = -1$ if $\ell(S) = [(1, 0, 2, \ldots, d)]_{\equiv_{d+1}}$, and $N_d[\langle S \rangle] = 0$ otherwise, and then extend it by linearity into a homomorphism $N_d : C_d(K; R) \to R$.*

Sperner's Lemma [23] has been generalized in several ways. The following statement from [25] is also a straightforward consequence of results of [11].

Theorem 1 (Sperner's Lemma [23,11,25]). *Let K be a simplicial complex of dimension d, let ℓ be a labeling of K, and let R be a ring. For an element C of $C_d(K; R)$, we have $N_d[C] = (-1)^d N_{d-1}[\partial_d C]$.*

Using Fact 2, we translate Theorem 1 into terms of pseudo-manifolds.

Theorem 2 (Sperner's Lemma on pseudo-manifolds). *Let d be an integer, let $\mathcal{M}$ be a pseudo d-manifold, and let ℓ be a labeling of $\mathcal{M}$. Then $N_d[\widehat{\mathcal{M}}] = (-1)^d N_{d-1}[\partial \widehat{\mathcal{M}}]$ where $\widehat{\mathcal{M}} \in C_d(\mathcal{M}, \mathbb{Z}/(2))$, $\partial \widehat{\mathcal{M}} \in C_{d-1}(\partial \mathcal{M}, \mathbb{Z}/(2))$ if $\mathcal{M}$ is not oriented, and $\widehat{\mathcal{M}} \in C_d(\mathcal{M}, \mathbb{Z})$, $\partial \widehat{\mathcal{M}} \in C_{d-1}(\partial \mathcal{M}, \mathbb{Z})$ if $\mathcal{M}$ is oriented.*

This version of Sperner's lemma can be viewed, from a physicist's point of view, as a result equivalent to a global conservation law of a flow. If there is a source for the flow and the space is bounded then there must be a sink for that flow. More concretely, the lines of flow can be drawn over d-simplices, that goes from one d-simplex to another if they share a $(d-1)$-face that has all possible labels in $\{0, \ldots, d-1\}$. The sources and sinks of the flow are the fully labeled d-simplices. The lemma basically says that if the amount of flow entering the manifold at the boundary is larger than the exiting flow, then there must exist sinks inside. The local conservation is stated by the fact that if there is an ingoing edge, there will not be two outgoing edges, and conversely. Formally, we have the following.

Fact 3. *Let (S, τ_S) be an oriented d-simplex. Then at most two of its oriented $(d-1)$-faces have a non-zero image by N_{d-1}. Moreover, if there are exactly two $(d-1)$-faces $(S', \tau_{S'}^S)$ and $(S'', \tau_{S''}^S)$ that have non-zero image by N_{d-1}, then $N_d[\langle (S, \tau_S) \rangle] = 0$ and $N_{d-1}[\langle (S', \tau_{S'}^S) \rangle] = -N_{d-1}[\langle (S'', \tau_{S''}^S) \rangle]$.*

This gives a relation between the problem of finding fully labeled d-simplices and the natural complete problems for the parity argument classes. We can consider an oriented d-simplex (S, τ_S) with $N_d[\langle (S, \tau_S) \rangle] = 1$ as a source for the flow, and $(S', \tau_{S'})$ with $N_d[\langle (S', \tau_{S'}) \rangle] = -1$ as a sink.

We now state the non-oriented black-box Sperner problems we will consider. The statement of d-**OSPM**, the general oriented problem can be found in the full paper.

Sperner on Pseudo d-Manifolds (d-SPM)

Input: a pseudo d-manifold $\mathcal{M}$, and $S \in \mathcal{M}_d$.
Oracle input: a labeling $\ell : \mathcal{M}_0 \rightarrow \{0, 1, \ldots, d\}$.
Promise: one of the two conditions holds, with $R = \mathbb{Z}/(2)$:
   a) $N_{d-1}[\widehat{\partial \mathcal{M}}] = 1$,
   b) $N_{d-1}[\widehat{\partial \mathcal{M}}] = 0$ and $N_d[\langle S \rangle] = 1$.
Output: $S' \in \mathcal{M}_d$ such that $N_d[\langle S' \rangle] = 1$, with $S \neq S'$ for case b.

We will deal in particular with the following important special case of 2-**SPM**. Let $V_m = \{(i, j) \in \mathbb{N}^2 \mid 0 \leq i + j \leq m\}$. Observe that $|V_m| = \binom{m+2}{2}$.

Regular Sperner (REGULAR 2-SPM)

Input: $n = \binom{m+2}{2}$ for some integer m.
Oracle input: a labeling $\ell : V_m \rightarrow \{0, 1, 2\}$.
Promise: for $0 \leq k \leq m$, $\ell(0, k) \neq 1$, $\ell(k, 0) \neq 0$, and $\ell(k, m - k) \neq 2$.
Output: p, p' and $p'' \in V$, such that $p' = p + (\varepsilon, 0)$, $p'' = p + (0, \varepsilon)$ for some $\varepsilon \in \{-1, 1\}$, and $\{\ell(p), \ell(p'), \ell(p'')\} = \{0, 1, 2\}$.

In fact, **REGULAR** 2-**SPM** on input $n = \binom{m+2}{2}$ is the instance of d-**SPM** on the regular m-subdivision of an elementary 2-simplex. Theorem 2 states that both d-SPM and d-OSPM have always a solution. The solution is not necessarily unique as it can be easily checked on simple instances. Thus the problems are not functional oracle problems.

5   Black-Box Algorithms for Pseudo d-Manifolds

The purpose of this section is to give a black-box algorithm for d-**SPM**. The corresponding algorithm for d-**OSPM** can be found in the full paper. To solve these problems, we adopt a divide and conquer approach. This kind of approach was successfully used in [16,15] and [20], to study the query complexity of the oracle version of the Local Search problem. However, the success of the divide and conquer paradigm for Sperner problems relies heavily on the use of the very strong statement of Sperner's Lemma that is given in Theorem 2. The usual, simpler version of Sperner's Lemma, like the one given in [19] does not appear to be strong enough for this purpose. Observe that though the standard proof of Sperner's Lemma is constructive, it yields only an algorithm of complexity $O(n)$. In our algorithms the division of the pseudo d-manifold $\mathcal{M}$ will be done according to the combinatorial properties of its skeleton graph. The particular parameter we will need is its *iterated separation number* that we introduce now for general graphs.

Definition 9. *Let $G = (V, E)$ be a graph. If A and C are subsets of V such that $V = A \cup C$, and that there is no edge between $A \setminus C$ and $C \setminus A$, then (A, C) is* *said to be a* separation *of the graph* G, *in notation* $(A, C) \prec G$. *The set* $A \cap C$ *is called a* separator *of the graph* G.

The iterated separation number *is defined by induction on the size of the graph* G *by* $s(G) = \min_{(A,C) \prec G} \{|A \cap C| + \max(s(G[A \setminus C]), s(G[C \setminus A]))\}$. *A pair* $(A, C) \prec G$ *such that* $s(G) = |A \cap C| + \max(s(G[A \setminus C]), s(G[C \setminus A]))$ *is called a* best separation *of* G.

The iterated separation number of a graph is equal to the *value of the separation game* on the graph G, which was introduced in [16]. In that article, that value was defined as the gain of a player in a certain game. Notice, also, that the iterated separation number is at most $\log |V|$ times the *separation number* as defined in [20]. Before giving the algorithms, and their analyses, we still need a few observations.

Lemma 2. *Let* $\mathcal{A}$ *and* $\mathcal{B}$ *be two pseudo d-manifolds, such that* $\mathcal{A} \cup \mathcal{B}$ *is also a pseudo d-manifold. Let* ℓ *be a labeling of* $\mathcal{A} \cup \mathcal{B}$. *If* $\mathcal{A}$ *and* $\mathcal{B}$ *have no d-simplex in their intersection, then* $N_d[\widehat{\mathcal{A} \cup \mathcal{B}}] = N_d[\widehat{\mathcal{A}}] + N_d[\widehat{\mathcal{B}}]$.

Lemma 3. *Let* $\mathcal{M}$ *be a pseudo d-manifold, and* $\mathcal{M}'$ *be a union of elementary d-complexes such that* $\mathcal{M}' \subseteq \mathcal{M}$. *Then* $\mathcal{M}'$ *is a pseudo d-manifold.*

Theorem 3. *Let* $\mathcal{M}$ *be a pseudo d-manifold,* H *a subset of* $\mathcal{M}_0$, *and* ℓ *be a labeling of the vertices of* $\mathcal{M}$. *Let* $(A, C) \prec G_{\mathcal{M}}[\mathcal{M}_0 \setminus H]$, $B = H \cup (A \cap C)$, *and* $M' = A \setminus C$ *and* $M'' = C \setminus A$. *Denote by* $\mathcal{B}$ *the set of elementary d-complexes of* $\mathcal{M}$ *whose vertices are all in* B, *and by* $\mathcal{M}'$ *(resp.* $\mathcal{M}''$*) the set of elementary d-complexes of which at least one of the vertices belongs to* M' *(resp.* M''*). Denote also by* $\mathcal{B}'$ *the set of elementary* $(d-1)$-*complexes of* $\mathcal{M}$ *whose vertices are all in* B. *Then,*

(i) $\mathcal{B}$, $\mathcal{M}'$, $\mathcal{M}''$ *and* $\mathcal{M}' \cup \mathcal{M}''$ *are pseudo d-manifolds,*

(ii) *if* $H \neq \mathcal{M}_0$ *then* $\mathcal{B}$, $\mathcal{M}'$ *and* $\mathcal{M}''$ *are proper subsets of* $\mathcal{M}$,

(iii) $N_d[\widehat{\mathcal{M}}] = N_d[\widehat{\mathcal{B}}] + N_d[\widehat{\mathcal{M}'}] + N_d[\widehat{\mathcal{M}''}]$,

(iv) *the inclusions* $\partial \mathcal{M}' \subseteq (\partial \mathcal{M}) \cup \mathcal{B}'$ *and* $\partial \mathcal{M}'' \subseteq (\partial \mathcal{M}) \cup \mathcal{B}'$ *hold,*

We are now ready to state Algorithm 1 which solves d-**SPM** when the labels of the 0-faces of $\partial \mathcal{M}$ are also known. We next give the result which states the correctness of our algorithm and specifies its complexity.

Lemma 4. *If* $\mathcal{M}$ *and* S *satisfy the promises of d-SPM, then Algorithms 1 returns a solution and uses at most* $s(G_{\mathcal{M}}[\mathcal{M}_0 \setminus H])$ *queries.*

Theorem 4. $\mathsf{DQC}(d\text{-}\mathbf{SPM}) = O(s(G_{\mathcal{M}}[\mathcal{M}_0 \setminus (\partial \mathcal{M})_0])) + |(\partial \mathcal{M})_0|$ *and* $\mathsf{DQC}(d\text{-}\mathbf{OSPM}) = O(s(G_{\mathcal{M}}[\mathcal{M}_0 \setminus (\partial \mathcal{M})_0])) + |(\partial \mathcal{M})_0|$.

Proof. The algorithms consist in querying the labels of the vertices of $\partial \mathcal{M}$ and then running Algorithm 1 with the initial choice $H = (\partial \mathcal{M})_0$. For the oriented case, an appropriate modification of Algorithm 1 works.

Algorithm 1. Main routine for solving d-**SPM**

Input: A pseudo d-manifold $\mathcal{M}$, $S \in \mathcal{M}_d$, a set $H \supseteq (\partial \mathcal{M})_0$ together with the labels of its elements.

Let $(A, C) \prec G_{\mathcal{M}}[\mathcal{M}_0 \setminus H]$ be a best separation, and $B = H \cup (A \cap C)$.
Let the complexes $\mathcal{B}$, $\mathcal{M}'$ and $\mathcal{M}''$ be defined as in Theorem 3.
Query the labels of the vertices in $A \cap C$.
if $\mathcal{B}$ contains a fully labeled elementary d-complex **then**
    Return the corresponding oriented d-simplex.
end if
Evaluate $N_{d-1}[\widehat{\partial \mathcal{B}}]$, $N_{d-1}[\widehat{\partial \mathcal{M}'}]$ and $N_{d-1}[\widehat{\partial \mathcal{M}''}]$.
if $N_{d-1}[\widehat{\partial K}] = 1$ for $K \in \{\mathcal{B}, \mathcal{M}', \mathcal{M}''\}$ **then**
    Iterate on K, any d-simplex $S \in K$, and B with the labels of its elements.
else
    Iterate on $K \in \{\mathcal{B}, \mathcal{M}', \mathcal{M}''\}$ containing S, S and B with the labels of its elements.
end if

To bound the complexity of our algorithms we need an upper-bound on the iterated separator number of the skeleton graph. The following theorem gives, for any graph, an upper bound on the size of a balancing separator, whose deletion leaves the graph with two roughly equal size components. The bound depends on the genus and the number of vertices of the graph.

Theorem 5 (Gilbert, Hutchinson, Tarjan [13]). *A graph of genus g with n vertices has a set of at most $6\sqrt{g \cdot n} + 2\sqrt{2n} + 1$ vertices whose removal leaves no component with more than $2n/3$ vertices.*

For our purposes we can immediately derive an upper bound on the iterated separation number.

Corollary 1. *For graphs $G = (V, E)$ of size n and genus g we have $s(G) \leq \lambda(6\sqrt{g \cdot n} + 2\sqrt{2n}) + \log_{3/2} n$, where λ is solution of $\lambda = 1 + \lambda\sqrt{2/3}$.*

In general, there is no immediate relationship between the genus of a pseudo d-manifold and the genus of its skeleton graph. However, if the pseudo d-manifold $\mathcal{M}$ is a triangulated oriented surface, then the genus of the graph is equal to the genus of $\mathcal{M}$. Used in conjunction with Corollary 1, Theorem 4 gives an effective upper bound for pseudo d-manifolds.

Corollary 2. *Let $\mathcal{M}$ be a pseudo d-manifold such that $G_{\mathcal{M}}$ is of size n and of genus g. Then,* $\mathrm{DQC}(d\text{-}\mathbf{SPM}) = O(\sqrt{g}) \cdot \sqrt{n} + |(\partial \mathcal{M})_0|$ *and* $\mathrm{DQC}(d\text{-}\mathbf{OSPM}) = O(\sqrt{g}) \cdot \sqrt{n} + |(\partial \mathcal{M})_0|$.

Since the skeleton graph of the underlying pseudo 2-manifold of **REGULAR** 2-**SPM** is planar, it has genus 0. Thus we get:

Theorem 6. $\mathrm{DQC}(\mathbf{REGULAR}\ 2\text{-}\mathbf{SPM}) = O(\sqrt{n})$.

In the next section, we show nontrivial lower bounds on the randomized and the quantum query complexity of the **REGULAR** 2-**SPM** problem. Observe that for some general instances of the 2-**SPM** over the same pseudo 2-manifold we can easily derive exact lower bounds from the known complexity of Grover's search problem [6]. For example, if a labeling is 2 everywhere, except on two consecutive vertices on the boundary where it takes respectively the values 0 and 1, then finding a fully labeled 2-simplex is of the same complexity as finding a distinguished element on the boundary.

6   Lower Bounds for REGULAR 2-SPM

We denote by **UNIQUE-SPERNER** all those instances of **REGULAR** 2-**SPM** for which there exists a unique fully labeled triangle. There exist several equivalent adversary methods for proving quantum lower bounds in the query model [24]. Here, we will use the weighted adversary method [2,3,14].

Theorem 7. *Let Σ be a finite set, let $n \geq 1$ be an integer, and let $S \subseteq \Sigma^n$ and S' be sets. Let $f : S \to S'$. Let Γ be an arbitrary $S \times S$ nonnegative symmetric matrix that satisfies $\Gamma[x,y] = 0$ whenever $f(x) = f(y)$. For $1 \leq k \leq n$, let Γ_k be the matrix such that $\Gamma_k[x,y] = 0$ if $x_k = y_k$, and $\Gamma_k[x,y] = \Gamma[x,y]$ otherwise. For all $S \times S$ matrix M and $x \in S$, let $\sigma(M,x) = \sum_{y \in S} M[x,y]$. Then*

$$\mathsf{QQC}(f) = \Omega\left(\min_{\Gamma[x,y]\neq 0, x_k \neq y_k} \sqrt{\frac{\sigma(\Gamma,x)\sigma(\Gamma,y)}{\sigma(\Gamma_k,x)\sigma(\Gamma_k,y)}}\right),$$

$$\mathsf{RQC}(f) = \Omega\left(\min_{\Gamma[x,y]\neq 0, x_k \neq y_k} \max\left(\frac{\sigma(\Gamma,x)}{\sigma(\Gamma_k,x)}, \frac{\sigma(\Gamma,y)}{\sigma(\Gamma_k,y)}\right)\right).$$

For the lower bound we will consider specific instances of **REGULAR** 2-**SPM**. For that, we need a few definitions. For any binary sequence b, let $|b|$ denote the length of the sequence b, and for $i = 0, 1$ let $w_i(b)$ be the number of bits i in b. For $0 \leq t \leq |b|$, let $b^t = b_1 \ldots b_t$ denote the prefix of length t of b.

The instances of **REGULAR** 2-**SPM** we will consider are those whose oracle inputs C_b are induced by binary sequences $b = b_1 \ldots b_{m-2}$ of length $m-2$ as follows:

$$C_b(i,j) = \begin{cases} 1 & \text{if } j = 0 \text{ and } i \neq 0, \\ 2 & \text{if } i = 0 \text{ and } j \neq m, \\ 0 & \text{if } i + j = m \text{ and } j \neq 0, \\ 1 & \text{if there exists } 0 \leq t \leq m-2 \text{ with } (i,j) = (w_0(b^t)+1, w_1(b^t)), \\ 2 & \text{if there exists } 0 \leq t \leq m-2 \text{ with } (i,j) = (w_0(b^t), w_1(b^t)+1), \\ 0 & \text{otherwise.} \end{cases}$$

Notice that the first and fourth (*resp.* second and fifth) conditions can be simultaneously satisfied, but the labeling definition is consistent. Also observe that, for any b, there is a unique fully labeled triangle, whose coordinates are $\{(w_0(b)+1, w_1(b)), (w_0(b), w_1(b)+1), (w_0(b)+1, w_1(b)+1)\}$. Therefore C_b is an instance of **UNIQUE-SPERNER**. We illustrate an instance of C_b in Figure 1.

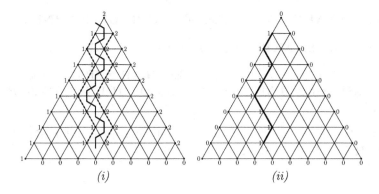

Fig. 1. In the coordinates system of the Figure, the point $(0,0)$ is the highest corner of the triangles, the x coordinates increase by going down and left, and the y coordinates increase by going down and right. On sub-figure *(i)*, the labeling C_b corresponds to the binary sequence $b = 0100110$. On sub-figure *(ii)*, the labeling O_b corresponds to the same sequence b. The unmarked vertices are all labeled 0.

It turns out that technically it will be easier to prove the lower bound for a problem which is closely related to the above instances of **REGULAR** 2-SPM, that we call **SNAKE**. Recall that $V_m = \{(i,j) \in \mathbb{N}^2 \mid 0 \leq i+j \leq m\}$. For every binary sequence $b = b_1 \ldots b_{m-2}$, we denote by O_b the function $V_m \to \{0,1\}$ defined for $p \in V_m$ by

$$O_b(p) = \begin{cases} 1 & \text{if there exists } 0 \leq t \leq m-2 \text{ with } (i,j) = (w_0(b^t)+1, w_1(b^t)), \\ 0 & \text{otherwise.} \end{cases}$$

See again Figure 1 for an example.

SNAKE
Input: $n = \binom{m}{2}$ for some integer m.
Oracle input: a function $f : V_m \to \{0,1\}$.
Promise: there exists a binary sequence $b = b_1 \ldots b_{m-2}$ such that $f = O_b$.
Output: $(w_0(b), w_1(b))$.

We recall here the definition of [20] of c-query reducibility between black-box problems, which we will use to prove our lower bound.

Definition 10. *For an integer $c > 0$, a functional oracle problem $A : S_1 \to T_1$ with $S_1 \subseteq \Sigma_1^n$ is c-query reducible to a functional oracle problem $B : S_2 \to T_2$ with $S_2 \subseteq \Sigma_2^{n'}$ if the following two conditions hold:*

(i) *$\exists \alpha : S_1 \to S_2$, $\exists \beta : T_2 \to T_1$, such that $\forall x \in S_1$, $A(x) = \beta(B(\alpha(x)))$,*
(ii) *$\exists \gamma_1, \ldots, \gamma_c : \{1, \ldots, n'\} \to \{1, \ldots, n\}$ and $\gamma : \{1, \ldots, n'\} \times \Sigma_1^c \to \Sigma_2$ such that $\forall x \in S_1$, $k \in \{1, \ldots, n'\}$, $\alpha(x)(k) = \gamma(k, x_{\gamma_1(k)}, \ldots, x_{\gamma_c(k)})$.*

Lemma 5 ([20]). *If A is c-query reducible to B then $\mathsf{QQC}(B) \geq \mathsf{QQC}(A)/2c$, and $\mathsf{RQC}(B) \geq \mathsf{RQC}(A)/c$.*

Lemma 6. SNAKE *is 3-query reducible to* **UNIQUE-SPERNER**.

Lemma 7. RQC(**SNAKE**) = $\Omega(\sqrt[4]{n})$ *and* QQC(**SNAKE**) = $\Omega(\sqrt[8]{n})$.

Theorem 8. *The query complexity of* **REGULAR** 2-**SPM** *satisfies*
RQC(**REGULAR** 2-**SPM**) = $\Omega(\sqrt[4]{n})$ *and* QQC(**REGULAR** 2-**SPM**) = $\Omega(\sqrt[8]{n})$.

Proof. By Lemma 5 and 6, the lower bounds of Lemma 7 for **SNAKE** also apply to **REGULAR** 2-**SPM**.

References

1. S. Aaronson. Quantum lower bound for the collision problem. In *34th STOC*, pp. 635–642, 2002.
2. S. Aaronson. Lower bounds for local search by quantum arguments. In *36th STOC*, pp. 465–474, 2004.
3. A. Ambainis. Polynomial degree vs. quantum query complexity. In *44th FOCS*, pp. 230–239, 2003.
4. R. Beals, H. Buhrman, R. Cleve, M. Mosca and R. de Wolf. Quantum lower bounds by polynomials, *J. of the ACM* (48):4, pp. 778–797, 2001.
5. P. Beame, S. Cook, J. Edmonds, R. Impagliazzo and T. Pitassi. The relative complexity of NP search problems. *J. Comput. System Sci.*, 57(1):3–19, 1998.
6. C. Bennett, E. Bernstein, G. Brassard and U. Vazirani. Strength and weakness of quantum computing. *SIAM J. on Computing*, 26(5):1510–1523, 1997.
7. E. Bloch. Mod 2 degree and a generalized no retraction Theorem. To appear in *Mathematische Nachrichten*.
8. J. Buresh-Oppenheim and T. Morioka. Relativized NP search problems and propositional proof systems. In *19th Conference on Computational Complexity*, pp. 54–67, 2004.
9. P. Crescenzi and R. Silvestri. Sperner's lemma and robust machines. *Comput. Complexity*, 7(2):163–173, 1998.
10. D. Deutsch and R. Jozsa. Rapid solution of problems by quantum computation. *Proc. of the Royal Society A*, volume 439,1985.
11. K. Fan. Simplicial maps from an orientable n-pseudomanifold into S^m with the octahedral triangulation. *J. Combinatorial Theory*, 2:588–602, 1967.
12. K. Friedl, Gábor Ivanyos, Miklos Santha and Yves Verhoeven. On the black-box complexity of Sperner's Lemma. http://xxx.lanl.gov/abs/quant-ph/0505185.
13. J. Gilbert, J. Hutchinson and R. Tarjan. A separator theorem for graphs of bounded genus. *J. Algorithms*, 5(3):391–407, 1984.
14. S. Laplante and F. Magniez. Lower bounds for randomized and quantum query complexity using kolmogorov arguments. In *19th Conference on Computational Complexity*, pp. 294–304, 2004.
15. D. Llewellyn and C. Tovey. Dividing and conquering the square. *Discrete Appl. Math.*, 43(2):131–153, 1993.
16. D. Llewellyn, C. Tovey and M. Trick. Local optimization on graphs. *Discrete Appl. Math.*, 23(2):157–178, 1989.
17. N. Megiddo and C. Papadimitriou. On total functions, existence theorems and computational complexity. *Theoret. Comput. Sci.*, 81:317–324, 1991.

18. C. Papadimitriou. On graph-theoretic lemmata and complexity classes. In *31st FOCS*, pp. 794–801, 1990.

19. C. Papadimitriou. On the complexity of the parity argument and other inefficient proofs of existence. *J. Comput. System Sci.*, 48(3):498–532, 1994.

20. M. Santha and M. Szegedy. Quantum and classical query complexities of local search are polynomially related. In *36th STOC*, pp. 494–501, 2004.

21. Y. Shi. Quantum lower bounds for the collision and the element distinctness problems. In *43rd FOCS*, pp. 513–519, 2002.

22. D. Simon. On the power of quantum computation. *SIAM J. on Computing* (26):5, pp. 1474–1783, 1997.

23. E. Sperner. Neuer Beweis für die Invarianz der Dimensionzahl und des Gebietes. *Abh. Math. Sem. Hamburg Univ.* 6:265–272, 1928.

24. R. Špalek and M. Szegedy. All quantum adversary methods are equivalent. http://xxx.lanl.gov/abs/quant-ph/0409116.

25. L. Taylor. Sperner's Lemma, Brouwer's Fixed Point Theorem, The Fundamental Theorem of Algebra. http://www.cs.csubak.edu/~larry/math/sperner.pdf.

26. A. Thomason. Hamilton cycles and uniquely edge colourable graphs. *Ann. Discrete Math.* 3: 259–268, 1978.

Property Testing and the Branching Program Size of Boolean Functions

(Extended Abstract)

Beate Bollig

FB 15, Johann-Wolfgang Goethe-Univ. Frankfurt am Main,
60054 Frankfurt, Germany
bollig@informatik.uni-frankfurt.de

Abstract. Combinatorial property testing, initiated formally by Goldreich, Goldwasser, and Ron (1998) and inspired by Rubinfeld and Sudan (1996), deals with the relaxation of decision problems. Given a property P the aim is to decide whether a given input satisfies the property P or is *far* from having the property. For a family of boolean functions $f = (f_n)$ the associated property is the set of 1-inputs of f. Newman (2002) has proved that properties characterized by oblivious read-once branching programs of constant width are testable, i.e., a number of queries that is independent of the input size is sufficient. We show that Newman's result cannot be generalized to oblivious read-once branching programs of almost linear size. Moreover, we present a property identified by restricted oblivious read-twice branching programs of constant width and by CNFs with a linear number of clauses, where almost all clauses have constant length, but for which the query complexity is $\Omega(n^{1/4})$.

1 Introduction and Results

1.1 Property Testing

Property testing is a field in computational theory that deals with the information that can be deduced from the input, when the number of allowable queries (reads from the input) is significantly smaller than its size. Applications could be practical situations in which the input is so large that even taking linear time in its size to provide an answer is too much. Given a particular set, called property, and an input x one wants to decide whether x has the property or is *far* from it. Far usually means that many characters of the input have to be modified to obtain an element in the set. The definition of property testing is a relaxation of the standard definition of a decision problem in the sense that the tester is allowed arbitrary behavior when the object does not have the property and yet is close to an object having the property.

Now we make this idea a little bit more precise. Let P be a property, i.e., a non-empty family of binary words. A word w of length n is called ϵn-far from satisfying P if no word w' of the same length, which differs in at most ϵn places (Hamming distance), satisfies P. An ϵ-test for P is a randomized algorithm,

M. Liśkiewicz and R. Reischuk (Eds.): FCT 2005, LNCS 3623, pp. 258–269, 2005.

which queries the quantity n, has the ability to make queries about the value of any desired bit of an input word w of length n, and distinguishes with probability $2/3$ between the case of $w \in P$ and the case of w being ϵn-far from satisfying P. (The choice of the success probability $2/3$ is not crucial since any constant strictly greater than $1/2$ is sufficient because of probability amplification.) A property P is said to be $(\epsilon, q(\epsilon, n))$-testable if there is an ϵ-test that for every input x of size n queries at most $q(\epsilon, n)$ bits of the input string. If a property P is $(\epsilon, q(\epsilon, n))$-testable with $q = q(\epsilon)$ (i.e., q is a function of ϵ only, and is independent of n), P is said to be ϵ-testable. Finally, we say that property P is testable if P is ϵ-testable for every fixed $\epsilon > 0$.

Properties can be identified with the collection of their characteristic boolean functions, i.e., a property $P \subseteq \{0,1\}^*$ is identified with a family of boolean function $f = (f_n)$, where $f_n : \{0,1\}^n \rightarrow \{0,1\}$ so that $f_n(x) = 1$ iff $x \in P$ and $|x| = n$. Let $x, y \in \{0,1\}^n$. If $g : \{0,1\}^n \rightarrow \{0,1\}$ and g is not the constant function 0, we define $dist(x, g) = \min\{H(x,y)|y \in g^{-1}(1)\}|$, where $H(x,y)$ is the Hamming distance. An input x is ϵn-close to a function g iff $dist(x, g) \leq \epsilon n$. Otherwise the input x is ϵn-far.

The general notion of property testing was first formulated by Rubinfeld and Sudan [18] and first studied for combinatorial objects by Goldreich, Goldwasser, and Ron [14]. These investigations were motivated by the notion of testing serving as a new notion of approximation and by some related questions that arise in computational learning theory. Recently, it has become quite an active research area, see e.g. [1] - [6], [8] - [13], [15] for an incomplete list and [7], [17] for excellent surveys on the topic.

1.2 Branching Programs

A *branching program* (BP) on the variable set $X_n = \{x_1, \ldots, x_n\}$ is a directed acyclic graph with one source and two sinks labeled by the constants 0 and 1. Each non-sink node (or decision node) is labeled by a boolean variable and has two outgoing edges, one labeled by 0 and the other by 1.

An input $a \in \{0,1\}^n$ activates all edges consistent with a, i.e., the edges labeled by a_i which leave nodes labeled by x_i. The computation path for an input a in a BP G is the path of edges activated by a which leads from the source to a sink. A computation path for an input a is called accepting if it reaches the 1-sink.

Let B_n denote the set of all boolean functions $f : \{0,1\}^n \rightarrow \{0,1\}$. The BP G represents the function $f \in B_n$ for which $f(a) = 1$ iff the computation path for the input a is accepting. The size of a branching program G is the number of its nodes. The branching program size of a boolean function f is the size of the smallest BP representing f. The length of a branching program is the maximum length of a path.

A branching program is called (syntactically) *read k times* (BPk) if each variable is tested on each path at most k times.

A branching program is called *s-oblivious*, for a sequence of variables $s = (s_1, \ldots, s_l)$, $s_i \in X_n$, if the set of decision nodes can be partitioned into disjoint sets V_i, $1 \leq i \leq l$, such that all nodes from V_i are labeled by s_i and the edges which leave V_i-nodes reach only V_{i+1}-nodes. The level i, $1 \leq i \leq l$, contains the V_i-nodes. The level $l + 1$ contains the 0- and the 1-sink. If the sequence of variables is unimportant we call an s-oblivious branching program an oblivious branching program. An oblivious branching program is of width w if its largest level contains w nodes.

An oblivious read k times branching program for a sequence $s = (s_1, \ldots, s_{kn})$ is called k-IBDD if s can be partitioned into k subsequences $(s_{(i-1)n+1}, \ldots, s_{in})$, $1 \leq i \leq k$, for which $\{s_{(i-1)n+1}, \ldots, s_{in}\} = X_n$.

1.3   Property Testing and the Branching Program Size of Boolean Functions

There are properties that are hard to decide but are testable such as 3-colorability [14] and properties that are easy to decide but are hard to test, e.g. in [2] examples of NC^1 functions have been presented that require $\Theta(n^{1/2})$ queries. Since a logical characterization of the properties testable with a constant number of queries is far from achieved, one goal is to identify whole classes of properties (instead of individual properties) that are testable and to formulate sufficient conditions for problems to be testable. Alon, Krivelevich, Newman, and Szegedy [2] have proved that membership in any regular language is testable, hence obtaining a general result identifying a non-trivial class of properties each being testable.

Newman [15] has extended the result described in [2] asserting that regular languages are testable by considering a non-uniform counterpart of the notion of a finite automaton. He has proved the following result. If $g = (g_n)$ where $g_n : \{0, 1\}^n \to \{0, 1\}$ is a family of boolean functions representable by oblivious read-once branching programs of width w then for every n and $\epsilon > 0$ there is a randomized algorithm that always accepts every $x \in \{0, 1\}^n$ if $g_n(x) = 1$ and rejects it with high probability if at least ϵn bits of x should be modified to obtain some $x' \in g_n^{-1}(1)$. His algorithm makes $(2^w/\epsilon)^{O(w)}$ queries. Therefore, for constant ϵ and w the query complexity is $O(1)$.

Recently, Fischer, Newman, and Sgall [13] put a bound on the branching program complexity of boolean functions that still guarantees testability by presenting a property identified by an oblivious read-twice branching program of width 3 for which at least $\Omega(n^{1/3})$ queries are necessary. Newman's result can be generalized to nonoblivious read-once branching programs of constant width [16]. Now it is quite natural to make one step further and to investigate whether Newman's result can be generalized to boolean functions representable by (oblivious) read-once branching programs of superlinear size. Bollig and Wegener [6] have shown that functions representable by read-once branching programs of quadratic size are not necessarily testable. Nevertheless, the presented lower bound on the query complexity is very small. Here we present a boolean function that can be represented by read-once branching programs of quadratic size but for which $o(n^{1/2})$ queries are insufficient in the sense of property testing. This is the best known lower bound on the query complexity of a property identified by boolean functions representable by restricted branching programs of small size. As a corollary we present boolean functions that can be represented by (oblivious) read-once branching programs of almost linear size which are not testable. This result is astonishing since the boolean functions representable by oblivious read-once branching programs of almost linear size are very simple.

1.4   Testing CNF Properties

A boolean formula is in conjunctive normal form (CNF) if it is a conjunction of clauses, where every clause is a disjunction of literals. (A literal is a boolean variable or a negated boolean variable.) The size of a CNF is the number of its clauses. If all clauses contain at most k literals, the formula is a kCNF. A boolean function f is said to have $O(1)$ size 0-witnesses if it can be represented by a kCNF, where $k = O(1)$. For a long time all properties that were known to be hard for two-sided error testing were functions whose 0-witnesses were large. E.g., the linear lower bound of Bogdanov, Obata, and Trevisan [5] capitalizes on the existence of inputs that are far from having the property, yet any local view of a constant fraction of them can be extended to an input having the property. If the property is defined by a kCNF, $k = O(1)$, this cannot happen. For each input that does not satisfy the property, there exists a set of k queries that witnesses the fact that the input does not have the property. Ben-Sasson, Harsha, and Raskhodnikova [4] have shown the existence of families of 3CNF formulas that require a linear number of queries. Fischer, Newman, and Sgall [13] have proved that there exists a property with $O(1)$ size 0-witnesses that can be represented by a width 3 oblivious read-twice branching program but for which a $5/8 \cdot 10^{-7}$-test requires $\Omega(n^{1/10})$ queries. Like the result in [4] the existence of the property involves a probabilistic argument and the proof is not constructive. In Section 3 we present a boolean function that can be represented by restricted oblivious read-twice branching programs of constant width and by CNFs of linear size, where almost all clauses have constant length, but for which the query complexity is $\Omega(n^{1/4})$ for any ϵ-test, $\epsilon \leq 1/8 - \delta'$ and δ' a constant.

Table 1 presents some lower bounds on the query complexity of boolean functions representable by CNFs of small size.

Table 1. Some results for boolean functions representable by CNFs of small size

Source	Proof constructive	Representable by 2IBDDs of constant width	Query complexity
[4]	no	no	$\gamma n,\ 0 < \gamma < 1$
[13]	no	yes	$\Omega(n^{1/10})$
New	yes	yes	$\Omega(n^{1/4})$

2 Non-testability of Functions with Almost Linear (Oblivious) Read-Once Branching Program Size

Bollig and Wegener [6] have already proved that boolean functions representable by read-once branching programs of quadratic size are not necessarily testable. Here we investigate a function which has some similarities to the function described in [6] but for the lower bound on the query complexity we have to use a proof which is more complicated since the number of allowable queries is not bounded by a constant.

W.l.o.g. let ℓ be an even number and $m := 2^\ell$. The function MP_n^* (*mixed pairs*) is defined on $n := 2m + \ell \cdot 2^{\ell/2}$ a-, x-, and y-variables, namely

$$a_0, \ldots, a_{\ell \cdot 2^{\ell/2}-1}, x_0, \ldots, x_{m-1}, y_0, \ldots, y_{m-1}.$$

The $\ell \cdot 2^{\ell/2}$ a-variables serve as address variables in the following way. First, we define the auxiliary h-variables as

$$h_i := a_{i \cdot 2^{\ell/2}} \oplus \cdots \oplus a_{(i+1) \cdot 2^{\ell/2}-1},$$

where the sum is computed mod 2 and $0 \leq i \leq \ell - 1$. Then, the address value is defined as $val(a) := 2^{\ell-1} \cdot h_{\ell-1} + \cdots + 2^0 \cdot h_0$ and

$$\mathrm{MP}_n^*(a, x, y) = \bigwedge_{0 \leq i \leq m-1} (x_i \oplus y_{(i+val(a))}),$$

where all indices are considered mod m.

The idea for the construction of the address value is the following one. On the one hand there are not too many a-variables in comparison to the number of x- and y-variables which serve as data variables. Moreover, we only have to modify at most ℓ a-variables to change a given address value to a different one. On the other hand we have to know at least $2^{\ell/2}$ a-variables to fix one of the auxiliary h-variables.

Theorem 1. *The function* MP_n^* *can be represented by read-once branching programs of quadratic size but for any ϵ-test, $\epsilon \leq 1/4 - \delta$, δ a constant, $o(n^{1/2})$ queries are insufficient.*

Proof. First, we prove the upper bound on the size of read-once branching programs representing MP_n^*. In the following all indices are considered mod m. We start with a complete binary tree of the h-variables, afterwards we replace each h_i-node, $0 \leq i \leq \ell - 1$, by a read-once branching program representing $a_{i \cdot 2^{\ell/2}} \oplus \cdots \oplus a_{(i+1) \cdot 2^{\ell/2}-1}$. Altogether the size is

$$(2^\ell - 1) \cdot (2 \cdot 2^{\ell/2} - 1) = 2^{(3/2)\ell+1} - 2^{\ell/2+1} - 2^\ell + 1.$$

If $val(a)$ is fixed, the function $f_{val(a)} := \bigwedge_{0 \leq i \leq m-1} (x_i \oplus y_{(i+val(a))})$ can be computed by a BP1 in size $3m$, where the variables are tested in the order

$$x_0, y_{(0+val(a))}, x_1, y_{(1+val(a))}, \ldots, x_{m-1} y_{(m-1+val(a))}.$$

Altogether the function MP_n^* can be represented by a read-once branching program of size $2^{(3/2)\ell+1} - 2^{\ell/2+1} + 1 + 2^\ell \cdot (3m - 1) = O(m^2) = O(n^2)$.

For the non-testability result we use Yao's *minimax* principle [19] which says that to show a lower bound on the complexity of a randomized test, it is enough to present an input distribution on which any deterministic test with that complexity is likely to fail. Positive instances are generated according to the distribution P which is the uniform distribution on all 1-inputs of MP_n^*. This can be realized as follows. Because of the symmetric properties of MP_n^*, the a-variables or, equivalently, the value $val(a)$ can be chosen according to the uniform distribution. Also the x-bits are chosen according to the uniform distribution. Afterwards, it is necessary to set $y_i := 1 - x_{i-val(a)}$. Negative instances are generated according to the distribution N which is the uniform distribution on the set of all inputs which are ϵn-far from MP_n^*. The probability distribution D over all inputs is now defined by choosing with probability $1/2$ positive instances according to P and with probability $1/2$ negative instances according to N.

The following claims which are similar to claims that have already been proved in [6] are helpful in order to show that any deterministic algorithm will fail with high probability.

Let U be the uniform distribution on all inputs.

Claim 1. *For each $\epsilon \leq 1/4 - \delta$, $\delta > 0$ a constant, the probability that an instance generated according to U is ϵn-close to MP_n^* is exponentially small with respect to n.*

Claim 2. *A pair (x_i, y_j) is called k-pair if $j \equiv i + k$ mod m. Let S be a set of d variables from $\{x_0, \ldots, x_{m-1}, y_0, \ldots, y_{m-1}\}$ and let w be an instance generated according to P (or to N). Let k be the address value of w. The probability that S contains a k-pair is bounded above by $d^2/(4m)$ which is $o(1)$ if $d = o(m^{1/2}) = o(n^{1/2})$.*

Now the idea is to prove that even an adaptive deterministic algorithm that queries at most $d = o(n^{1/2})$ bits has an error probability of more than $1/6$ on inputs taken from the distribution D. By probability amplification we can conclude that every adaptive deterministic algorithm that queries at most $d/9 = o(n^{1/2})$ bits must have an error probability of more than $1/3$.

Let $\mathcal{T}$ be an adaptive algorithm that queries at most d bits. Every leaf in the decision tree that represents $\mathcal{T}$ is labeled by either accept or reject. We may assume that $\mathcal{T}$ queries each input bit at most once for a randomly chosen input w according to D and that the decision tree is a complete binary tree of depth d, because we can transform each decision tree in such a tree without increasing the error probability. Let L be the set of all leaves that are labeled by reject. Let $B(L)$ be the event to reach a leaf from the set L. We assume that $\text{Prob}_N(B(L)) \geq 2/3$ as otherwise the algorithm errs on inputs which are ϵn-far from MP_n^* with probability of more than $1/6$. Our aim is to prove that from this assumption it follows that $\text{Prob}_P(B(L)) \geq (1 - o(1))2/3 > 1/3$ which implies that the algorithm errs by rejecting positive inputs with probability of more than $1/6$.

Every leaf $\alpha \in L$ corresponds to a set of variables S_α that were queried along the way to α and an assignment b_{S_α} to these variables. The algorithm reaches for an input w the leaf α if the assignment to the variables in S_α is consistent with b_{S_α}. The assignment to variables that are queried in other branches of the decision tree is irrelevant.

Let $B(\alpha)$ be the event that the algorithm reaches the leaf α, M_ϵ the event that an input is ϵn-far, and $A(S_\alpha)$ the event that for an input w the set S_α contains no k-pair, where k is the address value of w. The distribution N equals the uniform distribution U on all inputs with the restriction that the input is ϵn-far from MP_n^*. Using Claim 1 we know that $\mathrm{Prob}_U(M_\epsilon) \geq 1 - o(1)$ if n is large enough. Furthermore, if the input is chosen according to the uniform distribution on all inputs all leaves are reached with the same probability. Therefore,

$$\mathrm{Prob}_N(B(\alpha)) = \mathrm{Prob}_U(B(\alpha) \mid M_\epsilon) = \mathrm{Prob}_U(B(\alpha) \cap M_\epsilon)/\mathrm{Prob}_U(M_\epsilon)$$
$$\leq \mathrm{Prob}_U(B(\alpha))/\mathrm{Prob}_U(M_\epsilon)$$
$$\leq (1 + o(1))2^{-d}.$$

Since there are at most $o(n^{1/2}) = o(2^{\ell/2})$ a-variables in S_α and because of the definition of $val(a)$, all address values are still possible with equal probability. Using Claim 2 we know that

$$\mathrm{Prob}_P(B(\alpha)) \geq \mathrm{Prob}_P(B(\alpha) \cap A(S_\alpha))$$
$$= \mathrm{Prob}_P(B(\alpha) \mid A(S_\alpha)) \cdot \mathrm{Prob}_P(A(S_\alpha))$$
$$\geq (1 - o(1))2^{-d}.$$

Using the fact that $(1 - o(1))(1 - o(1))$ is equal to $(1 - o(1))$ we can conclude that

$$\mathrm{Prob}_P(B(L)) = \sum_{\alpha \in L} \mathrm{Prob}_P(B(\alpha)) \geq (1 - o(1)) \sum_{\alpha \in L} \mathrm{Prob}_N(B(\alpha))$$
$$= (1 - o(1))\mathrm{Prob}_N(B(L)) \geq (1 - o(1))2/3$$
$$> 1/3$$

if n is large enough. This completes the proof of Theorem 1. $\square$

The lower bound on the query complexity is determined by the number of different boolean functions f_i if we choose the number of a-variables in the right way. Using n^γ different boolean functions f_i, $0 \leq i \leq n^\gamma - 1$, and $0 < \gamma < 1$, we get the following result.

Corollary 1. *There exist boolean functions that can be represented by read-once branching programs of size $O(n^{1+\gamma})$, γ an arbitrary constant with $0 < \gamma < 1$, but for any ϵ-test, $\epsilon \leq 1/4 - \delta$, δ a constant, $o(n^{\gamma/2})$ queries are insufficient.*

For oblivious read-once branching programs we cannot choose different variable order for different address values. Therefore, it is impossible to represent a large number of different boolean functions f_i in small size. Using $\lceil \log n^{\gamma/2} \rceil$ different boolean functions we get the following result.

Corollary 2. *There exist boolean functions that can be represented by oblivious read-once branching programs of size $O(n^{1+\gamma})$, γ an arbitrary constant with $0 < \gamma < 1$, but for any ϵ-test, $\epsilon \leq 1/4 - \delta$, δ a constant, $o(\log^{1/2} n)$ queries are insufficient.*

3   Large Query Complexity for a Function with Constant Width 2IBDDs and CNFs of Small Size

In this section we construct a property identified by a family of boolean functions $g = (g_n)$ that can be represented by 2IBDDs of constant width and by CNFs of linear size, where even most of the clauses have constant length, but for which any ϵ-test requires n^δ queries for some $0 < \epsilon < 1$ and $0 < \delta < 1$. First, we reinvestigate a function $f_n : \{0, 1, 2\}^{n^2} \to \{0, 1\}$ that has already been considered by Fischer and Newman [11] to present a $\forall\exists$ property that is not testable but we use a different proof to present a larger lower bound on the query complexity. Afterwards we consider a boolean encoding of that function. Since our aim is to construct a boolean function that can be represented by 2IBDDs of constant width and by CNFs of linear size, we have to use a different approach as described in [11].

The property *symmetric permutation* uses the alphabet $\{0, 1, 2\}$. We say that a matrix satisfies the property *symmetric permutation* if it is a row permutation of a symmetric matrix with all 2's on its primary diagonal, and no 2's anywhere else. Obviously, this requirement is equivalent to the following two conditions:

1. In every row and in every column there exists exactly one 2-entry.
2. The matrix contains none of the following 2×2 matrices as a submatrix (to ensure that the original matrix was symmetric):

$$\begin{pmatrix} 2 & 0 \\ 1 & 2 \end{pmatrix}, \begin{pmatrix} 2 & 1 \\ 0 & 2 \end{pmatrix}, \begin{pmatrix} 0 & 2 \\ 2 & 1 \end{pmatrix}, \begin{pmatrix} 1 & 2 \\ 2 & 0 \end{pmatrix}.$$

As usual we identify the property *symmetric permutation* with its characteristic function $f = (f_n)$, where f_n is defined on $n \times n$ matrices M on the variables m_{ij}, $1 \leq i, j \leq n$.

Proposition 1. *For any ϵ-test for f_n, $\epsilon \leq 1/4 - \delta$ and δ a constant, $o(n)$ queries are insufficient, where n^2 is the number of variables of f_n.*

Sketch of proof. The proof follows the lines of the proof of Theorem 1 but we have to take into consideration that f_n is a non-boolean function and therefore the decision tree that represents an adaptive algorithm for f_n is a ternary tree. Furthermore, negative instances are generated according to the distribution N which is the uniform distribution on the set of all boolean $n \times n$ matrices that are ϵn^2-far from f_n (and not on the set of all inputs that are ϵn^2-far from f_n). Moreover, we need the following two claims.

Claim 3. *Let U be the uniform distribution on all boolean $n \times n$ matrices. For each $\epsilon \leq 1/4 - \delta$, δ a constant, the probability that an instance generated according to U is ϵn^2-close to f_n is $o(1)$.*

A pair $(m_{ij}, m_{i'j'})$, $i \neq i'$ and $j \neq j'$, is called a σ-pair for a permutation $\sigma \in S_n$ if $\sigma(i) = j'$ and $\sigma(i') = j$.

Claim 4. *Let S be a set of $d = o(n)$ variables from $\{m_{11}, \ldots, m_{nn}\}$ and M be an instance generated according to P, the uniform distribution on $f_n^{-1}(1)$. Let σ_M be the corresponding permutation to M. Let $A(S)$ be the event that S contains no σ_M-pair and no variable m_{ij} for which $\sigma_M(i) = j$. Then $\mathrm{Prob}_P(A(S)) \geq 1 - o(1)$.* □

The function $f_n^{A,B} : \{0,1\}^{2n^2} \to \{0,1\}$ is a boolean encoding of the function f_n and is defined on two $n \times n$ boolean matrices A and B on the variables a_{ij} and b_{ij}, $1 \leq i, j \leq n$. The function $f_n^{A,B}$ outputs 1 iff the following conditions are fulfilled:

i) A is a permutation matrix, i.e., there exists exactly one 1-entry in each row and one 1-entry in each column.

ii) If a_{ij} and $a_{i'j'}$ are equal to 1 then $b_{ij'}$ is equal to $b_{i'j}$.

Using a so-called distance preserving reduction we can prove the following claim.

Claim 5. *If there exists an ϵ-test with $q(\epsilon, 2n^2)$ queries for $f_n^{A,B}$ then there exists a 2ϵ-test with the same number of queries for f_n.*

Proof. For brevity we denote an ϵ-test with $q(\epsilon, n)$ queries by $(\epsilon, q(\epsilon, n))$-test in the following. Assume that there exists an $(\epsilon, q(\epsilon, 2n^2))$-test T for $f_n^{A,B}$. Our aim is to construct an $(2\epsilon, q(\epsilon, 2n^2))$-test for f_n. For this reason we define a mapping p from inputs M to f_n to inputs (A, B) for $f_n^{A,B}$. For every $M = (m_{ij})_{1 \leq i,j \leq n} \in \{0, 1, 2\}^{n^2}$ let $p(M)$ be defined as follows:

- $a_{ij} := m_{ij}(m_{ij} - 1)/2$,
- $b_{ij} := m_{ij}(m_{ij} - 1)/2 - (m_{ij} - 2)m_{ij}$.

If $m_{ij} = 0$ then $a_{ij} = b_{ij} = 0$, if $m_{ij} = 1$ then $a_{ij} = 0$ and $b_{ij} = 1$, and if $m_{ij} = 2$ then $a_{ij} = 1$ and $b_{ij} = 1$. Obviously, $M \in f_n^{-1}(1)$ implies $f_n^{A,B}(p(M)) = 1$. In order to obtain an input from $f_n^{-1}(1)$ there have to be at least as many bit positions in M to be changed as bit positions in $p(M)$ in order to get a 1-input from $f_n^{A,B}$. Therefore, if $dist(p(M), f_n^{A,B}) \leq \epsilon(2n^2)$ then $dist(M, f_n) \leq 2\epsilon n^2$.

To $(2\epsilon, q(\epsilon, 2n^2))$-test f_n on an input M, we perform the $(\epsilon, q(\epsilon, 2n^2))$-test T on $p(M)$. The only difference is that each time that a_{ij} (b_{ij}) is queried for $p(M)$ we just query m_{ij} and compute a_{ij} (b_{ij}). Since the reduction is distance preserving we can inherit the result of T without changing the error probability on 1-inputs and inputs that are $2\epsilon n^2$-far from f_n. □

Now we transform the function $f_n^{A,B}$ to a boolean function g_n that can be represented by 2IBDDs of constant width and by CNFs of linear size, where most of the clauses have constant length. The idea for the construction of g_n is the following one. We use the same number of copies for the variables in the matrices A and B. For every original variable a_{ij} (b_{ij}) we generate $(n-1)^2$ new variables $a_{ij}^{i'j'}$ ($b_{ij}^{i'j'}$), where $i \neq i'$ and $j \neq j'$. Then we add for each a_{ij} (b_{ij}) two new variables a_{ij}^r and a_{ij}^c (b_{ij}^r and b_{ij}^c). Altogether the function g_n is defined on $N := 2n^2((n-1)^2 + 2)$ variables and outputs 1 iff the following conditions are fulfilled:

a) For $1 \leq i, j \leq n$, all variables $a_{ij}^{i'j'}$, $i \neq i'$ and $j \neq j'$, a_{ij}^r, and a_{ij}^c have the same value. The same holds for the b_{ij}-variables.
b) For each row i, $1 \leq i \leq n$, there exists exactly one variable a_{ij}^r, $1 \leq j \leq n$, that is set to 1.
c) For each column j, $1 \leq j \leq n$, there exists exactly one variable a_{ij}^c, $1 \leq i \leq n$, that is set to 1.
d) If $a_{ij}^{i'j'} = a_{i'j'}^{ij} = 1$ then $b_{ij}^{i'j'}$ and $b_{i'j}^{ij'}$ are equal.

Claim 6. *If there exists an $(\epsilon, q(\epsilon, N))$-test for g_n then there exists an $(\epsilon, q(\epsilon, N))$-test for $f_n^{A,B}$.*

The proof of Claim 6 is similar to the proof of Claim 5.

Claim 7. *The function g_n can be represented by 2IBDDs of width 3 and by CNFs of linear size.*

Proof. In the first part of the 2IBDD we verify the requirements b), c), and d) one after another. Since the requirements are defined on different sets of variables we obtain an oblivious read-once branching program for these requirements by glueing the oblivious read-once branching programs together, i.e., the 1-sink of one branching program is replaced by the source of the next one and so on. The requirements b) and c) can be verified by an oblivious read-once branching program of width 3. The requirement d) can be checked for each group of four variables $a_{ij}^{i'j'}$, $a_{i'j'}^{ij}$, $b_{ij}^{i'j'}$, and $b_{i'j}^{ij'}$ by an oblivious read-once branching program of width 3 realising the function

$$\neg((a_{ij}^{i'j'} \wedge a_{i'j'}^{ij}) \wedge ((b_{ij}^{i'j'} \wedge \neg b_{i'j}^{ij'}) \vee (\neg b_{ij}^{i'j'} \wedge b_{i'j}^{ij'}))).$$

In the second part of the oblivious 2IBDD we just check the requirement a), i.e., if all copies of the same variable of the original function $f_n^{A,B}$ have the same value. All copies of the same variable are tested one after another. Width 3 is sufficient.

The resulting CNF is a conjunction of CNFs checking the requirements a)-d) separately. The requirement that for each row i (column j) there exists exactly one variable a_{ij}^r, $1 \leq j \leq n$, (a_{ij}^c, $1 \leq i \leq n$) that is set to 1 is equivalent to the requirement that there exists for each row i (column j) at least one variable that is set to 1 and that there do not exist two variables in the same row (column) that are set to 1. Hence, we obtain

$$(a_{i1} \lor a_{i2} \lor \cdots \lor a_{in}) \bigwedge_{1 \leq j_1 < j_2 \leq n} (\neg a_{ij_1} \lor \neg a_{ij_2}).$$

Altogether there are $2n \cdot n(n-1)/2 = n^3 - n^2$ clauses of length 2 and $2n$ clauses of length n to check the requirements b) and c). The requirement d) can be tested for each group of variables by a 4CNF with 2 clauses

$$(\neg a_{ij}^{i'j'} \lor \neg a_{i'j'}^{ij} \lor \neg b_{ij'}^{ij'} \lor b_{i'j}^{ij'}) \land (\neg a_{ij}^{i'j'} \lor \neg a_{i'j'}^{ij} \lor b_{ij'}^{i'j} \lor \neg b_{i'j}^{ij'}).$$

Altogether there are $n^2 \cdot 2(n-1)^2$ clauses of length 4 to verify condition d). Finally, the test whether some variables have the same value can be done by pairwise checking that two of them are equal or in a more clever way by checking whether the first one is equal to the second one, the second one to the third one and so on. Therefore, the requirement a) can be tested by a 2CNF with $2n^2 \cdot 2((n-1)^2 + 1)$ clauses. Summarizing, the function g_n has a CNF representation with $4n^4 - 7n^3 + 7n^2$ clauses of length 2, $2n^4 - 4n^3 + 2n^2$ clauses of length 4 and $2n$ clauses of length n, where $N = \Theta(n^4)$ is the number of variables. $\qquad \square$

Combining Proposition 1 and Claims 5 - 7 we obtain the following main result.

Theorem 2. *The function g_n can be represented by 2IBDDs of width 3 and by CNFs of linear size but for any ϵ-test, $\epsilon \leq 1/8 - \delta'$, δ' a constant, $o(N^{1/4})$ queries are insufficient, where N is the number of variables of g_n.*

4   Concluding Remarks

As we have seen Newman's result [15] that properties representable by read-once branching programs of constant width are testable cannot be generalized to functions of almost linear (oblivious) read-once branching program size. We have improved the best known lower bound on the query complexity of properties identified by boolean functions representable by very restricted branching programs up to $\Omega(n^{1/2})$. The question whether there exist linear lower bounds on the query complexity of properties identified by some restricted branching programs of small size or whether there exist sublinear ϵ-tests remains unsolved. Furthermore, we have presented a boolean function with 2IBDDs of constant width and CNFs with a linear number of clauses, where almost all clauses have constant length, that has query complexity $\Omega(n^{1/4})$, where n is the input length. A constructive proof that there exists a non-testable boolean function with 2IBDDs of constant width and constant size 0-witnesses, i.e., where not almost all but all clauses of its CNF representation have constant length, would be interesting.

Acknowledgement

I would like to thank Ingo Wegener for several valuable hints and fruitful discussions on the subject of the paper.

References

1. Alon, N., Fischer, E., Krivelevich, M., and Szegedy, M. (2000). Efficient testing of large graphs. Combinatorica 20, 451–476.
2. Alon, N., Krivelevich, M., Newman, I., and Szegedy, M. (2000). Regular languages are testable with a constant number of queries. SIAM Journal on Computing 30, 1842–1862.
3. Alon, N. and Shapira, A. (2005). Every monotone graph property is testable. To appear in Proc. of 37th STOC.
4. Ben-Sasson, E., Harsha, P., and Raskhodnikova, S. (2003). Some 3CNF properties are hard to test. Proc. of 35th STOC, 345–354.
5. Bogdanov, A., Obata, K., and Trevisan, L. (2002). A lower bound for testing 3-colorability in bounded-degree graphs. Proc. of 43rd FOCS, 93–102.
6. Bollig, B. and Wegener, I. (2003). Functions that have read-once branching programs of quadratic size are not necessarily testable. Information Processing Letters 87, 25–29.
7. Fischer, E. (2001). The art of uninformed decisions: a primer to property testing. Bulletin of the European Association for Theoretical Computer Science 75 (Oct. 2001), 97–126. The Computational Complexity Column.
8. Fischer, E. (2004). On the strength of comparisons in property testing. Information and Computation 189, 107–116.
9. Fischer, E. (2004a). The difficulty of testing for isomorphism against a graph that is given in advance. Proc. of 36th STOC, 391–397.
10. Fischer, E., Lehman, E., Newman, I., Reskhodnikova, S., Rubinfeld, R., and Samorodnitsky (2002). Monotonicity testing over general poset domains. Proc. of 34th STOC, 73–79.
11. Fischer, E. and Newman, I. (2001). Testing of matrix properties. Proc. of 33rd STOC, 286–295.
12. Fischer, E. and Newman, I. (2005). Testing versus estimation of graph properties. To appear in Proc. of 37th STOC.
13. Fischer, E., Newman, I., and Sgall, J. (2004). Functions that have read-twice constant width branching programs are not necessarily testable. Random Structures and Algorithms 24(2), 175–193.
14. Goldreich, O., Goldwasser, S., and Ron, D. (1998). Property testing and its connection to learning and approximation. Journal of the ACM 45, 653–750.
15. Newman, I. (2002). Testing membership in languages that have small width branching programs. SIAM Journal of Computing 31(5), 1557–1570.
16. Newman, I. (2002a). Private communication.
17. Ron, D. (2001). Property testing (a tutorial). In Handbook of Randomized Computing, Rajasekaran, S., Pardalos, P.M., Reif, J.H., and Rolim, J.D., Eds., vol. 9 of Combinatorial Optimization, Kluwer Academic Publishers, 597–649.
18. Rubinfeld, R. and Sudan, M. (1996). Robust characterization of polynomials with applications to program testing. SIAM Journal of Computing 25(2), 252–271.
19. Yao, A. C. (1977). Probabilistic computation, towards a unified measure of complexity. Proc. of 18th FOCS, 222–227.

Almost Optimal Explicit Selectors

Bogdan S. Chlebus[1,*] and Dariusz R. Kowalski[2, 3,**]

[1] Department of Computer Science and Eng., UCDHSC, Denver,
CO 80217, USA
[2] Department of Computer Science, University of Liverpool,
Liverpool L69 7ZF, UK
[3] Instytut Informatyki, Uniwersytet Warszawski, Banacha 2,
Warszawa, Poland

Abstract. We understand *selection by intersection* as distinguishing a single element of a set by the uniqueness of its occurrence in some other set. More precisely, given two sets A and B, if $A \cap B = \{z\}$, then element $z \in A$ is *selected* by set B. Selectors are such families $\mathcal{S}$ of sets B of some domain that allow to select many elements from sufficiently small subsets A of the domain. Selectors are used in communication protocols for the multiple-access channel, in implementations of distributed-computing primitives in radio networks, and in algorithms for group testing. We give new explicit (n, k, r)-selectors of size $\mathcal{O}(\min\left[n, \frac{k^2}{k-r+1} \text{ polylog } n\right])$, for any parameters $r \leq k \leq n$. We establish a lower bound $\Omega(\min\left[n, \frac{k^2}{k-r+1} \cdot \frac{\log(n/k)}{\log(k/(k-r+1))}\right])$ on the length of (n, k, r)-selectors, which demonstrates that our construction is within a polylog n factor close to optimal. The new selectors are applied to develop explicit implementations of selection resolution on the multiple-access channel, gossiping in radio networks and an algorithm for group testing with inhibitors.

1   Introduction

Selection by intersection means distinguishing a single element of a set as the only element of some other set. More precisely, given a subset $A \subseteq X$ of a finite domain X, element $z \in A$ is *selected* by a set $B \subseteq X$ when $A \cap B = \{z\}$.

The power of such a selection is often considered in quantitative terms, which translate into efficiency in applications. A natural parameter to consider is the size of sets A from which we select. A family $\mathcal{S}$ of subsets of X is called *k-selective*, following Chlebus *et al.* [4], if we can select an element from *any* subset $A \subseteq X$ of size $|A| \leq k$ by a set in $\mathcal{S}$. Families $\mathcal{S}$ that are useful in application, because of their selection-related properties, are typically parametrized by the size n of the domain X, and we want the number k to be close to n, while keeping the size of $\mathcal{S}$ small. Additionally, we may want to have many elements $z \in A$ to be selected, for any $A \subseteq X$ of size k.

---

* The work of this author is supported by the NSF Grant 0310503.
** The work of this author supported in part by the KBN Grant 4T11C04425.

M. Liśkiewicz and R. Reischuk (Eds.): FCT 2005, LNCS 3623, pp. 270–280, 2005.

This is captured by the following definition. Let n, k and r be positive integers so that $r \leq k \leq n$. Let $\mathcal{S}$ be a family of subsets of $[n] = [1..n]$. We say that $\mathcal{S}$ is an (n, k, r)-selector if, for each set $A \subseteq [n]$ of size $|A| = k$, there are at least r elements in A that can be selected from A by sets in $\mathcal{S}$.

The name "selectors" was coined by De Bonis, Gąsieniec and Vaccaro [13] in the context of their work on group testing. Their definition of selectors is in terms of binary matrices and corresponds to certain generalized superimposed codes. The notion of a selector generalizes many popular combinatorial structures. Among them there are (n, k)-selective families, introduced by Chlebus et al. [4], which are $(n, k, 1)$-selectors in selector terminology. Objects called simply (n, k)-selectors by Chrobak, Gąsieniec and Rytter [7] correspond to $(n, 2k, 3k/2)$-selectors. Finally, (n, k)-strongly-selective families introduced by Clementi, Monti and Silvestri [11] are nothing but (n, k, k)-selectors. Such (n, k, k)-selectors are closely related to $(k - 1)$-cover-free families, in the hypergraph terminology [23], and to superimposed codes [17,24]. See Section 2 for an overview of the related combinatorics and matrix representations.

Selection by intersection is a notion that occurs in many disguises in combinatorial settings and in algorithmic and communication applications. Selectors can be applied in deterministic conflict resolution in multiple-access channels [3,25], in broadcasting and gossiping algorithms for ad-hoc radio networks [1,4,7,12,26] and in deterministic algorithms for group testing [13,14,15,16]. Related combinatorial structures called radio synchronizers are directly applicable in algorithms for waking up radio networks [5,6,20] and to implement distributed-computing primitives in radio networks [5,6], like leader election and synchronization of local clocks.

Combinatorial structures used in implementations of algorithms as part of their code are said to be *explicit* when there are algorithms that produce them in time that is polynomial in the size of the output.

Our Results. The contributions are summarized as follows.

I. We construct explicit (n, k, r)-selectors of size $\mathcal{O}(\min\left[n, \frac{k^2}{k-r+1} \text{ polylog } n \right])$, for any configuration of parameters $r \leq k \leq n$. The design involves explicit dispersers. This result extends the ranges of two previously known explicit constructions. One is that of explicit superimposed codes of n codewords of length $\mathcal{O}(\min[n, k^2 \log^2 n])$ that are k-disjunct. This is the classical design by Kautz and Singleton [24]. The codes can be interpreted as (n, k, k)-selectors of size given by the length of codewords. The other is that of explicit $(n, k, 3k/4)$-selectors of size $\mathcal{O}(k \text{ polylog } n)$ given by Indyk [22].

II. We show that the length of an (n, k, r)-selector has to be $\Omega(\min\left[n, \frac{k^2}{k-r+1} \cdot \frac{\log(n/k)}{\log(k/(k-r+1))} \right])$. This demonstrates that the above mentioned explicit construction is within a polylog n factor close to optimal.

III. The new selectors are applied to obtain the following specific applications: (i) an explicit oblivious solution to a variant of a static selection problem for the multiple-access channel, (ii) an explicit implementation of gossiping in radio networks, and (iii) an algorithm for group testing with inhibitors.

Previous Work. Selectors generalize many kinds of families of finite sets, and the work on special cases of selectors has been motivated by either purely combinatorial interests, as in the case of cover-free families, or by applications of combinatorics, as in group testing and in communication in the multiple-access channel and ad-hoc radio networks. We summarize briefly the known facts about upper and lower bounds on the size of selectors, and on explicitness of known selectors.
Existence of small selectors: Komlós and Greenberg [25] showed that there are $(n, k, 1)$ selectors of size $\mathcal{O}(k \log(n/k))$. Dyachkov and Rykov [18] showed that there exist (n, k, k)-selectors of size $\mathcal{O}(k^2 \log n)$; see [19,23] for a simple proof and also [16] for a detailed account of existential upper bounds for superimposed codes. De Bonis, Gąsieniec and Vaccaro [13] showed that there exist (n, k, r)-selectors of size $\mathcal{O}\left(\frac{k^2}{k-r+1} \log(n/k)\right)$.
Lower bounds on size of selectors: Clementi, Monti and Silvestri [11] showed that $(n, k, 1)$-selectors have to be of size $\Omega(k \log(n/k))$. Lower bounds on (n, k, r)-selectors with r close to k are stronger. In particular, (n, k, k)-selectors obey a lower bound $\Omega(\min[n, k^2 \log n / \log k])$ on their size. The first component n in this bound follows from the observation that a family of all n singletons is an (n, k, k)-selector for any $k \leq n$. This lower bound was first showed by Dyachkov and Rykov [17] in a slightly weaker form $\Omega(c_k \cdot n)$, where $c_k = \Theta(k^2 / \log k)$. It was rediscovered by Chaudhuri and Radhakrishnan [2] in a stronger form $\frac{k^2 \ln n}{100 \ln k}$ for $k \leq n^{1/3}$, which was later improved by Clementi, Monti and Silvestri [11] who showed that the constant 100 can be replaced by 16, for $k \leq \sqrt{2n}$. See also [16] for a detailed account of lower bounds for superimposed codes. De Bonis, Gąsieniec and Vaccaro [13] gave a general lower bound $\Omega\left(\min\left[n, \frac{(r-1)^2}{k-r+1} \cdot \frac{\log(n/(k-r+1))}{\log((r-1)/(k-r+1))}\right]\right)$ on the size of (n, k, r)-selectors.
Explicit constructions: Explicit (n, k, k)-selectors of size $\mathcal{O}(\min[n, k^2 \log^2 n])$ were given by Kautz and Singleton [24]. Indyk [22] was the first to observe a relation between selectors and dispersers. He gave explicit $(n, k, 3k/4)$-selectors of size $\mathcal{O}(\min[n, k \text{ polylog } n])$. Clementi et. al [10] explicitly constructed $(n, k, 1)$-selectors of size $\mathcal{O}(\min[n, k \log k \log(n/k)])$.

Explicit graphs with good expansion properties, on which we rely in our constructions, were given by Ta-Shma, Umans and Zuckerman [28].

Structure of This Document. Section 2 discusses interrelations between selection, in the sense of obtaining singleton sets as intersections, and superimposed coding. Section 3 describes the construction of explicit selectors with a matching lower bound. Section 4 discusses applications in the areas of multiple-access channel, radio networks and group testing. We conclude with a discussion in Section 5.

2   Selection and Superimposed Coding

Given a finite domain of size ℓ, or simply $[1..\ell]$, a subset A can be uniquely represented by its binary characteristic vector of length ℓ: an occurrence of 1 in position i means that number i belongs to A. This allows to represent families of subsets of a finite domain as binary two-dimensional arrays. This may be defined in two ways, depending on the role of rows and columns. A representation is called *primal* when rows represent elements of the domain and columns represent subsets. A representation is called *dual* when columns represent elements of the domain and rows represent subsets. In the literature on selection in families of sets and on superimposed codes, the primal representation is typically used.

2.1 Selection by Intersection

Selection of elements of a finite set can be defined in terms of binary matrices as follows: for a subset A of the domain, represented as a set of rows, row $z \in A$ is *selected* by a column if there is exactly one occurrence of 1 in this column among the rows in A and this occurrence is at row z.

Let $\mathcal{B}$ be an $n \times m$ binary array. It represents, in a primal way, m subsets of set $[1..n]$. Array $\mathcal{B}$ is an $(n, k, 1)$-selector of size m if for any set $A \subseteq [1..n]$ of rows of $\mathcal{B}$, where $|A| = k$, there is a column with exactly one occurrence of 1 among the rows in A. In general, array $\mathcal{B}$ is an (n, k, r)-selector of size m if for any set $A \subseteq [1..n]$ of rows of $\mathcal{B}$, where $|A| = k$, there are at least r columns with exactly one occurrence of 1 among the rows in A.

A dynamic adversarial component, in binary arrays representing families of subsets of a finite domain, is added by a possibility to have rows shifted. By this we mean that the distance of a shift is at most the original number of columns and the obtained array has new entries filled with zeroes. We say that an array $\mathcal{B}$ has good *synchronization properties* when, for any set A of rows of a sufficiently small size, some column selects a single row among the rows in A after these rows have been shifted by arbitrary distances. When we want to be able to select against such adversaries from all sets A of size $|A| = k$, then $\mathcal{B}$ could be called *k-synchronizing*, following [5,6,20].

This synchronization terminology is motivated by the application in the multiple-access channel with collision detection we describe next. It was first considered by Gąsieniec, Pelc and Peleg [20]; see [3] for a detailed exposition of this model of communication. The model has the following properties. A single transmission by an attached station is heard by all stations. More than one simultaneous transmissions interfere with one another, and none can be heard by the stations, but the stations receive a feedback notifying them of the interference. Suppose there are n stations, some k of which wake up spontaneously and immediately start attempts to broadcast a message to all. The first successful transmission wakes up the whole network and allows to synchronize local clocks. A schedule of transmissions, for a station, is specified as a binary sequence. An occurrence of 1 as the i-th bit represents a transmission in the i-th step according to the local clock.

We say that a binary $n \times m$ array $\mathcal{B}$ is a (n, k)-synchronizer of length m if for any nonempty set $A \subseteq [1..n]$ of rows of size at most k, and for any shifts of rows in A, there is a column that selects exactly one (shifted) row in A. Such synchronizers were defined by Chrobak, Gąsieniec and Kowalski [6] in the context of their work on the problems of wake-up, leader election and synchronization of local clocks in multi-hop radio networks. This notion was also implicitly used by Gąsieniec, Pelc and Peleg [20] in their work on waking up a multiple-access channel. The fastest known algorithm to wake up a multi-hop radio networks, given by Chlebus and Kowalski [5], uses *universal synchronizers*, which are arrays with properties stronger than those of radio synchronizers.

A construction of a (n, n)-synchronizers of length $\mathcal{O}(n^{1+\varepsilon})$, for any constant $\varepsilon > 0$, was given by Indyk [22]; it can be performed in a quasi-polynomial time $\mathcal{O}(2^{\text{polylog } n})$. Chlebus and Kowalski [5] described explicit (n, k)-synchronizers of a length $\mathcal{O}(k^2 \text{ polylog } n)$.

Radio synchronizers have the properties of selective families. It follows that (n, k)-synchronizers have to be of lengths $\Omega(k \log(n/k))$. Using the probabilistic method, Gąsieniec, Pelc and Peleg [20] showed that there are (n, n)-synchronizers of a length $\mathcal{O}(n \log^2 n)$, and Chrobak, Gąsieniec and Kowalski [6] showed that there are (n, k)-synchronizers of a length $\mathcal{O}(k^2 \log n)$.

2.2 Superimposed Coding

Superimposed codes are typically represented as binary arrays, with columns used as binary codewords. Take an $a \times b$ binary array with the property that no boolean sum of columns in any set D of $d = |D|$ columns can cover a column not in D. This is a superimposed code of b binary codewords of length a each that is *d-disjunct*. When columns are representing sets, then d-disjunctness means that no union of up to d sets in any family of sets D could cover a set outside D.

A book by Du and Hwang [16] provides a contemporary exposition of super-imposed coding and its relevance to nonadaptive group testing. There is a natural correspondence between such codes and strongly selective families, which we give for completeness sake. Using this correspondence, the explicit superimposed codes given by Kautz and Singleton [24] can be interpreted as (n, k, k)-selectors of size $\mathcal{O}(k^2 \log^2 n)$.

The correspondence is obtained by using the representations, primal and dual, of families of sets as boolean arrays. Take a (n, k)-strongly-selective family $\mathcal{S}$, that is, an (n, k, k)-selector, of some length m. This means there are m sets in $\mathcal{S}$, and the domain is of size n. A dual boolean representation of $\mathcal{S}$ is an $m \times n$ binary array $\mathcal{A}$. Let us interpret this array in the primal way. This representation yields a superimposed code: it consists of n codewords of length m each. Observe that this code is $(k - 1)$-disjunct. To show this, suppose, to the contrary, that some $(k - 1)$ columns of a set C of columns can cover column x of $\mathcal{A}$. Then the columns in $C \cup \{x\}$ represent a subset of $[1..n]$ of size k. By the property of $\mathcal{S}$ being a strongly selective family, there is a row in $\mathcal{A}$ with an occurrence of 1 in column x and only occurrences of 0 in columns in C. This means that column x is not covered by the columns in C, which is a contradiction. A reasoning in the opposite direction is similar.

Generalizations of superimposed codes can be proposed, which correspond to (n, k, r)-selectors being a generalization of (n, k)-strongly-selective families. This was already done by Dyachkov and Rykov [17]. De Bonis and Vaccaro [14,15] considered such generalizations in the context of their work on group testing.

Similar, but more restricted, generalized superimposed codes were considered by Chu, Colbourn and Syrotiuk [8,9] in their work on distributed communication in ad-hoc multi-hop radio networks.

3   Explicit Selectors

We show how to construct (n, k, r)-selectors of size $\mathcal{O}(\min[\,n, \frac{k^2}{k-r+1}\, \text{polylog}\, n\,])$, for *any* configuration of parameters $r \leq k \leq n$, in time polynomial in n. The construction is by combining strongly selective families with dispersers. Strongly selective families are (n, k, k)-selectors. We show how to use dispersers to decrease the third parameter r in (n, k, r)-selectors while also gracefully decreasing the size of the family of sets.

If $r \leq 3k/4$ then we can use the construction of an $(n, k, 3k/4)$-selector given by Indyk [22]. Assume that $r > 3k/4$. Let $0 < \varepsilon < 1/2$ be a constant.

A bipartite graph $H = (V, W, E)$, with set V of inputs and set W of outputs and set E of edges, is a (ℓ, d, ε)-*disperser* if it has the following two properties:

Dispersion: for each $A \subseteq V$ such that $|A| \geq \ell$, the set of neighbors of A
            is of size at least $(1 - \varepsilon)|W|$.
Regularity: H is d-left-regular.

Let graph $G = (V, W, E)$, where $|V| = n$, $|W| = \Theta((k - r + 1)d/\delta)$, be a $(k - r + 1, d, \varepsilon)$-disperser, for some numbers d and δ. (The amount $\log \delta$ is called the *entropy loss* of this disperser.) An explicit construction of such graphs, that is, in time polynomial in n, was given by Ta-Shma, Umans and Zuckerman [28], for any $n \geq k \geq r$, and some $\delta = \mathcal{O}(\log^3 n)$, where $d = \mathcal{O}(\text{polylog}\, n)$ is a bound on the left-degrees.

Let $\mathcal{M} = \{M_1, \ldots, M_m\}$ be an explicit $(n, c\delta\frac{k}{k-r+1})$-strongly-selective family, for a sufficiently large constant $c > 0$ that will be fixed later, of size $m = \mathcal{O}(\min[\,n, \delta^2(\frac{k}{k-r+1})^2 \log^2 n\,])$, as constructed by Kautz and Singleton [24].

We define an (n, k, r)-selector $\mathcal{S}(n, k, r)$ of size $\min[n, m|W|]$, which consists of sets $F(i)$, for $1 \leq i \leq \min[n, m|W|]$. There are two cases to consider, depending on the relation between n and $m|W|$. The case of $n \leq m|W|$ is simple: take the singleton containing only the i-th element of V as $F(i)$. Consider a more interesting case when $n > m|W|$. For $i = am + b \leq m|W|$, where a and b are non-negative integers satisfying $a + b > 0$, let $F(i)$ contain all the nodes $v \in V$ such that v is a neighbor of the a-th node in W and $v \in M_b$.

Theorem 1. *The family $\mathcal{S}(n, k, r)$ is an (n, k, r)-selector of size*

$$\mathcal{O}(\min\left[\,n, \frac{k^2}{k - r + 1}\, \text{polylog}\, n\,\right]) \,.$$

Proof. First we show that $\mathcal{S}(n, k, r)$ is an (n, k, r)-selector. The case $n \leq m|W|$ is clear, since each node in a set A of size k occurs as a singleton in some set $F(i)$. Consider the case $n > m|W|$. Let set $A \subseteq V$ be of size k. Suppose, to the contrary, that there is a set $C \subseteq A$ of size $k - r + 1$ so that none among the elements in C is selected by sets from $\mathcal{S}(n, k, r)$, that is, $F(i) \cap A \neq \{v\}$, for each $v \in C$ and $1 \leq i \leq m|W|$.

Claim: Every $w \in N_G(C)$ has more than $c\delta\frac{k}{k-r+1}$ neighbors in A.

The proof is by contradiction. Assume, for simplicity of notation, that $w \in W$ is the w-th element of set W. Suppose, to the contrary, that there is $w \in N_G(C)$ which has at most $c\delta\frac{k}{k-r+1}$ neighbors in A, that is, $|N_G(w) \cap A| \leq c\delta\frac{k}{k-r+1}$. By the fact that $\mathcal{M}$ is a $(n, c\delta\frac{k}{k-r+1})$-strongly-selective family we have that, for every $v \in N_G(w) \cap A$, the equalities

$$F(w \cdot m + b) \cap A = (M_b \cap N_G(w)) \cap A = M_b \cap (N_G(w) \cap A) = \{v\}$$

hold, for some $1 \leq b \leq m$. This holds in particular for every $v \in C \cap N_G(w) \cap A$. There is at least one such $v \in C \cap N_G(w) \cap A$ because set $C \cap N_G(w) \cap A$ is nonempty since $w \in N_G(C)$ and $C \subseteq A$. The existence of such v is in contradiction with the choice of C. Namely, C contains only elements which are not selected by sets from $\mathcal{S}(n, k, r)$ but $v \in C \cap N_G(w) \cap A$ is selected by some set $F(w \cdot m + b)$. This makes the proof of Claim complete.

Recall that $|C| = k - r + 1$. By dispersion, the set $N_G(C)$ is of size larger than $(1 - \varepsilon)|W|$, hence, by the Claim above, the total number of edges between the nodes in A and $N_G(C)$ in graph G is larger than

$$(1 - \varepsilon)|W| \cdot c\delta\frac{k}{k-r+1} = (1 - \varepsilon)\Theta((k-r+1)d/\delta) \cdot c\delta\frac{k}{k-r+1} > kd,$$

for a sufficiently large constant c. This is a contradiction, since the total number of edges incident to nodes in A is at most $|A|d = kd$. It follows that $\mathcal{S}(n, k, r)$ is an (n, k, r)-selector.

The size of this selector is

$$\min[n, m|W|] = \mathcal{O}(\min\left[n, \delta^2(\frac{k}{k-r+1})^2 \log^2 n \cdot (k-r+1)d/\delta\right])$$

$$= \mathcal{O}(\min\left[n, d\delta\frac{k^2}{k-r+1} \log^2 n\right])$$

$$= \mathcal{O}(\min\left[n, \frac{k^2}{k-r+1} \text{ polylog } n\right]),$$

since $d = \mathcal{O}(\text{ polylog } n)$ and $\delta = \mathcal{O}(\log^3 n)$.

Indyk [22] gave an explicit construction of $(n, k, 3k/4 + 1)$-selectors of size $\mathcal{O}(k \text{ polylog } n)$. His method does not appear to be directly adaptable to produce (n, k, r)-selectors in the case when $k - r$ is significantly smaller than k.

Theorem 2. *The length of an (n, k, r)-selector has to be*

$$\Omega(\min\left[n, \frac{k^2}{k-r+1} \cdot \frac{\log(n/k)}{\log(k/(k-r+1))}\right]).$$

Proof. We show that this fact follows from the lower bounds given in [11] and [13]. We may assume that $\frac{k^2}{k-r+1} \cdot \frac{\log(n/k)}{\log(k/(k-r+1))} = o(n)$, because otherwise it is sufficient to take a family of n singletons to obtain a selector of size n.

A bound on the size of $(n, k, 1)$-selectors given in [11] is $\Omega(k \log(n/k))$; we call it CMS.

A bound on the size of (n, k, r)-selectors given in [13] is $\Omega(\min\left[n, \frac{(r-1)^2}{k-r+1} \cdot \frac{\log(n/(k-r+1))}{\log((r-1)/(k-r+1))}\right])$; we call it DGV.

Suppose the parameters k and r are functions of n. If $k = \mathcal{O}(1)$, then the size of an (n, k, r)-selector is $\Omega(\log n)$, which is consistent with the three bounds mentioned. Suppose $k = \omega(1)$. We consider two cases.

Case $1 \le r \le k/2$:

Apply the CMS bound. Observe that $\frac{k^2}{k-r+1} = \Theta(k)$ since $k/(k-r+1)$ is $\Theta(1)$.

Case $k/2 < r \le k$:

Apply the DGV bound. Observe that $(r-1) = \Theta(k)$ and $\log(n/(k-r+1)) = \Omega(\log(n/k))$.

This completes the proof.

4 Applications

Theorem 2 demonstrates that the construction of Theorem 1 is close to optimal within a polylog n factor. It follows that any algorithmic application of selectors can be made explicitly instantiated with only an additional poly-logarithmic overhead factor in performance. We describe three such applications.

4.1 Multiple Access Channel

There are n stations attached to a multiple-access channel. A transmission performed by exactly one station is heard by every station, while more simultaneous transmissions interfere with one another, which prevents hearing any of the transmitted messages. The channel is said to be *with collision detection* if each station receives a feedback notifying about an interference of many messages sent simultaneously. We consider the weaker channel *without* collision detection.

The problem of *k-selection* is defined as follows. Suppose each among some k of the stations stores its own input value, and the goal is to make at least one such a value heard on the channel. This problem can be solved deterministically in time $\mathcal{O}(\log n)$ applying the binary-search paradigm. It requires expected time $\Omega(\log n)$, as was shown by Kushilevitz and Mansour [27]. This selection problem can be generalized to (k, r)-selection as follows: we want to hear at least r values from among k held by the stations.

Corollary 1. *The (k, r)-selection problem for n stations, where $r \le k \le n$, can be solved deterministically by an explicitly instantiated oblivious algorithm in the multiple-access channel without collision detection in time*

$$\mathcal{O}(\min\left[n, \frac{k^2}{k-r+1}\, \mathrm{polylog}\, n\right]).$$

Proof. An (n, k, r)-selector $\mathcal{S}$ can be used to provide an oblivious deterministic solution to the selection problem as follows. The sets in $\mathcal{S}$ are ordered, and station i performs a transmission if i is in the i-th set of $\mathcal{S}$. The performance bound follows from Theorem 1.

4.2   Gossiping in Radio Networks

The fastest known distributed algorithm for gossiping in directed ad-hoc multi-hop radio networks, given by Gąsieniec, Radzik and Xin [21], employs general (n, k, r)-selectors. The bound $\mathcal{O}(n^{4/3} \log^4 n)$ on time obtained in [21] relies on existence of (n, k, r)-selectors of size $\mathcal{O}\big(\frac{k^2}{k-r+1} \log(n/k)\big)$ shown in [13].

Corollary 2. *Gossiping in directed ad-hoc radio networks of n nodes can be performed in time $\mathcal{O}(n^{4/3} \text{ polylog } n)$ by an explicitly instantiated distributed algorithm.*

Proof. Use our explicit selectors in the algorithm of [21], instead of those known to exist only, to make the algorithm explicit. The performance bound follows from the estimates in [21] and Theorem 1. The additional overhead is of order polylog n.

4.3   Group Testing with Inhibitors

There is a set of n objects, some k of which are categorized as *positive*. The task of group testing it to determine all positive elements by asking queries of the following form: does the given subset of objects contain at least one positive element? The efficiency is measured by the number of queries.

The c-stage group testing consists of partitioning all objects c times into disjoint pools and testing the pools separately in parallel in each among c stages. Groups testing with inhibitors allows a category of some r objects, called *inhibitors*, so that a presence of such an element in a query hides the presence of a positive item. De Bonis, Gąsieniec and Vaccaro [13] showed how to implement 4-stage group testing with inhibitors relying on (n, k, r)-selectors.

Corollary 3. *There is an explicit implementation of a 4-stage group testing on a set of n objects with k positive items and r inhibitors, which consist of $\mathcal{O}(\min\big[n, \frac{k^2}{k-r+1} \text{ polylog } n\big])$ queries, if only $k < n - 2r$.*

Proof. Instantiate the scheme of tests developed in [13] with our explicit selectors. The bound on the number of tests follows from the estimates given in [13] and from Theorem 1. The additional overhead for explicitness is of order polylog n.

5   Conclusion

We showed how to construct (n, k, r)-selectors in time that is polynomial in n, for any configuration $r \le k \le n$ of parameters. The obtained selectors are close to optimal, in terms of size, within a polylog n factor. Our construction is by way of combining explicit dispersers with explicit superimposed codes to obtain a family of a prescribed size with the desired degree of selectiveness.

This construction has a number of applications, as exemplified in Section 4. Such applications are fairly direct in the case of selection in multiple-access channel. A general scheme of application works by using any algorithm relying on selectors and making it explicit by plugging in the explicit selectors given in Theorem 1. We presented this for gossiping in radio networks and group testing with inhibitors. Since our construction is within a polylog-n factor from optimal, the additional overhead factor in efficiency is always of order polylog n.

Synchronizers are closely related to selectors. They are more robust, in that they exhibit selection-related properties even if rows of arrays representing them are shifted arbitrarily. The best know explicit (n, k)-synchronizers of a length $\mathcal{O}(k^2 \text{ polylog } n)$ were given in [5]. It is an open problem if explicit synchronizers of length $\mathcal{O}(k \text{ polylog } n)$ can be developed.

Known explicit constructions of dispersers, of a quality we need in construction of almost optimal selectors, are fairly complex. Simpler explicit dispersers applicable to obtain close to optimal selectors would be interesting to construct.

Exploring further a connection between selectors and graphs with expansion properties is an interesting topic of research.

References

1. R. Bar-Yehuda, O. Goldreich, and A. Itai, On the time complexity of broadcast in radio networks: an exponential gap between determinism and randomization, *Journal of Computer and System Sciences*, 45 (1992) 104 - 126.
2. S. Chaudhuri, and J. Radhakrishnan, Deterministic restrictions in circuit complexity, in *Proc., 28th ACM Symposium on Theory of Computing (STOC)*, 1996, pp. 30 - 36.
3. B.S. Chlebus, Randomized communication in radio networks, in *"Handbook of Randomized Computing,"* P.M. Pardalos, S. Rajasekaran, J. Reif, and J.D.P. Rolim (Eds.), Kluwer Academic Publishers, 2001, Vol. I, pp. 401 - 456.
4. B.S. Chlebus, L. Gąsieniec, A. Gibbons, A. Pelc, and W. Rytter, Deterministic broadcasting in unknown radio networks, *Distributed Computing*, 15 (2002) 27 - 38.
5. B.S. Chlebus, and D.R. Kowalski, A better wake-up in radio networks, in *Proc., 23rd ACM Symposium on Principles of Distributed Computing (PODC)*, 2004, pp. 266 - 274.
6. M. Chrobak, L. Gąsieniec, and D.R. Kowalski, The wake-up problem in multi-hop radio networks, in *Proc., 15th ACM-SIAM Symposium on Discrete Algorithms (SODA)*, 2004, pp. 985 - 993.
7. M. Chrobak, L. Gąsieniec, and W. Rytter, Fast broadcasting and gossiping in radio networks, *Journal of Algorithms*, 43 (2002) 177 - 189.
8. W. Chu, C.J. Colbourn, and V.R. Syrotiuk, Slot synchronized topology-transparent scheduling for sensor networks, *Computer Communications*, to appear.
9. W. Chu, C.J. Colbourn, and V.R. Syrotiuk, Topology transparent scheduling, synchronization, and maximum delay, in *Proc., 18th International Parallel and Distributed Processing Symposium (IPDPS)*, 2004, pp. 223 - 228.

10. A.E.F. Clementi, P. Crescenzi, A. Monti, P. Penna, and R. Silvestri, On computing ad-hoc selective families, in *Proc., 5th International Workshop on Randomization and Approximation Techniques in Computer Science (RANDOM-APPROX)*, 2001, LNCS 2129, pp. 211 - 222.

11. A.E.F. Clementi, A. Monti, and R. Silvestri, Distributed broadcast in radio networks of unknown topology, *Theoretical Computer Science*, 302 (2003) 337 - 364.

12. A. Czumaj, and W. Rytter, Broadcasting algorithms in radio networks with unknown topology, in *Proc., 44th IEEE Symposium on Foundations of Computer Science (FOCS)*, 2003, pp. 492 - 501.

13. A. De Bonis, L. Gąsieniec, and U. Vaccaro, Generalized framework for selectors with applications in optimal group testing, in *Proc., 30th International Colloquium on Automata, Languages and Programming (ICALP)*, 2003, LNCS 2719, pp. 81 - 96.

14. A. De Bonis, and U. Vaccaro, Constructions of generalized superimposed codes with applications to group testing and conflict resolution in multiple access channels, *Theoretical Computer Science*, 306 (2003) 223 - 243.

15. A. De Bonis, and U. Vaccaro, Improved algorithms for group testing with inhibitors, *Information Processing Letters*, 67 (1998) 57 - 64.

16. D.Z. Du, and F.K. Hwang, *"Combinatorial Group Testing and its Applications,"* World Scientific, 2000.

17. A.G. Dyachkov, and V.V. Rykov, A survey of superimposed code theory, *Problems of Control and Information Theory*, 12 (1983) 229 - 244.

18. A.G. Dyachkov, and V.V. Rykov, Bounds on the length of disjunctive codes, *Problemy Peredachi Informatsii*, 18 (1982) 7 - 13.

19. Z. Füredi, On r-cover free families, *Journal of Combinatorial Theory (A)*, 73 (1996) 172 - 173.

20. L. Gąsieniec, A. Pelc, and D. Peleg, The wakeup problem in synchronous broadcast systems, *SIAM Journal on Discrete Mathematics*, 14 (2001) 207 - 222.

21. L. Gąsieniec, T. Radzik, and Q. Xin, Faster deterministic gossiping in directed ad-hoc radio networks, in *Proc., 9th Scandinavian Workshop on Algorithm Theory (SWAT)*, 2004, LNCS 3111, pp. 397 - 407.

22. P. Indyk, Explicit constructions of selectors and related combinatorial structures, with applications, in *Proc., 13th ACM-SIAM Symposium on Discrete Algorithms (SODA)*, 2002, pp. 697 - 704.

23. S. Jukna, *"Extremal Combinatorics,"* Springer-Verlag, 2001.

24. W.H. Kautz, and R.R.C. Singleton, Nonrandom binary superimposed codes, *IEEE Transactions on Information Theory*, 10 (1964) 363 - 377.

25. J. Komlós, and A.G. Greenberg, An asymptotically nonadaptive algorithm for conflict resolution in multiple-access channels, *IEEE Transactions on Information Theory*, 31 (1985) 303 - 306.

26. D.R. Kowalski, and A. Pelc, Time of deterministic broadcasting in radio networks with local knowledge, *SIAM Journal on Computing*, 33 (2004) 870 - 891.

27. E. Kushilevitz and Y. Mansour, An $\Omega(D \log(N/D))$ lower bound for broadcast in radio networks, *SIAM Journal on Computing*, 27 (1998) 702 - 712.

28. A. Ta-Shma, C. Umans, and D. Zuckerman, Loss-less condensers, unbalanced expanders, and extractors, in *Proc., 33rd ACM Symposium on Theory of Computing (STOC)*, 2001, pp. 143 - 152.

The Delayed k-Server Problem

Wolfgang W. Bein[1], Kazuo Iwama[2], Lawrence L. Larmore[1], and John Noga[3]

[1] School of Computer Science, University of Nevada,
Las Vegas, Nevada 89154, USA*
bein@cs.unlv.edu larmore@cs.unlv.edu
[2] School of Informatics, Kyoto University,
Kyoto 606-8501, Japan
iwama@kuis.kyoto-u.ac.jp
[3] Department of Computer Science, California State University Northridge,
Northridge, CA 91330, USA
jnoga@csun.edu

Abstract. We introduce a new version of the server problem: the *delayed server problem*. In this problem, once a server moves to serve a request, it must wait for one round to move again, but could serve a repeated request to the same point. We show that the delayed k-server problem is equivalent to the $(k-1)$-server problem in the uniform case, but not in general.

Keywords: Design and analysis of algorithms; approximation and randomized algorithms.

1 Introduction

The *k-server problem* is defined as follows: We are given $k \geq 2$ mobile servers that reside in a metric space M. A sequence of requests is issued, where each request is specified by a point $r \in M$. To *service* this request, one of the servers must be moved to r, at a cost equal to the distance moved. The goal is to minimize the total cost over all requests. $\mathcal{A}$ is said to be *online* if it must decide which server, or servers, to move without the knowledge of future requests.

We say that an online algorithm $\mathcal{A}$ for any online problem is *C-competitive* if the cost incurred by $\mathcal{A}$ for any input sequence is at most C times the optimal (offline) cost for that same input sequence, plus possibly an additive constant independent of the input. The *competitive ratio of* $\mathcal{A}$ is the smallest C for which $\mathcal{A}$ is C-competitive. The *competitiveness* of any online problem is then defined to be the smallest competitive ratio of any online algorithm for that problem.

The competitive ratio is frequently used to study the performance of online algorithms for the k-server problem, as well as other optimization problems. We refer the reader to the book of Borodin and El-Yaniv [2] for a comprehensive discussion of competitive analysis.

---

* Research of these authors supported by NSF grant CCR-0312093.

M. Liśkiewicz and R. Reischuk (Eds.): FCT 2005, LNCS 3623, pp. 281–292, 2005.
© Springer-Verlag Berlin Heidelberg 2005

The k-server problem is originally given by Manasse, McGeoch and Sleator [11], who prove that no online algorithm for the k server problem in a metric space M has competitive ratio smaller than k, if M has at least $k+1$ points. They also present an algorithm for the 2-server problem which is 2-competitive, and thus optimal, for any metric space. They furthermore state the *k-server conjecture*, namely that for each k, there exists an online algorithm for the k server problem which is k-competitive in any metric space. For $k > 2$, this conjecture has been settled only in a number of special cases, including trees and spaces with at most $k+2$ points [4,5,10]. Bein *et al.* [1] have shown that the work function algorithm for the 3-server problem is 3-competitive in the Manhattan plane, while Koutsoupias and Papadimitriou have shown that the work function algorithm for the k-server problem is $(2k-1)$-competitive in arbitrary metric spaces [8,9].

We study here a modified problem, the *delayed k-server problem*. Informally, in the delayed server problem, once a server serves a request, it must wait for one round. This problem is motivated by applications where there are latencies to be considered or service periods to be scheduled.

More precisely, we consider the following two versions of the problem. Let $r^1, r^2, \ldots$, be a given request sequence.

(a): If a server serves the request at time t, it must stay at that request point until time $t+2$. However, if $r^{t+1} = r^t$, the server may serve the request at time $t+1$.

(b): If a server is used to serve the request at time t, it cannot be used to serve the request at time $t+1$.

In practice, the difference between these two versions is that, in Version (b), it may be necessary to have two servers at the same point, while in Version (a) that is never necessary.

We refer to the server which served the last request as *frozen*. We will assume an initial configuration of servers, one of which will be initially designated to be frozen.

Lemma 1.1. *Let $C_{a,M,k}$ and $C_{b,M,k}$ to be the competitiveness of the delayed k-server problem in a metric space M, for Versions (a) and (b), respectively. Then $C_{a,M,k} \leq C_{b,M,k}$.*

Proof. Let $\mathcal{A}$ be a C-competitive online algorithm for Version (b). We can then construct a C-competitive online algorithm $\mathcal{A}'$ for Version (a), as follows. For any request sequence ϱ, let ϱ' be the request sequence obtained from ϱ by deleting consecutive duplicate requests. Then $\mathcal{A}'$ services ϱ by emulating $\mathcal{A}$ on the requests of ϱ', except that $\mathcal{A}'$ services any consecutive duplicate request at zero cost and then forgets that it happened.

Let OPT be the optimal offline algorithm for Version (b), and let OPT$'$ be the optimal offline algorithm for Version (a). Note that the optimal cost to service a request sequence with no consecutive duplicates is the same for both versions. We know that there exists a constant K such that $cost_{\mathcal{A}} \leq C \cdot cost_{\text{OPT}} + K$ for every request sequence. Thus, for every request sequence ϱ,

$$cost_{\mathcal{A}'}(\varrho) - C \cdot cost_{\mathrm{OPT}'}(\varrho) =$$
$$cost_{\mathcal{A}'}(\varrho') - C \cdot cost_{\mathrm{OPT}'}(\varrho') =$$
$$cost_{\mathcal{A}}(\varrho') - C \cdot cost_{\mathrm{OPT}}(\varrho') \leq K$$

We remark that, trivially, the delayed k-server problem is only defined for $k \geq 2$ and that it is 1-competitive if $k = 2$ in all situations.

In the *cache problem*, we consider a two-level memory system, consisting of fast memory (the *cache*), which can hold k memory units (commonly called *pages*) and an area of slow memory capable of holding a much larger number of pages. For fixed k, we refer to the cache problem as the k-*cache problem* if the cache size is k.

In the most basic model, if a page in slow memory is needed in fast memory this is called a *page request*. Such a request causes a *hit* if the page is already in the cache at the time of the request. But in the case of a *fault*, *i.e.* when the page is not in the cache, the requested page must be brought into the cache – we assume unit cost for such a move – while a page in the cache must be evicted to make room for the new page. An online paging algorithm must make decisions about such evictions as the request sequence is presented to the algorithm. The k-cache problem is equivalent to the k-server problem in a uniform metric space and thus the definition of the delayed server problem implies the following two versions of the *delayed k-cache problem*:

(a): If a page in the cache is read at time t, it cannot be ejected at time $t + 1$.
(b): If a page in the cache is read at time t, it can neither be read nor ejected at time $t + 1$. (That implies that duplication of pages in the cache must be allowed.)

We remark that Version (a) of the delayed k-cache problem is equivalent to Version (a) of the delayed k-server problem in uniform spaces, while Version (b) of the delayed k-cache problem is equivalent to Version (b) of the delayed k-server problem in uniform spaces.

We refer to the cache location which was read in the previous step as *frozen*. We will assume an initial cache configuration, and one of the cache locations will be initially frozen.

Henceforth, we shall consider only Version (a) of the delayed k-server problem. We claim that it makes more sense for applications, such as the cache problem described above, but we have included a short section (Section 5, at the end of the paper) which discusses Version (b). In the Section 2 we show that the classic online algorithm LRU can be adapted to the delayed k-cache problem, and is $(k-1)$-competitive. More generally, we show that the (randomized or deterministic) delayed k-cache problem is equivalent to the (randomized or deterministic) $(k-1)$-cache problem. This implies that the (randomized or deterministic) delayed k-server problem in uniform spaces is equivalent to the (randomized or deterministic) $(k-1)$-server problem in uniform spaces. This result might prompt the conjecture that the delayed k-server problem is equivalent to the $(k-1)$-server problem in all metric spaces. This is however not the case, as we show in Section 3. In Section 4, we give a k-competitive algorithm for the delayed k-server problem in a tree. Finally, we discuss future work in Section 6.

2   The Delayed k-Cache Problem

2.1   LRU is $(k-1)$-Competitive

It is well-known that *least recently used* (LRU) a deterministic algorithm for the k-cache problem (and hence for the k-server problem in a uniform space) is k-competitive. At each fault, the least recently used page is ejected. (In the terminology of the server problem this means that at each step the least recently used server is moved to serve the request, if the request is at a point which does not already have a server.)

Theorem 2.1. *The algorithm LRU is $(k-1)$-competitive for the delayed k-cache problem.*

Proof. Let a request sequence $\varrho = r^1 \ldots r^n$ be given. Let OPT be an optimal offline algorithm. The most recently used page is the frozen page. We insist that LRU eject all other initial pages before it ejects the initially frozen page.

We partition ϱ into *phases* $\sigma^0, \sigma^1, \ldots$ We will show that OPT faults once during the phase σ^t for each $t > 0$, and that LRU faults at most $k - 1$ times during any phase. The result follows.

Define σ^0 to be the (possibly empty) sequence of requests that precedes the first fault of OPT. For $t > 0$, let σ^t consists of all requests starting with the tth fault of OPT, up to but not including the $(t + 1)^{\text{st}}$ fault of OPT.

Clearly, OPT faults at most once during each phase. We need to prove that LRU faults at most $k - 1$ times during a phase.

Initially, mark all pages in the cache. If a page p is requested, and if OPT does not fault on that request, mark p, unless it is already marked. If OPT faults on that that request, unmark all pages except the frozen page, and then mark p.

We observe that OPT never ejects a marked page, because the only time OPT faults is when a phase begins, and at that time all pages are unmarked except for the frozen page, which cannot be ejected. Thus, at any given step, all marked pages are in OPT's cache. LRU also never ejects a marked page. If there are unmarked pages in the cache, each of the marked pages has been used more recently than any of the unmarked pages, and if all of LRU's cache pages are marked and LRU faults, then OPT must have those same pages in the cache, so OPT must fault also, ending the phase, and unmarking the least recently used page before it is ejected. Thus, at any given step, all marked pages are in LRU's cache.

During the phase σ^t, for $t > 0$, each time LRU faults, the number of marks increases by 1. Since σ^t begins with one page marked, LRU faults at most $k - 1$ times during that phase.

2.2   Equivalence of the Delayed $(k-1)$-Cache Problem and the k-Cache Problem

We now generalize Theorem 2.1 by showing that the delayed k-cache problem and the $(k-1)$-cache problem are equivalent in a very strong sense, and thus have the same competitiveness. This equivalence is valid for both the deterministic and the randomized cases. This result implies that the delayed k-server problem for uniform spaces is equivalent to the $(k-1)$-server problem for uniform spaces. In particular, given any algorithm for the k-cache problem we construct an algorithm for the $(k-1)$-cache problem, and vice versa.

To formally construct the equivalence, it helps to assign standard names to all pages. Let $\{p_1, p_2, \ldots\}$ be the set of pages for the $(k-1)$-cache problem, and let $Q = \{q_0, q_1, q_2, \ldots\}$ be the set of pages for the delayed k-cache problem. To simplify our construction below, it helps if both caches are the same size. Thus, we introduce a fictitious page p_0 for the $(k-1)$-cache problem, which is never requested, and which is always in a fictitious cache location which can never be used. Write $P = \{p_0, p_1, p_2, \ldots\}$. We will designate q_0 to be the initially frozen page.

Without loss of generality, a request sequence for the delayed k-cache problem has no duplicate consecutive request, since such a request can be serviced at zero cost by doing nothing. Let $\mathcal{R}$ be the set of all request sequences for the $(k-1)$-cache problem, and $\mathcal{S}$ the set of all request sequences with no consecutive duplicate requests for the delayed k-cache problem. We will first construct a 1-1 onto mapping $f : \mathcal{S} \to \mathcal{R}$. Let $s^0 = p_0$. Given any $\varsigma = s^1 s^2 \ldots s^n \in \mathcal{S}$, we inductively construct a sequence of one-to-one and onto functions $f^t : Q \to P$, for $0 \le t < n$, as follows:

1. $f^0(q_i) = p_i$ for all $i \ge 0$.

2. If $t > 0$, then $f^t(q_i) = \begin{cases} p_0 & \text{if } q_i = s^t \\ f^{t-1}(s^t) & \text{if } q_i = s^{t-1} \\ f^{t-1}(q_i) & \text{otherwise} \end{cases}$

Now, define $f(\varsigma) = \varrho = r^1 r^2 \ldots r^n$, where $r^t = f^{t-1}(s^t)$. Note that ϱ can be defined online, i.e., r^t is determined by $s^1 \ldots s^u$.

If $\mathcal{A}$ is an algorithm for the $(k-1)$-cache problem, we construct an algorithm $\mathcal{B} = F(\mathcal{A})$ for the delayed k-cache problem, as follows. Given a request sequence $\varsigma \in \mathcal{S}$ for the delayed k-cache problem, let $\varrho = f(\varsigma)$ and $f^t : \mathcal{S} \to \mathcal{R}$ be defined as above. For each step t, let $r^t = f^{t-1}(s^t)$. If $\mathcal{A}$ ejects p_i at step t with request r^t, then $\mathcal{B}$ ejects $f^{t-1}(p_i)$ at step t with request s^t. The following two lemmas can be verified inductively:

Lemma 2.2. Let $\mathcal{A}$ be an algorithm for the $(k-1)$-cache problem, and let $\mathcal{B} = F(\mathcal{A})$. Then

1. for any $i \ge 0$ and any $t \ge 0$, q_i is in $\mathcal{B}$'s cache after t steps if and only if either $f^t(q_i) = p_0$ or $f^t(q_i)$ is in $\mathcal{A}$'s cache after t steps,
2. for any $t \ge 0$, $cost_{\mathcal{B}}^t(\varsigma) = cost_{\mathcal{A}}^t(\varrho)$.

Lemma 2.3. *If $\mathcal{B}$ is an algorithm for the delayed k-cache problem, then there is a unique algorithm $\mathcal{A}$ for the $(k-1)$-cache problem such that $F(\mathcal{A}) = \mathcal{B}$.*

The following theorem reduces the delayed cache problem to the cache problem.

Theorem 2.4. *Let $k \geq 2$. There is a C-competitive online algorithm, deterministic or randomized, for the $(k-1)$-cache problem, if and only if there is a C-competitive online algorithm, deterministic or randomized, respectively, for the delayed k-cache problem.*

Proof. We will only give the proof for the deterministic case, and in only one direction, as the converse has a similar proof, and the randomized case has the same proof where cost is replaced by expected value of cost.

Suppose that $\mathcal{A}$ is a C-competitive online algorithm for the $(k-1)$-cache problem. Let OPT be an optimal offline algorithm for the $(k-1)$-cache problem, and let OPT' be an optimal offline algorithm for the delayed k-cache problem. Let $\mathcal{B} = F(\mathcal{A})$. Note that $\mathcal{B}$ is online. By Lemma 2.3, there exists an offline algorithm $\mathcal{D}$ for the $(k-1)$-cache problem such that $F(\mathcal{D}) = $ OPT'.

We know there is some constant K such that $cost_{\mathcal{A}}(\varrho) \leq C \cdot cost_{\text{OPT}}(\varrho) + K$ for any $\varrho \in \mathcal{R}$. If $\varsigma \in \mathcal{S}$, let $\varrho = f(\varsigma)$. Then, by Lemma 2.2,

$$cost_{\mathcal{B}}(\varsigma) - C \cdot cost_{\text{OPT}'}(\varsigma) =$$
$$cost_{\mathcal{A}}(\varrho) - C \cdot cost_{\mathcal{D}}(\varrho) \leq$$
$$cost_{\mathcal{A}}(\varrho) - C \cdot cost_{\text{OPT}}(\varrho) \leq K$$

2.3   An Example

Figure 1 illustrates the equivalence of the delayed 3-cache problem and the 2-cache problem in an example. Suppose that $\varsigma = q_3 q_2 q_3 q_1 q_0$. Then $\varrho = f(\varsigma) =$

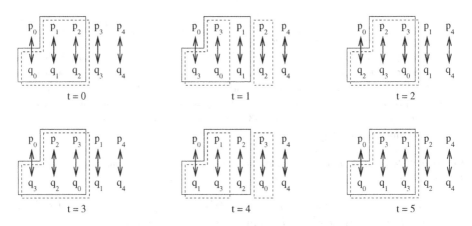

Fig. 1. Equivalence of 2-Cache and Delayed 3-Cache

$p_3p_2p_2p_1p_3$. The vertical arrows show the one-to-one correspondence f^t for each $0 \le t \le 5$.

Let $\mathcal{A}$ be LRU for the 2-cache problem, and OPT an optimal offline algorithm for the 2-cache problem. Then the contents of $\mathcal{A}$'s cache and $F(\mathcal{A})$'s cache, after each number of steps, are enclosed in solid lines. The contents of OPT's cache and $F(\text{OPT})$'s cache are enclosed in dotted lines. Note that $cost_{\mathcal{A}}(\varrho) = cost_{F(\mathcal{A})}(\varsigma) = 4$ and $cost_{\text{OPT}}(\varrho) = cost_{F(\text{OPT})}(\varsigma) = 2$.

3   Lower Bounds for the Delayed k-Server Problem

In this section we show lower bounds for the delayed k-server problem. Our first theorem shows that a lower bound of $(k-1)$ holds for arbitrary metric spaces of $k+1$ or more points.

Theorem 3.1. *In any metric space M of at least $k+1$ points, the competitiveness of the delayed k-server problem is at least $k-1$.*

Proof. Pick a set $X = \{x_1, \ldots, x_{k+1}\} \subseteq M$. Initially, all servers are in X, and, without loss of generality, there is a server at x_{k+1}. We then consider the following adversary:

- For each odd t, $r^t = x_{k+1}$.
- For each even t, r^t is the point in X where the algorithm does not have a server.

Note that the server which was initially at x_{k+1} will serve all odd requests, but will never move, yielding a cost of zero for all requests at odd-numbered steps, for both the online algorithm and the optimal algorithm. Thus, we can ignore all requests for odd t, and consider the sequence $r^2, r^4, \ldots, r^{2t}, \ldots$ in the metric space $M^* = M - x_{k+1}$, which is the sequence created by the cruel adversary for the $(k-1)$-server problem in M^*. The remainder of the proof is the same as the classic proof that the cruel adversary gives a lower bound of $k-1$ for the $(k-1)$-server problem in any space with at least k points. (See, for example, [11].) $\square$

From this, the results of Section 2, and from the well-known k-server conjecture [11], we might be tempted to conjecture that the competitiveness of the delayed k-server problem is $k-1$. That conjecture can be immediately shown to be false, however. In fact, we give a proof of a lower bound of 3 for the delayed 3-server problem.

Let M be the metric space with four points, $\{a, b, c, d\}$, where $ab = bc = cd = ad = 1$ and $ac = bd = 2$. We call this metric space the *square*.

Theorem 3.2. *The competitiveness of the delayed 3-server problem in the square is at least 3.*

Proof. Let $\mathcal{A}$ be an online algorithm for the delayed 3-server problem in the square. We show that the adversary can construct a request sequence consisting

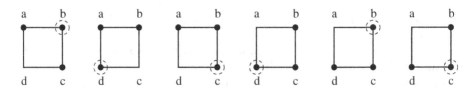

Fig. 2. One Phase Costs 3 for the Square, Optimal Cost is 1

of *phases*, where in each phase, $\mathcal{A}$ is forced to pay 3, while the optimal offline algorithm pays 1.

At the beginning and end of each phase, the servers are at three points, and the middle point is frozen. Without loss of generality, the servers are at $\{a, b, c\}$ and the server at b is frozen.

The adversary then constructs the phase as follows:

A1. The adversary requests d.
B1. Without loss of generality, $\mathcal{A}$ serves from c. $\mathcal{A}$'s servers are now at $\{a, b, d\}$.
A2. The adversary requests $c(dbc)^N$ for sufficiently large N.
B2. The algorithm responds. No possible response costs less than 2. The phase ends when the algorithm's servers are at $\{b, c, d\}$, if ever.

We now analyze the cost. The optimal cost to serve the phase is 1, since the optimal offline algorithm moves a server from a to d at A1. The optimal algorithm's servers will now be at $\{b, c, d\}$, and move A2 is free.

If $\mathcal{A}$ ever moves its servers to $\{b, c, d\}$, it pays 1 for B1, and at least 2 for B2; the configuration is symmetric to the initial configuration, and the next phase begins. If $\mathcal{A}$ never moves its servers to $\{b, c, d\}$, the phase never ends, but $\mathcal{A}$ pays unbounded cost.

Figure 2 shows one phase where $\mathcal{A}$ pays 3. The frozen server is enclosed by a dotted circle at each step.

We note that M cannot be embedded in a Euclidean space. However, if one models M with an ordinary square in the Euclidean plane, where the diagonal distance is $\sqrt{2}$ instead of 2, the above request sequence gives a lower bound of $1 + \sqrt{2}$ for the competitiveness of the delayed 3-server problem in the Euclidean plane.

4 k-Competitiveness for Trees

In this section we prove that the deterministic competitiveness of the delayed k-server problem is at most k for all continuous trees. The proof is similar to that for the regular k-server problem given in [5].

We define a *continuous tree* to be a metric space where there is exactly one continuous simple path between any two points. For example, the line is a continuous tree. We note that a continuous tree is called simply a *tree* in [5].

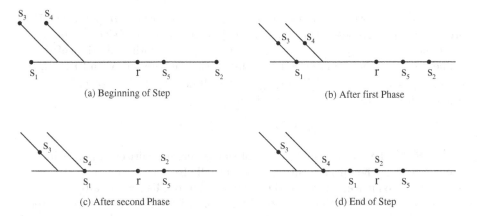

(a) Beginning of Step

(b) After first Phase

(c) After second Phase

(d) End of Step

Fig. 3. One Step of Tree Algorithm Showing Three Phases

Let T be a continuous tree. We define an online algorithm, $\mathcal{A}$, which we call the *tree algorithm*, for the delayed k-server problem in T. $\mathcal{A}$ is very similar to the algorithm given in [5].

Suppose that $s_1, \ldots, s_k$ are the positions of $\mathcal{A}$'s servers in T, and r is the request point. Without loss of generality, s_k is the frozen page.

In response to a request, $\mathcal{A}$ moves servers continuously toward r according to a protocol given below, stopping when one of them reaches r. We call this movement a *service step*. The movement consists of *phases*, where during each phase, the number of servers moving remains constant, and all of the moving servers move toward r. At the end of each phase, one or more servers stop. Once a server stops moving, it does not start up again during that service step.

We say that a server s_i is *blocked* by a server s_j if s_j is unfrozen and s_j lies on the simple path from s_i to r. (In the special case that s_i and s_j are at the same point, we say that s_j blocks s_i if $i > j$, but not if $i < j$.) During each phase, all unblocked servers move toward r at the same speed. If one of the moving servers becomes blocked, the phase ends.

Figure 3 shows an example of a step of the tree algorithm. In the figure, $k = 5$, and initially, in (a), s_5 is frozen, and none of the other servers are blocked. During the first phase, s_1, s_2, s_3, and s_4 move toward r. When s_1 blocks s_3, as shown in (b), the second phase begins, during which s_1, s_2, and s_4 move toward r. When s_1 and s_4 reach the same point, s_1 blocks s_4, as shown in (c), starting the third phase, during which s_1 and s_2 move towards r. Finally, s_2 reaches r and services the request, ending the step, as shown in (d).

Theorem 4.1. *The tree algorithm is k-competitive for the delayed k-server problem in a tree T.*

Proof. Let OPT be an optimal offline algorithm. We need to prove that, for some constant K,

$$cost_\mathcal{A}(\varrho) \le k \cdot cost_{\text{OPT}}(\varrho) + K \tag{1}$$

Just as in [5], we use the *Coppersmith-Doyle-Raghavan-Snir* potential [7] to prove k-competitiveness. If $X = \{x_1, \ldots x_k\}$ is any multiset of size k of points in T, define $\Sigma X = \sum_{1 \leq i < j \leq k} x_i x_j$. If $Y = \{y_1, \ldots y_k\}$ is another multiset of size k, define $||X, Y||$ to be the minimum matching distance between X and Y, i.e., the smallest possible value of $\sum_{i=1}^{k} x_i y_{\pi(i)}$, over all permutations π of $\{1, \ldots, k\}$. If S and O are multisets of size k, we define

$$\Phi(S, O) = \Sigma S + k||S, O||$$

Let S^0 be the initial configuration of the servers, a multiset of points in T of size k. Let r^0 be the initial position of the initially frozen server. If $\varrho = r^1 r^2 \ldots r^n$ is a request sequence, let S^t be the configuration of $\mathcal{A}$'s servers after servicing $r^0 \ldots r^t$, and let O^t be the multiset of positions of OPT's servers after t steps, and let $\Phi^t = \Phi(S^t, O^t)$, the potential after t steps. Let $cost_{\mathcal{A}}^t$ be the cost of $\mathcal{A}$ during Step t, and let $cost_{\text{OPT}}^t$ be the cost of OPT during Step t. Then $cost_{\mathcal{A}}(\varrho) = \sum_{t=1}^{n} cost_{\mathcal{A}}^t$, and $cost_{\text{OPT}}(\varrho) = \sum_{t=1}^{n} cost_{\text{OPT}}^t$.

Let $S^{t,i}$ be the multiset of positions of $\mathcal{A}$'s servers after i phases of Step t; for example, $S^{t,0} = S^{t-1}$ and $S^{t,m_t} = S^t$, where m_t is the number of phases of Step t. Let $p_{t,i}$ be the number of moving servers during the ith phase of Step t. and let $\ell_{t,i}$ be the distance that each of those servers moves. We verify the following sequence of equalities and inequalities.

$$||S^{t,i-1}, S^{t,i}|| = p_{t,i}\ell_{t,i} \quad \text{for all } t \text{ and all } 1 \leq i \leq m_t \tag{2}$$

$$\Sigma S^{t,i} - \Sigma S^{t,i-1} \leq (2k - (1+k)p_{t,i})\ell_{t,i} \tag{3}$$
$$\text{for all } t \text{ and all } 1 \leq i \leq m_t$$

$$||S^{t,i}, O^t|| - ||S^{t,i-1}, O^t|| \leq (p_{t,i} - 2)\ell_{t,i} \quad \text{for all } t \text{ and all } 1 \leq i \leq m_t \tag{4}$$

$$||S^{t,i-1}, S^{t,i}|| + \Phi(S^{t,i}, O^t) \leq \Phi(S^{t,i-1}, O^t) \quad \text{for all } t \text{ and all } 1 \leq i \leq m_t \tag{5}$$

$$\Phi(S^{t-1}, O^t) \leq k \cdot ||O^{t-1}, O^t|| + \Phi(S^{t-1}, O^{t-1}) \quad \text{for all } t \tag{6}$$

$$||S^{t-1}, S^t|| + \Phi(S^t, O^t) \leq \Phi(S^{t-1}, O^t) \quad \text{for all } t \tag{7}$$

$$cost_{\mathcal{A}}^t + \Phi^t =$$
$$||S^{t-1}, S^t|| + \Phi(S^t, O^t) \leq k \cdot ||O^{t-1}, O^t|| + \Phi(S^{t-1}, O^{t-1}) \tag{8}$$
$$= k \cdot cost_{\text{OPT}}^t + \Phi^{t-1} \quad \text{for all } t$$

(2) follows from the fact that $p_{t,i}$ servers move a distance of $\ell_{t,i}$ each.

During Phase i of Step t, each stationary server s_j gets farther away from at most one server, namely the moving server, if any, that blocks it; and s_j gets closer to each other moving server. Furthermore, any two moving servers get closer to each other. (3) follows from routine calculation.

During Phase i of Step t, in the minimum matching of O^t with S, as S varies from $S^{t,i-1}$ to $S^{t,i}$, the server of OPT which served r^{t-1} can be matched with s_k, and the server of OPT which served r^t can be matched with one moving server, say s_j. Since s_j gets closer to its partner during the phase, and since, in the worst case, the other $p_{t,i} - 1$ moving servers get farther away from their partners, (4) holds. Then, (5) follows from Inequalities (2), (3) and (4).

By the triangle inequality for minimum matching,

$$k \cdot ||O^{t-1}, O^t|| + \Phi(S^{t-1}, O^{t-1}) - \Phi(S^{t-1}, O^t) =$$
$$k \cdot ||O^{t-1}, O^t|| + k \cdot ||S^{t-1}, O^{t-1}|| - k \cdot ||S^{t-1}, O^t|| \geq 0$$

which verifies (6). Then (7) follows from (5) for each phase, while (8) follows from (6) and (7). Note that $\Phi^n \geq 0$; summing (8) over all t and letting $K = \Phi^0$, we obtain (1).

We give the following lemma without proof:

Lemma 4.2. *If $M_1 \subset M_2$, then $C_{a,M_1,k} \leq C_{a,M_2,k}$ and $C_{b,M_1,k} \leq C_{b,M_2,k}$.*

From this we have:

Theorem 4.3. *If a metric space M can be embedded into a continuous tree T, then the delayed k-server problem is k-competitive in M.*

5 Version (b)

We remark that Theorem 2.1 holds if we use Version (b) of the delayed k-cache problem. In this case, we must be sure to define LRU properly; LRU keeps track of when each cache location was used, and in case of a fault, ejects the page in the cache that was least recently read. The proof of $(k - 1)$-competitiveness is very similar to that for Version (a). In the proof, if a page is read, the cache location of that page is marked. If there are two copies of the page in the cache, it is important not to mark both of them unless forced to do so. More precisely, if a page p is requested, and there is one copy of p in the cache, that copy is read and its location marked, unless it is frozen, in which case another copy is moved into the cache and its location marked. But if p is requested and there are already two copies in the cache and their locations are not both marked, only the location of the copy that was most recently moved into the cache is marked.

We remark that, using essentially the same proof as that of Theorem 4.3, we can show that the tree algorithm is k-competitive for Version (b).

6 Open Problems

We conjecture that there is a lower bound of k for the delayed k-server problem in general spaces. In fact, it could well be that a lower bound of k could be provable for tree metric spaces. We also conjecture that $C_{a,M,k} \leq C_{b,M,k} \leq C_{M,k}$ for any metric space M, where $C_{M,k}$ is the competitiveness of the k-server problem in M.

Further work is necessary to give a competitive algorithm for the delayed server problem in general spaces. We conjecture that a modification of the Work Function Algorithm (see, for example, [3,6,9]) could yield such an algorithm.

Acknowledgement

Wolfgang Bein thanks Kazuo Iwama for the generous support he received while he visited Kyoto University to work on this project during December 2004 and January 2005.

References

1. Wolfgang Bein, Marek Chrobak, and Lawrence L. Larmore. The 3-server problem in the plane. *Theoretical Computer Science*, 287(1):387–391, 2002.
2. Allan Borodin and Ran El-Yaniv. *Online Computation and Competitive Analysis*. Cambridge University Press, 1998.
3. William R. Burley. Traversing layered graphs using the work function algorithm. *Journal of Algorithms*, 20:479–511, 1996.
4. Marek Chrobak, Howard Karloff, Tom H. Payne, and Sundar Vishwanathan. New results on server problems. *SIAM Journal on Discrete Mathematics*, 4:172–181, 1991.
5. Marek Chrobak and Lawrence L. Larmore. An optimal online algorithm for k servers on trees. *SIAM Journal on Computing*, 20:144–148, 1991.
6. Marek Chrobak and Lawrence L. Larmore. Metrical task systems, the server problem, and the work function algorithm. In Amos Fiat and Gerhard J. Woeginger, editors, *Online Algorithms: The State of the Art*, pages 74–94. Springer, 1998.
7. Don Coppersmith, Peter G. Doyle, Prabhakar Raghavan, and Marc Snir. Random walks on weighted graphs and applications to online algorithms. In *Proc. 22nd Symp. Theory of Computing (STOC)*, pages 369–378. ACM, 1990.
8. Elias Koutsoupias and Christos Papadimitriou. On the k-server conjecture. In *Proc. 26th Symp. Theory of Computing (STOC)*, pages 507–511. ACM, 1994.
9. Elias Koutsoupias and Christos Papadimitriou. On the k-server conjecture. *Journal of the ACM*, 42:971–983, 1995.
10. Elias Koutsoupias and Christos Papadimitriou. The 2-evader problem. *Information Processing Letters*, 57:249–252, 1996.
11. Mark Manasse, Lyle A. McGeoch, and Daniel Sleator. Competitive algorithms for server problems. *Journal of Algorithms*, 11:208–230, 1990.

Leftist Grammars and the Chomsky Hierarchy

Tomasz Jurdziński and Krzysztof Loryś*

Institute of Computer Science, Wrocław University,
Przesmyckiego 20, 51151 Wrocław, Poland
{tju, lorys}@ii.uni.wroc.pl

Abstract. Leftist grammars can be characterized in terms of rules of
the form $a \rightarrow ba$ and $cd \rightarrow d$, without distinction between terminals and
nonterminals. They were introduced by Motwani et. al. [9], where the
accessibility problem for some general protection system was related to
the membership problem of these grammars. This protection system was
originally proposed in [3,10] in the context of Java virtual worlds. We
show that the set of languages defined by general leftist grammars is not
included in CFL, answering in negative a question from [9]. Moreover, we
relate some restricted but naturally defined variants of leftist grammars
to the language classes of the Chomsky hierarchy.

1 Introduction

Leftist grammars were introduced by Motwani et. al. [9] and related to the accessibility problem for some general protection system of computer systems. A
protection system is a set of policies that prescribes the ways in which *objects*
interact with each other. By objects we mean users, processes or other entities
and interactions can include access rights, information sharing privileges and so
on. The accessibility problem for the protection system is formulated in the form
"Can object p gain (illegal) access to object q by a series of legal moves (as prescribed by the policy)?". A formal treatment of accessibility was first presented
by Harrison, Ruzzo and Ullman [5] who showed that the accessibility problem
is undecidable for a general access-matrix model of object-resource interaction.
This result prompted a broad research on tradeoffs between expressibility and
verifiability in protection systems. The work on protection systems took place
mainly in the context of operating systems and currently, operating systems
have efficient protection mechanisms. However, these mechanisms often fail at
the scale necessary for today's Internet [1].

The protection system related to leftist grammars was originally proposed
in [3,10] in the context of Java virtual worlds. The model of this protection
system strictly generalizes grammatical protection systems [2,7] and the take
grant model [8], it is a special case of the general access-matrix model [5]. Its
advantage over the general access-matrix model is the fact that accessibility is
decidable for this model what was obtained by the reduction to the membership

---

* Partially supported by Komitet Badań Naukowych, grant 0 T00A 003 23D.

M. Liśkiewicz and R. Reischuk (Eds.): FCT 2005, LNCS 3623, pp. 293–304, 2005.

problem of leftist grammars [9]. For a formal definition of this protection systems and the accessibility problem we refer the reader to [9]. Intuitively, the resources of the computer system are represented by vertices of the graph with labels, and access rights are represented by edges. Some additional constraints define the rules which allow to add/remove objects and change access rights (i.e. modify the graph). The accessibility problem is, given two vertices p and q, to decide if there exists a set of allowed operations on the graph such that finally there exists an edge between p and q.

Leftist grammars can be characterized in terms of rules of the form $a \to ba$ and $cd \to d$ over the alphabet Σ (there is no distinction between terminals and nonterminals). A symbol $x \in \Sigma$ is called a final symbol and a word $w \in \Sigma^*$ belongs to the language defined by a grammar G iff there exists a derivation which starts at wx and ends at x. Intuitively, the rules of leftist grammars correspond to the rules of the evolution of the graphs describing the protection system. And the derivations correspond to changes on one path of such a graph.

As pointed out above, the membership problem for these grammars is decidable [9]. Moreover, the problem of emptiness of the intersection of the language defined by a leftist grammar and a regular language is decidable. This result implied decidability of the accessibility problem for the protection system from [9]. However, no efficient algorithm for the membership problem of leftist grammars is known. And, there are no nontrivial lower bounds for this problem. In particular, a question if leftist grammars can recognize languages which are not context-free was addressed in [9]. The lack of efficient algorithms motivates also exploration of some restricted variants of these grammars. This research direction is prompted also by the fact that slight generalizations of leftist grammars make the membership problem undecidable [9].

From language theoretic point of view, leftist grammars do not even satisfy restrictions of context-sensitivity, as they can have length-reducing rules and length-increasing rules simultaneously. On the other hand, the productions of these grammars are severely restricted, so one could expect that their complexity is restricted as well. Thus, the study of their expressiveness is motivated both by their connections to the complexity of the accessibility problem and by itself. As suggested in [9], a natural research topic is here the placement of these grammars into the Chomsky hierarchy.

We study relationships between language classes defined by various types of leftist grammars and classes of the Chomsky hierarchy. Our main technical contribution states that general leftist grammars recognize some languages which are not context free, what answers in negative the question from [9]. Moreover, we propose a natural classification of leftist grammars according to the restrictions on so called *delete graphs* and *insert graphs*. Though our classification does not correspond to the way in which protection systems are usually restricted, we think that this research direction may help to fix the complexity of the problem for general leftist grammars. We relate restricted classes of leftist grammars to the set of regular, deterministic context free and context free languages. Our results are summarized in the following table, where FIN, REG, CFL, and DCFL, denote the classes of finite, regular, context-free, and deterministic context-free languages, respectively.

Delete graph	Insert graph	Included in	Not included in
acyclic	arbitrary	REG	FIN
arbitrary	empty	DCFL	REG
arbitrary	acyclic	CFL	DCFL
arbitrary	arbitrary	recursive*	CFL

$\star$ – proved in [9]

In Section 2 we provide some basic definitions and notations. In Section 3, we exploit properties of so-called leftmost derivations. Next, in Section 4 we relate restricted variants of leftist grammars to the classes of the Chomsky hierarchy. Section 5 is devoted to the proof of the fact that the set of languages defined by general leftist grammars is not included in CFL. Due to limited space, we omit some proofs and technical details.

2   Definitions and Notations

Throughout the paper ε denotes the empty word, $\mathbb{N}$, $\mathbb{N}_+$ denote the set of non-negative and positive integers. For a word x, $|x|$ denotes its length, the ith symbol of x is denoted by $x[i]$ ($0 < i \leq |x|$), and $x[i, j]$ denotes the factor $x[i] \ldots x[j]$ for $0 < i \leq j \leq |x|$. Let $[i, j] = \{l \in \mathbb{N} \mid i \leq l \leq j\}$. Sometimes, we will identify regular expressions with regular languages defined by them. We refer the reader to [6,4] for basics from formal language theory.

A leftist grammar $G = (\Sigma, P, x)$ consists of a finite alphabet Σ, a final symbol $x \in \Sigma$, and a set of production rules P of the following two types,

$$ab \to b \ \text{(Delete Rule)} \qquad c \to dc \ \text{(Insert Rule)}$$

where $a, b, c, d \in \Sigma$.

We say that $u \Rightarrow v$ is a derivation step for $u, v \in \Sigma^*$, if $u = u_1 y u_2$ and $v = u_1 z u_2$ such that $y \to z$ is a production rule in P. As usual, $\Rightarrow^*$ denotes the reflexive and transitive closure of $\Rightarrow$. A sequence of derivation steps $u_1 \Rightarrow u_2 \Rightarrow \ldots \Rightarrow u_p$ is called a derivation. A word u_i for $i \in [1, p]$ is called a sentential form in this derivation. Finally, the language of G is defined to be

$$L(G) = \{w \in \Sigma^* \mid wx \Rightarrow^* x\}.$$

Throughout the paper, we will implicitly deal with symbols of sentential forms as objects which can insert/delete other symbols and can be inserted or deleted. So, we make distinction between different occurences of a symbol $a \in \Sigma$ in a sentential form. However, in order to simplify notations, we will often identify the occurence of the symbol a in a sentential form with its value a. It should be clear from the context whether we say about a symbol as an element of the alphabet or an element of a sentential form.

We say that the symbol b in the delete rule $ab \to b$ is *active*. Similarly, the symbol c is *active* in the insert rule $c \to dc$.

Let $u \Rightarrow v$, where $u = u_1 y u_2$ and $v = u_1 z u_2$ such that $y \to z$ is a production rule in P. We would like to say that a symbol which is active in the production rule $y \rightarrow z$ is also *active* in the derivation step $u \Rightarrow v$ (that is, the rightmost symbol of the prefix $u_1 y$). However, it is possible that, for fixed u, v, there are many factorizations $u = u_1 y u_2$ such that $v = u_1 z u_2$ and $y \rightarrow z$ is a production. It turns out that it is possible to avoid this ambiguity.

Lemma 1. *For each leftist grammar G, there exists a leftist grammar $G' = (\Sigma', P', x')$ such that $L(G) = L(G')$ and, for each possible derivation step $u \Rightarrow_{G'} v$, there exists only one factorization $u = u_1 y u_2$ such that $v = u_1 z u_2$ where $y \rightarrow z$ is a production of G'. In particular, there are no rules of type $a \rightarrow aa$ or $aa \rightarrow a$ in P' for $a \in \Sigma$.*

The idea of the proof of the above lemma is to show that after removing all rules of type $aa \rightarrow a$, $a \rightarrow aa$ for $a \in \Sigma$ from a grammar G, adding the extra "artificial" final symbol x' and some rules for it, we obtain a grammar G' which defines the language $L(G)$ and satisfies conditions of the lemma. All our results concern leftist grammars which satisfy these conditions.

Let us make the following observation.

Fact 1. *The set of languages generated by leftist grammars is disjoint with* $\mathsf{FIN}_{>0}$, *where* $\mathsf{FIN}_{>0}$ *denotes a set* $\mathsf{FIN} \setminus \{\emptyset, \{\varepsilon\}\}$.

Proof. Let G be a leftist grammar such that $L(G) \notin \{\emptyset, \{\varepsilon\}\}$. Let $w \neq \varepsilon$ be a word in $L(G)$. Then, there exists a derivation which starts at wx, ends at x and no insert rule is applied in which the leftmost symbol of w is active (indeed, there is nothing to delete to the left of the leftmost symbol, so it is not needed to insert any symbols to the left of it). Further, the symbol $w[1]$ is deleted at some derivation step. So, $w[1]^i w[2, |w|]x \Rightarrow^* x$ for each $i > 0$, what implies that the language $L(G)$ is infinite.   □

Now, we introduce notions of insert graphs and delete graphs and we obtain a taxonomy of leftist grammars based on these notions. Let $G = (\Sigma, P, x)$ be a leftist grammar, where $\Sigma = \{a_1, \ldots, a_p\}$. An *Insert Graph* of G has p vertices $v_1, \ldots, v_p$. There exists an edge (v_i, v_j) in this graph iff the grammar contains a rule $a_i \rightarrow a_j a_i$. Similarly, a *Delete Graph* of G has p vertices $v_1, \ldots, v_p$. There exists an edge (v_i, v_j) in this graph iff the grammar contains a rule $a_j a_i \rightarrow a_i$.

We consider the cases that the insert/delete graph is empty, acyclic, or arbitrary. These cases will be denoted by empty, acyclic and arb, respectively. Leftist grammars with delete graphs of type A and insert graphs of type B are denoted by $\mathsf{LG}(A, B)$. The construction of the proof of Lemma 1 ensures that, for each grammar G of type $\mathsf{LG}(A, B)$ where $A, B \in \{\mathsf{empty}, \mathsf{acyclic}, \mathsf{arb}\}$, there exists a leftist grammar $G' = (\Sigma', P', x')$ such that $L(G) = L(G')$, G' is of type $\mathsf{LG}(A, B)$ and G' satisfies the conditions from Lemma 1.

3   Leftmost Derivations and Their Properties

Let $u_1 \Rightarrow u_2 \Rightarrow \ldots \Rightarrow u_p$ be a derivation. A symbol $u_1[i]$ is *alive* in u_1 with respect to the derivation $u_1 \Rightarrow^* u_p$ if there exists $j \leq i$ such that $u_1[j]$ is active in any of the steps $u_1 \Rightarrow u_2 \Rightarrow \ldots \Rightarrow u_p$. A symbol which is not alive is *gone*.

A derivation $u_1 \Rightarrow u_2 \Rightarrow \ldots \Rightarrow u_p$ is the *leftmost derivation* if the leftmost alive symbol with respect to $u_i \Rightarrow^* u_p$ is active in $u_i \Rightarrow u_{i+1}$ for each $i \in [1, p-1]$.

Let $u \Rightarrow^* v$ be a leftmost derivation. Assume that $u[i]$ is gone with respect to $u \Rightarrow^* v$. Then, $u[i]$ is *firm* in u with respect to this derivation if it is not deleted until v. Otherwise, $u[i]$ is *useless* in u.

Proposition 1. *If there exists a derivation $u \Rightarrow^* v$ then there exists a leftmost derivation which starts at u and ends at v.*

Next, we investigate some useful properties of leftmost derivations.

Proposition 2. *Let G' be a leftist grammar. Then, each leftmost derivation $v \Rightarrow^* w$ satisfies the following condition: Each sentential form u in this derivation has a factorization $u = u_1 u_2 u_3$ such that all symbols in u_3 are alive, all symbols in u_2 are useless, and all symbols in u_1 are firm with respect to $u \Rightarrow^* w$.*

Proof. The fact that alive symbols form the suffix of a sentential form follows directly from the definition of alive symbols. For the sake of contradiction, assume that a useless symbol a is located directly to the left of a firm symbol b. However, as each firm symbol is not active nor deleted in a further derivation, it is not possible to delete a symbol located directly to the left of it. Contradiction, because a should be deleted (it is useless). $\qquad\square$

We introduce a notion which formally describes the way in which symbols in sentential forms were inserted. Let $U \equiv u_1 \Rightarrow u_2 \Rightarrow \ldots \Rightarrow u_p$ be a derivation. Let b, d be symbols which appear in some sentential forms of this derivation. We say that b is a *descendant* of d in U if (d, b) belongs to the reflexive and transitive closure of the relation $\{(x, y) \mid vxw \Rightarrow vyxw$ is a derivation step in U for some v and $w\}$.

Further, we define a *history* of each symbol which appears during the derivation $U \equiv u_1 \Rightarrow u_2 \Rightarrow \ldots \Rightarrow u_p$. A history of a symbol a which appears in u_1 is equal to a word $h(a) = a$. Further, let c be a symbol inserted in a derivation step $vbw \Rightarrow vcbw$ of U. Then, a history of (this copy of) c is equal to a word $h(c) = ch(b)$. So, the history of each symbol (except symbols which appear in the "initial" sentential form) is fixed at a moment when it is inserted.

Proposition 3. *Let*

$$u \Rightarrow^* y_1 a y_2 y_3 \Rightarrow y_1 b a y_2 y_3 \Rightarrow^* v$$

be a leftmost derivation, let a symbol b following y_1 be a descendant of the rightmost symbol of y_2. Then, a history of this symbol b is equal to bay_2.

Proposition 4. *Let $uav \Rightarrow^* w$ be a leftmost derivation. Then, the suffix v remains unchanged as long as the symbol a following u is alive with respect to this derivation. If the symbol a following the prefix u is useless with respect to the derivation $uav \Rightarrow^* w$, then the prefix ua remains unchanged, as long as the symbol a following u is not deleted.*

The statements of the above proposition follow immediately from the definitions of alive symbols, useless symbols and leftmost derivations.

4   Restricted Leftist Grammars

4.1   Grammars with Empty Insert Graphs

Theorem 1. *The set of languages defined by grammars* LG(arb, empty) *is included in* DCFL.

Proof. Assume that $uawx \Rightarrow^* x$ is the leftmost derivation for an input word uaw and a at the position $|u| + 1$ is the leftmost alive symbol with respect to this derivation. Thus, all symbols in the prefix u are gone, they are not active in any derivation step. Then, there exists a leftmost derivation $u'awx \Rightarrow^* x$ for each u' which is a subsequence of u. Indeed, as no symbol from u is active in the derivation, we obtain a leftmost derivation $u'aw \Rightarrow^* x$ by removing from the derivation $uawx \Rightarrow^* x$ all steps which delete symbols in u that do not appear in u'.

The above observation implies the following fact. Let $u \in L(G)$ for a grammar G of type LG(arb, empty), let $v = ux$. Let $i \in [2, n + 1]$ be minimal value such that G contains a production $v[i - 1]v[i] \to v[i]$, where $n = |u|$. Then, a word $u' = u[1, i - 2]u[i, n]$ belongs to $L(G)$ as well. Using this property, we obtain the following algorithm determining if $w \in \Sigma^*$ belongs to $L(G)$. In each step, the algorithm finds the leftmost position in the current sentential form, where an application of the (delete) production rule is possible. Then, an appropriate symbol is deleted from the sentential form. This process is repeated until we obtain x what means that $w \in L(G)$ or $y \neq x$ such that no application of any rule of G is possible in y, what implies that $w \notin L(G)$.

One can implement the above algorithm on deterministic pushdown automaton (DPDA), because the choice of the position of the active symbol in consecutive derivation steps is always deterministic, and, this position moves from left to right. So, DPDA can move on the pushdown the part of the sentential form which is to the left of the position of the symbol which is active in a "current" derivation step.   □

Theorem 2. *The set of languages defined by grammars* LG(arb, empty) *is not included in* REG.

Proof. We describe a grammar $G = (\Sigma, P, x)$ of type LG(arb, empty) which generates a non-regular language. Let $\Sigma = \{a_0, a_1, b_0, b_1, x\}$ and let P contain the following production rules: $\{a_i b_i \to b_i \mid i = 0, 1\} \cup \{b_{1-i} b_i \to b_i \mid i = 0, 1\} \cup \{b_1 x \to x\}$. Now, let $w = a_1 a_0$ and $u = b_0 b_1$. We show that $w^n u^m \in L(G) \iff m \geq n$. The implication $\Leftarrow$ is obvious. The second implication follows from two observations. First, a symbol a_i for $i = 0, 1$ can be deleted only by b_i. Second, each copy of b_i for $i = 0, 1$ in the input word from $w^* u^*$ is able to delete at most one copy of a_i.

So, the language $w^* u^* \cap L(G)$ is equal to a non-regular language $\{w^n u^m \mid m \geq n\}$. As the set of regular languages is closed under intersection, the language $L(G)$ is non-regular as well.   □

4.2 Grammars with Acyclic Insert Graphs

Next, we analyze grammars with acyclic insert graphs.

Proposition 5. *Let $G = (\Sigma, P, x)$ be a leftist grammar with acyclic insert graph. Let $u \Rightarrow^* v$ be a leftmost derivation in G'. Then, each symbol in each sentential form of this derivation has at most $|\Sigma|$ descendants which are alive.*

Proof. Let $way \Rightarrow wbay$ be a derivation step which inserts a symbol b. Then, by Proposition 3, $y = y_1y_2$ such that bay_1 is equal to the history of this b and b is the descendant of the rightmost symbol of y_1. As the insert graph is acyclic, this history is not longer than the depth of the insert graph, which is bounded by $|\Sigma|$. And, because we consider the leftmost derivation, all symbols to the left of this b are gone in $wbay$. Thus, all alive descendants of the rightmost symbol of y_1 are included in bay_1 and $|bay_1| \leq |\Sigma|$.   □

Proposition 6. *Let $G = (\Sigma, P, x)$ be a leftist grammar with acyclic insert graph. Then, for each leftmost derivation $w \Rightarrow^* w'$, there exists a leftmost derivation which starts at w, ends at w' and there is no sentential form in this derivation which contains two useless symbols with equal histories, descendants of the same symbol. In particular, each symbol has at most $|\Sigma|^{|\Sigma|}$ useless descendants in each sentential form of this derivation.*

Proof. Let $w \Rightarrow^* w'$ be a leftmost derivation. Assume that a sentential form $uavay$ appears in this derivation and copies of a located directly to the left of v and directly to the right of v are useless symbols that are descendants of the same symbol and have equal histories. We show how to shorten this derivation such that one of these a's does not appear in any sentential form. Moreover, we do not introduce any new derivation steps which would insert symbols. Let az be a history of both these copies of a. According to Propositions 3 and 4, the original derivation $w \Rightarrow^* w'$ contains a subderivation

$$w \Rightarrow^* u_1azz' \Rightarrow^* uazz' \Rightarrow^* uav_1azz' \Rightarrow^* uavazz' \Rightarrow^* uavay, \text{ where}$$

- u_1azz' is a sentential form obtained after a derivation step which inserts the left copy of a, preceding v (see Proposition 3);
- $uazz'$ is a first sentential form in which the left copy of a is gone (i.e. useless)
  - (see Propositions 4 and 3);
- uav_1azz' is a sentential form obtained in a derivation step which inserts the right copy of a, following v (see Propositions 3 and 4);
- $uavazz'$ is a first sentential form in which the right copy of a is gone and useless (see Proposition 4);

As the whole av following u is useless in the sentential form $uavazz'$ (what follows from Proposition 2 and the fact that the first and the last symbol in the factor ava are useless), the derivation $uavazz' \Rightarrow^* w'$ implies that there exists also a derivation $uazz' \Rightarrow^* w'$ obtained from the original derivation by deleting the subderivation $uazz' \Rightarrow^* uavazz'$, and all derivation steps which remove symbols from the factor av in the subderivation $uavazz' \Rightarrow^* w'$. Thus, we obtained a new (shorter) leftmost derivation $w \Rightarrow^* w'$ which avoids the sentential form $uavazz'$ with two useless symbols a that have equal histories and are descendants of the same symbol. In this way, each derivation $w \Rightarrow^* w'$ may be stepwise transformed into a derivation $w \Rightarrow^* w'$ which satisfies conditions of the proposition.

Finally, as the insert graph is acyclic, the number of different histories of symbols is not larger than $|\Sigma|^{|\Sigma|}$.   □

Theorem 3. *The set of languages recognized by grammars of type* LG(arb, acyclic) *is included in* CFL.

Proof. (Sketch) The idea is to construct a pushdown automaton (PDA) which simulates (exactly) all leftmost derivations which end at x and the number of descendants of each input symbol in each sentential form of these derivations is not larger than $s = |\Sigma| + |\Sigma|^{|\Sigma|}$. If it is possible, the result follows on base of Propositions 6 and 5. In order to simulate such computations (in step-by-step manner), the automaton stores on the pushdown all useless symbols, and stores in its finite control all descendants of the leftmost alive symbol which appeared in the input word (this symbol is equal to the symbol scanned currently by the input head).   □

Theorem 4. *The set of languages recognized by grammars* LG(arb, acyclic) *is not included in* DCFL.

Proof. We will define a grammar $G = (\Sigma, P, x)$ with an acyclic insert graph, such that the language $L(G)$ does not belong to DCFL. Let

$$\Sigma = \{a_i \mid i \in [0,3]\} \cup \{b_i, e_i, f_i \mid i \in [0,1]\} \cup \{c, d, x\}.$$

Production rules of G are following, where $i = 0, 1$, $j = 0, 1, 2, 3$:

$$
\begin{array}{ll}
b_i \rightarrow e_i b_i & b_1 c \rightarrow c \\
b_i \rightarrow f_i b_i & e_1 c \rightarrow c \\
b_{1-i} e_i \rightarrow e_i & b_1 d \rightarrow d \\
b_{1-i} f_i \rightarrow f_i & f_1 d \rightarrow d \\
e_{1-i} e_i \rightarrow e_i & cx \rightarrow x \\
f_{1-i} f_i \rightarrow f_i & dx \rightarrow x \\
a_j e_{j \bmod 2} \rightarrow e_{j \bmod 2} & \\
a_j f_{j \operatorname{div} 2} \rightarrow f_{j \operatorname{div} 2} &
\end{array}
$$

Let $w = a_3 a_2 a_1 a_0$, $u = b_0 b_1$. We show that

$$w^n u^m c \in L(G) \iff m \geq 2n.$$

The implication $\Leftarrow$ is simple. In fact, there exists a leftmost derivation in which each b_i inserts e_i. This e_i deletes b_{1-i}, e_{1-i} (if they occur directly to the left of e_i) and one element of $\{a_i, a_{i+2}\}$ (if exists). Finally, c deletes the rightmost b_1 and e_1, and x deletes c.

For the implication $\Rightarrow$, observe that a derivation which starts from a word in w^*u^*cx and ends at x cannot use symbols f_0, f_1, as these symbols can be deleted only by d and there is no insert rule which inserts d. On the other hand, descendants of each copy of b_i ($i = 0, 1$) in the input word can delete at most one element of $\{a_0, a_1, a_3, a_3\}$. It follows from the observation that a_i can insert e_i (but not e_{1-i}) which can delete a_i or a_{i+2} (but not $a_{i+1}, a_{(i+3)}$ mod 4). On the other hand, each two consecutive copies of a_i and a_{i+2} are separated by a_{i+1} or $a_{(i+3)}$ mod 4. Using similar arguments, one can show that $w^n u^m d \in L(G) \iff m \geq n$.

The above observations imply that $L(G) \cap (w^*u^*c \cup w^*u^*d)$ is equal to the language $L' = \{w^n u^m c \,|\, m \geq 2n\} \cup \{w^n u^m d \,|\, m \geq n\}$. As the language L' is not in DCFL, and DCFL is closed under intersection with regular languages, $L(G)$ is not in DCFL, either.   □

4.3   Grammars with Restricted Delete Graphs

In this section we sketch the proof of the fact that leftist grammars with acyclic delete graphs define only regular languages.

Let G be a grammar of type LG(acyclic, arb). Let G' be a "reversed" grammar with respect to G. That is, for each production $\alpha \to \beta$ in G, G' contains a production $\beta \to \alpha$. Certainly, $L(G)$ is equal to the set $\{w \,|\, x \Rightarrow^*_{G'} wx\}$. The delete graph of G is equal to the insert graph of G' and the insert graph of G is equal to the delete graph of G'. So, if the delete graph of G is acyclic then the insert graph of G' is acyclic as well.

Note that each symbol in each sentential form of the leftmost derivation $x \Rightarrow^* wx$ is a descendant of the rightmost ("initial") symbol x. So, by Propositions 5 and 6, if there exists a derivation $x \Rightarrow^* wx$ then there exists a leftmost derivation $x \Rightarrow^* wx$, such that the number of alive symbols in each sentential form is not larger than $|\Sigma|$ and the number of useless symbols in each sentential form of this derivation is not larger than $|\Sigma|^{|\Sigma|}$. On the other, the leftmost derivation $x \Rightarrow^* wx$ makes the symbols from the final word w firm from left to right. That is, first $w[1]$ becomes a status firm, then $w[2], w[3]$ and so on (see Proposition 2). Thus, one can design a nondeterministic one-way finite automaton A which simulates derivations with at most $|\Sigma| + |\Sigma|^{|\Sigma|}$ non-firm descendants of the rightmost x. First, this automaton simulates a subderivation until the leftmost symbol $w[1]$ is inserted, storing all alive and useless (i.e. non firm) symbols in its finite control. Then, A moves right its input head and continues the simulation until $w[2]$ is inserted. This process is continued until all symbols of w are inserted and there are no useless nor alive symbols – then A accepts. If A is not able to simulate such derivation, it rejects. Finally, we obtain the following theorem.

Theorem 5. *The set of languages recognized by grammars of type* LG(acyclic, arb) *is included in* REG.

5   General Leftist Grammars

In this section we describe a leftist grammar that defines a language which is not context-free. Let $G = (\Sigma, P, x)$ be a grammar with the alphabet

$$\Sigma = \{a_i, B_i, F_i \mid i = 0, 1\} \cup \{X_{i,j}, D_{i,j} \mid i, j = 0, 1\} \cup \{x\},$$

and the following set of productions, where $i, j, k \in \{0, 1\}$:

$$
\begin{array}{llll}
(10) & a_i \rightarrow B_i a_i & (60) & a_{1-i} B_i \rightarrow B_i \\
(20) & a_i \rightarrow X_{i,0} a_i & (70) & B_i D_{i,j} \rightarrow D_{i,j} \\
(30) & X_{i,j} \rightarrow Y_{i,j} X_{i,j} & (80) & D_{i,j} D_{i,1-j} \rightarrow D_{i,1-j} \\
(40) & Y_{i,j} \rightarrow D_{i,j} Y_{i,j} & (83) & X_{1-i,k} D_{i,0} \rightarrow D_{i,0} \\
(50) & Y_{i,j} \rightarrow X_{i,1-j} Y_{i,j} & (86) & Y_{1-i,k} D_{i,1} \rightarrow D_{i,1}
\end{array}
$$

$$
\begin{array}{ll}
(90) & a_0 F_0 \rightarrow F_0 \quad (140) \quad F_1 x \rightarrow x \\
(100) & X_{0,j} F_0 \rightarrow F_0 \quad (150) \quad D_{i,j} x \rightarrow x \\
(110) & Y_{0,j} F_1 \rightarrow F_1 \\
(120) & F_{1-i} F_i \rightarrow F_i
\end{array}
$$

Let

$$
\begin{array}{ll}
\mathcal{A} = \{a_0, a_1\} & \mathcal{X}_i = \{X_{i,0}, X_{i,1}\} \\
\mathcal{F} = \{F_0, F_1\} & \mathcal{Y}_i = \{Y_{i,0}, Y_{i,1}\} \\
\mathcal{D}_i = \{D_{i,0}, D_{i,1}\} & \mathcal{Z}_i = \{X_{i,0}, Y_{i,0}, X_{i,1}, Y_{i,1}\}
\end{array}
$$

where $i \in [0, 1]$. For sets of symbols $\mathcal{U}, \mathcal{V}$, by $\mathcal{U}\mathcal{V}$ we mean the set $\{uv \mid u \in \mathcal{U}$ and $v \in \mathcal{V}\}$.

Proposition 7. *Let $n, m \in \mathbb{N}$. If $n \geq 2^{2m-2}$, then a word $w = (a_1 a_0)^m (F_0 F_1)^n$ belongs to $L(G)$.*

Now, for an input word $w = (a_1 a_0)^m (F_0 F_1)^n$, we formulate conditions which are necessary in order to exist a derivation $wx \Rightarrow^* x$.

Proposition 8. *Let $(a_1 a_0)^m (F_0 F_1)^n x \Rightarrow^* x$ be a derivation in G. Then, $n \geq 2^{2m-2}$.*

Proof. First, we specify some necessary conditions satisfied by each derivation $(a_1 a_0)^m (F_0 F_1)^n x \Rightarrow^* x$. All conditions specified in claims stated below concern such derivations. Observe that no insert rule of G inserts a symbol x, so each sentential form of each derivation which starts from $(a_1 a_0)^m (F_0 F_1)^n x$ contains only one x, at its rightmost position.

Claim 1. For each $i \in [0, 1]$ and each copy of a_i in the word $(a_1 a_0)^m (F_0 F_1)^n x$ except the leftmost a_1, a_i inserts a symbol B_i which deletes a_{1-i} located directly to the left of it. Moreover, descendants of this a_i insert $D_{i,j}$ for $j \in [0, 1]$ which deletes B_i inserted by it. And, the rightmost element of $\mathcal{D}_i$ which is the descendant of this a_i is deleted by x.

Claim 2. For each $i \in [0, 1]$ and each copy of a_i in the word $(a_1 a_0)^m (F_0 F_1)^n x$ except the leftmost a_1, all symbols which belong to $\mathcal{Z}_{1-i}$ and are descendants of

a_{1-i} located directly to the left of this a_i, should be deleted by the descendants of this a_i.

Claim 3. Let a_i for $i \in [0, 1]$ be a symbol which appears in the input word. Then, a sequence of its descendants in each sentential form of the derivation which ends at x belongs to the set $(\{B_i\} \cup \mathcal{D}_i)^* \mathcal{Z}_i^* \{B_i\}^*$.

By Proposition 1, there exists a leftmost derivation $wx \Rightarrow^* x$ for each $w \in L(G)$. So, let us consider only leftmost derivations which start from $w = (a_1 a_0)^m (F_0 F_1)^n x$. Remind that a_i (for $i \in [0, 1]$) and its descendants are not able to insert symbols which could delete elements of $\mathcal{Z}_i$. By Claim 1, the factor $Y_{0,0} X_{0,0}$ should appear in the sequence of descendants of the leftmost a_0. Indeed, this a_0 has to insert an element of $\mathcal{D}_0$ (see Claim 1) and the only way to insert such an element is by inserting $X_{0,0}$ which inserts $Y_{0,0}$. Now, we show by induction that the sequence of descendants of the pth symbol from A (for $p > 2$), say a_i, contains (in some sentential form) a subsequence $(Y_{i,1} X_{i,1} Y_{i,0} X_{i,0})^{2^{p-3}}$ (not necessarily a subword!) which is not deleted as long as this a_i is not deleted. The latter statement follows from the fact that a_i and its descendants are not able to insert symbols which delete the elements of $\mathcal{Z}_i$. Let $p = 3$, that is, we consider the third symbol from A, the second a_1. As the first (leftmost) a_0 inserted the subsequence $Y_{0,0} X_{0,0}$, the descendants of the second a_1 should insert the symbols which delete $Y_{0,0} X_{0,0}$ (see Claim 2). That is, the subsequence $D_{1,0} D_{1,1}$ should appear among the descendants of the second a_1 in some sentential form (as $D_{1,0}$ is the only symbol which deletes elements of $\mathcal{X}_0$, $D_{1,1}$ is the only symbol which deletes elements of $\mathcal{Y}_0$ and $D_{1,1}$ is the only possible descendant of a_1, which deletes $D_{1,0}$). However, in order to insert this sequence to the left of all elements of $\mathcal{Z}_1$ (which are descendants of this a_i – see Claim 3), a subsequence

$$Y_{1,1} X_{1,1} Y_{1,0} X_{1,0} = (Y_{1,1} X_{1,1} Y_{1,0} X_{1,0})^{2^{3-3}}$$

should be inserted.

Now, assume that the statement is true for $p - 1 < 2m$. Thus, the pth element of A, say a_i, and its descendants should insert symbols which are able to delete the sequence $(Y_{1-i,1} X_{1-i,1} Y_{1-i,0} X_{1-i,0})^{2^{p-4}}$ (see Claim 2). Note that elements of $\mathcal{X}_{1-i}$ can be deleted only by $D_{i,0}$ or F_0. However, F_0 cannot be a descendant of a_i. Similarly, elements of $\mathcal{Y}_{1-i}$ can be deleted only by $D_{i,1}$ or F_1, but F_1 cannot be a descendant of a_i. As the descendants of the pth symbol from A (which is equal to a_i) have to delete

$$(Y_{1-i,1} X_{1-i,1} Y_{1-i,0} X_{1-i,0})^{2^{p-4}} \in (\mathcal{Y}_{1-i} \mathcal{X}_{1-i})^{2^{p-3}},$$

by Claim 2, the derivation should contain a subsequence of sentential forms $v_1, \ldots, v_{2^{p-2}}$ such that the leftmost descendant of the considered a_i in v_j is equal to $D_{i,(j-1) \bmod 2}$. Indeed, only the leftmost descendant of a symbol a is able to delete a symbol which is not a descendant of this a. Moreover, by Claim 3, only the leftmost descendants of a_i which belongs to $\mathcal{Z}_i$ is allowed to insert an element of $\mathcal{D}_i$. And, no descendant of a_i which belongs to $\mathcal{Z}_i$ is deleted until this a_i is deleted (because it is not possible that a descendant of a_i is able to delete an element of $\mathcal{Z}_i$). So, in order to obtain first $D_{i,0}$ (as the leftmost descendant of a_i in v_1), a subsequence $Y_{i,0}X_{i,0}$ should be inserted. Further, assume that $D_{i,(j-1) \bmod 2}$ is the leftmost descendant of a_i in v_j. In order to obtain $D_{i,j \bmod 2}$ as the leftmost descendant of a_i, it is needed that at least $Y_{i,(j-1) \bmod 2}$ which inserted $D_{i,(j-1) \bmod 2}$, inserts $X_{i,j \bmod 2}$ which in turn inserts $Y_{i,j \bmod 2}$ and it finally inserts $D_{i,j \bmod 2}$. However, it means that the pth element of A (and its descendants) inserts a subsequence $(Y_{i,1}X_{i,1}Y_{i,0}X_{i,0})^{2^{p-3}}$ which is not deleted by its descendants.

So, finally, the rightmost element of A, that is a_0 which is the $(2m)$th element, inserts a subsequence $(Y_{0,1}X_{0,1}Y_{0,0}X_{0,0})^{2^{2m-3}}$ that will not be deleted by its descendants. Similarly as in Theorem 2, at least one copy of $\mathcal{F}$ from the sequence $(F_0F_1)^n$ is needed to delete one symbol from the sequence $(Y_{0,1}X_{0,1}Y_{0,0}X_{0,0})^{2^{2m-3}}$ (as F_0 deletes only $X_{0,1}, X_{0,0}$ and F_1 deletes only $Y_{0,1}, Y_{0,0}$. Thus, it is required that $n \geq 2^{2m-2}$. $\square$

Theorem 6. *The language $L(G)$ is not context-free.*

Proof. As CFL is closed under intersection with regular languages, if $L(G)$ is context-free, then the language $L' = L \cap (a_1a_0)^+(F_0F_1)^+$ is a context-free language as well. However, by Propositions 7 and 8, L' is equal to the non context-free language $\{(a_1a_0)^m(F_0F_1)^n \mid n \geq 2^{2m-2}\}$. $\square$

References

1. M. Blaze, J. Feigenbaum, J. Ioannidis, A. Keromytis. The role of trust management in distributed security. In Proc. of Secure Internet Programming LNCS 1603, 185–210.
2. T. Budd. Safety in grammatical protection systems. *International Journal of Computer and Information Sciences*, 12(6):413–430, 1983.
3. O. Cheiner, V. Saraswat. Security Analysis of Matrix. Technical report, AT&T Shannon Laboratory, 1999.
4. M. Harrison. Introduction to Formal Language Theory. Addison-Wesley, 1978.
5. M. Harrison, W. Ruzzo, J. Ullman. Protection in operating systems. *Communications of the ACM*, 19(8):461–470, August 1976.
6. J. Hopcroft, R. Motwani, J. Ullman. Introduction to Automata Theory, Languages, and Computation. *Addison-Wesley*, 2000.
7. R. Lipton, T. Budd. On Classes of Protection Systems. In *Foundations of Secure Computation*, Academic Press, 1978, pp. 281–296.
8. R. Lipton, L. Snyder. A linear time algorithm for deciding subject security. *Journal of the ACM*, 24(3):455–464, July 1977.
9. R. Motwani, R. Panigrahy, V. Saraswat, S. Venkatasubramanian. On the decidability of accessibility problems (extended abstract). *STOC 2000*, 306–315.
10. V. Saraswat. The Matrix Design. Technical report, AT&T Laboratory, April 1997.

Shrinking Multi-pushdown Automata

Markus Holzer[1] and Friedrich Otto[2]

[1] Institut für Informatik, Technische Universität München,
Boltzmannstrasse 3, D-85748 Garching bei München
holzer@in.tum.de
[2] Fachbereich Mathematik/Informatik, Universität Kassel,
D-34109 Kassel
otto@theory.informatik.uni-kassel.de

Abstract. The shrinking two-pushdown automaton is known to char-
actize the class of growing context-sensitive languages, while its deterministic variant accepts the Church-Rosser languages. Here we study
the expressive power of shrinking pushdown automata with more than
two pushdown stores, obtaining a close correspondence to linear time-
bounded multi-tape Turing machines.

1 Introduction

The *pushdown automaton* is a by now classical machine model used to analyze
languages. It is known that the (deterministic) pushdown automaton corresponds
exactly to the (deterministic) context-free languages [9]. Although of great importance for practical applications as well as for theoretical considerations, the
context-free languages are not sufficiently powerful to express all those properties
of languages that one is interested in in applications. Therefore various methods have been proposed to extend the class of languages considered beyond the
context-free languages, but staying properly within the class of context-sensitive
languages.

Here we consider the *multi-pushdown automaton*, which is an automaton that
has a constant number $k \geq 2$ of pushdown stores. On each of them it operates
in exactly the same way as a classical pushdown automaton operates on its
pushdown store. However, instead of a separate one-way input tape, the input
to a multi-pushdown automaton M is provided as the initial content of the first
pushdown store. Without any restrictions already the two-pushdown automaton
is a universal computing device. Therefore, we only consider multi-pushdown
automata that are *shrinking*, that is, for which there exists a weight function ω
that assigns a positive integer weight to each symbol of the pushdown alphabet
and to each internal state such that each transition step decreases the overall
weight of the actual configuration. It follows that these automata are linear
time-bounded.

In fact, it is easily seen that a realtime multi-pushdown automaton, that is,
a multi-pushdown automaton that consumes an input symbol in each step of
its computations, is shrinking. On the other hand, a linear time-bounded multi-
pushdown automaton can in general execute a linear number of steps within a

M. Liśkiewicz and R. Reischuk (Eds.): FCT 2005, LNCS 3623, pp. 305–316, 2005.

computation without changing the weight of the configurations involved. Hence, the shrinking model can be seen as an intermediate stage between the multi-pushdown automaton that runs in realtime and the one that runs in linear time.

For the special case of two pushdown stores, this model has been introduced by Buntrock and Otto in [6], where it is shown that the shrinking variant of the two-pushdown automaton (2-PDA) characterizes the class GCSL of *growing context-sensitive languages* considered by Dahlhaus and Warmuth [7]. Further, based on the results of [6] it is shown in [11] that the deterministic variant of the shrinking 2-PDA characterizes the class CRL of *Church-Rosser languages* of McNaughton et. al. [10].

Here we investigate the expressive power of the shrinking multi-pushdown automaton with more than two pushdown stores. Clearly, a multi-pushdown automaton with three pushdown stores, that is, a 3-PDA, can simulate a *flip-pushdown automaton* M_{FP} [8] by using its first pushdown store as a read-only input tape, and by using its other two pushdown stores for simulating the pushdown store of M_{FP}, realizing a flip of the pushdown of M_{FP} by shifting the content of its second pushdown to its third pushdown or vice versa. Also a 3-PDA can simulate an *input-reversal pushdown automaton* M_{IP} [4] by using its first two pushdown stores for simulating the input tape of M_{IP}, realizing an input reversal of M_{IP} by shifting the content of its first pushdown to its second pushdown or vice versa, and by using its third pushdown to simulate the pushdown store of M_{IP}. Thus, the shrinking multi-pushdown automaton can also be seen as a common generalization of these automata.

We will see that the shrinking 3-PDA already characterizes the class of *quasi-realtime* languages Q, which coincides with the complexity class NTIME(lin) [3]. In the deterministic case, on the other hand, we will see that by increasing the number of pushdown stores, we obtain a strict infinite hierarchy of language classes that approximates the complexity class DTIME(lin). Thus, we obtain a close correspondence between shrinking multi-pushdown automata on the one hand and linear time-bounded multi-tape Turing machines on the other hand, both in the nondeterministic and in the deterministic case.

The paper is structured as follows. In Section 2 we give the formal definition of the shrinking multi-pushdown automaton, and by presenting a detailed example we show that the shrinking 3-PDA is already strictly more powerful than the shrinking 2-PDA. In Section 3 we present the characterization of Q by the shrinking 3-PDA, and in Section 4 we derive the announced results on the shrinking deterministic multi-pushdown automaton. We conclude with Section 5, where we address in short the generalization to shrinking alternating multi-pushdown automata and some other variants of our model.

2   Definition

Throughout the paper ε will denote the empty word, and for a word w, we will use w^R to denote the mirror image of w. Finally, for any type of automaton A, $\mathcal{L}(\mathsf{A})$ will denote the class of languages accepted by automata from that class.

Definition 1. *A k-pushdown automaton (k-PDA) is a nondeterministic automaton with k pushdown stores. Formally, it is defined by a 7-tuple $M = (Q, \Sigma, \Gamma, \bot, q_0, F, \delta)$, where*

- *Q is a finite set of internal states,*
- *Σ is a finite input alphabet,*
- *Γ is a finite tape alphabet containing Σ such that $\Gamma \cap Q = \emptyset$,*
- *$\bot \notin \Gamma$ is a special symbol used to mark the bottom of the pushdown stores,*
- *$q_0 \in Q$ is the initial state,*
- *$F \subseteq Q$ is the set of final states, and*
- *$\delta : Q \times (\Gamma \cup \{\bot\})^k \to \mathcal{P}_{fin}(Q \times (\Gamma^* \cup \Gamma^* \cdot \bot)^k)$ is the transition relation, where $\mathcal{P}_{fin}(Q \times (\Gamma^* \cup \Gamma^* \cdot \bot)^k)$ denotes the set of finite subsets of $Q \times (\Gamma^* \cup \Gamma^* \cdot \bot)^k$.*

The automaton M is a deterministic k-pushdown automaton (k-dPDA), if δ is a (partial) function from $Q \times (\Gamma \cup \{\bot\})^k$ into $Q \times (\Gamma^* \cup \Gamma^* \cdot \bot)^k$.

A multi-pushdown automaton, MPDA for short, is a k-PDA for some $k \geq 2$, and a deterministic multi-pushdown automaton, dMPDA for short, is a k-dPDA for some $k \geq 2$.

A configuration of a k-PDA is described by a $(k+1)$-tuple $(q, u_1, u_2, \ldots, u_k)$, where $q \in Q$ is the actual state, and $u_i \in (\Gamma^* \cdot \{\bot\}) \cup \{\varepsilon\}$ is the current content of the i-th pushdown store ($1 \leq i \leq k$). Here we assume that the first letter of u_i is at the top and the last letter of u_i is at the bottom of the pushdown store. For an input string $w \in \Sigma^*$, the corresponding *initial configuration* is $(q_0, w\bot, \bot, \ldots, \bot)$, that is, the input is given as the initial content of the first pushdown store, while all other pushdown stores just contain the bottom marker. The k-PDA M induces a computation relation $\vdash_M^*$ on the set of configurations, which is the reflexive transitive closure of the single-step computation relation $\vdash_M$. The k-PDA M accepts with empty pushdown stores, that is,

$$L(M) := \{\, w \in \Sigma^* \mid (q_0, w\bot, \bot, \ldots, \bot) \vdash_M^* (q, \varepsilon, \varepsilon, \ldots, \varepsilon) \text{ for some } q \in F \,\}$$

is the *language accepted by M*.

For our investigation the notion of *weight function* will be crucial. A *weight function* ω on an alphabet Γ is a mapping $\omega : \Gamma \to \mathbb{N}_+$. It is extended to a morphism $\omega : \Gamma^* \to \mathbb{N}$ by defining $\omega(\varepsilon) := 0$ and $\omega(wa) := \omega(w) + \omega(a)$ for all $w \in \Gamma^*$ and $a \in \Gamma$.

Definition 2. *A k-PDA M is called* shrinking *if there exists a weight function $\omega : Q \cup \Gamma \cup \{\bot\} \to \mathbb{N}_+$ such that, for all $q \in Q$ and $u_1, \ldots, u_k \in \Gamma \cup \{\bot\}$, if $(p, v_1, \ldots, v_k) \in \delta(q, u_1, \ldots, u_k)$, then $\omega(p) + \sum_{i=1}^k \omega(v_i) < \omega(q) + \sum_{i=1}^k \omega(u_i)$ holds. By k-sPDA we denote the corresponding class of shrinking automata.*

Analogously, the shrinking variant of the k-dPDA is defined, which is denoted by k-sdPDA. Without fixing the parameter k we obtain the corresponding classes sMPDA *and* sdMPDA.

The 2-sPDA coincides with the sTPDA of Buntrock and Otto [6], both in the deterministic and the nondeterministic case. It is known that the class of languages accepted by the sTPDA coincides with the class GCSL of growing context-sensitive languages, and that the sdTPDA characterizes the class CRL of Church-Rosser languages [11].

Example 1. We present a 3-sdPDA $M_{Gl} = (Q, \Sigma, \Gamma, \bot, q_0, F, \delta)$ for the *Gladkij language* $L_{Gl} := \{ w\#w^R\#w \mid w \in \{a, b\}^* \}$. This language is quasi-realtime, but it is not growing context-sensitive [2,5], that is, it is not accepted by any 2-sPDA.

We define M_{Gl} by taking $Q := \{q_0, q_1, q_2, q_3\}$, $\Sigma := \{a, b, \#\}$, $\Gamma := \Sigma \cup \{a', b'\}$, $F := \{q_3\}$, and δ is defined as follows:

$$
\begin{aligned}
(q_0, c, \bot, \bot) &\rightarrow (q_0, \varepsilon, c'\bot, \bot) &&\text{for all } c \in \{a, b\}; \\
(q_0, c, d', \bot) &\rightarrow (q_0, \varepsilon, c'd', \bot) &&\text{for all } c, d \in \{a, b\}; \\
(q_0, \#, d', \bot) &\rightarrow (q_1, \varepsilon, d', \bot) &&\text{for all } d' \in \{a', b', \bot\}; \\
(q_1, c, c', \bot) &\rightarrow (q_1, \varepsilon, \varepsilon, c'\bot) &&\text{for all } c \in \{a, b\}; \\
(q_1, c, c', d') &\rightarrow (q_1, \varepsilon, \varepsilon, c'd') &&\text{for all } c, d \in \{a, b\}; \\
(q_1, \#, \bot, d') &\rightarrow (q_2, \varepsilon, \bot, d') &&\text{for all } d' \in \{a', b', \bot\}; \\
(q_2, c, \bot, c') &\rightarrow (q_2, \varepsilon, \bot, \varepsilon) &&\text{for all } c \in \{a, b\}; \\
(q_2, \bot, \bot, \bot) &\rightarrow (q_3, \varepsilon, \varepsilon, \varepsilon).
\end{aligned}
$$

Let $w = c_1 c_2 \ldots c_n \in \{a, b\}^+$. On input $w\#w^R\#w$, M_{Gl} executes the following computation:

$$
\begin{aligned}
(q_0, w\#w^R\#w\bot, \bot, \bot) &\vdash_M (q_0, c_2 \ldots c_n\#w^R\#w\bot, c_1'\bot, \bot) \\
&\vdash_M (q_0, c_3 \ldots c_n\#w^R\#w\bot, c_2'c_1'\bot, \bot) \\
&\vdash_M^* (q_0, \#w^R\#w\bot, c_n' \ldots c_2'c_1'\bot, \bot) \\
&\vdash_M (q_1, w^R\#w\bot, c_n' \ldots c_2'c_1'\bot, \bot) \\
&\vdash_M (q_1, c_{n-1} \ldots c_2 c_1\#w\bot, c_{n-1}' \ldots c_2'c_1'\bot, c_n'\bot) \\
&\vdash_M^* (q_1, \#w\bot, \bot, c_1'c_2' \ldots c_n'\bot) \\
&\vdash_M (q_2, w\bot, \bot, c_1'c_2' \ldots c_n'\bot) \\
&\vdash_M (q_2, c_2 \ldots c_n\bot, \bot, c_2' \ldots c_n'\bot) \\
&\vdash_M^* (q_2, \bot, \bot, \bot) \\
&\vdash_M (q_3, \varepsilon, \varepsilon, \varepsilon).
\end{aligned}
$$

If $w = \varepsilon$, then on input $w\#w^R\#w$, M_{Gl} executes the computation

$$(q_0, \#\#\bot, \bot, \bot) \vdash_M (q_1, \#\bot, \bot, \bot) \vdash_M (q_2, \bot, \bot, \bot) \vdash_M (q_3, \varepsilon, \varepsilon, \varepsilon),$$

and if the given input is not of the form $w\#w^R\#w$, then M_{Gl} is easily seen to not accept. Thus, we have $L(M_{Gl}) = L_{Gl}$.

It remains to prove that M_{Gl} is shrinking. We define a weight function $\omega : Q \cup \Gamma \cup \{\bot\} \rightarrow \mathbb{N}_+$ as follows:

$$
\begin{aligned}
\omega(c) &:= 2 &&\text{for all } c \in \{a, b\}; \\
\omega(c') &:= 1 &&\text{for all } c' \in \{a', b'\}; \\
\omega(\#) &:= 1; \\
\omega(\bot) &:= 1; \\
\omega(q) &:= 1 &&\text{for all } q \in Q.
\end{aligned}
$$

Then it is easily checked that each transition step of M_{GI} is weight-reducing with respect to this weight function.

For $k \geq 1$, let $\mathsf{NTIME}_k(lin)$ denote the class of languages that are accepted by nondeterministic k-tape Turing machines in linear time. Thus, $L \in \mathsf{NTIME}_k(lin)$ if and only if there exist a k-tape nondeterministic Turing machine T and a constant $c \in \mathbb{N}_+$ such that, for all words w, $w \in L$ if and only if T accepts on input w within $c \cdot |w|$ many steps. Further, let

$$\mathsf{NTIME}(lin) := \bigcup_{k \geq 1} \mathsf{NTIME}_k(lin),$$

that is, $\mathsf{NTIME}(lin)$ is the class of languages accepted by nondeterministic Turing machines in linear time. With $\mathsf{DTIME}_k(lin)$ and $\mathsf{DTIME}(lin)$ we denote the corresponding deterministic classes.

Obviously, each k-sPDA (k-sdPDA) can be simulated by a nondeterministic (deterministic) k-tape Turing machine that runs in linear time. Hence, we have the following inclusions.

Corollary 1.
(a) $\mathsf{CRL} = \mathcal{L}(2\text{-sdPDA}) \subsetneq \mathcal{L}(3\text{-sdPDA}) \subseteq \mathcal{L}(\text{sdMPDA}) \subseteq \mathsf{DTIME}(lin)$.
(b) $\mathsf{GCSL} = \mathcal{L}(2\text{-sPDA}) \subsetneq \mathcal{L}(3\text{-sPDA}) \subseteq \mathcal{L}(\text{sMPDA}) \subseteq \mathsf{NTIME}(lin)$.

3 On the Expressive Power of 3-sPDA

Here we derive a characterization of the class $\mathcal{L}(3\text{-sPDA})$ in terms of language and complexity classes that have been studied in the literature, which proves that the 3-sPDA is quite expressive indeed.

By Q we denote the class of *quasi-realtime languages*. This is the class of languages that are accepted by nondeterministic multitape Turing machines in realtime, that is, $\mathsf{Q} = \mathsf{NTIME}(n)$. As shown by Book and Greibach, Q admits the following characterization.

Proposition 1. [3] *The following statements are equivalent for each language L:*

(a) *L is quasi-realtime, that is, $L \in \mathsf{Q}$;*
(b) *L is accepted by a nondeterministic multitape Turing machine in linear time, that is, $L \in \mathsf{NTIME}(lin)$;*
(c) *L is accepted in realtime by a 3-PDA with an additional one-way read-only input tape;*
(d) *L is the length-preserving homomorphic image of the intersection of three context-free languages, that is, there exist three context-free languages L_1, L_2, and L_3 and a length-preserving morphism h such that $L = h(L_1 \cap L_2 \cap L_3)$.*

Thus, we see that $\mathcal{L}(\text{sMPDA}) \subseteq \mathsf{Q}$ holds. Actually, we will show in the following that already $\mathcal{L}(3\text{-sPDA})$ coincides with the complexity class Q, which improves upon part (c) of the above proposition is as far as our 3-sPDA has no additional input tape. For doing so we will use the characterization of Q in Proposition 1(d).

Theorem 1. *The class $\mathcal{L}$(3-sPDA) is closed under intersection.*

Proof. For $i = 1, 2$, let $M_i = (Q_i, \Sigma, \Gamma_i, \bot, q_0^{(i)}, F_i, \delta_i)$ be a 3-sPDA that accepts the language $L_i \subseteq \Sigma^*$. Here we assume that $\Gamma_1 \cap \Gamma_2 = \Sigma$, and that $Q_1 \cap Q_2 = \emptyset$. Further, let ω_i be a weight function that is compatible with M_i, $i = 1, 2$. We construct a 3-sPDA M for the language $L := L_1 \cap L_2$ as follows.

Let $\hat{\Sigma}$ and $\tilde{\Sigma}$ be two new copies of the alphabet Σ, and let M_1' be the 3-sPDA that is obtained from M_1 by replacing the input alphabet Σ by $\hat{\Sigma}$, and let M_2' be the 3-sPDA that is obtained from M_2 by replacing the input alphabet Σ by $\tilde{\Sigma}$.

The 3-PDA M has input alphabet Σ and tape alphabet $\Gamma := \Sigma \cup \Gamma_1' \cup \Gamma_2' \cup \Sigma'$, where Γ_i' is the tape alphabet of M_i', $i = 1, 2$, and Σ' is another new copy of the input alphabet Σ. Further, its set of states is $Q := Q_1 \cup Q_2 \cup \{q_0, q_1\}$, where q_0 and q_1 are two new states. It proceeds as follows, starting with an initial configuration of the form $(q_0, w\bot, \bot, \bot)$, where $w \in \Sigma^*$. First it shifts the input word w to its second pushdown store, replacing w by the corresponding word $w' \in \Sigma'^*$, which yields a configuration of the form $(q_1, \bot, w'^R \bot, \bot)$. Then it shifts w' back to the first pushdown store, replacing each symbol a' by the corresponding symbol $\hat{a}$ ($a \in \Sigma$), and at the same time it copies w' to the third pushdown store, replacing each symbol a' by the corresponding symbol $\tilde{a}$ ($a \in \Sigma$). In this way a configuration of the form $(q_0^{(1)}, \hat{w}\bot, \bot, \tilde{w}\bot)$ is reached. Next M simulates the 3-sPDA M_1' on input $\hat{w}$, while treating the first symbol of $\tilde{w}$ on its third pushdown store as the bottom marker of that pushdown store. If $w \in L_1$, then M_1' eventually reaches a configuration of the form $(q', \bot, \bot, \tilde{w}\bot)$ for some $q' \in Q_1$, in which it would now pop the topmost symbols from all three pushdown stores, in this way accepting the input $\hat{w}$. Instead M moves to the configuration $(q_0^{(2)}, \bot, \bot, \tilde{w}\bot)$ and starts simulating the 3-sPDA M_2' on input $\tilde{w}$, interchanging the roles of its first and its third pushdown stores. Finally, M accepts if M_2' accepts. It follows easily that $L(M) = L_1 \cap L_2$.

Finally we obtain a compatible weight function ω for M as follows:

$$
\begin{aligned}
\omega(a) &:= \omega_1(a) + \omega_2(a) + 2 && \text{for all } a \in \Sigma; \\
\omega(a') &:= \omega_1(a) + \omega_2(a) + 1 && \text{for all } a \in \Sigma; \\
\omega(\hat{a}) &:= \omega_1(a) && \text{for all } a \in \Sigma; \\
\omega(\tilde{a}) &:= \omega_2(a) && \text{for all } a \in \Sigma; \\
\omega(A) &:= \omega_1(A) && \text{for all } A \in \Gamma_1' \setminus \hat{\Sigma}; \\
\omega(B) &:= \omega_2(B) && \text{for all } B \in \Gamma_2' \setminus \tilde{\Sigma}; \\
\omega(q_0) &:= \omega_1(q_0^{(1)}) + \omega_2(q_0^{(2)}) + 2; \\
\omega(q_1) &:= \omega_1(q_0^{(1)}) + \omega_2(q_0^{(2)}) + 1; \\
\omega(q^{(1)}) &:= \omega_1(q^{(1)}) + \omega_2(q_0^{(2)}) && \text{for all } q^{(1)} \in Q_1; \\
\omega(q^{(2)}) &:= \omega_2(q^{(2)}) && \text{for all } q^{(2)} \in Q_2.
\end{aligned}
$$

It is easily seen that M is weight-reducing with respect to this weight function. $\qquad\square$

In the proof above the 3-sPDA M for $L_1 \cap L_2$ is deterministic, if the 3-sPDA M_1 and M_2 for L_1 and L_2 are.

Theorem 2. *The class* $\mathcal{L}(3\text{-sPDA})$ *is closed under* ε*-free morphisms.*

Proof. Let $M = (Q, \Sigma, \Gamma, \bot, q_0, F, \delta)$ be a 3-sPDA that accepts the language $L \subseteq \Sigma^*$ and that is compatible with the weight function ω, and let $\varphi : \Sigma^* \to \Delta^*$ be an ε-free morphism. We claim that the language $\varphi(L) := \{ \varphi(w) \mid w \in L \}$ is accepted by a 3-sPDA M'.

The 3-sPDA M' proceeds as follows, where Δ' is a new alphabet that is in one-to-one correspondence to Δ, and $w \in \Delta^*$:

$$(q_0', w\bot, \bot, \bot) \vdash^*_{M'} (q_1', \bot, w'^R \bot, \bot)$$
$$\vdash^*_{M'} (q_0, x\bot, \bot, \bot) \quad \text{for some } x \in \varphi^{-1}(w),$$

where the pre-image $x \in \varphi^{-1}(w)$ is computed iteratively by popping a non-empty factor u'^R from the second pushdown store while pushing a letter $a \in \Sigma$ satisfying $\varphi(a) = u$ onto the first pushdown store. Observe that $\varphi(a) \neq \varepsilon$ for each letter $a \in \Sigma$, which means that only non-empty factors u' must be considered. Then the 3-sPDA M is being simulated on input x. Thus, M' accepts on input w if and only if $\varphi^{-1}(w) \cap L \neq \emptyset$, that is, $L(M') = \varphi(L)$.

As φ is an ε-free morphism, it is easily seen that there exists a weight function ω' that is compatible with M'. Essentially ω' agrees with ω on all symbols $A \in \Gamma$, and for all $c \in \Delta$, we simply take $\omega'(c) := 1 + \max\{ \omega(a) \mid a \in \Sigma \}$.   □

Based on the closure properties of $\mathcal{L}(3\text{-sPDA})$ established above, we obtain the following characterization.

Theorem 3. $\mathcal{L}(3\text{-sPDA}) = \mathsf{Q}$.

Proof. According to Proposition 1 a language L belongs to the class Q if and only if it is the image of the intersection of three context-free languages with respect to a length-preserving morphism. As $\mathsf{CFL} \subset \mathsf{GCSL} \subset \mathcal{L}(3\text{-sPDA})$, and as $\mathcal{L}(3\text{-sPDA})$ is closed under intersection, the intersection of three context-free languages is accepted by a 3-sPDA. Furthermore, a length-preserving morphism is obviously ε-free, and hence, the closure of $\mathcal{L}(3\text{-sPDA})$ under ε-free morphisms implies that $\mathsf{Q} \subseteq \mathcal{L}(3\text{-sPDA})$.

On the other hand, each 3-sPDA can be simulated by a nondeterministic three-tape Turing machine that runs in linear time. Thus, $\mathcal{L}(3\text{-sPDA}) \subseteq \mathsf{NTIME}_3(lin) \subseteq \mathsf{NTIME}(lin) = \mathsf{Q}$, implying that $\mathcal{L}(3\text{-sPDA}) = \mathsf{Q}$ holds.   □

As a consequence we obtain that adding more pushdown stores to shrinking multi-pushdown automata does not increase their expressive power. Thus, the chain of inclusions of Corollary 1(b) actually looks as follows.

Corollary 2.
$$\mathsf{GCSL} = \mathcal{L}(2\text{-sPDA}) \subsetneq \mathcal{L}(3\text{-sPDA}) = \mathcal{L}(\mathsf{sMPDA}) = \mathsf{Q} = \mathsf{NTIME}(lin).$$

As a special case we may include the class 1-sPDA, where an automaton from this class has a one-way read-only input tape in addition to its one pushdown store. Based on Greibach's normal form result, it is easily seen that each context-free language is accepted by a 1-sPDA, which yields the three-level hierarchy

$$\mathsf{CFL} = \mathcal{L}(1\text{-sPDA}) \subsetneq \mathsf{GCSL} = \mathcal{L}(2\text{-sPDA}) \subsetneq \mathsf{Q} = \mathcal{L}(3\text{-sPDA}).$$

4   An Infinite Hierarchy for Deterministic Shrinking Multi-pushdown Automata

Obviously, we have the chain of inclusions

$$(\,*\,)\ \mathcal{L}(\text{1-sdPDA}) \subseteq \cdots \subseteq \mathcal{L}(k\text{-sdPDA}) \subseteq \mathcal{L}((k+1)\text{-sdPDA}) \subseteq \cdots \subseteq \mathcal{L}(\text{sdMPDA}),$$

where each 1-sdPDA is equipped with an additional one-way read-only input tape in addition to its one pushdown store.

Proposition 2. $\text{DCFL} = \mathcal{L}(\text{1-sdPDA})$.

Proof. As each 1-sdPDA is a deterministic PDA, the inclusion from right to left is obvious.

Conversely, it is known that each deterministic context-free language can be accepted by a deterministic PDA for which the only ε-transitions pop symbols from the pushdown store (see [9] Ex. 10.2). Such a deterministic PDA has running time $O(n)$, and it is easily seen to be shrinking. Thus, we see that $\text{DCFL} = \mathcal{L}(\text{1-sdPDA})$ holds.   □

Next we relate Turing machines with linear running time to shrinking multi-pushdown automata.

Theorem 4. $\text{DTIME}_1(lin) \subseteq \mathcal{L}(\text{3-sdPDA})$.

Proof. Let $T = (Q_T, \Sigma, \Gamma_T, q_0^{(T)}, F_T, \delta_T)$ be a single-tape deterministic Turing machine with state set Q_T, input alphabet Σ, tape alphabet Γ_T containing Σ, initial state $q_0^{(T)} \in Q_T$, the set of final states $F_T \subseteq Q_T$, and the transition relation $\delta_T : Q_T \times \Gamma_T \to (Q \times \Gamma_T \times \{-1, 0, +1\})$. Further, let $c \in \mathbb{N}_+$ be a constant such that the running time of T is bounded from above by the function $c \cdot n$.

We define a 3-sdPDA $M = (Q, \Sigma, \Gamma, \bot, q_0, F, \delta)$ for simulating the Turing machine T as follows:

- $Q := \{q_0, q_1, q_a\} \cup Q_T$,
- $\Gamma := \Gamma_T \cup \Sigma' \cup \{\#\}$, where $\Sigma' := \{\, a' \mid a \in \Sigma \,\}$,
- $F := \{q_a\}$, and
- δ is defined as follows, where $d' := d$ for all symbols $d \in \Gamma_T \setminus \Sigma$ and $\square \in \Gamma_T$ denotes the blank symbol:

$$
\begin{aligned}
&(1)\ (q_0, \bot, \bot, \bot) && \to (q_a, \varepsilon, \varepsilon, \varepsilon) && \text{if } \varepsilon \in L(T); \\
&(2)\ (q_0, a, \bot, \bot) && \to (q_0, \varepsilon, a'\bot, \#^{c+2}\bot) && \text{for all } a \in \Sigma; \\
&(3)\ (q_0, a, b', \#) && \to (q_0, \varepsilon, a'b', \#^{c+2}) && \text{for all } a, b \in \Sigma; \\
&(4)\ (q_0, \bot, b', \#) && \to (q_1, b'\bot, \varepsilon, \varepsilon) && \text{for all } b \in \Sigma; \\
&(5)\ (q_1, a', b', \#) && \to (q_1, b'a', \varepsilon, \varepsilon) && \text{for all } a, b \in \Sigma; \\
&(6)\ (q_1, a', \bot, \#) && \to (q_0^{(T)}, a', \bot, \#) && \text{for all } a \in \Sigma; \\
&(7)\ (q^{(T)}, a', d, \#) && \to (p^{(T)}, b', d, \varepsilon) && \text{if } (p^{(T)}, b, 0) \in \delta_T(q^{(T)}, a); \\
&(8)\ (q^{(T)}, \bot, d, \#) && \to (p^{(T)}, b'\bot, d, \varepsilon) && \text{if } (p^{(T)}, b, 0) \in \delta_T(q^{(T)}, \square);
\end{aligned}
$$

(9) $(q^{(T)}, a', d, \#) \rightarrow (p^{(T)}, db', \varepsilon, \varepsilon)$ if $(p^{(T)}, b, -1) \in \delta_T(q^{(T)}, a)$, $d \neq \perp$;

(10) $(q^{(T)}, a', \perp, \#) \rightarrow (p^{(T)}, \Box b', \perp, \varepsilon)$ if $(p^{(T)}, b, -1) \in \delta_T(q^{(T)}, a)$;

(11) $(q^{(T)}, \perp, d, \#) \rightarrow (p^{(T)}, db'\perp, \varepsilon, \varepsilon)$ if $(p^{(T)}, b, -1) \in \delta_T(q^{(T)}, \Box)$, $d \neq \perp$;

(12) $(q^{(T)}, \perp, \perp, \#) \rightarrow (p^{(T)}, \Box b'\perp, \perp, \varepsilon)$ if $(p^{(T)}, b, -1) \in \delta_T(q^{(T)}, \Box)$;

(13) $(q^{(T)}, a', d, \#) \rightarrow (p^{(T)}, \varepsilon, b'd, \varepsilon, \varepsilon)$ if $(p^{(T)}, b, 1) \in \delta_T(q^{(T)}, a)$, $a' \neq \perp$;

(14) $(q^{(T)}, \perp, d, \#) \rightarrow (p^{(T)}, \perp, b'd, \varepsilon)$ if $(p^{(T)}, b, 1) \in \delta_T(q^{(T)}, \Box)$;

(15) $(q^{(T)}, a', b', \#) \rightarrow (q_a, a', b', \varepsilon)$ for $q^{(T)} \in F_T$;

(16) $(q_a, a', b', \#) \rightarrow (q_a, a', b', \varepsilon)$ for all $a', b' \in \Gamma_T \cup \Sigma'$;

(17) $(q_a, a', b', \perp) \rightarrow (q_a, a', \varepsilon, \perp)$ for all $b' \in \Gamma_T \cup \Sigma'$;

(18) $(q_a, a', \perp, \perp) \rightarrow (q_a, \varepsilon, \perp, \perp)$ for all $a' \in \Gamma_T \cup \Sigma'$;

(19) $(q_a, \perp, \perp, \perp) \rightarrow (q_a, \varepsilon, \varepsilon, \varepsilon)$.

Given an input $w = a_1 a_2 \ldots a_n \in \Sigma^n$ for some $n \geq 1$, M proceeds as follows. First M pushes $c \cdot n + 1$ copies of the special symbol $\#$ onto its third pushdown store:

$$
\begin{aligned}
(q_0, a_1 a_2 \ldots a_n \perp, \perp, \perp) &\vdash_M (q_0, a_2 \ldots a_n \perp, a_1' \perp, \#^{(c+1)+1} \perp) \\
&\vdash_M^* (q_0, \perp, a_n' \ldots a_1' \perp, \#^{(c+1) \cdot n + 1} \perp) \\
&\vdash_M (q_1, a_n' \perp, a_{n-1}' \ldots a_1' \perp, \#^{(c+1) \cdot n} \perp) \\
&\vdash_M^* (q_1, a_1' \ldots a_n' \perp, \perp, \#^{c \cdot n + 1} \perp).
\end{aligned}
$$

Next M simulates the computation of T step by step, using its first pushdown store for storing the suffix of the tape content of T starting from the position of T's head, and its second pushdown store for storing the prefix of the tape content of T to the left of the position of T's head. In each simulation step an occurrence of the symbol $\#$ is popped from the third pushdown store. Should M run out of $\#$-symbols before a final state of T is reached, then the computation halts without accepting. However, if $w \in L(T)$, then according to our assumption, there is an accepting computation of T of length at most $c \cdot n$, that is, by simulating this computation, M will eventually reach a configuration of the form $(q^{(T)}, v\perp, u\perp, \#^m \perp)$, where $q^{(T)} \in F_T$, $u, v \in \Gamma_T^*$, and $m \geq 1$. Then steps (15) to (19) make M enter its final state q_a and empty its pushdown stores. Thus, we see that $L(T) \subseteq L(M)$ holds.

Conversely, if M accepts an input $w \in \Sigma^n$, then each accepting computation of M consists of three phases. In the first phase $c \cdot n + 1$ copies of the symbol $\#$ are pushed onto the third pushdown store, in the second phase an accepting computation of T on input w is simulated, and then the pushdown stores are emptied. Hence, we actually have the equality $L(M) = L(T)$.

It remains to show that M is shrinking with respect to some weight function. We define a weight function $\omega : Q \cup \Gamma \cup \{\perp\} \rightarrow \mathbb{N}_+$ as follows:

$$
\begin{aligned}
\omega(a) &:= 3 \cdot c + 8 && \text{for all } a \in \Sigma; \\
\omega(b) &:= 1 && \text{for all } b \in \Gamma \setminus (\Sigma \cup \{\#\}); \\
\omega(q) &:= 1 && \text{for all } q \in Q_T \cup \{q_a, \perp\}; \\
\omega(\#) &:= 3; \\
\omega(q_i) &:= 2 && \text{for all } i \in \{0, 1\}.
\end{aligned}
$$

In instruction (2) an occurrence of a letter $a \in \Sigma$ is replaced by an occurrence of the corresponding letter $a' \in \Sigma'$ and $(c + 2)$ occurrences of the symbol #. As $\omega(a') + (c + 2) \cdot \omega(\#) = 1 + (c + 2) \cdot 3 = 3 \cdot c + 7 < \omega(a)$, we see that this step is actually shrinking with respect to ω.

In instruction (12) an occurrence of the symbol # is replaced by $\Box b'$ for some $b' \in \Gamma \smallsetminus (\Sigma \cup \{\#\})$. As $\omega(\#) = 3 > 2 = \omega(\Box) + \omega(b')$, also this instruction is shrinking with respect to ω.

All the other instructions are easily seen to be weight-reducing with respect to ω. Thus, M is indeed shrinking.

Thus, $\mathsf{DTIME}_1(lin) \subseteq \mathcal{L}(3\text{-sdPDA})$. $\hfill\square$

In the simulation above two pushdown stores are used to simulate the tape of the Turing machine, and the third pushdown store is used to make the simulation weight-reducing. The same technique can be used to simulate a k-tape Turing machine that is linear time-bounded. Of course, then we need $2k + 1$ pushdown stores. Further, let $\mathsf{DTIME}_{1,k}(lin)$ denote the class of languages that are accepted in linear time by deterministic *on-line* Turing machines with k work tapes. In addition to its k work tapes, such a Turing machine has a separate input tape that is one-way and read-only. For the simulation of such a machine, we need $2k + 2$ pushdown stores, as we need one pushdown store to play the role of the input tape.

Corollary 3. *For all $k \geq 1$,* (a) $\mathsf{DTIME}_k(lin) \subseteq \mathcal{L}((2k + 1)\text{-sdPDA})$.
(b) $\mathsf{DTIME}_{1,k}(lin) \subseteq \mathcal{L}((2k + 2)\text{-sdPDA})$.

Aanderaa has shown in [1] that

$$\mathsf{DTIME}_{1,k}(n) \subsetneq \mathsf{DTIME}_{1,k+1}(n)$$

holds for all $k \in \mathbb{N}_+$ by presenting a language L_k that is accepted by an on-line $(k + 1)$-tape Turing machine in realtime, but that is not accepted by any on-line k-tape Turing machine in realtime. Actually, L_k is not accepted by any on-line k-tape Turing machine with time-bound $t(n)$ satisfying $t(n) < n \cdot (\log n)^{1 \triangleleft (1+k)}$ [13]. Even more, the on-line $(k + 1)$-tape Turing machine for L_k just uses its $(k + 1)$ work tapes as pushdown stores. As realtime computations are obviously weight-reducing, it follows that

$$L_k \in \mathcal{L}((k + 2)\text{-sdPDA}) \smallsetminus \mathsf{DTIME}_{1,k}(lin).$$

Thus, for each $k \geq 2$, we have $\mathcal{L}(k\text{-sdPDA}) \subseteq \mathsf{DTIME}_k(lin) \subseteq \mathsf{DTIME}_{1,k}(lin)$, while $\mathcal{L}((k + 2)\text{-sdPDA}) \not\subseteq \mathsf{DTIME}_{1,k}(lin)$. It follows that

$$\mathcal{L}(k\text{-sdPDA}) \subsetneq \mathcal{L}((k + 2)\text{-sdPDA}).$$

This proves the following separation result.

Theorem 5. *The chain* $(*)$ *contains an infinite number of proper inclusions.*

Hence, the language classes $\mathcal{L}(k\text{-sdPDA})$ $(k \geq 1)$ form an infinite strict hierarchy within the class Q, with DCFL $= \mathcal{L}(1\text{-sdPDA})$ being the first level and CRL $= \mathcal{L}(2\text{-sdPDA})$ being the second level. Observe that

$$\bigcup_{k \geq 1} \mathcal{L}(k\text{-sdPDA}) = \mathcal{L}(\text{sdMPDA}) = \text{DTIME}(lin) \subsetneq \text{NTIME}(lin) = \text{Q},$$

where the fact that $\text{DTIME}(lin)$ is properly included in $\text{NTIME}(lin)$ is proved in [15].

It remains open at this point whether there exist any integers $k \geq 3$ such that the language classes $\mathcal{L}(k\text{-sdPDA})$ and $\mathcal{L}((k+1)\text{-sdPDA})$ coincide, or whether we have $\mathcal{L}(k\text{-sdPDA}) \subsetneq \mathcal{L}((k+1)\text{-sdPDA})$ for all $k \geq 3$.

5 Concluding Remarks

In [12] the shrinking *alternating* two-pushdown automaton (2-sAPDA) is considered, which is the alternating variant of the 2-sPDA. Obviously, we can generalize alternation to shrinking multi-pushdown automata with more than two pushdown stores. If we denote by $\text{ATIME}_k(lin)$ the class of languages that are accepted by *alternating k-tape Turing machines in linear time*, and $\text{ATIME}(lin) := \bigcup_{k \geq 1} \text{ATIME}_k(lin)$, then we have the equality $\text{ATIME}(lin) = \text{ATIME}_1(lin)$ according to [14]. As Theorem 4 easily extends to the case of alternating machines, this yields the following consequence.

Corollary 4. $\mathcal{L}(3\text{-sAPDA}) = \text{ATIME}(lin)$.

Together with the results of [12] this gives the following sequence of inclusions:

$$\text{GCSL} = \mathcal{L}(2\text{-sPDA}) \subsetneq \mathcal{L}(2\text{-sAPDA}) \subseteq \mathcal{L}(3\text{-sAPDA}) = \text{ATIME}(lin).$$

However, it remains open whether or not the 3-sAPDA is more expressive than the 2-sAPDA.

At least two other variants of the shrinking multi-pushdown automata considered here come to mind. First, there is the multi-pushdown automaton with an additional one-way read-only input tape. The 3-sPDA of Example 1 uses its first pushdown store just as an input tape. Hence, already the shrinking two-pushdown automaton with an additional input tape accepts the Gladkij language. Is the corresponding language class properly contained in the complexity class Q, or is this model as powerful as the 3-sPDA? Secondly, instead of considering pushdown stores one may consider *stacks* [9]. It is easily seen that a shrinking one-stack machine with an additional one-way input tape can accept the Gladkij language. Where does the resulting language class lie in relation to the language classes considered in this paper? What about shrinking two-stack automata with or without an additional one-way input tape? Obviously, the shrinking three-stack automaton yields another characterization of the class Q.

References

1. S.O. Aanderaa: On k-tape versus $(k-1)$-tape real time computation. In R.M. Karp (ed.): *Complexity of Computation, SIAM-AMS Symp. in Appl. Math.*, Vol. 7, 75–96. American Math. Society, Providence, R.I., 1974.

2. R.V. Book: Time-bounded grammars and their languages. *J. Comput. System Sci.*, 5:397–429, 1971.

3. R.V. Book and S.A. Greibach: Quasi-realtime languages. *Math. Systems Theory*, 4:97–111, 1970.

4. H. Bordihn, M. Holzer, and M. Kutrib: Input reversals and iterated pushdown automata: A new characterization of Khabbaz geometric hierachy of languages. In C.S. Calude, E. Calude, and M.J. Dinneen (eds.): *DLT 2004, Proc.*, LNCS 3340, 102–113. Springer, Berlin, 2004.

5. G. Buntrock: *Wachsende kontext-sensitive Sprachen.* Habilitationsschrift, Fakultät für Mathematik und Informatik, Universität Würzburg, 1996.

6. G. Buntrock and F. Otto: Growing context-sensitive languages and Church-Rosser languages. *Inform. and Comput.*, 141:1–36, 1998.

7. E. Dahlhaus and M. Warmuth: Membership for growing context-sensitive grammars is polynomial. *J. Comput. System Sci.*, 33:456–472, 1986.

8. M. Holzer and M. Kutrib: Flip-Pushdown Automata: $k+1$ Pushdown Reversals Are Better Than k. In J.C.M. Baeten, J.K. Lenstra, J. Parrow, and G.J. Woeginger (eds.): *ICALP 2003, Proc.*, LNCS 2719, 490–501. Springer, Berlin, 2003.

9. J.E. Hopcroft and J.D. Ullman: *Introduction to Automata Theory, Languages, and Computation.* Addison-Wesley, Reading, MA, 1979.

10. R. McNaughton, P. Narendran, and F. Otto: Church-Rosser Thue systems and formal languages. *J. Assoc. Comput. Mach.*, 35:324–344, 1988.

11. G. Niemann and F. Otto: The Church-Rosser languages are the deterministic variants of the growing context-sensitive languages. *Inform. and Comput.*, 197:1–21, 2005. An extended abstract appeared in M. Nivat (ed.): *FoSSaCS 1998, Proc.*, LNCS 1378, 243–257. Springer, Berlin, 1998.

12. F. Otto and E. Moriya: Shrinking alternating two-pushdown automata. *IEICE Trans. on Inform. and Systems*, E87-D (2004) 959–966.

13. W.J. Paul: On-line simulation of $k+1$ tapes by k tapes requires nonlinear time. *Inform. and Control*, 53:1–8, 1982.

14. W.J. Paul, E. Prauß, and R. Reischuk: On alternation. *Acta Inform.*, 14 (1980) 243–255.

15. W.J. Paul, N. Pippenger, E. Szemerédi, and W.T. Trotter: On determinism versus non-determinism and related problems. In *24th Annual Symposium on Foundations of Computer Science, Proc.*, 429–438. IEEE Computer Society, Los Angeles, 1983.

A Simple and Fast Min-cut Algorithm

Michael Brinkmeier

Technical University of Ilmenau,
Institute for Theoretical and Technical Computer Science
mbrinkme@tu-ilmenau.de

Abstract. We present an algorithm which calculates a minimum cut and its weight in an undirected graph with nonnegative real edge weights, n vertices and m edges, in time $O((\max(\log n, \min(m/n, \delta_G/\varepsilon)) n^2)$, where ε is the minimal edge weight, and δ_G the minimal weighted degree. For integer edge weights this time is further improved to $O(\delta_G n^2)$ and $O(\lambda_G n^2)$.

In both cases these bounds are improvements of the previously known best bounds of deterministic algorithms. These were $O(nm + \log n n^2)$ for real edge weights and $O(nM + n^2)$ and $O(M + \lambda_G n^2)$ for integer weights, where M is the sum of all edge weights.

1 Introduction

The problem of finding a minimum cut of a graph appears in many applications, for example, in network reliability, clustering, information retrieval and chip design. More detailed, a minimum cut of an undirected graph with edge weights, is a set of edges with minimum sum of weights, such that its removal would cause the graph to become unconnected. The total weight of edges in a minimum cut of G is denoted by λ_G and caled *(edge-)connectivity* of G.

In the literature many algorithms can be found applying various methods. One group of algorithms is based on the well-known result of Ford and Fulkerson [FF56] regarding the duality of maximum s-t-flows and minimum s-t-cuts for arbitrary vertices s and t. In [GH61] Gomory and Hu presented an algorithm which calculated $n - 1$ maximum s-t-flows from a given source s to all other vertices t. Hao and Orlin adapted the maximum flow algorithm of Goldberg and Tarjan [GT88] and were able to construct a minimum cut of a directed graph with nonnegative real numbers as edge weights in time $O(nm \log(n^2/m))$.

Nagamochi and Ibaraki [NI92a, NOI94, NII99] described an algorithm without using maximum flows. Instead they constructed spanning forests and iteratively contracted edges with high weights. This lead to an asymptotic runtime of $O(nm + n^2 \log n)$ on undirected graphs with nonnegative real edge weights. On undirected, unweighted multigraphs they obtained a runtime of $O(n(n + m))$. Translated to integer weighted graphs without parallel edges, this corresponds to a runtime of $O(n(n + M))$ where M is the sum of all edge weights. Using a 'searching' technique, they improved this to $O(M + \lambda_G n^2)$.

M. Liśkiewicz and R. Reischuk (Eds.): FCT 2005, LNCS 3623, pp. 317–328, 2005.

Their approach was refined in [SW97] by Stoer and Wagner. They replaced the construction of spanning forests with the construction of *Maximum Adjacency Orders* but the asymptotic runtime stayed unchanged.

Karger and Stein [KS96] used the contraction technique for a randomized algorithm calculating the minimum cut of undirected graphs in $O(n^2 \log^3 n)$ expected time. Later Karger [Kar96, Kar98] presented two related algorithms using expected time $O(m \log^3 n)$ and $O(n^2 \log n)$.

The last two algorithms are related to an approach of Gabow [Gab95] based on matroids. He presented an algorithm for the minimum cut of a directed, unweighted graph requiring $O(\lambda m \log(n^2/m))$ time and based on this an algorithm for undirected unweighted graphs with $O(m + \lambda^2 n \log(n/\lambda))$, where λ is the weight of a minimum cut.

In this paper we propose two changes of the algorithm of Stoer and Wagner [SW97], which calculates a *maximum adjacency order* on the vertices of an undirected graph and then contracts the last two vertices in this order. Repeating this $n - 1$ times, allows the construction of a minimum cut. Our first change leads to a reduced average runtime by contracting more than one pair of vertices if possible, reducing the asymptotic worst case runtime for real weighted edges to $O\left(\max\left(\log n, \min\left(m/n, \delta_G/\varepsilon\right)\right) n^2\right)$, where ε is the minimal edge weight. In fact the same idea was already used in later versions of the algorithms of Nagamochi and Ibaraki [NOI94], but for some reason did not find its way into the MA-order based algorithms (even those described by Nagamochi and Ibaraki) [SW97, NI02] and the analysis of the worst case runtime.

For integer weighted edges our algorithm allows an additional relaxation of the applied maximum adjacency orders, which in turn allows the usage of an alternative data structure for the construction of the order. These *priority queues with threshold* reduce the asymptotic runtime for undirected graphs with nonnegative integer weights to $O(\delta_G n^2)$, instead of $O(Mn + n^2)$ as obtained by Nagamochi and Ibaraki[1]. Applying the same 'search' as Nagamochi and Ibaraki, we obtain a time of $O(\lambda_G n^2)$ instead of $O(M + \lambda_G n^2)$.

Independent of the types of edge weights, the algorithm presented in this paper requires at most the same time as the algorithm of Stoer and Wagner [SW97].

2   The Problem and Notations

In the following let $G = (V, E)$ be an undirected Graph without multiple edges and self-loops. Let $n = |V|$ the number of vertices and $m = |E|$ the number of edges. The latter ones are weighted by positive real numbers or integers, given by a map $w \colon V \times V \to \mathbb{R}^+$ with $w(u, v) = w(v, u)$ and $w(u, v) = 0$ if and only if $(u, v) \notin E$. The *degree* $\deg(v)$ of a vertex is the sum of weights of incident edges, ie. $\deg(v) = \sum_{u \in V} w(u, v)$. The *minimal degree* over all vertices of G is denoted by δ_G. The sum of all edge weights is $M = \frac{1}{2} \sum_{v \in V} \deg(v)$.

---

[1] Remember that M is the sum of all edge weights and m the number of edges.

A *cut* of G is an (unordered) partition $(C, V \setminus C)$ of the vertex set into two parts. For shorter notation the cut will often be written as C. The *weight* $w(C)$ of the cut C is the total weight of "cutted" edges, ie.

$$w(C) = \sum_{\substack{(u,v) \in E \\ u \in C, v \notin C}} w(u, v).$$

For two vertices u and v a *u-v-cut* is a cut C such that $u \in C$ and $v \notin C$ or vice versa, ie. C *separates u and v*. A *minimum u-v-cut* is a u-v-cut, whose weight is minimal amongst all u-v-cuts of G. The weight of a minimum u-v-cut is denoted by $\lambda_G(u, v)$. A *minimum cut* is a cut C with minimal weight among all cuts of G. Its weight λ_G is called the *edge-connectivity* of G. Obviously we have

$$\lambda_G = \min \{\lambda_G(u, v) \mid u, v \in V\}.$$

For a subset $U \subseteq V$ of vertices $G[U]$ denotes the induced subgraph of G, ie. the graph consisting of the vertices in U and all edges of G between them. If $U = \{v_1, \ldots, v_m\}$ we also write $G[v_1, \ldots, v_m]$.

Like the ones of Nagamochi/Ibaraki and Stoer/Wagner our algorithm will be a *contraction algorithm*. Basically this means that it identifies one or more pairs of "critical" vertices and contracts (or merges) them, obtaining a graph with less nodes. Let u and v be two vertices of $G = (V, E)$. The graph $G/u \sim v$ is obtained from G by identifying u and v, this means that u and v are replaced by a new vertex $[u] = [v]$ and that the weights of the edges (x, u) and (x, v) are added, ie. $w(x, [u]) = w(x, u) + w(x, v)$.

Contraction algorithms rely on the following well-known theorem, relating the edge-connectivity of a graph with the one of a quotient graph.

Theorem 1 (Thm 2.1 of [SW97]). *Let u and v be two vertices of an undirected, weighted graph $G = (V, E)$. Then the edge-connectivity of G is the minimum of the weights of a minimum v-u-cut and the edge-connectivity of the graph $G/u \sim v$, obtained by the identification of u and v, ie.*

$$\lambda_G = \min \left(\lambda_G(u, v), \lambda_{G/u \sim v}\right).$$

Proof. We have to differentiate two cases. Either each minimum cut of G separates u and v, then $\lambda_G = \lambda_G(u, v)$ and $\lambda_{G/u \sim v} > \lambda_G$, or there exists at least one minimum cut not separating u and v and hence induces a minimum cut of $G/u \sim v$, leading to $\lambda_G = \lambda_{G/u \sim v} \leq \lambda_G(u, v)$.

3   Maximum Adjacency Orders

The key for the algorithms of Nagamochi/Ibaraki and Stoer/Wagner are *maximum adjacency orders* on the vertices of the graph. The vertices $v_1, \ldots v_n$ are arranged in a maximum adjacency order, if for each v_i, $i > 1$, the sum of weights from v_i to all preceeding vertices $v_1, \ldots v_{i-1}$ is maximal among all vertices v_k with $k \geq i$.

Definition 1 (Maximum Adjacency Order). *Let $G = (V, E)$ be an undirected, weighted graph. An order $v_1, v_2, \ldots, v_n$ on the vertices of G is an maximum adjacency order or MA-order, if*

$$w(v_1, \ldots, v_{i-1}; v_i) := \sum_{j=1}^{i-1} w(v_i, v_j) \geq \sum_{j=1}^{i-1} w(v_k, v_j) =: w(v_1, \ldots, v_{i-1}; v_k)$$

for all $k \geq i$. The values $w(v_1, \ldots, v_{i-1}; v_i)$ are called adjacencies.

The foundation of the algorithms of Nagamochi/Ibaraki and Stoer/Wagner is the observation that the degree of v_n in an MA-order is equal to the weight of a minimum v_n-v_{n-1}-cut in G.

Lemma 1 (Lemma 3.1 of [SW97]). *For each MA-order $v_1, \ldots, v_n$ of the undirected, weighted Graph $G = (V, E)$, the cut $(\{v_1, \ldots, v_{n-1}\}, \{v_n\})$ is a minimum v_n-v_{n-1}-cut.*

Our algorithm is additionally based on the following simple observation.

Lemma 2. *Let $v_1, \ldots, v_n$ be an MA-order of $G = (V, E)$. Then $v_1, \ldots, v_l$ is an MA-order of $G[v_1, \ldots, v_l]$.*

Proof. Since

$$\sum_{j=1}^{i-1} w(v_i, v_j) \geq \sum_{j=1}^{i-1} w(v_k, v_j)$$

for all $k \geq i$ for all $2 \leq i \leq l$, the same obviously holds for $i \leq k \leq l$.

Corollary 1. *For each MA-order $v_1, \ldots, v_n$ of $G = (V, E)$ we have*

$$\lambda_G(v_i, v_{i-1}) \geq \lambda_{G[v_1, \ldots, v_i]}(v_i, v_{i-1}) = w(v_1, \ldots, v_{i-1}; v_i).$$

Proof. The equality of $w(v_1, \ldots, v_{i-1}; v_i)$ and $\lambda_{G[v_1, \ldots, v_i]}(v_i, v_{i-1})$ is a direct consequence of Lemma 2 and 1. Since the weight of a minimal v_i-v_{i-1}-cut increases, if vertices and edges are added to the graph, the inequality is immediately clear.

For $\tau > 0$, an MA-order $v_1, \ldots, v_n$ of the vertices of $G = (V, E)$, and $1 \leq i \leq n$, let the graphs G_i^τ be defined by $G_1^\tau := G$ and $G_{i+1}^\tau := G_i^\tau$ if $w(v_1, \ldots, v_i; v_{i+1}) < \tau$, and

$$G_{i+1}^\tau \in \{G_i^\tau, G_i^\tau / [v_i] \sim v_{i+1}\}$$

if $w(v_1, \ldots, v_i; v_{i+1}) \geq \tau$ where $[v_i]$ is the class of vertices in G containing v_i. In other words, the graphs G_i^τ are obtained from G by iteratively contracting pairs of vertices v_i and v_{i+1}, such that the adjacency of v_{i+1} is greater than or equal to τ. The resulting vertices represent sets, denoted by $[v_i]$, of vertices in the original graph G. The last graph G_n^τ of this sequence is denoted by G^τ.

Theorem 2. *For an MA-order $v_1, \ldots, v_n$ of $G = (V, E)$ and $\tau > 0$ the following equation holds for $1 \leq i \leq n$*

$$\min(\lambda_G, \tau) = \min(\delta_G, \lambda_{G_i^\tau}, \tau).$$

Proof. We prove the theorem by induction over i. For $i = 1$ the statement is trivial, since $\delta_G \geq \lambda_G = \lambda_{G_1^\tau}$. Now assume that $\min(\lambda_G, \tau) = \min(\delta_G, \lambda_{G_{i-1}^\tau}, \tau)$ for $i > 1$. If $G_i^\tau = G_{i-1}^\tau$, then the statement obviously holds. Now assume $G_i^\tau = G_{i-1}^\tau/[v_{i-1}] \sim v_i$. Then Lemma 1 induces $\lambda_{G_{i-1}^\tau} = \min(\lambda_{G_{i-1}^\tau}([v_{i-1}], v_i), \lambda_{G_i^\tau})$ and hence

$$\min(\lambda_G, \tau) = \min(\delta_G, \lambda_{G_{i-1}^\tau}, \tau) = \min(\delta_G, \lambda_{G_{i-1}^\tau}([v_{i-1}], v_i), \lambda_{G_i^\tau}, \tau).$$

Since $[v_{i-1}]$ represents a set of vertices in G, each $[v_{i-1}]$-v_i-cut in G_{i-1}^τ induces a v_{i-1}-v_i-cut in G and hence, by definition of G_i^τ and Corollary 1

$$\lambda_{G_{i-1}^\tau}([v_{i-1}], v_i) \geq \lambda_G(v_{i-1}, v_i) \geq w(v_1, \ldots, v_{i-1}; v_i) \geq \tau,$$

thus

$$\min(\lambda_G, \tau) = \min(\delta_G, \lambda_{G_i^\tau}, \tau).$$

Even though it is not clear how much vertices G^τ contains, we can limit the total weight of all edges in G^τ.

Lemma 3. *For an MA-order $v_1, \ldots, v_n$ on $G = (V, E)$ and $\tau > 0$ the total weight of all edges of G^τ is less than $(n-1)\tau$.*

Proof. The lemma is proved by induction over the number n of vertices in G. It is obviously true for $n = 1$.

Since $v_1, \ldots, v_{n-1}$ is an MA-order of $G[v_1, \ldots, v_{n-1}]$, we know that the total edge weight of the induced subgraph of G^τ is less than $(n-2)\tau$. If the adjacency of v_n is less than τ, then G^τ obviously has total edge weight less than $(n-1)\tau$.

Now assume that the adjacency of v_n is at least τ. Let v_i be the first vertex in the MA-order, such that $w(v_1, \ldots, v_i; v_n) \geq \tau$, implying $w(v_1, \ldots, v_j; v_n) \geq \tau$ for all indices $i \leq j \leq n-1$. Hence, the adjacencies of $v_i, \ldots, v_{n-1}$ are at least τ and the complete sequence is contracted. Therefore, at most the edges between v_n and $v_1, \ldots, v_{i-1}$ 'survive' in G^τ and the total edge weight of G^τ is less than $(n-1)\tau$.

4   Lax Adjacency Orders

In this section we are going to use the threshold τ for the contraction in each round to relax the restrictions of the maximum adjacency order. The basic observation leading to this improvement is the fact, that the exact order inside a sequence of vertices with adjacency $\geq \tau$ does not matter. Hence, during the construction of the adjacency order, we may choose any vertex, as long as its adjacency is above the threshold (if possible). We only have to chose a vertex of maximum adjacency if it is below τ.

Definition 2 (Lax Adjacency Order). *Let* $G = (V, E)$ *be an undirected, weighted graph and* $\tau \geq 0$. *An order* $v_1, v_2, \ldots, v_n$ *on the vertices of* G *is an* lax adjacency order *or* LA-order *for threshold* τ, *if*

$$\min\left(\tau, w(v_1, \ldots, v_{i-1}; v_i)\right) \geq \min\left(\tau, w(v_1, \ldots, v_{i-1}; v_k)\right)$$

for all $k \geq i$.

Now we have to prove that a LA-order of threshold τ is an allowed replacement for an MA-order in the construction of G^τ.

Lemma 4 (LA- and MA-orders). *Let* $u_1, \ldots, u_n$ *be an* LA-*order of* $G = (V, E)$ *and* $i_1 < \cdots < i_k$ *the indices of all vertices with* $w(u_1, \ldots, u_{i_l-1}; u_{i_l}) < \tau$ *for* $1 \leq l \leq k$. *Then there exists an* MA-*order* $v_1, \ldots, v_n$ *with*

1. $v_{i_l} = u_{i_l}$ *for* $1 \leq l \leq k$ *and*
2. *if* $u_j = v_{\bar{j}}$ *and* $i_l < j < i_{l+1}$ *for* $1 \leq l < k$, *then* $i_l < \bar{j} < i_{l+1}$ *and*
3. *if* $u_j = v_{\bar{j}}$ *and* $i_k < j$, *then* $i_k < \bar{j}$ *and.*

In other words, for each LA-order there exists an MA-order, such that the vertices with adjacencies below the threshold are at the same positions, which we are going to call 'fix points'. Furthermore, vertices with adjacencies equal to or greater than τ may be permuted, as long as they stay between the same "fix points". The sequences between two subsequent fix points are called 'high adjacency sequences' and their members 'vertices of high adjacency'.

Proof (of Lemma 4). We are going to construct an MA-order $v_1, \ldots, v_n$ for each fix point i_l, such that the two conditions of the lemma hold for all indeces up to position i_l.

Obviously $i_1 = 1$ and hence, each MA-order beginning with u_1 satisfies the conditions up to position i_1.

Now assume that we have an MA-order $v_1, \ldots, v_n$ with $v_{i_k} = u_{i_k}$ for $1 \leq i \leq l$, and the high adjacency vertices up to position i_l are only permuted in their specific sequences.

If $i_l = i_k$, ie. if the last fixed point is reached, we are done.

Otherwise, observe that $w(u_1, \ldots, u_{i_{l+1}-1}; u_k) < \tau$ for each $i_{l+1} \leq k$, since

$$\tau > w(u_1, \ldots, u_{i_{l+1}-1}; u_{i_{l+1}}) \geq w(u_1, \ldots, u_{i_{l+1}-1}; u_k).$$

If $i_{l+1} = i_l + 1$, then we obviously may extend $v_1, \ldots, v_{i_l}$ by $u_{i_{l+1}}$ and obtain an MA-order, which respects the fixed points and high adjacency sequences up to position i_{l+1}.

If $i_{l+1} > i_l + 1$, then there obviously exists a vertex (eg. $u_{i_{l+1}}$) with an adjacency equal to or greater than τ. Since all vertices u_k with $k \geq i_{l+1}$ have low adjacency, the next vertex in each MA-order extending $v_1, \ldots, v_{i_l}$, has to be in the high adjacency sequence between i_l and i_{l+1}.

As long as vertices of the high adjacency sequence between i_l and i_{l+1} are not integrated into the MA-order, it does not matter which vertex is chosen. Simply choose the first unused vertex of the sequence. Since all vertices preceeding it in the LA-order are already used in the MA-order, its adjacency is at least τ, requiring the next vertex to be one member of the sequence (their adjacency is below τ as observed above).

Since the high adjacency sequences of an LA- and a corresponding MA-order are contracted to a single vertex, the resulting graph G^τ is the same.

Theorem 3. *For an LA-order $v_1, \ldots, v_n$ with threshold $\tau > 0$ on $G = (V, E)$ the following equation holds for $1 \le i \le n$*

$$\min(\lambda_G, \tau) = \min(\delta_G, \lambda_{G^\tau}, \tau).$$

5   The Algorithm

Thm. 3 provides us with a simple algorithm for the determination of a minimum cut. Let $\tau > 0$ be given.

1. While G contains two or more vertices, repeat
   (a) Determine a vertex $[v]$ of G with minimal degree.
   (b) If $\delta_G < \tau$, set cut $= [v]$.
   (c) $\tau = \min(\tau, \delta_G)$.
   (d) $G = G^\tau$ for some LA-order with threshold τ.
2. The cut is given by the vertices in $[v]$ and has weight τ.

First of all, the algorithm terminates, since the adjacency of the last vertex v_n is its degree and hence equal to or greater than δ_G and τ. Therefore G^τ has less vertices than G and the algorithm reduces the number of nodes in each round.

Let $\lambda(G, \tau)$ denote the resulting value τ of the algorithm with inputs G and τ. We claim $\lambda(G, \tau) = \min(\lambda_G, \tau)$.

If G has exactly two nodes, then the algorithm returns $\min(\tau, \delta_G)$ and one vertex as cut. If G has more than two nodes then the algorithm returns the same result as if it was started with G^τ and $\tau = \min(\delta_G, \tau)$, ie.

$$\lambda(G, \tau) = \lambda\left(G^{\min(\tau, \delta_G)}, \min(\tau, \delta_G)\right).$$

By induction over the number of nodes and thm 3 this leads to

$$\lambda(G, \tau) = \min\left(\lambda_{G^{\min(\tau, \delta_G)}}, \tau, \delta_G\right) = \min(\lambda_G, \tau).$$

For the calculation of G^τ we have to construct an LA-order and then iteratively contract subsequent vertices v_{i-1} and v_i with $w(v_1, \ldots, w_{i-1}; v_i) \ge \delta_G$, where δ_G is the minimum degree. One way to implement one of these rounds would be to order these nodes and then contract some of them. But in fact ordering and contraction may be done at the same time. Assume that $v_1, \ldots, v_n$ is an arbitrary order of the vertices. Then

$$w(v_1, \ldots, v_{i-1}; v_i) = w(v_1, \ldots, v_j \sim v_{j+1}, \ldots, v_{i-1}; v_i),$$

ie. contraction of preceeding vertices does not affect the adjacency of a vertex. Hence we may construct the LA-order and, as soon as the adjacency is greater or equal to δ_G, we may contract the newly added and the last vertex.

The LA-order may be build using a maximum priority queue. Assume that we have n' vertices and m' edges. Each vertex has to be inserted and extracted at most once, leading to n' inserts and extractions. The priorities are updated at most once for each edge, leading to m' operations, resulting in a runtime $O(n'T_{\text{insert}} + n'T_{\text{extractMax}} + m'T_{\text{increaseKey}})$ for queue operations per round.

For the first round we have $n' = n$ and $m' = m$. For the subsequent rounds, the total edge weight is bounded by $\delta_G n$ (Lemma 3). If $\varepsilon > 0$ is the minimal edge weight, this bounds the number of edges by $\delta_G n/\varepsilon$. Furthermore, we have at most m edges, implying a bound of $\min(m, \delta_G n/\varepsilon)$. This leads to a worst case runtime $O(nT_{\text{insert}} + nT_{\text{extractMax}} + \min(m, \frac{\delta_G}{\varepsilon}n)T_{\text{increaseKey}})$ in all subsequent rounds.

Together with the time of $O(n^2)$, required for the $n-1$ contractions of vertex pairs, the sum over at most $n-1$ rounds is

$$O\left(n^2 + n^2 T_{\text{insert}} + n^2 T_{\text{extractMax}} + \left(m + \min\left(\frac{m}{n}, \frac{\delta_G}{\varepsilon}\right)n^2\right)T_{\text{increaseKey}}\right).$$

Using Fibonacci Heaps, our algorithm has an amortized runtime of

$$O\left(n^2 + n^2 + n^2\log n + m + \frac{\delta_G}{\varepsilon}n^2\right) = O\left(m + n^2\left(\log n + \min\left(\frac{m}{n}, \frac{\delta_G}{\varepsilon}\right)\right)\right)$$

$$= O\left(\max\left(\log n, \min\left(\frac{m}{n}, \frac{\delta_G}{\varepsilon}\right)\right)n^2\right).$$

Theorem 4. *Let $G = (V, E)$ be a undirected graph with n vertices and m edges.*

1. *If its edges are weighted by nonnegative real numbers, Algorithm 1 calculates a minimum cut in $O\left(\left(\max\left(\log n, \min\left(\frac{m}{n}, \frac{\delta_G}{\varepsilon}\right)\right)\right)n^2\right)$ time, where ε is the minimal weight of an edge.*
2. *If its edges are weighted by nonnegative integers, algorithm 1 calculates a minimum cut in $O\left(\left(\max\left(\log n, \min\left(\frac{m}{n}, \delta_G\right)\right)\right)n^2\right)$ time.*

As the following examples show, our estimation of $n-1$ required rounds is strict (exp. 2), but in many cases the algorithm needs less rounds (examples 1 and 3), leading to a lower average runtime.

Example 1. If $G = (V, E)$ has vertices $v_1, \ldots, v_n$ and edges (v_i, v_{i+1}) for $1 \le i < n$, ie. it is a straight line, then $v_1, \ldots, v_n$ is a LA-order and v_1 is a vertex of minimum degree 1. Hence the algorithm contracts all vertices of G in the first and only round, requiring $O(n + m + \log n)$ time.

Example 2. If G is a circle with n vertices $v_1, \ldots, v_n$, ie. it has edges (v_i, v_{i+1}) for $1 \le i < n$ and (v_n, v_1). Then $v_1, \ldots, v_n$ is a LA-order and each vertex has minimum degree. Since the adjacencies of the v_i, except for v_n, are all 1, only v_{n-1} and v_n are contracted, leading to a circle with one vertex less. Hence the algorithm requires $(n-1)$ rounds, as does the original algorithm of Stoer and Wagner, leading to the same runtime.

Algorithm 1: The Minimum Cut Algorithm

Input: An undirected graph $G = (V, E)$ with weights w
Output: A set cut of vertices of G forming a minimal cut of weight τ.

cut $\leftarrow \emptyset$, $\tau \leftarrow \infty$

while $|V| >= 2$ **do**
    forall $v \in V$ **do**
        Q.insert$(v, 0)$
        if $\deg(v) < \tau$ **then** $\tau \leftarrow \deg(v)$, cut $\leftarrow [v]$
    end
    while Q *is not empty* **do**
        adj $\leftarrow$ Q.maxKey()
        $v \leftarrow$ Q.extractMax()
        forall $u \in V$ *with* $(v, u) \in E$ **do**
            if $u \in Q$ **then** Q.increaseKey$(u, Q.\text{key}(u) + w(v, u))$
        end
        if $\tau \leq$ *adj* **then**
            contract v into u
            $[u] \leftarrow [u] \cup [v]$
        end
        $u \leftarrow v$
    end
end

Example 3. On random graphs with n vertices and m edges the number of rounds required by the algorithm varies depending on n and m. For graphs with n between 1000 and 5000 and m between 1000 and $200n$. Figure 1 shows the number of rounds over the ratio m/n. The experiments showed that the minimal number of required rounds increases with the edge-vertex-ratio, ie. the average degree in the random graph.

Obviously an LA-order for threshold τ is an LA-order for a lower threshold $\tau' < \tau$. Hence we may decrease the weight during a round, possibly increasing the number of contractions. This observation may be used in two ways. First of all, the degree of the newly contracted node may be compared to the recent weight. If it is smaller, then the new node describes a cut of lower weight. Secondly, the set of all scanned vertices is a cut of the graph. Its weight can be updated easily, since for each cut C and each vertex $v \notin C$, the following holds:

$$w(C \cup \{v\}) = w(C) + \deg(v) - 2 \sum_{\substack{(v,u) \\ u \in C}} w(v, u).$$

Translated to our situation the weight of the cut increases by the degree of the added node minus twice its adjacency. If the resulting weight is lower than the currently known bound for the minimum cut, we may continue using the better value instead. Both observations may be used to improve Algorithm 1.

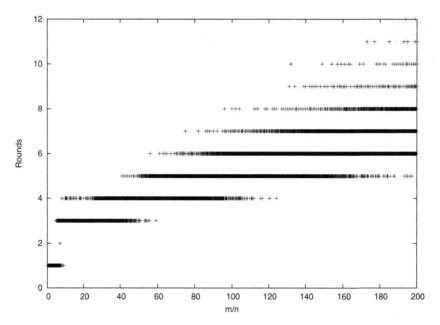

Fig. 1. The number of rounds required for random graphs

5.1   Priority Queues with Threshold

Up to this point we did not use the advantages of the threshold τ of lax adjacency orders. Since we do not have to differentiate between vertices with priorities above τ, we may use an alternative data structure, leading to a decreased worst case runtime $O(\delta_G n^2)$.

We are going to describe a *priority queue with threshold* τ, which allows the same operations as a priority queue, namely `insert`, `increaseKey` and `extractMax`. The two first operations behave as usual, but the third is changed slightly. `extractMax` returns an entry of maximal priority, as a regular priority queue does, if the maximal priority is lower than the threshold τ. Otherwise, it returns an entry with a not necessarily maximal priority $\geq \tau$.

A priority queue with threshold τ consists of $\tau + 1$ Buckets $B_0, \dots, B_\tau$. For $0 \leq j \leq \tau - 1$ the bucket B_j contains all entries (v, j) with value v and priority j. The last bucket, B_τ contains all entries (v, p) with $p \geq \tau$ in arbitrary order. In addition we maintain the maximal index of a nonempty bucket in $j_{\max}$. The operations are implemented in the following way:

- `insert(v, p)`: Insert (v, p) into $B_{\min(\tau, p)}$ and update $j_{\max}$ if necessary. This requires $O(1)$ time.
- `increaseKey((v, p), q)`: If $\min(\tau, q) > p$ the remove (v, p) from B_p and insert (v, q) into $B_{\min(\tau, q)}$ and update $j_{\max}$ if necessary. This takes $O(1)$ time.
- `extractMax`: Simply remove the first entry from bucket $B_{j_{\max}}$. If this was the last element in the bucket, search for the next nonempty bucket in decreasing

order starting at $j_{\max}-1$. In the worst case all τ buckets have to be examined, requiring time $O(\tau)$.

Applying the same estimations as for Theorem 4, we obtain the following result.

Theorem 5. *A minimum cut of an undirected graph $G = (V, E)$ with edges weighted by nonnegative integers, can be calculated in $O(\delta_G n^2 + m) = O(\delta_G n^2)$ time.*

Proof. As seen above, we require time

$$O\left(n^2 + n^2 T_{\text{insert}} + n^2 T_{\text{extractMax}} + \left(m + \min\left(\frac{m}{n}, \delta_G n^2\right)\right) T_{\text{increaseKey}}\right).$$

Using priority queues with thresholds, leads to a time of

$$O\left(n^2 + n^2 + \delta_G n^2 + \left(m + \min\left(\frac{m}{n}, \delta_G n^2\right)\right)\right) = O(\delta_G n^2).$$

By using a technique of Nagamochi and Ibaraki [NI92b] we may reduce the asymptotic runtime to $O(\lambda_G n^2)$.

Corollary 2. *Let $G = (V, E)$ an undirected, integer weighted graph.*

1. *Given $\tau > 0$ we may check in $O(\tau n^2)$ time, wether $\lambda_G < \tau$. If this is the case a minimum cut can be computed in the same time.*
2. *A minimum cut of G can be calculated in $O(\lambda_G n^2)$ time.*

Proof. Following Theorem 2, we have $\min(\lambda_G, \tau) = \min(\delta_G, \lambda_{G^\tau}, \tau)$. Hence we may calculate $\min(\lambda_G, \tau)$ by using our algorithm with a starting threshold of τ. This requires $O(\tau n^2)$ time. If $\lambda_G < \tau$, we obtain a minimum cut of G, and part one is proved.

We can use the first part for the second part. For increasing $i = 1, 2, \ldots$ check wether $\lambda_G < 2^i$ using the above algorithm. This is repeated until $2^{i-1} \le \lambda_G < 2^i$. In this case we obtain a minimum cut. The total runtime is

$$O(2n^2) + \ldots + O(2^{i-1}n^2) + O(2^i n^2)$$
$$= O((2 + \cdots + 2^i)n^2) = O(2^{i+1}n^2)$$
$$= O(\lambda_G n^2).$$

References

[CGK+97] Chandra Chekuri, Andrew V. Goldberg, David R. Karger, Matthew S. Levine, and Clifford Stein. Experimental study of minimum cut algorithms. In *Symposium on Discrete Algorithms*, pages 324–333, 1997.

[FF56]   L. R. Ford and D. R. Fulkerson. Maximal flow through a network. *Can. J. Math.*, 8:399–404, 1956.

[Gab95]   Harold N. Gabow. A matroid approach to finding edge connectivity and packing arborescences. *J. Comput. Syst. Sci.*, 50(2):259–273, 1995.

[GH61]   R. E. Gomory and T. C. Hu. Multi-terminal network flows. *J. SIAM*, 9:551–570, 1961.

[GT88]   A. V. Goldberg and R. E. Tarjan. A new approach to the maximum flow problem. *J. Assoc. Comput. Mach.*, 35:921–940, 1988.

[Kar96]   David R. Karger. Minimum cuts in near-linear time. In *STOC*, pages 56–63, 1996.

[Kar98]   David R. Karger. Minimum cuts in near-linear time. *CoRR*, cs.DS/9812007, 1998.

[KS96]   David R. Karger and Clifford Stein. A new approach to the minimum cut problem. *J. ACM*, 43(4):601–640, 1996.

[Mat93]   David W. Matula. A linear time 2+epsilon approximation algorithm for edge connectivity. In *SODA*, pages 500–504, 1993.

[NI92a]   Hiroshi Nagamochi and Toshihide Ibaraki. Computing edge-connectivity in multigraphs and capacitated graphs. *SIAM J. Disc. Math.*, 5(1):54–66, 1992.

[NI92b]   Hiroshi Nagamochi and Toshihide Ibaraki. A linear-time algorithm for finding a sparse k-connected spanning subgraph of a k-connected graph. *Algorithmica*, 7(5&6):583–596, 1992.

[NI02]   Hiroshi Nagamochi and Toshihide Ibaraki. Graph connectivity and its augmentation: applications of ma orderings. *Discrete Applied Mathematics*, 123(1-3):447–472, 2002.

[NII99]   Nagamochi, Ishii, and Ibaraki. A simple proof of a minimum cut algorithm and its applications. *TIEICE: IEICE Transactions on Communications/Electronics/Information and Systems*, 1999.

[NNI94]   Hiroshi Nagamochi, Kazuhiro Nishimura, and Toshihide Ibaraki. Computing all small cuts in undirected networks. In Ding-Zhu Du and Xiang-Sun Zhang, editors, *ISAAC*, volume 834 of *Lecture Notes in Computer Science*, pages 190–198. Springer, 1994.

[NOI94]   Hiroshi Nagamochi, Tadashi Ono, and Toshihide Ibaraki. Implementing an efficient minimum capacity cut algorithm. *Math. Program.*, 67:325–341, 1994.

[PR90]   M. Padberg and G. Rinaldi. An efficient algorithm for the minimum capacity cut problem. *Math. Program.*, 47:19–36, 1990.

[SW97]   Mechtild Stoer and Frank Wagner. A simple min-cut algorithm. *Journal of the ACM*, 44(4):585–591, July 1997.

(Non)-Approximability for the Multi-criteria $TSP(1, 2)$

Eric Angel[1], Evripidis Bampis[1], Laurent Gourvès[1], and Jérôme Monnot[2]

[1] LaMI, CNRS UMR 8042, Université d'Évry Val d'Essonne, France
{angel, bampis, lgourves}@lami.univ-evry.fr
[2] LAMSADE, CNRS UMR 7024, Université de Paris-Dauphine, France
monnot@lamsade.dauphine.fr

Abstract. Many papers deal with the approximability of multi-criteria optimization problems but only a small number of non-approximability results, which rely on NP-hardness, exist in the literature. In this paper, we provide a new way of proving non-approximability results which relies on the existence of a small size good approximating set (*i.e.* it holds even in the unlikely event of $P = NP$). This method may be used for several problems but here we illustrate it for a multi-criteria version of the traveling salesman problem with distances one and two ($TSP(1, 2)$). Following the article of Angel et al. (FCT 2003) who presented an approximation algorithm for the bi-criteria $TSP(1, 2)$, we extend and improve the result to any number k of criteria.

1 Introduction

Multi-criteria optimization refers to problems with two or more objective functions which are normally in conflict. Vilfredo Pareto stated in 1896 a concept (known today as "Pareto optimality") that constitutes the origin of research in this area. According to this concept, the solution to a multi-criteria optimization problem is normally not a single value, but instead a set of values (the so-called *Pareto curve*). From a computational point of view, this Pareto curve is problematic. Approximating it with a performance guarantee, *i.e.* computing an $\varepsilon - approximate\ Pareto\ curve$, motivated a lot of papers (see [1,8,11] among others). Up to our knowledge, non-approximability in the specific context of multi-criteria optimization has been investigated only from the point of view of NP-hardness [3,8]. In this paper, we aim to provide some negative results which are based on the non-existence of a small size approximating set: In multi-criteria optimization, one tries to approximate a set of solutions (the Pareto curve) with another set of solutions (the ε-approximate Pareto curve) and the more the ε-approximate Pareto curve contains solutions, the more accurate the approximation can be. Then, the best approximation ratio that could be achieved can be related to the size of the approximate Pareto curve. As a first attempt, we propose a way to get some negative results which works for several multi-criteria problems and we put it into practice on a special case of the multi-criteria traveling salesman problem.

M. Liśkiewicz and R. Reischuk (Eds.): FCT 2005, LNCS 3623, pp. 329–340, 2005.
© Springer-Verlag Berlin Heidelberg 2005

The traveling salesman problem is one of the most studied problem in the operations research community, see for instance [6]. The case where distances are either one or two (denoted by $TSP(1,2)$) was investigated by Papadimitriou and Yannakakis [9] who gave some positive and negative approximation results (see also [4]). Interestingly, this problem finds an application in a frequency assignment problem [5]. In this article, we deal with a generalization of the $TSP(1,2)$ where the distance is a vector of length k instead of a scalar: the k-criteria $TSP(1,2)$. Previously, Angel et al. [1] proposed a *local search* algorithm (called BLS) for the bi-criteria $TSP(1,2)$ which, with only two solutions generated in $\mathcal{O}(n^3)$, returns a $1/2$-approximate Pareto curve.

A question arises concerning the ability to improve the approximation ratio with an approximate Pareto curve containing two (or more) solutions. Conversely, given a fixed number of solutions, how accurate an approximate Pareto curve can be? More generally, given a multi-criteria problem, how many solutions are necessary to approximate the Pareto curve within a level of approximation? A second question arises concerning the ability to generalize BLS to any number of criteria. Indeed, a large part of the literature on multi-criteria optimization is devoted to bi-criteria problems and an algorithm which works for any number of criteria would be interesting.

The paper is organized as follows: In Section 2, we recall some definitions on exact and approximate Pareto curves. Section 3 is devoted to a method to derive some negatives results in the specific context of multi-criteria optimization. We use it for the k-criteria $TSP(1,2)$ but it works for several other problems. In Section 4, we study the approximability of the k-criteria $TSP(1,2)$. Instead of generalizing BLS, we adapt the classical *nearest neighbor* heuristic which is more manageable. This multi-criteria nearest neighbor heuristic works for any k and produces a $1/2$-approximate Pareto curve when $k \in \{1,2\}$ and a $(k-1)/(k+1)$-approximate Pareto curve when $k \geq 3$. This result extends for several reasons the one of Angel et al.. First, the new algorithm works for any $k \geq 2$, second the time complexity is decreased when $k = 2$.

2   Generalities

The Traveling Salesman Problem (TSP) is about to find in a complete graph $G = (V, E)$ a Hamiltonian cycle whose total distance is minimal. For the k-criteria TSP, each edge e has a *distance* $\boldsymbol{d}(e) = (\boldsymbol{d}_1(e), \ldots, \boldsymbol{d}_k(e))$ which is a vector of length k (instead of a scalar). The *total distance* of a tour T is also a vector $\boldsymbol{D}(T)$ where $\boldsymbol{D}_j(T) = \sum_{e \in T} \boldsymbol{d}_j(e)$ and $j = 1, \ldots, k$. In fact, a tour is evaluated with k objective functions. Given this, the goal of the optimization problem could be the following: Generating a feasible solution which simultaneously minimizes each coordinate. Unfortunately, such an ideal solution rarely exists since objective functions are normally in conflict. However a set of solutions representing all best possible trade-offs always exists (the so-called Pareto curve). Formally, a Pareto curve is a set of feasible solutions, each of them optimal in the sense of Pareto, which *dominates* all the other solutions. A tour T dominates another one T' (usually denoted by $T \leq T'$) iff $D_j(T) \leq D_j(T')$ for all $j = 1, \ldots, k$ and, for at least one coordinate j', one has $D_{j'}(T) < D_{j'}(T')$. A solution is optimal in the sense of Pareto if no solution dominates it.

From a computational point of view, Pareto curves are problematic [8,11]. Two of the main reasons are:

- the size of a Pareto curve which is often exponential with respect to the size of the corresponding problem;
- a multi-criteria optimization problem often generalizes a mono-criterion problem which is itself hard.

As a consequence, one tries to get a relaxation of this Pareto curve, *i.e.* an ε-*approximate Pareto curve* [8,11]. An ε-approximate Pareto curve P_ε is a set of solutions such that for every solution s of the instance, there is an s' in P_ε which satisfies $D_j(s') \leq (1 + \varepsilon)D_j(s)$ for all $j = 1, \ldots, k$.

In [8], Papadimitriou and Yannakakis prove that every multi-criteria problem has an ε-approximate Pareto curve that is polynomial in the size of the input, and $1/\varepsilon$, but exponential in the number k of criteria. The design of polynomial time algorithms which generate approximate Pareto curves with performance guarantee motivated a lot of recent papers. In this article we study the k-criteria $TSP(1, 2)$. In this problem, each edge e of the graph has a distance vector $d(e)$ of length k and $d_j(e) \in \{1, 2\}$ for all j between 1 and k.

3   Non-approximability Related to the Number of Generated Solutions

We propose in this section a new way to get some negative results which works for several multi-criteria problems and we put it into practice on the k-criteria $TSP(1, 2)$.

Usually, non-approximability results for mono-criterion problems bring thresholds of performance guarantee under which no polynomial time algorithm is likely to exist. Given a result of this kind for a mono-criterion problem Π, we directly get a negative result for a multi-criteria version of Π. Indeed, the multi-criteria version of Π generalizes Π. For example, hardness of inherent difficulty of the mono-criterion $TSP(1, 2)$ has been studied in [4,9] and the best known lower bound is $1 + 1/5380 - \delta$ (for all $\delta > 0$). Consequently, for all $\delta > 0$, no polynomial time algorithm can generate a $(1/5380 - \delta)$-approximate Pareto curve unless $P = NP$. However, the structure of the problem, namely the fact that several criteria are involved, is not taken into account.

In multi-criteria optimization, one tries to approximate a set of solutions (the Pareto curve) with another set of solutions (the ε-approximate Pareto curve) and the more the ε-approximate Pareto curve contains solutions, the more accurate the approximation can be. As a consequence, the best approximation ratio that could be achieved can be related to the size of the approximate Pareto curve. Formally, ε is a function of $|P_\varepsilon|$. If we consider instances for which the whole (or a large part of the) Pareto curve P is known and if we suppose that we approximate it with a set $P' \subset P$ such that $|P'| = x$ then the best approximation ratio ε such that P' is an ε-approximate Pareto curve is related to x. Indeed, there must be a solution in P' which approximates at least two (or more) solutions in P.

In the following, we explicitly give a family of instances (denoted by $I_{n,r}$) of the k-criteria $TSP(1,2)$ for which we know a lot of different Pareto optimal tours covering a large spectrum of the possible values.

We first consider an instance I_n with $n \geq 2k + 1$ vertices where distances belong to $\{(1,2,\ldots,2),(2,1,2,\ldots,2),\ldots,(2,\ldots,2,1)\}$. We suppose that for any $i = 1,\ldots,k$, the subgraph of I_n induced by the edges whose distance is 1 only on coordinate i is Hamiltonian (T_i denotes this tour). Using an old result [7], we know that K_n is Hamiltonian cycles decomposable into k disjoint tours if $n \geq 2k + 1$ and then, I_n exists.

We duplicate r times the instance I_n to get $I_{n,r}$. We denote by v_a^c the vertex v_a of the c-th copy of I_n. Between two copies with $1 \leq c_1 < c_2 \leq r$, we set $d([v_a^{c_1}, v_b^{c_2}]) = d([v_a, v_b])$ if $a \neq b$ and $d([v_a^{c_1}, v_a^{c_2}]) = (1, 2, \ldots, 2)$.

Lemma 1. *There are $\binom{r+k-1}{r}$ Pareto optimal tours in $I_{n,r}$ (denoted by $T_{c_1,\ldots,c_{k-1}}$ where c_i for $1 \leq i \leq k-1$ are $k-1$ indexes in $\{0,\ldots,r\}$) satisfying:*

(i) $\forall i = 1,\ldots,k-1$, $c_i \in \{0,\ldots,r\}$ and $\sum_{i=1}^{k-1} c_i \leq r$.
(ii) $\forall i = 1,\ldots,k-1$, $D_i(T_{c_1,\ldots,c_{k-1}}) = 2rn - c_in$ and $D_k(T_{c_1,\ldots,c_{k-1}}) = rn + n(\sum_{i=1}^{k-1} c_i)$.

Proof. Let $c_1, \ldots, c_{k-1}$ be integers satisfying (i), we build the tour $T_{c_1,\ldots,c_{k-1}}$ by applying the following process: On the c_1 first copies, we take the tour T_1, on the c_2 second copies, we take the tour T_2 and so on. Finally, for the $r - \sum_{i=1}^{k-1} c_i$ last copies, we take T_k. For any $1 \leq l_1 < l_2 \leq r$, and any tours T and T', we patch T on copy l_1 with T' on copy l_2 by replacing the edges $[v_i^{l_1}, v_j^{l_1}] \in T$ and $[v_j^{l_2}, v_m^{l_2}] \in T'$ by the edges $[v_i^{l_1}, v_j^{l_2}]$ and $[v_m^{l_2}, v_j^{l_1}]$. Observe that the resulting tour has a total distance $D(T') + D(T)$. So, by applying r times this process, we can obtain a tour $T_{c_1,\ldots,c_{k-1}}$ satisfying (ii). Moreover, the number of tours is equal to the number of choices of $k - 1$ elements among $r + (k - 1)$. □

Theorem 1. *For any $k \geq 2$, any ε-approximate Pareto curve with at most x solutions for the k-criteria $TSP(1,2)$ satisfies:*

$$\varepsilon \geq max_{i=2,\ldots,k}\{\frac{1}{(2i-1)r(i,x)-1}\}$$

$$\text{where } r(i,x) = min\{r|\ x \leq \binom{r+i-1}{r} - 1\}.$$

Proof. Let $r(k,x) = r$ be the smallest integer such that $x \leq \binom{r+k-1}{r} - 1$ and consider the instance $I_{n,r}$. Since $x \leq \binom{r+k-1}{r} - 1$, there exists two distinct tours $T_{c_1,\ldots,c_{k-1}}$ and $T_{c'_1,\ldots,c'_{k-1}}$ and a tour T in the approximate Pareto curve such that:

$$D(T) \leq (1+\varepsilon)D(T_{c_1,\ldots,c_{k-1}}) \text{ and } D(T) \leq (1+\varepsilon)D(T_{c'_1,\ldots,c'_{k-1}}) \quad (1)$$

Let $l_i = \max\{c_i, c_i'\}$ for $i = 1, \ldots, k-1$ and $l_k = \min\{\sum_{i=1}^{k-1} c_i, \sum_{i=1}^{k-1} c_i'\}$. By construction, we have $l_k \leq \sum_{i=1}^{k-1} l_i - 1$. Moreover, the total distance of T can be written $D_i(T) = 2rn - q_i$ for $i = 1, \ldots, k-1$ and $D_k(T) = rn + \sum_{i=1}^{k-1} q_i$ for some value of q_i (q_i is the number of edges of T where the distance has a 2 on coordinate i and 1 on the others). Thus, using inequalities (1), we deduce that for $i = 1, \ldots, k-1$, we have $2nr - q_i \leq (1+\varepsilon)(2rn - l_i n)$ which is equivalent to

$$q_i \geq l_i n(1+\varepsilon) - 2rn\varepsilon. \tag{2}$$

We also have $rn + \sum_{i=1}^{k-1} q_i \leq (1+\varepsilon)(rn + l_k n)$ which is equivalent to

$$\sum_{i=1}^{k-1} q_i \leq \varepsilon rn + l_k n(1+\varepsilon). \tag{3}$$

Adding inequalities (2) for $i = 1, \ldots, k-1$ and by using inequality (3) and $l_k \leq \sum_{i=1}^{k-1} l_i - 1$, we deduce:

$$\varepsilon \geq \frac{1}{(2k-1)r(k, x) - 1}. \tag{4}$$

Finally, since an ε-approximation for the k-criteria $TSP(1, 2)$ is also an ε-approximation for the i-criteria $TSP(1, 2)$ with $i = 2, \ldots, k-1$ (for the $k-i$ last coordinates, we get a factor 2), we can apply $k-1$ times the inequality (4) and the result follows. □

The following Table illustrates the Theorem 1 for some values of k and x:

$k \backslash x$	1	2	3	4	5	6	7	8	9
2	1.500	1.200	1.125	1.090	1.071	1.058	1.050	1.043	1.038
3	1.500	1.250	1.125	1.111	1.111	1.071	1.071	1.071	1.071
4	1.500	1.250	1.166	1.111	1.111	1.076	1.076	1.076	1.076

The method presented in this section can be applied to several other multi-criteria problems. For instance, it works with problems where all feasible solutions have the same size ($|V|$ for a Hamiltonian cycle, $|V| - 1$ for a spanning tree, etc).

4 Nearest Neighbor Heuristic for the k-Criteria $TSP(1, 2)$

Angel et al. present in [1] a *local search* algorithm (called BLS) for the bi-criteria $TSP(1, 2)$. This algorithm returns in time $\mathcal{O}(n^3)$ a 1/2-approximate Pareto curve. Since BLS works only for the bi-criteria $TSP(1, 2)$, an algorithm which works for any number of criteria would be interesting.

A generalization of BLS may exist but it is certainly done with difficulty. Since BLS uses the $2 - opt$ neighborhood, two neighboring solutions differ on two edges. Defining an order on each couple of possible distance vector is necessary to decide, among two neighboring solutions, which one is the best. When k grows, such an order is hard to handle.

In this section, we present a different algorithm which is more manageable. It works for any number of criteria and its time complexity is better than BLS's one for the bi-criteria $TSP(1, 2)$. We propose a *nearest neighbor* heuristic which computes in $\mathcal{O}(n^2 k!)$ time a $\frac{k-1}{k+1}$-approximate Pareto curve when $k \geq 3$ and a $1/2$-approximate Pareto curve when $k \in \{1, 2\}$. Let us observe here that the dependence of the time complexity on $k!$ is not surprising since the size of the approximate ε-Pareto curve is not necessarily polynomial on the number of the optimization criteria [8].

Traditionally, the nearest neighbor heuristic [10] consists in starting from a randomly chosen node and greedily insert non-visited vertices, chosen as the closest ones from the last inserted vertex. Adapting this heuristic to the k-criteria $TSP(1, 2)$ gives rise to two questions: How can we translate the notion of closeness when multiple objectives are considered? How many solutions must be generated to get an approximation of the Pareto curve? In the following, we propose a way which simultaneously brings an answer to both questions. Given the problem, the total distance of a Pareto optimal tour T^* is enclosed in a k-dimensional cost space. The way to generate a tour T which approximates T^*, and also the notion of closeness, depend on where $\boldsymbol{D}(T^*)$ is located in the cost space. The idea is to partition the cost space into a fixed number of parts. Then, with each part we associate an appropriate notion of closeness. Given a part and its proper notion of closeness, we can generate with the nearest neighbor rule a tour which approximates any Pareto optimal solution whose total distance is in the part. For any instance of the k-criteria $TSP(1, 2)$, we propose to divide the cost space into $k!$ parts as follows: Each part is identified by a permutation of $\{1, \ldots, k\}$. Given a permutation L of $\{1, \ldots, k\}$, a tour T is in the part identified by L if $\boldsymbol{D}_{L(1)}(T) \leq \ldots \leq \boldsymbol{D}_{L(k)}(T)$. For the notion of closeness, we introduce a preference relation over all possible distance vectors which looks like a lexicographic order. This preference relation which depends on L (denoted by $\prec_L$) is defined by using $k + 1$ sets $S_1, \ldots, S_{k+1}$:

$$S_q = \{\boldsymbol{a} \in \{1, 2\}^k \mid \forall j \leq k + 1 - q \quad \boldsymbol{a}_{L(j)} = 1\}, \quad \text{for } 1 \leq q \leq k$$
$$S_{k+1} = \{1, 2\}^k.$$

Definition 1. *For any edge e, we say that e is S_q-preferred (for $\prec_L$) if $\boldsymbol{d}(e) \in S_q \backslash S_{q-1}$ (where $S_0 = \emptyset$). For two edges e and e' such that e is S_q-preferred and e' is $S_{q'}$-preferred, we say that $\boldsymbol{d}(e)$ is preferred (resp., weakly preferred) to $\boldsymbol{d}(e')$ and we note $\boldsymbol{d}(e) \prec_L \boldsymbol{d}(e')$ (resp., $\boldsymbol{d}(e) \preccurlyeq_L \boldsymbol{d}(e')$) iff $q < q'$ (resp., $q \leq q'$).*

An example where $k = 3$ and L is the identity permutation is given in Figure 1.

The algorithm that we propose for the k-criteria $TSP(1, 2)$ is given in Table 1. Called KNN for k-criteria Nearest Neighbor, it is composed of $k!$ steps. A permutation L of $\{1, 2, \ldots, k\}$ is determined at each step. With a permutation

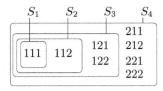

Fig. 1. One has $111 \prec_L 112 \prec_L 121 \preccurlyeq_L 122 \prec_L 211 \preccurlyeq_L 212 \preccurlyeq_L 221 \preccurlyeq_L 222$

Table 1. For $v \in V$ and p a tour, $p(v)$ denotes the node which immediately follows v in p

```
KNN: k-criteria Nearest Neighbor
P := ∅;
For each permutation L of {1 2 ... k} do
    Take arbitrarily v ∈ V ;
    W := {v} ; u := v ;
    While W ≠ V do
        Take r ∈ V\W s.t. r is the closest vertex to u by ≼L ;
        W := W ∪ {r} ;
        p(u) := r ; u := r ;
    End While ;
    p(r) := v ;
    P := P ∪ {p};
End do ;
Return P ;
```

L, we build a preference relation $\prec_L$ and finally, a solution is greedily generated with the nearest neighbor rule.

Theorem 2. KNN *returns a* $(k-1)/(k+1)$*-approximate Pareto curve for the* k*-criteria* $TSP(1,2)$ *when* $k \geq 3$ *and a* $1/2$*-approximate Pareto curve when* $k \in \{1,2\}$.

The proof of the theorem requires some notations and intermediate lemmata. In the following, we consider two particular tours p and p^*. We assume that p is the tour generated by KNN with the preference relation $\prec_L$ and that p^* is a Pareto optimal tour satisfying

$$D_{L(1)}(p^*) \leq D_{L(2)}(p^*) \leq \ldots \leq D_{L(k)}(p^*). \tag{5}$$

The set of all possible distance vectors $\{1,2\}^k$ is denoted by Ω. For all $j \leq k$, we introduce $U_j = \{a \in \Omega \mid a_j = 1\}$ and $\overline{U}_j = \{a \in \Omega \mid a_j = 2\}$. For $a \in \Omega$, we note $X_a = \{v \in V \mid d([v, p(v)]) = a\}$ and $X_a^* = \{v \in V \mid d([v, p^*(v)]) = a\}$. Finally, x_a (resp. x_a^*) denotes the cardinality of X_a (resp. X_a^*).

If n is the number of vertices then by construction we have $\sum_{a \in \Omega} x_a = \sum_{a \in \Omega} x_a^* = n$, $D_j(p) = 2n - \sum_{a \in U_j} x_a$ and $D_j(p^*) = 2n - \sum_{a \in U_j} x_a^*$.

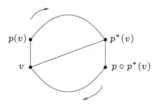

Fig. 2. The tour p generated by KNN. The edge $[v^c p^*(v)]$ belongs to p^*

Lemma 2. *The following holds for any $q \leq k$:*

$$2 \sum_{a \in \cap_{j=1}^{k+1-q} U_{L(j)}} x_a \geq \sum_{a \in \cap_{j=1}^{k+1-q} U_{L(j)}} x_a^*.$$

Proof. We define $F_q = \{v \in V \mid d([v,p(v)]) \in S_q\}$ and $F_q^* = \{v \in V \mid d([v,p^*(v)]) \in S_q\}$. Then, we have to prove that $2|F_q| \geq |F_q^*|$. The key result is to see that $p^*[F_q^* \backslash F_q] \subseteq F_q$ where $p^*[W] = \bigcup_{v \in W} \{p^*(v)\}$. Take a vertex v in $F_q^* \backslash F_q$ (see Figure 2). Then, $d([v,p^*(v)]) \in S_q$, $d([v,p(v)]) \in S_{q'}$ and $q' > q$. During the computation of p, suppose that v is the current node and that $p^*(v)$ is not already visited. We get a contradiction (the nearest neighbor rule is violated) since $p(v)$ immediately follows v in p and $d([v,p^*(v)]) \prec_L d([v,p(v)])$. Now, suppose $p^*(v)$ was already visited. It directly precedes $p \circ p^*(v)$ in p and then $d([p*(v),p \circ p^*(v)]) \preceq_L d([v,p^*(v)])$. As a consequence, $d([p^*(v),p \circ p^*(v)]) \in S_{q''}$ such that $q'' \leq q$ and $p^*(v) \in F_q$ since $S_{q''} \subseteq S_q$.

Since $|p^*[F_q^* \backslash F_q]| = |F_q^* \backslash F_q|$, $|F_q^*| = |F_q^* \backslash F_q| + |F_q^* \cap F_q|$ and $|F_q| \geq |F_q^* \cap F_q|$, we deduce $|F_q^*| = |p^*[F_q^* \backslash F_q]| + |F_q^* \cap F_q| \leq 2|F_q|$. Finally, since $\cap_{j=1}^{k+1-q} U_{L(j)} = S_q$, $|F_q| = \sum_{a \in S_q} x_a$ and $|F_q^*| = \sum_{a \in S_q} x_a^*$, the result follows. $\square$

The following inequality is equivalent to (5):

$$\sum_{a \in U_{L(1)}} x_a^* \geq \sum_{a \in U_{L(2)}} x_a^* \geq \ldots \geq \sum_{a \in U_{L(k)}} x_a^*.$$

We easily deduce that for any couple j_1, j_2 such that $j_1 < j_2$ we have:

$$\sum_{a \in U_{L(j_2)} \backslash U_{L(j_1)}} x_a^* \leq \sum_{a \in U_{L(j_1)} \backslash U_{L(j_2)}} x_a^*. \tag{6}$$

Let b_1, b_2, j and m be such that $b_1 \in \{1,2\}$, $b_2 \in \{1,2\}$, $1 \leq j \leq k$ and $1 \leq m < j$. Let $R(b_1,j,m,b_2)$ be the set of all $a \in \Omega$ such that $a_{L(j)} = b_1$ and there exists exactly m distinct coordinates of a among $\{a_{L(1)}, a_{L(2)}, \ldots, a_{L(j-1)}\}$ which are equal to b_2. Remark that $R(b_1,j,m,b_2) = R(b_1,j,j-1-m,\overline{b_2})$ where $\overline{b_2} = 3-b_2$.

Lemma 3. *For any $j \leq k$, one has:*

$$\sum_{q=1}^{j-1} \left(q \sum_{a \in R(1,j,q,2) \cup R(2,j,q,2)} x_a^* \right) \leq (j-1) \sum_{q=0}^{j-1} \left(\sum_{a \in R(2,j,q,1)} x_a^* \right).$$

Proof of Theorem 2

Proof. The proof is cut into 3 cases ($j = 1$, $j = 2$ and $j \geq 3$). In the following, we consider that L is any permutation of $\{1, \ldots, k\}$, p^* is a Pareto optimal tour satisfying (5) and p is built with the nearest neighbor rule and the preference relation $\prec_L$. Then, we have to show that:

(i) if $j = 1$ or 2 then $\boldsymbol{D}_{L(j)}(p) \leq (1 + \frac{1}{2})\boldsymbol{D}_{L(j)}(p^*)$,

(ii) if $j \geq 3$ then $\boldsymbol{D}_{L(j)}(p) \leq (1 + \frac{j-1}{j+1})\boldsymbol{D}_{L(j)}(p^*)$.

Case $j = 1$. $\boldsymbol{D}_{L(1)}(p) \leq \frac{3}{2}\boldsymbol{D}_{L(1)}(p^*)$ is equivalent to the following inequality:

$$2 \sum_{a \in U_{L(1)}} x_a - \sum_{a \in U_{L(1)}} x_a^* + 2 \sum_{a \in \overline{U}_{L(1)}} x_a^* \geq 0. \tag{7}$$

Indeed, $\boldsymbol{D}_{L(1)}(p) \leq \dfrac{3}{2}\boldsymbol{D}_{L(1)}(p^*) \Leftrightarrow 2\left(2n - \sum_{a \in U_{L(1)}} x_a\right) \leq 3\left(2n - \sum_{a \in U_{L(1)}} x_a^*\right)$

$$\Leftrightarrow -2 \sum_{a \in U_{L(1)}} x_a \leq 2n - 3 \sum_{a \in U_{L(1)}} x_a^*$$

Using $n = \sum_{a \in U_{L(1)}} x_a^* + \sum_{a \in \overline{U}_{L(1)}} x_a^*$, the equivalence follows. Thus, using Lemma 2 with $q = k$ and $\sum_{a \in \overline{U}_{L(1)}} x_a^* \geq 0$ (which is true since for all $a \in \Omega$, $x_a^* \geq 0$), inequality (7) follows.

Case $j = 2$. $\boldsymbol{D}_{L(2)}(p) \leq \frac{3}{2}\boldsymbol{D}_{L(2)}(p^*)$ is equivalent to the following inequality:

$$-2 \sum_{a \in U_{L(2)} \backslash U_{L(1)}} x_a - 2 \sum_{a \in U_{L(2)} \cap U_{L(1)}} x_a \leq 2 \sum_{a \in \overline{U}_{L(2)}} x_a^* - \sum_{a \in U_{L(2)} \backslash U_{L(1)}} x_a^* - \sum_{a \in U_{L(2)} \cap U_{L(1)}} x_a^*.$$
$$\tag{8}$$

Indeed, $\boldsymbol{D}_{L(2)}(p) \leq \dfrac{3}{2}\boldsymbol{D}_{L(2)}(p^*) \Leftrightarrow -2 \sum_{a \in U_{L(2)}} x_a \leq 2 \sum_{a \in \overline{U}_{L(2)}} x_a^* - \sum_{a \in U_{L(2)}} x_a^*.$

If we partition $U_{L(2)}$ into two subsets $U_{L(2)} \backslash U_{L(1)}$ and $U_{L(2)} \cap U_{L(1)}$ then the equivalence follows. By Lemma 2 with $q = k - 1$ we get:

$$2 \sum_{a \in U_{L(1)} \cap U_{L(2)}} x_a \geq \sum_{a \in U_{L(1)} \cap U_{L(2)}} x_a^*.$$

Then, using inequality (8), we have to prove:

$$-2 \sum_{a \in U_{L(2)} \backslash U_{L(1)}} x_a \leq 2 \sum_{a \in \overline{U}_{L(2)}} x_a^* - \sum_{a \in U_{L(2)} \backslash U_{L(1)}} x_a^*.$$

By inequality (6), when $j_1 = 1$ and $j_2 = 2$, we get:

$$- \sum_{a \in U_{L(1)} \setminus U_{L(2)}} x_a^* \leq - \sum_{a \in U_{L(2)} \setminus U_{L(1)}} x_a^*$$

Thus:

$$2 \sum_{a \in \overline{U}_{L(2)}} x_a^* - \sum_{a \in U_{L(1)} \setminus U_{L(2)}} x_a^* \leq 2 \sum_{a \in \overline{U}_{L(2)}} x_a^* - \sum_{a \in U_{L(2)} \setminus U_{L(1)}} x_a^*.$$

Since $U_{L(1)} \setminus U_{L(2)} \subseteq \overline{U}_{L(2)}$, we have:

$$-2 \sum_{a \in U_{L(2)} \setminus U_{L(1)}} x_a \leq 0 \leq 2 \sum_{a \in \overline{U}_{L(2)}} x_a^* - \sum_{a \in U_{L(1)} \setminus U_{L(2)}} x_a^*.$$

Case $j \geq 3$. $D_{L(j)}(p) \leq \frac{2j}{j+1} D_{L(j)}(p^*)$ holds if we have the following inequality:

$$-(j+1) \sum_{a \in U_{L(j)}} x_a \leq 2(j-1) \sum_{a \in \overline{U}_{L(j)}} x_a^* - 2 \sum_{a \in U_{L(j)}} x_a^*. \tag{9}$$

$$D_{L(j)}(p) \leq \frac{2j}{j+1} D_{L(j)}(p^*) \Leftrightarrow (j+1)\left(2n - \sum_{a \in U_{L(j)}} x_a\right) \leq 2j\left(2n - \sum_{a \in U_{L(j)}} x_a^*\right)$$

$$\Leftrightarrow -(j+1) \sum_{a \in U_{L(j)}} x_a \leq 2(j-1)n - 2j \sum_{a \in U_{L(j)}} x_a^*$$

$$\Leftrightarrow -(j+1) \sum_{a \in U_{L(j)}} x_a \leq 2(j-1) \sum_{a \in \overline{U}_{L(j)}} x_a^* - 2 \sum_{a \in U_{L(j)}} x_a^*,$$

using $n = \sum_{a \in U_{L(j)}} x_a^* + \sum_{a \in \overline{U}_{L(j)}} x_a^*$.

Let us denote by $\mathcal{A}$ and $\mathcal{B}$ the following quantities:

$$\sum_{a \in U_{L(j)}} x_a = \sum_{a \in U_{L(j)} \setminus (\cap_{m \leq j-1} U_{L(m)})} x_a + \sum_{a \in \cap_{m \leq j} U_{L(m)}} x_a = \mathcal{A}$$

$$\sum_{a \in U_{L(j)}} x_a^* = \sum_{a \in U_{L(j)} \setminus (\cap_{m \leq j-1} U_{L(m)})} x_a^* + \sum_{a \in \cap_{m \leq j} U_{L(m)}} x_a^* = \mathcal{B}.$$

Then, inequality (9) becomes:

$$-(j+1)\mathcal{A} \leq 2(j-1) \sum_{a \in \overline{U}_{L(j)}} x_a^* - 2\mathcal{B}. \tag{10}$$

To prove (10), we propose the following decomposition:

$$\mathcal{C} = 2(j-1) \sum_{a \in \overline{U}_{L(j)}} x_a^* - 2 \sum_{a \in U_{L(j)} \setminus \cap_{m \leq j-1} U_{L(m)}} x_a^* - 4 \sum_{a \in \cap_{m \leq j} U_{L(m)}} x_a \tag{11}$$

$$-(j+1)\mathcal{A} \leq \mathcal{C} \tag{12}$$

$$\mathcal{C} \leq 2(j-1) \sum_{a \in \overline{U}_{L(j)}} x_a^* - 2\mathcal{B} \tag{13}$$

Thus, (12) becomes:

$$-(j+1) \sum_{a \in U_{L(j)} \backslash \bigcap_{m \leq j-1} U_{L(m)}} x_a - (j-3) \sum_{a \in \bigcap_{m \leq j} U_{L(m)}} x_a \leq$$

$$\leq 2(j-1) \sum_{a \in \overline{U}_{L(j)}} x_a^* - 2 \sum_{a \in U_{L(j)} \backslash \bigcap_{m \leq j-1} U_{L(m)}} x_a^*$$

Since the left part of this inequality is negative, we want to prove that the right part is positive:

$$0 \leq 2(j-1) \sum_{a \in \overline{U}_{L(j)}} x_a^* - 2 \sum_{a \in U_{L(j)} \backslash \bigcap_{m \leq j-1} U_{L(m)}} x_a^* \tag{14}$$

$$\sum_{a \in U_{L(j)} \backslash \bigcap_{m \leq j-1} U_{L(m)}} x_a^* \leq (j-1) \sum_{a \in \overline{U}_{L(j)}} x_a^* \tag{15}$$

We also have:

$$\sum_{a \in U_{L(j)} \backslash \bigcap_{m \leq j-1} U_{L(m)}} x_a^* = \sum_{q=1}^{j-1} \left(\sum_{a \in R(1,j,q,2)} x_a^* \right) \text{ and}$$

$$(j-1) \sum_{a \in \overline{U}_{L(j)}} x_a^* = (j-1) \sum_{q=0}^{j-1} \left(\sum_{a \in R(2,j,q,1)} x_a^* \right).$$

The first equality follows from $U_{L(j)} \backslash \bigcap_{m \leq j-1} U_{L(m)} = \bigcup_{q=1}^{j-1} R(1,j,q,2)$ since $a \in U_{L(j)} \backslash \bigcap_{m \leq j-1} U_{L(m)}$ iff $a_{L(j)} = 1$ and there exists exactly q indexes $\{i_1, \ldots i_q\}$ such that $1 \leq q \leq j-1$ and $a_{L(i_1)} = a_{L(i_2)} = \ldots = a_{L(i_q)} = 2$, which is equivalent to $a \in R(1,j,q,2)$. The second equality follows from $\overline{U}_{L(j)} = \bigcup_{q=0}^{j-1} R(2,j,q,1)$ because $a \in \overline{U}_{L(j)}$ means $a_{L(j)} = 2$.

As a consequence, (15) becomes:

$$\sum_{q=1}^{j-1} \left(\sum_{a \in R(1,j,q,2)} x_a^* \right) \leq (j-1) \sum_{q=0}^{j-1} \left(\sum_{a \in R(2,j,q,1)} x_a^* \right).$$

With Lemma 3, we have:

$$\sum_{q=1}^{j-1} \left(q \sum_{a \in R(1,j,q,2) \cup R(2,j,q,2)} x_a^* \right) \leq (j-1) \sum_{q=0}^{j-1} \left(\sum_{a \in R(2,j,q,1)} x_a^* \right)$$

and (15) follows from

$$\sum_{q=1}^{j-1} \left(q \sum_{a \in R(1,j,q,2) \cup R(2,j,q,2)} x_a^* \right) \geq \sum_{q=1}^{j-1} \left(\sum_{a \in R(1,j,q,2)} x_a^* \right).$$

By Lemma 2 with $q = k + 1 - j$ we have:

$$2 \sum_{a \in \bigcap_{m \leq j} U_{L(m)}} x_a \geq \sum_{a \in \bigcap_{m \leq j} U_{L(m)}} x_a^*$$

which is exactly (13).   □

There is an instance which shows that the analysis is tight but, because of the restricted number of pages, it is given in the full version of this paper [2].

References

1. E. Angel, E. Bampis and L. Gourvès: Approximating the Pareto curve with local search for the bi-criteria TSP(1,2) problem. in Proc. of FCT'2003, LNCS 2751, 39-48, 2003.
2. E. Angel, E. Bampis, L. Gourvès and J. Monnot: (Non)-Approximability for the multi-criteria $TSP(1, 2)$. Technical Report, Université d'Évry Val d'Essonne, N°116-2005, 2005.
3. X. Deng, C.H. Papadimitriou and S. Safra: On the Complexity of Equilibria. in Proc. of STOC'2002, 67–71, 2002.
4. L. Engebretsen: An Explicit Lower Bound for TSP with Distances One and Two. Algorithmica, 35(4), 301-318, 2003.
5. D. Fotakis and P. Spirakis: A Hamiltonian Approach to the Assignment of Non-Reusable Frequencies. in Proc. of FCS&TCS'98, LNCS 1530, 18–29, 1998.
6. D.S. Johnson and C.H. Papadimitriou: Performance guarantees for heuristics. chapter in The Traveling Salesman Problem: a guided tour of Combinatorial Optimization, E.L Lawler, J.K. Lenstra, A.H.G. Rinnooy Kan and D.B. Shmoys (eds.), Wiley Chichester, 145–180, 1985.
7. D.E. Lucas: Récréations mathématiques. Vol. II, Gauthier Villars, Paris 1892.
8. C.H. Papadimitriou and M. Yannakakis: On the approximability of trade-offs and optimal access of web sources. in Proc. of FOCS'2000, 86–92, 2000.
9. C.H. Papadimitriou and M. Yannakakis: The traveling salesman problem with distances one and two. Mathematics of Operations Research, 18(1), 1–11, 1993.
10. D.J. Rosenkrantz, R.E. Stearns and P.M. Lewis II: An analysis of several heuristics for the traveling salesman problem. SIAM J. Comp., 6, 563–581, 1977.
11. A. Warburton: Approximation of Pareto optima in multiple-objective shortest path problems. Operations Research, 35(1), 70–79, 1987.

Completeness and Compactness
of Quantitative Domains

Paweł Waszkiewicz

Jagiellonian University, ul. Nawojki 11,
30-072 Kraków, Poland
pqw@ii.uj.edu.pl
http://www.ii.uj.edu.pl/~pqw

Abstract. In this paper we study the interplay between metric and order completeness of semantic domains equipped with generalised distances. We prove that for bounded complete posets directed-complete-ness and partial metric completeness are interdefinable. Moreover, we demonstrate that Lawson-compact, countably based domains are precisely the compact pmetric spaces that are continuously ordered.

1 Introduction

At the heart of Scott's denotational semantics for a, say, typed functional programming language, lies the idea that types are represented (denoted) by certain partially ordered and topologised sets, called domains, and terms are denoted by continuous functions between domains; in particular, the closed terms (programs) are then represented by elements of domains. The success of denotational semantics stems from the fact that it is able to attach a precise mathematical meaning to various involved syntactic constructs; questions about programs are then changed into conjectures about order properties of domains. For example, semantic equivalence of two syntactically different programs can be decided by showing that their denotations are exactly the same elements of domains. However, in classical domain theory some questions – the quantitative questions about programs – cannot be asked, simply because domains are *just* partial orders with no more refined ways of comparison between points.

The goal of *quantitative domain theory* (QDT) is to equip semantic domains with a distance and create denotational semantics that would allow to reason about quantitative properties of programs such as their complexity or speed of convergence. Research in this area focused on attempts in unifying order and metric structures and led to a number of concepts that generalise the notion of metric space to an ordered setting. Examples include: distances with values in spaces more general than the real line, so called $\mathcal{V}$-continuity spaces [FK97], and various refinements of these; quasimetrics, which by virtue of being asymmetric encode topology and order at the same time [Smy88]; partial metrics [Mat94], [Hec99]; measurements [Mar00a] and many more. The diversity of approaches and proposals without virtually any attempt towards creating a quantitative

M. Liśkiewicz and R. Reischuk (Eds.): FCT 2005, LNCS 3623, pp. 341–351, 2005.

denotational semantics caused QDT to be an esoteric and mostly unpopular research activity among computer scientists. Amusingly, any approach to QDT offers rich theory and challenging foundational questions that make some general topologists vitally interested.

On the way to understanding the menagerie of QDT, in [Was01], [Was03a], [Was03] we have unified the theories of partial metrics and measurements on domains. Since every partially metrized domain induces a measurement and a metric in the canonical way, one can regard domains as either partial metric spaces or metric spaces, or spaces equipped with a measurement. This point of view on domains is taken as the starting point of our paper, where we hope to take the second step and apply the unification results to study the interplay between various aspects of completeness and compactness that arise on a quantitative domain from both its order and metric structures. Our paper is a sequel to [Was]. We present a summary of relevant results from [Was] in the next section. In Section 3 main results of our present paper are stated: In short, we show that once one defines a distance on a bounded complete domain in an appropriate way, then the order properties and completeness (compactness) of the distance are fundamentally tied together. This facts demonstrate that partial metrization is a very natural way of equipping domains with the notion of distance.

Our paper is aimed at specialists in domain theory and/or at topologists interested in weakly separated spaces. These may find our general results useful and engaging, the others left bored to death. The tale to be read here is about how to speak about distance in the ordered setting – a sheer aesthetic pleasure of understanding this general problem rather than a quest for computer-oriented applications is our only motivation at the moment.

2   Background

We now give a brief review of basic definitions and results from domain theory and theories of partial metrics and measurements that are needed for our paper. Despite being elementary, this part of the paper is fairly concise and technical, and therefore can be hard to digest. It is thought as a quick reference; for a more comprehensive treatment of these issues, we refer the reader to the quoted literature.

2.1   Domain Theory

Our primary references in domain theory are [AJ94] together with [Gie03]. Let P be a poset. A subset $A \subseteq P$ of P is *directed* if it is nonempty and any pair of elements of A has an upper bound in A. If a directed set A has a supremum, it is denoted $\bigsqcup^\uparrow A$. A poset P in which every directed set has a supremum is called a *dcpo*. A dcpo P in which every nonempty subset has an infimum is *bounded complete*. We say that $x \in P$ *approximates* (*is way-below*) $y \in P$, and write $x \ll y$, if for all directed subsets A of P with $\bigsqcup^\uparrow A \in P$, $y \sqsubseteq \bigsqcup^\uparrow A$ implies $x \sqsubseteq a$ for some $a \in A$. Now, $\downarrow x$ is the set of all approximants of x and $\uparrow x$ is the set of all elements that x approximates. We say that a subset B of a poset P

is a (*domain-theoretic*) *basis* for P if for every element x of P, the set $\downarrow x \cap B$ is directed with supremum x. A poset is called *continuous* if it has a basis. A poset P is continuous if and only if $\downarrow x$ is directed with supremum x, for all $x \in P$. A poset is called a *domain* if it is a continuous dcpo. If a poset admits a countable basis, we say that it is ω-continuous, or countably based. The poset of nonnegative real numbers in the order $\sqsubseteq$ opposite to the natural one $\leq$ is denoted $[0, \infty)^o$ and is a domain without a least element.

A subset $U \subseteq P$ of a poset P is *upper* if $x \sqsupseteq y \in U$ implies $x \in U$. If for all directed sets D, $\bigsqcup^{\uparrow} D \in U \Rightarrow \exists d \in D$ $(d \in U)$, then U is called inaccessible by directed suprema. Upper sets inaccessible by directed suprema form a topology called the *Scott topology*; it is usually denoted $\sigma(P)$ here. A function $f \colon P \to Q$ between posets is Scott-continuous if and only if it preserves the order and existing suprema of directed subsets. The collection $\{\uparrow x \mid x \in P\}$ forms a basis for the Scott topology on a continuous poset P. The topology satisfies only weak separation axioms: It is always T_0 on a poset but T_1 only if the order is trivial. For an introduction to T_0 spaces, see [Hec90]. An excellent general reference on Topology is [Eng89]. The *weak topology* on a poset P, $\omega(P)$, is generated by a subbasis $\{P \setminus \uparrow x \mid x \in P\}$ and the *Lawson topology* $\lambda(P)$ is the join $\sigma(P) \vee \omega(P)$ in the lattice of all topologies on P.

2.2 Measurements

We will now give a brief summary of the main elements of Keye Martin's theory of measurements. Our main reference is [Mar00]. Let P be a poset. For a monotone mapping $\mu \colon P \to [0, \infty)^o$ and $A \subseteq P$ we define

$$\mu(x, \varepsilon) = \{y \in P \mid y \sqsubseteq x \wedge \mu y < \mu x + \varepsilon\}.$$

We say that $\mu(x, \varepsilon)$ is the set of elements of P which are ε-*close to* x. We say that μ *measures* P (or: μ *is a measurement on* P) if μ is Scott-continuous and

$$(\forall U \in \sigma)(\forall x \in U)(\exists \varepsilon > 0)\ \mu(x, \varepsilon) \subseteq U.$$

A measurement μ is *weakly modular* if for all $a, b, r \in P$ and for all $\varepsilon > 0$, if $a, b \sqsubseteq r$, then there exists $s \in P$ with $s \sqsubseteq a, b$ such that $\mu r + \mu s \leq \mu a + \mu b + \varepsilon$.

Any measurement $\mu \colon P \to [0, \infty)^o$ on a continuous poset P with a least element induces a distance function $p_\mu \colon P \times P \to [0, \infty)$ by $p_\mu(x, y) = \inf\{\mu z \mid z \ll x, y\}$. The map p_μ, when considered with codomain $[0, \infty)^o$, is Scott-continuous, encodes the order on the poset P: $x \sqsubseteq y$ if and only if $p_\mu(x, y) = p(x, x)$, and the Scott topology: the collection of p_μ-balls $\{B_{p_\mu}(x, \varepsilon) \mid x \in P, \varepsilon > 0\}$, where $B_{p_\mu}(x, \varepsilon) = \{y \in P \mid p_\mu(x, y) < \mu x + \varepsilon\}$, is a basis for the Scott topology on P. If the measurement μ is weakly modular, then the induced distance p_μ is a partial metric [Was01].

2.3 Partial Metrics

A *partial metric* (*pmetric*) on a set X is a map $p \colon X \times X \to [0, \infty)$ governed by the following axioms due to Steve Matthews [Mat94]:

- *small self-distances:* $p(x, x) \leq p(x, y)$,
- *symmetry:* $p(x, y) = p(y, x)$,
- T_0 *separation:* $p(x, y) = p(x, x) = p(y, y) \Rightarrow x = y$,
- *the sharp triangle inequality,* $\Delta^{\sharp}$: $p(x, y) \leq p(x, z) + p(z, y) - p(z, z)$.

that hold for all $x, y, z \in X$. The topology induced by p on X, denoted $\tau_p(X)$, is given by: $U \in \tau_p(X)$ iff for all $x \in U$ there exists $\varepsilon > 0$ such that $\{y \in X \mid p(x, y) < p(x, x) + \varepsilon\} \subseteq U$. Note that if the last inclusion is changed to $\{y \in X \mid p(x, y) < p(y, y) + \varepsilon\} \subseteq U$, then we are defining some other topology (different than $\tau_p(X)$ in general) called the *dual topology* and denoted $\tau_p^o(X)$. Every pmetric induces a metric $d_p \colon X \times X \to [0, \infty)$ by $d_p(x, y) = 2p(x, y) - p(x, x) - p(y, y)$. Its topology is denoted by $\tau_{d_p}(X)$. Finally, every pmetric induces a partial order by $x \sqsubseteq_p y$ if and only if $p(x, y) = p(x, x)$. If a poset $(P, \sqsubseteq)$ admits a pmetric $p \colon P \times P \to [0, \infty)$ with $\sqsubseteq = \sqsubseteq_p$, then we say that p is compatible with the order on P. If, in addition, $\tau_p(P)$-open sets are inaccessible by directed suprema, then we call p *order-consistent*.

2.4 Exactly Radially Convex Metrics on Posets

A metric $d \colon P \times P \to [0, \infty)$ on a poset P is *exactly radially convex (erc)* providing that $x \sqsubseteq y \sqsubseteq z$ if and only if $d(x, y) + d(y, z) = d(x, z)$. Clearly, if P has a least element $\bot$, then $x \sqsubseteq y$ if and only if $d(x, y) = d(\bot, y) - d(\bot, x)$. For example, let the product of ω copies of the unit interval, $\mathbb{I}^\omega$, be equipped with a metric $d(x, y) = \sum_{i \in \omega} 2^{-i} |x_i - y_i|$, where $x = (x_0, x_1, ...)$, $y = (y_0, y_1, ...)$. The metric d is erc with respect to the coordinatewise order.

From now on, we assume that P is a poset with a least element $\bot$. If $d \colon P \times P \to [0, \infty)$ is a bounded erc metric, then the map $p_d \colon P \times P \to [0, \infty)$ given by $p_d(x, y) = d(x, y) + \mu_d x + \mu_d y$, where $\mu_d \colon P \to [0, \infty)^o$ is defined as $\mu_d x = \sup\{d(\bot, z) \mid z \in P\} - d(\bot, x)$, is a pmetric (called the *induced partial metric*) compatible with the order, with the self-distance $2\mu_d$ [Was]. Moreover:

- If either $\tau_d(P) = \lambda(P)$, or $\sigma(P) \subseteq \tau_d(P)$, then μ_d has the measurement property: for all $x \in P$, $U \in \sigma(P)$, if $x \in U$, then there exists $\varepsilon > 0$ with $\mu_d(x, \varepsilon) \subseteq U$.
- The inclusion $\tau_{p_d}(P) \subseteq \sigma(P)$ is equivalent to Scott-continuity of μ_d, which, in turn, is equivalent to p_d being order-consistent.

Conversely, for every pmetric $p \colon P \times P \to [0, 1)$ which is compatible with the order, the induced metric $d_p \colon P \times P \to [0, \infty)$ is bounded, erc and satisfies $d_p(\bot, x) = \mu_p \bot - \mu x_p$ for every $x \in P$, where μ_p is the self-distance of p. (It is always possible to assume that the codomain of p is bounded by 1 by results of [KV94].)

3   Main Results

In the previous section we have observed that for partial orders existence of bounded erc metrics is determined by existence of partial metrics compatible

Table 1. Domains with their partial metrics

Domain	Pmetric p
Closed, nonempty intervals of $\mathbb{R}$ with reverse inclusion	$p([a,b],[c,d]) = \max\{b,d\} - \min\{a,c\}$
The powerset of natural numbers with inclusion	$p(x,y) = \sum_{n \notin x \cap y} 2^{-(n+1)}$
Words over $\{0,1\}$, prefix order	length of common prefix
Plotkin's $\mathbb{T}^\omega$ $\{(P,N) \subseteq \omega \times \omega \mid P \cap N = \emptyset\}$ with coordinatewise inclusion	$p((A,B),(C,D)) = \sum_{n \notin (A \cap C) \cup (B \cap D)} 2^{-(n+1)}$
Any ω-continuous poset	$p(x,y) = 1 - \sum_{\{n \mid x,y \in U_n\}} 2^{-(n+1)}$ $\{U_n \mid n \in \omega\}$ - basis of open filters

with the order and *vice versa*. Therefore in our presentation of quantitative domains we can choose to work with the distances that are most suitable for current purposes, usually assuming the existence of both at the same time.

Let us discuss the relevance of assumptions that are made in the statements of the main results below: We often assume that a pmetric p on a continuous poset P is compatible with the order and generates topology weaker than the Scott topology, that is $\tau_p(P) \subseteq \sigma(P)$. This guarantees that the partial metric is order-consistent and its self-distance $\mu \colon P \to [0,\infty)^o$ is Scott-continuous.

Additionally, we assume that the Scott topology of P is weaker than the metric topology, $\sigma(P) \subseteq \tau_{d_p}(P)$, so the map μ is a measurement. This situation is very common, examples include all domains where the partial metric topology *is* the Scott topology, see Table 1.

It is important to realise that the assumptions made above are by no means restrictive: the statement $\tau_p(P) \subseteq \sigma(P) \subseteq \tau_{d_p}(P)$ for p compatible with the order, is a minimal requirement for a partial metric that would describe existing order properties of the underlying poset.

Let us start with an easy observation of how a partial metric encodes countability of the order:

Lemma 1. *If P is a domain equipped with a partial metric compatible with the order and such that $\tau_p(P) \subseteq \sigma(P) \subseteq \tau_{d_p}(P)$, then the following are equivalent:*

1. *P is ω-continuous,*
2. *$\sigma(P)$ is second-countable,*
3. *$\tau_{d_p}(P)$ is second-countable.*

Proof. The equivalence of 1 and 2 is observed in [AJ94]. For 2$\Rightarrow$3 note that $\tau_{d_p}(P)$ is separable, hence second-countable. The remaining implication follows immediately from the fact that $\sigma(P)$ is coarser that $\tau_{d_p}(P)$.

3.1   Completeness

In order to motivate suitable definition of completeness for partial metric spaces, observe that these spaces are examples of *syntopologies* and as such are *com-*

pletable in the sense of M.B. Smyth who gives an appropriate construction in [Smy94]. This completion is nowadays known as *S-completion* or *Smyth-completion*. In [Kün93] Künzi noted that in fact each partial metric induces a Smyth-completable quasi-uniform space. Subsequently, Sünderhauf [Sün95] showed that for S-completable spaces, Smyth-completion reduces to a so called *bicompletion*. In order to avoid introducing unnecessary terminology, we just note that in our setting this means precisely that the Smyth-completion of a partial metric space is the usual completion of the induced metric and hence a pmetric $p\colon P \times P \to [0, \infty)$ is complete if its induced metric d_p is complete. It follows from metric space theory that such a completion always exists, see for example [Wil70], Theorem 24.4, page 176.

In this paper we are interested in the relationship between completeness of the induced metric and the structure of the underlying domain. First of the main results states that existence of a complete partial metric on a continuous poset P implies directed-completeness of the order.

Theorem 1. *Let* $p\colon P \times P \to [0, \infty)$ *be a pmetric on a continuous poset* P *with a least element* $\perp$ *which is compatible with the order and such that* $\tau_p(P) \subseteq \sigma(P) \subseteq \tau_{d_p}(P)$. *If* p *is complete, then* P *is a dcpo.*

Proof. Every directed subset of P contains an increasing sequence with the same upper bounds [Hec99]. Let $\{x_n\}$ be an ω-chain, increasing. Then the sequence $\{\mu_p x_n\}$ in $[0, \infty)$ is monotone, decreasing, bounded by 0. Hence it is convergent and Cauchy. Therefore $\lim d_p(x_n, x_{n+1}) = \lim(\mu_p x_n - \mu_p x_{n+1}) = 0$, which proves that $\{x_n\}$ is Cauchy. But d_p is complete, so there is $z \in P$ with $\lim d_p(x_n, z) = 0$. We claim that $z = \bigsqcup^\uparrow x_n$.

Clearly, for all $n \in \omega$, $\{x_n\} \to_{\tau_p(P)} x_n$, so every $\tau_p(P)$-open set containing element x_n, contains the sequence $\{x_n\}$ cofinally. Now, we will show that if $\{x_n\} \to_{\tau_p(P)} a \in P$, then $a \sqsubseteq z$. For any $\varepsilon > 0$ there is $N \in \omega$ such that for all $n \geq N$ we have $p(a, x_n) < p(a, a) + \varepsilon/2$ and $d_p(x_n, z) < \varepsilon/2$. Hence for $n \geq N$, $p(a, a) \leq p(a, z) \leq p(a, x_n) + p(x_n, z) - p(x_n, x_n) \leq p(a, x_n) + d_p(x_n, z) < p(a, a) + \varepsilon$. Thus $p(a, a) = p(a, z)$, which gives $a \sqsubseteq_p z$, that is, $a \sqsubseteq z$. We conclude that $x_n \sqsubseteq z$.

Suppose that $c \in P$ is another upper bound for $\{x_n\}$. Then $p(z, z) \leq p(z, c) \leq p(z, x_n) + p(x_n, c) - p(x_n, x_n) = d(z, x_n) + p(z, z) \to_{n \to \infty} p(z, z)$. Hence $p(z, c) = p(z, z)$, which proves that $z \sqsubseteq c$ and consequently $\bigsqcup^\uparrow x_n = z$.

For bounded complete domains we have a sort of converse as well: order-completeness of the domain implies that the erc metric topology $\tau_{d_p}(P)$ is completely metrizable, which is our second main result. In order to prove it, we need to recall a result from general topology: A Tychonoff space X is Čech-complete if and only if there is a countable collection $\{\mathcal{R}_n \mid n \in \omega\}$ of open covers of X, which has the following property: for any collection $\mathcal{F}$ of closed sets of X with the finite intersection property and such that

$$\forall k \in \omega.\ \exists F_k \in \mathcal{F}.\ \exists U_k \in \mathcal{R}_k.\ F_k \subseteq U_k,$$

we have $\bigcap \mathcal{F} \neq \emptyset$.

Theorem 2. *Let P be a bounded complete domain equipped with a partial metric $p \colon P \times P \to [0, \infty)$ which is compatible with the order and such that $\tau_p(P) \subseteq \sigma(P) \subseteq \tau_{d_p}(P)$. Then the induced metric topology is completely metrizable.*

Proof. In order to show complete metrizability of $\tau_{d_p}(P)$, it is enough to prove its Čech-completeness. Let $\{\mathcal{G}_n \mid n \in \omega\}$ be a development for $\tau_{d_p}(P)$ (meaning by definition that $\{\mathcal{G}_n \mid n \in \omega\}$ is a sequence of $\tau_{d_p}(P)$-open covers of P such that if $x \in U \in \tau_{d_p}(P)$, then there exists $n_0 \in \omega$ with $x \in \mathbf{St}(x, \mathcal{G}_{n_0}) \subseteq U$, where $\mathbf{St}(x, \mathcal{G}_{n_0}) = \bigcup\{V \in \mathcal{G}_{n_0} \mid x \in V\}$). Let $\{A_i \mid i \in I\}$ be a basis for the dual topology $\tau_p^o(P)$ and observe that it consists of Scott-closed sets. Define a countable collection of $\tau_{d_p}(P)$-open covers $\{\mathcal{R}_n \mid n \in \omega\}$ of P by

$$\mathcal{R}_n := \{\, \Uparrow b \cap A_i \mid \exists U \in \mathcal{G}_n. \; \Uparrow b \cap A_i \subseteq U \text{ for some } b \in P \text{ and } i \in I \,\}.$$

The family $\{\mathcal{R}_n\}$ is well-defined due to the fact that the sets of the form $\Uparrow b \cap A_i$ form a basis for $\tau_{d_p}(P)$. Indeed, let $U \in \tau_{d_p}(P)$ and $x \in U$. Let $\{B_j \mid j \in J\}$ be a basis for $\tau_p(P)$. Since $\tau_{d_p}(P) = \tau_p(P) \vee \tau_p^o(P)$, the collection $\{A_i \cap B_j \mid i \in I, j \in J\}$ is a basis for $\tau_{d_p}(P)$. This means that there are $i \in I, j \in J$ such that $x \in A_i \cap B_j \subseteq U$. However $B_j \in \sigma(P)$, thus there exists $b \in P$ with $x \in \Uparrow b \subseteq B_j$. One observes that $\Uparrow b \in \tau_{d_p}(P)$ by assumption, and hence $\Uparrow b \cap A_i \in \tau_{d_p}(P)$. This proves the claim.

Let $\mathcal{F}$ be a collection of $\tau_{d_p}(P)$-closed sets with the finite intersection property such that

$$\forall k \in \omega. \; \exists F_k \in \mathcal{F}. \; \exists R \in \mathcal{R}_k. \; F_k \subseteq R. \tag{1}$$

We must demonstrate that $\bigcap \mathcal{F} \neq \emptyset$. Define for every $n \in \omega$ a set $C_n = \bigcap_{k=1}^{n} F_k$ and choose $x_n \in C_n$. By formula (1) and definition of $\{R_n\}$, we have

$$\forall n \in \omega. \; x_n \in C_n \subseteq \Uparrow b_n \cap A_n \subseteq U_n, \tag{2}$$

where $U_n \in \mathcal{G}_n$. Next, define $y_n = \bigcap_{i=n}^{\infty} x_i$. Since the sequence $\{C_n\}$ is descending,

$$\forall n. \; \forall k \geq n. \; x_k \in \Uparrow b_n \cap A_n$$

and so $b_n \sqsubseteq y_k \sqsubseteq x_k \in A_n$. Since each A_n is downward-closed,

$$\forall n. \; \forall k \geq n. \; y_k \in \Uparrow b_n \cap A_n. \tag{3}$$

Now, since P is a dcpo, the lub of the chain $\{y_n\}$ exists. Put $z = \bigsqcup^{\uparrow} y_n$. By formula (3) and the fact that each A_n is closed under directed suprema,

$$\forall n. \; \forall k \geq n. \; z \in \Uparrow b_n \cap A_n$$

and hence by formula (2) we have

$$\forall n \in \omega. \; z \in U_n \in \mathcal{G}_n. \tag{4}$$

We claim that $\{x_n\} \to_{\tau_{d_p}(P)} z$. Let V be any $\tau_{d_p}(P)$-open set around z. Then there exists $m \in \omega$ such that $z \in \mathbf{St}(z, \mathcal{G}_m) \subseteq V$. Consequently, by (2),(3) and (4),

$$\forall k \geq m. \; x_k \in C_m \subseteq U_m \subseteq \mathbf{St}(z, \mathcal{G}_m) \subseteq V,$$

and this proves that $\{x_n\} \to_{\tau_{d_p}(P)} z$.

Since each set C_n is $\tau_{d_p}(P)$-closed and the sequence $\{x_n\}$ is cofinally in each of C_n, we infer that $z \in C_n$ and so $z \in F_n$ for all $n \in \omega$. Since $\{\mathbf{St}(z, \mathcal{G}_n)\}_{n \in \omega}$ is a basis at z and $\tau_{d_p}(P)$ is Hausdorff,

$$\{z\} \subseteq \bigcap_n F_n \subseteq \bigcap_n \mathbf{St}(z, \mathcal{G}_n) = \{z\}.$$

Therefore, $\bigcap_n F_n = \{z\}$.

Now, for any $F \in \mathcal{F}$ we have that $\{F \cap F_n \mid n \in \omega\}$ is again a nonempty sequence of $\tau_{d_p}(P)$-closed sets satisfying equation (1). Hence, the same argument proves that its intersection is nonempty. We have

$$\emptyset \neq \bigcap_n (F \cap F_n) = F \cap \bigcap_n F_n = F \cap \{z\}$$

and then $z \in F$. We conclude that $z \in \bigcap \mathcal{F}$. That is, $\bigcap \mathcal{F} \neq \emptyset$, as required.

Finally, we describe an important special case where completeness of a partial metric (equivalently: completeness of the induced erc metric) can be detected by inspecting properties of the self-distance of the pmetric.

Theorem 3. *Let P be a bounded complete domain equipped with a partial metric $p: P \times P \to [0, \infty)$ which is compatible with the order and such that $\tau_p(P) \subseteq \sigma(P) \subseteq \tau_{d_p}(P)$ and such that the self-distance $\mu_p: P \to [0, \infty)^\circ$ of p is weakly modular. The following are equivalent:*

1. *The partial metric p is complete,*
2. *The self-distance μ_p preserves infima of filtered subsets of P.*

Proof. Before embarking on the proof of the equivalence in the hypothesis, observe that since μ_p is weakly modular, p is of the form $p(x, y) = \inf\{\mu_p z \mid z \sqsubseteq x, y\}$ for all $x, y \in P$ [Was01]. Moreover, the fact that μ_p is a measurement implies that the partial metric topology $\tau_p(P)$ is the same as the Scott topology of P.

For (2)$\Rightarrow$(1), let (x_n) be a Cauchy sequence in P with respect to the induced metric d_p. Define $y_n = \bigsqcap_{i=n+1}^{\infty} x_i$. Since P is a dcpo, the chain $\{y_n\}$ has a lub, say, $z = \bigsqcup^\uparrow y_n$. We will show that z is the metric limit of $\{x_n\}$. Since the sequence $\{x_n\}$ is Cauchy, for all $\varepsilon > 0$ there is $n_0 \in \omega$ such that for all $n \geq n_0$ we have $d_p(x_n, x_{n+1}) < \varepsilon$. The special form of p yields $\mu_p(x_n \sqcap x_{n+1}) - \mu_p x_n < \varepsilon$. Therefore $\inf_n \mu_p(x_n, x_{n+1}) = \inf_n \mu_p x_n$. It follows easily by induction that for a fixed $k \in \omega$ we have

$$\inf_n \mu_p \left(\bigsqcap_{i=n+1}^{n+1+k} x_i \right) = \mu_p x_n. \tag{5}$$

Now, we claim that

$$\lim_n p(z, x_n) = \mu_p z \tag{6}$$

and the limit on the left-hand side exists. For, $\mu_p z \leq p(z, x_{n+1}) \leq p(z, y_n) + p(y_n, x_{n+1}) - \mu_p y_n = p(y_n, x_{n+1}) = \mu_p y_n$ and since $\lim_n \mu_p y_n = \mu_p z$, we have $\lim_n p(z, x_n) = \mu_p z$.

Next, we calculate

$$\lim_n \mu_p y_n = \inf_n \mu_p(\bigcap_{i=n+1}^{\infty} x_i) = \inf_n \mu_p(\bigcap_{k \in \omega} \bigcap_{i=n+1}^{n+1+k} x_i)$$

$$= \inf_n \sup_k \mu_p(\bigcap_{i=n+1}^{n+1+k} x_i) = \sup_k \inf_n \mu_p(\bigcap_{i=n+1}^{n+1+k} x_i)$$

$$\overset{(5)}{=} \sup_k \inf_n \mu_p x_n = \inf_n \mu_p x_n.$$

Therefore, $\lim(\mu_p y_n - \mu_p x_n)$ exists and equals 0. Combining this observation with (5) we get $\lim(p(z, x_n) - \mu_p z) + \lim(p(z, x_n) - \mu_p y_n) + \lim(\mu_p y_n - \mu_p x_n) = 0$, which is equivalent to saying that $\lim_n d_p(z, x_n) = 0$.

The proof of (1)$\Rightarrow$(2) follows from Lemma 3.4 in [ONeill95]. We sketch the construction: let F be a filtered subset of P. By bounded completeness of P, F has infimum $x \in P$. Since $F \sqsupseteq x$, we have $\mu_p(F) \le \mu_p x$ and so $\alpha = \lim \mu_p(F)$ exists in $[0, \infty)$. Next, we find a descending chain $\{z_n\}$ in F with $\lim \mu_p z_n = \alpha$. Since $\{z_n\}$ is Cauchy, it has a metric limit $c \in P$. By showing that $c = x$ (using the sharp triangle inequality of p), we complete the proof.

3.2 Compactness

Similarly as in the definition of completeness, we define a partial metric space (X, p) to be *compact* if the induced metric topology $\tau_{d_p}(X)$ is compact. Existence of a compact pmetric on a poset determines much of its structure.

Theorem 4. *Let P be a continuous poset with a least element $\bot$. Suppose that P admits a compact pmetric $p \colon P \times P \to [0, \infty)$ which is compatible with the order and such that $\tau_p(P) \subseteq \sigma(P)$ and $\lambda(P) \subseteq \tau_{d_p}(P)$. Then:*

1. *P is a dcpo,*
2. *P is countably based,*
3. *P is Lawson-compact.*

Proof. Since every compact pmetric is complete, Theorem 1 applies and hence P is a dcpo. Moreover, $\tau_{d_p}(P)$ is second-countable, since it is a compact metric space. Lemma 1 implies that P is countably based. Suppose now that C is a closed subset of $\tau_{d_p}(P)$. Then it is compact and since by assumption $\lambda(P) \subseteq \tau_{d_p}(P)$, the subset C is compact, hence closed in the Lawson topology $\lambda(P)$. Therefore $\tau_{d_p}(P) = \lambda(P)$ and thus P is Lawson-compact.

Let us discuss some special cases of the theorem above. If the pmetric topology is the Scott topology on P, then the assumption $\lambda(P) \subseteq \tau_{d_p}(P)$ holds already and the conclusions of the theorem remain valid. (For example, if P is bounded complete and the self-distance of the compact pmetric is weakly modular, then $\tau_p(P) = \sigma(P)$ and the theorem above applies.) However, from results of [Was] we can infer more:

Corollary 1. *Let P be a continuous poset with a least element. The following are equivalent:*

1. *P is an ω-continuous, Lawson-compact dcpo,*
2. *P admits a compact pmetric $p\colon P \times P \to [0,\infty)$ compatible with the order with $\tau_p(P) \subseteq \sigma(P)$ and $\lambda(P) \subseteq \tau_{d_p}(P)$ and such that for all $x \in P$ we have $p(x,x) = 0$ if and only if x is maximal in P.*

Proof. The unproved direction, that is, the construction of the pmetric is described in detail in Theorem 6.5 of [Was].

References

[AJ94]   Abramsky, S. and Jung, A. (1994) Domain Theory. In S.Abramsky, D.M. Gabbay and T.S.E. Maibaum, editors, *Handbook of Logic in Computer Science* **3**, 1–168, Oxford University Press.

[Eng89]   Engelking, R. (1989) *General Topology.* Sigma Series in Pure Mathematics. Heldermann Verlag.

[FK97]   Flagg, R.C. and Kopperman, R. (1997) Continuity spaces: reconciling domains and metric spaces. *Theoretical Computer Science* **177** (1), 111–138.

[Gie03]   Gierz, G., Hofmann, K.H., Keimel, K., Lawson, J.D., Mislove, M. and Scott, D.S. (2003) Continuous lattices and domains. Volume 93 of *Encyclopedia of mathematics and its applications*, Cambridge University Press.

[Hec90]   Heckmann, R. (1990) Power Domain Constructions (Potenzbereich-Konstruktionen). PhD Thesis, Universität des Saarlandes.

[Hec99]   Heckmann, R. (1999) Approximation of metric spaces by partial metric spaces. *Applied Categorical Structures* **7**, 71–83.

[Kün93]   Künzi, H.-P. (1993) Nonsymmetric topology. In *Bolyai Society of Mathematical Studies*, **4**, 303–338, Szekszárd, Hungary (Budapest 1995).

[KV94]   Künzi, H.-P. and Vajner, V. (1994) Weighted quasi-metrics. *Papers on General Topology and Applications: Proceedings of the Eighth Summer Conference on General Topology and Its Applications.* In *Ann. New York Acad. Sci.* **728**, 64–77.

[Mar00]   Martin, K. (2000) A foundation for computation. PhD Thesis, Tulane University, New Orleans LA 70118.

[Mar00a]   Martin, K. (2000) The measurement process in domain theory. Proceedings of the 27th ICALP. In *Lecture Notes in Computer Science* **1853**, Springer-Verlag.

[Mat94]   Matthews, S.G.. (1994) Partial metric topology. *Papers on General Topology and Applications: Proceedings of the Eighth Summer Conference on General Topology and Its Applications.* In *Ann. New York Acad. Sci.* **728**, 64–77.

[ONeill95] O'Neill, S.J. (1995) Partial metrics, valuations and domain theory. Research Report CS-RR-293, Department of Computer Science, University of Warwick, Coventry, UK.

[Smy88]   Smyth, M.B. (1988) Quasi-uniformities: reconciling domains and metric spaces. *Lecture Notes in Computer Science*, **298**, 236–253.

[Smy94]   Smyth, M.B. (1994) Completeness of quasi-uniform and syntopological spaces. *Journal of the London Mathematical Society*, **49**, 385–400.

[Sün95]   Sünderhauf, P. (1995) Quasi-uniform completeness in terms of Cauchy nets. *Acta Mathematica Hungarica*, **69**, 47–54.

[Was01]   Waszkiewicz, P. (2001) Distance and measurement in domain theory. In S. Brookes and M. Mislove, editors, 17th Conference on the Mathematical Foundations of Programming Semantics, volume **45** of *Electronic Notes in Theoretical Computer Science*, Elsevier Science Publishers.

[Was03]   Waszkiewicz, P. (2003) Quantitative continuous domains. *Applied Categorical Structures* **11**, 41–67.

[Was03a]   Waszkiewicz, P. (2003a) The local triangle axiom in topology and domain theory. *Applied General Topology* **4** (1), 47–70.

[Was]   Waszkiewicz, P. Partial Metrizability of Continuous Posets. *Submitted* Available at: http://www.ii.uj.edu.pl/~ pqw

[Wil70]   Willard, S. (1970) General Topology. Addison-Wesley.

A Self-dependency Constraint in the Simply Typed Lambda Calculus

Aleksy Schubert*

Institute of Informatics, Warsaw University,
ul. Banacha 2, 02–097 Warsaw, Poland

Abstract. We consider a class of terms in the simply typed lambda calculus in which initially, during reduction, and for the potential terms it is never allowed that a variable x is applied to an argument containing x. The terms form a wide class which includes linear terms. We show that corresponding variant of the dual interpolation problem (i.e. the problem in which all expressions can be restricted to terms of this kind) is decidable. Thus the model for this kind of expressions can admit fully abstract semantics and the higher-order matching problem is decidable.

Classification: Semantics, logic in computer science.

1 Introduction

The higher-order matching problem consists in solving certain equations in the simply typed λ-calculus. The equations have the form $M \doteq N$ where the unknowns occur only in the term M. The higher-order matching problem can be applied in automated theorem proving (see e.g. [Har96]) as a special case of the higher-order unification. The higher-order matching problem has also connections with semantics of the simply typed λ-calculus [Sta82]. It is strongly connected with the problem of λ-definability. The problem of λ-definability is to decide whether a given function from a function space over a finite base can be defined by a λ-term. The problem of λ-definability is undecidable which corresponds to the fact that there is no way to effectively describe a fully-abstract model for the simply typed λ-calculus [Loa01]. The higher-order matching problem defines a stronger notion of definability the decidability of which is still unknown. Semantical investigations in simply typed lambda calculus have some practical impact as they are strongly connected to the area of programme transformations (see e.g. [dMS01]). In particular, a research like this can lead to procedures that find programme transformations for certain kinds of programmes in a complete way.

The higher-order matching problem has a long history. The problem was posed by G. Huet in his PhD thesis [Hue76]. In the last years. V. Padovani presented a new approach to higher-order matching. He proved that higher-order

---

* This work was partly supported by KBN grant 3 T11C 002 27; part of the work was conducted while the author was a winner of the fellowship of The Foundation for Polish Science (FNP).

M. Liśkiewicz and R. Reischuk (Eds.): FCT 2005, LNCS 3623, pp. 352–364, 2005.

matching is decidable without restriction on the order but with a restriction on the term N — it must be a single constant of a base type [Pad95a]. In addition, R. Loader presented [Loa03] the undecidability of unrestricted higher-order matching, but for β-equality (the standard formulation requires $\beta\eta$-equality). This was supplemented by results that take into account restrictions on domains of solutions [dG00, DW02, SSS02, SS03].

This paper expands significantly the technique used in [Pad95a] and proposes a new interesting class of terms. Informally, a term M is self-independent when in each self-independent context and for each subterm N of the term if the evaluation of the subterm N is executed somewhere in the reduction then all its arguments may not contain a copy of N. For example the term

$$\lambda x : (\alpha \to \alpha) \to \alpha \to \alpha.x(\lambda z_1.fz_1z_1)a,$$

where f is a constant of the type $\iota \to \iota \to \iota$ and a is a constant of the type ι, is self-independent since the only self-independent terms that can be substituted for x are the identity, and the constant term and these can only result in forgetting of the most of the term or in copying a to z_1. This class contains linear terms and, as the example shows, is wider.

The main result of the paper is the decidability for the higher-order matching problem when the instances and solutions are restricted to self-independent terms. Schmidt-Schauß proposed a refined technique to solve the higher-order matching problem which is based on strictness analysis [SS99]. The proof in this paper is based on the earlier Padovani's approach which gives an explicit syntactical characterisation of the introduced class. A presentation of this kind is obscured when Schmidt-Schauß's technique is used. What is more, such an explicit presentation eases potential implementation since it shows the structure of terms to be find.

In sections 2 and 3, the paper defines the basic notions. In the section 4, the main technical development is sketched. The section 5 presents a sketch of the decidability proof. The paper is concluded with a discussion in the section 6.

2 Preliminaries

Assume we are given a set $\mathcal{B}$ of *base types*. Let $\mathcal{T}_\mathcal{B}$ be the set of all *simple types over* $\mathcal{B}$ defined as the smallest set containing $\mathcal{B}$ and closed on $\to$. We shall often omit the subscript $\mathcal{B}$ if $\mathcal{B}$ is clear from the context or unimportant. In order to simplify the presentation, we limit ourselves here to the case where $\mathcal{B}$ is a singleton $\{\iota\}$. We conjecture that the case with many base types can be developed without much problem using the same technique.

The set of simply typed terms is defined, as usual, based on the set of pre-terms. The set $\lambda^*_\to$ of *simply typed pre-terms* contains an infinite, countable set of variables V, a countable set of constants C (disjoint with V), and a map $T : V \cup C \to \mathcal{T}_\mathcal{B}$ that indicates types of symbols. We assume that there exists an infinite number of variables of a type α. This set is also closed on the application and λ-abstraction operations. We will usually write $s : \tau$ to denote the fact that a pre-term s has a type τ. As usual, we deal with α-equivalence. The symbol $[s]_\alpha$ denotes the α-equivalence class of s. The set $\lambda_\rightarrow$ is defined as a quotient of $\lambda_\rightarrow^*$ by α-equivalence. The elements of $\lambda_\rightarrow$ are denoted usually by $M, M_1, \ldots, N, N_1 \ldots$ etc. as well as the elements of $\lambda_\rightarrow^*$ are denoted by $s, s_1, \ldots, t, t_1, \ldots$ etc. The notion of a closed term is defined as usual, similarly the set $\mathrm{FV}(M)$ of free variables in M. The symbol Const_M denotes the set of constants occurring in M. We assume here that all the terms and pre-terms are properly typed according to the rules of the Church-style simply typed lambda calculus.

We denote by $T(\tau, C)$ the set of all closed terms of type τ built-up of constants from the set C. The notion of *sorted set* is of some usefulness here. The family of sets $T^C = \{T(\tau, C)\}_{\tau \in T_B}$ is an example of a sorted set. We adopt standard $\in$ notation to sorted sets.

The term Const_M denotes the set of constants that occur in M. This notation is extended to sets (and other structures) of terms.

The *order* of a type τ, denoted by $\mathrm{ord}(\tau)$ is defined inductively as
— $\mathrm{ord}(\iota) = 0$ for $\iota \in \mathcal{B}$;   — $\mathrm{ord}(\tau_1 \rightarrow \tau_2) = \max(\mathrm{ord}(\tau_1) + 1, \mathrm{ord}(\tau_2))$.
The notion of order extends to terms and pre-terms. We define $\mathrm{ord}(M) = \mathrm{ord}(\tau)$ or $\mathrm{ord}(t) = \mathrm{ord}(\tau)$, where τ is the type of M or t respectively.

The relation of $\beta\eta$-reduction is written $\rightarrow_{\beta\eta}^*$ its reflexive closure $=_{\beta\eta}$. The β-reduction is written $\rightarrow_\beta^*$. We write $M[x := N]$ to denote the substitution of N for x with usual renaming of bound variables. We sometimes write substitutions in the prefix mode as in $S(N)$. The β-*normal, η-long form* for a simply typed term M is a β-normal term M' such that $M =_{\beta\eta} M'$ and which cannot be η-expanded i.e. there is no simply typed M'' such that $M'' \rightarrow_\eta M'$.

We call *an instance of the higher-order matching problem* each pair of simply typed λ-terms $\langle M, N \rangle$. We usually denote them as $M \doteq N$. A *solution* of such an instance is a substitution S such that $S(M) =_{\beta\eta} N$. We often restrict ourselves to the case when N has no free variables. This restriction is not essential.

Definition 1. (the higher-order matching problem)
The higher-order matching problem is a decision problem — given an instance $M \doteq N$ of the higher-order matching problem whether there exists a solution of $M \doteq N$.

An *interpolation equation* is a pair of terms, usually written $xN_1 \cdots N_k \doteq N$, such that x is the only free variable in the left-hand side of the equation and ≥ 0.

Let $\langle E, E' \rangle$ be a pair of sets of interpolation equations such that there exists a variable x which occurs free in the left-hand side of each equation in $E \cup E'$. We call $\langle E, E' \rangle$ *an instance of the dual interpolation problem*. We also call $\langle E, E' \rangle$ a *dual set*.

A *solution* of such an instance is a term P such that for each equation $[xN_1 \cdots N_k \doteq N] \in E$ we have $PN_1 \cdots N_k =_{\beta\eta} N$, and for each equation $[xN_1 \cdots N_k \doteq N] \in E'$ we have $PN_1 \cdots N_k \neq_{\beta\eta} N$.

Definition 2. (the dual interpolation problem)
The dual interpolation problem is a decision problem — given an instance $\langle E, E' \rangle$ of the dual interpolation problem whether there exists a solution of $\langle E, E' \rangle$.

The dual interpolation problem and the higher-order matching problem are connected in the following way:

Theorem 1. *The problem of higher-order matching reduces to the dual interpolation problem.*

Proof. See [Pad95b, Pad96] or [Sch01].

2.1 Self-independent Terms

We define self-independent terms by means of a marked reduction:

$$(\lambda x.M)N \to_{\beta\#} M[x := N] \qquad (M)^{\#}N \to_{\beta\#} MN.$$

where the substitution $[x := N]$ is defined so that $(P)^{\#}[x := N] = (P[x := N])^{\#}$.

A *self-dependency* is a term of the form $(M)^{\#}N_1 \cdots N_k$ where either M or one of N_i contains the mark $\#$. For each term M we define the set $\mathtt{M}^{\#}(M)$ as

$$\mathtt{M}^{\#}(M) = \{C[x^{\#}N_1 \cdots N_k] \mid M = C[xN_1 \cdots N_k]\}$$

($C[\cdot]$ denotes a term with a single hole usually called context, substitution for $[\cdot]$ is not capture avoiding.)

All normal forms of a base type are self-independent. We say that a term M is self-independent if for each sequence $N_1, \ldots, N_n$ of self-independent terms the outermost-leftmost reduction of each term $M'N_1 \cdots N_n : \iota$ where ι is a base type and $M' \in \mathtt{M}^{\#}(M)$ does not include a term with a subterm being a self-dependency.

By a routine inductive argument we can obtain that all the linear terms are self-independent. Clearly, Church numerals greater than zero are not self-independent, because their normal forms are self-dependencies. The class of self-independent terms is wider, though. It contains for instance terms $\lambda xy.xyy$ (where y is a variable of the base type) and $\lambda y.fyy$ (where f is a constant of the type $\iota \to \iota \to \iota$).

Note that when the term $\lambda xy.xyy$ is applied to the term $\lambda v_1 v_2 u.v_1(v_2 u)$ then we obtain $\lambda yu.y(yu)$ which is a self-dependency. Thus we cannot restrict the self-dependency check to normal forms.

3   Tools and Definitions

This section contains definitions specific to the dual interpolation problem. The first two subsections present some notational conventions. The subsection 3.3 presents a notion that allows to decompose sets of dual interpolation equations and the subsection 3.4 presents a notion which identifies unnecessary parts of a hypothetical solution.

3.1 Addresses in Terms

Let t be a pre-term. A sequence of numbers γ *points out* to a subterm u when $\gamma = \varepsilon$ and u is the greatest subterm of t that does not start with λ or when $\gamma = i \cdot \gamma'$, and $t = \lambda x_1 \ldots x_m.u_0 u_1 \cdots u_n$ while γ' points out to u in u_i for $0 \le i \le n$. A sequence that points to a subterm in a term t is called *an address in t*. The prefix order on addresses is written as $\gamma \preceq \gamma'$. The strict version is denoted as $\gamma \prec \gamma'$. The set of all addresses in a pre-term t is written $\mathrm{Addr}(t)$.

A *graft* of a pre-term u in t at an address γ is a term t'

- equal to u when $\gamma = \varepsilon$, or
- equal to $\lambda x_1 \ldots x_m.u_0 u_1 \cdots u_i' \cdots u_n$ when $t = \lambda x_1 \ldots x_n.u_0 u_1 \cdots u_n$, u_i' is a graft of u in u_i at γ' with $\gamma = i \cdot \gamma'$.

The term t' is denoted by $t[\gamma \leftarrow u]$. Let c be a constant $t[c \leftarrow u]$ denotes the result of grafting of the term u at every occurrence of the constant c.

Let C be a set of fresh constants. We say that a pre-term u is a C-pruning of a pre-term t iff

- $u = t$, or
- $u = u'[\gamma \leftarrow c]$ where $c \in C$, γ is an address in u' and u' is a C-pruning of t.

Let t' and t'' be C-prunings of a term t. The prefix order extends to terms so that $t' \preceq t''$ iff $\mathrm{Addr}(t') \subseteq \mathrm{Addr}(t'')$. The above-mentioned notions extend to λ-terms, sets of λ-terms and sets of pre-terms in a standard way. We sometimes use natural numbers as constants in prunings.

3.2 Matrix Notation

Consider a set of equations $\{x\overline{\mathsf{M}}_1^1 \cdots \overline{\mathsf{M}}_n^1 \doteq W^1, \cdots, x\overline{\mathsf{M}}_1^m \cdots \overline{\mathsf{M}}_n^m \doteq W^m\}$. The terms $\overline{\mathsf{M}}_j^i$ form a matrix $\overline{\mathsf{M}}$. $\overline{\mathsf{M}}_i$ denotes the i-th column in $\overline{\mathsf{M}}$ while $\overline{\mathsf{M}}^j$ the j-th row. The terms $W_1, \ldots, W_m$ form a *result column*. We denote the set of equations as $[x\overline{\mathsf{M}} \doteq W]$. The notions $\mathrm{NFL}(M)$, Const_M extend to matrices, columns and rows in a natural way. The concatenation of rows and columns is defined as the concatenation of the respective sequences. Columns are denoted by V, W etc. while rows by R, Q etc. The terms *height* and *width* denote the number of elements in a column or in a row respectively and for a particular row R and column V are written as $|R|$ and $|V|$. If all the elements in a column V are the same then the column is *constant*. A *type of a column* V is the (unique) type of any of its elements. We say that a pair W, W' of columns is *constant* iff W is constant as a column and $W'^i \ne W^1$ for all i.

3.3 Approximations

The notions of approximation allow us to semantically characterise minimal building blocks of solutions for equations.

An approximation of a pair of columns W_1, W_2 in a dual set $\mathcal{E} = \langle [x\overline{\mathsf{M}}_1 \doteq W_1], [x\overline{\mathsf{M}}_2 \doteq W_2] \rangle$ of equations for the solution M is any pair of columns $\widetilde{W_1}, \widetilde{W_2}$ of the heights $|W_1|, |W_2|$ such that

- there exists an $\{a, b, 0, \ldots, l\}$-pruning M' of M such that for each equation $[x\overline{M}_{i,1}^k \cdots \overline{M}_{i,r}^k \doteq W_i^k] \in \mathcal{E}$ we have $\mathrm{NFL}(M'\overline{M}_{i,1}^k \cdots \overline{M1}_{i,r}^k) = \widetilde{W}_i^k$,

- for $c \in \{a, b\}$ if c occurs in $\widetilde{W}_j^i$ then $\widetilde{W}_j^i = c$,

- for each constant $c \in \mathbb{N}$ if some $\widetilde{W}_j^i = c$ then there exists $\gamma \succ \varepsilon$ and k, l such that $\gamma(\widetilde{W}_l^k) = c$ (we say c is *guarded*).

For the sake of notational convenience we assume that for each approximation, the set of numeric constants that occur in it forms an initial connected subset of $\mathbb{N}$, e.g. $\{0, 1, 2\}$ or $\{0, 1\}$.

Let $\widetilde{W}_1, \widetilde{W}_1'$ and $\widetilde{W}_2, \widetilde{W}_2'$ be approximations of a pair of columns W, W'. We write $\widetilde{W}_1, \widetilde{W}_1' \preceq \widetilde{W}_2, \widetilde{W}_2'$ when for each k we have $\widetilde{W}_1^k \preceq \widetilde{W}_2^k$ and $\widetilde{W}_2'^k \preceq \widetilde{W}_2'^k$. It is easily verified that $\trianglelefteq$ is a pre-order. Its strict version is denoted by $\triangleleft$.

A pair of columns W, W' is *trivial* iff for each i, j we have $W^i = W'^j = c$, where c is either a or b.

Let $\langle[x\overline{M} \doteq W], [x\overline{M'} \doteq W']\rangle$ be a dual set and M its solution. We say that $\widetilde{W}, \widetilde{W}'$ *is a minimal approximation* of W, W' iff there is no non-trivial $\widetilde{W}_1\widetilde{W}_1' \triangleleft \widetilde{W}, \widetilde{W}'$.

We say that a term M' is a *minimal pruning for minimal approximation* $\widetilde{W}, \widetilde{W}'$ iff $\widetilde{W}, \widetilde{W}'$ is a minimal approximation of W, W' and there is no $M'' \prec M'$ such that $\mathrm{NFL}(M''\overline{M}) = \widetilde{W}$ and $\mathrm{NFL}(M''\overline{M'}) = \widetilde{W}'$.

Let W, W' be a pair of columns with a type $\tau_1 \to \cdots \to \tau_n \to \iota$. We say that a pair of columns V, V' is a *splitting pair of columns* for W, W' iff there exist terms $M_1, \ldots, M_n$ such that the pair of columns $\mathrm{NFL}(VM_1 \cdots M_n)$, $\mathrm{NFL}(V'M_1 \cdots M_n)$ is an approximation of W, W'.

3.4   Accessibility

The notions of accessibility allow us to identify the portions of terms which are not necessary and to remove them. Let $E = [x\overline{M} \doteq W]$ be a set of interpolation equations. We say that an address γ is *accessible* in a term M wrt. E iff

- γ is an address in M,
- there is an equation $[x\overline{M}_1^i \cdots \overline{M}_n^i \doteq W^i] \in E$ such that

$$\mathrm{NF}(M[\gamma \leftarrow c]\overline{M}_1^i \cdots \overline{M}_n^i) \tag{1}$$

has an occurrence of c, where c is a fresh constant of a base type.

We say that an address is *totally accessible* iff for each equation in E the condition (1) holds. We say that an address is *totally head accessible* iff for each equation in E we have $\mathrm{NF}(M[\gamma \leftarrow c]\overline{M}_1^i \cdots \overline{M}_n^i) = c$, where c is a fresh constant. W.l.o.g. we may assume that c is a constant of the base type. Let $\langle E, E'\rangle$ be a dual set and M its solution. We say that an occurrence γ is *totally head accessible* wrt. the dual set iff it is totally head accessible wrt. E and E'.

3.5   Observational Equivalence

We introduce a notion of the observational equivalence. This notion is closely related to the dual interpolation problem. Solutions of an instance of the dual interpolation problem form an equivalence class in this relation.

Let $\mathcal{R} = \{\mathcal{R}_\tau\}_{\tau \in \mathcal{T}}$ be an indexed family of sets containing λ-terms, and satisfying conditions (1) all terms in $\mathcal{R}$ are in β-normal, η-long form, and (2) for each term $M \in \mathcal{R}$, there exists an α-representant s_M of M such that for each subterm t of the pre-term s_M we have $[t]_\alpha \in \mathcal{R}$. Such a set is called an *observable*.

The notion of an observable gives rise to a variation of the dual interpolation problem and the higher-order matching problem.

An instance of the dual interpolation problem for an observable $\mathcal{R}$ is a pair of sets $\mathcal{E} = \langle E, E' \rangle$ of interpolation equations such that there exists a variable x which occurs free in the left-hand side of each equation in $E \cup E'$ and all right-hand sides of $E \cup E'$ belong to $\mathcal{R}$.

The dual interpolation problem for an observable $\mathcal{R}$ is to decide whether an instance of the dual interpolation problem for the observable $\mathcal{R}$ has a solution.

Definition 3. (the higher-order matching problem for an observable)
An instance of the higher-order matching problem for an observable $\mathcal{R}$ *is a pair of simply typed λ-terms $\langle M, N \rangle$ where $N \in \mathcal{R}$.*

The higher-order matching problem for an observable $\mathcal{R}$ is to decide whether a given instance of the higher-order matching problem for the observable $\mathcal{R}$ has a solution.

The notion of an observable is also a base for a pre-order and an equivalence relation which, in turn, allow us to define a semantic structure for the simply typed λ-calculus.

Definition 4. (observational pre-order)
For each observable $\mathcal{R}$ we define an observational pre-order with respect to $\mathcal{R}$ as the relation $\sqsubseteq_\mathcal{R}$ on λ-terms such that

> $M \sqsubseteq_\mathcal{R} M'$ *iff M and M' have both the type $\sigma \to \tau$, and for each sequence of terms $N_1, \dots, N_n$ with appropriate types and $n \geq 1$ we have that if $\mathrm{NFL}(MN_1 \cdots N_n) \in \mathcal{R}$ then $MN_1 \cdots N_n =_{\beta\eta} M'N_1 \cdots N_n$ or*
> M *and M' are both of a base type and if $\mathrm{NFL}(M) \in \mathcal{R}$ then $M =_{\beta\eta} M'$*

Definition 5. (observational equivalence)
For each observable $\mathcal{R}$ we define an observational equivalence with respect to this observable as the relation on closed terms $\approx_\mathcal{R}$ such that

$$M \approx_\mathcal{R} M' \text{ iff } M \sqsubseteq_\mathcal{R} M' \text{ and } M' \sqsubseteq_\mathcal{R} M$$

More details concerning this equivalence are in [Sch01]. In particular, we have the following theorem

Theorem 2. *For each solvable instance $\mathcal{E}$ of dual interpolation for an observable $\mathcal{R}$ there exists an equivalence class of $\approx_\mathcal{R}$ such that all its elements are solutions of $\mathcal{E}$.*

For each class A of the relation $\approx_\mathcal{R}$ there exists an instance $\mathcal{E}$ of the dual interpolation problem for the observable $\mathcal{R}$ such that terms from A solve $\mathcal{E}$.

There is an algorithm which for a type $\tau = \tau_1 \to \cdots \tau_n \to \iota$ and complete sets of representants of $\approx_\mathcal{R}$ classes for types $\tau_1, \ldots, \tau_n$ generates a set C of instances such that for each $\approx_\mathcal{R}$ class A in the type τ there is an instance $\mathcal{E}_A$ such that terms of A are the only solutions of $\mathcal{E}_A$.

Proof. Easy conclusion from [Sch01]. ∎

3.6   Transferring Terms

The proof in this paper is based on the following schema: We define *transferring terms.* (The name comes from the fact that these terms govern the transfer of the symbols from the original places to the solution.) We show that each solvable dual interpolation instance has a solution of this form. The form can be further simplified to pseudo transferring terms which is done in Section 5. The latter form is so simple that we are able to enumerate them and thus we obtain a proof that the dual interpolation problem is decidable.

This section contains the crucial definition of transferring terms. The definition is difficult, but we explain some of its elements in this section later on.

Definition 6. (transferring terms)
Let $n, m \in \mathbb{N}$ and let C be a set of constants. A closed term $M : \tau_1 \to \cdots \to \tau_p \to \iota$ is an (n, m, C)-transferring term iff

1. $M = \lambda y_1 \ldots y_p.M'$ *and M' is a term over C without any occurrence of y_i where $i = 1, \ldots, p$, or*
2. $M = \lambda y_1 \ldots y_p.f M_1 \cdots M_k$, *where*
   - *$f \in C$ has a type $\sigma_1 \to \cdots \to \sigma_k \to \iota$, and*
   - *$M_i = \lambda z_1 \ldots z_r.M_i'$ where M_i' does not begin with λ, and*
   - *$\lambda y_1 \ldots y_p.M_i'$ are (n_i, m, C_i)-transferring terms, and*
   - *$n_i < n$, and*
   - *$C_i = C \cup \{z_1, \ldots, z_r\}$, or*
3. $M = \lambda y_1 \ldots y_p.y_i M_1 \cdots M_k[a \leftarrow N_a \boldsymbol{y}, b \leftarrow N_b \boldsymbol{y}, 0 \leftarrow N_0 \boldsymbol{y} \boldsymbol{x_0}, \ldots, l \leftarrow N_l \boldsymbol{y} \boldsymbol{x_l}]$, *where*
   - $\boldsymbol{y} = y_1, \ldots, y_p$;
   - *each M_i is a closed term over $C \cup \{a, b, 0, \ldots, l\}$;*
   - *each constant $0, \ldots, l$ occurs only once in $y_i M_1 \cdots M_k$;*
   - *N_a and N_b are (n_a, m_a, C)-transferring and (n_b, m_b, C)-transferring respectively;*
   - *N_j is (n_j, m_j, C)-transferring for each j;*
   - *$n_a, n_b \leq n$;*
   - *$m_a + m_b \leq m$ and $m_a, m_b \neq m$;*
   - *$n_j < n$ and $m_j \leq n$.*

Sometimes when (n, m, C) are unimportant or clearly seen from the context, we shorten the name to transferring.

Suppose that case (3) is abridged so that there are no constants from $\mathbb{N}$. The resulting term M may be presented as $M = \lambda \boldsymbol{y}.C_{y_k}[y_k M_1 \cdots (\lambda z z_j \boldsymbol{z}'.M_i) \cdots M_r]$. Consider $\lambda z z_j \boldsymbol{z}'.M_i$. In this term, none of the occurrences of z_j in M_i is in a subterm beginning with y_l from $\boldsymbol{y}$.

When this constraint is applied, the number of occurrences of variables from $\boldsymbol{y}$ in an (n, m, C)-transferring term M is bounded by m.

This simple picture is contaminated by the presence of constants from $\mathbb{N}$. We allow these constants to occur in subterms of M beginning with variables y_l from $\boldsymbol{y}$. This adds some flexibility and, consequently, expressive power to our terms. This flexibility is restricted, though. We have to pay for blocks that declare these variables with coins kept in n.

4   Transferring Terms and Dual Interpolation

Here is a theorem that relates the dual interpolation problem to transferring representatives. This is the central theorem of the paper.

Theorem 3. *Let $\mathcal{E}$ be an instance of the dual interpolation problem. If M is a solution of $\mathcal{E}$ then there exist $n, m \in \mathbb{N}$ together with a set of constants C and an (n, m, C)-transferring term M' such that M' is a solution of $\mathcal{E}$, too.*
Moreover, n, m, C depend recursively on $\mathcal{E}$.

In order to prove this theorem we need a lemma that enables cleanup of sometimes excessively complicated terms.

4.1   Skipping Unimportant Variables

The induction step in the proof of Theorem 3 consists in splitting a dual set and finding transferring solutions for the results of the split. These solutions are based on a term M that solves the whole set at the very beginning. This term can be too complicated to be a compact solution of the split sets. During the process of construction of transferring solutions for them, we have to compact M in several different ways. This section is devoted to the major step of the compactification.

Below, we use C-prunings for $C = \{a, b\} \cup \mathbb{N}$. Thus we shorten the name C-pruning to pruning here. We impose a constraint on the shape of prunings performed on solutions. *Constants from $\mathbb{N}$ may occur only once.*

First of all, we have to determine what we want to skip.

Definition 7. (an unimportant occurrence)
Let $M = \lambda \boldsymbol{y}.y_i M_1 \cdots M_k$ be a term that solves a dual set $\mathcal{E} = \langle [x\overline{\mathrm{M}} \doteq W], [x\overline{\mathrm{M}} \doteq W'] \rangle$, and let M' be a pruning of M that gives a minimal approximation of W, W'. An occurrence $\gamma \cdot 0$ of a variable z in M' is unimportant *iff*

- γ is totally head accessible, and
- there is $\gamma' \neq 0$ such that $\gamma \cdot \gamma'$ is an occurrence of a variable bound on γ and $\gamma \cdot \gamma'$ is totally head accessible.

Note that unimportant occurrence has always the shape $\gamma \cdot 0$. We say that an unimportant occurrence $\gamma \cdot 0$ is maximal if there is no unimportant occurrence $\gamma \cdot \gamma' \cdot 0$ where $\gamma' \succ \varepsilon$.

Lemma 1. Let $M = \lambda \boldsymbol{y}.zM_1 \cdots M_k$ be a term that solves a set of dual equations $\mathcal{E} = \langle [x\overline{\mathsf{M}} \doteq W], [x\overline{\mathsf{M}'} \doteq W'] \rangle$, where W, W' is a non-constant pair of columns. If $M' = \lambda \boldsymbol{y}.z\widehat{M_1} \cdots \widehat{M_k}$ is a minimal pruning of M for minimal approximation of W, W' wrt. $\mathcal{E}$, then

- either there exists j such that $\overline{\mathsf{M}}_j, \overline{\mathsf{M}}'_j$ is a splitting pair of columns for W, W',
- or there is an address γ in M that points to a constant $f \in \mathrm{Const}_M$ that is totally head accessible.

Proof. We get rid of the unimportant occurrences. This is done by a partial evaluation. The results of evaluation can be controlled since self-independent terms are used.

Proof of Theorem 3:
The aforementioned lemma allows to prove the main theorem of the paper. This is done by splitting dual interpolation instances. The way the split is done is controlled by the shape of minimal approximation for the hypothetical solution. □

5 Decidability

We have to provide yet another characterisation of solutions for the higher-order matching problem in the case of self-independent terms. The inductive definition of transferring terms requires in some places any terms instead of the transferring ones. In the proofs here, we replace these terms by equivalent pseudo transferring representatives which describe the behavioural equivalence in a more precise way than transferring ones.

Definition 8. (pseudo transferring terms)
Let $\mathcal{R}$ be a finite observable, $n, m \in \mathbb{N}$, and C be a set of constants. A term $M : \tau_1 \rightarrow \cdots \rightarrow \tau_p \rightarrow \iota$ is a pseudo (n, m, C)-transferring term *for the observable* $\mathcal{R}$ *iff*

1. $M = \lambda y_1 \ldots y_p.M'$ and M' is a term over constants from C without any occurrence of y_i,
2. $M = \lambda y_1 \ldots y_p.fM_1 \cdots M_k$, where $f \in C$ has a type $\sigma_1 \rightarrow \cdots \rightarrow \sigma_k \rightarrow \iota$ and $\lambda y_1 \ldots y_p.M_i$ are pseudo (n_i, m, C_i)-transferring terms for the observable $\mathcal{R}$ with $n_i < n$ and $C_i = C \cup \{z_1, \ldots, z_r\}$ where $M_i = \lambda z_1 \ldots z_r.M'_i$ and M'_i does not begin with λ.

3. $M = \lambda y_1 \ldots y_p.y_i M_1 \cdots M_k[a \leftarrow N_a \boldsymbol{y}, b \leftarrow N_b \boldsymbol{y}, z_0 := N_0 \boldsymbol{y}, \ldots, z_l := N_l \boldsymbol{y}]$, where

- $\boldsymbol{y} = y_1, \ldots, y_p$;
- $z_0, \ldots, z_l$ have order less than the order of y_i;
- $\lambda z_{i_1} \ldots z_{i_k}.M_i$ are pseudo $(n_i, m_i, C \cup \{a, b\})$-transferring terms for the observable $\mathcal{R}_{\{a,b\}}$, where n_i is the maximal number of equations in a dual set $\mathcal{E}$ characterising a single equivalence class as in Theorem 2, and m_i is the maximal total size of right-hand sides in such an $\mathcal{E}$
- N_a and N_b are pseudo (n_a, m_a, C)-transferring and pseudo (n_b, m_b, C)-transferring respectively for the observable $\mathcal{R}$;
- T is a set of terms such that $\mathcal{R}^T \subsetneq \mathcal{R}$;
- for each j the term N_j is pseudo (n_j, m_j, C)-transferring for the observable $\mathcal{R}^T$.
- $n_a, n_b \leq n$;
- $m_a + m_b \leq m$ and $m_a, m_b \neq m$;
- n_j is the number of equations in a dual set characterising an equivalence class of observational equivalence for $\mathcal{R}^T$ and m_j is the total size of right-hand sides in the aforementioned dual set.

Sometimes when (n, m, C) are unimportant or clearly seen from the context, we shorten the name to pseudo transferring.

We have the following theorem:

Theorem 4. *If a dual interpolation set $\mathcal{E}$ for an observable $\mathcal{R}$ has an (n,m,C)-transferring solution M then it has a pseudo (n,m,C)-transferring solution, too.*

We are now ready to conclude with the following theorem.

Theorem 5. *The dual interpolation problem for self-independent terms is decidable so the higher-order matching problem for these terms is decidable.*

Proof. The decision procedure consists in enumerating all pseudo transferring terms for the type of the unknown in an instance of the dual interpolation problem. This is done by recursion on the form of pseudo transferring terms.

6   Discussion and Further Research

The presented construction works for a single base type. The case with many base types requires an appropriate change in the definition of the transferring terms. I believe that the proof can be extended to cover this case, too, but the notation will be more difficult to handle.

A successful solution of the higher-order matching problem requires a thorough study of contexts in the simply typed lambda calculus. Similarly, a conscious application of programme transformation techniques and prover tactics based on higher-order matching demands a fine-grained map of different decidable cases. It is desirable to have procedures of this kind for various classes of contexts since they give there a precisely described partial completeness. The notion of self-independent terms relies on the outermost-leftmost reduction. A further research should be conducted in order to provide a similar characterisation for other reduction strategies. In particular the call-by-value strategy leads to a stronger notion of self-independent terms. The presented study shows that the higher-order matching problem poses much difficulty already in the case with a single base type.

It is interesting to see the connections between the self-independent terms and the linear logic. A possible direction of studies is to investigate if the higher-order matching problem is decidable in variants of the linear logic with ! operation. This could also give more classes of terms for which the higher-order matching problem is decidable.

Acknowledgements

I would like to thank Prof. Pawel Urzyczyn, Prof. Jerzy Tiuryn, Gilles Dowek, Vincent Padovani, Jurek Marcinkowski, and ToMasz Wierzbicki for discussions on the higher-order matching problem. The remarks of anonymous referees and my wife Joanna helped in the preparation of the paper. I am grateful for that.

References

[dG00]   Philippe de Groote. Linear Higher-Order Matching Is NP-Complete. In Leo Bachmair, editor, *Proceedings of 11th International Conference on Rewriting Techniques and Applications*, number 1833 in LNCS, pages 127–140. Springer-Verlag, 2000.

[dMS01]   Oege de Moor and Ganesh Sittampalam. Higher-order matching for program transformation. *Theoretical Computer Science*, 269(1–2):135–162, 2001.

[DW02]   Daniel J. Dougherty and Tomasz Wierzbicki. A decidable variant of higher order matching. In Sophie Tison, editor, *Proc. Thirteenth Intl. Conf. on Rewriting Techniques and Applications (RTA)*, pages 340–351, 2002.

[Har96]   John Harrison. HOL light: a tutorial introduction. In Mandayam Srivas and Albert Camilleri, editors, *Proceedings of the First International Conference on Formal Methods in Computer-Aided Design*, pages 265–269, 1996.

[Hue76]   G. Huet. *Résolution d'Équations dans les Langages d'Ordre $1, 2, \ldots, \omega$*. PhD thesis, Université de Paris VII, 1976.

[Loa01]   Ralph Loader. Finitary PCF is not decidable. *Theoretical Computer Science*, 1-2(266):341–364, 2001.

[Loa03]   Ralph Loader. Higher Order β Matching is Undecidable. *Logic Journal of the IGPL*, 11(1):51–68, January 2003.

[Pad95a]   Vincent Padovani. Decidability of all minimal models. In Stefano Berardi and Mario Coppo, editors, *Types for Proofs and Programs, International Workshop TYPES'95*, number 1158 in LNCS, pages 201–215. Springer Verlag, 1995.

[Pad95b]   Vincent Padovani. On equivalence classes of interpolation equations. In M. Dezani-Ciancaglini and G. Plotkin, editors, *Typed Lambda Calculi and Applications*, number 902 in LNCS, pages 335–349, 1995.

[Pad96]   Vincent Padovani. *Filtrage d'Ordre superieur*. PhD thesis, Université Paris VII, January 1996.

[Sch01]   Aleksy Schubert.   A note on observational equivalence in the simply typed lambda calculus.   Technical Report TR 02-01 (266), Institute of Informatics, Warsaw University, 2001.   Can be found at `http://www.mimuw.edu.pl/~alx/ftp-public/domains.ps.gz`.

[SS99]   Manfred Schmidt-Schauß. Decidability of behavioural equivalence in unary PCF. *Theor. Comput. Sci.*, 216(1-2):363–373, 1999.

[SS03]   Manfred Schmidt-Schauß.   Decidability of arity-bounded higher-order matching. In Franz Baader, editor, *CADE-19*, number 2741 in LNCS, pages 488–502. Springer, 2003.

[SSS02]   Manfred Schmidt-Schauß and Klaus U. Schulz. Decidability of bounded higher-order unification. In *CSL 2002*, LNCS 2471, pages 522–536. Springer-Verlag, 2002.

[Sta82]   Richard Statman. Completness, invariance and λ-definability. *Journal of Symbolic Logic*, 1(47), 1982.

A Type System for Computationally Secure Information Flow*

Peeter Laud and Varmo Vene

Tartu University
{peeter_l, varmo}@ut.ee

Abstract. The paper presents a novel type system for checking the security of information flow in programs containing operations of symmetric encryption. The type system is correct with respect to the complexity-theoretic security definitions of the encryption primitive.

Topics: Semantics, cryptography.

1 Introduction and Related Work

Suppose that you have received a program that purports to help you to organize some kind of your personal data. The program has functionality to make that data available over a network, but only in encrypted form, so that only people designated by you may have access to it. How can you be sure that the program really does what it claims to do, and does not leak your data to someone not entitled to it?

Here we have an instance of the problem of *secure information flow* (SIF). The security here is *not absolute*, though. If the program acts as promised then it indeed leaks your personal data — someone that is able to break the encryption can recover it. At stake here is *computational* security — if we assume that the encryption cannot be broken with realistic resources, does that also mean that your data is safe against all realistic adversaries?

When blindly trusting the source of the program is not an option, we have to verify it somehow. One possibility for such a verification is typing the program with a type system that ensures that correctly typed programs have SIF. Using static analysis for certification of SIF was pioneered by Denning and Denning [8,9]. The correctness of the analysis was not proven directly from the semantics of the program, though. Volpano et al. [23] gave a definition of SIF without using any instrumentations. They also gave a type system which could check whether programs satisfy this definition. In subsequent papers [21,25] they have extended their approach to richer sets of programming constructions, and have also attempted to handle operations that provide non-information-theoretical security — namely one-way functions [26,22]. The work in this paper can be seen as a significant extension to their approach. The interest in security types is motivated by the relative ease of incorporating them into existing programming languages [18]. A good overview of research on language-based information flow security is given by Sabelfeld and Myers [19].

---

* Partially supported by Estonian Science Foundation, grant #6095.

M. Liśkiewicz and R. Reischuk (Eds.): FCT 2005, LNCS 3623, pp. 365–377, 2005.

Type systems for handling cryptographic operations were first investigated by Abadi [1]; a more recent paper is [2]. In his approach, the type of each piece of data expresses the intended uses of that data; the type of keys shows what kind of data it is intended to encrypt so as to make it safe to communicate over public channels. Our approach is somewhat different — our types reflect the source of the data; communicating data over public channels is safe if the sources are not sensitive. The main difference between their and our approaches is however the computational setting — their type systems are correct with respect to the Dolev-Yao model [10] while ours is correct with respect to complexity-theoretic definitions of security [27].

Automatic or computer-assisted handling of cryptographic operations while remaining true to the complexity-theoretic security definitions has also attracted research but the emphasis or results of the approaches have been more or less different from the current paper. Universal composability [7,20] is a very general approach that strives to abstract away from the complexity-theoretic details of cryptographic primitives so that Dolev-Yao-style, but computationally justified arguments become possible. Tools for manipulating these abstractions have unfortunately failed to appear until now. Also, the current best abstraction of symmetric encryption [6] has some restrictions. Lincoln et al. [15,16] have given probabilistic semantics to spi-calculus [3], stated some protocols and proved them correct with respect to this semantics and the complexity-theoretic security definitions, but there is no tool support. Abadi et al. [5,4] have given an automatic means to decide when certain reasonable computational interpretations of two messages in Dolev-Yao model are indistinguishable. We [11,13] have proposed program analyses for checking SIF in presence of encryption in programs.

In this paper we present a type system that can also gracefully handle symmetric encryptions as operations in the program, and, as the first of its kind, is correct with respect to the *complexity-theoretic* security definitions of the cryptographic primitive, instead of being based on the Dolev-Yao model. The paper has the following structure. In Sec. 2 we state the preliminaries — we define our program language, its semantics, and the security of both information flow and encryption systems. Sec. 3 presents the type system and Sec. 4 states its correctness theorem and gives an overview of its proof. Sec. 5 concludes.

2   The Settings

We consider programs in a simple imperative language (the WHILE-language) whose expressions E and programs P are defined by the following grammar:

$$P ::= x := E \mid skip \mid P_1; P_2 \mid if\ b\ then\ P_1\ else\ P_2 \mid while\ b\ do\ P_1$$
$$E ::= o(x_1, \ldots, x_k)$$

Here $x, x_1, \ldots, x_k, b$ are variables from the set **Var** and o is an operator from the set of operators **Op**. The set **Op** has to contain two special operators — a nullary operator $\mathcal{G}en$ that denotes the generation of new encryption keys, and a binary operator $\mathcal{E}nc$ denoting the symmetric encryption. Our type system does not handle decryption specifically, therefore we do not mention it here.

Semantics. As our type system is proved correct mostly by showing the bisimilarity of certain programs, we use (small-step) structural operational semantics as the main method to describe what a program does. The encryption system that gives the semantics to the operators $\mathcal{G}en$ and $\mathcal{E}nc$ must be probabilistic, otherwise it cannot satisfy the requirements we are going to put on it. Hence the semantics of the programs must be probabilistic as well. Let $\mathcal{D}(X)$ denote the set of all probability distributions over the set X. For $x \in X$ let $\eta(x) \in \mathcal{D}(X)$ be the probability distribution that puts all its weight onto x. Let $x \leftarrow D$ denote that the random variable x is picked according to the probability distribution D. Let $\{\!|E : C|\!\}$ denote the distribution of E under the conditions C. A state of the program is a mapping from **Var** to the set of values $\mathbf{Val} = \{0, 1\}^*$. A program configuration is a pair of a program (yet to be executed) and a state. A probabilistic program configuration is a pair of a program and a probability distribution over states. The semantics is a relation from program configurations to probabilistic program configurations and probability distributions over states. We assume that each k-ary operator $o \in \mathbf{Op}$ has been given a semantics $[\![o]\!] : \mathbf{Val}^k \to \mathcal{D}(\mathbf{Val})$. The semantics of programs is given in Fig. 1, where $\mathbf{Val} = \text{true} \,\dot\cup\, \text{false}$ is a fixed partition. Note that $\longrightarrow$ is a function.

$$\langle x := o(x_1, \ldots, x_k), S \rangle \longrightarrow \{\!| S[x \mapsto v] : v \leftarrow [\![o]\!](S(x_1), \ldots, S(x_k)) |\!\} \tag{1}$$

$$\langle skip, S \rangle \longrightarrow \eta(S) \tag{2}$$

$$\frac{\langle \mathsf{P}_1, S \rangle \longrightarrow D}{\langle \mathsf{P}_1; \mathsf{P}_2, S \rangle \longrightarrow \langle \mathsf{P}_2, D \rangle} \tag{3}$$

$$\frac{\langle \mathsf{P}_1, S \rangle \longrightarrow \langle \mathsf{P}_1', D \rangle}{\langle \mathsf{P}_1; \mathsf{P}_2, S \rangle \longrightarrow \langle \mathsf{P}_1'; \mathsf{P}_2, D \rangle} \tag{4}$$

$$\frac{S(b) \in \text{true}}{\langle if\ b\ then\ \mathsf{P}_1\ else\ \mathsf{P}_2, S \rangle \longrightarrow \langle \mathsf{P}_1, \eta(S) \rangle} \tag{5}$$

$$\frac{S(b) \in \text{false}}{\langle if\ b\ then\ \mathsf{P}_1\ else\ \mathsf{P}_2, S \rangle \longrightarrow \langle \mathsf{P}_2, \eta(S) \rangle} \tag{6}$$

$$\frac{S(b) \in \text{true}}{\langle while\ b\ do\ \mathsf{P}_1, S \rangle \longrightarrow \langle \mathsf{P}_1; while\ b\ do\ \mathsf{P}_1, \eta(S) \rangle} \tag{7}$$

$$\frac{S(b) \in \text{false}}{\langle while\ b\ do\ \mathsf{P}_1, S \rangle \longrightarrow \eta(S)} \tag{8}$$

Fig. 1. The operational semantics of programs

For defining security we also have to state what the outcome of the program is. A *program run* is a sequence $C_0 \xrightarrow{p_1} C_1 \xrightarrow{p_2} \cdots \xrightarrow{p_n} C_n$ where $C_0, \ldots, C_{n-1}$ are program configurations and C_n is a program state. If $C_{i-1} = \langle \mathsf{P}_{i-1}, S_{i-1} \rangle \longrightarrow \langle \mathsf{P}_i, D_i \rangle$ then

C_i must be equal to $\langle P_i, S_i \rangle$ for some state S_i and $p_i = D_i(S_i)$ (if $i = n$ then we have just D_n instead of $\langle P_i, D_i \rangle$). The *probability* of a run is the product of all p_i that it contains. Let $\mathbf{State}_\perp = \mathbf{State} \,\dot\cup\, \{\perp\}$ where $\perp$ denotes nontermination. If $\langle P, S \rangle$ is a configuration and $D \in \mathcal{D}(\mathbf{State}_\perp)$ then we write $\langle P, S \rangle \Longrightarrow D$ if for all $S \in \mathbf{State}$, $D(S)$ equals the sum of the probabilities of all runs starting with $\langle P, S \rangle$ and ending with S. The relation $\Longrightarrow$ defines the result of running a program on an initial state.

Encryption Systems. An *encryption system* is a triple of algorithms $(\mathcal{G}, \mathcal{E}, \mathcal{D})$. They all must have running times polynomial to the length of their arguments. The algorithm $\mathcal{G}$ is the *key-generation algorithm*. It is invoked to create new encryption keys. The algorithm $\mathcal{G}$ takes one argument — the *security parameter* $n \in \mathbb{N}$ (represented in unary, because of the comment about the running times of algorithms) which determines the security of the system — more concretely, it determines the length of the keys. Larger security parameter means longer keys. The *encryption algorithm* takes as its arguments the security parameter, a key returned by $\mathcal{G}(1^n)$ (actually, we could assume that the security parameter is contained in that key but this is the usual presentation), and a plaintext — a bit-string. It returns the corresponding ciphertext. The arguments and the return value of the *decryption algorithm* are similar, only the places of plaintext and ciphertext are reversed. The key generation algorithm is obviously probabilistic, the decryption algorithm is deterministic. The encryption algorithm may either be deterministic or probabilistic but for satisfying the security requirements stated below it has to be probabilistic. It is required that the decryption of an encryption of a bit-string is equal to that bit-string.

The security requirement we put on the encryption system is the same as Abadi and Rogaway [5] used. We want the encryption to conceal the identity of both plaintexts and encryption keys and we want it also to hide the length of the plaintexts. Formally, for all probabilistic polynomial-time (PPT) algorithms $\mathcal{A}$ (with access to two oracles) the difference

$$\mathbf{P}[\mathcal{A}^{\mathcal{E}(1^n, k, \cdot), \mathcal{E}(1^n, k', \cdot)}(1^n) = 1 \mid k, k' \leftarrow \mathcal{G}(1^n)] -$$
$$\mathbf{P}[\mathcal{A}^{\mathcal{E}(1^n, k, 0), \mathcal{E}(1^n, k, 0)}(1^n) = 1 \mid k \leftarrow \mathcal{G}(1^n)] \quad (9)$$

must be a negligible function in n where $\mathbf{0}$ is a fixed bit-string. Here $\mathcal{E}(1^n, k, \cdot)$ is an oracle that encrypts its inputs with key k, and $\mathcal{E}(1^n, k, \mathbf{0})$ is an oracle that discards its inputs and returns encryptions of $\mathbf{0}$ (under key k) instead. A function is negligible if it is asymptotically smaller than the reciprocal of any polynomial.

We see that an encryption system does not just define a nullary and two binary algorithms, but it defines an entire family (indexed by $n \in \mathbb{N}$) of them. The semantics of programs therefore also has to be indexed by the security parameter n. Instead of a single relation $\longrightarrow$ we have a family $\{\xrightarrow{n}\}_{n \in \mathbb{N}}$. The semantics of a k-ary operation o is a family of probabilistic functions $[\![o]\!]_n : \mathbf{Val}^k \to \mathcal{D}(\mathbf{Val})$. We require that $[\![\mathcal{G}en]\!]_n = \mathcal{G}(1^n)$ and $[\![\mathcal{E}nc]\!]_n = \mathcal{E}(1^n, \cdot, \cdot)$ for some encryption system $(\mathcal{G}, \mathcal{E}, \mathcal{D})$ satisfying (9).

Secure Information Flow. We have to fix what are the secret inputs and the public outputs of a program. For simplicity, let there be two fixed subsets $\mathbf{Var}_S, \mathbf{Var}_P \subseteq \mathbf{Var}$.

The secret inputs are the initial values of the variables in $\mathbf{Var_S}$, and the public outputs are the final values of the variables in $\mathbf{Var_P}$.

Our definition of security is termination-insensitive. Sensitivity to termination is orthogonal to other issues and can be easily added as an afterthought [24]. Actually, non-termination cannot be detected at all, but running for a too long time can. In our setting, superpolynomial (in the security parameter) running time is definitely too long because encryption is secure only against polynomial-time adversaries. We say that a program P runs in *expected polynomial time* if there exists a polynomial q and a negligible function α, such that the sum of the probabilities of all program runs (for the semantics $\xrightarrow{}_{n}$) of length at most $q(n)$ is at least $1 - \alpha(n)$.

The inputs of the program have to come from somewhere; when the program is run then its input state is picked from some probability distribution over program states. The nature of that distribution can have a profound effect on the security of the program. If the family of input distributions (indexed by the security parameter) D is not polynomial-time samplable (i.e. there exists no PPT algorithm $\mathcal{A}$ whose outputs on input 1^n are distributed as D_n) then some effects not achievable in polynomial time may happen during the program run and we no longer can be sure that the encryption is secure. Another consideration is, that we are interested whether *the program* leaks the secret inputs or not, so we want to exclude the cases where the secrets have already been leaked before running the program. We say that a family of distributions D over program states *isolates the secrets* if the values of the variables in $\mathbf{Var_S}$ are *computationally independent* of the values of the rest of the variables. I.e. the families of probability distributions

$$\{(S_n|_{\mathbf{Var_S}}, S_n|_{\mathbf{Var} \setminus \mathbf{Var_S}}) \ : \ S_n \leftarrow D_n\} \tag{10}$$

and

$$\{(S_n|_{\mathbf{Var_S}}, S'_n|_{\mathbf{Var} \setminus \mathbf{Var_S}}) \ : \ S_n, S'_n \leftarrow D_n\} \tag{11}$$

have to be indistinguishable, i.e. no PPT algorithm that is given either a sample of (10) or (11) can tell with probability non-negligibly higher than $1/2$ which of these two distributions the sample was taken from.

A program P that runs in expected polynomial time has *computationally secure information flow* (CSIF) if the secret inputs and public outputs of the program are computationally independent. I.e. for all polynomial-time samplable families of probability distributions D that isolate the secrets, the families of probability distributions

$$\{(S_n|_{\mathbf{Var_S}}, S'_n|_{\mathbf{Var_P}}) \ : \ S_n \leftarrow D_n, \langle \mathsf{P}, S_n \rangle \xrightarrow{}_{n} D'_n, S'_n \leftarrow D'_n\} \tag{12}$$

and

$$\{(S_n|_{\mathbf{Var_S}}, S'_n|_{\mathbf{Var_P}}) \ : \ S_n, S''_n \leftarrow D_n, \langle \mathsf{P}, S''_n \rangle \xrightarrow{}_{n} D'_n, S'_n \leftarrow D'_n\} \tag{13}$$

have to be indistinguishable.

3   Type System

A *typing* assigns types to variables. The type of a variable indicates what kind of information is allowed to influence its value. The type also indicates whether the value of the variable is a valid encryption key. I.e. the type of a variable is a pair of an *information type* and a *usage type*. *Inference rules* allow to deduce the type of the program from the types of its variables. A typing is *valid* if the program has a type.

The "basic" secrets that the program operates on are its secret inputs and also the encryption keys that the program operates on; we definitely have to keep track where they are flowing. We have to distinguish between different keys; if we did not then each key would potentially be able to decrypt any ciphertext. In our current approach we distinguish the keys statically — let $\mathcal{G}$ be a (finite) set whose elements we use to label the key generation statements $x := \mathcal{G}en$ in the program. We can distinguish two keys if they have been generated at statements with different labels. Let $\mathcal{T}_0 = \{h\} \cup \mathcal{G}$; it is the set of types for basic secrets. Here h denotes the type of secret inputs.

Let $\mathcal{T}_1 = \{\{t\}_N \mid t \in \mathcal{T}_0, N \subseteq \mathcal{G}\}$. The type $\{t\}_N$ means that the information of type t has been encrypted with keys of the type N. To recover the information, one needs at least one key for each type that is contained in N. The set $\mathcal{T}_1$ is ordered — $\{t\}_N \leq \{t'\}_{N'}$ iff $t = t'$ and $N \supseteq N'$. We get less information out of something that is protected by more keys.

Let $\mathcal{T}_2 = \mathcal{P}(\mathcal{T}_1)$ (the power set). The elements of $\mathcal{T}_2$ denote the merging of information from several types in $\mathcal{T}_1$. The information types are basically elements of $\mathcal{T}_2$. The element $\emptyset \in \mathcal{T}_2$ denotes public data. The set $\mathcal{T}_2$ is ordered: $T \leq T'$ if $\forall \{t\}_N \in T \exists \{t'\}_{N'} \in T' : \{t\}_N \leq \{t'\}_{N'}$. The relation $\leq$ is a preorder (reflexive and transitive), so we identify T and T' whenever $T \leq T'$ and $T' \leq T$. In practice this amounts to the deletion of all non-maximal elements from $T \in \mathcal{T}_2$.

Besides the equivalence $\leq \cap \geq$ on $\mathcal{T}_2$ there is another one that corresponds to the usage of keys that are not protected by other keys. For example, if $T = \{\{h\}_{\{1\}}, \{1\}_{\emptyset}\}$ then we the possible knowledge of a key generated at $1 \in \mathcal{G}$ allows us to recover h — T is equivalent to $T \cup \{h\}$ (denote $T \equiv T \cup \{h\}$). To formally define the relation $\equiv$ we introduce the sets T^N for $T \in \mathcal{T}_2$ and $N \subseteq \mathcal{G}$. The set T^N corresponds to the information that may be recovered if we have the information in T and possibility to decrypt information with keys in N. They are defined as the least sets satisfying

$$
\begin{aligned}
&1. & T &\subseteq T^{\emptyset} \\
&2. & N \subseteq N' &\Rightarrow T^N \subseteq T^{N'} \\
&3. & \{t\}_M \in T^N &\Rightarrow \{t\}_{M \setminus \{i\}} \in T^{N \cup \{i\}} \\
&4. & \{t\}_M \in T^{N \cup \{i\}} \wedge \{i\}_{\emptyset} \in T^N &\Rightarrow \{t\}_M \in T^N \\
&5. & \{i\}_{\emptyset} \in T^{N \cup \{i\}} &\Rightarrow \{i\}_{\emptyset} \in T^N .
\end{aligned}
\tag{14}
$$

Finally we define that $T \equiv T^{\emptyset}$. The items 1.-4. specify just the abilities of a Dolev-Yao attacker. The 5. item is used to break encryption cycles (see [5] for a more thorough discussion on them). This "Dolev-Yao attacker with the ability to break encryption cycles" first appeared in [12].

There is a more direct way of computing $T^{\emptyset}$ than iterating the rules in (14). First we want to determine the set I of all key labels $i \in \mathcal{G}$ that occur as $\{i\}_{\emptyset}$ in $T^{\emptyset}$. We let $I \subseteq \mathcal{G}$ be the *largest* set satisfying

$$
\left(\forall \{t\}_N \in T : t \neq i \vee N \not\subseteq I\right) \Rightarrow i \notin I
\tag{15}
$$

for all $i \in \mathcal{G}$. Such an I can be found by initializing I with $\mathcal{G}$ and then iterating (15).

Proposition 1. *Let $T \in \mathcal{T}_2$ and let I be defined as in (15). Then $T^\emptyset = \{\{t\}_{N\setminus I} \mid \{t\}_N \in T\}$ (up to the relation $\le \cap \ge$).*

See the full version [14] for a proof. In the following, when we talk about the elements of $\mathcal{T}_2$ then we mean the equivalence classes with respect to $(\le \cap \ge) \sqcup \equiv$.

The set of *usage types* is $\mathcal{U} = \{\mathsf{Data}\} \cup \{\mathsf{Key}_N \mid N \subseteq \mathcal{G}\}$. If the usage type of variable is Data then its value is not usable as an encryption key. The value of a variable with the usage type Key_N is an encryption key generated at a key generation statement labeled with an element of N.

We can now state the actual sets of types for expressions, variables and programs. Define

$$\mathcal{T}_E := \{\langle T, J, U\rangle \mid T, J \in \mathcal{T}_2, U \in \mathcal{U}, T \ge J\}$$
$$\mathcal{T}_V := \{\langle T, U\rangle \ var \mid T \in \mathcal{T}_2, U \in \mathcal{U}\}$$
$$\mathcal{T}_C := \{T \ cmd \mid T \in \mathcal{T}_2\} \ .$$

For a program type $T \ cmd$ the type T is a lower bound on the information types of variables that are assigned to in this program. A variable type shows both the kinds of information that may be contained in that variable, and whether it may be used as an encryption key. The components T and U in an expression type $\langle T, J, U\rangle$ have the same meaning (for that expression). Additionally, J is an upper bound on the information that may control whether this expression is evaluated. A typing γ is a mapping from **Var** to $\mathcal{T}_V$. Compared to the type system of Volpano et al. [23] the new details (besides the much richer set of information types) are the usage types for variables and expressions and the extra component in the expression types recording the information through implicit flow.

There is an order defined on $\mathcal{T}_E$:

$$T \le T' \wedge J \le J' \Rightarrow \langle T, J, \mathsf{Data}\rangle \le \langle T', J', \mathsf{Data}\rangle$$
$$T \le T' \wedge J \le J' \wedge N \subseteq N' \Rightarrow \langle T, J, \mathsf{Key}_N\rangle \le \langle T', J', \mathsf{Key}_{N'}\rangle$$
$$J \le J' \wedge T \cup \{\{i\}_\emptyset \mid i \in N\} \le T' \Rightarrow \langle T, J, \mathsf{Key}_N\rangle \le \langle T', J', \mathsf{Data}\rangle \ .$$

The set $\mathcal{T}_C$ is ordered as well: $T \ cmd \le T' \ cmd$ if $T \ge T'$. The set $\mathcal{T}_V$ is unordered. Let $\top$ be the greatest element of $\mathcal{T}_2$, i.e. $\top = \{\{t\}_\emptyset \mid t \in \mathcal{T}_0\}$. The rules for typing expressions and programs are given in Fig. 2. Note that the rule (18) is a "general" rule for typing expressions and it may also applied to encryptions and key generations.

Recall that we called a typing γ *valid* (for the program P) if $\gamma \vdash \mathsf{P} : T \ cmd$ is derivable for some $T \in \mathcal{T}_2$. The rules in Fig. 2 put certain constraints on γ. To further explain these rules, let us state these constraints explicitly. Each assignment statement $x := o(x_1, \ldots, x_k)$ in the program introduces a set of constraints. Let $b_1, \ldots, b_m$ be the variables controlling whether this assignment is executed (i.e. these are the conditional variables occurring in the *if*- and *while*-statements enclosing this assignment). Stating the constraints is simpler if we also introduce an order on $\mathcal{T}_V$ by defining

$$T \le T' \Rightarrow \langle T, \mathsf{Data}\rangle \ var \le \langle T', \mathsf{Data}\rangle \ var$$
$$T \le T' \wedge N \subseteq N' \Rightarrow \langle T, \mathsf{Key}_N\rangle \ var \le \langle T', \mathsf{Key}_{N'}\rangle \ var$$
$$T \cup \{\{i\}_\emptyset \mid i \in N\} \le T' \Rightarrow \langle T, \mathsf{Key}_N\rangle \ var \le \langle T', \mathsf{Data}\rangle \ var$$

and define $\gamma_{\mathsf{Data}}(x) \ge \gamma(x)$ to be the smallest type whose usage component equals Data.

$$\frac{\gamma \vdash e : \langle T', J', U' \rangle \quad \langle T', J', U' \rangle \leq \langle T, J, U \rangle}{\gamma \vdash e : \langle T, J, U \rangle} \tag{16}$$

$$\frac{\gamma \vdash \mathsf{P} : T'\ cmd \quad T'\ cmd \leq T\ cmd}{\gamma \vdash \mathsf{P} : T\ cmd} \tag{17}$$

$$\frac{\gamma \vdash e_i : \langle T, J, \mathsf{Data} \rangle}{\gamma \vdash o(e_1, \ldots, e_k) : \langle T, J, \mathsf{Data} \rangle} \tag{18}$$

$$\gamma \vdash \mathcal{G}en^i : \langle \emptyset, \emptyset, \mathsf{Key}_{\{i\}} \rangle \tag{19}$$

$$\frac{\gamma \vdash y : \langle T, J, \mathsf{Data} \rangle \quad \gamma \vdash k : \langle T, J, \mathsf{Key}_N \rangle}{\gamma \vdash \mathcal{E}nc(k, y) : \langle \{ \{t\}_{M \cup \{i\}} \mid \{t\}_M \in T, i \in N \} \cup J, J, \mathsf{Data} \rangle} \tag{20}$$

$$\frac{\gamma(x) = \langle T, U \rangle\ var}{\gamma \vdash x : \langle T, \emptyset, U \rangle} \tag{21}$$

$$\frac{\gamma \vdash e : \langle T, J, \mathsf{Data} \rangle \quad \gamma(x) = \langle T, \mathsf{Data} \rangle\ var}{\gamma \vdash x := e : J\ cmd} \tag{22}$$

$$\frac{\gamma \vdash e : \langle T, J, \mathsf{Key}_N \rangle \quad \gamma(x) = \langle T, \mathsf{Key}_N \rangle\ var}{\gamma \vdash x := e : J\ cmd} \tag{23}$$

$$\gamma \vdash skip : \top\ cmd \tag{24}$$

$$\frac{\gamma \vdash \mathsf{P}_1 : T\ cmd \quad \gamma \vdash \mathsf{P}_2 : T\ cmd}{\gamma \vdash \mathsf{P}_1; \mathsf{P}_2 : T\ cmd} \tag{25}$$

$$\frac{\gamma \vdash e : \langle T, J, \mathsf{Data} \rangle \quad \gamma \vdash \mathsf{P}_1 : T\ cmd \quad \gamma \vdash \mathsf{P}_2 : T\ cmd}{\gamma \vdash if\ e\ then\ \mathsf{P}_1\ else\ \mathsf{P}_2 : T\ cmd} \tag{26}$$

$$\frac{\gamma \vdash e : \langle T, J, \mathsf{Data} \rangle \quad \gamma \vdash \mathsf{P} : T\ cmd}{\gamma \vdash while\ e\ do\ \mathsf{P} : T\ cmd} \tag{27}$$

Fig. 2. Typing rules

An assignment $x := o(x_1, \ldots, x_k)$ simply introduces the constraints $\gamma(x) \geq \gamma(x_i)$ and $\gamma(x) \geq \gamma_{\mathsf{Data}}(b_j)$ for all i and j. Also, the usage component of $\gamma(x)$ must be Data. For the special kinds of assignments, there is a choice between the set of constraints we just stated and a set of constraints that depends of the statement.

For $x := \mathcal{E}nc(k, y)$ the alternative set of constraints is the following. Let $\gamma(x) = \langle T_x, \mathsf{Data} \rangle\ var$, $\gamma(k) = \langle T_k, \mathsf{Key}_N \rangle\ var$, $\gamma_{\mathsf{Data}}(y) = \langle T_y, \mathsf{Data} \rangle\ var$, and $\gamma_{\mathsf{Data}}(b_j) = \langle B_j, \mathsf{Data} \rangle\ var$. The constraints $T_x \geq \{ \{t\}_{M \cup \{i\}} \mid \{t\}_M \in T_y \cup T_k, i \in N \}$ and $T_x \geq B_j$ must then hold. Such different handling of information flowing from k and y vs. the information flowing from b_j-s was the reason of introducing the second information-type component to the expression types. Notice that typing rule (20) is the only rule that handles two information-type components differently.

For $x := \mathcal{G}en^i$, the alternative set of constraints is the following. Let $\gamma(x) = \langle T_x, \mathsf{Key}_N \rangle \ var$ and $\gamma_{\mathsf{Data}}(b_j) = \langle B_j, \mathsf{Data} \rangle \ var$. Then $i \in N$ and $T_x \geq B_j$.

For $x := y$, where x and y are both keys, we have the following alternative set of constraints. Let $\gamma(x) = \langle T_x, \mathsf{Key}_{N_x} \rangle \ var$, $\gamma(y) = \langle T_y, \mathsf{Key}_{N_y} \rangle \ var$ and $\gamma_{\mathsf{Data}}(b_j) = \langle B_j, \mathsf{Data} \rangle \ var$. Then $T_x \geq T_y$, $T_x \geq B_j$ and $N_x \supseteq N_y$.

These constraints can be used to automatically infer typings for programs. The next proposition is proved in the full version of this paper [14].

Proposition 2. *A typing γ of a program P is valid iff it satisfies the constraints given above.*

4 Correctness of the Type System

For stating the correctness theorem we have to define which variables actually constitute the inputs of the program and which are merely used for storing the intermediate results or outputs. We say that $x \in \mathbf{Var}$ is an *input variable* if there is a path through the program where a read of x precedes the first write to x. If x is not an input variable then its initial value has no effect on the computation. Let $\mathbf{Var_I} \subseteq \mathbf{Var}$ be the set of all input variables. The set $\mathbf{Var_I}$ can be found using the same methods as for determining the potentially uninitialized variables in Java methods [17]. For a given γ let $\gamma_I(x) = T$ where $\langle T, U \rangle \ var = \gamma_{\mathsf{Data}}(x)$.

Theorem 1. *Let P be a program running in expected polynomial time with the set of variables* **Var.** *Let* **Var$_S$** *and* **Var$_P$** *be fixed. If P has a valid typing γ, such that $\gamma(x) \geq \langle \{\{h\}_\emptyset\}, \mathsf{Data} \rangle \ var$ for all $x \in$ **Var$_S$**, $\gamma(x) \geq \langle \emptyset, \mathsf{Data} \rangle \ var$ for all $x \in$ **Var$_I$** $\cup$ **Var$_P$**, and $\bigvee_{x \in \mathbf{Var_P}} \gamma_I(x) \not\geq \{\{h\}_\emptyset\}$ then P has secure information flow.*

We see that for applying this theorem the inputs and outputs of the program may not be keys. If we had allowed the inputs to be keys then the theorem would have had to demand that they really are valid keys. I.e. the values of corresponding variables must have been distributed indistinguishably to keys and two variables that are keys would have to be either equal or independent. In any case they would have to be independent of non-keys. We believe that the restriction that the theorem has in its current wording is not a major one. If one wants to consider keys as inputs as well, then one could just prepend the program with commands to generate those keys. If the public output of the program would have been a key then we can just assign it to a different (new) variable.

Theorem 1 follows by some simple manipulation of probability distributions [14] from the following lemma basically stating that the public outputs of a program satisfying the premises of Thm. 1 can be computed without ever accessing the secret inputs.

Lemma 1 (Simulation Lemma). *Let P,* **Var,** **Var$_S$,** **Var$_P$,** **Var$_I$** *satisfy the premises of Theorem 1. Then there exists a program P′ running in expected polynomial time with the set of variables* **Var′** *and the set of input variables* **Var$_I$′,** *such that* **Var$_S$** $\cup$ **Var$_P$** $\subseteq$ **Var′,** **Var$_I$′** $\subseteq$ **Var$_I$,** *P′ does not access the variables in* **Var$_S$,** *and for every* *polynomial-time samplable family of probability distributions D over program states the families $\{(S_n|_{\mathbf{Var_S}}, S'_n|_{\mathbf{Var_P}}) : S_n \leftarrow D_n, \langle \mathsf{P}, S_n \rangle \Rrightarrow D'_n, S'_n \leftarrow D'_n \}$ and $\{(S_n|_{\mathbf{Var_S}}, S'_n|_{\mathbf{Var_P}}) : S_n \leftarrow D_n, \langle \mathsf{P'}, S_n \rangle \Rrightarrow D'_n, S'_n \leftarrow D'_n \}$ are indistinguishable.*

Let us give a short description of the proof of the Simulation Lemma (full proof can be found in [14]). The main tools for constructing the program $\mathsf{P'}$ and showing that its public outputs are indistinguishable from those of P are *probabilistic bisimulations* and the definition of secure encryption (9). The definition states that sometimes the encryption $\mathcal{E}nc(k, y)$ may be replaced with $\mathcal{E}nc(k_\mathbf{n}, \mathbf{0})$ where $\mathbf{0}$ is a constant and $k_\mathbf{n}$ is a fixed key (generated somewhere in the beginning of the program).

Let $T \in \mathcal{T}_2$ be an information type. We say that some $i \in \{\mathsf{h}\} \cup \mathcal{G}$ occurs in T as *data* if $\{i\}_N \in T$ for some N. We say that $i \in \mathcal{G}$ *occurs in T as a key* if $\{t\}_{N \cup \{i\}} \in T$ for some t and N. We say that j *encrypts i in T* if $\{i\}_{N \cup \{j\}} \in T$ for some N. A key label i is *free* in T if occurs in T only as a key. If h does not occur in $\bigvee_{x \in \mathbf{Var_P}} \gamma_I(x)$ then the program $\mathsf{P'}$ mentioned in the Simulation Lemma can be constructed by just deleting all statements that access variables of types where h occurs. These statements are assignments to variables whose information type is at least $\{\mathsf{h}\}_\mathcal{G}$ and *if*- and *while*-statements whose guards have types with the same property. We can show that P and $\mathsf{P'}$ are bisimilar with respect to a bisimulation that requires the equality of public variables.

If h occurs in $\bigvee_{x \in \mathbf{Var_P}} \gamma_I(x)$ then we repeatedly use the indistinguishability (9) to construct programs that use certain keys to encrypt only public data (actually, the constant $\mathbf{0}$), not secrets. The behavior of these programs is indistinguishable from the original program if we only look at public variables. The typing γ is also a valid typing for these programs, but they also have more permissive typings. A valid typing γ'' of the last constructed program $\mathsf{P''}$ is such, that h does not occur in $\bigvee_{x \in \mathbf{Var_P}} \gamma''_I(x)$. From the program $\mathsf{P''}$ one can construct $\mathsf{P'}$ as before.

The replacement of encryptions may only be done if the encryption key is only used in ways that is possible in (9) — it may only be used for encryption. To better account for the flow of different keys, we first separate the keys with different labels. For this we introduce to our programming language a new expression

$$\mathcal{C}\mathcal{E}nc(k^{(t)} \| i_1, k_1, y_1 | i_2, k_2, y_2 | \ldots | i_n, k_n, y_n), \tag{28}$$

where $i_1, \ldots, i_n \in \mathcal{G}$ and the rest are variables. The semantics of the expression compares the value $k^{(t)}$ with $i_1, \ldots, i_n$ (it is guaranteed to match one; say i_j) and returns $\mathcal{E}nc(k_j, y_j)$. Such expressions occur in the intermediate steps of transformation but not in the final program $\mathsf{P'}$. The typing rule for (28) is

$$\frac{\gamma \vdash y_j : \langle T_j, J, \mathsf{Data} \rangle \quad \gamma \vdash k^{(t)} : \langle T, J, \mathsf{Data} \rangle \quad T \leq T_j \quad \gamma \vdash k_j : \langle T_j, J, \mathsf{Key}_{\{i_j\}} \rangle}{\gamma \vdash \mathcal{C}\mathcal{E}nc(k^{(t)} \| i'_1, k_1, y_1 | \ldots | i'_n, k_n, y_n) : \langle \bigcup_{j=1}^{n} \{\{t\}_{N \cup \{i_j\}} \mid \{t\}_N \in T_j\}, J, \mathsf{Data} \rangle}$$

In the set of variables $\mathbf{Var}$ we replace each k where $\gamma(k) = \langle T, \mathsf{Key}_{\{i_1, \ldots, i_n\}} \rangle$ *var* by the variables $k^{(t)}$ and $k^{(i_1)}, \ldots, k^{(i_n)}$. The variable $k^{(t)}$ gets the type $\langle T, \mathsf{Data} \rangle$ *var*; its value chooses which one of the variables $k^{(i_j)}$ is to be used. The type of $k^{(i_j)}$ will be $\langle T, \mathsf{Key}_{\{i_j\}} \rangle$ *var*. Let γ_0 be the new typing. In the program we have to change key generation statements (we assign the key to the right $k^{(i_j)}$ and i_j to $k^{(t)}$), the assignments of a key to a key (we copy $k^{(t)}$ and all variables $k^{(i_j)}$), the uses of a key as an encryption key (we use the expression (28)) and the uses of a key in other ways (we use nested if-statements to check for different values of $k^{(t)}$ and use the right $k^{(i_j)}$ inside). The resulting program P_0 is bisimilar to P, it types according to γ_0 and the values of all variables of P can be recovered from the values of variables of P_0.

We introduce a new variable $k_\mathfrak{n}$ to the set of variables of P_0 and prepend P_0 with the statement $k_\mathfrak{n} := \mathcal{G}en^\mathfrak{n}$; here $\mathfrak{n}$ is a new key label. The key $k_\mathfrak{n}$ will be used by the transformed programs instead of keys that have been processed. In short, $k_\mathfrak{n}$ will play the role of k at the right hand side of (9).

We will now describe one iteration of replacing the encryptions with encryptions under the key $k_\mathfrak{n}$. Let P_o be the current program and γ_o the current typing; initially $\mathsf{P}_o = \mathsf{P}_0$ and $\gamma_o = \gamma_0$. Consider again the type $T_P = \bigvee_{x \in \mathbf{Var}_P} \gamma_{oI}(x)$. The type T_P satisfies certain invariants — they are satisfied for P_0 and remain satisfied during the iterations. First — the key label $\mathfrak{n}$ either does not occur in T_P (this is the case for P_0) or is free in it. Second — if $\mathfrak{n}$ encrypts some $i \in \{\mathsf{h}\} \cup \mathcal{G}$ in T_P then some other key label encrypts i in T_P as well. Therefore, if any key labels occur in T_P as a key (if no key labels occur as a key then h does not occur in T_P; then we are done, see above) then some $i \neq \mathfrak{n}$ is free in T_P. This follows from the fact that each $T \in \mathcal{T}_2$ (in normal form) contains a free key label, if any key labels occur in T as a key. If we delete from T_P all elements $\{t\}_N$ where $\mathfrak{n} \in N$ then we get another element of $\mathcal{T}_2$ where some key labels still occur as a key.

We delete from P_o all statements that access variables of types where i occurs as data; this deletion is identical to the deletion of h above. The deletion does not change the values of public variables because their types are not larger than T_P and i does not occur in T_P as data. In the resulting program keys generated at statements $x := \mathcal{G}en^i$ are only used for encryption. We then replace triples $|i, k, y|$ in $\mathcal{C}\mathcal{E}nc$-expressions, where $\gamma_o(k) = \langle T, \mathsf{Key}_{\{i\}} \rangle$ var, with $|i, k_\mathfrak{n}, \mathbf{0}|$. After that, if the triples of a $\mathcal{C}\mathcal{E}nc$-expression all have $k_\mathfrak{n}$ and $\mathbf{0}$ as their second and third components, we replace the entire expression with $\mathcal{E}nc(k_\mathfrak{n}, \mathbf{0})$. In this way we get rid of encrypted *secrets* — the type of $\mathcal{E}nc(k_\mathfrak{n}, \mathbf{0})$ is $\langle \emptyset, \emptyset, \mathsf{Data} \rangle$. The resulting program $\mathsf{P}_{o'}$ is the input to the next iteration.

The typing γ_o changes as well. All variables in whose types i occurred as data will be deleted. For the rest of the variables x we get $\gamma_{o'}(x)$ from $\gamma_o(x) = \langle T_x, U_x \rangle$ var in the following way. The usage type U_x remains the same. The information type will contain all such $\{t\}_N \in T_x$ where $i \notin N$. If $\{t\}_{N \cup \{i\}} \in T_x$ (here $i \notin N$) then the information type according to $\gamma_{o'}$ may or may not contain $\{t\}_{N \cup \{\mathfrak{n}\}}$. We choose the least $\gamma_{o'}$ satisfying these conditions that is a valid typing of $\mathsf{P}_{o'}$.

5   Conclusions and Future Work

We have presented a type system for computationally secure information flow that should be simple enough to be integrated into existing programming languages and used by software engineers.

The presented type system could definitely be developed further. It could be used for programs containing procedures; the existing data flow analyses [13] cannot cope with them. Extending the programming language with procedures probably requires some form of key label polymorphism.

It is also important to get rid of the constraint that two keys generated at the same program point cannot be distinguished. Here some form of key relabeling during the program run could be useful.

References

1. Abadi, M.: Secrecy by Typing in Security Protocols. Journal of the ACM **46** (1999) 749–786
2. Abadi, M., Blanchet, B.: Secrecy types for asymmetric communication. Theoretical Computer Science **298** (2003) 387–415
3. Abadi, M., Gordon, A.: A Calculus for Cryptographic Protocols: The Spi Calculus. Information and Computation **148** (1999) 1–70
4. Abadi, M., Jürjens, J.: Formal Eavesdropping and Its Computational Interpretation. In proc. of TACS 2001 (LNCS 2215), pages 82–94
5. Abadi, M., Rogaway, P.: Reconciling Two Views of Cryptography (The Computational Soundness of Formal Encryption). In proc. of the International Conference IFIP TCS 2000 (LNCS 1872), pages 3–22
6. Backes, M., Pfitzmann, B.: Symmetric Encryption in a Simulatable Dolev-Yao Style Cryptographic Library. In proc. of CSFW 2004, pages 204–218
7. Canetti, R.: Universally Composable Security: A New Paradigm for Cryptographic Protocols. In proc. of FOCS '01, pages 136–145
8. Denning, D.: A Lattice Model of Secure Information Flow. Communications of the ACM **19** (1976) 236–243
9. Denning, D., Denning, P.: Certification of Programs for Secure Information Flow. Communications of the ACM **20** (1977) 504–513
10. Dolev, D., Yao, A.: On the security of public key protocols. IEEE Transactions on Information Theory **IT-29** (1983) 198–208
11. Laud, P.: Semantics and Program Analysis of Computationally Secure Information Flow. In proc of. ESOP 2001 (LNCS 2028), pages 77–91
12. Laud, P.: Encryption Cycles and Two Views of Cryptography. In proc. of Nordsec 2002, pages 85–100
13. Laud, P.: Handling Encryption in Analyses for Secure Information Flow. In proc. of ESOP 2003 (LNCS 2618), pages 159–173
14. Laud, P., Vene, V.: A Type System for Computationally Secure Information Flow. Tech. Report IT-LU-O-043-050307, Cybernetica AS, March 7th 2005.
15. Lincoln, P., Mitchell, J., Mitchell, M., Scedrov, A.: A Probabilistic Poly-Time Framework for Protocol Analysis. In proc. of ACM CCS '98, pages 112–121
16. Lincoln, P., Mitchell, J., Mitchell, M., Scedrov, A.: Probabilistic Polynomial-Time Equivalence and Security Analysis. In proc. of the World Congress on Formal Methods in the Development of Computing Systems '99 (LNCS 1708), pages 776–793
17. Lindholm, T., Yellin, F.: The Java Virtual Machine Specification. Addison-Wesley (1999)
18. Myers, A.C.: JFlow: Practical Mostly-Static Information Flow Control. In proc. of POPL '99, pages 228–241
19. Sabelfeld, A., Myers, A.C.: Language-Based Information-Flow Security. IEEE Journal on Selected Areas in Communications **21** (2003) 5–19
20. Pfitzmann, B., Waidner, M.: A Model for Asynchronous Reactive Systems and its Application to Secure Message Transmission. In proc. of IEEE S&P 2001, pages 184–200

21. Smith, G., Volpano, D.: Secure Information Flow in a Multi-threaded Imperative Language. In proc. of POPL '98, pages 355–364
22. Volpano, D.: Secure Introduction of One-way Functions. In proc. of CSFW '00, pages 246–254
23. Volpano, D., Smith, G., Irvine, C.: A Sound Type System for Secure Flow Analysis. Journal of Computer Security **4** (1996) 167–187
24. Volpano, D.M., Smith, G.: Eliminating Covert Flows with Minimum Typings. In proc. of CSFW '97, pages 156–169
25. Volpano, D., Smith, G.: Probabilistic Noninterference in a Concurrent Language. In proc. of CSFW '98, pages 34–43
26. Volpano, D., Smith, G.: Verifying Secrets and Relative Secrecy. In: proc. of POPL 2000, pages 268–276
27. Yao, A.: Theory and applications of trapdoor functions (extended abstract). In proc. of FOCS '82, pages 80–91

Algorithms for Graphs Embeddable with Few Crossings Per Edge

Alexander Grigoriev[1] and Hans L. Bodlaender[2]

[1] Maastricht University, Faculty of Economics and Business Administration,
Quantitative Economics, P.O. Box 616,
6200 MD Maastricht, The Netherlands
a.grigoriev@ke.unimaas.nl
[2] Institute of Information and Computing Sciences, Utrecht University,
Padualaan 14, De Uithof, P.O. Box 80089,
3508 TB Utrecht, The Netherlands
hansb@cs.uu.nl

Abstract. We consider graphs that can be embedded on a surface of bounded genus such that each edge has a bounded number of crossings. We prove that many optimization problems, including maximum independent set, minimum vertex cover, minimum dominating set and many others, admit polynomial time approximation schemes when restricted to such graphs. This extends previous results by Baker [1] and Eppstein [7] to a much broader class of graphs.

1 Introduction

Already more than two decades ago, Baker [1] showed that the *maximum independent set* and many other NP-hard optimization problems on graphs admit polynomial time approximation schemes (PTAS) when restricted to planar graphs. The basic idea of Baker's algorithm was to remove the vertices in every kth level of a breadth first search tree (BFS) and to solve the problem on the remaining components by a dynamic programming algorithm. Baker proved that from k ways of choosing which set of levels to remove there is at least one which only decreases the size of the maximum independent set by a factor of at most $(k-1)/k$. Moreover, remaining components after levels deletion are k-outerplanar graphs, and dynamic programming can solve the problem on these components efficiently.

Recently, Eppstein in [7] observed that the results by Baker [1] can be extended to any minor-closed family of graphs satisfying so-called diameter-treewidth property. This implies that the problem admits a PTAS if restricted to bounded-genus graphs. This result has been generalized to other minor-closed classes; in particular, Grohe gave PTAS's for several problems, for any minor-closed family that does not contain all graphs [6].

Nowadays, there is a growing body of work, mainly developed by Demaine and Hajiaghayi, based on the concept of "bidimensionality" and presenting directions for generalizations of the Baker-Eppstein ideas of using a diameter like

M. Liśkiewicz and R. Reischuk (Eds.): FCT 2005, LNCS 3623, pp. 378–387, 2005.

parameter to bound treewidth and thus yielding polynomial time approximation schemes for problems on even more general classes of graphs; see Demaine et al. in [2,3,4].

In this paper we continue the line of investigations — in which way can Baker's technique be further extended? Revisiting Eppstein [7] result, we observe that the restriction that the class of graphs must be minor-closed can be relaxed. By moving from the input graph to an auxiliary graph obtained by replacing each crossing by a vertex and back, we can obtain Baker-type PTAS's for several problems on graphs that are embeddable on a surface of bounded genus (e.g., the plane, the torus) with a bounded number of crossings per edge. We emphasize on the fact that all known results, also in Demaine et al. [2,3,4], work only under assumption that the graph family is minor-closed. In contrast, in this paper, we introduce the graph families on which Baker-Eppstein techniques work perfectly but actually any graph is a minor of sufficiently large graph of the considering families.

In the end of the paper we present several additional results which provide an insight on the graphs with few crossings per edge.

2 Problem and Definitions

We illustrate the basic ideas of the PTAS on the maximum independent set problem. Given a graph $G = (V, E)$, we look for a maximum cardinality independent set in G, i.e., a vertex subset $V' \subseteq V$ such that no two vertices from V' are adjacent by an edge from E. This problem is known to be NP-hard even for planar graphs. The problem admits a PTAS if restricted to planar graphs [1] and even to bounded-genus graphs [7]. Let $n = |V|$.

Definition 1 (Good embedding). *We call an embedding of graph G on a surface S of genus g a* good *embedding if it satisfies the following conditions: (i) all vertices of the graph are given as distinct points in S; (ii) no two edge crossings happen in the same point in S; (iii) for any edge no vertex of the graph, except the endpoints of the edge, is situated on the edge.*

Definition 2 (Crossing parameter). *Let the* crossing parameter φ *of a graph (on surface S) be the minimum over all good embeddings on S of the maximum over all edges e of the number of edge crossings of e.*

Through this paper we assume that a good embedding of G is given and both the crossing parameter φ and the genus g of S are bounded by some constants. Clearly, the graph is planar if $g = 0$ and $\varphi = 0$.

Definition 3 (Tree decomposition). *A* tree decomposition $(\{X_i \mid i \in I\}, T = (I, F))$ *of a graph $G = (V, E)$ is a pair, with $\{X_i \mid i \in I\}$ a collection of subsets of V (called* bags*), such that*

- $\bigcup_{i \in I} X_i = V$.
- *For all $\{v, w\} \in E$, there is an $i \in I$ with $v, w \in X_i$.*

- For each $v \in V$, the set $T_v = \{i \in I \mid v \in X_i\}$ forms a connected subtree of T.

Definition 4 (Treewidth). The width of a tree decomposition $\{X_i \mid i \in I\}$ is $\max_{i \in I} |X_i| - 1$. The treewidth of a graph G is the minimum width over all tree decompositions of G.

3   The Polynomial Time Approximation Scheme

We now describe our polynomial time approximation scheme for the maximum independent set problem on graphs with bounded crossing parameter on bounded genus. We assume the embedding is given. Consider the following algorithm $\mathcal{A}$ which is a revised version of the algorithms by Baker [1] and Eppstein [7].

Input: Graph G, parameter k (without loss of generality, let $\varphi < k$).

Algorithm $\mathcal{A}$:

1. Construct the graph $G' = (V', E')$ obtained from G by replacing each edge crossing by a vertex. This can be done by the following recursive procedure. Starting from graph G, find in the graph embedding a pair of crossing edges. Let (v_1, u_1) and (v_2, u_2) be such edges. Redefine the graph introducing at the crossing point a new vertex w and replacing edges (v_1, u_1) and (v_2, u_2) by edges (v_1, w), (v_2, w), (u_1, w), and (u_2, w). Recurse on the new graph unless there is no edge crossings.
2. Build a breadth first search tree T of G', with an arbitrary root v_0, and consider the levels of the tree (i.e., vertex sets with equal distance to v_0).
3. For all i, $0 \le i \le k$, we perform the following procedure.
   (a) Remove from G' all levels of T congruent to $i (\mathrm{mod}\ k)$ together with their φ successive levels. This decomposes G' into a collection of subgraphs $\mathcal{H} = \{H_1, H_2, \ldots, H_r\}$ where each subgraph $H_t = (V_t, E_t)$ is induced by $k - \varphi - 1$ consecutive levels in T of G'.
   (b) Consider a subgraph G_t of G induced by vertices $V_t \cap V$. Since the number of crossings per edge is at most φ and we removed $\varphi + 1$ consecutive levels from G', we have that after deletion of levels there is no edge $e \in E$ such that its two endpoints belong to two different subgraphs $G_{t'}$ and $G_{t''}$. Therefore, for each i, $0 \le i \le k$, we have a subgraph of G formed by a collection of disconnected subgraphs $G_1, G_2, \ldots, G_r$. By Lemma 1 below, the treewidth of G_t is bounded by $O(k)$ for all $t = 1, 2, \ldots, r$. Hence, the maximum independent set for G_t can be found in time $O(n2^{O(k)})$ by a dynamic programming algorithm, using standard treewidth techniques; see, e.g., Telle and Proskurowski [12].
   (c) Let S_i be a union of the maximum independent sets of all G_t, $t = 1, 2, \ldots, r$.
4. Define $S_{\max}$ by a maximum cardinality set over all $S_i, 0 \le i \le k$.

Output: Return $S_{\max}$.

The following lemma is a key for algorithm $\mathcal{A}$ and for the main result of the paper.

Lemma 1. *The treewidth of G_t is bounded by $O(k)$ for all $t = 1, \ldots, r$.*

Proof. Consider a subgraph H_t induced by levels $r + 1, r + 2, \ldots, r + s$ in T of G' where $s = k - \varphi - 1 = O(k)$. Consider a minor of G' obtained by contraction of the first r levels in T to a single vertex and deletion of all levels above $r + s$. Clearly, this minor is a graph of genus g. Moreover, it has a diameter of at most $2(k - \varphi - 1) = O(k)$. By Eppstein [7] the treewidth of such a minor is $O(gk)$. Therefore, H_t as a subgraph of such a minor has the treewidth of at most $O(gk)$ as well.

Now, let us estimate how much the treewidth of G_t and H_t can differ. Construct a graph H'_t from H_t by replacing each vertex v in H_t that represents an edge crossing, say e_1 and e_2, by two adjacent vertices v_1 and v_2 representing e_1 and e_2 respectively. Let v_1 be adjacent to all vertices corresponding to the neighborhood of v representing e_1, and let v_2 be adjacent to all vertices corresponding to the neighborhood of v representing e_2. A tree decomposition of H_t of treewidth d can be turned into a tree decomposition of H'_t of treewidth at most $2d + 1$, by replacing each occurrence of an vertex that represents a crossing of two edges in a bag by the corresponding two vertices; this gives a tree decomposition of H'_t whose maximum bag size is at most doubled. One can also observe that we can select for each edge in G_t a path in H'_t between its endpoints, such that these paths do not have internal vertices in common. Thus, G_t is a minor of H'_t and hence the treewidth of G_t is at most twice the treewidth of H_t plus one, and thus $O(gk) = O(k)$ as required. □

Now, we are ready to summarize the main results of the paper in the following theorem and corollary.

Theorem 1. *Algorithm $\mathcal{A}$ outputs an independent set of graph G of size at least $1 - O(1/k)$ times the optimum in time $O(kn2^{O(k)})$, and thus, there is a PTAS for maximum independent set for graphs given with an embedding on a surface of bounded genus and with bounded crossing parameter.*

Proof. Since for all i, $0 \leq i \leq k$, set S_i is a union of independent sets of disconnected subgraphs of G, Algorithm $\mathcal{A}$ returns an independent set of graph G.

As in Baker [1], there is at least one i, $0 \leq i \leq k$, such that at most $(\varphi + 1)/k$ of the nodes in the optimal solution are at the levels congruent to i, $i + 1$, $\ldots, i + \varphi \pmod{k}$, otherwise we would have a contradiction to maximality of the solution. This implies that $|S_{\max}|$ is approximating the optimum within a factor $(k - \varphi - 1)/k = 1 - O(1/k)$.

Notice that the most time consuming operation in Algorithm $\mathcal{A}$ is the dynamic programming used in step 3. As we already noticed above, this dynamic programming requires $O(n2^{O(k)})$ time. Since we run step 3 for all choices of i, $0 \leq i \leq k$, the total running time of Algorithm $\mathcal{A}$ is $O(kn2^{O(k)})$. □

Corollary 1. *For each of the following problems (and many others) there is a PTAS for graphs embeddable on a surface of bounded genus with bounded crossing parameter:*

- *minimum vertex cover;*
- *minimum dominating set;*
- *minimum edge dominating set;*
- *minimum triangle matching;*
- *maximum H-matching;*
- *maximum tile salvage.*

Proof. This can be proven in the same way as Theorem 1, using techniques similar to those of Baker [1].   □

4   More on the Crossing Parameter

Graphs with bounded crossing parameter were investigated by several authors in the context of graph drawing; see, e.g., Pach and Toth [11]. However, before this article nothing was known on the recognition complexity of the graphs with small crossing parameter. To give the reader more insight on the graphs with few crossings per edge, in this section we present some results on the computational complexity of the crossing parameter and some other useful properties of the class of graphs with bounded crossing parameter.

Theorem 2. *Crossing parameter 1 recognition in the plane is NP-complete.*

Proof. We prove the theorem by reduction of the well known strongly NP-complete problem 3-PARTITION; see Garey and Johnson [5]: Given a set A of $3m$ elements, a bound $B \in \mathbb{Z}^+$, and a size $s(a) \in \mathbb{Z}^+$ for each $a \in A$ such that $B/4 < s(a) < B/2$ and such that $\sum_{a \in A} s(a) = mB$, can A be partitioned into m disjoint sets $A_1, A_2, \ldots, A_m$ such that for $1 \leq i \leq m$, $\sum_{a \in A_i} s(a) = B$?

Before starting the reduction, we would like to mention several properties of the complete graph on six vertices: (i) by Guy's conjecture proven for complete graphs on up to 10 vertices, see, e.g., [13], K_6 has crossing number 3 (i.e., the minimum possible number of edge crossings with which K_6 can be drawn in the plane is 3); (ii) K_6 can be drawn in the plane with three crossings, at most one crossing per edge, and two vertices in the exterior, see Figure 1; (iii) in any drawing of K_6 having at most one crossing per edge for any two vertices of the graph there is a path between those two vertices such that all edges of the path are crossed. For a proof of (iii) see Theorem 4 in Appendix. Taking into account properties (i)-(iii), we can use graph K_6 as an edge that cannot be crossed. In the figures below thick edges are graphs K_6.

Now, we reduce 3-PARTITION to the crossing parameter 1 recognition. Given an instance of 3-PARTITION, we construct the graph for the crossing parameter 1 recognition as follows. For each element $a \in A$ we introduce a gadget P_a called *splitter* which is a simple star having $s(a) + 1$ edges. We also introduce two special gadgets called *transmitter* and *collector*. Both these gadgets have a "double"-wheel form with $3m$ thick radials for the transmitter and Bm thick radials for the collector; see Figure 2.

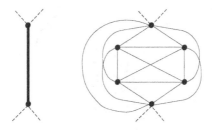

Fig. 1. Thick edge is a graph K_6

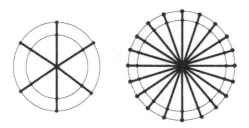

Fig. 2. $m = 2$ transmitter and $B = 10$; $m = 2$ collector

We finish construction by adding the following edges:

- We connect the transmitter center to a degree 1 vertex of each splitter $P_a, a \in A$.
- We connect remaining $s(a)$ degree 1 vertices of each splitter $P_a, a \in A$, to the collector center.
- Let a cycle $[t_1, t_2, \ldots, t_{3m}]$ be an exterior circuit of the transmitter and a cycle $[c_1, c_2, \ldots, c_{Bm}]$ be an exterior circuit of the collector. For all $i \in \{1, 2, \ldots, m\}$, connect vertex t_{3i} to vertex c_{Bi} by a thick edge; for illustration see Figure 3.

Let us refer to the obtained graph as to G. Now, we claim that G is embeddable with at most one crossing per edge if and only if the instance of 3-PARTITION has an affirmative answer.

Part "IF" of the claim is rather straightforward. We illustrate this with an instance of 3-PARTITION having 6 elements of weights 2,3,3,3,4,5. This instance has a required partition (3+3+4=10 and 2+3+5=10) and the corresponding graph G can be drawn with at most one crossing per edge as on Figure 3. In general, we draw graph G as follows. First we draw transmitter and collector such that both are placed in the exterior face of each other. Then, we connect by thick edges in the exterior each third vertex of the transmitter to each Bth vertex of the collector, creating m distinct faces. We assign the splitters to the faces according to the partition. Since the total size of each triple in the partition is B and in each of the m faces the collector has B edges in the exterior circle,

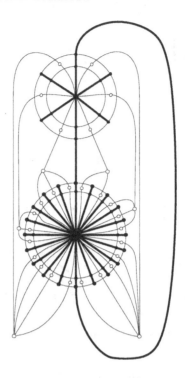

Fig. 3. Instance with $A = \P 2^c\, 3^c\, 3^c\, 3^c\, 4^c\, 5\Diamond$

we can assign the edges of splitters to the sectors of the collector such that each edge will be crossed only once.

Now we prove part "ONLY IF" of the statement. Consider a drawing of G having at most one crossing per edge and let us construct the corresponding partition for 3-PARTITION.

First, let us analyze the possible ways of drawing G. It is convenient to consider the possible drawings on a globe. It is well known that a sphere drawing has an equivalent planar representation with respect to the edge crossings; see [8, Proposition 8.3.1]. Without loss of generality we can assume that the transmitter center is a North Pole and the collector center is a South Pole of the globe. By construction, the globe is partitioned by the thick non-crossable meridian paths into m distinct faces $F_1, F_2, \ldots F_m$. Moreover, since these meridian paths are non-crossable, the ordering of the meridian paths on the globe and the ordering of the faces on the globe correspond to the vertex ordering in the exterior circuits of the transmitter and collector.

Now, let us find out how can we draw the thick paths of the transmitter adjacent to the North Pole but not participating in the meridian paths. Consider four consequent vertices of the exterior circuit of the transmitter, for instance, t_{3m}, t_1, t_2, t_3. By construction and observation above, t_{3m} and t_3 are the vertices on two consequent meridian paths. These two meridian paths form one of the distinct faces, say F_1. Vertices t_1 and t_2 in the drawing must be placed in F_1

otherwise at least one of the meridian paths will be crossed. Moreover, since the transmitter is a "double"-wheel, the ordering of the thick paths adjacent to the North Pole and ending in t_1 and t_2 must be consistent with the ordering of the vertices in the exterior circuit of the transmitter. The same arguments work for all other consequent four-tuples of the exterior circuit of transmitter. This implies that there is a unique way of drawing the transmitter around the North Pole. Similarly, there is a unique way of drawing the collector around the South Pole. We also notice, that since we can not cross thick edges and other edges can be crossed at most once, we do not have any intersections between transmitter and collector.

Now, consider a drawing of the splitters. In any face F_i, $i \in \{1, 2, \ldots, m\}$, we can place at most 3 splitters, otherwise one of the edges of the transmitter will be crossed more than once. The center of a splitter must be placed in the exterior of the transmitter and collector, otherwise one of the splitter edges will be crossed more than once. Hence, for each face F_i, $i \in \{1, 2, \ldots, m\}$, the number of paths between the South Pole and the centers of three splitter assigned to F_i is at most B. Since we have in total Bm such paths, each face contains exactly three splitters with exactly B paths between the splitter centers and the South Pole.

Consider a partition of set A correspondent to the assignment of splitters to the faces. By observation above, each triple of numbers correspondent to three splitters assigned to a face sums to B. Therefore, A has a required partition. It remains to notice that 3-PARTITION is strongly NP-complete and we are allowed to use unary encoding to describe the inputs of the problems. Hence, the reduction was polynomial.   □

Corollary 2. *When $P \neq NP$, there does not exist a polynomial time 2-approximation algorithm for finding the crossing parameter of a graph on the plane.* □

Notice, however, that several natural classes of graphs have a bounded crossing parameter on the plane. For instance, graphs of intersections of objects in the plane with bounded objects density (disk graphs with bounded density are special case of these); graphs with bounded degree and bounded tree width; planar graphs.

Observation 3. *The class of graphs with an embedding on the plane with crossing parameter 1 is not closed under taking minors. In fact, every graph is a minor of a graph with crossing parameter 1: take any good embedding, and then add a new vertex of degree two between every two successive crossings.*

From work on the crossing *number* of graphs (the minimum total number of crossings in a planar embedding), we can also obtain bounds on the crossing parameter (on the plane). E.g., the crossing number of a complete graph with n vertices is $\Theta(n^4)$ [9], hence its crossing parameter is $\Theta(n^2)$.

5   Conclusions and Open Problems

For several classes of graphs, it is now known that there are polynomial time approximation schemes for a large collection of problems. Each of these build upon the work by Baker [1]. In this paper, we gave a new class of graphs where the same approach can be used. An interesting question is whether there is a general notion under which the different results of the type can be unified.

A disadvantage of our algorithm is that an embedding with bounded crossings per edge is requested as part of the input. As discussed earlier, for some applications, we indeed get such an embedding. However, it would be interesting if "robust" versions of the algorithms can be designed, i.e., algorithms that do not need the embedding as part of the input. Note that such a robust PTAS has been designed by Nieberg et al. for the dominating set problem on unit disk graphs [10].

Recent work (see e.g., [2]) shows that there is a PTAS for the connected dominating set problem and other related problems on planar graphs and generalizations of it. It would be interesting to see if these results carry over to graphs with bounded crossing parameter.

Acknowledgements

This research is a part of the project "Tree Width and Combinatorial Optimization (TACO)" supported by the Netherlands Organization for Scientific Research NWO.

References

1. Baker, B.: Approximation algorithms for NP-complete problems on planar graphs. JACM **41** (1994) 153–180.
2. Demaine, E., Hajiaghayi, M.: Bidimensionality: New Connections between FPT Algorithms and PTASs. In Proc. of the 16th Annual ACM-SIAM Symposium on Discrete Algorithms (SODA 2005), Vancouver, British Columbia, Canada (2005) 590–601.
3. Demaine, E., Hajiaghayi, M.: Graphs Excluding a Fixed Minor have Grids as Large as Treewidth, with Combinatorial and Algorithmic Applications through Bidimensionality. In Proc. of the 16th Annual ACM-SIAM Symposium on Discrete Algorithms (SODA 2005), Vancouver, British Columbia, Canada (2005) 682–689.
4. Demaine, E., Nishimura, N., Hajiaghayi, M., Ragde, P., Thilikos, D.M.: Approximation algorithms for classes of graphs excluding single-crossing graphs as minors. J. Comp. System Sc. **69** (2004) 166–195.
5. Garey, M.R., Johnson, D.S.: Computers and intractability: A guide to the theory of NP-completeness. W.H. Freeman, San Francisco (1979).
6. Grohe, M.: Local tree-width, excluded minors, and approximation algorithms. Combinatorica **23** (2003) 613-632.
7. Eppstein, D.: Diameter and treewidth in minor-closed graph families. Algorithmica **27** (2000) 275–291.

8. Gross, J., Yellen, J.: Graph Theory and Its Applications. New York, CRC Press (1999).
9. Leighton, F.T.: New lower bound techniques for VLSI. Math. Systems Theory **17** (1984) 47–70.
10. Nieberg, T., Hurink, J.L., Kern, W.: A robust PTAS for Maximum Independent Sets in unit disk graphs. In Proc. of the 30th Workshop on Graph Theoretic Concepts in Computer Science, Bad Honnef, (WG 2004), Germany (2004) 214–221.
11. Pach, J., Toth, G.: Graphs drawn with few crossings per edge. Combinatorica **17** (1997) 427–439.
12. Telle, J.A., Proskurowski, A.: Algorithms for vertex partitioning problems on partial k-trees. SIAM J. Disc. Math. **17** (1997) 529–550.
13. World Wide Web: http://mathworld.wolfram.com/GraphCrossing Number.html

Appendix

Theorem 4. *In any planar embedding of K_6 having at most one crossing per edge, between any two vertices there exists a path such that all edges in that path are crossed.*

Proof. First we prove that each vertex is contained in at least two distinct crossed edges. Assume this is not true and there is an embedding of K_6 such that for some vertex v_1 edges $e_1 = (v_1, v_2)$, $e_2 = (v_1, v_3)$, $e_3 = (v_1, v_4)$, $e_4 = (v_1, v_5)$ are not crossed by any edge. Without loss of generality, we assume that edges e_1, e_2, e_3, e_4 are drawn clockwise in this particular order. Since graph is complete, there is a simple cycle formed by edges e_1, e_3 and $e_5 = (v_2, v_4)$ with only one edge e_5 that can be crossed. Vertices v_3 and v_5 belong to the different faces formed by that cycle. Therefore edge (v_3, v_5) crosses e_5. On the other hand, there is a vertex v_6 that must be connected to both v_3 and v_5. Hence e_5 is crossed at least twice that leads to the contradiction.

Now, we prove that crossed edges form a connected graph. For a contradiction we assume that there are 2 or more connectivity components. Since every vertex is contained in two distinct crossed edges, each connectivity component has at least 3 vertices. Therefore, we can have only two components with 3 vertices each. Moreover, each component forms a triangle (cycle). Hence, the question is whether we can cross two triangles with curved sides such that each side of each triangle will be crossed exactly once? Take a side (v_1, v_2) of triangle 1. Vertices v_1 and v_2 belong to different faces formed by triangle 2. The third vertex v_3 of triangle 1 will share the face either with v_1 or with v_2. Therefore, either edge (v_3, v_1) or edge (v_3, v_2) will cross the boundary of triangle 2 even number of times which contradicts to the requirement that each edge is crossed once. Therefore, crossed edges in K_6 form a connected graph as required.   □

The reader may even verify that, when K_6 is drawn with at most one crossing per edge, the crossing edges form a Hamiltonian circuit. This observation is out of the scope of this article and we leave it without a proof.

Approximation Results for the Weighted P_4 Partition Problems

Jérôme Monnot[1] and Sophie Toulouse[2]

[1] Université Paris Dauphine, LAMSADE, CNRS UMR 7024,
75016 Paris, France
monnot@lamsade.dauphine.fr
[2] Université Paris 13, Institut Galilée LIPN, CNRS UMR 7030,
93430 Villetaneuse, France
sophie.toulouse@lipn.univ-paris13.fr

Abstract. We present several new standard and differential approximation results for P_4-partition problem by using the algorithm proposed in
Hassin and Rubinstein (Information Processing Letters, 63: 63-67, 1997),
for both minimization and maximization versions of the problem. However, the main point of this paper is the robustness of this algorithm,
since it provides good solutions, whatever version of the problem we deal
with, whatever the approximation framework within which we estimate
its approximate solutions.

Keywords: Graph partition, 3-length chain, approximation algorithms,
performance ratio, standard approximation, differential approximation.

1 Introduction

Consider an instance I of an **NP**-hard optimization problem Π and a polynomial time algorithm A computing feasible solutions for Π. Denote by $\mathrm{apx}_{\Pi}(I)$
the value of a solution computed by A on I, by $\mathrm{opt}_{\Pi}(I)$, the value of an optimal solution and by $\mathrm{wor}_{\Pi}(I)$ the value of a worst solution (that corresponds
to the optimal value when reversing the optimization goal). The quality of A is
expressed by the way of approximation ratios that somehow compare the approximate value to the optimal one. So far, two measures stand out from the
literature: the *standard* ratio [2] (the most widely used) and the *differential* ratio
[3,4,5,8]. The standard ratio is defined by $\rho_{\Pi}(I, \mathrm{A}) = \frac{\mathrm{apx}_{\Pi}(I)}{\mathrm{opt}_{\Pi}(I)}$ if Π is a maximization problem, by $\rho_{\Pi}(I, \mathrm{A}) = \frac{\mathrm{opt}_{\Pi}(I)}{\mathrm{apx}_{\Pi}(I)}$ otherwise, whereas the differential ratio is
defined by $\delta_{\Pi}(I, \mathrm{A}) = \frac{\mathrm{wor}_{\Pi}(I) - \mathrm{apx}_{\Pi}(I)}{\mathrm{wor}_{\Pi}(I) - \mathrm{opt}_{\Pi}(I)}$. Instead of dividing the approximate value
by the optimum value, this latter measure divides the distance from a worst so-
lution to the approximate value by the instance diameter. Within the worst case
analysis framework and given a universal constant $\varepsilon \leq 1$ (*resp.*, $\varepsilon \geq 1$), an algorithm A is said to be an ε-standard approximation for a maximization (*resp.*
a minimization) problem Π if $\rho_{\mathrm{A}_{\Pi}}(I) \geq \varepsilon \ \forall I$ (*resp.*, $\rho_{\mathrm{A}_{\Pi}}(I) \leq \varepsilon \ \forall I$). According to differential ratio, A is said to be an ε-differential approximation for Π

M. Liśkiewicz and R. Reischuk (Eds.): FCT 2005, LNCS 3623, pp. 388–396, 2005.

if $\delta_{\mathbf{A}_{II}}(I) \geq \varepsilon$, $\forall I$, for a universal constant $\varepsilon \leq 1$. Alternatively, any solution being a convex combination of the two values $\mathrm{wor}_{II}(I)$ and $\mathrm{opt}_{II}(I)$, an approximate value $apx_{II}(I)$ will be an ε-differential approximation if for any instance I, $apx_{II}(I) \geq \varepsilon \times \mathrm{opt}_{II}(I) + (1 - \varepsilon) \times \mathrm{wor}_{II}(I)$ (for the maximization case). Within the worst case analysis framework according to both standard and differential ratios, we focus on a special problem, the weighted P_4-partition problem (P_4P in short). Furthermore, we study the performance of a single algorithm on various versions of this problem. Precisely, we show that this algorithm is efficient for P_4P, proving approximation ratios for both standard and differential measures, for both maximization and minimization versions of the problem.

In the weighted P_4-partition problem, we are given a complete graph K_{4n} together with a distance function $d : E \to \mathbb{R}^+$ on its edges. A P_4 is an induced path of length 3 (or, equivalently, an induced path on 4 vertices); for any instance $I = (K_{4n}, d)$, the cost of a path P is given by the sum of its edges weight, and the goal is to find a partition $T^* = \{P_1^*, \ldots, P_n^*\}$ of n vertex-disjoint chains of length 3 (that we call a P_4-partition) such that $d(T^*) = \sum_{i=1}^q d(P_i^*)$ is optimum, i.e., of maximum distance if the goal is to maximize ($\mathrm{MAXP_4P}$), of minimum distance otherwise ($\mathrm{MINP_4P}$). For the minimization version, we will more often assume that the distance function satisfies the triangular inequality, i.e., $d(x, y) \leq d(x, z) + d(z, y), \forall x, y, z$; $\mathrm{MINMETRICP_4P}$ will refer to this restriction. Finally, we also deal with a special case of metric instances where the distance function is worth either 1 or 2; the corresponding problems will be denoted by $\mathrm{MAXP_4P_{1,2}}$ and $\mathrm{MINP_4P_{1,2}}$. All these problems are known to be **NP**-hard, [7,14]; nevertheless, $\mathrm{MAXP_4P}$ is standard-approximable within ratio 3/4, Hassin and Rubinstein, [9], whereas (to our knowledge), no approximation rate has been established for $\mathrm{MINP_4P}$ yet. Concerning $\mathrm{MINP_kP}$ (the general version), it cannot be approximated within $2^{p(n)}$ for any polynomial p and for any k; this is due to the fact that deciding whether a graph admits a P_k-partition of its vertex set is **NP**-complete, [7,13,14]. The P_k-partition problem consists in, given a simple graph $G = (V, E)$, deciding if there exists a collection $\mathcal{P}$ of vertex-disjoint k-length paths $\{P_1, \ldots, P_\ell\}$ such that any vertex from V belongs to exactly one path P_i from $\mathcal{P}$. Finally, note that $\mathrm{MINMETRICP_kP}$ is very close to the k-vehicle routing problem when restricting the route of each vehicle to at most k intermediate stops, [1,6].

In the first section, we study the relationship between TSP and P_kP under differential ratio, showing how a differential approximation for TSP allows a differential approximation for P_kP (where P_kP seeks to determine a partition of a vertex set into k-length paths of optimum weight). In the second section, that contains the main result of this paper, we propose a complete analysis, from both a standard and a differential point of view, of an algorithm proposed by Hassin and Rubinstein [9]. We prove that, with respect to the standard ratio, this algorithm provides new approximation rates for $\mathrm{METRICP_4P}$, that is to say: the approximate solution respectively achieves a $3/2$-, a $7/6$- and a $9/10$-standard approximation for $\mathrm{MINMETRICP_4P}$, $\mathrm{MINP_4P_{1,2}}$ and $\mathrm{MAXP_4P_{1,2}}$. Under differential ratio, the approximate solution is a $1/2$-approximation for general P_4P, a $2/3$-approximation for $P_4P_{a,b}$. The gap between differential and standard ratios for maximization version of a problem come from the fact that within differential framework, the approximate value has to be located into $[\mathrm{wor}(I), \mathrm{opt}(I)]$, instead of $[0, \mathrm{opt}(I)]$ within the standard one. That is the point of differential measure: the reference it does to $\mathrm{wor}(I)$ makes it more precise and makes robust the approximation level (since minimizing and maximizing are equivalent and more generally, differential ratio is stable under affine transformation of the objective function). In addition to the new approximation bounds that we provide, the main result of this paper is the robustness of the algorithm we study, since this latter provides good solutions, whatever version of the problem we deal with, whatever the approximation framework within which we estimate its approximate solutions.

2   From Traveling Salesman Problem to P_kP

A common technic to obtain an approximate solution for MAXP_kP from a Hamiltonian cycle is called the *deleting and turning around* method, see [9,10,6]. This method consists, starting from a tour, in producing k solutions of MAXP_kP by turning around the cycle and deleting 1 edge upon k; then it picks the best solution. Obviously, the quality of the output T' depends on the quality of the initial tour. Hence, it is proved in [9,10], that any ε-standard approximation of MAXTSP provides a $\frac{k-1}{k}\varepsilon$-standard approximation for MAXP_kP. Under a differential point of view, things are less optimistic, since already for $k = 4$ there exists an instance family $(I_n)_{n \geq 1}$ verifying $\mathrm{apx}(I_n) = \frac{1}{2}\mathrm{opt}_{\mathrm{MAXP}_4P}(I_n) + \frac{1}{2}\mathrm{wor}_{\mathrm{MAXP}_4P}(I_n)$. For $n \geq 1$, the instance is $I_n = (K_{8n}, d)$ where the vertex set can be partitioned into two sets $L = \{\ell_1, \ldots, \ell_{4n}\}$ and $R = \{r_1, \ldots, r_{4n}\}$; the associated distance function d equals 0 on L^2, 2 on R^2, and 1 on $L \times R$. We have for any $n \geq 1$:

Property 1. $\mathrm{apx}(I_n) = 6n$, $\mathrm{opt}_{\mathrm{MAXP}_4P}(I_n) = 8n$ and $\mathrm{wor}_{\mathrm{MAXP}_4P}(I_n) = 4n$.

Nevertheless, using this method one can obtain:

Lemma 1. *From an ε-differential approximation of MAXTSP, one can polynomially compute a $\frac{\varepsilon}{k}$-differential approximation of MAXP_kP*

Thus, from [8,11], we deduce a $\frac{2}{3k}$-differential approximation for MAXP_kP. Finally, observe that even if we consider $\mathrm{MINMETRICP}_4P$, we are not able to obtain a result as good as standard approximation for MAX_4P. Consider the instances $I'_n = (K_{8n}, d')$ built from I_n where we modify the distance d by $d'(\ell_i, \ell_j) = d'(r_i, r_j) = 1$ and $d'(\ell_i, r_j) = n^2 + 1$ for all i, j. We thus have $\mathrm{opt}_{\mathrm{MINMETRICTSP}}(I'_n) = 2n^2 + 8n$ and $\mathrm{opt}_{\mathrm{MINMETRICP}_4P}(I'_n) = 6n$.

3   Approximating P_4P by the Way of Optimal Matchings

Here starts the analysis, both on a standard and a differential point of view, of an algorithm proposed by Hassin and Rubinstein [9]; the authors show that the approximate solution is a 3/4-standard approximation for MAXP_4P. First, dealing with the standard ratio, we prove that this algorithm provides a 3/2-approximation for $\text{MINMETRICP}_4\text{P}$ and respectively a 7/6 and a 9/10-approximation for $\text{MINP}_4\text{P}_{1,2}$ and $\text{MAXP}_4\text{P}_{1,2}$ when restricting us to 1, 2-valuated graphs. As corollary of a general result, we also obtain an alternate proof of the result of [9].

We then prove that, with respect to the differential measure, the approximate solution achieves a 1/2-approximation in general graphs, for both maximization and minimization versions of the problem; this latter ratio is raised up to 2/3 in bi-valuated graphs.

3.1 Description of the Algorithm

The algorithm proposed in [9] runs in two stages: first, it computes an optimum weight perfect matching $M_{T'}$ on (K_{4n}, d); then, it builds a second optimum weight perfect matching $R_{T'}$ on edges $M_{T'}$ in order to complete the solution (note that "optimum weight" signifies "*maximum weight*" if the goal of P_4P is to maximize, "*minimum weight*" if the goal is to minimize). Precisely, we define the instance (K_{2n}, d') (to any edge $e_v \in M_{T'}$ corresponds a vertex v in K_{2n}), where the distance function d' is defined as follows: for any edge $[v_1, v_2]$, $d'(v_1, v_2)$ is set to the weight of the heaviest edge linking e_{v_1} and e_{v_2}, that is, if v_1 represents $e_{v_1} = [x_1, y_1]$ and v_2 represents $e_{v_2} = [x_2, y_2]$, then $d'(v_1, v_2)$ is set to $\max\{d(x_1, x_2), d(x_1, y_2), d(y_1, x_2), d(y_1, y_2)\}$ (when dealing with the minimum version of the problem, set the weight to the lightest). On (K_{2n}, d'), we build an optimum weight matching $R_{T'}$ that is transposed to (K_{4n}, d) by selecting the edge that realizes the same cost. Since the computation of an optimum weight perfect matching is polynomial, the whole algorithm runs in polynomial time for both versions.

3.2 General P_4P Within the Standard Framework

For any solution T, we denote respectively by M_T and R_T the set of the final edges and the set of middle edges of its chains. Moreover, for any edge $e \in T$, we denote by $P_T(e)$ the P_4 from T that contains e, by $C_T(e)$ the cycle of length 4 containing $P_T(e)$ ($C_T(e)$ is somehow the closure of $P_T(e)$). Finally, we define $\overline{R_T} = \cup_{e \in T}(C_T(e) \setminus P_T(e))$; note that $R_T \cup \overline{R_T}$ is a perfect matching.

Lemma 2. *For any instance $I = (K_{4n}, d)$, if T is a feasible solution and T^* an optimal solution, then there exist 3 pairwise disjoint edge sets A, B and C verifying:*

(i) $A \cup B = T^*$ and $C \subseteq \overline{R_{T^*}}$.
(ii) $A \cup C$ and $B \cup (\overline{R_{T^*}} \setminus C)$ are both perfect matchings on I.
(iii) $A \cup C \cup M_T$ is a 2-matching on I of which cycles are of length a multiple of 4.

Proof. Let $T^* = M_{T^*} \cup R_{T^*}$ be an optimal solution; we apply the following process:

---

1 Set $A = M_{T^*}$, $B = R_{T^*}$, $C = \emptyset$ and $G' = (V, A \cup M_T)$;
2 While there exists a connected component V' of G' having an edge $e = [x, y] \in R_{T^*}$ with $x \in V'$ and $y \notin V'$, do:
  2.1 $A \leftarrow (A \setminus P_{T^*}(e)) \cup \{e\}$, $B \leftarrow (B \cup P_{T^*}(e)) \setminus \{e\}$, $C \leftarrow C \cup (C_{T^*}(e) \setminus P_{T^*}(e))$;
  2.2 $G' \leftarrow (V, A \cup M_T)$;
3 output A, B and C;

At the initialization stage, the connected components of the partial graph induced by $(A \cup C \cup M_T)$ are either cycles alternating edges from $(A \cup C)$ and M_T, or isolated edges from $M_{T^*} \cap M_T$. During step 2, at each iteration, the process merges together two cycles, or a cycle and an isolated edge, or two isolated edges, into a single cycle; an illustration of the process is proposed in Figure 1. Note that all along the process, the sets A, B and C remain pairwise disjoints.

For (i): Immediate from definition of the process.

For (ii): Sets C and $\overline{R_{T^*}} \setminus C$ are both matchings by construction; let us now prove that A and B also are matchings. Two edges $e \neq e'$ from A (*resp.*, B) are adjacent one to each other if and only if they belong to the same chain P_{T^*}. However, according to steps 1 and 2.1, for any chain P_{T^*}, either $P_{T^*} \cap R_{T^*}$ or $P_{T^*} \cap M_{T^*}$ may belong to A (*resp.*, B), that contradicts the fact that e and e' could be adjacent.

Moreover, before processing step 2, $A \cup C = M_{T^*}$ is a perfect matching; now, at each iteration of step 2, we swap two couples of edges $C_{T^*} \cap M_{T^*}$ and $C_{T^*} \setminus M_{T^*}$ (the former edges are removed from A, the latter are added to $A \cup C$); thus, the resulting edge set $A \cup C$ remains a perfect matching. Finally, since the equality $B \cup (\overline{R_{T^*}} \setminus C) = (T^* \cup \overline{R_{T^*}}) \setminus (A \cup C)$ holds, the set $B \cup (\overline{R_{T^*}} \setminus C)$ also is a perfect matching.

For (iii): At the end of process, $(A \cup C) \cap M_T = \emptyset$ and thus, $A \cup C \cup M_T$ is a perfect 2-matching. Now consider a cycle Γ of $(V, A \cup C \cup M_T)$; by definition

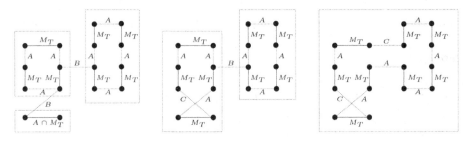

Fig. 1. The construction of sets A and C

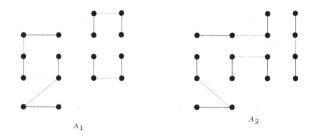

Fig. 2. The sets $A_1 \cup M_{T'}$ and $A_2 \cup M_{T'}$

of step 2, for any edge $e = [x, y] \in \Gamma \cap R_{T^*}$, its begin and end points both belong to $V(\Gamma)$, which means that $\Gamma \setminus M_T$ is a subset of T^* and therefore, we get $|V(\Gamma)| = 4k$.

Theorem 1. *The solution T' provided by the algorithm achieves a $\frac{3}{2}$-standard approximation for* MINMETRICP$_4$P *and this ratio is tight.*

Proof. Let T^* be an optimal solution on $I = (K_{4n}, d)$, we consider 3 pairwise disjoint sets A, B and C according to the application of Lemma 2 to T'; according to property (iii), we can split $A \cup C$ into two sets A_1 and A_2 in such a way that $A_i \cup M_{T'}$ $(i = 1, 2)$ is a P$_4$-partition (see Figure 2 for an illustration). Thus, we get from the optimality of $R_{T'}$ the inequality $d(A_i) \geq d(R_{T'})$ and deduce:

$$2d(R_{T'}) \leq d(A) + d(C) \tag{1}$$

From property (ii) of Lemma 2 which states that $B \cup (\overline{R_{T^*}} \setminus C)$ is a perfect matching, and because $M_{T'}$ is optimal, we get:

$$d(M_{T'}) \leq d(B) + d(\overline{R_{T^*}} \setminus C) \tag{2}$$

Adding inequalities (1) and (2), and since I satisfies the triangular inequality, we obtain:

$$d(M_{T'}) + 2d(R_{T'}) \leq 2\mathrm{opt}_{\mathrm{MINMETRICP_4P}}(I) \tag{3}$$

(Note that this latter inequality is only true when minimizing.) Considering $d(M_{T'}) \leq \mathrm{opt}_{\mathrm{MINMETRICP_4P}}(I)$, the proof is complete. The tightness is given by the instances $I_n = (K_{8n}, d)$ described in Property 1.

Based upon Lemma 2, one can also obtain an alternate proof of the result given in [9].

Theorem 2. *The solution T' provided by the algorithm achieves a $\frac{3}{4}$-standard approximation for* MAXP$_4$P.

Proof. The inequality (3) becomes

$$d(M_{T'}) + 2d(R_{T'}) \geq \mathrm{opt}_{\mathrm{MAXP_4P}}(I) + d(\overline{R_{T^*}}) \tag{4}$$

Considering this time that $2 \times d(M_{T'}) \geq \mathrm{opt}_{\mathrm{MAXP_4P}}(I) + d(\overline{R_{T^*}})$, we deduce $\mathrm{apx}_{\mathrm{MAXP_4P}}(I) \geq \frac{3}{4}\left(\mathrm{opt}_{\mathrm{MAXP_4P}}(I) + d(\overline{R_{T^*}})\right)$.

3.3   General P_4P Within the Differential Framework

Dealing with differential ratio, $MAXP_4P$ $MINP_4P$ and $MINMETRICP_4P$ are equivalent to approximate ; this is generally true for any couple of problems that only differ by an affine transformation of their objective function.

Theorem 3. *The solution T' provided by the algorithm achieves a $\frac{1}{2}$-differential approximation for P_4P and this ratio is tight.*

Proof. We prove the result for the maximization version. First, observe that $\overline{R_{T^*}}$ is a n-cardinality matching; hence, considering any perfect matching M of I such that $M \cup \overline{R_{T^*}}$ do form a P_4-partition, we get:

$$d(M) + d(\overline{R_{T^*}}) \geq \text{wor}_{\text{MAXP}_4\text{P}}(I) \tag{5}$$

Adding inequalities (4) and (5), we obtain:

$$2\text{apx}_{\text{MAXP}_4\text{P}}(I) \geq d(M_{T'}) + 2d(R_{T'}) + d(M) \geq \text{wor}_{\text{MAXP}_4\text{P}}(I) + \text{opt}_{\text{MAXP}_4\text{P}}(I)$$

To show the tightness of this ratio, we refer to Property 1.

3.4   Bi-valuated Metric P_4P with Weights 1 & 2 Within the Standard Framework

As done in [12] for $MINTSP$, we now focus on instances where any edge is worth either 1 or 2; indeed, such an analysis enables a keener comprehension of a given algorithm. Moreover, since the P_4-partition problem is **NP**-complete, the problems $MAXP_4P_{1,2}$ and $MINP_4P_{1,2}$ are **NP**-hard.

Theorem 4. *The solution T' provided by the algorithm achieves a $\frac{9}{10}$-standard approximation for $MAXP_4P_{1,2}$ and a $\frac{7}{6}$-standard approximation for $MINP_4P_{1,2}$. These ratios are tight.*

Proof. We only prove the maximization case. Let $I = (K_{4n}, d)$ be an instance of $MAXP_4P_{1,2}$ with $d(e) \in \{1, 2\}$. We will denote by $M_{T',2}$ (*resp.*, $R_{T',2}$) the set of $M_{T'}$ edges of weight 2, by p (*resp.*, q) its cardinality. Trivially, cardinalities p and q verify: $p \leq 2n$, $d(M_{T'}) = 2n + p$ and $q \leq n$, $d(R_{T'}) = n + q$. Similarly, let us denote by $M_{T^*,2} = \{e \in M_{T^*} | d(e) = 2\}$ with size p^* and by $R_{T^*,2} = \{e \in R_{T^*} | d(e) = 2\}$ with size q^*. We have $\text{apx}_{\text{MAXP}_4\text{P}_{1,2}}(I) = 3n + p + q$ and $\text{opt}_{\text{MAXP}_4\text{P}_{1,2}}(I) = 3n + p^* + q^*$. Wlog., we may assume that the following property always holds for T^*:

Property 2. For any 3-length chain $P \in T^*$, $|P \cap M_{T^*,2}| \geq |P \cap R_{T^*,2}|$.

Otherwise, T^* would contain a chain $P = \{[x, y], [y, z], [z, t]\}$ verifying $d(x, y) = d(z, t) = 1$ and $d(y, z) = 2$; thus, by swapping P and $P' = \{[y, z], [z, t], [t, x]\}$ into T^*, one does generate another optimal solution.

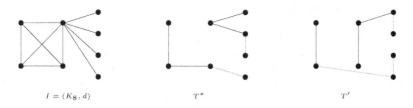

Fig. 3. Instance $I = (K_8, d)$ establishing the tightness for $\text{MaxP}_4\text{P}_{1,2}$

Property 3. $q \geq \frac{q^*}{2}$.

Let G' be the multi-graph induced by $M_{T'} \cup R_{T^*,2}$ (edges from $M_{T'} \cap R_{T^*,2}$ appear twice). This graph is constituted of chains, cycles and isolated edges: its chains alternate edges from $M_{T'}$ and $R_{T^*,2}$, with the particularity that their extremal edges all belong to $M_{T'}$; its cycles also alternate edges from $M_{T'}$ and $R_{T^*,2}$ and the 2-length cycles correspond to the edge set $R_{T^*,2} \cap M_{T'}$; finally, its isolated edges belong to $M_{T'} \setminus R_{T^*,2}$. For any cycle Γ on G', there exists an edge $e(\Gamma) = [x, y]$ with $e(\Gamma) \in M_{T^*,2}$, $x \in V(\Gamma)$, and y is an endpoint of a chain; we denote by $e'(\Gamma)$ the edge from $\Gamma \cap R_{T^*,2}$ to which $e(\Gamma)$ is incident. The existence of $e(\Gamma)$ follows from the Property 2; moreover, vertex y cannot be incident to any edge from $R_{T^*,2}$, since otherwise $e(\Gamma) \in M_{T^*}$ would link two edges of T^*, contradiction! Thus, setting $A = R_{T^*,2}$, and replacing in A for any cycle Γ on G' the edge $e'(\Gamma)$ by the edge $e(\Gamma)$, we build an edge set A satisfying $|A| = |R_{T^*,2}| = q^*$ and $(V, M_{T'} \cup A)$ is a simple graph made of pairwise disjoint chains. Like we did while proving Theorem 1, we split A into two sets A_1 and A_2 in such a way that the partial graph induced by $M_{T'} \cup A_i$ for $i = 1, 2$ is a set of at-most-3-length chains. We arbitrarily complete A_i by the way of an edge set B_i in order to obtain a P_4-partition $A_i \cup B_i \cup M_{T'}$. Obviously, $|A_i| + |B_i| = n$ and $d(B_1) + d(B_2) \geq |B_1| + |B_2| = (n - |A_1|) + (n - |A_2|) = 2n - q^*$. Moreover, $d(A_i \cup B_i) \leq d(R_{T'})$ due to $R_{T'}$ optimality. Since $d(A) = 2q^*$, we deduce $2n + q^* \leq d(B_1) + d(B_2) + d(A) \leq 2d(R_{T'}) = 2n + 2q$ and Property 3 is established. Thanks to this latter, we obtain:

$$\text{apx}_{\text{MaxP}_4\text{P}_{1,2}}(I) \geq 3n + p + \frac{q^*}{2} \qquad (6)$$

On the other hand, since $M_{T',2}$ is a matching containing a maximum number of 2-edges, we have in particular $p \geq p^*$ and deduce:

$$\text{opt}_{\text{MaxP}_4\text{P}_{1,2}}(I) \leq 3n + p + q^* \qquad (7)$$

On the behalf of inequalities $p \geq q^*$, $n \geq q^*$, (6) and (7), we obtain the expected result $\text{apx}_{\text{MaxP}_4\text{P}_{1,2}}(I) \geq \frac{9}{10}\text{opt}_{\text{MaxP}_4\text{P}_{1,2}}(I)$.

The tightness comes from the instance $I = (K_8, d)$ depicted in Figure 3, where edges of distance 2 are drawn in continuous line and edges of distance 1 of T^* and T' are drawn in dotted line; other edges are not drawn. One can easily see $\text{opt}_{\text{MaxP}_4\text{P}_{1,2}}(I) = 10$ and $\text{apx}_{\text{MaxP}_4\text{P}_{1,2}}(I) = 9$.

3.5   Bi-valuated Metric P_4P with Weights $a\&b$ Within the Differential Framework

As we mentioned earlier, the differential measure is stable under affine transformation; now, any instance from $\text{MAX}P_4P_{a,b}$ may be mapped into an instance of $\text{MAX}P_4P_{1,2}$ or $\text{MIN}P_4P_{a,b}$ by the way of such a transformation. Thus, proving $\text{MAX}P_4P_{1,2}$ is ε-differential approximable actually establishes that $\text{MIN}P_4P_{a,b}$ and $\text{MAX}P_4P_{a,b}$ are ε-differential approximable for any couple of real values $a < b$.

Theorem 5. *The solution T' provided by the algorithm achieves a $\frac{2}{3}$-differential approximation for $P_4P_{a,b}$ and this ratio is tight.*

References

1. Arkin, E. M., Hassin, R. and Levin, A.: Approximations for minimum and min-max vehicle routing problems. J. of Algorithms, (article in press)
2. Ausiello, G., Crescenzi, P., Gambosi, G., Kann, V., Marchetti-Spaccamela, A., Protasi, M.:Complexity and Approximation (Combinatorial Optimization Problems and Their Approximability Properties). *Springer, Berlin* (1999)
3. Ausiello, G., D'Atri, A., Protasi, M.: Structure preserving reductions among convex optimization problems. J. Comput. System Sci. **21** (1980) 136–153
4. Bellare, M., Rogaway, P.: The complexity of approximating a nonlinear program. Mathematical Programming **69** (1995) 429–441
5. Demange, M., Paschos, V. Th.: On an approximation measure founded on the links between optimization and polynomial approximation theory. Theoretical Computer Science **158** (1996) 117–141
6. Frederickson, G. N., Hecht, M. S., Kim, C. E.: Approximation algorithms for some routing problems. SIAM J. on Computing **7** (1978) 178–193
7. Garey, M. R., Johnson, D. S.: Computers and intractability. a guide to the theory of NP-completeness. *CA, Freeman* (1979)
8. Hassin, R., Khuller, S.: z-approximations. Journal of Algorithms **41** (2001) 429–442
9. Hassin, R., Rubinstein, S.: An Approximation Algorithm for Maximum Packing of 3-Edge Paths. Information Processing Letters **63** (1997) 63–67
10. Hassin, R., Rubinstein, S.: An Approximation Algorithm for Maximum Triangle Packing. ESA, LNCS **3221** (2004) 403–413
11. Monnot, J.: Differential approximation results for the traveling salesman and related problems. Information Processing Letters **82** (2002) 229–235
12. Papadimitriou, C., Yannakakis, M.: The traveling salesman problem with distances one and two. Mathematics of Operations Research **18** (1993) 1–11
13. Sahni, S., Gonzalez, T.: P-complete approximation problems. Journal of the Association for Computing Machinery **23** (1976) 555–565
14. Steiner, G.: On the k-path partition of graphs. Theoretical Computer Science **290** (2003) 2147–2155

The Maximum Resource Bin Packing Problem*

Joan Boyar[1], Leah Epstein[2], Lene M. Favrholdt[1], Jens S. Kohrt[1], Kim S. Larsen[1],
Morten Monrad Pedersen[1], and Sanne Wøhlk[3]

[1] Dept. of Math. and Computer Science, University of Southern Denmark
¶joan, lenem, svalle, kslarsen, mortenm◊@imada.sdu.dk
[2] Dept. of Mathematics, University of Haifa, Israel
lea@math.haifa.ac.il
[3] Dept. of Organization and Management, University of Southern Denmark
swo@sam.sdu.dk

Abstract. Usually, for bin packing problems, we try to minimize the number of bins used or in the case of the dual bin packing problem, maximize the number or total size of accepted items. This paper presents results for the opposite problems, where we would like to maximize the number of bins used or minimize the number or total size of accepted items. We consider off-line and on-line variants of the problems.

For the off-line variant, we require that there be an ordering of the bins, so that no item in a later bin fits in an earlier bin. We find the approximation ratios of two natural approximation algorithms, First-Fit-Increasing and First-Fit-Decreasing for the maximum resource variant of classical bin packing.

For the on-line variant, we define maximum resource variants of classical and dual bin packing. For dual bin packing, no on-line algorithm is competitive. For classical bin packing, we find the competitive ratio of various natural algorithms.

We study the general versions of the problems as well as the parameterized versions where there is an upper bound of $\frac{1}{k}$ on the item sizes, for some integer k.

1 Introduction

Many optimization problems involve some resource, and the task for algorithm designers is typically to get the job done using the minimum amount of resources. Below, we give some examples.

Bin packing is the problem of packing items of sizes between zero and one in the smallest possible number of bins of unit size. Here, the bins are the resources. The traveling salesperson problem is the problem of finding a tour which visits each vertex in a weighted graph while minimizing the total weight of visited edges. Here the weight is the resource. Scheduling jobs on a fixed number of machines is the problem of minimizing the completion time of the last job. Here time is the resource.

Each of these problems come in many variations and there are many more entirely different optimization problems. Since these problems are computationally hard, the

---

* The work of Boyar, Favrholdt, Kohrt, and Larsen was supported in part by the Danish Natural Science Research Council (SNF). The work of Epstein was supported in part by the Israel Science Foundation (ISF).

M. Liśkiewicz and R. Reischuk (Eds.): FCT 2005, LNCS 3623, pp. 397–408, 2005.

optimal solution can usually not be computed in reasonable time for large instances, so polynomial time approximation algorithms are devised. For many of these problems, there are interesting variants where the entire instance is not known when the computation must commence. The area of on-line algorithms deals with this problem scenario. For detailed descriptions of many of these problems and their solutions in terms of approximation or on-line algorithms, see [4,10,13], for instance.

For all of these problems, minimizing the resources used seems to be the obvious goal. However, if the resource is not owned by the problem solver, but is owned by another party who profits from selling the resource, there is no longer agreement about the objective, since the owner of the resource wants to maximize the resources used, presumably under some constraints which could be outlined in a contract. Thus, many of the classical problems are interesting also when considered from the reverse perspective of trying to maximize the amount of resources that are used.

In [1], the Lazy Bureaucrat Scheduling Problem is considered. Here, tasks must be scheduled and processed by an office worker. The authors consider various constraints and objective functions. The flavor of the constraints is that the office worker cannot sit idle if there is work that can be done, and the office worker's objective is to schedule tasks under these constraints so as to minimize the work carried out; either total work, arranging to leave work as early as possible, or a similar goal. Though it is presented as a toy problem, it is an important view on some optimization problems, and many other problems are interesting in this perspective, provided that the constraints imposed on the problem are natural. Also other problems have been investigated in this reverse perspective, e.g., longest path [16], maximum traveling salesperson problem [12] and lazy online interval coloring [9].

Maximum Resource Bin Packing (MRBP). In this paper, we consider bin packing from the maximum resource perspective. We consider it as an approximation problem, but we also investigate two on-line variants of the problem. To our knowledge, this is the first time one of these reverse problems has been considered in an on-line setting. Note that the complexity status of the off-line problems studied in this paper is open.

The abstract problem of packing items of a given size into bins has numerous concrete applications, and for many of these, when the resource must be purchased, the reverse problem becomes interesting for one of the parties involved. We use the following concrete problem for motivation.

Assume that we hire a company to move some items by truck from one site, the origin, to another, the destination. Say that the price we must pay is proportional to the number of trucks used. Some companies may try to maximize the number of trucks used instead of trying to get the items packed in few trucks. To prevent the company from cheating us, the following constraint has been placed on the packing procedure:

Constraint 1: When a truck leaves the origin, none of the unpacked items remaining at the origin should fit into that truck.

In the off-line variant, *Off-Line MRBP*, we are given an unlimited number of unit sized bins and a sequence of items with sizes in $(0, 1]$, and the goal is to maximize the number of bins used to pack all the items subject to Constraint 1. A set of items fits in a bin if the sum of the sizes of the items is at most one. In the off-line variant, there must be an ordering of the bins such that no item in a later bin fits in an earlier bin. Explained using the motivating example, Constraint 1 can be illustrated as follows: Trucks arrive at the origin one at a time. A truck is loaded, and may leave for its destination when none of the remaining items can fit into the truck. At this time, the next truck may arrive.

On-Line MRBP is similar to the off-line version. However, the problem is on-line, meaning that items are revealed one at a time, and each item must be processed before the next item becomes available. Because of the on-line nature of the problem, instead of Constraint 1, the following modified constraint is used:

Constraint 2: The company is not allowed to begin using a new truck if the current item fits in a truck already being used.

Thus, the on-line algorithm is allowed to open a new bin every time the next item to be processed does not fit in any of the previous bins. The objective is still to use as many bins as possible. Thus, all partly loaded trucks are available all the time, and whenever an item does not fit, a new truck may pull up to join the others.

We also consider another on-line problem, *On-Line Dual MRBP*. Here, the number of available bins is fixed. For each item, an algorithm has to accept it and place it in one of its bins, if it is possible to do so. Thus, here a fixed number of trucks have been ordered. In this case, neither Constraint 1 nor Constraint 2 is used; the objective is not to maximize the number of trucks used, since this number is fixed. There are two possible objective functions: the number of accepted items or the total size of the accepted items. In both cases, the objective is to minimize this value. Thus, the truck company wants to pack as few items or as little total size as possible into the trucks, minimizing the fuel needed for each truck, or maybe hoping to get a new order of trucks for the items which do not fit into the fixed number of trucks which have been ordered.

For all three problems, we study the general version as well as the parameterized version where there is an upper bound of $\frac{1}{k}$ on the item sizes, for some integer k.

A closely related problem is the Bin Covering Problem. In this problem, the algorithm is given a sequence of items and has to place them in bins, while trying to maximize the number of bins that contain items with a total size of at least one. This is quite similar to Off-Line MRBP with bins twice as large and Constraint 1 replaced by the following weaker constraint:

Constraint 3: No pair of trucks leaving the origin may have a total load of items that could have been packed in one truck.

The problem is NP-complete but has an asymptotic fully polynomial time approximation scheme (AFPTAS) [14]. Further results on that problem can be found in [2,7,8].

Our Results. For Off-Line MRBP, we show that no algorithm has an approximation ratio of more than $\frac{17}{10}$. For the parameterized version, the upper bound is $1 + \frac{1}{k}$ for $k \geq 2$. The algorithm First-Fit-Decreasing is worst possible in the sense that it meets this upper bound. First-Fit-Increasing is better; it has a competitive ratio of $\frac{6}{5}$ and a parameterized competitive ratio of $1 + \frac{k-1}{k^2+1}$ for $k \geq 2$. See Section 2 for a definition of the algorithms.

For On-Line MRBP, we prove a general lower bound of $\frac{3}{2}$ on the parameterized competitive ratio for $k \leq 3$ and $1 + \frac{1}{k-1}$ for $k \geq 3$. We prove a general upper bound of 2 for $k \leq 2$ and $1 + \frac{1}{k-1}$ for $k \geq 2$. Hence, for $k \geq 3$, all algorithms have the same parameterized competitive ratio. We prove that First-Fit, Best-Fit, and Last-Fit all meet the general upper bound.

For On-Line Dual MRBP, we prove that if the objective function is the total *number* of items packed, no deterministic algorithm is competitive; this also holds for any value of k for the parameterized problem. If the objective function is the total *size* of the packed items, no algorithm for the general problem is competitive. For the parameterized version, we prove general lower and upper bounds of $1 + \frac{1}{e(k-1)}$ and $1 + \frac{1}{k-1}$, respectively.

The proof of Theorem 3, below, showing that for Off-Line MRBP, the approximation ratio of First-Fit-Increasing is $\frac{6}{5}$, uses a new variant of the standard weighting argument. That result also gives a connection between Off-Line MRBP and the relative worst order ratio for on-line algorithms for the classical bin packing problem. The relative worst order ratio [5] is a new measure for the quality of on-line algorithms. Theorem 3 has been used to prove the upper bound on a result comparing First-Fit to Harmonic(k) using the relative worst order ratio [6]. Perhaps other "reverse" problems will have similar connections to the relative worst order ratio.

2   Notation and Algorithms

The input is a sequence of items, $I = \langle s_1, s_2, \ldots, s_n \rangle$. For convenience, we identify an item with its size and require that item s_i has size $0 < s_i \leq 1$ (or $0 < s_i \leq \frac{1}{k}$, for some integer k, for the parameterized problem). Items have to be placed in bins of size one.

For any input sequence I, let $ALG(I)$ be both the packing produced when running ALG on this input sequence and the number of bins used for this packing. In particular, let OPT be an algorithm which produces an optimal packing, and let $OPT(I)$ be both this packing and the number of bins used. For the off-line variant, let $SMALL(I)$ be a packing using the minimum number of bins that the items from I can be packed in without putting items with sizes totaling more than one in any bin, and let $SMALL$ be an algorithm that creates this packing. Note that $SMALL$ is an optimal algorithm from the classical bin packing problem, but for MRBP, it is a worst possible algorithm.

An approximation algorithm ALG is a *c-approximation algorithm*, $c \geq 1$, if there is a constant b such that for all possible input sequences I, $OPT(I) \leq c\,ALG(I) + b$. The infimum of all such c is called the *approximation ratio* of the algorithm, $\mathcal{R}_{ALG}$. For the parameterized problem, we consider the *parameterized approximation ratio*, $\mathcal{R}_{ALG}(k)$, which is the approximation ratio in the case where all items have size at most $\frac{1}{k}$ for some integer k.

An important algorithm in this context is First-Fit (*FF*), which places an item in the first bin in which it fits. In this paper, we investigate two well known off-line variants of *FF* in detail: *First-Fit-Increasing* (*FFI*) handles items in non-decreasing order with respect to their sizes, placing them using First-Fit. *First-Fit-Decreasing* (*FFD*) handles items in non-increasing order with respect to their sizes, also placing them using First-Fit.

In the on-line variants of the problem, the algorithms receive the input, i.e., the items, one at a time and have to decide where to pack the item before the next item (if any) is revealed. Similarly to the approximation ratio for approximation algorithms, the performance of deterministic on-line algorithms is measured in comparison with the optimal off-line algorithm OPT [11,17,18]. An on-line algorithm ALG is *c-competitive*, $c \geq 1$, if there is a constant b such that for all possible input sequences I, $OPT(I) \leq c\,ALG(I) + b$. The infimum of all such c is called the *competitive ratio* of the algorithm, $\mathcal{C}_{ALG}$. For the parameterized problem, we consider the *parameterized competitive ratio*, $\mathcal{C}_{ALG}(k)$, which is the competitive ratio in the case where all items have size at most $\frac{1}{k}$ for some integer k.

For the on-line variants, we consider the following natural algorithms, all of which, except for Last-Fit, have been well studied in other contexts: *First-Fit (FF)* as defined previously. *Last-Fit (LF)* is the opposite of *FF*, i.e., it places a new item in the last opened bin in which it fits. *Best-Fit (BF)* places the item in a feasible bin which is as full as possible, i.e., a feasible bin with least free space. *Worst-Fit (WF)* is the opposite of *BF*, i.e., it places an item in a feasible bin with most free space.

Omitted proofs can be found in the full version of this paper.

3   Off-line Maximum Resource Bin Packing

For Off-Line MRBP, the goal is to maximize the number of bins used, subject to Constraint 1, so there must be an ordering of the bins such that no item placed in a later bin fits in an earlier bin. We show that no algorithm for the problem has an approximation ratio worse than $\frac{17}{10}$. Using the proof that for classical Bin Packing, First-Fit's approximation ratio is $\frac{17}{10}$ [15] we prove that *FFD* has this worst possible approximation ratio. Finally we show that *FFI* has a better ratio of $\frac{6}{5}$.

Theorem 1 (General upper bound). *Any algorithm ALG for Off-Line MRBP has*
$$\mathcal{R}_{ALG}(k) \leq \tfrac{17}{10}, \text{for } k = 1, \text{ and } \mathcal{R}_{ALG}(k) \leq 1 + \tfrac{1}{k}, \text{for } k \geq 2.$$

Proof. Consider any multiset of requests, I. The minimum number of bins, m, ALG could use on I is no less than the number of bins used by *SMALL*.

Consider OPT's packing of I, and create an ordered list I' containing the items in I, starting with the items OPT packed in the first bin, followed by those in the second bin, etc., until all items have been included.

By the restrictions on what an algorithm may do, First-Fit packs the items in I' exactly as OPT packed the items of I. By [15], the number of bins, n, used by First-Fit is at most $\frac{17}{10}m + 2$, if $k = 1$, and at most $(1 + \frac{1}{k})m + 2$, if $k \geq 2$. Thus, OPT uses at most $\frac{17}{10}m + 2$ bins, if $k = 1$, and at most $(1 + \frac{1}{k})m + 2$ bins, if $k \geq 2$, giving the stated ratio. □

This result is tight since First-Fit-Decreasing has this approximation ratio.

Theorem 2. *For Off-Line MRBP,*
$$\mathcal{R}_{FFD}(k) = \tfrac{17}{10}, \text{for } k = 1, \text{ and } \mathcal{R}_{FFD}(k) = 1 + \tfrac{1}{k} \text{ for } k \geq 2.$$

We now turn to the better algorithm, *FFI*.

Theorem 3. *For Off-Line MRBP,*
$$\mathcal{R}_{FFI}(k) = \tfrac{6}{5}, \text{ if } k = 1, \text{ and } \mathcal{R}_{FFI}(k) = \tfrac{k^2+k}{k^2+1}, \text{ if } k \geq 2.$$

Proof. Note that $\mathcal{R}_{FFI}(1) = \mathcal{R}_{FFI}(2) = \mathcal{R}_{FFI}(3)$. We prove the lower bound first. For $k \leq 2$ we use the following input: n items of size $\tfrac{1}{2}$ and n items of size $\tfrac{1}{3}$, where n is a large integer divisible by 6. The optimal packing is to put one item of size $\tfrac{1}{2}$ and one item of size $\tfrac{1}{3}$ in each bin. This makes OPT use n bins, each with a fraction of $\tfrac{1}{6}$ empty space. On the other hand, FFI packs $\tfrac{n}{3}$ bins, each containing three elements of size $\tfrac{1}{3}$, followed by $\tfrac{n}{2}$ bins, each with two items of size $\tfrac{1}{2}$. Hence, in total, FFI uses $\tfrac{5n}{6}$ bins, and the ratio follows.

For $k \geq 3$ we use a slightly more complicated sequence. Let n be a large integer. The input contains $n(k^2 - 1)$ items of size $\tfrac{1}{k+1}$ and $n(k + 1)$ items of size $\tfrac{1}{k}$. FFI uses $n(k - 1)$ bins for the smaller items and $n(k + 1)/k$ bins for the larger ones, which is $n(k^2 + 1)/k$ in total. All the bins are completely full. An optimal packing would be to combine one larger item with $k - 1$ smaller ones, using $n(k + 1)$ bins. Each bin is thus full by a fraction of $\tfrac{1}{k} + \tfrac{k-1}{k+1} > \tfrac{k}{k+1}$, which makes the packing valid. The approximation ratio for this sequence is thus exactly $\tfrac{k^2+k}{k^2+1}$.

Note that in this case, FFI's packing is actually the same as the packing made by $SMALL$. This is not always the case, though.

Next, we prove the upper bound. We first prove the case $k \leq 3$ which is slightly different from the other cases and has to be treated separately. For this part of the proof, we do not assume an upper bound on the item sizes.

We assign weights to items in the following way. For all items in the interval $(0, \tfrac{1}{6}]$ (small items), the weight is defined to be equal to the size. An item which belongs to an interval $(\tfrac{1}{i+1}, \tfrac{1}{i}]$ for some $i = 1, 2, 3, 4, 5$ (large items), is assigned the weight $\tfrac{1}{i}$.

The intuition for this weighting comes from considering a bin in the packing made by FFI that contains only items from a single interval $(\tfrac{1}{i+1}, \tfrac{1}{i}]$, $i \in \{1, 2, 3, 4, 5\}$. This bin contains at most i items, and therefore each item in such a bin can be thought of as contributing $\tfrac{1}{i}$ to the total size of items plus empty space in FFI's packing.

Let W be the total weight of the items in a given input sequence. We prove that $FFI + 5 \geq W \geq \tfrac{5}{6}(OPT - 5)$, which implies the upper bound.

Consider first the optimal solution OPT. We show that the total weight of items is at least $W \geq \tfrac{5}{6}(OPT - 5)$. To show that, we claim that all bins in OPT, except for at most five bins, have items of weight at least $\tfrac{5}{6}$. First, consider the bins containing at least one small item. Due to Constraint 1, there is at most one such bin whose total sum of item sizes is less than $\tfrac{5}{6}$. Note that the weight of an item is at least its size. A bin which contains items of total size of at least $\tfrac{5}{6}$ has weight at least that amount. A bin which contains an item larger than $\tfrac{1}{2}$ has weight at least 1. Therefore, we only need to consider bins containing only items in $(\tfrac{1}{6}, \tfrac{1}{2}]$. We define a pattern to be a multiset of numbers in $\tfrac{1}{2}, \tfrac{1}{3}, \tfrac{1}{4}, \tfrac{1}{5}$ whose sum is at most 1. The type of a pattern is the inverse of the smallest number in it. A pattern P of type j is a maximal pattern if adding another instance of $\tfrac{1}{j}$ to P, would not result in a pattern, i.e. $P \cup \{\tfrac{1}{j}\}$ is not a pattern. The pattern of a bin is the multiset of the weights of its items.

For each $j = 2, 3, 4, 5$, the packing has at most one bin whose pattern is of type j but is not maximal. We show that a bin of any maximal pattern has weight at least

$\frac{5}{6}$ [1]. The only maximal pattern of type 2 is $\{\frac{1}{2}, \frac{1}{2}\}$. The maximal patterns of type 3 are $\{\frac{1}{3}, \frac{1}{3}, \frac{1}{3}\}$ and $\{\frac{1}{2}, \frac{1}{3}\}$. Consider a maximal pattern of type 4. We need to show that the sum of elements in the pattern is at least $\frac{5}{6}$. Let a, b, c be the amounts of $\frac{1}{2}, \frac{1}{3}$, and $\frac{1}{4}$ in the pattern. If the sum is less than $\frac{5}{6}$, we have $\frac{3}{4} < \frac{a}{2} + \frac{b}{3} + \frac{c}{4} < \frac{5}{6}$. This gives $9 < 6a + 4b + 3c < 10$. Since a, b, c are integers, this is impossible. Similarly, consider a maximal pattern of type 5. Let a, b, c, d be the amounts of $\frac{1}{2}, \frac{1}{3}, \frac{1}{4}$, and $\frac{1}{5}$ in the pattern. If the sum is less than $\frac{5}{6}$, we have $\frac{4}{5} < \frac{a}{2} + \frac{b}{3} + \frac{c}{4} + \frac{d}{5} < \frac{5}{6}$. This gives $48 < 30a + 20b + 15c + 12d < 50$ or $30a + 20b + 15c + 12d = 49$. Since a, b, c, d are non-negative integers, this combination is impossible.

Consider now the packing of FFI. We show that the total weight of items is at most $W \le FFI + 5$. Note that the algorithm actually acts as Next Fit Increasing and never assigns an item to an old bin once a new bin is opened. A bin of FFI is called a transition bin if it has both at least one small item and at least one other item, or it has only large items, but it contains items of distinct weights. The last case means that the algorithm is done packing all items of weight $\frac{1}{j}$ for some $5 \ge j \ge 2$ and has started packing items of weight $\frac{1}{j-1}$. Therefore, there are at most five transition bins. In any other bin, the sum of the weights of the items is at most one; this is clear if there are only small items whose weights are equal to their sizes. For other items, there are j items of weight $\frac{1}{j}$ in a bin containing only such items. As for the transition bins, the total weight of items whose size is at most one can be at most 2, so $W \le FFI + 5$. Hence $OPT \le \frac{6}{5}FFI + 11$.

We now prove the upper bound for $k \ge 4$. We slightly revise the definitions. Items are small if they are in the interval $(0, \frac{1}{k+3}]$. The weight of a small item is its size. An item which belongs to an interval $(\frac{1}{i+1}, \frac{1}{i}]$ for some $i = k, k+1, k+2$, is assigned the weight $\frac{1}{i}$.

Consider the optimal solution OPT. We show that the total weight of items is at least $W \ge (k^2 + 1)(OPT - 4)/(k^2 + k)$. To show that, we claim that all bins except for at most four bins have items of weight at least $(k^2 + 1)/(k^2 + k)$. First, consider the bins containing at least one small item. Due to Constraint 1, there is at most one such bin whose total sum of items is less than $1 - \frac{1}{k+3} = \frac{k+2}{k+3} \ge \frac{k^2+1}{k^2+k}$ (which holds for all $k \ge 3$).

Note that the weight of an item is at least its size. A bin which contains items of total size at least $\frac{k^2+1}{k^2+k}$ has weight at least that amount. We need to consider bins containing only items in $(\frac{1}{k+3}, \frac{1}{k}]$. We define a pattern to be a multiset of numbers in $\frac{1}{k+2}, \frac{1}{k+1}, \frac{1}{k}$ whose sum is at most 1. Again, we define the type of a pattern as the inverse of the smallest item in it.

For each $j = k, k+1, k+2$, the packing has at most one bin whose pattern is of type j but is not maximal. We show that a bin of any maximal pattern has weight at least $\frac{k^2+1}{k^2+k}$. The only maximal pattern of type k is $\{\frac{1}{k}, \ldots, \frac{1}{k}\}$.

Consider a maximal pattern of type $k+1$. We need to show that the sum of elements in the pattern is at least $\frac{k^2+1}{k^2+k}$. Let a and b be the amounts of $\frac{1}{k}$ and $\frac{1}{k+1}$ in the pattern. If the sum is less than $\frac{k^2+1}{k^2+k}$, we have $\frac{k}{k+1} < \frac{a}{k} + \frac{b}{k+1} < \frac{k^2+1}{k^2+k}$. This gives $k^2 < (k+1)a + kb < k^2 + 1$. Since a and b are integers, this is impossible. Similarly,

---

[1] The paper [3] showed a similar result on patterns for a different purpose.

consider a maximal pattern of type $k + 2$. Let a, b, and c be the amounts of $\frac{1}{k}$, $\frac{1}{k+1}$, and $\frac{1}{k+2}$ in the pattern. If the sum is less than $\frac{k^2+1}{k^2+k}$, we have $\frac{k+1}{k+2} < \frac{a}{k} + \frac{b}{k+1} + \frac{c}{k+2} < \frac{k^2+1}{k^2+k}$. This gives $k(k+1)^2 < (k+2)(k+1)a + k(k+2)b + k(k+1)c < (k^2+1)(k+2)$ or $a(k^2 + 3k + 2) + b(k^2 + 2k) + c(k^2 + k) = k(k+1)^2 + 1$. We need to exclude the existence of an integer solution for a, b, and c. Assume that such a solution exists. Note that $a + b + c < k + 2$, since otherwise the left hand side is at least $k(k+1)(k+2) > k(k+1)^2 + 1$. Also note that $a + b + c > k - 1$, since otherwise the left hand side is at most $(k+1)(k+2)(k-1) < k(k+1)^2 + 1$. If $a + b + c = k + 1$ we get $(a + b + c)(k^2 + k) + 2a(k+1) + kb = k(k+1)^2 + 1$. Simplifying, we have $2a(k+1) + kb = 1$, which is clearly impossible. If $a + b + c = k$, we need that $2a(k+1) + kb = k^2 + k + 1$. If $a = 0$, there is no solution since $k^2 + k + 1$ is not divisible by k. Otherwise, $2a - 1$ must be divisible by k. Since $0 < a \leq k$ and a is an integer, the only value it can have is $\frac{k+1}{2}$, but this gives $b = -1$ which is impossible.

Now consider the packing of *FFI*. The total weight of items is at most $W \leq FFI + 4$ since there can be only four transition bins. This implies the upper bound on the approximation ratio. $\Box$

4   On-line Maximum Resource Bin Packing

For *On-Line Maximum Resource Bin Packing*, the goal is to maximize the number of bins used subject to Constraint 2, so the algorithm is only allowed to open a new bin if the current item does not fit in any open bin. We have matching lower and upper bounds for most algorithms, and we conjecture that the algorithm Worst-Fit (*WF*) has an optimal competitive ratio of $\frac{3}{2}$.

Theorem 4 (General upper bound). *Any algorithm ALG for On-Line MRBP has*
$$\mathcal{C}_{ALG}(k) \leq 2, \text{ for } k = 1, \text{ and } \mathcal{C}_{ALG}(k) \leq \frac{k}{k-1}, \text{ for } k \geq 2.$$

Proof. For $k = 1$, this is proven using the fact that for any algorithm, all bins, except possibly one, are at least half full. For $k \geq 2$, we use the fact that all bins, except possibly one, are full to at least $\frac{k-1}{k}$. $\Box$

For $k \geq 3$, Theorem 4 is tight for deterministic algorithms. The following theorem follows from Lemmas 1 and 2 handling the cases $k \leq 2$ and $k \geq 3$ respectively.

Theorem 5 (General lower bound). *Any deterministic algorithm ALG for On-Line MRBP has $\mathcal{C}_{ALG}(k) \geq \frac{3}{2}$, for $k \leq 2$, and $\mathcal{C}_{ALG}(k) \geq \frac{k}{k-1}$, for $k \geq 3$.*

Lemma 1. *There exists a family of sequences I_n with items no larger than $\frac{1}{2}$ such that $OPT(I_n) \to \infty$ for $n \to \infty$ and, for any deterministic on-line algorithm ALG, $OPT(I_n) \geq \frac{3}{2}ALG(I_n)$.*

Proof. The sequence is given in phases. Each phase begins with $\langle \frac{1}{2}, \varepsilon, \frac{1}{2}, \varepsilon \rangle$, where $\varepsilon < \frac{1}{12}$. For this sequence, there are two possible packings. In the first one, there are two bins with one item of each size. In the second one, each bin has one item of size $\frac{1}{2}$, and both small items are in the first bin.

If the on-line algorithm chooses the first packing, the sequence continues with $\langle 2\varepsilon, \frac{1}{2} - \varepsilon, \frac{1}{2} - 3\varepsilon \rangle$, filling up the two on-line bins. An optimal off-line algorithm chooses the second packing, places the 2ε-item in the second bin, and opens a new bin for the last two items. Thus, OPT uses three bins, and has all bins filled to at least $\frac{1}{2} + 2\varepsilon$.

If the on-line algorithm chooses the second packing, the sequence continues with $\langle \frac{1}{2}, \frac{1}{2} - 4\varepsilon, \varepsilon, \varepsilon \rangle$. In this case, an optimal off-line algorithm chooses the first packing, and thus opens a new bin for the first two of the last four items. The last two items are placed in the first two bins. Again, OPT uses three bins and has all bins filled to at least $\frac{1}{2} + 2\varepsilon$.

Since each on-line bin is filled completely and each off-line bin is filled to at least $\frac{1}{2} + 2\varepsilon$, this can be repeated arbitrarily many times, with the result that ALG uses two bins per phase and OPT uses three bins per phase. □

Lemma 2. *For $k \geq 3$, there exists a family of sequences I_n with items no larger than $\frac{1}{k}$ such that $OPT(I_n) \rightarrow \infty$ for $n \rightarrow \infty$ and, for any deterministic on-line algorithm ALG, $OPT(I_n) \geq \frac{k}{k-1}ALG(I_n) - \frac{1}{k-1}$.*

First-Fit, Best-Fit, and Last-Fit are worst possible:

Theorem 6. *For On-Line MRBP,*
$$C_{FF}(k) = C_{BF}(k) = 2, \text{ for } k = 1, \text{ and } C_{FF}(k) = C_{BF}(k) = \frac{k}{k-1}, \text{ for } k \geq 2.$$

Proof. The upper bound follows from Theorem 4, and the lower bound for $k \geq 3$ follows from Theorem 5. Thus, we only need to prove the lower bound of 2 for $k \leq 2$. To this end, consider the sequence $\langle \frac{1}{2}, \varepsilon \rangle^n$, where n is a large odd integer and $\varepsilon \leq \frac{1}{2n}$. FF as well as BF puts all the small items in the first bin, using $1 + \frac{n-1}{2}$ bins in total. OPT on the other hand distributes the small items one per bin, using n bins. This gives a ratio arbitrarily close to 2 for n arbitrarily large. □

Theorem 7. *For On-Line MRBP, $C_{LF}(k) = 2$, for $k = 1$, and $C_{LF}(k) = \frac{k}{k-1}$, for $k \geq 2$.*

Investigation of Worst-Fit seems to indicate that it works very well in comparison with the other algorithms studied here. However, the gap between the lower bound of $\frac{3}{2}$ and the upper bound of 2 remains. Based on our investigation, we conjecture the following:

Conjecture 1. For On-Line MRBP, the competitive ratio of WF is $\frac{3}{2}$.

5   On-line Maximum Resource Dual Bin Packing

For this problem, there are exactly n bins. An item cannot be rejected if it fits in some bin, but there are no constraints as to which bins the algorithm may use, except that no bin may be filled to more than 1. We have two different cases corresponding to the two different objective functions. For both objective functions, no deterministic algorithm is competitive in the general case with no upper bound less than 1 on item sizes.

Theorem 8. *For On-Line Dual MRBP with accepted total size as cost function, no deterministic algorithm is competitive in general.*

Proof. We describe only the adversarial sequence. Let $n \geq 2$ be the number of available bins, and let *ALG* be any deterministic algorithm. The input sequence is constructed in up to n rounds. In round i, for $1 \leq i \leq n-1$, n items of size ε are given, for some small $\varepsilon > 0$. If, after the ith round, *ALG* has one or more bins with fewer than i items, then an item of size $1 - \varepsilon(i - 1)$ is given. If, after $n - 1$ rounds, *ALG* has $n - 1$ items in each of its n bins, we give an item of size $n\varepsilon$, and then $n - 1$ items of size $1 - \varepsilon(n - 1)$. $\square$

We note that the situation for the parameterized problem for $k > 1$ is very different from the situation for the general problem. For every $k > 1$, it is not hard to show that any algorithm has competitive ratio of at most $k/(k - 1)$. The reason for this is that if *OPT* rejected any item at all, then its bins are full up to at least $1 - 1/k$.

The following lower bound tends to $1 + \frac{1}{e(k-1)}$ as n tends to infinity and is at least $1 + \frac{1}{3(k-1)}$ for any $n \geq 2$.

Theorem 9. *Consider On-Line Dual MRBP with accepted total size as cost function For $k \geq 2$, any deterministic algorithm ALG has*
$$C_{ALG}(k) \geq 1 + \frac{m}{n(k-1)}, \text{ where } m = \max\left\{ j \;\Big|\; \sum_{i=j}^{n} \frac{1}{i} > 1 \right\} \in \left\{ \left\lfloor \frac{n}{e} \right\rfloor, \left\lceil \frac{n}{e} \right\rceil \right\}.$$

Proof. Let n be the number of bins and let $\varepsilon > 0$ be a very small constant. Let m be the largest number $1 \leq m < n$ such that $\sum_{i=m}^{n} \frac{1}{i} > 1$. Since $1/x$ is a monotonically decreasing function for $x > 0$, we get a lower bound of $\left\lfloor \frac{n}{e} \right\rfloor$ on m: $\sum_{i=\left\lfloor \frac{n}{e} \right\rfloor}^{n} \frac{1}{i} > \int_{\left\lfloor \frac{n}{e} \right\rfloor}^{n+1} \frac{1}{x} dx > 1$. For the upper bound, m can be at most $\left\lceil \frac{n}{e} \right\rceil$, since $\sum_{i=\left\lceil \frac{n}{e} \right\rceil+1}^{n} \frac{1}{i} < \int_{\left\lceil \frac{n}{e} \right\rceil}^{n} \frac{1}{x} dx < 1$. Hence, depending on the exact value of n, m is either $\left\lfloor \frac{n}{e} \right\rfloor$ or $\left\lceil \frac{n}{e} \right\rceil$.

The initial input is $n!$ items of size ε. We first prove that, for any packing of these items, there exists an integer i, $m \leq i \leq n$, such that at least i bins receive strictly less than $\frac{n!}{n+m-i}$ items. Assume for the purpose of contradiction that this is not the case. Then, at least one bin has at least $\frac{n!}{m}$ small items, and for each $i = n - 1, n - 2, \ldots, m$, there is at least one additional bin that receives at least $\frac{n!}{n+m-i}$ items. Since the total number of items is $n!$, we get that $\sum_{i=m}^{n} \frac{n!}{n+m-i} \leq n!$, which is equivalent to $\sum_{i=m}^{n} \frac{1}{i} \leq 1$. By the definition of m, this cannot be the case.

Now, pick an i, $m \leq i \leq n$, such that at least i bins receive strictly less than $\frac{n!}{n+m-i}$ items in *ALG*'s packing. Give $\left\langle \frac{n!}{n+m-i} \varepsilon \right\rangle^{i-m}$, $\left\langle \frac{1}{k} \right\rangle^{n(k-1)}$, $\left\langle \frac{1}{k} - \frac{n!-1}{n+m-i} \varepsilon \right\rangle^{m}$.

After packing the first $i - m$ of these items, *ALG* still has at least m bins filled to strictly less than $\frac{n!}{n+m-i} \varepsilon$. Let r be the number of *ALG*'s bins which are completely empty. Since all bins are less than $\frac{1}{k}$ full, there is room for exactly $n(k - 1) + r$ items of size $\frac{1}{k}$ and at least $\max\{m - r, 0\}$ items of size $\frac{1}{k} - \frac{n!-1}{n+m-i} \varepsilon$. Thus, *ALG* is able to pack the remaining items as well, giving a total size of $n!\varepsilon + (i - m) \frac{n!}{n+m-i} \varepsilon + \frac{n(k-1)+m}{k} - \frac{n!-1}{n+m-i} m\varepsilon$.

Before the arrival of the size $\frac{1}{k}$ items, *OPT* can pack the items from phase one in one bin each and distribute the initial $n!$ items in the remaining bins to fill all bins up to exactly $\frac{n!}{n+m-i}\,\varepsilon$. Each bin gets $k-1$ items of size $\frac{1}{k}$, and no further items can be packed. The total size packed by OPT is $n!\varepsilon + (i-m)\,\frac{n!}{n+m-i}\,\varepsilon + \frac{n(k-1)}{k}$. As ε decreases, the ratio converges to $1 + \frac{m}{n(k-1)}$. □

The lowest possible value of $\frac{m}{n}$ is $\frac{1}{3}$ which is obtained when n equals 3, 6 or 9.

For the case where the objective function is the number of accepted items, the situation is even worse.

Theorem 10. *For On-Line Parameterized MRBP with the number of accepted items as cost function, no deterministic algorithm is competitive, for any k.*

Proof. We describe only the adversarial sequence. The input sequence begins with $2kn-2$ items of size $\frac{1}{2k}$. ALG fills all but at most two bins completely, and the remaining two bins are either both filled to $1 - \frac{1}{2k}$, or one is filled completely and the other to $1 - \frac{1}{k}$. In the first case, the sequence continues with 1 item of size $\frac{1}{k}$ and $\lfloor \frac{1}{k\varepsilon} \rfloor$ items of size ε. In the second case, the sequence continues with one item of size $\frac{1}{2k} + \varepsilon$, two items of size $\frac{1}{2k}$, and $\lfloor \frac{1}{2k\varepsilon} \rfloor - 1$ items of size ε. □

6 Concluding Remarks

The most interesting open problem is to prove that the off-line maximum resource bin packing problem is NP-hard (or to find a polynomial time algorithm for it).

For the off-line version of the problem, we have investigated First-Fit-Decreasing, which is worst possible, and First-Fit-Increasing, which performs better and obtains an approximation ratio of $\frac{6}{5}$. It would be interesting to establish a general lower bound on the problem, and, if it is lower than $\frac{6}{5}$, to determine the optimal algorithm for the problem. Does there exist a polynomial time approximation scheme for the off-line version?

For the on-line version, we have considered the two standard bin packing problems from the literature. For dual bin packing, no algorithm is competitive in general, independent of whether the cost measure is the total size or the total number of accepted items. With the total accepted size as cost function, the situation is completely different for the parameterized version; for $k \geq 2$, any algorithm has a parameterized competitive ratio between $1 + \frac{1}{k-1}$ and about $1 + \frac{1}{e(k-1)}$.

For the classic variant of on-line bin packing, we have established general upper and lower bounds and proved that First-Fit, Best-Fit, and Last-Fit perform worst possible. The behavior of Worst-Fit seems very promising, but we leave it as an open problem to determine its competitive ratio.

Acknowledgments

First of all, we would like to thank Michael Bender for suggesting the problem and for interesting initial discussions. We would also like to thank Morten Hegner Nielsen for interesting discussions at the 3rd NOGAPS, where the work on the problem was initiated. In addition, we are thankful to Gerhard Woeginger for comments on an earlier version of this paper.

References

1. E.M. Arkin, M.A. Bender, J.S.B. Mitchell, and S. Skiena. The Lazy Bureaucrat scheduling problem. *Information and Computation*, 184(1):129–146, 2003.
2. S.F. Assman, D.S. Johnson, D.J. Kleitman, and J.Y-T. Leung. On a dual version of the one-dimensional bin packing problem. *J. Alg.*, 5(4):502–525, 1984.
3. A. Bar-Noy, R. E. Ladner, and T. Tamir. Windows scheduling as a restricted version of bin packing. In *Proc. 12th Annual ACM-SIAM Symposium on Discrete Algorithms*, pages 224–233, 2004.
4. A. Borodin and R. El-Yaniv. *Online Computation and Competitive Analysis*. Cambridge University Press, 1998.
5. J. Boyar and L.M. Favrholdt. The relative worst order ratio for on-line algorithms. In *Algorithms and Complexity: 5th Italian Conference, LNCS 2653*, pages 58–69, 2003.
6. J. Boyar and L.M. Favrholdt. The relative worst order ratio for on-line bin packing algorithms. Technical report PP–2003–13, Department of Mathematics and Computer Science, University of Southern Denmark, 2003.
7. J. Csirik, J.B.G. Frenk, M. Labbé, and S. Zhang. Two simple algorithms for bin covering. *Acta Cybernetica*, 14(1):13–25, 1999.
8. J. Csirik, D.S. Johnson, and C. Kenyon. Better approximation algorithms for bin covering. In *Proc. 12th Annual ACM-SIAM Symposium on Discrete Algorithms*, pages 557–566, 2001.
9. L. Epstein and M. Levy. Online interval coloring and variants. In *Proc. 32nd International Colloquium on Automata, Languages and Programming*, 2005. To appear.
10. M.R. Garey and D.S. Johnson. *Computers and Intractability – A Guide to the Theory of NP-Completeness*. W. H. Freeman and Company, 1979.
11. R.L. Graham. Bounds for certain multiprocessing anomalies. *Bell Systems Technical Journal*, 45:1563–1581, 1966.
12. R. Hassin and S. Rubinstein. An approximation algorithm for the maximum traveling salesman problem. *Information Processing Letters*, 67(3):125–130, 1998.
13. D.S. Hochbaum, editor. *Approximation Algorithms for NP-Hard Problems*. PWS Publishing Company, 1997.
14. K. Jansen and R. Solis-Oba. An asymptotic fully polynomial time approximation scheme for bin covering. *Theoretical Computer Science*, 306(1–3):543–551, 2003.
15. D.S. Johnson, A. Demers, J.D. Ullman, M.R. Garey, and R.L. Graham. Worst-case performance bounds for simple one-dimensional packing algorithms. *SIAM J. Comp.*, 3:299–325, 1974.
16. D.R. Karger, R. Motwani, and G.D.S. Ramkumar. On approximating the longest path in a graph. *Algorithmica*, 18(1):82–98, 1997.
17. A.R. Karlin, M.S. Manasse, L. Rudolph, and D.D. Sleator. Competitive snoopy caching. *Algorithmica*, 3(1):79–119, 1988.
18. D.D. Sleator and R.E. Tarjan. Amortized efficiency of list update and paging rules. *Communications of the ACM*, 28(2):202–208, 1985.

Average-Case Non-approximability
of Optimisation Problems

Birgit Schelm

Technische Universität Berlin,
Fakultät IV - Elektrotechnik und Informatik,
10587 Berlin, Germany
bts@cs.tu-berlin.de

Abstract. Both average-case complexity and the study of the approximability of NP optimisation problems are well established and active fields of research. Many results concerning the average behaviour of approximation algorithms for NP optimisation problems exist, both with respect to their running time and their performance ratio, but a theoretical framework to examine their structural properties with respect to their average-case approximability is not yet established. With this paper, we hope to fill the gap and provide not only the necessary definitions, but show that

1. The class of NP optimisation problems with p-computable input distributions has complete problems with respect to an average-approximability preserving reduction.
2. The average-time variants of worst-case approximation classes form a strict hierarchy if NP is not easy on average. By linking average-ratio approximation algorithms to p-time algorithms schemes, we can prove similar hierarchy results for the average-ratio versions of the approximation classes. This is done under the premise that not all NP-problems with p-computable input distributions have p-time algorithm schemes.
3. The question whether NP is easy on average is equivalent to the question whether every NP optimisation problem with a p-computable input distribution has an average p-time approximation scheme.

Classification: Computational and structural complexity.

1 Introduction

There are several ways of dealing with the intractability of many computational problems. One way is to consider algorithms whose running time is polynomial on most inputs, though it is super-polynomial on some instances that rarely occur. Another way, applicable to optimisation problems, is to search for p-time algorithms that do not compute an optimal solution, but one that comes close to the optimum, e.g. up to a constant factor.

There are already a number of results on the probabilistic behaviour of approximation algorithms, see for example [14,22,20] for results on the performance

M. Liśkiewicz and R. Reischuk (Eds.): FCT 2005, LNCS 3623, pp. 409–421, 2005.

ratio of greedy algorithms, and [8] for approximation algorithms that run in average p-time. In this paper we introduce a framework that allows us to examine the structural properties of optimisation problems with respect to such results. To do so we give formal definitions of average-case approximation classes. Reflecting the two parameters of interest, namely the running time and the quality of the solution found, there are three possibilities of relaxing worst-case requirements to average-case requirements.

The notion of average p-time for decision problems is well established, see [29] for a survey. Recent results like [7] attest to the relevance of this field of research. We apply this notion to the running time of approximation algorithms and examine average p-time versions of the worst-case approximation classes PTAS, APX, poly-APX and exp-APX and also adapt it to the performance ratio of approximation algorithms.

Introducing a reduction that preserves approximability for the majority of the average-case approximation classes in Section 3, we show that DistNPO, the class of NP optimisation problems with p-computable input distributions, has complete problems with respect to it.

We examine the inclusion structure of the average-case approximation classes in Section 4. We show that if NP is not easy on average, the average p-time approximation classes form a strict hierarchy. The average-ratio versions of the approximation classes also form a strict hierarchy, but under the premise that there are DistNP-problems that do not have a p-time algorithm scheme. An algorithm scheme is a deterministic algorithm that is allowed to make errors for some inputs with probability that is bounded by an additional input parameter.

Differences between the worst-case and average-case setting become apparent in Section 5. There we look at the average-case variant of the worst-case relation "P = NP iff every NP optimisation problem is solvable in polynomial time". The average-case equivalent of this question is already addressed in [28]. It is shown that computing an optimal solution for an NP opt-problem on average is equivalent to solving membership for problems in Δ_2^p on average, and that if every NP problem with a P^{NP}-samplable input distribution is easy on average, then every NP opt-problem with a P^{NP}-samplable input distribution is solvable in average p-time as well. Instead of trying to compute an optimal solution in average p-time, we focus on computing an approximate solution in average p-time and show that the question whether NP is easy on average is equivalent to the question whether every NP opt-problem with p-computable input distributions has an average p-time approximation scheme.

2   Preliminaries

We follow standard definitions of complexity theory, see e.g. [24,3,11]. We consider only nonempty words over $\Sigma = \{0,1\}$. By $|x|$ we denote the length of a word x, by Σ^n the set of all words of length n, by $\leq$ the standard ordering on Σ^+ – words are ordered lengthwise, words of equal lengths lexicographically – and by χ_L the characteristic function of $L \subseteq \Sigma^+$, that is $\chi_L(x) = 1$ iff $x \in L$.

Optimisation and Approximation Problems: (see [2] for further details) An NP *opt-problem* F over Σ is given as a tuple $(I, S, m, type)$ with:

1. $I \subseteq \Sigma^+$ the set of input instances; $I \in P$.
2. $S \subseteq I \times \Sigma^+$ a relation of inputs and their feasible solutions. $S(x)$ denotes the set of feasible solutions for any $x \in I$. We require $S \in P$ and $|y| \leq q(|x|)$ for all $y \in S(x)$ and some polynomial q.
3. $m : S \to \mathbb{Q}^+$ the objective function; $m \in FP$.
4. *type* the information whether F is a max- or min-problem.

NPO denotes the class of all NP opt-problems. Since we consider only NPO problems in this paper, we omit to explicitly mention NPO membership for optimisation problems. For all $x \in I$, the optimal value of the objective function is $\mathrm{opt}(x) = \max\{m(x,y) \mid y \in S(x)\}$ for max- and $\mathrm{opt}(x) = \min\{m(x,y) \mid y \in S(x)\}$ for min-problems.

If not stated otherwise, an *approximation algorithm* is a deterministic p-time algorithm. By $A(x)$ we denote the solution computed by an approximation algorithm A on input x. Clearly, $A(x) \in S(x)$ has to hold for $x \in I$ if $S(x) \neq \emptyset$.

For $x \in I$ with $S(x) \neq \emptyset$, the *performance ratio* R of a solution $y \in S(x)$ is defined as $R(x,y) = \max\{\mathrm{opt}(x)/m(x,y), m(x,y)/\mathrm{opt}(x)\}$. So $R(x,y) \geq 1$ for all $x \in I$ and $y \in S(x)$. If $R(x,y) = 1$, then y is optimal. We write $R^A(x) = R(x, A(x))$ for the ratio of an approximation algorithm A. Let $r : \mathbb{N} \to [1, \infty)$ be a function. A computes an *r-approximate solution* for $F \in NPO$ if $R^A(x) \leq r$ for every $x \in I$. Then A is an *r-approximate algorithm* and F is *r-approximable*.

A problem $F \in NPO$ is in *APX* if F is r-approximable for a constant $r > 1$. It is in *poly-APX* if F is $p(n)$-approximable for a polynomial p, and it is in *exp-APX* if F is $2^{p(n)}$-approximable for a polynomial p. F is in *PTAS* if an approximation algorithm exists that on input (x, r), where $x \in I$ and $r > 1$, returns an r-approximate solution for x in time polynomial in $|x|$.

The notion of an approximation preserving reduction, on which our definition of an average-case approximation preserving reduction is based, was first introduced in [9], see also [2]. Let $F_1, F_2 \in NPO$. F_1 *AP-reduces* to F_2, ($F_1 \leq_{AP} F_2$), if there exist functions $f : \Sigma^+ \times (1, \infty) \to \Sigma^+$, $g : \Sigma^+ \times \Sigma^* \times (1, \infty) \to \Sigma^*$, computable in time polynomial in $|x|$ for $r \in (1, \infty)$ fixed, and a constant $\alpha \geq 1$ such that for every $x \in I_{F_1}$ and for every $r > 1$:

1. $f(x,r) \in I_{F_2}$.
2. If $S_{F_1}(x) \neq \emptyset$, then $S_{F_2}(f(x,r)) \neq \emptyset$.
3. $g(x,y,r) \in S_{F_1}(x)$ for every $y \in S_{F_2}(f(x,r))$.
4. $R_{F_2}(f(x,r),y) \leq r \Rightarrow R_{F_1}(x,g(x,y,r)) \leq 1 + \alpha(r-1)$ for all $y \in S_{F_2}(f(x,r))$.

The additional parameter r is only required to preserve membership in PTAS. If the quality of the required solution is not needed for the reduction, the additional parameter is omitted. The AP-reduction is transitive, and PTAS, APX, poly-APX and exp-APX are closed under it.

Average-Case Complexity: A *probability function* on Σ^+ is a function μ : $\Sigma^+ \to [0,1]$ s.t. $\sum_x \mu(x) = 1$. Its *distribution* is $\mu^*(x) = \Sigma_{y \leq x}\mu(y)$. A distribution ν *dominates* a distribution μ if a polynomial p exists s.t. $\nu(x) \geq \mu(x)/p(|x|)$ for all $x \in \Sigma^+$. ν *dominates* μ *wrt. a function* $f \colon \Sigma^+ \to \Sigma^+$, if a constant $l \geq 0$ exists s.t. $\nu(y) \geq |y|^{-l} \sum_{x.f(x)=y} \mu(x)$ for every $y \in \text{range}(f)$. A function f *transforms* μ *into* ν if $\nu(y) = \sum_{x.f(x)=y} \mu(x)$ for all $y \in \text{range}(f)$.

A function $f : \Sigma^+ \to [0,1]$ is *p-computable* if a DTM M exists s.t. $|M(x,k) - f(x)| \leq 2^{-k}$ for all $x \in \Sigma^+, k \geq 1$, and M is time-bounded by a polynomial in $|x|$ and k. If a distribution μ^* is p-computable, then μ is p-computable as well. The converse is not true unless P = NP, see [16], so a probability function μ is called p-computable if both μ and μ^* are, thus following standard notation used in the literature.

Fact 1 (Gurevich [16]). *For every p-computable prob. function μ on Σ^+ a p-computable prob. function ν exists s.t. the length of the binary representation of $\nu(x)$ is bounded by $4 + 2|x|$, and $\mu(x) \leq 4\nu(x)$ for all $x \in \Sigma^+$.*

For *standard prob. functions* we use $\mu_\mathbb{N}(n) = c_0/n^2$ with $c_0 = 6/\pi^2$ for positive integers. The standard prob. function μ_{Σ^+} on Σ^+ is obtained by first selecting a length according to $\mu_\mathbb{N}$, then a word uniformly from Σ^n, i.e. $\mu_{\Sigma^+}(x) = \mu_\mathbb{N}(|x|) \cdot 2^{-|x|}$. For more details on such prob. functions see [16].

Since we deal with average-case properties of opt-problems, we cannot separate the problems from their input distributions. The pairs of opt-problem or language and prob. function are called *distributional opt-problems* and *distributional problems* respectively. *DistNP* is the class of distr. problems (L, μ), where $L \in \text{NP}$ and μ is p-computable. Likewise, *DistNPO* is the class of all distr. opt-problems (F, μ), where $F \in \text{NPO}$ and μ is p-computable. *DistMinNPO* is the class of all distr. min-problems in DistNPO, *DistMaxNPO* the class of all distr. max-problems in DistNPO.

A robust definition of efficiency on average was given by Levin [21]. A distr. problem (L, μ) is efficiently solvable on average if it is solvable in time t that is polynomial on μ-average. A function $t : \Sigma^+ \to \mathbb{Q}$ is *polynomial on μ-average*, if constants $k, c > 0$ exist s.t. $\sum_x t^{1/k}(x)|x|^{-1}\mu(x) \leq c$. We say that t is polynomial on μ-average *via constants* $k, c > 0$ to indicate that these constants are used to verify that t satisfies the definition above. Detailed discussions of the benefits of this definition can be found in [16], [6], and [29].

Fact 2 (Gurevich [16]). *1. If functions f, g are polynomial on μ-average, then $\max\{f, g\}, f \times g, f+g$, and f^c for any constant $c > 0$ are polynomial on μ-average as well.*

2. If a function f is polynomial on ν-average and ν dominates a prob. distribution μ, then f is polynomial on μ-average as well.

A distr. problem (L, μ) is in *AvgP* if L is decidable by a deterministic algorithm whose running time is polynomial on μ-average. The result DistNP $\not\subseteq$ AvgP unless NE = E [6] links the question whether NP is easy on average to a proposition about worst-case classes that is generally assumed not to hold.

A distributional problem (L_1, μ_1) many-one reduces to (L_2, μ_2), ($(L_1, \mu_1) \leq_m^p$ (L_2, μ_2)), if a function f exists s.t. $L_1 \leq_m^p L_2$ via f and μ_2 dominates μ_1 with respect to f. AvgP is closed under this reduction, and DistNP contains complete problems with respect to it [21].

Algorithm Schemes for distributional problems were first introduced in [17] though they were already used in [25] under a different name.

A *p-time algorithm scheme* for a distributional problem (L, μ) is a deterministic algorithm A with input $x \in \Sigma^+$, $m \in \mathbb{N}^+$ and the properties:

1. $\Pr_\mu \left[A(x, m) \neq \chi_L(x) \right] < 1/m$ for all $m \in \mathbb{N}^+$,
2. The running time t_A is polynomial in $|x|$ and m.

A distributional problem (L, μ) belongs to *HP* (heuristic p-time) if it has a p-time algorithm scheme. In [23] it was shown that extending the notion of p-time algorithm schemes to average p-time algorithm schemes does not increase their computational power.

P-time algorithm schemes that return "?" instead of a wrong value for χ_L characterise AvgP, which was shown in [17]. From this characterisation AvgP $\subseteq$ HP follows immediately. The inclusion was shown to be strict in [23].

Schapire [25] introduced the class ApproxP, a restriction of HP to problems with p-computable prob. functions. His proof that if DistNP $\subseteq$ ApproxP, then NE = E also shows that if DistNP $\subseteq$ HP, then NE = E. He also shows that HP is closed under many-one reduction.

Average-Case Approximation Classes: Extending the notions of average-case complexity to distributional opt-problems yields the following definition of average-case approximation classes. The parameter over which the average is taken is denoted by the index.

Definition 1. *Let* (F, μ) *be a distributional opt-problem, A an approximation algorithm for F with ratio R and time that is polynomial on μ-average.*
$(F, \mu) \in AvgPO$ *if* $R^A(x) = 1$ *for every* $x \in I$.
$(F, \mu) \in Avg_t\text{-}APX$ *if a constant* $r > 1$ *exists s.t.* $R^A(x) \leq r$ *for every* $x \in I$.
$(F, \mu) \in Avg_t\text{-}poly\text{-}APX$ *if a polynomial p exists s.t.* $R^A(x) \leq p(|x|)$ *for every* $x \in I$.
$(F, \mu) \in Avg_t\text{-}exp\text{-}APX$ *if a polynomial p exists s.t.* $R^A(x) \leq 2^{p(|x|)}$ *for every* $x \in I$.
$(F, \mu) \in Avg_t\text{-}PTAS$ *if an approx. algorithm A exists s.t. on input* (x, r), $x \in I$, $r > 1$, *A returns an r-approximate solution for x in time* t_A, *and for every* $r > 1$ *constants* $k_r, c_r > 0$ *exist s.t.* $\sum_x t_A^{1/k_r}(x, r)|x|^{-1}\mu(x) \leq c_r$.

A function $f \colon \Sigma^+ \to \mathbb{Q}^+$ is called *constant on μ-average*, if a constant $r > 0$ exists such that $\sum_{x \in \Sigma^{\geq n}} f(x)\mu(x) < r$ for all $n > 0$. We also say that f *is constant on μ-average via r.* For a detailed discussion of this definition, see [26]. With this notion we can define average-ratio approximation classes.

Definition 2. *Let* (F, μ) *be a distributional opt-problem, A an approximation algorithm for F with performance ratio R and running time t_A.*

$(F, \mu) \in Avg_r\text{-}APX$ if R is constant on μ-average and t_A is bounded by a polynomial.

$(F, \mu) \in Avg_{r,t}\text{-}APX$ if R is constant on μ-average and t_A is polynomial on μ-average.

$(F, \mu) \in Avg_r\text{-}poly\text{-}APX$ if R is polynomial on μ-average and t_A is bounded by a polynomial.

$(F, \mu) \in Avg_{r,t}\text{-}poly\text{-}APX$ if R and t_A are polynomial on μ-average.

An average-ratio version of exp-APX is of little use, since the performance ratio of any solution of an NP opt-problem is exponentially bounded.

3 Reductions and DistNPO-Completeness

In order to define DistNPO-completeness, we present an average-case approximability preserving reduction, a distributional version of the AP-reduction.

Definition 3. *Let* (F_1, μ_1) *and* (F_2, μ_2) *be distr. opt-problems. Then* (F_1, μ_1) *AP-reduces to* (F_2, μ_2), *(* $(F_1, \mu_1) \leq_{AP} (F_2, \mu_2)$ *), if functions* $f : \Sigma^+ \times (1, \infty) \to \Sigma^+$, $g : \Sigma^+ \times \Sigma^* \times (1, \infty) \to \Sigma^*$ *exist, computable in time polynomial in* $|x|$ *for every fixed* $r \in (1, \infty)$, *and a constant* $\alpha \geq 1$ *s.t.:*

Reducibility: F_1 *AP-reduces to* F_2 *via* (f, g, α), *and*

Dominance: μ_2 *dominates* μ_1 *wrt.* f *for every fixed value of* r, *i.e. for every* $r > 1$ *a constant* l_r *exists s.t. for every* $y \in \mathrm{range}(f)$:

$$\mu_2(y) \geq |y|^{-l_r} \sum_{x \in I_{F_1} \cdot f(x,r) = y} \mu_1(x).$$

An AP-reduction (f, g, α) is *honest*, if a constant $c_r > 0$ exists for every $r > 1$ s.t. $|f(x, r)| \geq |x|^{c_r}$ for all $x \in \Sigma^+$. The following lemma can be easily shown by adapting standard techniques from average-case complexity and the theory of approximability. The full proof can be found in [26].

Lemma 1. *1. Honest AP-reductions for distr. opt-problems are transitive.*

2. $Avg_t\text{-}PTAS$, $Avg_t\text{-}APX$, $Avg_t\text{-}poly\text{-}APX$, $Avg_t\text{-}exp\text{-}APX$, $Avg_r\text{-}poly\text{-}APX$ *and* $Avg_{r,t}\text{-}poly\text{-}APX$ *are closed under AP-reduction.*

Definition 4. *Let* $\mathcal{C}$ *be a class of distr. opt-problems. A distr. opt-problem* (F, μ) *is* hard *for* $\mathcal{C}$ *wrt. AP-reduction, if* $(F', \mu') \leq_{AP} (F, \mu)$ *for all* $(F', \mu') \in \mathcal{C}$. *It is* complete *for* $\mathcal{C}$, *if it is hard for* $\mathcal{C}$ *and* $(F, \mu) \in \mathcal{C}$.

This section's main contribution is that DistNPO has complete problems. This is done by proving that distributional versions of the Universal Max-Problem MaxU and the Universal Min-Problem MinU, see [19] and [10], are complete for DistNPO. A standard prob. function on the inputs is used for both problems.

Definition 5 (Distr. Universal Max-Problem $(\mathrm{MaxU}, \mu_{\mathrm{MaxU}}))$**.**

Input: $(M, x, 1^k)$; M *an NTM with an output tape,* x *an input for* M, $k \in \mathbb{N}^+$.

Solution: *Any sequence $y \in \{0,1\}^*$ of nondeterministic choices of M that corresponds to a halting computation path of at most k steps.*
Objective function: $M(x,y)$, *the value M computes on input x on path y.*
Prob. function: $\mu_{\mathrm{MaxU}}(M, x, 1^k) = \mu_{\Sigma^+}(M) \cdot \mu_{\Sigma^+}(x) \cdot \mu_{\mathrm{N}}(k)$

$(\mathrm{MinU}, \mu_{\mathrm{MinU}})$ is the minimisation version of $(\mathrm{MaxU}, \mu_{\mathrm{MaxU}})$ with $\mu_{\mathrm{MinU}} = \mu_{\mathrm{MaxU}}$. Both MaxU and MinU are NPO-complete, see [10].

The DistNPO-completeness of $(\mathrm{MaxU}, \mu_{\mathrm{MaxU}})$ and $(\mathrm{MinU}, \mu_{\mathrm{MinU}})$ is shown in two parts. First we show that $(\mathrm{MaxU}, \mu_{\mathrm{MaxU}})$ is complete for DistMaxNPO and $(\mathrm{MinU}, \mu_{\mathrm{MinU}})$ is complete for DistMinNPO. Then $(\mathrm{MaxU}, \mu_{\mathrm{MaxU}})$ is reduced to $(\mathrm{MinU}, \mu_{\mathrm{MinU}})$ and vice versa, thus proving that $(\mathrm{MaxU}, \mu_{\mathrm{MaxU}})$ is hard for DistMinNPO and $(\mathrm{MinU}, \mu_{\mathrm{MinU}})$ is hard for DistMaxNPO.

The main problem that arises in the first part is that *any* p-computable prob. function has to be reduced to μ_{MaxU} and μ_{MinU} such that the dominance requirement of the AP-reduction is satisfied. The same problem has to be solved when proving completeness for DistNP. The key to solving this problem is an efficiently computable encoding of words that reflects – with respect to a given prob. function – the probability with which the encoded word is chosen. This approach was first employed in [15]. Ben-David et al. subsumed the existence of such an encoding function as well as its properties in a technical lemma, the *Coding Lemma* [6]. A modified version of this Lemma, the *Distribution Controlling Lemma*, was used in [5].

Coding Lemma. *Let μ be a p-computable prob. function such that the binary representation of $\mu(x)$ has length polynomial in $|x|$ for all $x \in \Sigma^+$. Then a coding function C_μ exists that satisfies the following conditions:*

Compression: $|C_\mu(x)| \le 1 + \min\{|x|, \log_2 1/\mu(x)\}$ *for all $x \in \Sigma^+$.*
Efficient Encoding: $C_\mu(x)$ *is computable in time polynomial in $|x|$.*
Unique Decoding: C_μ *is one-to-one (i.e. $C_\mu(x) = C_\mu(x')$ implies $x = x'$).*
Efficient Decoding: $C_\mu^{-1}(w) = x$ *is computable in time polynomial in $|x|$ for every $w \in \mathrm{range}(C_\mu)$.*

A function that has those properties is $C_{\mu(x)} = 0x$ if $\mu(x) \le 2^{-|x|}$, and $C_{\mu(x)} = 1z$ otherwise, where z is the shortest binary string s.t. $\mu^*(x^-) < 0.z1 \le \mu^*(x)$. For the proof, see [6] and [4]. Fact 1 allows restricting the probability functions in the Coding Lemma to probability functions with rational values.

The proofs are omitted due to space restrictions, but can be found in [26].

Theorem 1. $(\mathrm{MaxU}, \mu_{\mathrm{MaxU}})$ *is DistMaxNPO complete, and* $(\mathrm{MinU}, \mu_{\mathrm{MinU}})$ *is DistMinNPO complete.*

Main Theorem 1. DistNPO *has complete problems with respect to* $\le_{\mathrm{AP}}$.

4   Non-approximability in the Average Case

In this chapter we take a closer look at the inclusion structure of the average-case approximation classes introduced in Definition 1 and 2.

The inclusions $\text{AvgPO} \subseteq \text{Avg}_t\text{-PTAS} \subseteq \text{Avg}_t\text{-APX} \subseteq \text{Avg}_t\text{-poly-APX} \subseteq$ $\text{Avg}_t\text{-exp-APX}$, and $\text{Avg}_r\text{-APX} \subseteq \text{Avg}_r\text{-poly-APX}$ as well as $\text{Avg}_{r,t}\text{-APX} \subseteq$ $\text{Avg}_{r,t}\text{-poly-APX}$ follow immediately. Now we show that there are distr. opt-problems that are not approximable within an exponential factor in average p-time if $\text{DistNP} \not\subseteq \text{AvgP}$, and that under that premise the inclusions for the average-p-time approximation classes are strict. The inclusions $\text{Avg}_r\text{-APX} \subseteq$ $\text{Avg}_r\text{-poly-APX}$ and $\text{Avg}_{r,t}\text{-APX} \subseteq \text{Avg}_{r,t}\text{-poly-APX}$ are strict if $\text{DistNP} \not\subseteq \text{HP}$. We can also construct distr. opt-problems that are not in $\text{Avg}_r\text{-poly-APX}$ and $\text{Avg}_{r,t}\text{-poly-APX}$ under this premise.

4.1 Non-approximability Proofs for Distr. Optimisation Problems

To separate the average-case approximation classes, we use a technique similar to the gap-technique from worst-case non-approximability proofs that was first used by Garey and Johnson [13] to show that if $P \neq NP$, then Minimum Graph Colouring is not approximable with ratio less than 2, see also [2]. This technique requires reducing instances for an NP-complete problem L to instances for an opt-problem F via a function f, such that the question whether $x \in L$ or not can be reduced to the question whether the optimal value of $f(x)$ is greater or smaller than a certain threshold that depends on x alone. The latter question can then be decided with an appropriate approximation algorithm for F.

In the average-case setting we have to consider not only the opt-problems but also their distributions. The differences of average p-time and average-ratio approximation algorithms and the respective premises require different adaptations of the gap technique to capture the peculiarities.

Average-Polynomial Time: Since the basic concept of the gap-technique is that of a reduction, we use the same ideas like for the definition of the AP-reduction for distr. opt-problems, namely combining the notion of reducibility with the notion of dominance of the distributions involved. See [26] for the proof.

Lemma 2. *Let* (L, μ) *be complete for* DistNP, $L \subseteq \Sigma^+$ *and* (F, ν) *a distr. min-problem. Suppose p-time computable functions* $f : \Sigma^+ \to I_F$, $c : \Sigma^+ \to \mathbb{N}^+$ *and* $r : \mathbb{N}^+ \to [1, \infty)$ *exist s.t.:*
1. ν *dominates* μ *with respect to* f, *and*
2. *for any* $x \in \Sigma^+$

$$x \in L \Rightarrow \text{opt}_F(f(x)) = c(x)$$
$$x \notin L \Rightarrow \text{opt}_F(f(x)) > r(|f(x)|)c(x).$$

Then no r-*approximate algorithm that runs in time polynomial on* ν-*average can exist for* (F, ν) *if* $\text{DistNP} \not\subseteq \text{AvgP}$.

A similar lemma holds for max-problems. It can be derived from Lemma 2 by changing the requirement for the case $x \notin L$ to $\text{opt}_F(f(x)) < c(x)/r(|f(x)|)$.

Average Ratio: Linking non-approximability results for the average-ratio setting to the premise $\text{DistNP} \not\subseteq \text{AvgP}$ does not seem to work as there is no general way of telling whether a solution computed by an average-ratio approximation algorithm is good enough or not. But the property that a "good" solution is computed with high probability relates nicely to the notion of algorithm schemes.

In contrast to the average-time setting, average-constant ratio and average-polynomial ratio approximation classes differ in their robustness with respect to dominance of the prob. functions involved. So we need different lemmas that explain the general structure of non-approximability proofs for both classes.

The main difference to the gap technique for the average p-time case is that here an additional parameter is needed for the function that transforms instances for the DistNP-complete decision problem into instances for the distr. opt-problem. To show non-approximability within a ratio that is constant on average, we need a more restrictive relationship between the prob. functions. This restricted relation of the prob. functions as well as the influence of the additional parameter is expressed by the next definition.

Definition 6. Let $f \colon \Sigma^+ \times \mathbb{N}^+ \to \Sigma^+$ be a function, μ, ν prob. functions on Σ^+. Then f transforms μ into ν for every $m \in \mathbb{N}^+$ if there is a constant $l > 1$ such that $\nu(y) = m^{-l} \sum_{x \colon f(x,m)=y} \mu(x)$ for all $y \in \mathrm{range}(f)$ and all $m \in \mathbb{N}^+$.

This definition is an adaptation of the notion that a function transforms μ into ν to functions with an additional, non-random parameter. It ensures that if μ is a prob. function, then ν is one as well, because $\sum_y \nu(y) = \sum_m m^{-l} \sum_x \mu(x) = \sum_m m^{-l} \leq 1$. We also say that f transforms μ into ν for every $m \in \mathbb{N}^+$ via l, where l is the constant from the definition.

Lemma 3. Let (L, μ) be complete for DistNP, $L \subseteq \Sigma^+$, and (F, ν) a distr. min-problem. If p-time computable functions $f \colon \Sigma^+ \times \mathbb{N}^+ \to I_F$ and $c \colon \Sigma^+ \times \mathbb{N}^+ \to \mathbb{N}^+$, and a constant $r > 1$ exist s.t.:

1. f transforms μ into ν for every constant $m \in \mathbb{N}^+$ via a constant $l > 1$, and
2. for every $x \in \Sigma^+$ and $m \in \mathbb{N}^+$

$$x \in L \Rightarrow \mathrm{opt}_F(f(x,m)) = c(x,m),$$
$$x \notin L \Rightarrow \mathrm{opt}_F(f(x,m)) > rm^{l+1}c(x,m).$$

Then no p-time approximation algorithm with ratio that is constant on ν-average via r and no approximation algorithm with ratio that is constant on ν-average via r and running time that is polynomial on ν-average can exist if DistNP $\not\subseteq$ HP.

Proof. Suppose an approx. algorithm A for (F, ν) with ratio R that is constant on ν-average via a constant r exists. Then $\mathrm{E}\left[R^A(x)\right] = \sum_x R^A(x)\nu(x) < r$.

Let A' be an algorithm scheme that, on input (x, m), accepts if $A\big(f(x,m)\big) \leq rm^{l+1}c(x,m)$. If A' accepts, then $x \in L$, so A' works correctly on those inputs. If A' rejects, either $x \notin L$ or the result of A was too far away from the optimum. So we can link the error probability of A' to the probability that A computes a "bad" solution. Applying the Markov Inequality then yields

$$\mathrm{Pr}_\mu\left[A'(x,m) \neq \chi_L(x)\right] \leq m^l \, \mathrm{Pr}_\nu\left[R^A(y) > rm^{l+1}\right] \leq \mathrm{E}\left[R^A(y)\right](rm)^{-1} < m^{-1}.$$

The running time $t_{A'}$ of A' is $t_{A'}(x,m) = t_c(x,m) + t_f(x,m) + t_A\big(f(x,m)\big)$, where t_A is the running time of A, and t_f and t_c are the times needed to compute

$f(x,m)$ and $c(x,m)$, which are bounded by polynomials in $|x|$ and m. If t_A is bounded by a polynomial, then $t_{A'}$ is polynomial in $|x|$ and m. So A' is a p-time algorithm scheme for (L,μ), which contradicts the premise DistNP $\not\subseteq$ HP.

Else, if t_A is polynomial on ν-average, it is easy to show that $t_{A'}$ is polynomial on μ-average in $|x|$ and m by using standard techniques from average-case complexity. With the result from [23] this yields a p-time algorithm scheme, a contradiction to the premise. $\square$

Changing the requirements for $x \notin L$ to $\text{opt}_F\big(f(x)\big) < c(x,m)/(rm^{l+1})$ yields the respective lemma for max-problems. To apply this lemma, we essentially need a reduction technique that allows us to create a constant gap for any constant.

To adapt the Lemma to the polynomial case, we can relax the dominance requirement to the existence of constants $l > 1$, $s_m \geq 0$ such that $\nu(y) \geq |y|^{-s_m} m^{-l} \sum_{x.f(x,m)=y} \mu(x)$ for all $y \in \text{range}(f)$, $m \in \mathbb{N}^+$. Set $\text{opt}_F(f(x,m)) > (rm|f(x,m)|^{s_m+2})^k c(x,m)$ for the bound on the optimum for the case $x \notin L$. Here the variable error bound m affects the polynomial gap in two ways, so we need a reduction that preserves the gap for every polynomial. For max-problems $\text{opt}_F\big(f(x)\big) < c(x,m)/\big(rm|f(x,m)|^{s_m+1}\big)^k$ has to hold for the case $x \notin L$ so that the lemma can be applied. The proof is similar to that of Lemma 3.

4.2   Application of the Non-approximability Proofs

One of the difficulties when applying the gap-technique for distr. opt-problems is to find suitable DistNP-complete problems. Due to the difficulties described in Section 3, the list of problems with natural input distributions known to be complete for DistNP is rather short, see [30] for a compilation, and none of them relate to opt-problems in a natural way.

It is not difficult to verify that for every NP-complete problem L a prob. function ν exists s.t. (L,ν) is hard for DistNP: reduce a known DistNP-complete problem (A,μ) to L via a reduction function f (such a function exists since L and A are NP-complete), and define ν such that f transforms μ into ν. Since many-one reduction functions are not necessarily one-to-one and monotone wrt. the standard order on Σ^+, we cannot guarantee that f transforms the p-computable prob. function μ into a p-computable one. Using a padded version of an NP-complete problem allows us to overcome this difficulty. This technique was first used by Köbler and Schuler in [18].

Definition 7. *Let* $\text{pad}(L) = \{1^{|x|}0xy10^i \mid x \in \{0,1\}^*, y \in L, i \in \mathbb{N}\}$ *for* $L \subseteq \Sigma^+$.

Lemma 4. *For every* NP-*complete problem* L *a p-computable prob. function* ν *exists s.t.* $(\text{pad}(L),\nu)$ *is complete for* DistNP.

See [26] for the proof. With this tool we can state our second main theorem.

Main Theorem 2.

1. $\text{AvgP} \subsetneq \text{Avg}_t\text{-PTAS} \subsetneq \text{Avg}_t\text{-APX} \subsetneq \text{Avg}_t\text{-poly-APX} \subsetneq \text{Avg}_t\text{-exp-APX}$ *if* $\text{DistNP} \not\subseteq \text{AvgP}$.

2. $\text{Avg}_r\text{-APX} \subsetneq \text{Avg}_r\text{-poly-APX}$, $\text{Avg}_t\text{-exp-APX} \not\subseteq \text{Avg}_r\text{-poly-APX}$,
   $\text{Avg}_{r,t}\text{-APX} \subsetneq \text{Avg}_{r,t}\text{-poly-APX}$ *and* $\text{Avg}_t\text{-exp-APX} \not\subseteq \text{Avg}_{r,t}\text{-poly-APX}$ *if*
   DistNP $\not\subseteq$ HP.

Proof. (Sketch) We take the same decision and opt-problems used to separate
worst-case approximation classes (see [2]). These are (with $x \in \{r, t, (r, t)\}$):

- Knapsack and MaxKnapsack to separate AvgPO and $\text{Avg}_t\text{-PTAS}$,
- Partition and MinBinPacking to separate $\text{Avg}_t\text{-PTAS}$ and $\text{Avg}_t\text{-APX}$,
- Any NP-complete problem L and MaxIndSet to separate $\text{Avg}_x\text{-APX}$ and
  $\text{Avg}_x\text{-poly-APX}$,
- HamiltonCircuit and MinTravellingSalesman to separate $\text{Avg}_x\text{-poly-APX}$ and
  $\text{Avg}_t\text{-exp-APX}$.

We construct DistNP complete versions of those problems by applying Lemma 4.
The DistNP complete problems reduce to the respective opt-problems in a gap
preserving way, inducing a prob. function on the instances of the opt-problem.
The opt-problems with those prob. functions form separating problems for the
respective classes. They are approximable within the bounds of the larger class
in the worst case, hence approximable withing the same bounds in the average
case wrt. any prob. function.

For the details of the gap-preserving reductions see [26].   □

By applying the padding technique to those separating opt-problems, we can
ensure that they lie in DistNPO, which shows:

Corollary 1.
- *Let $\mathcal{C} \in \{\text{AvgPO}, \text{Avg}_t\text{-PTAS}, \text{Avg}_t\text{-APX}, \text{Avg}_t\text{-poly-APX}, \text{Avg}_t\text{-exp-APX}\}$.
  If* DistNPO $\subseteq \mathcal{C}$, *then* DistNP $\subseteq$ AvgP.
- *Let $\mathcal{C} \in \{\text{Avg}_r\text{-APX}, \text{Avg}_r\text{-poly-APX}\} \cup \{\text{Avg}_{r,t}\text{-APX}, \text{Avg}_{r,t}\text{-poly-APX}\}$. If*
  DistNPO $\subseteq \mathcal{C}$, *then* DistNP $\subseteq$ HP.

By using a construction similar to the reduction in the proof of Theorem 1
(see [26]), we can reduce a DistNP-problem (L, μ) to $(\text{MaxU}, \mu_{\text{MaxU}})$ such that
$x \in L$ iff the instance of MaxU has at least one solution. This implies

Theorem 2. *If* DistNPO $\subseteq \text{Avg}_t\text{-exp-APX}$, *then* DistNP $\subseteq$ AvgP.

5 Average-Case Approximability

In contrast to the results of the previous section, we examine the average-case
equivalent of the worst-case relation P = NP iff PO = NPO here. If DistNP $\subseteq$
AvgP, combining a result from [28] with a recent result from [7] shows that then
every DistNPO-problem is in $\text{Avg}_t\text{-PTAS}$. The full proof is given in [26].

Theorem 3. *If* DistNP $\subseteq$ AvgP *then* DistNPO $\subseteq \text{Avg}_t\text{-PTAS}$.

Combining this theorem with Corollary 1 proves our third main result.

Main Theorem 3. DistNP $\subseteq$ AvgP $\Leftrightarrow$ DistNPO $\subseteq \text{Avg}_t\text{-PTAS}$

References

1. S. Arora, C. Lund, R. Motwani, M. Sudan, M. Szegedy. Proof Verification and Hardness of Approximation Problems. *33rd FOCS.* IEEE, 1992.
2. G. Ausiello, P. Crescenzi, G. Gambosi, V. Kann, A. Marchetti-Spaccamela, and M. Protasi. *Complexity and Approximation.* Springer, 1999.
3. J. Balcázar, J. Díaz, J. Gabarró. *Structural Complexity I.* Springer, 2nd ed., 1995.
4. J. Belanger, J. Wang. On average-P vs. average-NP. In K. Ambos-Spies, S. Homer, U. Schöning, eds., *Complexity Theory – Current Research.* Cambridge University Press, 1993.
5. J. Belanger, J. Wang. On the NP-isomorphism with respect to random instances. *JCSS,* 50:151–164, 1995.
6. S. Ben-David, B. Chor, O. Goldreich, M. Luby. On the theory of average case complexity. *JCSS,* 44:193–219, 1992.
7. H. Buhrman, L. Fortnow, A. Pavan. Some results on derandomization. *20th STACS.* Springer, 2003.
8. A. Coja-Oghlan, A. Taraz. Colouring random graphs in expected polynomial time. *20th STACS.* Springer, 2003.
9. P. Crescenzi, V. Kann, R. Silvestri, L. Trevisan. Structure in approximation classes. *SIAM J. Comp.,* 28(5):1759–1782, 1999.
10. P. Crescenzi, A. Panconesi. Completeness in approximation classes. *Inf. and Comp.,* 93:241–262, 1991.
11. D.-Z. Du, K. Ko. *Theory of Computational Complexity.* John Wiley & Sons, 2000.
12. U. Feige, S. Goldwasser, L. Lovász, S. Safra, M. Szegedy. Approximating Clique is almost NP-complete. *32nd FOCS.* IEEE, 1991.
13. M. Garey, D. Johnson. The complexity of near optimal graph coloring. *J. ACM,* 23:43–49, 1976.
14. G. Grimmet, C. McDiarmid. On colouring random graphs. *Math. Proc. Camb. Phil. Soc.,* 77:313–324, 1975.
15. Y. Gurevich. Complete and incomplete randomized NP problems. *28th FOCS.* IEEE, 1987.
16. Y. Gurevich. Average case completeness. *JCSS,* 42:346–398, 1991.
17. R. Impagliazzo. A personal view of average-case complexity. *10th Structure.* IEEE, 1995.
18. J. Köbler, R. Schuler. Average-case intractability vs. worst-case intractability. *22nd MFCS.* Springer, 1998.
19. M. Krentel. The complexity of optimization problems. *JCSS,* 36:490–509, 1988.
20. B. Kreuter, T. Nierhoff. Greedily approximating the r-independent set and k-center problems on random instances. *Rand. and Approx. Techniques in Comp. Science.* Springer, 1997.
21. L. Levin. Problems, complete in "average" instance. *16th STOC.* ACM, 1984.
22. C. McDiarmid. Colouring random graphs. *Ann. Operations Res.,* 1:183–200, 1984.
23. A. Nickelsen, B. Schelm. Average-case computations – Comparing AvgP, HP, and Nearly-P. *CCC 2005, to appear,* 2005.
24. C. Papadimitriou. *Computational Complexity.* Addison Wesley, 1994.
25. R. Schapire. The emerging theory of average-case complexity. Technical Report MIT/LCS/TM-431, MIT Laboratory of Computer Science, 1990.
26. B. Schelm. *Average-Case Approximability of Optimisation Problems.* Doctoral Thesis, TU Berlin, Fak. f. Elektrotechnik und Informatik, 2004. http://edocs.tu-berlin.de/diss/2004/schelm_birgit.pdf.

27. C. Schindelhauer. *Average- und Median-Komplexitätsklassen.* Doctoral Thesis, Universität Lübeck, 1996.
28. R. Schuler, O. Watanabe. Toward average-case complexity analysis of NP optimization problems. *10th Structure.* IEEE, 1995.
29. J. Wang. Average-case computational complexity theory. In L. Hemaspaandra, A. Selman, eds., *Complexity Theory Retrospective II.* Springer, 1997.
30. J. Wang. Average case intractable NP-problems. In D.-Z. Du, K. Ko, eds., *Advances in Complexity and Algorithms.* Kluwer, 1997.

Relations Between Average-Case
and Worst-Case Complexity

A. Pavan[1,*] and N.V. Vinodchandran[2,**]

[1] Department of Computer Science,
Iowa State University
pavan@cs.iastate.edu
[2] Department of Computer Science and Engineering,
University of Nebraska-Lincoln
vinod@cse.unl.edu

Abstract. The consequences of the worst-case assumption NP = P are very well understood. On the other hand, we only know a few consequences of the average-case assumption "NP is easy on average." In this paper we establish several new results on the *worst-case complexity* of Arthur-Merlin games (the class AM) under the *average-case complexity* assumption "NP is easy on average."

We first consider a stronger notion of "NP is easy an average" namely NP is easy on average with respect to distributions that are computable by a polynomial size circuit family. Under this assumption we show:
- AM ⊆ NSUBEXP.

Under the assumption that NP is easy on average with respect to polynomial-time computable distributions, we show:
- AME = E where AME is the exponential version of AM. This improves an earlier known result that if NP is easy on average then NE = E.
- For every $c > 0$, AM ⊆ [io-pseudo$_{\mathrm{NTIME}(n^c)}$]−NP. Roughly this means that for any language L in AM there is a language L' in NP so that it is computationally infeasible to distinguish L from L'.

We use results from several sub-areas of complexity theory, including derandomization, for establishing our results.

1   Introduction

Can an average-case complexity collapse lead to the collapse of worst-case complexity classes? We explore this question in the framework introduced by Levin [Lev86].

In Levin's framework, a distributional problem is a decision problem A with a probability distribution on the instances of A. Such a distributional problem

---
* Research supported in part by NSF grant CCF-0430807.
** Research supported in part by NSF grants CCF-0430991, CCR-0311577 and a Big 12 Faculty Fellowship.

M. Liśkiewicz and R. Reischuk (Eds.): FCT 2005, LNCS 3623, pp. 422–432, 2005.
© Springer-Verlag Berlin Heidelberg 2005

is in Average-P if there is a deterministic algorithm that solves A in polynomial time on the average. DistNP is the set of pair (A, μ) where $A \in$ NP and μ is a polynomial-time computable distribution. The question analogous to NP $\overset{?}{=}$ P here is whether DistNP $\subseteq$ Average-P (in words NP is easy on average). Levin [Lev86] showed that there are distributional problems that are complete for DistNP in the sense that NP is easy on average if and only if the complete problem is solvable in Average-P. We refer the reader to [Gur91, Wan97] for pointers to many nice results in this area.

Understanding relations between average-case complexity and worst-case complexity is one of the most fundamental problems in complexity theory. In general we would like to answer the following question: "If a class $\mathcal{C}$ is easy on average, then what happens in the worst-case world?" This question has been studied extensively for the class E under various notions of average-case complexity. For example, it is now known that if E can be approximated by small circuits, then E indeed has small circuits [BFNW91, IW97]. Some problems such as the Permanent are shown to have a remarkable property—their worst-case complexity is essentially the same as their average-case complexity [Lip91, GS92, FL92]. Such proofs are based on the random-self reducibility of the Permanent. Similarly, random-self reducibility of PSPACE-complete problems can be used to show that if PSPACE is easy on average, then PSPACE = P [KS04].

Consequences of the assumption "NP is easy on average" have also been studied a lot. Ben-David, Chor, Goldreich, and Luby [BDCGL92] showed that if NP is easy on average, then E = NE. Impagliazzo [Imp95] showed that if NP is easy on average, then BPP = ZPP.

Köbler and Schuler [KS04] explored this connection further and showed that the assumption derandomizes MA to NP. Later Arvind and Köbler [AK02] showed in fact the assumption implies AM $\cap$ co-AM = NP $\cap$ co-NP. Buhrman, Fortnow, and Pavan [BFP03] showed that if NP is easy on average, then pseudorandom generators exist and BPP = P. They used this fact to show that a witness of any NE machine can be computed in E time. But it is still open whether the assumption that NP is easy on average implies derandomization of AM to NP. In this paper we explore this possibility.

We first consider a stronger notion of NP being easy on average. We show that if NP is easy on average with respect to distributions that are computable by polynomial-size circuits, then AM can be derandomized to nondeterministic subexponential time.

Under the assumption NP is easy with respect to polynomial-time computable distributions, we show that AME = E where AME is the exponential version of AM. This improves the earlier mentioned collapse result DistNP $\subseteq$ Average-P $\Rightarrow$ NE = E [BDCGL92]. In the polynomial-time range we show that DistNP $\subseteq$ Average-P implies AM is almost NP: for any language $L \in$ AM there is a language L' in NP so that L and L' are computationally indistinguishable infinitely often.

We use results from several subareas of complexity theory including derandomization to prove our results.

2 Preliminaries

We assume familiarity with definitions of standard complexity classes. Please refer to [BDG88, Pap94] for definitions of standard complexity classes.

An Arthur-Merlin game is a combinatorial game, played by Arthur–a probabilistic polynomial-time machine (with public coins), and Merlin–a computationally unbounded Turing machine.

Given a string x, Merlin tries to convince Arthur that x belongs to some language L. The game consists of a finite number of moves. The moves alternate between Arthur and Merlin. In each move Arthur (or Merlin) prints a finite string on a read-write communication tape. Arthur's moves can depend on his random bits. In the end Arthur either accepts or rejects x.

Babai and Moran [Bab85, BM88] define a language L to be in AM as follows. For every string x of length n, the game consists of a random move by Arthur and a reply by Merlin. If $x \in L$, then the probability that there exists a move by Merlin that leads to acceptance by Arthur is at least $\frac{3}{4}$; on the other hand, if $x \notin L$, then the probability that there exists a move by Merlin that leads to acceptance by Arthur is at most $\frac{1}{4}$.

AMTIME$(l(n))$ denotes the class of languages accepted by a 2-round Arthur-Merlin interactive protocol with maximum message length $l(n)$ by both Arthur and Merlin. Let NLIN = NTIME$(O(n))$.

We need the exponential version of these complexity classes: E = DTIME$(2^{O(n)})$, EXP = DTIME$(2^{n^{O(1)}})$, NEXP = DTIME$(2^{n^{O(1)}})$, AME = AM$(2^{O(n)})$. For the nondeterministic case we will also need the sub-exponential version NSUBEXP = $\cap_{\epsilon>0}$NTIME(2^{n^ϵ}).

Let $\Sigma = \{0,1\}$ and $A^n = \Sigma^n \cap A$. For any complexity class $\mathcal{C}$, the class $_{io}\mathcal{C}$ is the class of languages $\{A \mid \exists B \in \mathcal{C}$ such that for infinitely many n, $A^n = B^n\}$.

Lemma 1. *For every $k > 0$, E $\not\subseteq$ $_{io}$DTIME(2^{kn}).*

2.1 Average-Case Complexity

We review some basic definitions from average-case complexity. We recommend Wang's survey [Wan97] for an excellent introduction to the area.

A *distributional problem* is a pair (A, μ) so that $A \subseteq \Sigma^*$ and μ is a probability distribution on Σ^*. DistNP is the class of distributional problems (A, μ) so that $A \in$ NP and μ^* is polynomial time computable where μ^* denotes the distribution function of μ. Recall that $\mu^*(x) = \sum_{y \leq x} \mu(y)$.

We need parameterized versions of the definitions of various notions in average-case complexity. In particular we will need the parameterized subclass of DistNP we call DistNLIN

Definition 1.

- DistNLIN = $\{(A, \mu) \mid A \in$ NLIN *and* μ^* *is linear-time computable*$\}$, where μ^* denotes the distribution function of μ.

- *Let μ be a distribution on Σ^*. Then a function $f : \Sigma^* \to \mathbf{N}$ is said to be n^k on μ-average [Lev86] if*

$$\sum_{x \in \Sigma^*} \frac{f^{1/k}(x)\mu(x)}{|x|} < \infty.$$

- *A distributional problem (A, μ) is in $\mathrm{AvgTIME}(n^k)$ if there exists a deterministic Turing machine M that accepts A so that the running time of M is n^k on μ-average. (A, μ) is in Average-P if there exists a k so that (A, μ) is in $\mathrm{AvgTIME}(n^k)$.*

By " NP is easy on average" we mean $\mathrm{DistNP} \subseteq \mathrm{Average\text{-}P}$. Thus if $\mathrm{DistNP} \subseteq$ Average-P, then for every language L in NP, and every polynomial-time computable distribution μ, $(L, \mu) \in$ Average-P.

Remark 1. Instead of working with distributions on Σ^*, we can also consider an ensemble of distributions $\mu = (\mu_1, \mu_2, \cdots)$, where μ_i is a distribution on Σ^i. An ensemble is polynomial-time computable, if there is a polynomial-time computable machine that on inputs i and x outputs $\mu_i^*(x)$. Impagliazzo [Imp95] defined a notion of average polynomial-time with respect to an ensemble of distributions. Though we state our results using distributions over Σ^*, all our results carry over to the case of ensemble of distributions.

Levin [Lev86] introduced the notion of reductions between distributional problems and defined DistNP-completeness. There are distributional problems that are DistNP-complete in the sense that NP is easy on average if and only if the complete problem is in Average-P. Proving distributional completeness is much more challenging than proving usual NP-completeness since the reductions must satisfy certain additional *domination* properties.

Definition 2. *([Lev86]) Let μ and ν be two distributions on Σ^*.*

- *We say that ν dominates μ within n^k if for all x, $\mu(x) \leq |x|^k \nu(x)$.*
- *Let $f : \Sigma^* \to \Sigma^*$. Then we say that ν dominates μ within n^k via f if there is a distribution μ_1 such that μ_1 dominates μ within n^k, and for all y in the range of f $\nu(y) = \sum_{y=f(x)} \mu_1(x)$.*
- *(A, μ) reduces to (B, ν) if there is a polynomial time computable many-one reduction f from A to B so that for some k, ν dominates μ within n^k via f*

Gurevich [Gur91] showed that the *distributional halting problem* (K, μ_K) is complete for DistNP, where $K = \{ \langle i, x, 0^n \rangle \mid N_i$ accepts x in n steps $\}$, and $\mu_k(\langle i, x, 0^n \rangle) = \frac{1}{2^{|i|}|i|^2} \frac{1}{2^{|x|}|x|^2} \frac{1}{n^2}$. Here N_i denotes the ith nondeterministic Turing machine in some fixed enumeration. We denote distributional halting problem by DH in this paper.

The following statement is easy to prove.

Observation 1. *If $(L, U) \in \mathrm{AvgTIME}(n^k)$, then there is a deterministic Turing machine M that decides L and for all but finitely many n, there exist at most $2^n/n^2$ strings of length n, on which M takes more than n^{4k} time. Here U is the standard uniform distribution on Σ^*.*

Definition 3. *Two languages L and L' are* ae-NTIME(n^c)-*distinguishable, if there is a n^c-time bounded nondeterministic machine N such that for all but finitely many n, $M(0^n)$ outputs a string from $L \Delta L'$ along every accepting path. We say L and L' are* io-NTIME(n^c)-*indistinguishable, if they are not* ae-NTIME(n^c)-*distinguishable.*

Definition 4. *We say* AM $\subseteq$ [io-pseudo$_{\mathrm{NTIME}(n^c)}$]$-$NP, *if for every language L in* AM *there exists a language L' in* NP *such that L and L' are* io-NTIME(n^c)-*indistinguishable.*

3   Arthur-Merlin Games

We show several results on the worst-case complexity of Arthur-Merlin games under the assumption that NP is easy on average. First we consider a slightly stronger notion of NP is easy on average and prove that under this assumption AM $\subseteq$ NSUBEXP.

Let DistNP$_{\mathrm{P/poly}}$ denote the class of distributional problems (A, μ) so that $A \in$ NP and the distribution function μ^* is computable by a polynomial-size circuit family. We show that if DistNP$_{\mathrm{P/poly}} \subseteq$ Average-P, then AM $\subseteq$ NSUBEXP. We first show that the hypothesis implies E $\not\subseteq$ io(NP/poly). The result then follows from the following result of Shaltiel and Umans [SU01].

Theorem 1 ([SU01]). *If* E $\not\subseteq$ io(NP/poly), *then* AM $\subseteq$ NSUBEXP.

The chain of arguments that we use is as follows. If E $\subseteq$ NP/poly and NP is easy on average then E $\subseteq$ Average-P/poly with respect to certain distribution (which is nonuniformly computable). Since there are random-self-reducible complete problems for E, we will actually get E $\subseteq$ P/poly. From [BFNW91] we have that E $\subseteq$ P/poly $\Rightarrow$ E = MA. Since NP is easy on average we have MA = NP [KS04] and NE = E [BDCGL92]. So we will finally get NE = NP which is a contradiction to the nondeterministic time hierarchy theorem. We now present a more formal argument.

We will need a non-uniform version of Average-P.

Definition 5. *A distributional problem (L, μ) is in* Average-P/poly, *if there is a Turing machine M and a polynomial-bounded function $a : \mathbb{N} \to \Sigma^*$, such that $x \in L \Leftrightarrow M$ accepts $\langle x, a(|x|) \rangle$, and there exists a constant $k > 0$, such that*

$$\sum_x \frac{T_M^{1/k}(\langle x, a(|x|) \rangle)}{|x|} \mu(x) < \infty.$$

where $T_M(y)$ denotes the running time of M on input y.

Note that the above definition requires the running time of M to be efficient on average only on correct advice.

We need the following lemma regarding random-self-reducibility of E complete problems.

Lemma 2 ([BFNW91]). *There exists a random-self-reducible complete language $L \in \mathrm{E}$ so that if there is a polynomial-size circuit C such that*

$$\forall n, \Pr_x[C(x) = L(x)] \geq 1 - 1/n^2,$$

then $L \in \mathrm{P}/\mathtt{poly}$.

We now prove the result we need to derandomize AM (actually we need an i.o. version of the following theorem. But we state and prove a cleaner non-i.o. version).

Theorem 2. *If $\mathrm{DistNP}_{\mathrm{P}/\mathrm{poly}} \subseteq \mathrm{Average\text{-}P}$ then $\mathrm{E} \not\subseteq \mathrm{NP}/\mathtt{poly}$.*

Proof. (*sketch*) Assume $\mathrm{E} \subseteq \mathrm{NP}/\mathtt{poly}$. Let L be a complete language for E that is random-self-reducible (provided by Lemma 2). Since $L \in \mathrm{NP}/\mathtt{poly}$, there is a language L' in NP and a polynomial-bounded advice function $a : \mathbb{N} \to \Sigma^*$, such that

$$\forall x, x \in L \Leftrightarrow \langle x, a(|x|) \rangle \in L'$$

Consider the following distribution μ

$$\mu(\langle x, y \rangle) = \begin{cases} U(x) & \text{if } y = a(|x|) \\ 0 & \text{otherwise} \end{cases}$$

Here U is the uniform distribution on Σ^*, i.e., $U(x) = \frac{1}{n^2}\frac{1}{2^n}$, where $n = |x|$. It is clear that μ is P/poly-computable. Since $\mathrm{DistNP}_{\mathrm{P}/\mathrm{poly}} \subseteq \mathrm{Average\text{-}P}$, $(L', \mu) \in \mathrm{Average\text{-}P}$. Consider the following reduction from (L, U) to (L', μ): $f(x) = \langle x, a(|x|) \rangle$. It is clear that f is P/poly-computable and satisfies the dominance condition. Thus $(L, U) \in \mathrm{Average\text{-}P}/\mathtt{poly}$.

Thus there is a Turing machine M and an advice function $a : \mathbb{N} \to \Sigma^*$, such that $x \in L \Leftrightarrow M$ accepts $\langle x, a(|x|) \rangle$, and there exists a constant l, such that

$$\sum_x \frac{T_M^{1/l}(\langle x, a(|x|) \rangle)}{|x|} \mu(x) < \infty.$$

By Observation 1, there is a constant k, such that for every n, on at least $(1 - 1/n^2)$ fraction of strings of the form $\langle x, a(|x|) \rangle$, M halts within n^k steps, and M accepts $\langle x, a(|x|) \rangle$ if and only if $x \in L$.

We now claim that this implies $L \in \mathrm{P}/\mathtt{poly}$. Define a new machine M' as follows: M' on input $\langle x, y \rangle$ runs M on $\langle x, y \rangle$ for n^k steps. If M does not halt within n^k steps, then M' rejects its input. If M halts within n^k steps, then M' accepts $\langle x, y \rangle$ if and only if M accepts $\langle x, y \rangle$.

By converting M' into a circuit and hardwiring $a(|x|)$ in it, we obtain a polynomial-size circuit C such that

$$\forall n, \Pr[C(x) = L(x)] \geq 1 - 1/n^2.$$

Since L is random-self-reducible, by Lemma 2, $L \in \mathrm{P}/\mathtt{poly}$. Since L is complete for E, $\mathrm{E} \subseteq \mathrm{P}/\mathtt{poly}$. By [BFNW91], if $\mathrm{E} \subseteq \mathrm{P}/\mathtt{poly}$, then $\mathrm{E} \subseteq \mathrm{MA}$. If $\mathrm{DistNP} \subseteq$

Average-P, then MA = NP [KS04] and also NE = E [BDCGL92]. Thus we have NE = NP. A contradiction follows from the nondeterministic time hierarchy theorem [Ž83].

Theorem 3. *If* $\text{DistNP}_{\text{P/poly}} \subseteq$ Average-P, *then* AM $\subseteq$ NSUBEXP.

Proof. (Sketch) Essentially the same proof will show that if $\text{DistNP}_{\text{P/poly}} \subseteq$ Average-P then E $\not\subseteq$ io(NP/poly). To see this, if E $\not\subseteq$ io(NP/poly) then we will get that E can be approximated by a polynomial size circuit on infinitely many input lengths. As argued in [BFNW91] we then get that E $\subseteq$ io$(\text{P/poly}) \subseteq$ ioMA. If DistNP $\subseteq$ Average-P, then MA = NP [KS04], Thus we get E $\subseteq$ ioNP. Since a straightforward application of nondeterministic time hierarchy theorem is not sufficient to separate NE from ioNP we need to argue slightly differently. By [BDCGL92], if DistNP $\subseteq$ Average-P, then E = NE. By a result of Impagliazzo, Kabanets, and Wigderson, if E = NE, then there is a fixed constant k such that $\text{NTIME}(2^n) \subseteq \text{DTIME}(2^{kn})$. Thus we obtain E $\subseteq$ ioNP $\subseteq$ ioNTIME$(2^n) \subseteq$ ioDTIME(2^{kn}). This contradicts Lemma 1.

Ben-David, Chor, Goldreich, and Luby [BDCGL92] showed that if DistNP $\subseteq$ Average-P, then E = NE. Buhrman, Fortnow, and Pavan showed that in fact a stronger conclusion follows, that is if DistNP $\subseteq$ Average-P, then for every NE-predicate $R(x, y)$, there is an E machine M such that $M(x)$ outputs a y such that $R(x, y)$ holds. Buhrman [Buh93] showed that if E = NE, and FSAT $\in \text{PF}_{tt}^{\text{NP}}$, then witnesses of NE predicates can be computed in E-time. Here FSAT denotes the problem of computing satisfying assignments of propositional formulas, and $\text{PF}_{tt}^{\text{NP}}$ is the class of functions that can be computed by polynomial-time machines that can make nonadaptive queries to an NP-oracle. This raises the following question: "If DistNP $\subseteq$ Average-P, is FSAT $\in \text{PF}_{tt}^{\text{NP}}$?". We have a partial answer.

Theorem 4. *If* $\text{DistNP}_{\text{P/poly}} \subseteq$ Average-P, *then* FSAT $\in \text{SUBEXP}_{tt}^{\text{NP}}$.

Proof. By Theorem 3, the hypothesis implies that E $\not\subseteq$ io(NP/poly). Very recently, Shaltiel and Umans [SU04] showed that E $\subseteq$ NP/poly if and only if E $\subseteq \text{P}_{tt}^{\text{NP}}/\text{poly}$. Thus the hypothesis implies the existence of a language in E that does not have $\text{P}_{tt}^{\text{NP}}$-circuits. Klivans and van Melkebeek [KvM02] showed that there exist pseudo-random generators with polynomial stretch that are secure against $\text{P}_{tt}^{\text{NP}}$-circuits. These pseudo-random generators can be used to derandomize $\text{BPP}_{tt}^{\text{NP}}$ to $\text{SUBEXP}_{tt}^{\text{NP}}$. Since FSAT $\in \text{BPP}_{tt}^{\text{NP}}$ [VV85], we have the conclusion.

Now we show that, under the assumption DistNP $\subseteq$ Average-P, AME = E, and AM $\subseteq$ [io-pseudo$_{\text{NP}}$]−NP. We also give an alternate proof of the result that under the assumption AM $\cap$ co-AM = NP $\cap$ co-NP due to Arvind and Köbler [AK02].

We use uniform derandomization results for AM and AM $\cap$ co-AM due to Gutfreud, Shaltiel, and Ta-Shma [GSTS03], and the fact that pseudorandom generators exist if NP is easy on average [BFP03] to prove our results.

Theorem 5 ([GSTS03]). *If* $E \not\subseteq AMTIME(2^{\beta n})$ *for some constant* β *then for every* $c > 0$, $AM \subseteq [io\text{-pseudo}_{NTIME(n^c)}] - NP$.

Theorem 6 ([BFP03]). *If* $DistNP \subseteq Average\text{-}P$ *then there is an algorithm that on input* 1^n *outputs a pseudorandom set for circuits of size* n. *The algorithm runs in time* n^c *for a fixed* c.

Recall that a set S is a pseudorandom set for circuits of size $s(n)$, if for every circuit C of size $s(n)$ if $\Pr_{x \in \Sigma^n}[C(x) = 1]$ is close to $\Pr_{x \in S}[C(x) = 1]$.

Our approach is as follows. Under the assumption $DistNP \subseteq Average\text{-}P$, we show that $AMTIME(2^n)$ is a subset of $DTIME(2^{kn})$ for a fixed k. The same argument will also show that $_{io}AMTIME(2^n) \subseteq {}_{io}DTIME(2^{kn})$. Since for any fixed k, $E \not\subseteq {}_{io}DTIME(2^{kn})$ we get that $E \not\subseteq {}_{io}AMTIME(2^n)$ and therefore $AM \subseteq Pseudo_{NTIME(n^c)} - NP$ from Theorem 5.

For showing that $AMTIME(2^n) \subseteq DTIME(2^{kn})$ for a fixed k, we first observe a general result that if $DistNP \subseteq Average\text{-}P$ then $DistNLIN$ is in $AvgTIME(n^k)$ for a fixed k. We use this result to first show that every tally language in $AMTIME(n)$ is in $BPTIME(n^k)$ for a fixed k. Finally we use the pseudorandom generator from Theorem 6 to get $BPTIME(n^k) \subseteq DTIME(n^l)$. A standard padding gives the collapse in the exponential level.

We know that if $NP = P$ then for any k there is an l so that $NTIME(n^k) \subseteq DTIME(n^l)$. One way to see this is to observe that any problem in $NTIME(T(n))$ is $T^2(n)$ time reducible to SAT and the result follows since under the assumption SAT is in $DTIME(n^{k'})$ for a fixed k'. Such a theorem exists in the average-case setting also. We use the completeness of the Distributional Halting problem [Gur91] to prove an analogous result in the average-case setting. An accessible proof can be found in [Wan97].

Lemma 3 ([Lev86, Gur91]). *Let* (A, μ) *and* (B, ν) *be two distributional problems and let* f *be a reduction from A to B so that:*

- *f is computable in time n^l*
- *ν dominates μ within n^r via f*

Then if $B \in AvgTIME(n^k)$ *then* $A \in AvgTIME(n^{2klr})$.

Lemma 4 ([Lev86, Gur91]). *Every distributional problem* $(A, \mu) \in DistNLIN$ *is reducible to the* $DistNP$ *complete problem* $DH = (K, \mu_K)$ *via a reduction f_A so that:*

- *f_A is computable in time n^3*
- *μ_K dominates μ within n^3 via f_A*

Using the above two lemmas we get the following theorem.

Theorem 7. *If* $DistNP \subseteq Average\text{-}P$ *then there exists a k so that* $DistNLIN \subseteq AvgTIME(n^k)$.

Now we prove our main Lemma.

Lemma 5. *If* DistNP $\subseteq$ Average-P *then there exists a k so that* AMTIME(2^n) $\subseteq$ DTIME(2^{kn}).

Proof. (*sketch*) We show that under the assumption any tally set in AMTIME(n) is in DTIME(n^l) for a fixed l. The Lemma follows from the fact that for any language $L \in$ AMTIME(2^n), the tally version is in AMTIME(n).

Let L be a tally language in AMTIME(n). Then there is a language $B \in$ NLIN so that

$$0^n \in L \Rightarrow \Pr_{y \in \Sigma^n}[\langle 0^n, y \rangle \in B] = 1$$
$$0^n \notin L \Rightarrow \Pr_{y \in \Sigma^n}[\langle 0^n, y \rangle \in B] < 1/4$$

Consider the linear-time computable distribution μ which on a string $\langle 0^n, y \rangle$ has a probability of $\frac{1}{n^2 2^{|y|}}$. The distributional problem (B, μ) is in DistNLIN and hence is in AvgTIME(n^k) for a fixed k. Let M be the deterministic machine that witnesses this fact. Now consider a n^{4k} time deterministic machine M' that simulates M for n^{4k} steps. An easy counting argument shows that for every n, M' correctly decides B on at least $1 - 1/n^2$ fraction of strings of the form $\langle 0^n, y \rangle$. From this it easily follows that $L \in$ BPTIME(n^{4k}).

By Theorem 6, if DistNP $\subseteq$ Average-P, there is an n^c time-bounded algorithm that outputs a pseudorandom set for circuits of size n. Using this we can derandomize BPTIME(n^{4k}) to DTIME($n^{8k^2 c}$). Thus L is in DTIME(n^l) for $l = 8k^2 c$.

Our theorems follow from the above Lemma.

Theorem 8. *If* DistNP $\subseteq$ Average-P, *then* AME = E.

Theorem 9. *If* DistNP $\subseteq$ Average-P *then,* AM $\subseteq$ [io-pseudo$_{\text{NP}}$]$-$NP *for every* $c > 0$.

Proof. Proof follows from Lemma 5 and Theorem 5.

Our approach gives a different proof of the following result due to Arvind and Köbler [AK02].

Theorem 10 ([AK02]). *If* DistNP $\subseteq$ Average-P, *then* AM $\cap$ co-AM = NP $\cap$ co-NP.

Proof. The same argument as in Lemma 5 shows that if DistNP $\subseteq$ Average-P, then $_{\text{io}}$AMTIME(2^n) $\subseteq$ $_{\text{io}}$DTIME(2^{kn}) for a fixed k. By Lemma 1, E $\not\subseteq$ $_{\text{io}}$DTIME(2^{kn}) for any fixed k. Therefore E $\not\subseteq$ $_{\text{io}}$AMTIME(2^n). Gutfreud, Shaltiel, and Ta-shma [GSTS03] showed that this implies AM $\cap$ co-AM = NP $\cap$ co-NP.

Acknowledgments

We thank V. Arvind and Johannes Köbler for helpful discussions. We thank Chris Bourke and the FCT referees for useful comments.

References

[AK02]   V. Arvind and J. Köbler. New lowness results for ZPP^{NP} and other complexity classes. *Journal of Computer and System Sciences*, 65(2):257–277, 2002.

[Bab85]   L. Babai. Trading group theory for randomness. In *Proc. 17th Annual ACM Symp. on Theory of Computing*, pages 421–429, 1985.

[BDCGL92]   S. Ben-David, B. Chor, O. Goldreich, and M. Luby. On the theory of average case complexity. *Journal of Computer and System Sciences*, 44(2):193–219, 1992.

[BDG88]   J. Balcázar, J. Diaz, and J. Gabarró. *Structural Complexity I*. Springer-Verlag, Berlin, 1988.

[BFNW91]   L. Babai, L. Fortnow, N. Nisan, and A. Wigderson. *BPP* has subexponential time simulations unless *exptime* has publishable proofs. In *Proceedings of the 6th Annual Conference on Structure in Complexity Theory, 1991*, pages 213–219, 1991.

[BFP03]   H. Buhrman, L. Fortnow, and A. Pavan. Some results on derandomization. In *Proceedings of the 20th Annual Symposium on Theoretical Aspects of Computer Science*, volume LNCS 2607, pages 212–222, 2003.

[BM88]   L. Babai and S. Moran. Arthur-merlin games: a randomized proof system, and a hierarchy of complexity class. *Journal of Computer and System Sciences*, 36(2):254–276, 1988.

[Buh93]   H. Buhrman. *Resource bounded reductions*. PhD thesis, University of Amsterdam, 1993.

[FL92]   U. Feige and C. Lund. On the hardness of computing permanent of random matrices. In *Proceedings of 24th Annual ACM Symposium on Theory of Computing*, pages 643–654, 1992.

[GS92]   P. Gemmel and M. Sudan. Higly resilient correctors for polynomials. *Information Processing Letters*, 43:169–174, May 1992.

[GSTS03]   D. Gutfreund, R. Shaltiel, and A. Ta-Shma. Uniform hardness vs. randomness tradeoffs for Arth ur-Merlin games. *Computational Complexity*, 12:85–130, 2003.

[Gur91]   Y. Gurevich. Average case completeness. *Journal of Computer and System Sciences*, 42:346–398, 1991.

[Imp95]   R. Impagliazzo. A personal view of average-case complexity theory. In *Proceedings of the 10th Annual Conference on Structure in Complexity Theory*, pages 134–147. IEEE Computer Society Press, 1995.

[IW97]   R. Impagliazzo and A. Wigderson. P = BPP if E requires exponential circuits: Derandomizing the XOR lemma. In *Proceedings of the 29th ACM Symposium on Theory of Computing*, pages 220–229, 1997.

[KS04]   J. Köbler and R. Schuler. Average-case intractability vs. worst-case intractability. *Information and Computation*, 190(1):1–17, 2004.

[KvM02]   A. Klivans and D. van Melkebeek. Graph nonisomorphism has subexponential size proofs unless the polynomial-time hierarchy collapses. *SIAM Journal on Computing*, 31:1501–1526, 2002.

[Lev86]   L. Levin. Average case complete problems. *SIAM Journal of Computing*, 15:285–286, 1986.

[Lip91]   R. Lipton. New directions in testing. In *Distributed Computing and Cryptography*, volume 2 of *DIMACS Series in Discrete Mathematics and Theoretical Computer Science*, pages 191–202. American Mathematics Society, 1991.

[Pap94]   C. Papadimitriou. *Computational Complexity*. Addison-Wesley, 1994.

[SU01]   R. Shaltiel and C. Umans. Simple extractors for all min-entropies and a new pseudo-random generator. In *42nd IEEE Symposium on Foundations of Computer Science*, pages 648–657, 2001.

[SU04]   R. Shaltiel and C. Umans. Pseudorandomness for approximate counting and sampling. Technical Report TR 04-086, ECCC, 2004.

[VV85]   L. Valiant and V. Vazirani. NP is as easy as detecting unique solutions. In *Proc. 17th ACM Symp. Theory of Computing*, pages 458–463, 1985.

[Ž83]   S. Žák. A Turing machine time hierarchy. *Theor. Computer Science*, 26:327–333, 1983.

[Wan97]   J. Wang. Average-case computational complexity theory. In L. Hemaspaandra and A. Selman, editors, *Complexity Theory Retrospective II*, pages 295–328. Springer-Verlag, 1997.

Reconstructing Many Partitions Using Spectral Techniques*

Joachim Giesen and Dieter Mitsche

Institute for Theoretical Computer Science,
ETH Zürich, CH-8092 Zürich
{giesen, dmitsche}@inf.ethz.ch

Abstract. A partitioning of a set of n items is a grouping of these items into k disjoint, equally sized classes. Any partition can be modeled as a graph. The items become the vertices of the graph and two vertices are connected by an edge if and only if the associated items belong to the same class. In a planted partition model a graph that models a partition is given, which is obscured by random noise, i.e., edges within a class can get removed and edges between classes can get inserted. The task is to reconstruct the planted partition from this graph. In the model that we study the number k of classes controls the difficulty of the task. We design a spectral partitioning algorithm that asymptotically almost surely reconstructs up to $k = c\sqrt{n}$ partitions, where c is a small constant, in time $C^k \operatorname{poly}(n)$, where C is another constant.

1 Introduction

The partition reconstruction problem, which we study in this paper, is related to the k-partition problem. In the latter problem the task is to partition the vertices of a given graph into k equally sized classes such that the number of edges between the classes is minimized. This problem is already NP-hard for $k = 2$, i.e.1, in the graph bisection case [6]. Thus researchers, see for example [4,2] and the references therein, started to analyze the problem in specialized but from an application point of view (e.g., parallel scheduling or mesh partitioning) still meaningful, graph families - especially families of random graphs. The random graph families typically assume a given partition of the vertices of the graph (planted partition), which is obscured by random noise. The families are parameterized by a set of parameters, e.g., the number of vertices n and classes k. The goal now becomes to assess the ability of a partitioning algorithm to reconstruct the planted classes. The most prominent such measure is the probability that the algorithm can reconstruct the planted partition.

The best studied random graph family for the partition reconstruction problem is the following: an edge in the graph appears with probability p if its two incident vertices belong to the same planted class and with probability $q < p$

---

* Partly supported by the Swiss National Science Foundation under the grant "Nonlinear manifold learning".

M. Liśkiewicz and R. Reischuk (Eds.): FCT 2005, LNCS 3623, pp. 433–444, 2005.

otherwise, independently from all other edges. In general the probabilities p and q can depend on the number n of vertices in the graph and on the number k of classes of the planted partition.

Related Work. The partitioning problem in the planted partition model that we have described above gets more difficult if the difference $p - q$ gets small and/or k gets large. If we assume that p and q are fixed the only parameter left to control the difficulty of the problem is k. The algorithm of Shamir and Tsur [11] which builds on ideas of Condon and Karp [4] can with high probability reconstruct correctly up to $k = O(\sqrt{n/\log n})$ planted classes. The same guarantees can be given for an algorithm due to McSherry [9]. Both algorithms are polynomial in time and even allow the classes to differ in size (only a lower bound on the size of the classes is needed), i.e., they deal with the more general planted clustering problem.

The algorithm of McSherry falls in the category of spectral clustering algorithms. The use of spectral methods for clustering has become increasingly popular in recent years. The vast majority of the literature points out the experimental success of spectral methods, see for example the review by Meila et al. [10]. On the theoretical side much less is known about the reasons why spectral algorithms perform well. In 1987 Boppana [3] presented a spectral algorithm for recovering the optimal bisection of a graph. Much later Alon et al. [1] showed how the entries in the second eigenvector of the adjacency matrix of a graph can be used to find a hidden clique of size $\Omega(\sqrt{n})$ in a random graph. Spielman and Teng [12] showed how bounded degree planar graphs and d-dimensional meshes can be partitioned using the signs of the entries in the second eigenvector of the adjacency matrix of the graph or mesh, respectively.

In [7] we designed an efficient (polynomial in n) spectral algorithm. We cannot prove that this algorithm with high probability reconstructs the planted partition, but we can prove that for this algorithm the relative number of misclassifications (with a suited definition of misclassification) for $k = o(\sqrt{n})$ goes to zero with high probability when n goes to infinity.

Our Result. We design a spectral algorithm that runs in time $C^{k/2} \text{poly}(n)$, for some constant C. We prove that this algorithm asymptotically almost surely reconstructs a planted partition for $k \leq c\sqrt{n}$, where c is another sufficiently small constant.

2   Planted Partitions

In this section we introduce the planted partition reconstruction problem. We first define the $A(\varphi, p, q)$ distribution, see also McSherry [9].

$A(\varphi, p, q)$ **Distribution.** Given a surjective function $\varphi : \{1, \ldots, n\} \to \{1, \ldots, k\}$ and probabilities $p, q \in (0, 1)$ with $p > q$. The $A(\varphi, p, q)$ distribution is a distribution on the set of $n \times n$ symmetric, 0-1 matrices with zero trace. Let $\hat{A} = (\hat{a}_{ij})$ be a matrix drawn from this distribution. It is $\hat{a}_{ij} = 0$ if $i = j$ and for $i \neq j$,

$$P(\hat{a}_{ij} = 1) = p \quad \text{if } \varphi(i) = \varphi(j)$$
$$P(\hat{a}_{ij} = 0) = 1 - p \quad \text{if } \varphi(i) = \varphi(j)$$
$$P(\hat{a}_{ij} = 1) = q \quad \text{if } \varphi(i) \neq \varphi(j)$$
$$P(\hat{a}_{ij} = 0) = 1 - q \quad \text{if } \varphi(i) \neq \varphi(j),$$

independently. The *matrix of expectations* $A = (a_{ij})$ corresponding to the $A(\varphi, p, q)$ distribution is given as

$$a_{ij} = 0 \quad \text{if } i = j$$
$$a_{ij} = p \quad \text{if } \varphi(i) = \varphi(j) \text{ and } i \neq j$$
$$a_{ij} = q \quad \text{if } \varphi(i) \neq \varphi(j)$$

Lemma 1 (Füredi and Komlós [5], Krivelevich and Vu [8]). *Let $\hat{A}$ be a matrix drawn from the $A(\varphi, p, q)$ distribution and A be the matrix of expectations corresponding to this distribution. Let $c' = \min\{p(1-p), q(1-q)\}$ and assume that $c'^2 \gg (\log n)^6/n$. Then*

$$|A - \hat{A}| \leq \sqrt{n}$$

with probability at least $1 - 2e^{-c'^2 n/8}$. Here $|\cdot|$ denotes the L_2 matrix norm, i.e., $|B| = \max_{|x|=1} |Bx|$. □

Planted Partition Reconstruction Problem. Given a matrix $\hat{A}$ drawn from from the $A(\varphi, p, q)$ distribution. Assume that all classes $C_l = \varphi^{-1}(l), l \in \{1, \ldots, k\}$ have the same size n/k. Then the function φ is called a *partition function*. The planted partition reconstruction problem asks to reconstruct φ up to a permutation of $\{1, \ldots, k\}$ only from $\hat{A}$ (up to permutations of of $\{1, \ldots, k\}$).

3   Spectral Properties

Any real symmetric $n \times n$ matrix has n real eigenvalues and $\mathbb{R}^n$ has a corresponding eigenbasis. Here we are concerned with two types of real symmetric matrices. First, any matrix $\hat{A}$ drawn from an $A(\varphi, p, q)$ distribution. Second, the matrix A of expectations corresponding to the distribution $A(\varphi, p, q)$.

We want to denote the eigenvalues of $\hat{A}$ by $\hat{\lambda}_1 \geq \hat{\lambda}_2 \geq \ldots \geq \hat{\lambda}_n$ and the vectors of a corresponding orthonormal eigenbasis of $\mathbb{R}^n$ by $v_1, \ldots, v_n$, i.e., it is $\hat{A}v_i = \hat{\lambda}_i v_i$, $v_i^T v_j = 0$ if $i \neq j$ and $v_i^T v_i = 1$, and the $v_1, \ldots, v_n$ span the whole $\mathbb{R}^n$.

For the sake of analysis we want to assume here without loss of generality that the matrix A of expectations has a block diagonal structure, i.e., the elements in the i-th class have indices from $\frac{n}{k}(i-1)+1$ to $\frac{n}{k}i$ in $\{1, \ldots, n\}$. It is easy to verify that the eigenvalues $\lambda_1 \geq \ldots \geq \lambda_n$ of A are $(\frac{n}{k}-1)p+(n-\frac{n}{k})q$, $\frac{n}{k}(p-q)-p$ and $-p$ with corresponding multiplicities 1, $k-1$ and $n-k$, respectively. A possible orthonormal basis of the eigenspace corresponding to the k largest eigenvalues of A is u_i, $i = 1, \ldots, k$, whose j-th coordinates are given as follows,

$$u_{ij} = \begin{cases} \sqrt{\frac{k}{n}}, & j \in \{\frac{n}{k}(i-1) + 1, \ldots, \frac{n}{k}i\} \\ 0, & \text{else.} \end{cases}$$

Spectral Separation. The *spectral separation* $\delta_k(A)$ of the eigenspace of the matrix A of expectations corresponding to its k largest eigenvalues from its complement is defined as the absolute difference between the k-th and the $(k+1)$-th eigenvalue, i.e., it is $\delta_k(A) = \frac{n}{k}(p-q)$.

Projection Matrix. The matrix $\hat{P}$ that projects any vector in $\mathbb{R}^n$ to the eigenspace corresponding to the k largest eigenvalues of a matrix $\hat{A}$ drawn from the distribution $A(\varphi, p, q)$, i.e., the projection onto the space spanned by the vectors $v_1, \ldots, v_k$, is given as

$$\hat{P} = \sum_{i=1}^{k} v_i v_i^T.$$

The matrix P that projects any vector in $\mathbb{R}^n$ to the eigenspace corresponding to the k largest eigenvalues of the matrix A of expectations can be characterized even more explicitly. Its entries are given as

$$p_{ij} = \begin{cases} \frac{k}{n}, & \varphi(i) = \varphi(j) \\ 0, & \varphi(i) \neq \varphi(j) \end{cases}$$

Lemma 2. *All the k largest eigenvalues of $\hat{A}$ are larger than $\sqrt{n}$ and all the $n - k$ smallest eigenvalues of $\hat{A}$ are smaller than $\sqrt{n}$ with probability at least $1 - 2e^{-c'^2 n/8}$ provided that n is sufficiently large and $k < \frac{p-q}{4}\sqrt{n}$.*

Proof. Will appear in the full version of the paper.   □

Theorem 1 (Stewart [13]). *Let $\hat{P}$ and P be the projection matrices as defined above. It holds*

$$|P - \hat{P}| \leq \frac{2|A - \hat{A}|}{\delta_k(A) - 2|A - \hat{A}|}$$

if $\delta_k(A) > 4|A - \hat{A}|$ where $|\cdot|$ is the L_2 matrix norm.

4   A Spectral Algorithm

Now we have all prerequisites at hand that we need to describe our spectral algorithm to solve the planted partition reconstruction problem. We assume that the input to the algorithm is a $4n \times 4n$ matrix $\hat{A}$ drawn from the $A(\varphi, p, q)$, where φ is a surjective function from $\{1, \ldots, 4n\} \to \{1, \ldots, k\}$.

GRIDRECONSTRUCT($\hat{A}$)
1    $k' :=$ number of eigenvalues $\hat{\lambda}_i$ of $\hat{A}$ that are larger than $2\sqrt{n}$.
2    $\alpha := c_0/\sqrt{k}$.

3   Randomly partition $\{1, \ldots, 4n\}$ into four subsets I_{11}, I_{12}, I_{21} and I_{22}
    of equal size.
4   **for** $i, j = 1, 2$ **do**
5      $\hat{A}_{ij} :=$ restriction of $\hat{A}$ to index set I_{ij}.
6   **end for**

In the first six lines of the algorithm we do some pre-processing. We compute
the eigenvalues of $\hat{A}$ and use them in line 1 to estimate the number of planted
partitions. According to Lemma 2 our estimate k' is with high probability the
correct number k of planted partitions. Thus we will in the following always use
k instead of k'. In line 2 we use the value of k to set the value α, which is an
essential parameter of the algorithm; c_0 is a small constant > 0.

In line 3 we randomly partition the set $\{1, \ldots, 4n\}$ into four equally sized
subsets. One way to compute such a random partition is to compute a random
permutation π of $\{1, \ldots, 4n\}$ and assign

$$I_{11} = \{\pi(1), \ldots, \pi(n)\}, \qquad I_{12} = \{\pi(n+1), \ldots, \pi(2n)\},$$
$$I_{21} = \{\pi(2n+1), \ldots, \pi(3n)\}, \quad I_{22} = \{\pi(3n+1), \ldots, \pi(4n)\}.$$

For the sake of the analysis we will use a slightly different method: we first put
every vertex into one of the four parts with equal probability $\frac{1}{4}$ (independently
from all other vertices) and later redistribute randomly chosen elements to make
the partition sizes equal. In Lemma 4 below we formalize this.

The only reason to partition $\{1, \ldots, 4n\}$ into four sets is for the sake of
analysis where we need independence at some point. The main idea behind our
algorithm needs only a partitioning into two sets.

In lines 4 to 6 we compute the restrictions of the matrix $\hat{A}$ to the index sets
I_{ij}. Note that the $\hat{A}_{ij}$ are $n \times n$ matrices.

The following lines 7 to 30 make up the main part of the algorithm.

7   **for** $i, j = 1, 2$ **do**
8     $l := 1$
9     $\{v_1, \ldots, v_k\} :=$ orthonormal eigenbasis of the eigenspace $(\subset \mathbb{R}^n)$ of
        $\hat{A}_{ij}$ that corresponds to the k largest eigenvalues.
10   **for all** $(\lambda_1, \ldots, \lambda_k) \in (\alpha\mathbb{Z})^k$ with $\sum_{s=1}^{k} \lambda_s^2 \leq 1$ **do**
11      $\hat{C}_l^{ij} := \emptyset$
12      $v := \sum_{s=1}^{k} \lambda_s v_s / \left| \sum_{s=1}^{k} \lambda_s v_s \right|$
13      $I :=$ subset of the index set I_{ij} that corresponds to the
        n/k largest coordinates of v (break ties arbitrarily).
14   **for all** $t \in I_{i(j \bmod 2 + 1)}$ **do**
15   **if** $\sum_{s \in I} \hat{a}_{st} \geq t(n, k, \hat{\lambda}_1, \hat{\lambda}_2)$ **do**
16        $\hat{C}_l^{ij} := \hat{C}_l^{ij} \cup \{t\}$
17   **end if**
18   **end for**
19   **if** $|\hat{C}_l^{ij}| \leq \frac{3}{4} \left(1 + 3(kn^{-3/4})^{1/3}\right) \frac{n}{k}$ **do**

```
20          l := l − 1
21      else
22          for 1 ≤ l′ < l do
23              if Ĉ_l^{ij} ∩ Ĉ_{l′}^{ij} ≠ ∅ do
24                  Ĉ_{l′}^{ij} := Ĉ_{l′}^{ij} ∪ Ĉ_l^{ij}; l := l − 1; break
25              end if
26          end for
27      end if
28      end for
29      l := l + 1
30  end for
```

The idea in the main part of the algorithm is to sample the unit sphere of the eigenspace of a matrix $\hat{A}_{ij}$ on a grid with grid spacing α. Every vector in this sample is used to form a class $\hat{C}_l^{ij}$ of indices in $I_{i(j \bmod 2+1)}$. That is, for the algorithm we pair the index sets I_{11} with I_{12} and I_{21} with I_{22}. Thus the sampled eigenvectors of the matrix A_{ij}, which corresponds to the index set I_{ij}, are used to reconstruct the classes in the partner index set $I_{i(j \bmod 2+1)}$. The unit sphere is sampled in line 12 and elements are taken into class $\hat{C}_l^{ij}$ in lines 15 to 17 using a threshold test. Note that the threshold value $t(n, k, \hat{\lambda}_1, \hat{\lambda}_2)$ is a function of values that all can be computed from $\hat{A}$.

This way we would form too many classes and elements that belong to one class would be scattered over several classes in the reconstruction. To prevent this we reject a class $\hat{C}_l^{ij}$ in lines 19 to 21 if it does contain too few elements. We know that the correct reconstruction has to contain roughly n/k elements. In lines 22 to 26 we check if the set $\hat{C}_l^{ij}$ is a (partial) reconstruction that already has been partially reconstructed. If this is the case then there should exist $\hat{C}_{l′}^{ij}$ with $l′ < l$ such that $\hat{C}_l^{ij} \cap \hat{C}_{l′}^{ij} \neq \emptyset$. We combine the partial reconstructions in line 24 and store the result in the set $\hat{C}_{l′}^{ij}$. With the break statement in line 24 we leave the for-loop enclosed by lines 22 and 26.

In lines 31 to 49 we post-process the reconstructions that we got in the main part of the algorithm.

```
31  for all l ∈ {1, …, k} do
32      Ĉ_l := Ĉ_l^{11}
33      for all l_1 ∈ {1, …, k} do
34          if ∑_{i∈Ĉ_l^{11}} ∑_{j∈Ĉ_{l_1}^{21}} â_{ij} > s(n, k, λ̂_1, λ̂_2) do
35              Ĉ_l := Ĉ_l ∪ Ĉ_{l_1}^{21}
36              for all l_2 ∈ {1, …, k} do
37                  if ∑_{i∈C_{l_1}^{21}} ∑_{j∈Ĉ_{l_2}^{12}} â_{ij} > s(n, k, λ̂_1, λ̂_2) do
38                      Ĉ_l := Ĉ_l ∪ Ĉ_{l_2}^{12}
39                      for all l_3 ∈ {1, …, k} do
40                          if ∑_{i∈C_{l_2}^{12}} ∑_{j∈Ĉ_{l_3}^{22}} â_{ij} > s(n, k, λ̂_1, λ̂_2) do
```

41         $\hat{C}_l := \hat{C}_l \cup \hat{C}_{l_3}^{22}$
42   **end if**
43   **end for**
44   **end if**
45   **end for**
46   **end if**
47   **end for**
48 **end for**
49 **return** all $\hat{C}_l$

After the main part of the algorithm the reconstruction of a class C_l is distributed into four sets corresponding to the four index sets I_{ij}. The purpose of the post-processing is to unite these four parts. This is again done using thresholding with threshold value $s(n, k, \hat{\lambda}_1, \hat{\lambda}_2)$, which again can be computed from $\hat{A}$. In line 49 we finally return the computed reconstructions.

5   Running Time Analysis

The running time of the algorithm GRIDRECONSTRUCT is essentially bounded by the number of points $(\lambda_1, \ldots, \lambda_k) \in (\alpha\mathbb{Z})^k$ with $\sum_{s=1}^k \lambda_s^2 \leq 1$.

Lemma 3. *The number of points* $(\lambda_1, \ldots, \lambda_k) \in (\alpha\mathbb{Z})^k$ *with* $\sum_{s=1}^k \lambda_s^2 \leq 1$ *is asymptotically bounded by* $\frac{1}{\sqrt{\pi k}} \left(\frac{\pi 2 e}{c_0^2} \right)^{k/2}$.

Proof. We want to bound the number of points $(\lambda_1, \ldots, \lambda_k) \in (\alpha\mathbb{Z})^k$ that are contained in the k-dimensional ball centered at the origin with radius 1. This number is asymptotically the number of such points in the cube $[-1, 1]^k$ times the volume of the k-dimensional unit ball divided by the volume of the cube, which is 2^k. The volume of the k-dimensional unit ball is $\frac{\pi^{k/2}}{k/2\,\Gamma(k/2)}$ and the number of points $(\lambda_1, \ldots, \lambda_k) \in (\alpha\mathbb{Z})^k$ that are contained in $[-1, 1]^k$ is $(2/\alpha)^k$. Plugging in $\alpha = c_0/\sqrt{k}$ and using Stirling's formula ($\Gamma(x) \sim \sqrt{2\pi}e^{-x}x^{x-1/2}$) asymptotically gives for the number of points in the unit sphere

$$\frac{\pi^{k/2}}{k/2\,\Gamma(k/2)\,2^k} \frac{2^k k^{k/2}}{c_0^k} = \frac{(k\pi)^{k/2}}{k/2\,\Gamma(k/2)\,c_0^k} \sim \frac{1}{\sqrt{2\pi}} \frac{(2\pi e k)^{k/2}}{\sqrt{k/2}\,(kc_0^2)^{k/2}}$$

$$= \frac{1}{\sqrt{\pi k}} \left(\frac{2\pi e}{c_0^2} \right)^{k/2}. \qquad \square$$

Remark 1. The time needed for the pre-processing in the algorithm is polynomially bounded in n. The same holds for the post-processing. The time we have to spend on the main part is polynomial in n for every point in the intersection of $(\alpha\mathbb{Z})^k$ with the k-dimensional unit sphere. That is, the running time of the whole algorithm is asymptotically bounded by $\frac{1}{\sqrt{\pi k}} \left(\frac{2\pi e}{c_0^2} \right)^{k/2} \text{poly}(n) = C^{k/2}\,\text{poly}(n)$, for some constant $C > 0$.

6   Correctness Proof

Lemma 4. *Let $l \in \{1, \ldots, k\}$. The size of $C_l^{ij} = C_l \cap I_{ij}$ is contained in the interval*

$$\left[\left(1 - 3(kn^{-3/4})^{1/3}\right) \frac{n}{k}, \left(1 + 3(kn^{-3/4})^{1/3}\right) \frac{n}{k} \right]$$

with probability at least $1 - e^{-c'' n^{1/4}}$ for some $c'' > 0$.

Proof. Will appear in the full version of the paper. $\square$

In the following A_{ij} and $\hat{A}_{ij}$ always refer to the restrictions of the matrices A and $\hat{A}$, respectively, to the index set I_{ij}. Also C_l^{ij} is the restriction of C_l to I_{ij}.

Lemma 5. *The spectral separation $\delta_k(A_{ij})$ is at least*

$$\left(1 - 3(kn^{-3/4})^{1/3}\right) \frac{n}{k}(p - q)$$

with probability at least $1 - ke^{-c'' n^{1/4}}$.

Proof. Will appear in the full version of the paper. $\square$

Lemma 6. *For every unit vector $v \in \mathbb{R}^n$ with $P_{ij}v = v$ the angle θ between v and $\hat{P}_{ij}v$ is bounded by*

$$\theta < \arccos\left(\sqrt{1 - \varepsilon}\right), \quad \text{and} \quad \varepsilon = \frac{2\sqrt{n}}{\left(1 - 3(kn^{-3/4})^{1/3}\right) \frac{n}{k}(p - q) - 2\sqrt{n}}.$$

with probability at least $\left(1 - 2e^{-c'^2 n/8}\right) \left(1 - ke^{-c'' n^{1/4}}\right)$.

Proof. With probability at least $\left(1 - 2e^{-c'^2 n/8}\right) \left(1 - ke^{-c'' n^{1/4}}\right)$ we have $|(P_{ij} - \hat{P}_{ij})v|^2 < \varepsilon$ by Stewart's theorem, the theorem of Füredi and Komlós and Lemma 5. It follows $\varepsilon > 1 + |\hat{P}_{ij}v|^2 - 2v^T\hat{P}_{ij}v = 1 + |\hat{P}_{ij}v|^2 - 2|\hat{P}_{ij}v|\cos\theta$, which in turn gives $\cos\theta > \frac{1 + |\hat{P}_{ij}v|^2 - \varepsilon}{2|\hat{P}_{ij}v|}$. As a function of $|\hat{P}_{ij}v|$ the cosine of θ is minimized at $|\hat{P}_{ij}v| = \sqrt{1 - \varepsilon}$. Thus we have $\cos\theta > \sqrt{1 - \varepsilon}$, which gives that the stated bound on θ holds with the stated probability. $\square$

Lemma 7. *For every vector w in the image of the projector $\hat{P}_{ij}$ there is a vector v computed in line 12 of the algorithm* GRIDRECONSTRUCT *such that the angle θ between w and v is bounded by $\theta < \arccos\left(1 - \alpha\sqrt{k}/2\right)$.*

Proof. Will appear in the full version of the paper. $\square$

Lemma 8. *For any* $l \in \{1, \ldots, k\}$ *there is a vector* v *computed in line 12 of the algorithm* GRIDRECONSTRUCT *such that with probability at least* $(1 - 2e^{-c'^2 n/8}) \left(1 - ke^{-c'' n^{1/4}}\right)$ *at least*

$$\left(1 - 4(1 - \cos \beta) \left(1 + 3(kn^{-3/4})^{1/3}\right)\right) \frac{n}{k}$$

with $\beta = \arccos\left(1 - \alpha\sqrt{k}/2\right) + \arccos\left(\sqrt{1 - \varepsilon}\right)$ *and* ε *as in lemma 6, of the indices corresponding to the* n/k *largest entries in* v *are mapped to* l *by* φ.

Proof. Let $c_l \in \mathbb{R}^n$ be the normalized characteristic vector of the class C_l^{ij}. By construction it holds $P_{ij} c_l = c_l$. Thus the angle between c_l and $\hat{P}_{ij} c_l$ is bounded by $\arccos(\sqrt{1 - \varepsilon})$ with probability at least $(1 - 2e^{-c'^2 n/8}) \left(1 - ke^{-c'' n^{1/4}}\right)$ by Lemma 6. For the vector $\hat{P}_{ij} c_l$ there exists by Lemma 7 a vector as constructed in line 12 of the algorithm such that the angle between $\hat{P}_{ij} c_l$ and v is bounded by $\arccos\left(1 - \alpha\sqrt{k}/2\right)$. Using the triangle inequality for angles we thus get

$$c_l^T v \geq \cos\left(\arccos\left(1 - \alpha\sqrt{k}/2\right) + \arccos\left(\sqrt{1 - \varepsilon}\right)\right) = \cos \beta.$$

Since c_l and v are both unit vectors we can get an upper bound on the length of $|c_l - v|$ from the lower bound on the dot product $c_l^T v$. First we decompose v into the projection of v onto c_l and the orthogonal complement $v^\perp$ of this projection. Since v is a unit vector we have $1 = (c_l^T v)^2 + |v^\perp|^2$. Thus $|v^\perp|^2$ is upperbounded by $1 - (\cos \beta)^2$. Also, $|(c_l^T v) c_l - c_l|^2$ is upper bounded by $(1 - \cos \beta)^2$ since c_l is a unit vector. Combining the two inequalities we get

$$|v - c_l|^2 = |v^\perp|^2 + |(c_l^T v) c_l - c_l|^2 \leq 1 - (\cos \beta)^2 + (1 - \cos \beta)^2 = 2(1 - \cos \beta).$$

Let $x = |C_l^{ij}|$ be the size of C_l^{ij} and let y be the number of indices whose corresponding entries in the v are among the x largest, but that are not mapped to l by φ. The number y is maximized under the upper bound on $|v - c_l|^2$ if the entries that are "large" in c_l but are "small" in v have a value just smaller than $\frac{1}{2}\sqrt{1/x}$ in v and if the entries whose value is 0 in c_l, but "large" in v, have a value just larger than $\frac{1}{2}\sqrt{1/x}$ in v and if all other entries coincide. For such a vector v it follows $|v - c_l|^2 = \frac{y}{2x}$, which implies $y \leq 4x(1 - \cos \beta)$. Since by Lemma 4 it holds $x \leq \left(1 + 3(kn^{-3/4})^{1/3}\right) \frac{n}{k}$ with probability at least $1 - e^{-c'' n^{1/4}}$ we have $y \leq 4(1 - \cos \beta) \left(1 + 3(kn^{-3/4})^{1/3}\right) \frac{n}{k}$ with probability at least $1 - ke^{-c'' n^{1/4}}$ (taking a union bound). This is also an upper bound on the number of indices whose corresponding entries in the vector v are among the n/k largest, but that are not mapped to l by φ. $\square$

Remark 2. If $k \leq c\sqrt{n}$ and $\alpha = c_0/\sqrt{n}$ with sufficiently small constants c and $c_0 = c_0(c)$ then for large enough n

$$\left(1 - 4(1 - \cos \beta) \left(1 + 3(kn^{-3/4})^{1/3}\right)\right) \geq \frac{3}{4}.$$

That is, asymptotically almost surely at least $3/4$ of the indices corresponding to the n/k largest entries in v are mapped to l by φ.

Lemma 9. *Let v_{ij} be a unit vector constructed in round (i, j) in line 12 of the algorithm* GRIDRECONSTRUCT. *Let I be subset of the index set I_{ij} that corresponds to the n/k largest entries in v.*

If at least $\nu n/k$ of the indices in I are mapped to the same element $l \in \{1, \ldots, k\}$ by φ then for $t \in C_l^{i(j \bmod 2+1)}$ it holds

$$\sum_{s \in I} \hat{a}_{st} \geq (\nu p + (1 - \nu)q) \frac{n}{k}(1 - \delta)$$

with probability at least $e^{-\frac{\delta^2}{2}(\nu p + (1-\nu)q)\frac{n}{k}}$. And if at most $\mu n/k$ of the indices in I are mapped to the same element $l \in \{1, \ldots, k\}$ by φ then for $t \in C_l^{i(j \bmod 2+1)}$ it holds

$$\sum_{s \in I} \hat{a}_{st} \leq (\mu p + (1 - \mu)q) \frac{n}{k}(1 + \delta)$$

with probability at least $e^{-\frac{q\delta^2}{4}\frac{n}{k}}$.

Proof. Will appear in the full version of the paper.   □

Remark 3. If we choose $\nu = 3/4$ and $\mu = 2/3$ and let $\delta < \frac{(\nu-\mu)(p+q)}{(\nu+\mu)(p-q)+2q}$ then

$$(\nu p + (1 - \nu)q)\frac{n}{k}(1 - \delta) > (\mu p + (1 - \mu)q)\frac{n}{k}(1 + \delta)$$

asymptotically almost surely. That is, if we choose the threshold in line 15 of the algorithm GRIDRECONSTRUCT in the interior of the interval

$$\left[\mu p + (1 - \mu)q)\frac{n}{k}(1 + \delta), (\nu p + (1 - \nu)q)\frac{n}{k}(1 - \delta)\right]$$

then with high probability the test in line 15 is only passed for vectors v (as constructed in line 12 of the algorithm) that have at least $\frac{2}{3}\frac{n}{k}$ indices corresponding to the n/k largest entries in v that are mapped to the same element by φ. Assume this element is $l \in \{1, \ldots, k\}$. The elements that pass the test (and are subsequently put into a class) are all mapped to l by φ. The only problem is that it is possible that for some vector v only some of the elements that are mapped to l by φ and that take the test also pass it. But from Remark 2 we know that for every $l \in \{1, \ldots, k\}$ there is a vector v such that with high probability (taking a union bound) all elements that are mapped to l by φ and that take the test also pass it. That is the reason for the post-processing in lines 19 to 28 of the algorithm. It remains to show how a good threshold value can be found. Only obstacle to that is that we do not know the values of p and q when running the algorithm.

Lemma 10. *With probability at least* $1 - 2e^{-c'^2 n/8}$,

$$q \leq \frac{k}{k-1} \frac{\hat{\lambda}_1}{n} + \frac{k}{k-1} \frac{1}{\sqrt{n}} =: q^+ \quad and \quad q \geq \frac{k}{k-1} \frac{\hat{\lambda}_1}{n} - \frac{k}{k-1} \left(\frac{1}{\sqrt{n}} + \frac{1}{k} \right) := q^-.$$

and with the same probability

$$p \leq \frac{k\hat{\lambda}_2 + \frac{k}{k-1}\hat{\lambda}_1}{n-k} + \frac{k\sqrt{n}}{(n-k)(k-1)} + \frac{k\sqrt{n}}{n-k} := p^+ \quad and$$

$$p \geq \frac{k\hat{\lambda}_2 + \frac{k}{k-1}\hat{\lambda}_1}{n-k} - \frac{k\sqrt{n}}{(n-k)(k-1)} - \frac{n}{(n-k)(k-1)} - \frac{k\sqrt{n}}{n-k} := p^-.$$

Proof. Will appear in the full version of the paper.   □

Remark 4. With high probability we can approximate q arbitrarily well by $\frac{k}{k-1} \frac{\hat{\lambda}_1}{n}$ for growing n if $k \in \omega(1)$. That is not the case for p. If $k = c\sqrt{n}$ then we can approximate p asymptotically by $\frac{k\hat{\lambda}_2 + \frac{k}{k-1}\hat{\lambda}_1}{n-k}$ only up to a constant that depends on c. But for sufficiently small c if we choose

$$\delta < \frac{\left(\frac{3}{4}p^- + \frac{1}{4}q^-\right) - \left(\frac{2}{3}p^+ + \frac{1}{3}q^+\right)}{\left(\frac{3}{4}p^- + \frac{1}{4}q^-\right) + \left(\frac{2}{3}p^+ + \frac{1}{3}q^+\right)},$$

which is positive for sufficiently large n, the algorithm GRIDRECONSTRUCT asymptotically almost surely reconstructs for $k \leq c\sqrt{n}$ (and $k \in \omega(1)$) all the classes C_l^{ij} for all $l \in \{1, \ldots, l\}$ (up to a permutation of $\{1, \ldots, k\}$) if we choose the threshold $t(n, k, \hat{\lambda}_1, \hat{\lambda}_2) = \left(\frac{2}{3}p^+ + \frac{1}{3}q^+\right) \frac{n}{k}(1 + \delta)$ in line 15 of the algorithm.

Lemma 11. *Let* $\delta > 0$ *be a constant. It holds asymptotically almost surely*

$$\sum_{i \in \hat{C}_l^{11}} \sum_{j \in \hat{C}_{l'}^{21}} \hat{a}_{ij} \geq \left(1 - 3(kn^{-3/4})^{1/3}\right)^2 \frac{n^2}{k^2}(1 - \delta)p$$

if $\varphi(l) = \varphi(l')$, *and it holds asymptotically almost surely*

$$\sum_{i \in \hat{C}_l^{11}} \hat{\sum}_{j \in \hat{C}_{l'}^{21}} \leq \left(1 + 3(kn^{-3/4})^{1/3}\right)^2 \frac{n^2}{k^2}(1 + \delta)q$$

if $\varphi(l) \neq \varphi(l')$.

Proof. Will appear in the full version of the paper.   □

Remark 5. Analogous results hold for the index sets $\hat{C}_l^{11}$ and $\hat{C}_{l'}^{22}$ with $l, l' \in \{1, \ldots, k\}$ and for the index sets $\hat{C}_l^{12}$ and $\hat{C}_{l'}^{21}$ with $l, l' \in \{1, \ldots, k\}$.

Remark 6. If $\delta < (p-q)/(p+q)$ and n sufficiently large then

$$\left(1 - 3(kn^{-3/4})^{1/3}\right)^2 \frac{n^2}{k^2}(1-\delta)p > \left(1 + 3(kn^{-3/4})^{1/3}\right)^2 \frac{n^2}{k^2}(1+\delta)q.$$

Again in order to use this result to derive a computable threshold value we have to approximate the unknown probabilities p and q by $p^{\pm}$ and $q^{\pm}$, which are functions of the known (or almost surely known) quantities $n, k, \hat{\lambda}_1$ and $\hat{\lambda}_2$ (see Lemma 10). If we choose $\delta < (p^- - q^+)/(p^+ + q^+)$ and the threshold $s(n, k, \hat{\lambda}_1, \hat{\lambda}_2) = \left(1 + 3(kn^{-3/4})^{1/3}\right)^2 \frac{n^2}{k^2}(1 + \delta)q^+$ then the algorithm GRIDRECONSTRUCT asymptotically almost surely finds the correct reconstruction of the planted partition (up to a permutation of $\{1, \ldots, k\}$).

References

1. N. Alon, M. Krivelevich, and B. Sudakov. Finding a large hidden clique in a random graph. *Random Structures and Algorithms*, 13:457–466, 1998.
2. B. Bollobas and A.D. Scott. Max cut for random graphs with a planted partition. *Combinatorics, Probability and Computing*, 13:451–474, 2004.
3. R. B. Boppana. Eigenvalues and graph bisection: An average-case analysis. *Proceedings of 28th IEEE Symposium on Foundations on Computer Science*, pages 280–285, 1987.
4. A. Condon and R. Karp. Algorithms for graph partitioning on the planted partition model. *Random Structures and Algorithms 8*, 2:116–140, 1999.
5. Z. Füredi and J. Komlós. The eigenvalues of random symmetric matrices. *Combinatorica I*, 3:233–241, 1981.
6. M. R. Garey, D. S. Johnson, and L. Stockmeyer. Some simplified NP-complete graph problems. *Theoretical Computer Science*, 1:237–267, 1976.
7. J. Giesen and D. Mitsche. Bounding the misclassification error in spectral partitioning in the planted partition model. *Proceedings of the 31st International Workshop on Graph-Theoretic Concepts in Computer Science*, 2005.
8. M. Krivelevich and V. H. Vu. On the concentration of eigenvalues of random symmetric matrices. *Microsoft Technical Report*, 60, 2000.
9. F. McSherry. Spectral partitioning of random graphs. *Proceedings of 42nd IEEE Symosium on Foundations of Computer Science*, pages 529–537, 2001.
10. M. Meila and D. Verma. A comparison of spectral clustering algorithms. *UW CSE Technical report 03-05-01*.
11. R. Shamir and D. Tsur. Improved algorithms for the random cluster graph model. *Proceedings 7th Scandinavian Workshop on Algorithm Theory*, pages 230–259, 2002.
12. D. Spielman and S.-H. Teng. Spectral partitioning works: Planar graphs and finite element meshes. *Proceedings of 37th IEEE Symposium on Foundations on Computer Science*, pages 96–105, 1996.
13. G. Stewart and J. Sun. *Matrix perturbation theory*. Academic Press, Boston, 1990.

Constant Time Generation of Linear Extensions

(Extended Abstract)

Akimitsu Ono and Shin-ichi Nakano

Gunma University, Kiryu-Shi 376-8515, Japan
nakano@cs.gunma-u.ac.jp

Abstract. Given a poset $\mathcal{P}$, several algorithms have been proposed for generating all linear extensions of $\mathcal{P}$. The fastest known algorithm generates each linear extension in constant time "on average". In this paper we give a simple algorithm which generates each linear extension in constant time "in worst case". The known algorithm generates each linear extension exactly twice and output one of them, while our algorithm generates each linear extension exactly once.

1 Introduction

A linear extension of a given poset $\mathcal{P}$ is one of the most important notion associated with $\mathcal{P}$. An example of a poset is shown in Fig. 1, and its linear extension is shown in Fig. 2. Many scheduling problems with precedence constraints are modeled by a linear extension of a poset, or equivalently a topological sort[C01] of an acyclic digraph [PR94]. Even though many such scheduling problems are NP-complete, one can solve the problem by first generating all linear extensions of a given poset and then picking the best one [PR94]. Linear extensions are also of interest to combinatorists, because of their relation to counting problems [St97].

Let $\mathcal{P} = (S, R)$ be a poset with a set S and a binary relation R on S. We write $n = |S|$ and $m = |R|$. It is known one can find a linear extension of a given poset $\mathcal{P}$ in $O(m + n)$ time [C01, p.550].

Many algorithms to generate a particular class of objects, without repetition, are already known [LN01, LR99, M98, N02, R78]. Many excellent textbooks have been published on the subject [G93, KS98, W89]. Given a poset $\mathcal{P}$, three algorithms to generate all linear extensions of $\mathcal{P}$ are explained in [KV83]. The best algorithm among them generates the first linear extension in $O(m+n)$ time, then generates each linear extension in $O(n)$ time.

Generally, generating algorithms produce huge outputs, and the outputs dominate the running time of the generating algorithms. So if we can compress the outputs, then it considerably improves the efficiency of the algorithms. Therefore many generating algorithms output objects in an order such that each object differs from the preceding one by a very small amount, and output each object as the "difference" from the preceding one. Such orderings of objects are known as *Gray codes* [J80, R93, R00, S97].

M. Liśkiewicz and R. Reischuk (Eds.): FCT 2005, LNCS 3623, pp. 445–453, 2005.

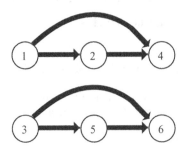

Fig. 1. An example of a poset $\mathcal{P}$

Let G be a graph, where each vertex corresponds to each object and each edge connects two similar objects. Then the Gray code corresponds to a Hamiltonian path of G. For the set $LE(\mathcal{P})$ of all linear extensions of a given poset $\mathcal{P}$ we can also define such a graph G. However, the graph G may not have a Hamiltonian path. Therefore the algorithm in [PR94] first constructs a new set S' so that if $x \in LE(\mathcal{P})$ then $+x, -x \in S'$, then prove that the graph G' corresponding to S' always has a Hamiltonian path. Based on this idea, the algorithm in [PR94] generates the first linear extension in $O(m + n)$ time, then generates each linear extension in only $O(1)$ time on average along a Hamiltonian path of G'. Note that the algorithm generates each linear extension exactly twice but output exactly one of them.

The paper [PR94] proposed the following question. Is there any algorithm to generate each linear extension in $O(1)$ time "in the worst case"? In this paper we answer the question affirmatively.

In this paper we give an algorithm to generate all linear extensions of $\mathcal{P}$. Our algorithm is simple and generates each linear extension in constant time in worst case (not on average). Our algorithm also outputs each linear extension as the difference from the preceding one. Thus our algorithm also generates a Gray code for linear extensions of a given poset.

The main idea of the algorithm is as follows. We first define a rooted tree (See Fig. 3.) such that each vertex corresponds to a linear extension of $\mathcal{P}$, and each edge corresponds to a relation between two linear extensions. Then by traversing the tree we generate all linear extensions of $\mathcal{P}$. With a similar technique we have already solved some generation problems for graphs[LN01, N02, NU03] and combinatorics[KN05]. In this paper we apply the technique for linear extensions.

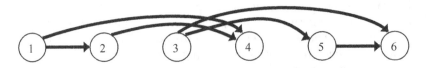

Fig. 2. A linear extension of a poset $\mathcal{P}$

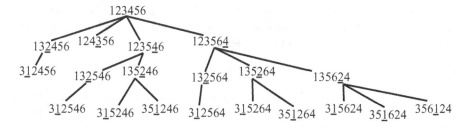

Fig. 3. The family tree T_P of $L(\mathcal{P})$

The rest of the paper is organized as follows. Section 2 gives some definitions. Section 3 introduces the family tree. Section 4 presents our first algorithm. The algorithm generates each linear extension of a given poset $\mathcal{P}$ in $O(1)$ time on average. In Section 5 we improve the algorithm so that it generates each linear extension in $O(1)$ time in worst case. Finally Section 6 is a conclusion.

2   Preliminaries

A *poset* P is a set S with a binary relation R which is reflexive, antisymmetric and transitive. Note that R is a partial ordering on S. We denote $n = |S|$. We say x precedes y if xRy. We regard a poset P as a directed graph D such that (1) each vertex corresponds to an element in S, and (2) direct edge (x, y) exists iff xRy. (See an example in Fig. 1.)

Given a poset $\mathcal{P} = (S, R)$, a *linear extension* of $\mathcal{P}$ is a permutation $(x_1, x_2, \cdots, x_n)$ of S, such that if $x_i R x_j$ then $i \leq j$. Intuitively, if we draw the directed graph corresponding to $\mathcal{P}$ so that $x_1, x_2, \cdots, x_n$ appear along a horizontal line from left to right in this order, then all directed edges go from left to right. (See an example in Fig. 2.)

3   The Family Tree

Let $LE(\mathcal{P})$ be the set of linear extensions of a given poset $\mathcal{P}$. In this section we define a tree structure among linear extensions in $LE(\mathcal{P})$.

Given a poset $\mathcal{P} = (\mathcal{S}, \mathcal{R})$, choose a linear extension $L_r \in LE(\mathcal{P})$. Without loss of generality we can assume that $S = \{1, 2, \cdots, n\}$ and $L = (1, 2, \cdots, n)$. (Otherwise, we rename the elements in S.) We call $L_r = (1, 2, \cdots, n)$ *the root linear extension* of $\mathcal{P}$.

Then we define *the parent $P(L)$* for each linear extension L in $LE(\mathcal{P})$ (except for L_r) as follows. Let $L = (x_1, x_2, \cdots, x_n)$ be a linear extension of $\mathcal{P}$, and assume that $L \neq L_r$. Let k be the minimum integer such that $x_k \neq k$. Since $L \neq L_r$ such k always exists. We define *the level* of L by k. For example the level of $(1, 2, 4, 3, 5, 6)$ is 3. For convenience we regard the level of L_r to be $n + 1$.

By removing k from L then inserting the k into the immediately before x_k of L, we have a different permutation L'. Note that k has moved to the left, and so $L' \neq L$. Now we have the following lemma.

Lemma 1. L' *is also a linear extensions of* $\mathcal{P}$ *in* $LE(\mathcal{P})$.

Proof. Assume otherwise. Then there must exists x_i such that (1) $x_i R k$ and (2) $i > k$. However this contradicts the fact that $L_r = (1, 2, \cdots, n)$ is a linear extension of $\mathcal{P}$.   □

We say that L' is *the parent linear extension* of L, and write $P(L) = L'$. If $P(L)$ is the parent linear extension of L then we say L is *a child linear extension* of $P(L')$. Note that L has the unique parent linear extension $P(L)$, while $P(L)$ may have many child linear extensions.

We also have the following lemma.

Lemma 2. *Let* ℓ *and* ℓ_p *be the levels of a linear extension* L *and its parent linear extension* $P(L)$. *Then* $\ell < \ell_p$.

Proof. Immediately from the definition of the parent.

The two lemmas above means the following. Given a linear extension L in $LE(\mathcal{P})$ where $L \neq L_r$, by repeatedly finding the parent linear extension of the derived linear extension, we have the unique sequence $L, P(L), P(P(L)), \cdots$ of linear extensions in $LE(\mathcal{P})$, which eventually ends with the root linear extension L_r. Since the level is always increased, $L, P(L), P(P(L)), \cdots$ never lead into a cycle.

By merging these sequences we have *the family tree* of $LE(\mathcal{P})$, denoted by $T_\mathcal{P}$, such that the vertices of $T_\mathcal{P}$ correspond to the linear extensions in $LE(\mathcal{P})$, and each edge corresponds to each relation between some L and $P(L)$. This proves that every linear extension appears in the tree as a vertex. For instance, $T_\mathcal{P}$ for a poset in Fig. 1 is shown in Fig. 3. In Fig. 3, the k for each linear extension is underlined. By removing the k, then inserting it into the k-th position the parent linear extension is obtained.

4   Algorithm

In this section we give our generation algorithm. Given a poset $\mathcal{P}$, our algorithm traverses the family tree $T_\mathcal{P}$ and generates all linear extensions of $\mathcal{P}$.

If we can generate all child linear extensions of a given linear extension in $LE(\mathcal{P})$, then in a recursive manner we can construct $T_\mathcal{P}$, and generate all linear extensions $LE(\mathcal{P})$ of $\mathcal{P}$. How can we generate all child linear extensions of a given linear extension?

Given a linear extension $L = (p_1, p_2, p_3, \cdots, p_n)$ in $LE(\mathcal{P})$, let ℓ_p be the level of L. Let $C = (c_1, c_2, c_3, \cdots, c_n)$ be a child linear extension of L, and let ℓ_c be the level of C. By Lemma 2, the level of C is smaller than the level of L. Thus $\ell_c < \ell_p$ holds. Therefore, for each $i = 1, 2, \cdots, \ell_p - 1$, if we generate all child linear extension of L "having level i", then by merging those child linear extensions we can generate all child linear extensions of L.

Now, given i, $1 \leq i \leq \ell_p - 1$, we are going to generate all child linear extensions of $L = (p_1, p_2, \cdots, p_n)$ having level i. We can generate such child linear extensions by deleting $p_i = i$ from L then insert it somewhere in $(p_{i+1}, p_{i+2}, \cdots, p_n)$ so that the resulting permutation is again a linear extension.

For example, see Fig. 3 for a poset $\mathcal{P}$ in Fig. 1. The last child $L = (1, 2, 3, 5, 6, 4)$ of the root linear extension has level 4. Thus each child linear extension has level either $1, 2$ or 3. For level 1, no child linear extensions of L having level 1 exists, since $1R2$ means we cannot move 1 to the right. For level 2, child linear extensions of L having level 2 are $L = (1, 3, 2, 5, 6, 4)$, $L = (1, 3, 5, 2, 6, 4)$ and $L = (1, 3, 5, 6, 2, 4)$. Note that $L = (1, 3, 5, 6, 4, 2)$ is not a linear extension because of $2R4$. For level 3, no child linear extensions of L having level 3 exists, since $3R5$ means we cannot move 3 to the right.

We have the following algorithm.

Procedure find-all-children$(L = (p_1 p_2 \cdots p_n), \ell_p)$
{ L is the current linear extension of $\mathcal{P}$.}
begin
01   Output L { Output the difference from the preceding one.}
02   **for** $i = 1$ **to** ℓ_p - 1
03   **begin** { generate children with level i }
04   j = i
05   **while** $(p_j, p_{j+1}) \notin R$ **do**
06   **begin**
07   swap p_j and p_{j+1}
08   **find-all-children**$(L = (p_1 p_2 \cdots p_n), i)$
09          $j = j + 1$
10   **end**
11   insert p_j into immediately after p_{i-1}
12   { Now $p_i = i$ again holds, and L is restored as it was.}
13   **end**
end

Algorithm find-all-linear-extensions$(\mathcal{P} = (S, R))$
begin
    Find a linear extension L_r
    find-all-children$(L_r, \text{n+1})$
end

For example, see Fig. 3. The last child $L = (1, 2, 3, 5, 6, 4)$ of the root linear extension has level 4. Assume we are going to generate child linear extensions having level 2. Since $p_2 = 2$, $p_3 = 3$ and $(2, 3) \notin R$, so we generate $L = (1, 3, 2, 5, 6, 4)$ by swapping p_2 and p_3. Then, since $p_3 = 2$, $p_4 = 5$ and $(2, 5) \notin R$, so we generate $L = (1, 3, 5, 2, 6, 4)$. Then, since $p_4 = 2$, $p_5 = 6$ and $(2, 6) \notin R$, so we generate $L = (1, 3, 5, 6, 2, 4)$. Then, since $p_5 = 2$, $p_6 = 4$ and $(2, 4) \in R$, so we do not generate $L = (1, 3, 5, 6, 4, 2)$.

Note that if $(p_i, p_{i+1}) \in R$, then L has no child linear extensions having level i. Therefore if (1) $(p_a, p_{a+1}), (p_{a+1}, p_{a+2}), \cdots, (p_b, p_{b+1}) \in R$, and (2) the level of L is $\ell_p > b$, then L has no child linear extension with level $a, a + 1, \cdots, b$. Then even if we execute Line 02 of the algorithm **find-all-children** several times, no linear extension is generated. Thus we cannot generate k child linear extensions in $O(k)$ time.

However, we can preprocess the root linear extension and provide a simple list to solve this problem, as follows. First let $LIST = L_r = (1, 2, \cdots, n)$. For each $i = 1, 2, \cdots, n - 1$, if $(p_i, p_{i+1}) \in R$, then we remove p_i from $LIST$. Then the resulting $LIST$ tell us all levels at which at least one child linear extension exists. Using $LIST$ we can skip the levels at which no child linear extension exists.

For instance, see T_P in Fig. 3 for a poset in Fig. 1. For the root linear extension $L_r = (1, 2, 3, 4, 5, 6)$, $LIST = (2, 3, 4, 6)$. The last child $L = (1, 2, 3, 5, 6, 4)$ of L_r has level 4.

Insted of generating all child linear extension at level i for $i = 1, 2, \cdots, \ell_p - 1$ by the **for** loop in Line 02, we generate all child linear extensions at level i only for each integer i in $LIST$ up to $\ell_p - 1$. Thus now we can generate k child linear extensions in $O(k)$ time.

Theorem 1. *The algorithm uses $O(n)$ space and runs in $O(|LE(\mathcal{P})|)$ time.*

Proof. Since we traverse the family tree T_P and output each linear extension at each corresponding vertex of T_P, we can generate all linear extensions in $LE(\mathcal{P})$ without repetition.

Since we trace each edge of the family tree in constant time, the algorithm runs in $O(|LE(\mathcal{P})|)$ time.

The argument L of the recursive call in Line 08 is passed by reference. Note that we restore L as it was when return occurs.

The algorithm outputs each linear extension as only the difference from the preceding one. For each recursive call we need a constant amount of space, and the depth of the recursive call is bounded by n. Thus the algorithm uses $O(n)$ space.

Note that if $\mathcal{P}$ is given as an adjacency matrix then we can check Line 05 in constant time. Although if $\mathcal{P}$ is given as adjacency lists we can still construct the adjacency matrix in $O(n+m)$ preprocessing time with the technique in [A74, p.71], and check Line 05 in constant time. □

5   Modification

The algorithm in Section 4 generates all linear extensions in $LE(\mathcal{P})$ in $O(|LE(\mathcal{P})|)$ time. Thus the algorithm generates each linear extension in $O(1)$ time "on average". However, after generating a linear extension corresponding to the last vertex in a large subtree of $T_\mathcal{P}$, we have to merely return from the deep recursive call without outputting any linear extension. This may take much time. Therefore, we cannot generate each linear extension in $O(1)$ time in worst case.

However, a simple modification [NU03] improves the algorithm to generate each linear extension in $O(1)$ time. The algorithm is as follows.

Procedure find-all-children2(L, *depth*)
{ L is the current sequence and *depth* is the depth of the recursive call.}
begin
01   **if** *depth* is even
02   **then** Output L { before outputting its child.}
03   Generate child linear extensions $L_1, L_2, \cdots, L_x$ by the method in Section 4, and
04   recursively call **find-all-children2** for each child linear extension.
05   **if** *depth* is odd
06   **then** Output L { after outputting its child.}
end

One can observe that the algorithm generates all linear extensions so that each sequence can be obtained from the preceding one by tracing at most three edges of $T_\mathcal{P}$. Note that if L corresponds to a vertex v in $T_\mathcal{P}$ with odd depth, then we may need to trace three edges to generate the next linear extension. Otherwise, we need to trace at most two edges to generate the next linear extension. Note that each linear extension is similar to the preceding one, since it can be obtained with at most three (delete then insert) operations. Thus, we can regard the derived sequence of the linear extensions as a combinatorial Gray code [J80, S97, R93, W89] for linear extensions.

6   Conclusion

In this paper we gave a simple algorithm to generate all linear extensions of a given poset. The algorithm is simple and generates each linear extension in constant time in worst case. This solve an open question in [PR94].

We have another choice for the definition of the family tree for $LE(\mathcal{P})$ as follows. Given a linear extension $L = (c_1, c_2, \cdots, c_n) \neq L_r$ in $\mathcal{P}$, let k be the level of L. Then let i be the index such that $c_i = k$. By definition $i > k$ holds. Now by swapping c_i with its left neighbour we obtain another linear extension $P(L)$, and we say $P(L)$ is the parent of L. Based on this parent-child relation we can define another family tree for $LE(\mathcal{P})$. For instance see Fig. 4. Based on this family tree, in a similar manner, we can design another simple algorithm to generate all linear extensions of a given poset. Note that each linear extension is also similar to the preceding one. The next linear extension can be obtained with at most three "adjacent transposition" operations.

452   A. Ono and S.-i. Nakano

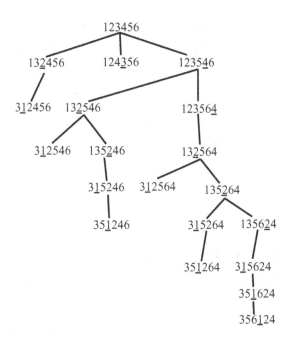

Fig. 4. Another family tree of $L(\mathcal{P})$

References

[A74]   A. Aho, J. Hopcroft and J. Ullman, *The Design and Analysis of Computer Algorithms*, Addison-Wesley, (1974).

[C01]   T. H. Cormen, C. E. Leiserson, R. L. Rivest and C. Stein, *Introduction to Algorithms*, MIT Press, (2001).

[G93]   L. A. Goldberg, *Efficient Algorithms for Listing Combinatorial Structures*, Cambridge University Press, New York, (1993).

[J80]   J. T. Joichi, D. E. White and S. G. Williamson, *Combinatorial Gray Codes*, SIAM J. on Computing, 9, (1980), pp.130-141.

[KS98]   D. L. Kreher and D. R. Stinson, *Combinatorial Algorithms*, CRC Press, Boca Raton, (1998).

[KV83]   A. D. Kalvin and Y. L. Varol, *On the Generation of All Topological Sortings*, J. of Algorithms, 4, (1983), pp.150–162.

[KN05]   S. Kawano and S. Nakano, *Constant Time Generation of Set Partitions*, IEICE Trans. Fundamentals, accepted, to appear, (2005).

[LN01]   Z. Li and S. Nakano, *Efficient Generation of Plane Triangulations without Repetitions*, Proc. ICALP2001, LNCS 2076, (2001), pp.433–443.

[LR99]   G. Li and F. Ruskey, *The Advantage of Forward Thinking in Generating Rooted and Free Trees*, Proc. 10th Annual ACM-SIAM Symp. on Discrete Algorithms, (1999), pp.939–940.

[M98]   B. D. McKay, *Isomorph-free Exhaustive Generation*, J. of Algorithms, 26, (1998), pp.306-324.

[N02]   S. Nakano, *Efficient Generation of Plane Trees*, Information Processing Letters, 84, (2002), pp.167–172.

[NU03]   S. Nakano and T. Uno, *A Simple Constant Time Enumeration Algo-rithm for Free Trees*, IPSJ Technical Report, 2003-AL-91-2, (2003). (http://www.ipsj.or.jp/members/SIGNotes/Eng/16/2003/091/article002.html)

[PR94]   G. Pruesse and F. Ruskey, *Generating Linear Extensions Fast*, SIAM Journal on Computing, 23, (1994), pp.373–386.

[R78]   R. C. Read, *How to Avoid Isomorphism Search When Cataloguing Combinatorial Configurations*, Annals of Discrete Mathematics, 2, (1978), pp.107–120.

[R93]   F. Ruskey, *Simple Combinatorial Gray Codes Constructed by Reversing Sublists*, Proc. ISAAC93, LNCS 762, (1993), pp.201–208.

[R00]   K. H. Rosen (Eds.), *Handbook of Discrete and Combinatorial Mathematics*, CRC Press, Boca Raton, (2000).

[S97]   C. Savage, *A Survey of Combinatorial Gray Codes*, SIAM Review, 39, (1997) pp. 605-629.

[St97]   R. Stanley, *Enumerative Combinatorics Volume 1*, Cambridge University Press, (1997).

[W89]   H. S. Wilf, *Combinatorial Algorithms : An Update*, SIAM, (1989).

On Approximating Real-World Halting Problems

Sven Köhler[2], Christian Schindelhauer[1], and Martin Ziegler[3]

[1] Heinz Nixdorf Institute, University of Paderborn
schindel@uni-paderborn.de
[2] University of Paderborn
skoehler@uni-paderborn.de
[3] University of Southern Denmark
ziegler@imada.sdu.dk

Abstract. No algorithm can of course solve the Halting Problem, that is, decide within finite time always correctly whether a given program halts on a certain given input. It might however be able to give correct answers for 'most' instances and thus solve it at least approximately. Whether and how well such approximations are feasible highly depends on the underlying encodings and in particular the Gödelization (programming system) which in practice usually arises from some programming language.

We consider BrainF*ck (BF), a simple yet Turing-complete real-world programming language over an eight letter alphabet, and prove that the natural enumeration of its syntactically correct sources codes induces a both efficient and dense Gödelization in the sense of [Jakoby&Schindelhauer'99]. It follows that any algorithm M approximating the Halting Problem for BF errs on at least a constant fraction $\varepsilon_M > 0$ of all instances of size n for infinitely many n.

Next we improve this result by showing that, in every dense Gödelization, this constant lower bound ε to be independent of M; while, the other hand, the Halting Problem does admit approximation up to arbitrary fraction $\delta > 0$ by an appropriate algorithm M_δ handling instances of size n for infinitely many n. The last two results complement work by [Lynch'74].

1 Introduction

In 1931, the logician KURT GÖDEL constructed a mathematical predicate which could neither be proven nor falsified. In 1936, ALAN TURING introduced and showed the Halting Problem H to be undecidable by a Turing machine. This was considered a strengthening of Gödel's result regarding that, at this time and preceding AIKEN's Mark I and ZUSE's Z3, the Turing machine was meant as an idealization of an average mathematician.

Nowadays the Halting Problem is usually seen from a quite different perspective. Indeed with increasing reliance on high speed digital computers and huge software systems running on them, source code verification or at least the detection of stalling behaviour becomes even more important. In fact, by RICE's Theorem, this is equivalent to many other real-world problems arising from goals like automatized software engineering, optimizing compilers, formal proof systems and so on. Thus, the Halting problem is a very practical one which has to be dealt with some way or another.

M. Liśkiewicz and R. Reischuk (Eds.): FCT 2005, LNCS 3623, pp. 454–466, 2005.

One direction of research considered and investigated the capabilities of extended Turing machines equipped with some kind of external device solving the Halting problem. While the physical realizability of such kinds of super-Turing computers is questionable and in fact denied by the Church-Turing Hypothesis, the current field of Hypercomputation puts in turn this hypothesis into question. On the theoretical side, these considerations led to the notion of relativized computability and the Arithmetical Hierarchy which have become standard topics in Recursion Theory [Soar87].

1.1 Approximate Problem Solving

Another approach weakens the usual notion of algorithmic solution from strict to approximate or from worst-case to average case. The first arises from the fact that many optimization problems are $\mathcal{NP}$-complete only if requiring the solution to exactly attain the, say, minimum whereas they become computationally much easier when asking for a solution only within a certain factor of the optimum.

Regarding decision problems, a notion of approximate solution has been established in Property Testing [Gold97]. Here for input $x \in \Sigma^n$, the answer "$x \in L$" is considered acceptable even for $x \notin L$ *provided* that $y \in L$ holds for some $y \in \Sigma^n$ with (edit or Hamming) distance $d(x, y) \leq \varepsilon n$. Observe that this notion of approximation strictly speaking refers to the arguments x to the problem rather than the problem L itself. Also, any program source x is within constant distance from the terminating one y obtained by changing the first command(s) in x by a halt instruction.

Average case analysis is an approach based on the observation that the hard instances which make a certain problem difficult might occur only rarely in practice whereas most 'typical' instances might turn out as easy. So, although for example $\mathcal{NP}$-complete, an algorithm would be able to correctly and efficiently solve this problem in, say, 99.9% of all cases while possibly failing on some few and unimportant others. In this example, $\varepsilon = 1/1000$ is called the error rate of the problem under consideration with respect to a certain probability distribution or encoding of its instances.

Such approaches have previously been mainly applied in order to deal with important problems where the practitioner cannot be silenced by simply remarking that they are $\mathcal{NP}$-complete, that is, within complexity theory. However the same makes sense, too, for important undecidable problems such as Halting: even when possibly erring on, say, every 10th instance, detecting the other 90% of stalling programs would have prevented many buggy versions of a certain operation system from being released prematurely.

1.2 The Error Complexity

So instead of deciding some (e.g., hard or even non-recursive) problem L, one is satisfied with solving some problem S which approximates L in the sense that the symmetric[1] set difference $A := L \triangle S := (L \setminus S) \cup (S \setminus L)$ is 'small'. For $L \subseteq \Sigma^*$ (with an at least two-letter alphabet Σ) this is formalized, analogously to the error complexity from average analysis, [JaSc99, DEFINITION 1] as the asymptotic behavior of $\mu\{\bar{x} \in (L \triangle S) | \bar{x} \in \Sigma^n\}$ for a fixed probability measure $\mu : \Sigma^* \to [0, 1]$; if this quantity

---

[1] The error to the Halting problem can in fact be made one-sided, see Corollary 22 below.

tends to 0 as $n \to \infty$ it basically means that, for (μ-) average instances, S ultimately equals L. In the case of μ denoting the counting measure, this amounts [Papa95, §14.2, p.336],[RoUl63] to the following

Definition 1. *For* $A \subseteq \Sigma^*$, *let* $\text{density}(A, =n) := \#(A \cap \Sigma^n)/\#\Sigma^n$ *and* $\text{density}(A, <n) := \#(A \cap \Sigma^{<n})/\#(\Sigma^{<n})$ *where* $\Sigma^{<n} = \bigcup_{j=0}^{n-1} \Sigma^j$. *For* $A \subseteq \mathbb{N}$, *let* $\text{Density}(A, N) := \#(A \cap \{0, \ldots, N-1\})/N$.

The latter formalization has been considered independently in [RoUl63, Lync74][2] for approximating decision problems $L \subseteq \mathbb{N}$. The notions are related as follows:

Lemma 2. *For* $x \in \mathbb{N}$, *let* $\bar{x}$ *denote the x-th string in* Σ^* *ordered with respect to length (ties broken arbitrarily). For* $A \subseteq \mathbb{N}$, $\tilde{A} := \{\bar{x} : x \in A\}$, *and* $0 \le \varepsilon \le 1$ *it holds:*
a) $\text{density}(\tilde{A}, =n) \le \varepsilon \ \forall n$ $\Rightarrow$ $\text{density}(\tilde{A}, <n) \le \varepsilon \ \forall n$.
b) $\text{Density}(A, N) \le \varepsilon \ \forall N$ $\Rightarrow$ $\text{density}(\tilde{A}, <n) \le \varepsilon \ \forall n$.
c) $\text{density}(\tilde{A}, <n) \le \varepsilon \ \forall n$ $\Rightarrow$ $\text{Density}(A, N) \le \varepsilon' \ \forall N$ *where* $\varepsilon' := \varepsilon \cdot (2 - \varepsilon)$.
Taking complements yields similar claims for reversed inequalities "$\ge \varepsilon$".

Since $0 < \varepsilon' < 1$ whenever $0 < \varepsilon < 1$ in b+c), both densities are essentially equivalent up to constants in that one tends to 0/1 iff so does the other. This allows us to deliberately switch in the sequel between $A \subseteq \Sigma^*$ encoded over one alphabet Σ (say, the decimals $\{0, 1, \ldots, 9\}$) and its re-coding over some other (e.g., binary or hexadecimal) finite Σ'.

Proof (Lemma 2). a) is obvious; for b) observe $\text{density}(\tilde{A}, <n) = \text{Density}\left(A, \#(\Sigma^{<n})\right)$. This also establishes c) in case $N = \#(\Sigma^{<n}) = \frac{\#\Sigma^n - 1}{\#\Sigma - 1}$ with $\#(A \cap \{0, \ldots, N-1\}) \le \varepsilon \cdot \#(\Sigma^{<n})$, whereas the worst-case occurs for $N = \#(\Sigma^{<n}) + \varepsilon \cdot \#\Sigma^n$ with $\#(A \cap \{0, \ldots, N-1\}) = \varepsilon \cdot \#(\Sigma^{<n}) + \varepsilon \cdot \#\Sigma^n$. Then and thus,

$$\text{Density}(A, N) \le \frac{\varepsilon \cdot \frac{\#\Sigma^{n+1} - 1}{\#\Sigma - 1}}{\frac{\#\Sigma^n - 1}{\#\Sigma - 1} + \varepsilon \cdot \#\Sigma^n} = \varepsilon \cdot \left(1 + (1 - \varepsilon) \cdot \frac{\#\Sigma - 1}{\#\Sigma - \frac{1}{\#\Sigma^n}}\right) \le \varepsilon \cdot \left(1 + (1 - \varepsilon)\right)$$

$\square$

For a good approximation S of L, one wants the density of $A = L \triangle S$ to eventually drop below some prescribed ε; that is satisfy, e.g., $\exists n_0 \ \forall n \ge n_0 : \text{density}(A, n) \le \varepsilon$.

Definition 3.   *An inequality "$f(n) \le g(n)$" depending on* $n \in \mathbb{N}$ *holds **almost everywhere**, denoted by "$f(n) \le_{ae} g(n)$", iff* $\exists n_0 \forall n \ge n_0 : f(n) \le g(n)$. *It holds **infinitely often** ("$f(n) \le_{io} g(n)$") iff* $\forall n_0 \exists n \ge n_0 : f(n) \le g(n)$.

So if "$\text{density}(A, n) \le_{ae} \varepsilon$" fails, one may try for the weaker "$\text{density}(A, n) \le_{io} \varepsilon$".

1.3   The Halting Problem

The halting problem is defined with respect to an (often implicitly chosen) programming system. Here we follow the notation of [Roge67, Soar87, Smit94].

Definition 4. *A **Gödelization** φ is a sequence of all partial recursive functions s.t.*
  – there exists a partial universal program u with $\varphi_u(\langle i, x \rangle) = \varphi_i(x)$ *(UTM)*

---
[2] We are considerably grateful to an anonymous referee for pointing out the work of N. LYNCH.

– *and a total program* s *with* $\varphi_{s(\langle i,x\rangle)}(y) = \varphi_i(\langle x,y\rangle)$ (*SMN*)
– *for a bijective computable function* $\langle\cdot,\cdot\rangle : \Sigma^* \times \Sigma^* \to \Sigma^*$ *or* $\langle\cdot,\cdot\rangle : \mathbb{N} \times \mathbb{N} \to \mathbb{N}$.
called **pairing function**. *The* **Halting problem** *for* φ *is* $H_\varphi = \{\langle i,x\rangle : x \in \mathrm{dom}(\varphi_i)\}$.

The Halting problem is sometimes alternatively defined as the task $\tilde{H}_\varphi$ of deciding whether a given program i terminates on the empty input, that is, whether $\lambda \in \mathrm{dom}(\varphi_i)$; or the question whether $i \in \mathrm{dom}(\varphi_i)$. Based on RICE's Theorem, all three versions can be reduced to one another and are thus equivalent from the point of view of strict computability but in generally *not* concerning approximations; see Example 24.

Similarly, strict undecidability of H_φ holds independently of the underlying programming system whereas a change in φ may sensitively affect its error complexity. In fact one can artificially 'blow up and pad' any Gödelization to obtain one where already a constant answer yields exponentially small error to the Halting problem [Lync74, PROPOSITION 1]. While the *Padding Lemma* of Recursion Theory asserts any programming system to repeat each computable function an infinite number of times, these repetitions should occur in a 'balanced' way for the Gödelization to be reasonable.

Definition 5. *Gödelization* φ *is* **dense** *iff* $\forall i \exists c > 0:$ density $(\{j : \varphi_i = \varphi_j\}, n) \geq_{ae} c$.

Another influence to the complexity of the Halting problem arises from the pairing function under consideration. Again, in order to avoid trivial approximations, we restrict to pairing functions which are **pair-fair** in the sense of [JaSc99, DEFINITION 5] and recall that for instance the standard pairing $\langle x,y\rangle = x + \frac{(x+y)(x+y+1)}{2}$ satisfies this condition. It has been proven that, under these natural restrictions, every heuristic claiming to solve the Halting problem makes at least a constant fraction of errors:

Theorem 6 ([JaSc99, THEOREM 4]). *Let* $\mathcal{REC}$ *denote the class of recursive languages and* φ *a dense Gödelization. Then* $\forall S \in \mathcal{REC}\ \exists \varepsilon > 0:$ density$(H_\varphi \triangle S, n)$ $\geq_{ae} \varepsilon$.

1.4 Own and Related Contributions

Observe that the lower approximation bound ε in Theorem 6 may in general depend on S; it seems thus still conceivable that H_φ admits an approximation *scheme* in the sense that better and better algorithms achieve smaller and smaller error densities. In fact the question whether or not there exists a *universal* constant lower bound was open for half a decade [JaSc99, bottom of p. 402].

The present paper gives both a positive and a negative answer to this question:

Theorem 7. *For any dense Gödelization* φ *it holds*
a) $\exists \varepsilon > 0 \ \forall S \in \mathcal{REC}:$ density$(H_\varphi \triangle S, n) \geq_{io} \varepsilon$.
b) $\forall \varepsilon > 0 \ \exists S \in \mathcal{REC}:$ density$(H_\varphi \triangle S, n) \leq_{io} \varepsilon$.

This complements [Lync74, PROPOSITION 6][5] where *ae-* rather than *io*-approximation is considered. In addition, our work differs from [Lync74] in treating the Halting problem H_φ *with* inputs as opposed to $\tilde{H}_\varphi$; see the discussion following Definition 4. Thirdly, we consider *dense* programming systems whereas [Lync74, p.147] requires them to be *optimal* in the sense of [Schn75] — a strictly[6] stronger condition:

Lemma 8. *Any optimal Gödelization* φ *is dense according to Definition 5.*

Proof. Start with some dense Gödelization φ'. In φ, fix an arbitrary index $i \in \mathbb{N}$. Thus $\varphi_i = \varphi'_{i'}$ for some $i' \in \mathbb{N}$. φ' being dense, the set $J' := \{ j' : \varphi'_{i'} = \varphi'_{j'} \}$ has Density(J',N) $\geq_{ae} c$ for some $c > 0$. By definition of optimality there exists $C \in \mathbb{N}$ and to each j' some $j \leq C \cdot j'$ such that $\varphi_j = \varphi'_{j'}$. Hence, Density$(\{ j : \varphi_i = \varphi_j \},N) \geq_{ae} c/C$. □

The above differences (H_φ rather than $\bar{H}_\varphi$, dense rather than optimal Gödel numberings) to [Lync74] are due to our interest in the Halting problem as arising *in practice*, that is, for real programming languages; see Sections 1.5 and 2. We focus on mere computability of according approximations; in particular our work is not related to the *restricted* Halting problem — *Given (i,t), does Turing machine #i terminate after $\leq t$ steps?* — considered in [Mach78, SECTION 6.1] for complexity purposes.

1.5 Omega Numbers

Approximations to the Halting problem have been treated by encoding H into a single real $r \in \mathbb{R}$ and then considering computational approximations to this r. A first encoding goes back to [Spec49] in terms of the number $x[H] = \sum_{n \in H} 2^{-n}$ whose binary digits are obviously not decidable but semi-decidable, i.e., any 1 can be verified within finite time.

CHAITIN's Omega-Number [Chai87, LiVi97] gives another way of encoding the entire Halting problem into a single real $\Omega_U = \sum_{\bar{x} \in \text{dom}(U)} 2^{-|\bar{x}|}$ where U denotes a universal Turing machine which is required to be self-delimiting. This implies by KRAFT's inequality that $\Omega_U \leq 1$ can be interpreted as the probability for U to terminate upon input of a random program. Ω_U is considered 'denser' and more difficult to approximate than $x[H]$ because its binary digits are not even semi-decidable; see, e.g., [LiVi97, CHKW01]. At first, it has therefore received noticable attention when [CDS01] did succeed in determining the first 64 bits of Ω_U.

However this approximation had been significantly simplified by the observation that, for the specific U considered in [CDS01], an overwhelming fraction of all instances do not contain a `halt` instruction at all and thus stall trivially. In other words, those program sources arising in practice form only a very sparse subset within the programming system treated there. In order to avoid such trivialities and instead obtain meaningful results about the possibility or impossibility of approximations to the Halting problem, we now present:

2 A Particularly Compact, Practical Dense Programming System

Concerning the applicability of Theorem 6, its prerequisite is satisfied by every Turing-complete programming language over alphabet Σ with some kind of *end-of-string* (eof) indicator. More generally it holds:

Example 9. Let $\varphi = (\varphi_{\bar{x}})_{\bar{x} \in \Sigma^*}$ denote a Gödelization which is *self-delimiting* in the sense that, whenever $\varphi_{\bar{x}}$ does not identically diverge, it holds $\varphi_{\bar{x}} = \varphi_{\bar{x} \circ \bar{y}}$ for all $\bar{y}$. Then, φ is dense. This includes, for arbitary Gödelization $\psi = (\psi_n)_{n \in \mathbb{N}}$, the 'tally'[3] re-coding

---

[3] See [Book74]. Also, this dense Gödelization is obviously non-*optimal* in the sense of [Schn75].

$$\phi_{1^n0} := \psi_n, \qquad \phi_{1^n0\bar{x}} := \psi_n \text{ for } \bar{x} \in \{0,1\}^*, \qquad \phi_{\bar{x}} :\equiv \perp \text{ for } \bar{x} \in \{1\}^* \ .$$

While a special symbol $\langle \text{eof} \rangle$ may always be added to Σ, we consider this cheating. Also from the practical side, compilers for programming languages nowadays rely on the end-of-file being indicated by the operating system (e.g., via $\texttt{feof}$) as opposed to the out-dated detection of characters like nul, ^D, or ^Z. In the present section we analyze and establish a practical, *non*-self delimiting programming language to be dense.

2.1 The BF Programming Language

BF ('BrainF*ck') was designed in 1993 by URBAN MÜLLER and has since then spread the Internet for its shrewd simplicity [Wiki05]. It is a Turing-complete programming language over the eight letter alphabet $\Sigma_{\text{BF}} = \{ \boxed{<}, \boxed{>}, \boxed{+}, \boxed{-}, \boxed{,}, \boxed{.}, \boxed{[}, \boxed{]} \}$. The first six characters represent commands, the remaining two brackets are used to construct simple loops.

A BF-program stores data on a tape similar to that of a Turing-Machine. Each cell of the tape may contain an integer between 0 and 255, that is, one byte. The current cell may be incremented using $\boxed{+}$ and decremented with $\boxed{-}$; (incrementing 255 will result in 0, decrementing 0 will result in 255). Other cells can be accessed by moving the read/write-head either to the left $\boxed{<}$ or to the right $\boxed{>}$. Initially, all cells are set to 0. The two commands $\boxed{,}$ and $\boxed{.}$ are for input and output: $\boxed{,}$ will fetch a byte from the input stream and store it into the current cell; $\boxed{.}$ appends the byte in the current cell to the output stream.

Loops are formed by putting commands inbetween the two bracket symbols $\boxed{[}$ and $\boxed{]}$. Each time the loop is about to be executed, the current cell is checked whether it contains a value other than 0. If so, the loop is executed again. The commands in the loop are skipped, if the current cell was 0. Note, that after each round of the loop, another cell could have been made current by the commands within the loop.

Definition 10. *Let $\mathcal{BF}_n \subseteq \Sigma_{\text{BF}}^n$ denote the set of strings $\bar{p}$ of length n representing a syntactically correct BF source code and $\mathcal{BF} = \bigcup_n \mathcal{BF}_n$.*

Observe that the syntax of this programming language is quite simple, the only requirement being that opening and closing brackets are nested correctly.

Remark 11. BF is sometimes refered to with a fixed tape of 30.000 cells size. However the level of standardization is not very advanced yet. In order to obtain a Turing-complete system, we shall assume an unbounded tape.

2.2 Naive Encoding of BF

This straight-forward idea takes BF source codes as Gödel indices:

Definition 12. *For $\bar{p} \in \Sigma_{\text{BF}}^*$, let $\psi_{\bar{p}}$ denote the function obtained by interpreting $\bar{p}$ as source code for some BF program; $\psi_{\bar{p}} :\equiv \perp$ in case $\bar{p}$ lacks syntactical correctness.*

However, closer analysis reveals that this programming system is *not* dense:

Theorem 13. *Let* $\mathrm{BF}_n := |\mathcal{BF}_n|$ *denote the number of correct BF-sources of length n.*

a) $\mathrm{BF}_{n+1} = 6 \cdot \mathrm{BF}_n + \sum_{i=0}^{n-1} \mathrm{BF}_i \cdot \mathrm{BF}_{n-1-i}.$

b) $\mathrm{BF}_n = \sum_{k=0}^{n/2} C_k \cdot \binom{n}{2k} \cdot 6^{n-2k}$, *where* $C_k = \frac{1}{k+1} \cdot \binom{2k}{k}$ *denotes* CATALAN*'s number.*

c) $(n+3) \cdot \mathrm{BF}_{n+1} = (12n+18) \cdot \mathrm{BF}_n - 32n \cdot \mathrm{BF}_{n-1}.$

d) $\mathrm{BF}_{n+1} \leq 8 \cdot \mathrm{BF}_n.$

e) $\mathrm{BF}_n \leq O(8^n/\sqrt{n}).$

Thus among all 8^n strings $\bar{p} \in \Sigma_{\mathrm{BF}}^n$, the fraction of syntactically correct sources tends to zero, permitting for H_ψ a trivial $O(\frac{1}{\sqrt{n}})$-approximation.

Proof. a) A syntactically correct BF program of length $n+1$ either consists of one (out of 6 possible) non-loop character followed by an, again syntactically correct, program of length n; or, in case it begins with the loop character $\boxed{[}$, it consists of a loop (whose body is a syntactically correct program of length i for some $i < n$) followed by some other syntactically correct source of length $n - 1 - i$.

b) Consider the collection $\mathcal{BF}_{n,k} \subseteq \mathcal{BF}_n$ of BF programs $\bar{p} \in \Sigma_{\mathrm{BF}}^n$ of length n with $0 \leq k \leq n/2$ occurrences of $\boxed{[}$ or, equivalently, of $\boxed{]}$. Then, C_k equals the number of correct ways of nesting $2k$ brackets [Bail96]. Any $\bar{p} \in \mathcal{BF}_{n,k}$ can be obtained in a unique way by choosing $2k$ out of n positions in $\bar{p}$ for placing these brackets and by filling each of the remaining $n - 2k$ positions independently with one out of the 6 non-bracket characters in Σ_{BF}.

c) follows from a) and b) by induction.

d) Claim c) immediately yields $\mathrm{BF}_{n+1} \leq c_0 \cdot \mathrm{BF}_n$ by induction, where $c_0 := 12$. Repeated application of c) establishes a sequence of improved bounds $\mathrm{BF}_{n+1} \leq c_k \cdot \mathrm{BF}_n \; \forall n$ with (c_k) decreasing down to 8.

e) Combining c+d), obtain $BF_{n+1} \leq 8 \cdot \dfrac{n + \frac{9}{4}}{n+3} \cdot \mathrm{BF}_n \leq 8^n \cdot \displaystyle\prod_{i=1}^{n} \dfrac{i + \frac{9}{4}}{i+3},$

$$\prod_{i=-2}^{n} \frac{i+\frac{9}{4}}{i+3} = \prod_{j=1}^{n+3} \frac{j-\frac{3}{4}}{j} \leq \prod_{j=1}^{n+3} \frac{j-\frac{1}{2}}{j}, \quad \Big(\prod_{j=1}^{n} \frac{2j-1}{2j}\Big)^2 \leq \Big(\prod_{j=1}^{n} \frac{2j-1}{2j}\Big) \cdot \Big(\prod_{j=1}^{n} \frac{2j}{2j+1}\Big) = \frac{1}{2n+1}$$

$\square$

2.3   Compact Encoding of BF

Regarding Theorem 13, a dense programming system based on BF better avoids enumerating syntactically incorrect sources. This leads to the following

Definition 14. *Define* φ_N *to denote the function computed by the N-th syntactically correct BF program* $\bar{p}_N$. *More formally, let* $\mathcal{BF}$ *be ordered primarily with respect to length n and secondarily according to the enumeration given by recursive application of Theorem 13a), that is, by first listing the* $6 \cdot \mathrm{BF}_n$ *programs starting with no loop and then listing, recursively and for each* $i = 0 \ldots n-1$, *the loop bodies and loop tails as BF sources of length i and* $n - 1 - i$, *respectively.*

Although this programming system is *not* self-delimiting, it holds:

Theorem 15. *The Gödelization* φ *from Definition 14 is dense.*

We emphasize that this is by no means a consequence of syntactical correctness alone!

Proof. Fix a partial recursive function computed by some BF source $\bar{p} \in \mathcal{BF}_m$ of length $m = |\bar{p}|$. For $n \geq m+2$, we construct BF_{n-m-2} equivalent programs $\bar{p}' \in \mathcal{BF}_n$. To this end preced $\bar{p}$ with a loop, i.e., let $\bar{p}' := \boxed{[} \circ \bar{q} \circ \boxed{]} \circ \bar{p}$ for an arbitrary syntactically correct source $\bar{q}$ of length $n - m - 2$. Since, upon start of execution, the current cell is initialized to 0, this loop gets skipped anyway and $\bar{p}'$ thus behaves like $\bar{p}$, indeed. The thus obtained sources $\bar{p}'$ constitute, in relation to BF_n and by Theorem 13d), a fraction

$$\frac{\mathrm{BF}_{n-m-2}}{\mathrm{BF}_n} \geq \frac{\mathrm{BF}_{n-m-2}}{8^{m+2} \cdot \mathrm{BF}_{n-m-2}} = 8^{-m-2} =: c > 0$$

among all programs of length $n \geq n_0 := m + 2$. Now proceed as in Lemma 2c). $\quad\square$

Conversion between BF sources and Gödel indices is a central part of efficient (rather than merely computable) SMN- and UTM-properties according to Definition 4. A naive approach enumerates *all* strings $\bar{p} \in \Sigma_{\mathrm{BF}}^n$ and counts the syntactically correct ones in order to obtain $\bar{p}_N$. This, however, gives rise to exponential time in $\log N$. The following result improves to running time polynomial in the input size, that is, $|\bar{p}|$ or $\log N$:

Theorem 16. *Given a program $\bar{p} = \bar{p}_N \in \mathcal{BF}_n$ of length n, one can calculate its index $N \in \mathbb{N}$ according to Definition 14 within time $O(n^3 \cdot \log n \cdot \log\log n)$. Conversely, from N, the according $\bar{p}_N$ is computable using $O(n^3 \cdot \log n \cdot \log\log n)$ steps where $n = \log N$. Both algorithms use memory of size $O(n^2)$. See also* http://www.upb.de/cs/bf

To conclude, the Gödelization introduced in this section is practical, efficient, and dense. It even seems plausible to satisfy the stronger condition of *optimality*; recall Lemma 8. To this end one might establish a sparse SMN-property for BF as required in

Lemma 17. *Let $\varphi = \left(\varphi_{\bar{p}}\right)_{\bar{p} \in \Sigma^*}$ denote a Gödelization and SMN-function $s : \Sigma^* \times \Sigma^* \to \Sigma^*$ according to Definition 4 satisfying $|s(\bar{p}, \bar{x})| \leq c(\bar{p}) + |\bar{x}|$ for all $\bar{p}, \bar{x} \in \Sigma^*$ with arbitrary $c : \Sigma^* \to \mathbb{N}$. Then, φ is optimal in the sense of [Schn75].*

Proof. Fix some other Gödelization Φ. Consider its UTM-function Φ_U and let u' denote the index of Φ_U in φ; i.e., $\forall \bar{x} \in \Sigma^* : \quad \Phi_{\bar{P}}(\bar{x}) = \Phi_U(\langle \bar{P}, \bar{x} \rangle) = \varphi_{u'}(\langle \bar{P}, \bar{x} \rangle) = \varphi_{\bar{p}}(\bar{x})$ where $\bar{p} := s(u', \bar{P})$ has by prerequisite length $|\bar{p}| \leq c(u') + |\bar{P}| = c_0 + |\bar{P}|$. $\quad\square$

3   The Error Complexity of Dense Programming Systems

In the last section we showed that a natural encoding of BF is dense. From Theorem 6 it follows that every algorithm A trying to solve the Halting problem of such a dense programming system errs on at least a constant fraction $\varepsilon_A > 0$. This constant fraction $\varepsilon_A > 0$ may depend on the algorithm A and can be arbitrarily small. In this section we will show that there is a universal constant $\varepsilon_0 > 0$ lower bounds the error made by any heuristic trying to approximate the Halting problem for a dense Gödelization.

3.1   Halting Ratio

A straight-forward implication of Theorem 6 is that neither nearly all programs halt nor do nearly all of them stall. This is formalized as follows:

Definition 18. *Call* $h_\varphi : N \mapsto \text{Density}\left(\{\langle i,x \rangle : x \in \text{dom}(\varphi_i)\}, N\right)$ *the* **halting ratio.**

Like Ω_U (see Section 1.5), h_φ describes a probability for a random instance to halt.

Corollary 19. *For every dense Gödelization* φ, $\exists c > 0 : c \leq_{ae} h_\varphi \leq_{ae} 1 - c$.

Proof. Consider two indices i, j with $\text{dom}(\varphi_i) = \Sigma^*$ and $\text{dom}(\varphi_j) = \emptyset$. Because of the dense programming system and the pair-fair pairing, these indices alone induce a constant fraction of halting and non-halting indices.   □

It seems desirable, again similarly to Section 1.5, to investigate the real number $r_\varphi :=$ $\lim_{n \to \infty} h_\varphi(n)$. However in many cases h_φ fails to converge:

Example 20. Take any programming system $\psi = (\psi_i)_{i \in \mathbb{N}}$ and define $\varphi = (\varphi_I)_{I \in \mathbb{N}}$ by

$$\varphi \;=\; (\psi_1,\;\; \psi_2, \psi_2, \psi_2, \psi_2, \;\ldots\ldots,\; \underbrace{\psi_i, \psi_i, \psi_i, \ldots, \psi_i, \psi_i,}_{i^i \text{ times}} \;\ldots\ldots\ldots) \; .$$

Obviously φ_I behaves identically for all I within a block arising from the same ψ_i. Since the size i^i of such a block dominates by far those of all previous blocks together, namely $N_i = 1 + 4 + \ldots + (i-1)^{i-1} \leq (i-1)^0 + (i-1)^1 + \ldots + (i-1)^{i-1} = \frac{(i-1)^i - 1}{i-2} \leq \frac{1}{i-2} \cdot i^i$,

a) termination of ψ_i determines whether $h_\varphi(N_i)$ is (arbitrarily close to) 1 or 0. In particular, h_φ is an oscillating function and fails to converge for $N \to \infty$.

b) As infinitely many instances of H_ψ are undecidable, so is almost every entire block of H_φ. In particular, $\forall \varepsilon > 0 \;\forall S \in \mathcal{REC} :$ Density$(H_\varphi \triangle S, N) \geq_{io} 1 - \varepsilon$.

c) On the other hand, $S := \emptyset \in \mathcal{REC}$ satisfies $\forall \varepsilon > 0 :$ Density$(H_\varphi \triangle S, N) \leq_{io} \varepsilon$. Indeed, each of the infinitely many i corresponding to stalling instances of H_ψ yields an entire block of them in H_φ, dominating Density$(H_\varphi, N_i + i^i) \leq \frac{1}{i-2}$ as above.   □

With the last two properties, this specific Gödelization concretizes the REMARK on top of p.147 in [Lync74]. Compare them to *io*-approximations of arbitrary dense programming systems according to Theorem 7.

3.2   Relation Between Two Approximations

Consider the question of approximating the function $h_\varphi : \mathbb{N} \to \mathbb{Q}$. This is related to the approximation of the Halting problem in the sense of Section 1.2 as follows:

Lemma 21. *Fix Gödelization* φ *with Halting problem* $H = H_\varphi$ *and halting ratio* $h = h_\varphi$.

a) *Given* $N \in \mathbb{N}$, $\varepsilon \in \mathbb{Q}$, *and* $b \in \mathbb{Q}$ *with* $|b - h(N)| \leq \varepsilon$, *one can compute a list* $H'_N \subseteq H \cap \{0, 1, \ldots, N - 1\}$ *of halting instances satisfying* Density$(H \triangle H'_N, N) \leq \varepsilon$.

b) *Let* $S \subseteq \mathbb{N}$ *be arbitrary. Given* $N \in \mathbb{N}$ *and* $\varepsilon \in \mathbb{Q}$ *such that* Density$(H \triangle S, N) \leq \varepsilon$, *an* S-oracle machine *can compute* $b \in \mathbb{Q}$ *with* $|b - h(N)| \leq \varepsilon$.

In particular for this φ *and any* $\varepsilon > 0$, *the Halting problem* H_φ *can* ae *be* ε-approximated iff *the halting ratio* h_φ *can* ae *be* ε-approximated; analogously for approximating io.

Proof. a) Recursively enumerate elements $x \in H \cap \{0, 1, \ldots, N - 1\}$ until having obtained a collection $H'_N \subseteq H$ of cardinality $\#H'_N \geq (b - \varepsilon) \cdot N$.

b) By repeatedly quering the oracle S for all finitely many $x \in \{0, 1, \ldots, N-1\}$, calculate the number $b := \#(S \cap \{0, 1, \ldots, N-1\})/N$. □

Corollary 22. *Fix computable $f : \mathbb{N} \to \mathbb{Q}$ and recursively enumerable $L \subseteq \mathbb{N}$ admitting (ae/io) an $f(N)$-approximation with two-sided error. Then L can (ae/io) be $f(N)$-approximated with one-sided error.*

Proof. W.l.o.g. $L = H_\varphi = H$ for some φ. Let $S \subseteq \mathbb{N}$ denote a recursive two-sided $f(N)$-approximation of H. Upon input of x, compute $\varepsilon := f(N)$, $N \geq |x|$; then obtain an approximation $b \in \mathbb{Q}$ for $h(N)$ by virtue of Lemma 21b), observing that oracle queries to S can be decided by presumption. Then apply Lemma 21a) to get a some $H'_N \subseteq H \cap \{0, \ldots, N-1\}$ with $\mathrm{Density}(H \bigtriangleup H'_N, N) \leq \varepsilon$, i.e., one-sided ε-approximation. □

3.3 Approximating the Halting Ratio

We now reveal that the Halting ratio of a dense programming system infinitely often admits a well approximation and infinitely often it does not.

Lemma 23. *Fix a dense programming system φ.*
a) *For all $\varepsilon > 0$, there exists a TM M such that* $|M(n) - h_\varphi(n)| \leq_{io} \varepsilon$.
b) *There exists $\varepsilon > 0$ such that all TMs M have* $|M(n) - h_\varphi(n)| \geq_{io} \varepsilon$.

Proof. a) For fixed $\varepsilon > 0$ consider $k := \lceil 1/\varepsilon \rceil$ and the $k+1$ constant (trivially computable) functions $0, \frac{1}{k}, \frac{2}{k}, \frac{3}{k}, \ldots, 1$. For every input length n, at least one of these values differs from $h_\varphi(n) \in [0, 1]$ by at most ε. A second application of pidgeon-hole's principle yields that some of these constant functions is close to h_φ for infinitely many n.

b) For a fixed rational $\varepsilon > 0$ (whose actual value we determine later) we assume $\varphi_i(n)$ for some $i \in \mathbb{N}$ is a candidate for computing $h_\varphi(n)$.
Now an algorithm A computes on input $z \in \Sigma^n$ the following. First it computes $b = \varphi_n(n)$ as an approximation to $h_\varphi(n)$ on input $z = \langle i, x \rangle \in \Sigma^n$. Let $f_1(z) = i$ and $f_2(z) = x$ be the decoding functions of $z = \langle i, x \rangle$. Let i^* be an index of A. According to the Recursion Theorem, A may know its own index and $\varepsilon \in \mathbb{Q}$. Then the algorithm simulates all inputs to H_φ of length n in parallel step by step until $(b - \varepsilon)|\Sigma|^n$ strings $y \in \Sigma^n$ have been found with $f_2(y) \in \mathrm{dom}(\varphi_{f_1(y)})$ and $f_1(y) \neq i^*$. Let s denote this number of halting inputs $\langle i, x \rangle \in \Sigma^n$ with $i \neq i^*$ found by A. If $s \geq (b - \varepsilon)|\Sigma|^n$ then the algorithm halts, else the algorithm does not halt.
There is a chance that the algorithm does not halt before this last condition, which means that $n \notin \mathrm{dom}(\varphi_n)$ or less than $(b - \varepsilon)|\Sigma|^n$ strings of length n corresponding to halting instances exist. In both cases φ_n was proven not to compute h_φ within the error margin ε.
Recall that i^* is an index for algorithm A and that the repetition rate of i^* is constant. The diagonalization argument is that all inputs $\langle i^*, x \rangle$ will result in $a(i^*, n) := |\{\langle i^*, x \rangle \in \Sigma^n\}|$ additional halting inputs of length n compared to $(b - \varepsilon)|\Sigma|^n$ where φ_n predicted at most $(b + \varepsilon)|\Sigma|^n$ halting instances. For large enough n, this number $a(i^*, n)$ is lower bounded by $\Omega(|\Sigma|^n)$, i.e. $\exists c > 0 : \forall_{ae} n : a(i^*, n) > c|\Sigma|^n$, because of the pair-fair property of $\langle \cdot, \cdot \rangle$.

Now if $s \geq (b - \varepsilon)|\Sigma|^n$ then there are at least $(b - \varepsilon + c)|\Sigma|^n$ halting instances for almost all input lengths n where $\varphi_n(n)$ is a candidate for $h_\varphi(n)$. Note that there are infinite many equivalent machines $n_1, n_2, \ldots$, with $\varphi_{n_i} = \varphi_n$. For $\varepsilon < c/2$ this implies that φ_n errs infinitely often on these inputs of length n_i with an error margin of at least ε (which can be determined independent from n).

Therefore for all machines $M = \varphi_n$ there are infinitely many input lengths n such that $M(n)$ does not approximate $h_\varphi(n)$ by an additional error term of ε. $\square$

3.4   The Halting Problem is *ae*-hard and *io*-easy

Combining Lemma 21 and Lemma 23 establishes the already announced

Theorem 7. *For any dense Gödelization* φ *it holds*
a) $\exists \varepsilon > 0 \ \forall S \in \mathcal{REC}:$ density$(H_\varphi \triangle S, n) \geq_{io} \varepsilon$.
b) $\forall \varepsilon > 0 \ \exists S \in \mathcal{REC}:$ density$(H_\varphi \triangle S, n) \leq_{io} \varepsilon$.

This solves the open problem stated in [JaSc99] for the case of dense programming system. In particular, it shows that the dense encoding of BF from Section 2.3 provides a natural hard problem which cannot be approximated better than up to a constant factor.

Furthermore, Theorem 7 nicely complements [Lync74, PROPOSITION 6]. Observe that Claim b) there only seems to be stronger than our Theorem 7a) because of the more restrictive presumption that the Gödelization φ under consideration be *optimal* in the sense of [Schn75] rather just dense.

In addition, [Lync74] refers to the Halting problem as termination of φ_i on the special input i (that is, in our notation, to $\tilde{H}_\varphi$; see Definition 4) whereas we treat the more general and practically relevant H_φ, i.e., termination of φ_i on given input x. Although both problems are equivalent with respect to exact computability, their behaviour concerning approximations differs significantly. This can be observed already in the proof of Lemma 23b) which heavily relies on the described algorithm A's behaviour to depend on (the length of) its input. More explicitly, we have the following

Example 24. Consider the dense tally Gödelization φ in Example 9. There, any ψ_n gives rise to an asymptotic 2^{-n-1}-fraction of equivalent instances $\varphi_{\bar{x}}$. Thus, storing the solutions to H_ψ for the first N inputs $\psi_1, \ldots, \psi_N$ allows for *ae* answering correctly a fraction $\varepsilon_N = \sum_{n=1}^N n \cdot 2^{-n-1}$ of instances to $\tilde{H}_\varphi$ with $\varepsilon_N \to 1$ as $N \to \infty$.

4   Conclusion

Since the Halting problem is of practical importance yet cannot be solved in the strict sense, we considered the possibility of approximating it. Similarly to the average-case theory of complexity, this depends crucially on the encoding of the problem, that is here, the programming system under consideration.

Many practical programming languages lacking density in fact do admit such an approximation with asymptotically vanishing relative error for the simple reason that the fraction of syntactically incorrect instances tends to 1. This was exemplified by a combinatorial analysis of the Turing-complete formal language BF. Here and in similar cases, the question for approximation the Halting problem is equivalent to a mere syntax check and thus becomes trivial and vain.

On the other hand, considering only syntactically correct sources was established to yield an efficient *and* dense programming system in the case of BF. For any such system, we proved a universal constant lower bound on relative approximations to the Halting problem even in the weak *io*-sense. Our third contribution establishes that, conversely, any constant relative error $\varepsilon > 0$ is *io* feasible by an appropriate machine M.

Question 25. Is there some optimal (but necessarily non-dense) programming system φ whose Halting problem H_φ satisfies the following even stronger inapproximability property similar to [Lync74, PROPOSITION 2]

$$\forall S \in \mathcal{REC} \quad \forall \varepsilon > 0: \quad \text{density}(H_\varphi \bigtriangleup S, n) \geq_{io} 1 - \varepsilon \quad \text{or even} \quad \geq_{ae} 1 - \varepsilon \ ?$$

Observe that [Lync74, PROPOSITION 6] reveals the answer to be negative concerning the Halting problem $\tilde{H}_\varphi$ *without* input which, regarding Example 24, tends to be strictly easier to approximate than H_φ anyway.

Another open problem, it remains whether BF leads in Section 2.3 to an even optimal (rather than just dense) Gödelization; cf. Lemma 8. Furthermore it is conceivable — although by no means obvious — that the programming system Jot by C. BARKER is dense as well; see http://ling.ucsd.edu/~barker/Iota/#Goedel.

References

[Bail96]   D.F. BAILEY: "Counting Arrangements of 1's and -1's", pp.128–131 in *Mathematics Magazine* vol.**69** (1996).

[Book74]   R.V. BOOK: "Telly languages and complexity classes", pp.186–193 in *Information and Control* vol.**26:2** (1974).

[CDS01]   C.S. CALUDE, M.J. DINNEEN, C.-K. SHU: "Computing a Glimpse of Randomness", pp.361–370 in *Experimental Mathematics* vol.**11:3** (2001).

[CHKW01] C.S. CALUDE, P. HERTLING, B. KHOUSSAINOV, Y. WANG: "Recursively enumerable reals and Chaitin Ω numbers", pp.125–149 in *Theoretical Computer Science* vol.**255** (2001).

[Chai87]   G.J. CHAITIN: *Algorithmic Information Theory*, Cambridge University Press (1987).

[Gold97]   O. GOLDREICH: "Combinatorial property testing (a survey)", pp.45–59 in *Proc. DIMACS Workshop in Randomized Methods in Algorithm Design* (1997).

[JaSc99]   A. JAKOBY, C. SCHINDELHAUER: "The Non-Recursive Power of Erroneous Computation", pp.394–406 in *Foundations of Software Technology and Theoretical Computer Science* (FSTTCS 1999), Springer LNCS vol.**1738**.

[Koeh04]   S. KÖHLER: "*Zur Approximierbarkeit des Halteproblems in einer praktischen Gödelisierung*", Bachelor's Thesis, University of Paderborn (2004).

[LiVi97]   M. LI, P. VITÁNI: *An Introduction to Kolmogorov Complexity and its Application*, 2nd Edition, Springer (1997).

[Lync74]   N. LYNCH: *Approximations to the Halting Problem*, pp.143–150 in *J. Computer and System Sciences* vol.**9** (1974).

[Mach78]   M. MACHTEY, P. YOUNG: *An Introduction to the General Theory of Algorithms*, The Computer Science Library (1978).

[Papa95]   C.H. PAPADIMITRIOU: *Computational Complexity*, Addison-Wesley (1995).

[Roge67]   H. ROGERS JR: *Theory of Recursive Functions and Effective Computability*, Mc-Graw Hill (1967).

[RoUl63]   G.F. ROSE, J.S. ULLIAN: "Approximation of Functions on the Integers", pp.693–701 in *Pacific Journal of Mathematics* vol.**13:2** (1963).

[Schn75]   C.P. SCHNORR: "Optimal Enumerations and Optimal Gödel Numberings", pp.182–191 in *Mathematical Systems Theory* vol.**8:2** (1975).

[Smit94]   C. SMITH: *A Recursive Introduction to the Theory of Computation*, Springer (1994).

[Soar87]   R.I. SOARE: *Recursively Enumerable Sets and Degrees*, Springer (1987).

[Spec49]   E. SPECKER: "Nicht konstruktiv beweisbare Sätze der Analysis", pp.145–158 in *J. Symbolic Logic* vol.**14:3** (1949).

[Wiki05]   http://wikipedia.org/wiki/BrainFuck; Wikipedia, the free encyclopedia (2005)

An Explicit Solution to Post's Problem over the Reals

Klaus Meer and Martin Ziegler

Department of Mathematics and Computer Science,
Syddansk Universitet, Campusvej 55, 5230 Odense M,
FAX: 0045 6593 2691, Denmark
{meer, ziegler}@imada.sdu.dk

Abstract. In the BSS model of real number computations we prove a
concrete and explicit semi-decidable language to be undecidable yet not
reducible from (and thus strictly easier than) the real Halting Language.
This solution to Post's Problem over the reals significantly differs from its
classical, discrete variant where advanced diagonalization techniques are
only known to yield the existence of such intermediate Turing degrees.

Then we strengthen the above result and show as well the existence of
an uncountable number of incomparable semi-decidable Turing degrees
below the real Halting problem in the BSS model. Again, our proof will
give concrete such problems representing these different degrees.

1 Introduction

Is every super-Turing computer capable of solving the discrete Halting Problem H? More formally, does each undecidable, recursively enumerable language $P \subseteq \mathbb{N}$, when serving as oracle to some appropriate Turing Machine M, enable this M^P to decide H? That question of E.L. POST from 1944 was answered to the negative in 1956/57 independently by MUCHNIK and FRIEDBERG [8][1]. Devising the *finite injury priority* sophistication of diagonalization, they proved the existence of r.e. Turing degrees strictly between those of $\emptyset$ and $\emptyset' = H$; cf. [21, CHAPTERS V to VII].

While the diagonal language is also based on a mere existence proof, its reduction to H reveals this as well as many other explicit and practical problems in automatized software verification undecidable. In contrast, problems like P are until nowadays only known to exist but have resisted any explicit — not to mention intuitive — description.

It turns out that for real number problems the situation is quite different. More precisely, for the $\mathbb{R}$-machine model due to BLUM, SHUB, and SMALE [2,3], we *explicitly* present a semi-decidable language (specifically, the set $\mathbb{Q}$ of rationals) and prove it to neither be reducible from the real Halting Problem $\mathbb{H}_\mathbb{R}$

---

[1] The existence of intermediate Turing degrees that need not to be r.e. follows from a result by KLEENE and POST from 1954, see [21, CHAPTER VI].

M. Liśkiewicz and R. Reischuk (Eds.): FCT 2005, LNCS 3623, pp. 467–478, 2005.

nor from the set $\mathbb{A}$ of algebraic reals. The proof exploits that real computability theory, apart from logic as in the discrete case, has also algebraic and topological aspects.

Section 1.2 recalls the basics of real number computation in the BSS model as well as the recursion-theoretic notions of reducibility and degrees; Section 2 contains the first main result of our work; we show $\mathbb{Q} \precsim \mathbb{A}$, i.e. the real algebraic numbers cannot be decided using a BSS oracle machine which has access to the (undecidable!) set of rationals as oracle set. Section 2.1 proves the '$\preceq$'-part, Section 2.2 the '$\precsim$'-part. In Section 3 the results are generalized in order to get an uncountable number of incomparable semi-decidable problems below the real Halting problem. We conclude in Section 4 with some general remarks on hypercomputation.

1.1 Related Work

Our contribution adds to other results, indicating that many (separation-) problems which seem to require non-constructive (e.g., diagonalization) techniques in the discrete case, admit an explicit solution over the reals. For instance, a problem neither in $\mathcal{VP}$ nor $\mathcal{VNP}$-complete (provided that $\mathcal{VP} \neq \mathcal{VNP}$, of course) was presented explicitly in [4, SECTION 5.5].

CUCKER's work [7] is about the Arithmetic Hierarchy over $\mathbb{R}$, that is, degrees beyond the real Halting Problem $\mathbb{H}_{\mathbb{R}}$.

HAMKINS and LEWIS considered Post's Problem over the reals for *Infinite Time* Turing Machines, that is, with respect to arguments $x \in \mathbb{R}$ given by their binary expansion and for hypercomputers performing an ordinal number of steps like $1, 2, 3, \ldots, n, \ldots, \omega, \omega + 1, \ldots, 2\omega, \ldots$ They showed in [9] that in this model,

- for sets of reals the answer is "no" just like in the classical discrete case.
- for *single* real numbers x on the other hand, considered as sets $L_x \subseteq \mathbb{N}$ of those indices where the binary expansion of x has a 1, there is no undecidable degree below that of the Halting Problem (of Infinite Time Machines). POST's Problem therefore is to be answered to the *positive* in this latter setting!

The existence of different *complexity degrees* below $\mathcal{NP}$ in the BSS model both for real and for complex numbers was studied in a series of papers [1,6,16] and related to classical results (cf. [13,20]) for the Turing model.

1.2 The BSS Model of Real Number Computation

This section summarizes very briefly the main ideas of real number computability theory. For a more detailed presentation see [3].

Essentially a (real) BSS-machine can be considered as a Random Access Machine over $\mathbb{R}$ which is able to perform the basic arithmetic operations at unit cost and which registers can hold arbitrary real numbers.

Definition 1. *([2])*

a) *Let $Y \subseteq \mathbb{R}^\infty := \bigoplus_{k \in \mathbb{N}} \mathbb{R}^k$, i.e. the set of finite sequences of real numbers. A BSS-machine M over $\mathbb{R}$ with admissible input set Y is given by a finite set*

I of instructions labeled by $1, \ldots, N$. A configuration of M is a quadruple $(n, i, j, x) \in I \times \mathbb{N} \times \mathbb{N} \times \mathbb{R}^{\infty}$. Here, n denotes the currently executed instruction, i and j are used as addresses (copy-registers) and x is the actual content of the registers of M. The initial configuration of $M's$ computation on input $y \in Y$ is $(1, 1, 1, y)$. If $n = N$ and the actual configuration is (N, i, j, x), the computation stops with output x.

The instructions M is allowed to perform are of the following types :

computation: $n : x_s \leftarrow x_k \circ_n x_l$, where $\circ_n \in \{+, -, \times, \div\}$ or $n : x_s \leftarrow \alpha$ for some constant $\alpha \in \mathbb{R}$.

The register x_s will get the value $x_k \circ_n x_l$ or α, respectively. All other register-entries remain unchanged. The next instruction will be $n + 1$; moreover, the copy-register i is either incremented by one, replaced by 0, or remains unchanged. The same holds for copy-register j.

branch: n: if $x_0 \geq 0$ **goto** $\beta(n)$ **else goto** $n + 1$. According to the answer of the test the next instruction is determined (where $\beta(n) \in I$). All other registers are not changed.

copy: $n : x_i \leftarrow x_j$, i.e. the content of the "read"-register is copied into the "write"-register. The next instruction is $n + 1$; all other registers remain unchanged.

b) A set $A \subseteq \mathbb{R}^{\infty}$ is a *decision problem* or a *language*. We call a function $f : A \to \mathbb{R}^{\infty}$ *(BSS-) computable* iff it is realized by a BSS machine over admissible input set A. Similarly, a set $A \subseteq \mathbb{R}^{\infty}$ is *decidable* in $\mathbb{R}^{\infty}$ iff its characteristic function is computable. It is *semi-decidable* (synonymously: *r.e.*) iff there is a BSS algorithm which takes inputs from $\mathbb{R}^{\infty}$ and halts precisely on the elements belonging to A.

c) A *BSS oracle machine* using an *oracle* set $B \subseteq \mathbb{R}^{\infty}$ is a BSS machine with an additional type of node called oracle node. Entering such a node the machine can ask the oracle whether a previously computed element $x \in \mathbb{R}^{\infty}$ belongs to B. The oracle gives the correct answer at unit cost.

Several further concepts and notions now emerge as in the discrete setting.

Definition 2. *The* real Halting Problem $\mathbb{H}_{\mathbb{R}}$ *is the following decision problem. Given the code* $c_M \in \mathbb{R}^{\infty}$ *of a BSS machine* M *together with an* $x \in \mathbb{R}^{\infty}$, *does* M *terminate its computation on input* x?

Both the existence of such a coding for BSS machines and the undecidability of $\mathbb{H}_{\mathbb{R}}$ in the BSS model were shown in [2].

Next, oracle reductions are defined as usual.

Definition 3. *a) A real number decision problem* A *is* reducible *to another decision problem* B *if there is a BSS oracle machine that decides membership in* A *by using* B *as oracle set. We denote this reducibility by* $A \preceq B$ *and write* $A \precnsim B$ *when* A *is reducible to* B, *but* B *is not reducible to* A.

b) *If* A *is reducible to* B *and vice versa, we write* $A \equiv B$. *This defines equivalence classes* $\{B : A \equiv B\}$ *among real number decision problems called* (real) Turing degrees *or* BSS degrees.

c) *If none of two problems is reducible to the other, they are said to be incomparable.*

The main question treated in this paper is: Are there incomparable Turing degrees strictly between the degree $\emptyset$ of decidable problems in $\mathbb{R}^\infty$ and the degree $\emptyset'$ of the real Halting problem $\mathbb{H}_\mathbb{R}$?

2   Explicit Solution to Post's Problem over the Reals

Consider the sets $\mathbb{Q}$ of all rational numbers and $\mathbb{A}$ of all algebraic reals, that is, of real zeros of polynomials with rational coefficients, only. $\mathbb{Q}$ is obviously semi-decidable (upon input of $x \in \mathbb{R}$, simply check for all pairs of integers $r, s \in \mathbb{Z}$ whether $x = r/s$) but well-known not to be decidable [10,17]. In fact the same holds for $\mathbb{A}$: Given $x \in \mathbb{R}$, try for all polynomials $p \in \mathbb{Q}[X]$ whether $p(x) = 0$.

Our first main result states that, even given oracle access to $\mathbb{Q}$, $\mathbb{A}$ remains undecidable: $\mathbb{A} \not\preceq \mathbb{Q}$. Since oracle access to the Halting Problem $\mathbb{H}_\mathbb{R}$ of BSS machines allows to decide $\mathbb{A}$ by querying whether the above search for $p \in \mathbb{Q}[X]$ terminates, $\mathbb{Q}$ thus constitutes an explicit example of a real BSS degree strictly between the decidable one and that of the Halting Problem.

We also show $\mathbb{Q} \preceq \mathbb{A}$.

Theorem 4. *In the BSS model of real number computation it holds $\mathbb{Q} \lesssim \mathbb{A}$. In particular, transcendence is not semi-decidable even when using $\mathbb{Q}$ as an oracle.*

This result is, in spite of the notational resemblance to $\mathbb{Q} \subsetneq \mathbb{A}$, by no means obvious.

2.1   Deciding $\mathbb{Q}$ in $\mathbb{R}$ by Means of an $\mathbb{A}$–Oracle

In this section, we prove

Lemma 5. $\mathbb{Q} \preceq \mathbb{A}$.

Proof. Consider some input $x \in \mathbb{R}$. By querying the $\mathbb{A}$-oracle, identify and rule out the case that x is not in $\mathbb{A}$ (and hence not in $\mathbb{Q}$ either). So it remains to distinguish $x \in \mathbb{Q}$ from $x \in \mathbb{A} \setminus \mathbb{Q}$. To this end, calculate $d := \deg(x)$ according to Lemma 6 below and test whether $d = 1$ $(x \in \mathbb{Q})$ or $d \geq 2$ $(x \notin \mathbb{Q})$.   □

Recall that the *degree* of an algebraic $a \in \mathbb{R}$ is defined to be

$$\deg(a) = \dim_\mathbb{Q} \mathbb{Q}(a) = [\mathbb{Q}(a) : \mathbb{Q}],$$

that is, the dimension of the rational extension field generated by a. It is well known, for example in [14, PROPOSITION V.§1.2], that finite field extensions $M \subset K \subset L$ satisfy

$$[L : M] = [L : K] \cdot [K : M] . \tag{1}$$

A non-algebraic number is *transcendental*, the set of which we denote by $\mathbb{T}$.

Lemma 6. *The function* $\deg : \mathbb{A} \to \mathbb{N}$, $a \mapsto \deg(a)$ *is BSS computable.*

We point out that the restriction of deg to algebraic numbers is essential here; in other words: While for reasons of mathematical convenience one can *define* $\deg(x) := \infty$ for transcendental x, a BSS machine cannot *compute* it.

Proof. Exploit that an alternative yet equivalent definition for $\deg(a)$ is given by the degree of a minimal polynomial of a, that is, of an irreducible non-zero $p \in \mathbb{Z}[X]$ with $p(a) = 0$ [14, PROPOSITION V.§1.4].

Therefore we enumerate all non-zero $p \in \mathbb{Z}[X]$ and, for each one, plug in a to test whether $p(a) = 0$. If so, check p for irreducibility — a property in classical $\mathcal{NP}$ by virtue of [5] and thus BSS decidable. If this test succeeds as well, return $\deg(p)$ and terminate; otherwise continue with the next $p \in \mathbb{Z}[X]$.   □

Remark 7. An alternative way for deciding irreducibility in $\mathbb{Z}[X]$ — although not within nondeterministic polynomial time — proceeds as follows:

Given $p \in \mathbb{Z}[X]$ of degree $n - 1$, choose some n arbitrary distinct arguments $x_1, \ldots, x_n \in \mathbb{Z}$ and multi-evaluate $y_i := p(x_i)$. Observe that, if $q \in \mathbb{Z}[X]$ is a non-trivial divisor of p, then $z_i := q(x_i)$ divides y_i for each $i = 1, \ldots, n$. This suggests to go through all (finitely many) choices for $(z_1, \ldots, z_n) \in \mathbb{Z}^n$ with $z_i | y_i$, to calculate the interpolation polynomial $q \in \mathbb{Q}[X]$ to data (x_i, z_i) and check whether its coefficients are integral.

2.2 Undecidability of $\mathbb{A}$ in $\mathbb{R}$ with Support of a $\mathbb{Q}$–Oracle

In this section, we prove $\mathbb{A} \not\leq \mathbb{Q}$.

The undecidability of $\mathbb{A}$ *without* further oracle assistance follows similarly to that of $\mathbb{Q}$ from a continuity argument, observing that each, $\mathbb{A}$ and $\mathbb{Q}$ as well as their complements, are dense in $\mathbb{R}$. In fact, algebraic numbers remain dense even when restricting to arbitrary high degree:

Lemma 8. *Let* $x \in \mathbb{R}$, $\varepsilon > 0$, *and* $N \in \mathbb{N}$.
Then, there exists an algebraic real a *of* $\deg(a) = N$ *with* $|x - a| < \varepsilon$.

Proof. Take some arbitrary algebraic real b of degree N, such as $b := 2^{1/N}$. Since $\mathbb{Q}$ is dense in $\mathbb{R} \ni y := x - b$, there exists some rational $r \in \mathbb{Q}$ with $|r - y| < \varepsilon$. Then $a := r + b$ has the desired property.   □

Of course, total discontinuity does not prevent a problem to be BSS decidable under the support of a $\mathbb{Q}$–oracle any more as, for example, $\mathbb{Q}$ now is decidable. More precisely a putative algorithm might try distinguishing algebraic from transcendental reals by mapping a given x through some rational function $f \in \mathbb{R}(X)$, then querying the oracle whether the value $f(x)$ is rational or not, and proceeding adaptively depending on the answer.

The following observation basically says that in any sensible such approach, for transcendental x, $f(x)$ will be irrational rather than rational.

Lemma 9. *Let* $f : \mathrm{dom}(f) \subseteq \mathbb{R} \to \mathbb{R}$ *be analytic and non-constant,* $T \subseteq \mathrm{dom}(f)$ *uncountable. Then,* f *maps some* $x \in T$ *to a transcendental, that is,* $f(x) \notin \mathbb{A}$.

Proof. Consider an arbitrary $y \in \mathbb{A}$; by uniqueness of analytic functions [19, THEOREM 10.18], f can map at most countably many different $x \in \mathrm{dom}(f)$ to that single value y. Hence, if $f(x) \in \mathbb{A}$ for all $x \in T$, $f^{-1}(\mathbb{A}) = \bigcup_{y \in \mathbb{A}} f^{-1}(\{y\})$ is a countable union of countable sets and thus countable, too — contradicting the prerequisite that $T \subseteq f^{-1}(\mathbb{A})$ is uncountable.   □

So it remains the case of an algorithm trying to map algebraic x to rationals $f(x)$ and transcendental x to irrational $f(x)$. The final ingredient formalizes the intuition that this approach cannot distinguish transcendentals from algebraic numbers of sufficiently high degree:

Proposition 10. *Let $f \in \mathbb{R}(X)$, $f = p/q$ with polynomials p, q of $\deg(p) < n$, $\deg(p) < m$. Let $a_1, \ldots, a_{n+m} \in \mathrm{dom}(f)$ be distinct real algebraic numbers with $f(a_1), \ldots, f(a_{n+m}) \in \mathbb{Q}$.*

a) There are co-prime polynomials $\tilde{p}, \tilde{q}$ of $\deg(\tilde{p}) < n$, $\deg(\tilde{q}) < m$ with co-efficients in the algebraic field extension $\mathbb{Q}(a_1, \ldots, a_{n+m})$ such that, for all $x \in \mathrm{dom}(f) = \{x : q(x) \neq 0\} \subseteq \mathbb{R}$, it holds $f(x) = \tilde{f}(x) := \tilde{p}(x)/\tilde{q}(x)$.

b) Let $d := \max_i \deg(a_i)$. Then $f(x) \notin \mathbb{Q}$ for all transcendental $x \in \mathrm{dom}(f)$ as well as for all $x \in \mathbb{A}$ of $\deg(x) > D := d^{n+m} \cdot \max\{n-1, m-1\}$.

Notice that p and q themselves in general do not satisfy claim a); e.g. $p = \pi \cdot \tilde{p}$ and $q = \pi \cdot \tilde{q}$.

Proof. a) Without loss of generality take p and q to be co-prime. Let $y_i := f(a_i)$. The idea is to solve the rational interpolation problem for (a_i, y_i). Already knowing that is *has* a solution (namely p, q) avoids many of the difficulties discussed in [15].

More precisely, observe that the coefficients $p_0, \ldots, p_{n-1}, q_0, \ldots, q_{m-1} \in \mathbb{R}$ of p and q satisfy the homogeneous $(n+m) \times (n+m)$-size system of linear equations

$$
\begin{pmatrix}
1 & a_1 & a_1^2 & \ldots & a_1^{n-1} & -y_1 & -y_1 a_1 & \ldots & -y_1 a_1^{m-1} \\
1 & a_2 & a_2^2 & \ldots & a_2^{n-1} & -y_2 & -y_2 a_2 & \ldots & -y_2 a_2^{m-1} \\
1 & a_3 & a_3^2 & \ldots & a_3^{n-1} & -y_3 & -y_3 a_3 & \ldots & -y_3 a_3^{m-1} \\
\vdots & \vdots & \vdots & \ddots & \vdots & \vdots & \vdots & \ddots & \vdots
\end{pmatrix}
\cdot
\begin{pmatrix}
p_0 \\
\vdots \\
p_{n-1} \\
q_0 \\
\vdots \\
q_{m-1}
\end{pmatrix}
= 0 .
$$

In particular, this system has $(p_0, \ldots, q_{m-1}) \in \mathbb{R}^{n+m}$ as non-zero solution.

The coefficients of the matrix live in $\mathbb{Q}(a_1, \ldots, a_{n+m})$. Therefore, GAUSSIAN Elimination yields a (possibly different) non-zero solution $(\bar{p}_0, \ldots, \bar{q}_{m-1})$, also with entries in $\mathbb{Q}(a_1, \ldots, a_{n+m})$. Now apply the EUCLIDEAN Algorithm to the thus obtained polynomials $\bar{p}, \bar{q}$ and calculate their greatest common divisor $\bar{h}$ which, again, has coefficients in $\mathbb{Q}(a_1, \ldots, a_{n+m})$.

Thus, $\tilde{p} := \bar{p}/\bar{h}$ and $\tilde{q} := \bar{q}/\bar{h}$ are co-prime polynomials over $\mathbb{Q}(a_1, \ldots, a_{n+m})$ of $\deg(\tilde{p}) < n$ and $\deg(\tilde{q}) < m$ such that $\tilde{p} \cdot q$ coincides with $p \cdot \tilde{q}$ on arguments $a_1, \ldots, a_{n+m}$. This implies the latter polynomials of degree less than $n + m$ to be identical: $\tilde{p} \cdot q = p \cdot \tilde{q}$.

It follows that q divides both sides; and co-primality of (p, q) in the factorial ring $\mathbb{R}[X]$ requires that q divides $\tilde{q}$. Similarly, $\tilde{q}$ divides q, yielding $\tilde{q} = \lambda q$ for some $\lambda \in \mathbb{R}$. Analogously, $\tilde{p} = \lambda p$ for the same λ.

b) Consider $x \in \mathbb{R}$ with $y := f(x) \in \mathbb{Q}$ and suppose x is algebraic of $\deg(x) > d^{n+m} \cdot \max\{n-1, m-1\}$ or transcendental. Being, by virtue of a), a zero of the polynomial $\tilde{p} - y \cdot \tilde{q}$ with coefficients from $\mathbb{Q}(a_1, \ldots, a_n)$, x lies in an algebraic extension of the latter field, hence ruling out the case that it is transcendental. More precisely, the degree of x over $\mathbb{Q}(a_1, \ldots, a_n)$ is bounded by $\deg(\tilde{p} - y \cdot \tilde{q})$; and $\deg(x)$, its degree over $\mathbb{Q}$, is at most $\deg(\tilde{p} - y \cdot \tilde{q}) \cdot \deg(a_1) \cdots \deg(a_{n+m}) \leq \max\{n-1, m-1\} \cdot d^{n+m}$ by Equation (1) — contradiction.   □

We are finally in the position for the

Proof (of Theorem 4). Suppose some BSS algorithm semi-decides $\mathbb{T}$ in $\mathbb{R}$ with oracle $\mathbb{Q}$ according to Definition 1; in other words, it proceeds by repeatedly evaluating a given $x \in \mathbb{R}$ at functions $f \in \mathbb{R}(X)$ and continuing adaptively according to whether $f(x)$ is positive/zero/negative and rational/irrational, such as to terminate iff $x \in \mathbb{T}$.

Consider this process unrolled into an (infinite yet countable) Decision Tree, each internal node u of which is labeled with an according $f_u \in \mathbb{R}(X)$ and has five successors according to the cases

$$\boxed{0 > f_u(x) \in \mathbb{Q}} \quad \boxed{0 > f_u(x) \notin \mathbb{Q}} \quad \boxed{0 = f_u(x)} \quad \boxed{0 < f_u(x) \in \mathbb{Q}} \quad \boxed{0 < f_u(x) \notin \mathbb{Q}}$$

with leafs corresponding to terminating computations, that is, to $x \in \mathbb{T}$. Observe that the sets T_v of $x \in \mathbb{T}$ terminating in leaf v give rise to a partition of $\mathbb{T}$. In fact, the at most countably many leafs — as opposed to $\mathbb{T}$ having cardinality of the continuum — require that T_v is uncountable for at least one v.

Consider the path leading from the root to that leaf. W.l.o.g. it contains no branches of type "$0 = f_u(x)$" nor of type "$f_u(x) \in \mathbb{Q}$" that are answered "yes"; for if it does, then the uncountable set T_v of transcendentals x passing through this branch implies that f_u is constant (Lemma 9) and node u thus is dispensable. By possibly changing from $+f_u$ to $-f_u$, we may finally suppose that every branch on the path to leaf v is of type $0 < f_u(x)$.

Summarizing, $T_v \neq \emptyset$ is the set of exactly those $x \in \mathbb{R}$ satisfying $0 < f_u(x) \notin \mathbb{Q}$ for the (finitely many) internal nodes u on the path from the root to v; in particular, $T_v \subseteq \mathrm{dom}(f_u)$. Now take some $t \in T_v \subseteq \mathbb{R}$. Due to continuity of rational functions, there exists $\varepsilon > 0$ such that $f_u(x) > 0$ on all nodes u on that path for any $x \in \mathbb{R}$ satisfying $|x - t| < \varepsilon$. In particular, $f_u(a) > 0$ holds for infinitely many algebraic numbers a of unbounded degree according to Lemma 8. Since by presumption, none of them completes the (terminating) computational path to leaf v, they must branch off somewhere, that is, satisfy $f_u(a) \in \mathbb{Q}$ for some of the finitely many nodes u. However by Proposition 10b), each single f_u can sort out only algebraics of degree up to some finite $D = D(u)$ — a contradiction.   □

3   More Undecidable and Incomparable Real Degrees

A further achievement of the works of FRIEDBERG and MUCHNIK was the proof of existence of incomparable r.e. degrees below the Halting problem. In this section, we extend the used techniques to establish in the real case explicit such problems.

More precisely, we shall construct natural incomparable subsets of $\mathbb{A}$. They are given as certain algebraic, infinite extensions of $\mathbb{Q}$ obtained by adjunction of all n-th roots of a fixed prime.

Here, we shall explicitly construct two incomparable problems, only. However, it will then be obvious from the presentation that an uncountable number of incomparable real Turing degrees exist.

3.1   Some Auxiliary Results from Algebra

Let $\mathbb{P} := \{2, 3, 5, 7, \ldots\}$ denote the set of prime numbers. We define the following infinite algebraic extensions of $\mathbb{Q}$:

$$\mathbb{Q}_{\sqrt{2}} := \mathbb{Q}(\{2^{\frac{1}{p}} | p \in \mathbb{P}\}) \quad \text{and} \quad \mathbb{Q}_{\sqrt{3}} := \mathbb{Q}(\{3^{\frac{1}{p}} | p \in \mathbb{P}\}),$$

where the corresponding roots are taken as the smallest positive real that is such a root. Thus, $\mathbb{Q}_{\sqrt{2}}$ results from $\mathbb{Q}$ by field adjunction of all p-th roots of $2, p \in \mathbb{P}$. It is easy to see that $[\mathbb{Q}_{\sqrt{2}} : \mathbb{Q}] = \infty$ and $[\mathbb{Q}_{\sqrt{3}} : \mathbb{Q}] = \infty$. In order to apply the techniques from Section 2 we need the following two results.

Theorem 11. *Let $n \in \mathbb{P}$ and let k be a field. If $a \in k$ is not the n-th power of an element in k, then the field extension $k(\sqrt[n]{a})$ has degree n over k.*

Proof. See [14], Chapter 6, Theorem 9.1.

We would like to guarantee that the elements $2^{\frac{1}{p}}, p \in \mathbb{P}$ do not only have degree p over $\mathbb{Q}$, but as well over $\mathbb{Q}_{\sqrt{3}}$ (and vice versa for elements $3^{\frac{1}{p}}$ and $\mathbb{Q}_{\sqrt{2}}$). In view of the previous theorem it thus suffices to show that $2^{\frac{1}{p}} \notin \mathbb{Q}_{\sqrt{3}}$. Though we strongly assume this to be known we could not find a suitable reference; therefore we add a proof.

Lemma 12. *For all $p \in \mathbb{P}$ it holds $2^{\frac{1}{p}} \notin \mathbb{Q}_{\sqrt{3}}$. Similarly, $3^{\frac{1}{p}} \notin \mathbb{Q}_{\sqrt{2}}$ for all $p \in \mathbb{P}$. Thus $[\mathbb{Q}_{\sqrt{3}}(2^{\frac{1}{p}}) : \mathbb{Q}_{\sqrt{3}}] = p$ and $[\mathbb{Q}_{\sqrt{2}}(3^{\frac{1}{p}}) : \mathbb{Q}_{\sqrt{2}}] = p$ for all $p \in \mathbb{P}$.*

Proof. Suppose to the opposite that $2^{\frac{1}{p}} \in \mathbb{Q}_{\sqrt{3}}$. Then $2^{\frac{1}{p}}$ is already element of a finite extension of $\mathbb{Q}$ with elements $3^{\frac{1}{2}}, 3^{\frac{1}{3}}, \ldots, 3^{\frac{1}{q}}$ for some $q \in \mathbb{P}$. Define $N := \prod_{\substack{i \in \mathbb{P} \\ i \leq q}} i$; it follows that $2^{\frac{1}{p}} \in \mathbb{Q}(3^{\frac{1}{N}})$. We can now proceed almost as in the classical proof of irrationality of $\sqrt{2}$. Suppose $2^{\frac{1}{p}}$ has a representation as $\frac{f(3^{\frac{1}{N}})}{g(3^{\frac{1}{N}})}$ for some polynomials $f, g \in \mathbb{Z}[x]$ such that the integer coefficients of f and g

have 1 as their joint greatest common divisor and such that the occurring powers of $3^{\frac{1}{N}}$ are non-integral. The usual arguments together with $p > 1$ result in the contradiction that 2 divides all those coefficients. The final claim now follows from Theorem 11. □

Since any rational can be incorporated into any minimal polynomial of an element in $\mathbb{Q}_{\sqrt{2}}$ over $\mathbb{Q}_{\sqrt{3}}$ it follows

Corollary 13. *Let $n \in \mathbb{N}$ and let $x \in \mathbb{Q}_{\sqrt{3}}$; for each $\epsilon > 0$ there are infinitely many $y \in \mathbb{Q}_{\sqrt{2}}$ of degree at least n over $\mathbb{Q}_{\sqrt{3}}$ such that $|x - y| < \epsilon$.*

3.2 Existence of Incomparable Degrees

The results from the previous subsection allow to generalize our results to obtain

Theorem 14. *The sets $\mathbb{Q}_{\sqrt{2}}$ and $\mathbb{Q}_{\sqrt{3}}$ are incomparable.*

To prove the theorem we need the following generalization of Proposition 10.

Proposition 15. *Let $f \in \mathbb{R}(X), f = \frac{p}{q}$ with polynomials p, q of degree less than n and m, respectively. Let $a_1, \ldots, a_{n+m} \in \mathbb{Q}_{\sqrt{2}} \cap \mathrm{dom}(f)$ be distinct with $f(a_i) \in \mathbb{Q}_{\sqrt{3}}$.*

a) *There are co-prime polynomials $\tilde{p}, \tilde{q}$ of $\deg(\tilde{p}) < n$, $\deg(\tilde{q}) < m$ with coefficients in the algebraic field extension $\mathbb{Q}_{\sqrt{3}}(a_1, \ldots, a_{n+m})$ such that, for all $x \in \mathrm{dom}(f) = \{x : q(x) \neq 0\} \subseteq \mathbb{R}$, it holds $f(x) = \tilde{f}(x) := \tilde{p}(x)/\tilde{q}(x)$.*
b) *Let d be the maximal degree of an a_i over the field $\mathbb{Q}_{\sqrt{3}}$. Then $f(x) \notin \mathbb{Q}_{\sqrt{3}}$ for all transcendental $x \in \mathrm{dom}(f)$ as well as for all $x \in \mathbb{Q}_{\sqrt{2}}$ of degree $> D := d^{n+m} \cdot \max\{n - 1, m - 1\}$ over $\mathbb{Q}_{\sqrt{3}}$.*

Proof. Follows the same way as Proposition 10. □

Proof (of Theorem 14). We only fill in the missing arguments for showing $\mathbb{Q}_{\sqrt{2}} \nleq \mathbb{Q}_{\sqrt{3}}$. Incomparability then follows obviously from that proof.

Assume M to be a machine semi-deciding $\mathbb{R} \setminus \mathbb{Q}_{\sqrt{2}}$ by means of an $\mathbb{Q}_{\sqrt{3}}$-oracle. Follow the proof of Theorem 4 to obtain in just the same way a leaf v together with the related path set $T_v \subseteq \mathbb{R} \setminus \mathbb{Q}_{\sqrt{2}}$. Since T_v is uncountable it contains a transcendental x and in each neighborhood of x by virtue of Lemma 12 and Corollary 13 elements of $\mathbb{Q}_{\sqrt{2}}$ of arbitrarily high degree over the field $\mathbb{Q}_{\sqrt{3}}$. Thus, applying Proposition 15 there exist elements in $\mathbb{Q}_{\sqrt{2}}$ that are branched along v, contradicting the assumed semi-decidability of $\mathbb{R} \setminus \mathbb{Q}_{\sqrt{2}}$. □

As an easy consequence of the above proof we obtain the related result for other extensions of $\mathbb{Q}$ such as $\mathbb{Q}_{\sqrt{p}}$ for $p \in \mathbb{P}$. Clearly, there exist uncountably many sequences of reals that we could attach to $\mathbb{Q}$ in order to get even more incomparable problems (which, however, may be less explicit than $\mathbb{Q}_{\sqrt{p}}$). Thus, we have proven

Theorem 16. *There are uncountably many real recursively enumerable Turing degrees below the real Halting problem.* □

3.3 Some Open Problems

The previous arguments lead to some other problems concerning the relation between some natural subsets of $\mathbb{R}$ that we consider to be interesting.

For $d \in \mathbb{N}$ let $\mathbb{A}_d := \{x \in \mathbb{A} : \deg(x) \leq d\} \subset \mathbb{R}$ denote the set of algebraic numbers that have degree at most d over $\mathbb{Q}$.

Problem 1. *Is it true that each step in the following chain is strict?*

$$\mathbb{Q} \preceq \mathbb{A}_2 \preceq \mathbb{A}_3 \preceq \ldots \preceq \mathbb{A} \preceq \mathbb{H}_{\mathbb{R}} \ ?$$

We have defined $\mathbb{A}_d$ to consist of numbers of degree *less or* equal to d but point out that considering, rather than $\mathbb{A}_2 =: \mathbb{A}_{\leq 2}$, the set $\mathbb{A}_{=2} := \{x \in \mathbb{A} : \deg(x) = 2\}$ of numbers of degree *exactly* 2, in fact makes no difference:

Lemma 17. *It holds* $\mathbb{A}_{=2} \equiv \mathbb{A}_{\leq 2}$.

Proof. Based on oracle access to $\mathbb{A}_{\leq 2}$, decide $\mathbb{A}_{=2}$ in $\mathbb{R}$ as follows: Upon input of $x \in \mathbb{R}$, query $\mathbb{A}_{\leq 2}$ to find out whether $\deg(x) \leq 2$. If not, reject; otherwise $x \in \mathbb{A}$ and we may apply Lemma 6 to compute $\deg(x)$.

Conversely, given $\mathbb{A}_{=2}$ as an oracle, decide whether $x \in \mathbb{A}_{\leq 2}$ by querying both x and $y := x + \sqrt{2}$. If at least one of them belongs to $\mathbb{A}_{=2}$, then x is surely algebraic and thus applicable to Lemma 6. If $x, y \in \mathbb{R} \setminus \mathbb{A}_{=2}$, we may reject immediately because $\deg(x) < 2$ would imply $x \in \mathbb{Q}$ and thus $y = x + \sqrt{2} \in \mathbb{A}_{=2}$. $\qquad \square$

But what about this question for general degrees $d \in \mathbb{N}$?

Problem 2. *Is it true that for all $d \geq 2$ it holds $\mathbb{A}_{=d} \equiv \mathbb{A}_{\leq d}$?*

Another interesting question kindly pointed out to us by an anonymous referee would yield, in addition to $\mathbb{Q}$, a vast number of further problems strictly below $\mathbb{H}_{\mathbb{R}}$.

Problem 3. *Does $\mathbb{H}_{\mathbb{R}} \preceq A \subseteq \mathbb{R}^{\infty}$ imply that A is uncountable?*

Currently we do not see such a proof. And even if, the stronger result $A \not\preceq \mathbb{Q}$ still would remain.

4 Conclusion

We have shown that oracle access to the set of rational numbers $\mathbb{Q}$ gives a BSS machine additional power but still prevents it from solving the real Halting Problem $\mathbb{H}_{\mathbb{R}}$ (of BSS machines). In addition we proved that there is an uncountable number of incomparable recursively enumerable degrees in the real number setting.

Our proofs do not rely on the ordering available over the real numbers. Thus with small corrections (for example a slightly changed definition of the characteristic path in a potential decision tree) it also yields corresponding results over the complex numbers.

We close with some (necessarily speculative) remarks concerning the raising field of super-Turing computation, that is, concerning hypercomputers of various sorts. While their realizability is questionable and in fact denied by the Church-Turing Hypothesis, recent works put in turn this hypothesis into question [22,11,12]. Since a super-Turing computer capable of solving some problem P can also solve any $L \preceq P$, hypercomputers for higher (Turing-) degrees are necessarily more difficult to realize than for lower ones. Therefore, rather than trying to solve the Halting Problem, it seems more promising to go for some strictly easier yet undecidable one for a start. FRIEDBERG and MUCHNIK's solution P to Post's Problem would be a candidate for this approach, were it not for its inherent non-constructivity. In contrast and over the reals, we have explicitly revealed $\mathbb{Q}$ as an undecidable problem strictly easier than the Halting problem.

One might object that, since 'Natura non facit saltus' according to LEIBNIZ, the discontinuity inherent in deciding $\mathbb{Q}$ in $\mathbb{R}$ (i.e., of distinguishing fractions from general reals) makes an according hypercomputing device physically impossible. However we point out that for example the Fractional Quantum Hall Effect (Nobel Prize Physics 1998) shows that nature does exhibit exactly this kind of discontinuous behavior.

Acknowledgments. K.MEER is partially supported by the IST Programme of the European Community, under the PASCAL Network of Excellence, IST-2002-506778 and by the Danish Natural Science Research Council (SNF). M.ZIEGLER's stay in Odense was made possible by project 21-04-0303 of SNF. We thank the unknown referees for some useful comments.

References

1. S. BEN-DAVID, K. MEER, C. MICHAUX: "A note on non-complete problems in $NP_\mathbb{R}$", pp.324–332 in *Journal of Complexity* vol. **16**, no. 1 (2000).
2. L. BLUM, M. SHUB, S. SMALE: "On a Theory of Computation and Complexity over the Real Numbers: $\mathcal{NP}$-Completeness, Recursive Functions, and Universal Machines", pp.1–46 in *Bulletin of the American Mathematical Society* (AMS Bulletin) vol.**21** (1989).
3. L. BLUM, F. CUCKER, M. SHUB, S. SMALE: "*Complexity and Real Computation*", Springer (1998).
4. P. BÜRGISSER: "*Completeness and Reduction in Algebraic Complexity Theory*", Springer (2000).
5. D.G. CANTOR: "Irreducible Polynomials with Integral Coefficients have Succinct Certificates", pp.385–392 in *J. Algorithms* vol.**2** (1981).
6. O. CHAPUIS, P. KOIRAN: "Saturation and stability in the theory of computation over the reals", pp.1–49 in *Annals of Pure and Applied Logic*, vol.**99** (1999).
7. F. CUCKER: "The arithmetical hierarchy over the reals", pp.375–395 in *Journal of Logic and Computation* vol.**2(3)** (1992).
8. R.M. FRIEDBERG: "Two recursively enumerable sets of incomparable degrees of unsolvability", pp.236–238 in *Proc. Natl. Acad. Sci.* vol.**43** (1957).
9. J.D. HAMKINS, A. LEWIS: "Post's Problem for supertasks has both positive and negative solutions", pp.507–523 in *Archive for Mathematical Logic* vol.**4(6)** (2002).

10. G.T. HERMAN, S.D. ISARD: "Computability over arbitrary fields", pp.73–79 in *J. London Math. Soc.* vol.**2** (1970).

11. M.L. HOGARTH: "Non-Turing Computers and Non-Turing Computability", pp.126–138 in *Proc. Philosophy of Science Association* vol.**1** (1994).

12. T. KIEU: "Hypercomputation with Quantum Adiabatic Processes", pp.93–104 in *Theoretical Computer Science* **317** (2004).

13. R. LADNER: "On the structure of polynomial time reducibility", pp.155–171 in *Journal of the ACM*, vol. **22** (1975).

14. S. LANG: "*Algebra*", 3rd Edition Addison-Wesley (1993).

15. N. MACON, D.E. DUPREE: "Existence and Uniqueness of Interpolating Rational Functions", pp.751–759 in *The American Mathematical Monthly* vol.**69** (1962).

16. G. MALAJOVICH, K. MEER: "On the Structure of $NP_{\mathbb{C}}$", pp.27–35 in *SIAM Journal on Computing*, vol. **28**, no.1 (1999).

17. K. MEER: "Real Number Models under Various Sets of Operations", pp.366–372 in *J. Complexity* vol.**9** (1993).

18. E.L. POST: "Recursively enumerable sets of positive integers and their decision problems", pp.284–316 in *Bull. Amer. Math. Soc.* vol.**50** (1944).

19. W. RUDIN: "*Real and Complex Analysis*", McGraw-Hill (1966).

20. U. SCHÖNING: "A uniform approach to obtain diagonal sets in complexity classes", pp.95–103 in *Theoretical Computer Science*, vol.**18** (1982).

21. R.I. SOARE: "*Recursively Enumerable Sets and Degrees*", Springer (1987).

22. A. C.-C. YAO: "Classical Physics and the Church-Turing Thesis", pp.100–105 in *J. ACM* vol.**50(1)** (2003).

The Complexity of Semilinear Problems in Succinct Representation
(Extended Abstract)*

Peter Bürgisser[1,**], Felipe Cucker[2,***], and Paulin Jacobé de Naurois

[1] Dept. of Mathematics, University of Paderborn,
D-33095 Paderborn, Germany
pbuerg@upb.de
[2] Department of Mathematics, City University of Hong Kong,
83 Tat Chee Avenue, Hong Kong, P.R. of China
macucker@math.cityu.edu.hk
[3] LORIA, 615 rue du Jardin Botanique, BP 101,
54602 Villers-lès-Nancy Cedex, Nancy, France
Paulin.De-Naurois@loria.fr

Abstract. We prove completeness results for twenty-three problems in semilinear geometry. These results involve semilinear sets given by additive circuits as input data. If arbitrary real constants are allowed in the circuit, the completeness results are for the Blum-Shub-Smale additive model of computation. If, in contrast, the circuit is constant-free, then the completeness results are for the Turing model of computation. One such result, the $\mathsf{P}^{\mathsf{NP}[\log]}$-completeness of deciding Zariski irreducibility, exhibits for the first time a problem with a geometric nature complete in this class.

1 Introduction and Main Results

A subset $S \subseteq \mathbb{R}^n$ is *semilinear* if it is a Boolean combination of closed half-spaces $\{x \in \mathbb{R}^n \mid a_1 x_1 + \ldots + a_n x_n \leq b\}$. That is, S is derived from closed half-spaces by taking a finite number of unions, intersections, and complements.

The geometry of semilinear sets and its algorithmics has been a subject of interest for a long time not the least because of its close relationship with linear programming and its applications. This relationship is at the heart of many algorithmic results on both semilinear geometry and linear programming. It is also a good starting point to motivate the results in this paper.

Consider the feasibility problem for linear programming. That is, the problem of deciding whether a system of linear equalities and inequalities has a solution. A celebrated result by Khachijan [8] states that if the coefficients of these equalities

---

* A full version of this paper can be obtained at http://www-math.upb.de/agpb
** Partially supported by DFG grant BU 1371 and Paderborn Institute for Scientific Computation (PaSCo).
*** Partially supported by City University SRG grant 7001558.

M. Liśkiewicz and R. Reischuk (Eds.): FCT 2005, LNCS 3623, pp. 479–490, 2005.
© Springer-Verlag Berlin Heidelberg 2005

and inequalities are integers then this problem can be solved in polynomial time in the Turing machine model; that is, it belongs to the class P. If the coefficients are not integers but arbitrary real numbers, the Turing machine model is no longer appropriate. Instead, we analyze this version of the problem using the machine model over the real numbers introduced by Blum, Shub and Smale (the BSS model in the following). While it is not difficult to show that the linear programming feasibility problem over $\mathbb{R}$ is in $\mathrm{NP}_{\mathbb{R}} \cap \mathrm{coNP}_{\mathbb{R}}$ (this is merely Farkas' Lemma), or even that it can be solved in average polynomial time, its membership to $\mathrm{P}_{\mathbb{R}}$ (i.e., its solvability in deterministic polynomial time in the BSS model) remains an open problem. This membership problem has even been proposed by Smale as one of the mathematical problems for the 21st century [17].

A situation intermediate between the two above is the one in which the inequalities $a_1 x_1 + \ldots + a_n x_n \leq b$ have integer coefficients a_i and real right hand side b. In this case, the appropriate model of computation is the *additive model*. This is a restriction of the BSS model over $\mathbb{R}$ where multiplications and divisions are excluded from the capabilities of the machine. Only additions, subtractions and comparisons may be performed. The rephrasing of a well known result by Tardos [18] shows that the feasibility problem for a system of linear inequalities of the above mixed type is solvable in $\mathrm{P}_{\mathrm{add}}$.[1]

Equalities and inequalities of the mixed type we just described are not as rare as they may appear at a first glance. They naturally occur in the defining equations of semilinear sets given in *succinct representation*. Here, a semilinear set is given by an additive decision circuit (a more precise development follows in Section 2): a point $x \in \mathbb{R}^n$ is in the set if and only if the circuit returns 1 with input x. Since additive circuits are natural input data for additive machines one may wonder about the complexity of the feasibility problem $\mathrm{CSAT}_{\mathrm{add}}$ for semilinear sets in succinct representation. This problem consists of deciding whether the semilinear set S given by an additive circuit is nonempty. As it turns out, this problem is $\mathrm{NP}_{\mathrm{add}}$-complete [2]. This is in contrast with the result by Tardos mentioned above and is explained by the fact that an additive circuit of size $\mathcal{O}(n)$ can describe a semilinear set defined with $\mathcal{O}(2^n)$ linear inequalities.

The completeness result for $\mathrm{CSAT}_{\mathrm{add}}$ is not an isolated fact. It was recently shown [3] that several other problems for semilinear sets in succinct representation are complete in some complexity class. Notably, to decide whether the dimension of such a set is at least a given number is also $\mathrm{NP}_{\mathrm{add}}$-complete, to compute its Euler characteristic is $\mathrm{FP}_{\mathrm{add}}^{\#\mathrm{P}_{\mathrm{add}}}$-complete, and to compute any of its Betti numbers is $\mathrm{FPAR}_{\mathrm{add}}$-complete.

One of the goals of this paper is to further expand the catalogue of complete problems in semilinear geometry. We will show completeness for twenty three problems in this domain. These results, together with the previous results mentioned above, draw an accurate landscape of the difficulty of different prob-

---

[1] The reader may have noticed that we use the subscript "add" for complexity classes in the additive model, the subscript "$\mathbb{R}$" for those in the unrestricted BSS model, and no subscript at all for those in the Turing model. In addition, to emphasize the latter, we use sanserif fonts.

lems in semilinear geometry providing, at the same time, examples of natural complete problems for many of the complexity classes defined in the additive model.

A final remark is relevant. If an additive circuit has no constant gates (other than those with associated constant 0 or 1) it is said to be *constant-free*. Such a circuit can be described by means of a binary string and thus be taken as input by ordinary Turing machines. In this way, all problems considered in this paper have a discrete version fitting the classical complexity setting.

By checking our proofs one can see that all our completeness results hold for these discrete versions with respect to the corresponding discrete complexity classes.

We next briefly describe our main results. The precise definition of some concepts (e.g., Zariski topology) will be given later on this paper. The following list should give, however, an idea of the results we obtain. We consider the following problems related to topological properties of semilinear sets:

EADH$_{add}$ (*Euclidean Adherence*)   Given a decision circuit $\mathscr{C}$ with n input gates and a point $x \in \mathbb{R}^n$, decide whether x belongs to the Euclidean closure of the semilinear set $S_{\mathscr{C}} \subseteq \mathbb{R}^n$ described by $\mathscr{C}$.

ECLOSED$_{add}$(*Euclidean Closed*)   Given a decision circuit $\mathscr{C}$, decide whether $S_{\mathscr{C}}$ is closed under the Euclidean topology.

EDENSE$_{add}$(*Euclidean Denseness*)   Given a decision circuit $\mathscr{C}$ with n input gates, decide whether $S_{\mathscr{C}}$ is dense in $\mathbb{R}^n$.

UNBOUNDED$_{add}$ (*Unboundedness*)   Given a decision circuit $\mathscr{C}$ with n input gates, decide whether $S_{\mathscr{C}}$ is unbounded in $\mathbb{R}^n$.

COMPACT$_{add}$ (*Compactness*)   Given a decision circuit $\mathscr{C}$, decide whether $S_{\mathscr{C}}$ is compact.

ISOLATED$_{add}$ (*Isolatedness*)   Given a decision circuit $\mathscr{C}$ with n input gates and a point $x \in \mathbb{R}^n$, decide whether x is isolated in $S_{\mathscr{C}}$.

EXISTISO$_{add}$ (*Existence of Isolated Points*)   Given a decision circuit $\mathscr{C}$ with n input gates, decide whether there exists $x \in \mathbb{R}^n$ isolated in $S_{\mathscr{C}}$.

#ISO$_{add}$ (*Counting Isolated Points*)   Given a decision circuit $\mathscr{C}$, count the number of isolated points in $S_{\mathscr{C}}$.

LOCDIM$_{add}$ (*Local Dimension*)   Given a decision circuit $\mathscr{C}$, a point $x \in S_{\mathscr{C}}$ and an integer $d \in \mathbb{N}$, decide whether $\dim_x S_{\mathscr{C}} \geq d$.

LOCCONT$_{add}$ (*Local Continuity*)   Given an additive circuit $\mathscr{C}$ with n input gates and a point $x \in \mathbb{R}^n$, decide whether the function $F_{\mathscr{C}}$ computed by $\mathscr{C}$ is continuous at x (for the Euclidean topology).

CONT$_{add}$ (*Continuity*)   Given an additive circuit $\mathscr{C}$, decide whether $F_{\mathscr{C}}$ is continuous (for the Euclidean topology).

SURJ$_{add}$ (*Surjectivity*)   Given an additive circuit $\mathscr{C}$, decide whether $F_{\mathscr{C}}$ is surjective.

#DISC$_{add}$ (*Counting Discontinuities*)   Given an additive circuit $\mathscr{C}$, count the number of points in $\mathbb{R}^n$ where $F_{\mathscr{C}}$ is not continuous for the Euclidean topology.

REACH$_{add}$ (*Reachability*)   Given a decision circuit $\mathscr{C}$ with n input gates, and two points s and t in $\mathbb{R}^n$, decide whether s and t belong to the same connected component of $S_{\mathscr{C}}$.

CONNECTED$_{add}$ (*Connectedness*)   Given a decision circuit $\mathscr{C}$, decide whether $S_{\mathscr{C}}$ is connected.

TORSION$_{\mathrm{add}}$ (*Torsion*)   Given a decision circuit $\mathscr{C}$, decide whether the homology of $S_{\mathscr{C}}$ is torsion free.

ZADH$_{\mathrm{add}}$ (*Zariski Adherence*)   Given a decision circuit $\mathscr{C}$ with n input gates and a point $x \in \mathbb{R}^n$, decide whether x belongs to the Zariski closure of $S_{\mathscr{C}}$.

ZCLOSED$_{\mathrm{add}}$(*Zariski Closed*)   Given a decision circuit $\mathscr{C}$, decide whether $S_{\mathscr{C}}$ is closed under the Zariski topology.

ZDENSE$_{\mathrm{add}}$(*Zariski Denseness*)   Given a decision circuit $\mathscr{C}$ with n input gates, decide whether $S_{\mathscr{C}}$ is Zariski dense in $\mathbb{R}^n$.

IRR$_{\mathrm{add}}$(*Zariski Irreducibility*)   Given a decision circuit $\mathscr{C}$, decide whether the Zariski closure of $S_{\mathscr{C}}$ is affine.

\#IRR$_{\mathrm{add}}$ (*Counting Irreducible Components*)   Given a decision circuit $\mathscr{C}$, count the number of irreducible components of $S_{\mathscr{C}}$.

\#IRR$_{\mathrm{add}}^{(d)}$ (*Counting Irreducible Components of Fixed Dimension*)   Given a decision circuit $\mathscr{C}$, count the number of irreducible components of $S_{\mathscr{C}}$ of dimension d.

\#IRR$_{\mathrm{add}}^{[c]}$ (*Counting Irreducible Components of Fixed Codimension*)   Given a decision circuit $\mathscr{C}$, count the number of irreducible components of $S_{\mathscr{C}}$ of codimension c.

\#IRR$_{\mathrm{add}}^{\{N\}}$ (*Counting Irreducible Components in Fixed Ambient Space*)   Given a decision circuit $\mathscr{C}$ with a fixed number N of input gates, count the number of irreducible components of $S_{\mathscr{C}}$.

Our main results can be summarized in the following table. Here (T) means that the hardness is for Turing reductions. In what follows, unless specified otherwise, completeness will always mean completeness with respect to many-one reductions.

Problems	Complete in	Discrete version complete in
EADH$_{\mathrm{add}}$, ZADH$_{\mathrm{add}}$	NP$_{\mathrm{add}}$	NP
ECLOSED$_{\mathrm{add}}$, ZCLOSED$_{\mathrm{add}}$	coNP$_{\mathrm{add}}$	coNP
EDENSE$_{\mathrm{add}}$	coNP$_{\mathrm{add}}$	coNP
ZDENSE$_{\mathrm{add}}$	NP$_{\mathrm{add}}$	NP
UNBOUNDED$_{\mathrm{add}}$	NP$_{\mathrm{add}}$	NP
COMPACT$_{\mathrm{add}}$	coNP$_{\mathrm{add}}$	coNP
ISOLATED$_{\mathrm{add}}$	coNP$_{\mathrm{add}}$	coNP
LOCDIM$_{\mathrm{add}}$	NP$_{\mathrm{add}}$	NP
LOCCONT$_{\mathrm{add}}$, CONT$_{\mathrm{add}}$	coNP$_{\mathrm{add}}$	coNP
IRR$_{\mathrm{add}}$	$\mathrm{P}_{\mathrm{add}}^{\mathrm{NP}_{\mathrm{add}}[\log]}$	$\mathrm{P}^{\mathrm{NP}[\log]}$
EXISTISO$_{\mathrm{add}}$	Σ_{add}^2	$\Sigma_2\mathrm{P}$
SURJ$_{\mathrm{add}}$	Π_{add}^2	$\Pi_2\mathrm{P}$
\#ISO$_{\mathrm{add}}$, \#DISC$_{\mathrm{add}}$	$\mathrm{FP}^{\#\mathrm{P}_{\mathrm{add}}}$ (T)	$\mathrm{FP}^{\#\mathrm{P}}$ (T)
\#IRR$_{\mathrm{add}}$, \#IRR$_{\mathrm{add}}^{(d)}$, \#IRR$_{\mathrm{add}}^{[c]}$, \#IRR$_{\mathrm{add}}^{\{N\}}$	$\mathrm{FP}^{\#\mathrm{P}_{\mathrm{add}}}$ (T)	$\mathrm{FP}^{\#\mathrm{P}}$ (T)
REACH$_{\mathrm{add}}$, CONNECTED$_{\mathrm{add}}$	PAR$_{\mathrm{add}}$ (T)	PSPACE

We remark that the Zariski topology and irreducible components are natural concepts studied in algebraic geometry [16]. In particular, we show that the problem to test irreducibility of a semilinear set given by a constant-free decision circuit is complete for the class $\mathrm{P}^{\mathrm{NP}[\log]}$. The latter class was first studied by Papadimitriou and Zachos [14] and consists of the decision problems that can be solved in polynomial time by $\mathcal{O}(\log n)$ queries to some NP language. Equivalently, $\mathsf{P}^{\mathsf{NP}[\log]}$ can also be characterized as the set of languages in P^{NP} whose queries are *non adaptive*, cf. [13, Th. 17.7]. This means that the input to any query does not depend on the oracle answer to previous queries, but only on the input of the machine. Several natural complete problems for $\mathsf{P}^{\mathsf{NP}[\log]}$ are known, see for instance [10,7].

For the problem TORSION$_{\mathrm{add}}$ we prove PAR$_{\mathrm{add}}$-hardness (with respect to Turing reductions) and membership in EXP$_{\mathrm{add}}$ (PSPACE-hardness and membership in EXP for its discrete version). This advances towards determining the complexity of TORSION$_{\mathrm{add}}$, a question left open in [3, §7]. Also, the PAR$_{\mathrm{add}}$-completeness of CONNECTED$_{\mathrm{add}}$ closes a question left open therein.

2 Preliminaries

We next review the notions which will be central in this paper, fixing notations at the same time. A basic reference (since this paper is an extension of it) is [3].

(1) The Euclidean norm in $\mathbb{R}^n$ induces a topology, called *Euclidean*, in $\mathbb{R}^n$. We will denote the closure of a subset $S \subseteq \mathbb{R}^n$ with respect to the Euclidean topology by $\overline{S}$. Following [16], we define another, coarser, topology in $\mathbb{R}^n$, hereby restricting us to semilinear sets.

Definition 1. *We call a semilinear set $S \subseteq \mathbb{R}^n$ Zariski closed if it is a finite union of affine subspaces of $\mathbb{R}^n$. The Zariski closure of a semilinear set $V \subseteq \mathbb{R}^n$, denoted by $\overline{V}^Z$, is the smallest Zariski-closed semilinear subset of $\mathbb{R}^n$ containing V.*

The use of the words "closed" or "closure" is appropriate: the semilinear Zariski-closed sets satisfy the axioms of the closed sets of a topology on $\mathbb{R}^n$.

We will use the sign functions $\mathsf{sg} : \mathbb{R} \to \{-1, 0, 1\}, \mathsf{pos} : \mathbb{R} \to \{0, 1\}$ defined by $\mathsf{sg}(x) = \mathsf{pos}(x) = 1$ if $x > 0$, $\mathsf{sg}(0) = 0, \mathsf{pos}(0) = 1$, and $\mathsf{sg}(x) = -1, \mathsf{pos}(x) = 0$ if $x < 0$. We extend these functions to $\mathbb{R}^n$ componentwise. A *quadrant* of $\mathbb{R}^n$ is an open subset of $\mathbb{R}^n$ of the form $\{x \in \mathbb{R}^n \mid \mathsf{sg}(x) = \sigma\}$ for some $\sigma \in \{-1, 1\}^n$.

(2) We next recall a few facts concerning additive circuits. Such circuits are defined in many places [2,3,9]. An *additive circuit* is a directed acyclic graph whose nodes are of one of the following types: input, output, constant, addition, substraction, and selection. The first four types of node have an obvious semantics; selection nodes have four inputs v, a, b, c and return a if $v > 0$, b if $v = 0$ and c otherwise.

An additive circuit $\mathscr{C}$ with n input nodes and m output nodes computes a function $F_{\mathscr{C}} : \mathbb{R}^n \to \mathbb{R}^m$. A *decision circuit* $\mathscr{C}$ is an additive circuit with exactly one output node that is preceded by a selection node with $a, b, c \in \{0, 1\}$. Such a circuit computes a function $F_{\mathscr{C}} : \mathbb{R}^n \to \{0, 1\}$ and decides the semilinear set $S_{\mathscr{C}} := \{x \in \mathbb{R}^n \mid F_{\mathscr{C}}(x_1, \ldots, x_n) = 1\}$. We say that $S_{\mathscr{C}}$ is given in *succinct representation*.

Definition 2. *Let $\mathscr{C}$ be a decision circuit with r selection gates and n input gates. A path γ of $\mathscr{C}$ is an element in $\{-1,0,1\}^r$. We say that $x \in \mathbb{R}^n$ follows a path γ of $\mathscr{C}$ if, on input x and for all j, the result of the test performed at the j-th selection gate is γ_j (i.e., $\gamma_j = -1$ if the tested value v satisfies $v < 0$, $\gamma_j = 0$ if $v = 0$, and $\gamma_j = 1$ if $v > 0$). The leaf set of a path γ is defined as*

$$D_\gamma = \{x \in \mathbb{R}^n \mid \text{ input } x \text{ follows the path } \gamma \text{ of } \mathscr{C}\}.$$

A path γ is accepting if and only if we have $F_\mathscr{C}(x) = 1$ for one (and hence for all) $x \in D_\gamma$. We denote by $\mathcal{A}_\mathscr{C}$ the set of accepting paths of the circuit $\mathscr{C}$.

(3)   We finally recall some notions of computation and complexity. In this paper we use additive machines (i.e., BSS machines over $\mathbb{R}$ which do not multiply or divide) as described in [2, Ch. 18] or in [9]. For these machines, versions of the usual complexity classes are defined yielding the classes $\mathrm{P_{add}}$, $\mathrm{NP_{add}}$, $\#\mathrm{P_{add}}$, $\mathrm{PAR_{add}}$, $\mathrm{EXP_{add}}$, and $\mathrm{FP_{add}}$ (note that the additive version of polynomial space requires instead polynomial parallel time). An overview of these classes and their properties can be found in [2, Ch. 18] and [3].

We already defined the problem $\mathrm{CSAT_{add}}$ and observed that it is $\mathrm{NP_{add}}$-complete. The following two problems are also $\mathrm{NP_{add}}$-complete:

$\mathrm{CBS_{add}}$ (*Circuit Boolean Satisfiability*)   Given a decision circuit $\mathscr{C}$ with n input gates, decide whether there exists $x \in \{0,1\}^n$ such that $\mathscr{C}(x) = 1$.

$\mathrm{DIM_{add}}$ (*Dimension*)   Given a decision circuit $\mathscr{C}$ with n input gates and $k \in \mathbb{N}$, decide whether the dimension of $S_\mathscr{C}$ is greater than or equal to k.

For $\mathrm{DIM_{add}}$ this follows easily from [3, Theorem 5.1] (there k is assumed to be fixed, but the proof carries over easily). Note that $\mathrm{CBS_{add}}$ deals with a digital form of nondeterminism since it requires the circuit to be satisfied by a point in $\{0,1\}^n$.

The $\mathrm{NP_{add}}$-completeness of $\mathrm{CBS_{add}}$ allows us to use a problem with a discrete flavor to prove completeness results in the additive setting. More generally, a series of results starting in [6], continued in [3], and relying on Meyer auf der Heide [11], allow us to use standard discrete problems as basis for reductions yielding Turing-hardness results in the additive setting.

We finish these preliminaries with a lemma gathering several facts which will be used later on in many proofs.

Lemma 1. *Given a decision circuit $\mathscr{C}$, two paths γ, γ' of $\mathscr{C}$, and a point $x \in \mathbb{R}^n$, the following tasks can be performed by an additive machine in time polynomial in the size of $\mathscr{C}$:*

  (i) *Decide whether D_γ is nonempty.*
  (ii) *Decide whether $x \in \overline{D_\gamma}$, or decide whether $x \in \overline{D_\gamma}^Z$.*
  (iii) *Compute $\dim D_\gamma$.*
  (iv) *Decide whether $\overline{D_\gamma}^Z \subseteq \overline{D_{\gamma'}}^Z$.*

3   Some Proofs

In this section we give some proofs to convey an idea of our techniques.

3.1   Basic Topology

Proposition 1. *The problem* $\mathrm{ZDENSE_{add}}$ *is* $\mathrm{NP_{add}}$*-complete.*

PROOF.   Note that $\overline{S_{\mathscr{C}}}^{Z} = \bigcup_{\gamma \in \mathcal{A}_{\mathscr{C}}} \overline{D_{\gamma}}^{Z}$. Therefore, $\overline{S_{\mathscr{C}}}^{Z} = \mathbb{R}^{n}$ if and only if there exists $\gamma \in \mathcal{A}_{\mathscr{C}}$ such that D_{γ} is Zariski dense in $\mathbb{R}^{n}$. Since $\overline{D_{\gamma}}^{Z}$ is the affine hull of D_{γ} (if $D_{\gamma} \neq \emptyset$), we see that D_{γ} is Zariski dense in $\mathbb{R}^{n}$ if and only if $\dim D_{\gamma} = n$. Hence, S is Zariski dense in $\mathbb{R}^{n}$ if and only if $\dim S = n$. The membership to $\mathrm{NP_{add}}$ now follows from the fact that $\mathrm{DIM_{add}}$ is in $\mathrm{NP_{add}}$.

For proving the hardness, we reduce $\mathrm{CBS_{add}}$ to $\mathrm{ZDENSE_{add}}$. Assume $\mathscr{C}$ is a decision circuit with n input gates. Consider a circuit $\mathscr{C}'$ computing the function

$$G_{\mathscr{C}} : \mathbb{R}^{n} \to \{0,1\}, \quad x \mapsto F_{\mathscr{C}}(\mathsf{pos}(x)). \tag{1}$$

The mapping $\mathscr{C} \mapsto (\mathscr{C}', 0)$ reduces $\mathrm{CBS_{add}}$ to $\mathrm{ZDENSE_{add}}$. Indeed, if $S_{\mathscr{C}} \cap \{0,1\}^{n} = \emptyset$ then $S_{\mathscr{C}'} = \emptyset$ as well and hence $0 \notin \overline{S_{\mathscr{C}'}}$. On the other hand, if $S_{\mathscr{C}} \cap \{0,1\}^{n} \neq \emptyset$ then $S_{\mathscr{C}'}$ contains at least one quadrant and hence $0 \in \overline{S_{\mathscr{C}'}}$. □

The following result is proved with similar arguments.

Proposition 2. *The problem* $\mathrm{EXISTISO_{add}}$ *is* $\Sigma^{2}_{\mathrm{add}}$*-complete.*

3.2   Zariski Irreducibility

Irreducibility is a natural concept in algebraic geometry [16]. For semilinear sets this notion can be defined as follows.

Definition 3. *A semilinear set* $S \subseteq \mathbb{R}^{n}$ *is* Zariski-irreducible *if its Zariski closure is an affine space. The Zariski closure of a semilinear set* $S \subseteq \mathbb{R}^{n}$ *is a nonredundant finite union of affine subspaces* $A_{1}, \ldots, A_{s}$ *of* $\mathbb{R}^{n}$*. We call* $A_{1}, \ldots, A_{s}$ the irreducible components of $\overline{S}^{Z}$ *and call the sets* $S \cap A_{i}$ the irreducible components *of* S.

We extend the definition of $\mathrm{P^{NP[log]}}$ to the additive setting in the obvious way thus obtaining the class $\mathrm{P_{add}^{NP_{add}[log]}}$. Again, it is not difficult to show that this class can also be characterized as the set of decision problems solvable in additive polynomial time with non adaptive queries to $\mathrm{NP_{add}}$.

The main result of this section is the following.

Theorem 1. *The problem* $\mathrm{IRR_{add}}$ *is* $\mathrm{P_{add}^{NP_{add}[log]}}$*-complete.*

We first prove the upper bound.

Lemma 2. *The problem* $\mathrm{IRR_{add}}$ *is in* $\mathrm{P_{add}^{NP_{add}[\log]}}$.

PROOF.   Consider the following algorithm:

> input $\mathscr{C}$ with n input gates
> for $k = -1, \ldots, n$ (independently) do
> > (i)   check whether $\dim S_{\mathscr{C}} \geq k$
> > (ii) check whether $\forall \gamma, \gamma' \in \mathcal{A}_{\mathscr{C}}$ $(\dim D_{\gamma'} = k \Rightarrow \overline{D_\gamma}^Z \subseteq \overline{D_{\gamma'}}^Z)$
> let $d = \max\{k : $ (i) holds $\}$
> if (ii) holds for $k = d$ then ACCEPT else REJECT

This algorithm decides whether $S_{\mathscr{C}}$ is Zariski irreducible. Indeed, the dimension d of $S_{\mathscr{C}}$ is computed, and the query (ii) for $k = d$ checks whether for all leaf sets $D_{\gamma'}$ of dimension d we have $\overline{S_{\mathscr{C}}}^Z = \overline{D_{\gamma'}}^Z$. This holds if and only if $S_{\mathscr{C}}$ is Zariski irreducible.

Since $\mathrm{DIM_{add}}$ is known to be in $\mathrm{NP_{add}}$ [3], (i) is a query to a problem in $\mathrm{NP_{add}}$. By Lemma 1, (ii) is a query to a problem in $\mathrm{coNP_{add}}$. Since the queries are nonadaptive and the algorithm runs in polynomial time, the set $\mathrm{IRR_{add}}$ is in $\mathrm{P_{add}^{NP_{add}[\log]}}$. $\square$

Lemma 3. **(i)** *Let $S_1 \subseteq \mathbb{R}^n$ and $S_2 \subseteq \mathbb{R}^m$ be two non-empty semilinear sets. Then, $S_1 \times S_2 \subseteq \mathbb{R}^{n+m}$ is irreducible if and only if both S_1 and S_2 are irreducible.*
(ii) *A nonempty union of reducible semilinear sets is reducible.* $\square$

We turn now to the proof of the lower bound in Theorem 1.

Lemma 4. *The problem* $\mathrm{IRR_{add}}$ *is* $\mathrm{P_{add}^{NP_{add}[\log]}}$*-hard under many-one reductions.*

PROOF.   Assume L is a problem in $\mathrm{P_{add}^{NP_{add}[\log]}}$. Then we may assume that L is decided by a polynomial time additive machine asking non adaptively a polynomial number of queries to the $\mathrm{NP_{add}}$-complete problem $\mathrm{ZDENSE_{add}}$. Hence, there exists a polynomial p and, for all $n \in \mathbb{N}$, a polynomial size circuit $\mathscr{C}^n$ with $n + p(n)$ input gates and a family of polynomial size circuits $\mathscr{C}_1^n, \ldots, \mathscr{C}_{p(n)}^n$ with n input gates, such that, for $x \in \mathbb{R}^n$, x is in L if and only if $F_{\mathscr{C}^n}(x, s) = 1$, where $s = (s_1, \ldots, s_{p(n)})$ denotes the *sequence of oracle answers* for the input x, that is $s_i = 1$ if the output of $\mathscr{C}_i^n$ on input x is in $\mathrm{ZDENSE_{add}}$ and $s_i = 0$ otherwise. Thus the circuits $\mathscr{C}_i^n$ compute the inputs to the oracle queries and $\mathscr{C}^n$ performs the final computation deciding the membership of x to L, given the sequence s of oracle answers.

The output $\mathcal{E}_i^n$ of $\mathscr{C}_i^n$ on input x is an input to $\mathrm{ZDENSE_{add}}$. Thus $\mathcal{E}_i^n$ is a (description of a) decision circuit defining a semilinear set, which we denote by $S_i \subseteq \mathbb{R}^{r(n)}$. (Without loss of generality, we may assume that all these sets lie in a Euclidean space of the same dimension $r(n) > 1$ and that all the circuits $\mathcal{E}_i^n$ use the same number of selection gates $q(n) > 1$.) We denote by $\mathcal{A}_i$ the set of accepting paths of $\mathcal{E}_i^n$. Moreover, for $\gamma \in \mathcal{A}_i$, we denote by $D_{\gamma i} \subseteq S_i$ the corresponding leaf set, and write $\partial D_{\gamma i}$ for its Euclidean boundary.

The reduction (1) from the proof of Proposition 1 that reduces CBS_{add} to $ZDENSE_{add}$ produces either a Zariski dense or an empty set. Moreover, the leaf sets produced by this reduction are, up to boundary points, quadrants of $\mathbb{R}^{r(n)}$. Taking this into account, we may therefore assume without loss of generality that S_i is either empty or Zariski dense in $\mathbb{R}^{r(n)}$, for all $x \in \mathbb{R}^n$ and all i. Moreover, we may assume that (recall $r(n) > 1$)

$$S_i \neq \emptyset \Longrightarrow \bigcup_{\gamma \in \mathcal{A}_i} \partial D_{\gamma i} \text{ is reducible.} \tag{2}$$

Our goal is to reduce L to IRR_{add}. Thus we have to compute from $x \in \mathbb{R}^n$, in polynomial time, a decision circuit defining a semilinear set Ω such that $x \in L$ iff Ω is irreducible. We will consider $x \in \mathbb{R}^n$ as fixed and suppress it notationally. To simplify notation, we will write $p := p(n), q := q(n), r := r(n)$ for fixed $x \in \mathbb{R}^n$.

The set Ω will be a set of tuples (u, y, a) in the Euclidean space $\Pi := \mathbb{R}^q \times (\mathbb{R}^r)^p \times \mathbb{R}^p$. To convey an idea of the intended meaning, we call $u \in \mathbb{R}^q$ *selection gate vector*, $y = (y_1, \ldots, y_p) \in (\mathbb{R}^r)^p$ *oracle vector*, and $a \in \mathbb{R}^p$ *oracle answer vector*. A selection gate vector u induces a discrete vector $\gamma := \mathsf{sg}(u) \in \{-1, 0, 1\}^q$, which describes a possible path of one of the circuits $\mathcal{E}_i^n$. An oracle answer vector a induces a bit vector $\alpha := \mathsf{pos}(a) \in \{0, 1\}^p$, which describes a possible sequence of oracle answers. The set Ω will be a finite union of *product sets* of the form $U \times Y_1 \times \cdots \times Y_p \times A \subseteq \Pi$, where $U \subseteq \mathbb{R}^q$, $Y_i \subseteq \mathbb{R}^r$, and $A \subseteq \mathbb{R}^p$ are semilinear sets. Note that, by Lemma 3, a nonempty product set is irreducible iff all U, Y_i, A are irreducible and nonempty.

Let z be a fixed point in $\mathbb{R}^r$ (for instance the origin). Recall that $s \in \{0, 1\}^p$ denotes the sequence of oracle answers for the fixed input x. We define the subsets $T_i := S_i \cup \{z\} \subseteq \mathbb{R}^r$, for which we make the following important observation:

$$\begin{aligned}
s_i = 1 &\Longleftrightarrow \overline{S_i}^Z = \mathbb{R}^r \Longleftrightarrow \overline{T_i}^Z = \mathbb{R}^r, \\
s_i = 0 &\Longleftrightarrow S_i = \emptyset \Longleftrightarrow \overline{T_i}^Z = \{z\}.
\end{aligned} \tag{3}$$

We define the set $\Omega \subseteq \Pi$ as the one accepted by the following algorithm:

> input $(u, y, a) \in \mathbb{R}^q \times (\mathbb{R}^r)^p \times \mathbb{R}^p$
> compute $\gamma := \mathsf{sg}(u) \in \{-1, 0, 1\}^q$, $\alpha := \mathsf{pos}(a) \in \{0, 1\}^p$
> (I)   case $(\forall i \ y_i \in T_i) \wedge (\exists i \ a_i = 0)$ ACCEPT
> (II) case $(F_{\mathscr{C}^n}(x, \alpha) = 1) \wedge (\forall i \ y_i \in T_i) \wedge \exists j \ (\alpha_j = 0 \wedge \gamma \in \mathcal{A}_j \wedge y_j \in \partial D_{\gamma j})$
>   ACCEPT
> (III) case $(F_{\mathscr{C}^n}(x, \alpha) = 1) \wedge \forall i \ ((\alpha_i = 0 \Longrightarrow y_i = z) \wedge (\alpha_i = 1 \Longrightarrow y_i \in S_i))$
>   ACCEPT
>   else REJECT.

It is easy to see that an additive circuit formalizing the above algorithm can be computed from the given $x \in \mathbb{R}^n$ in polynomial time by an additive machine. (Use that a description of the circuits $\mathscr{C}^n, \mathscr{C}_i^n$ can be computed from n by an additive machine in polynomial time.)

To prove the lemma, it is sufficient to show the following assertion:

$$x \in L \Longleftrightarrow \Omega \text{ is irreducible.} \tag{4}$$

In order to show this we are going to analyze the set Ω. We define

$$\Omega_{\mathrm{I}} = \{(u, y, a) \in \Pi \mid (u, y, a) \text{ satisfies Case (I)}\}$$

and similarly Ω_{II} and Ω_{III}. Note that Ω_{II} is not the set of (u, y, a) accepted by the step (II) of the algorithm. We have $\Omega = \Omega_{\mathrm{I}} \cup \Omega_{\mathrm{II}} \cup \Omega_{\mathrm{III}}$, but this union is not necessarily disjoint. It is obvious that Ω_{I} is reducible.

We introduce some more notation needed for analyzing the above algorithm. Consider the following subset

$$\mathcal{Y} := \{\alpha \in \{0, 1\}^p \mid F_{\mathscr{C}^n}(x, \alpha) = 1\}$$

of possible oracle answer sequences leading to acceptance. Note that $s \in \mathcal{Y}$ iff $x \in L$. Moreover, define for $\alpha \in \mathcal{Y}$ the following set of indices

$$J(\alpha) := \{j \mid \alpha_j = 0 \wedge s_j = 1\}$$

and for $j \in J(\alpha)$ let $\Omega_{\mathrm{II}}^j(\alpha)$ denote the set of $(u, y, a) \in \Pi$ that satisfy the condition of Case (II) with the α and j specified. Similarly, we define $\Omega_{\mathrm{III}}(\alpha)$. We have

$$\Omega = \Omega_{\mathrm{I}} \cup \bigcup_{\alpha \in \mathcal{Y}, j \in J(\alpha)} \left(\Omega_{\mathrm{II}}^j(\alpha) \cup \Omega_{\mathrm{III}}(\alpha)\right). \tag{5}$$

The following claim settles one direction of (4).

Claim A. If $x \in L$, then Ω is irreducible.

In order to prove this claim, note that $\Omega_{\mathrm{III}}(s) = \mathbb{R}^q \times F_1 \times \cdots \times F_p \times \mathsf{pos}^{-1}(s)$, where we have put $F_i := S_i$ if $s_i = 1$ and $F_i := \{z\}$ otherwise. This implies that

$$\overline{\Omega_{\mathrm{III}}(s)}^Z = \mathbb{R}^q \times \overline{T_1}^Z \times \cdots \times \overline{T_p}^Z \times \mathbb{R}^p =: \Theta,$$

since $\overline{\mathsf{pos}^{-1}(s)}^Z = \mathbb{R}^p$. The product set Θ is irreducible by Lemma 3(i) and (3). It is clear that $\Omega_{\mathrm{I}} \cup \Omega_{\mathrm{II}} \subseteq \Theta$. Moreover, we claim that $\Omega_{\mathrm{III}}(\alpha) \subseteq \Theta$ for all $\alpha \in \mathcal{Y}$. Indeed, assume $(u, y, a) \in \Omega_{\mathrm{III}}(\alpha)$. If we had $s_i = 0$ and $\alpha_i = 1$ for some i, then we would have $y_i \in S_i$, which contradicts the fact that $S_i = \emptyset$ due to $s_i = 0$. This shows that $(u, y, a) \in \Theta$.

Altogether, using (5), we have shown that $\Omega \subseteq \Theta$. Hence $\overline{\Omega}^Z = \Theta$, which finishes the proof of Claim A.

Claim B. For $\alpha \in \mathcal{Y} \setminus \{s\}$, $j \in J(\alpha)$, the set $\Omega_{\mathrm{II}}^j(\alpha) \cup \Omega_{\mathrm{III}}(\alpha)$ is reducible.

Claim B implies the other direction of the assertion (4). Indeed, assume $x \notin L$. Then $s \notin \mathcal{Y}$ and according to (5), Ω is a union of reducible sets and thus reducible.

It remains to prove Claim B. Let $\pi_j \colon \Pi \to \mathbb{R}^r$, $(u, y, a) \to y_j$ be the projection onto the jth factor. In order to show that a subset $\Omega' \subseteq \Pi$ is reducible, it is sufficient to prove that $\pi_j(\Omega')$ is reducible, since irreducibility is preserved by linear maps. Hence it is enough to show that $\pi_j\left(\Omega_{\mathrm{II}}^j(\alpha) \cup \Omega_{\mathrm{III}}(\alpha)\right)$ is reducible.

Taking into account (2) and the fact that $j \in J(\alpha)$ implies $S_j \neq \emptyset$, it suffices to prove that

$$\bigcup_{\gamma \in \mathcal{A}_j} \partial D_{\gamma j} \subseteq \pi_j \left(\Omega_{\mathrm{II}}^j(\alpha) \cup \Omega_{\mathrm{III}}(\alpha) \right) \subseteq \{z\} \cup \bigcup_{\gamma \in \mathcal{A}_j} \partial D_{\gamma j}.$$

The second inclusion is clear since $j \in J(\alpha)$ and thus $\alpha_j = 0$.

For the first inclusion, assume $y_j \in \partial D_{\gamma j}$ for some $\gamma \in \mathcal{A}_j$. Choose $a \in \mathbb{R}^p$ and $u \in \mathbb{R}^q$ such that $\mathsf{pos}(a) = \alpha$ and $\mathsf{sg}(u) = \gamma$. Then $(u, z, \ldots, z, y_j, z, \ldots, z, a) \in \Omega_{\mathrm{II}}^j(\alpha)$, where the y_j is at the jth position. Hence $y_j \in \pi_j \left(\Omega_{\mathrm{II}}^j(\alpha) \cup \Omega_{\mathrm{III}}(\alpha) \right)$. This finishes the proof of Claim B and completes the proof of the lemma.   □

3.3 Problems of Connectivity

The proof of the following result is inspired by a similar result for graphs in [4].

Theorem 2. *The problem* CONNECTED$_{\mathrm{add}}$ *is* PAR$_{\mathrm{add}}$-*complete under Turing reductions. The same holds when restricted to problems in* $\mathbb{R}^3$.

In [3] it was shown that, for all $k \in \mathbb{N}$, the problem to compute the kth Betti number of the semilinear set given by an additive circuit is FPAR$_{\mathrm{add}}$-complete and the question was raised whether this holds also for the problem of computing the torsion subgroup of the homology group $H_k(X; \mathbb{Z})$. We give a partial answer to this question by showing that this problem is in fact FPAR$_{\mathrm{add}}$-hard. Hereby we focus on the problem TORSION$_{\mathrm{add}}$ of deciding whether the torsion subgroups $T_k(S_{\mathscr{C}})$ of a semilinear set $S_{\mathscr{C}}$ given by a circuit vanish for all k, that is, whether all the homology groups $H_k(S_{\mathscr{C}}; \mathbb{Z})$ are free abelian groups. The question of the corresponding upper bound remains open, but at least we show that the problem is in EXP$_{\mathrm{add}}$.

Theorem 3. *The problem* TORSION$_{\mathrm{add}}$ *is* PAR$_{\mathrm{add}}$-*hard under Turing reductions and belongs to* EXP$_{\mathrm{add}}$.

For the lower bound proof, we start with the reduction in the proof of Theorem 2, which reduces any language L in PSPACE to CONNECTED$_{\mathrm{add}}$ by mapping a bit string x to a decision circuit describing a semilinear set $S'_n \subseteq \mathbb{R}^3$ such that S'_n is connected iff $x \in L$. Then we extend this construction by modifying the space $S'_n \times [0, 1]$ roughly by building in a Moebius strip and identifying the boundary lines of the resulting space.

4 Open Problems

Let the semilinear set $S_{\mathscr{C}}$ be given by a constant-free decision circuit $\mathscr{C}$. We remark that the problem to test simple connectivity of $S_{\mathscr{C}}$ is undecidable. This follows by reducing to it the group triviality problem, which is known to be undecidable [1,15].

We propose as open problems to determine the complexity of the following topological properties: Is $S_{\mathscr{C}}$ is a topological manifold? Is $S_{\mathscr{C}}$ contractible?

Acknowledgements

We thank the anonymous referees for clarifying an issue regarding the subtle difference between $P^{NP[\log]}$ and the corresponding function class, as well pointing out to us some references to natural problems for $P^{NP[\log]}$.

References

1. S.I. Adian. Unsolvability of certain algorithmic problems in the theory of groups (in Russian). *Trudy Moskov. Math. Obshch.*, 6:231–298, 1957.
2. L. Blum, F. Cucker, M. Shub, and S. Smale. *Complexity and Real Computation*. Springer-Verlag, 1998.
3. P. Bürgisser and F. Cucker. Counting complexity classes for numeric computations I: Semilinear sets. *SIAM J. Comp.*, 33:227–260, 2004.
4. A. Chandra, L. Stockmeyer, and U. Vishkin. Constant depth reducibility. *SIAM J. Comp.*, 13:423–439, 1984.
5. F. Cucker and P. Koiran. Computing over the reals with addition and order: Higher complexity classes. *Journal of Complexity*, 11:358–376, 1995.
6. H. Fournier and P. Koiran. Are lower bounds easier over the reals?, In *Proc. 30th ACM STOC*, pages 507–513, 1998.
7. E. Hemaspaandra and L.A. Hemaspaandra and J. Rothe. Exact Analysis of Dodgson Elections: Lewis Carroll's 1876 Voting System is Complete for Parallel Access to NP. *Journal of the ACM*, pages 806–825, 1997.
8. L.G. Khachijan. A polynomial algorithm in linear programming. *Dokl. Akad. Nauk SSSR*, 244:1093–1096, 1979. (In Russian, English translation in *Soviet Math. Dokl.*, 20:191–194, 1979.)
9. P. Koiran. Computing over the reals with addition and order. *Theoretical Computer Science*, 133:35–47, 1994.
10. M. W. Krentel. The complexity of optimization problems. In *Proc. 18th ACM Symp. on the Theory of Computing*, pages 79–86, 1986.
11. F. Meyer auf der Heide. A polynomial linear search algorithm for the n-dimensional knapsack problem. *J. ACM*, 31:668–676, 1984.
12. C.H. Papadimitriou. On the complexity of unique solutions. *J. ACM*, 31:392–400, 1984.
13. C.H. Papadimitriou. *Computational Complexity*. Addison-Wesley, 1994.
14. C.H. Papadimitriou and S. Zachos. Two remarks on the power of counting. *Proc. 6th GI conference in Theoretical Computer Science*, Lecture Notes in Computer Science 145, Springer Verlag, pages 269–276, 1983.
15. M. Rabin. Recursive unsolvability of group theoretic problems. *Ann. of Math.*, 67(2):172–194, 1958.
16. I.R. Shafarevich. *Basic Algebraic Geometry*. Springer Verlag, 1974.
17. S. Smale. Mathematical problems for the next century. *Mathematical Intelligencer*, 20:7–15, 1998.
18. E. Tardos. A strongly polynomial algorithm to solve combinatorial linear programs. *Oper. Res.*, 34:250–256, 1986.

On Finding Acyclic Subhypergraphs[*]

Kouichi Hirata[1], Megumi Kuwabara[2], and Masateru Harao[1]

[1] Department of Artificial Intelligence
[2] Graduate School of Computer Science and Systems Engineering,
Kyushu Institute of Technology,
Kawazu 680-4, Iizuka 820-8502, Japan
{hirata, harao}@ai.kyutech.ac.jp
megumik@dumbo.ai.kyutech.ac.jp

Abstract. In this paper, we investigate the problem of finding *acyclic* subhypergraphs in a hypergraph. First we show that the problem of determining whether or not a hypergraph has a *spanning connected acyclic subhypergraph* is NP-complete. Also we show that, for a given $K > 0$, the problem of determining whether or not a hypergraph has *an acyclic subhypergraph containing at least K hyperedges* is NP-complete. Next, we introduce a *maximal* acyclic subhypergraph, which is an acyclic subhypergraph that is cyclic if we add any hyperedge of the original hypergraph to it. Then, we design the linear-time algorithm *mas* to find it, which is based on the *acyclicity test algorithm* designed by Tarjan and Yannakakis (1984).

1 Introduction

Hypergraphs [2,3] have developed as one of the mathematical tools for characterizing the queries in Database Theory [1,7,14,15,16] in 1980's. Today, it is known that the hypergraphs are related to the several problems for Artificial Intelligence, for example, constraint satisfaction, clause subsumption, theory revision, abductive explanation, machine learning, data mining, and so on [5,6,9,11,12].

The *acyclicity*, which is not only defined on hypergraphs but also extended to several related problems, is the property that makes some intractable problems tractable [4,5,6,9,10]. For example, while the *evaluation problem for conjunctive queries* on databases is NP-complete in general, it is LOGCFL-complete if a conjunctive query is *acyclic* [9]. Furthermore, while the *clause subsumption* is NP-complete in general, it is also LOGCFL-complete if a clause is *acyclic* [9], and, while the *minimum condensation* is not polynomial-time approximable unless NP=ZPP, it is solvable in polynomial time if a clause is *acyclic* [4]. Here, the problem of determining whether or not a hypergraph (and the related concept) is acyclic is solvable in linear time [15,16] and in symmetric log space [9].

[*] This work is partially supported by Grand-in-Aid for Scientific Research 15700137 and 16016275 from the Ministry of Education, Culture, Sports, Science and Technology, Japan.

M. Liśkiewicz and R. Reischuk (Eds.): FCT 2005, LNCS 3623, pp. 491–503, 2005.

One direction to use the advantage for the acyclicity is to formulate the extended concept of the acyclicity. Hypergraphs with *bounded tree-width, bounded query-width* and *bounded hypertree-width* are examples of such a formulation [4,10]. In all cases, the acyclicity is corresponding to having the tree-width 1, the query-width 1 or the hypertree-width 1, and the related several problems on them are tractable with respect to the bounded width.

In this paper, we pay our attention to another direction to use the advantage for the acyclicity, that is, finding *acyclic subhypergraphs* in a connected and possibly cyclic hypergraph. If we can find them efficiently, then it is possible to give some approximate solution efficiently of the problem for the original hypergraph, by solving the problem for acyclic subhypergraphs. Unfortunately, in this paper, we show the following intractability of finding acyclic subhypergraphs.

For a hypergraph $H = (V, \mathcal{E})$ and a subhypergraph $H' = (V', \mathcal{E}')$ of H, we say that H' is *spanning* w.r.t. H if $V' = V$. Furthermore, we say that a hypergraph H is *connected* if, for each $u, v \in V$, there exists a *path* from u to v in H, that is, a sequence $v_0 E_1 v_1 \cdots E_n v_n$ such that $u = v_0$, $v = v_n$, $E_i \in \mathcal{E}$, $E_{i-1} \neq E_i$ and $v_{i-1}, v_i \in E_i$ for each i ($1 \leq i \leq n$). The spanning connected acyclic subhypergraph of a connected hypergraph is an extension of a spanning tree of a connected graph. While a connected graph always has a spanning tree and we can find it in polynomial time, we show that the problem of determining whether or not a connected hypergraph has a spanning connected acyclic subhypergraph is NP-complete. Furthermore, we show that, for a given $K > 0$, the problem of determining whether or not a hypergraph has an acyclic subhypergraph containing at least K hyperedges is NP-complete.

Next, in order to avoid such intractability, we introduce a *maximal* acyclic subhypergraph. Here, for a hypergraph $H = (V, \mathcal{E})$ and an acyclic subhypergraph $H' = (V', \mathcal{E}')$ of H, H' is *maximal* if there exists no hyperedge $E \in \mathcal{E} - \mathcal{E}'$ such that $H \cup \{E\}$ is acyclic.

Tarjan and Yannakakis [15] have designed the *acyclicity test algorithm* acy of determining whether or not a hypergraph is acyclic in linear time. Then, we can find a maximal acyclic subhypergraph by a generate-and-test method with acy. However, the time complexity of this method is beyond linear time, because this method calls acy whenever we add a hyperedge to a current subhypergraph.

Hence, in this paper, by modifying the algorithm acy, we design the algorithm mas to find a maximal acyclic subhypergraph in linear time.

2   Preliminaries

A *hypergraph* $H = (V, \mathcal{E})$ consists of a set V of vertices and a set $\mathcal{E} \subseteq 2^V$ of hyperedges. For $\mathcal{E} \subseteq 2^V$, $\|\mathcal{E}\|$ denotes the total size of $\mathcal{E}$, that is, $\|\mathcal{E}\| = \sum_{E \in \mathcal{E}} |E|$. For a vertex $v \in V$, the *degree* of v, denoted by $deg(v)$, is the number of hyperedges containing v, that is, $|\{E \in \mathcal{E} \mid v \in E\}|$. For a hyperedge $E \in \mathcal{E}$, $v(E)$ denotes the set of vertices contained by E. For a hypergraph $H = (V, \mathcal{E})$ and a hyperedge E, we sometimes denote $H \cup \{E\} = (V \cup v(E), \mathcal{E} \cup \{E\})$ and $H - \{E\} = (V, \mathcal{E} - \{E\})$. Furthermore, we sometimes denote H by $\mathcal{E}$ alone.

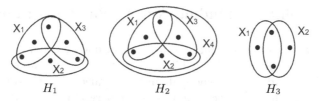

Fig. 1. Hypergraphs H_1, H_2 and H_3 in Example 1

For a hypergraph $H = (V, \mathcal{E})$ and $u, v \in V$, a *path* from u to v in H is a sequence $v_0 E_1 v_1 \cdots E_n v_n$ such that $u = v_0$, $v = v_n$, $E_i \in \mathcal{E}$, $E_{i-1} \neq E_i$ and $v_{i-1}, v_i \in E_i$ for each i ($1 \leq i \leq n$). A hypergraph H is *connected* if there exists a path from u to v for each $u, v \in V$.

Next, we introduce the *acyclicity* for hypergraphs.

Definition 1. For a hypergraph $H = (V, \mathcal{E})$, the *GYO-reduct* $GYO(H)$ [4,9,15] of H is the hypergraph obtained from H by repeatedly applying the following rules, called *GYO-reduction*, as long as possible:

1. Remove hyperedges that are empty or contained in other hyperedges.
2. Remove vertices that appear in ≤ 1 hyperedges.

Definition 2. We say that a hypergraph H is *acyclic* if $GYO(H)$ is an empty hypergraph, i.e., $GYO(H) = (\emptyset, \emptyset)$, and *cyclic* otherwise.

Definition 2 follows from one of the equivalent conditions that a hypergraph is acyclic [1]. Other equivalent conditions are presented in [1].

Example 1. In Figure 1, H_1 is cyclic but H_2 and H_3 are acyclic. In contrast to graphs, note that cycles contained completely in some hyperedge are removed by GYO-reduction (H_2) and a hypergraph with double hyperedges is acyclic (H_3).

Let $H = (V, \mathcal{E})$ and $H_1 = (V_1, \mathcal{E}_1)$ be hypergraphs. We say that H_1 is a *subhypergraph* of H if $V_1 \subseteq V$ and $\mathcal{E}_1 \subseteq \mathcal{E}$. Also we say that H_1 is and an *acyclic subhypergraph* of H if H_1 is a subhypergraph of H and H_1 is acyclic. Furthermore, we say that a subhypergraph H_1 of H is *spanning* if $V_1 = V$.

Finally, we introduce the following NP-complete problem.

MONOTONE 1-IN-3 3SAT [8]
INSTANCE: A set X of variables and a collection C of monotone 3-clauses over X.
QUESTION: Is there a truth assignment to X that makes exactly one literal of each clause in C true?

For an instance of MONOTONE 1-IN-3 3SAT, in this paper, we refer $\{x_1, \ldots, x_n\}$ to X, $\{c_1, \ldots, c_m\}$ to C and $x_{j1} \vee x_{j2} \vee x_{j3}$ to c_j for each j ($1 \leq j \leq m$).

3   Intractability of Finding Acyclic Subhypergraphs

In this section, first we investigate the problem of finding a spanning connected acyclic subhypergraph. Note that, while a connected graph always has a spanning tree and we can find it in polynomial time, a connected hypergraph does not always have a spanning connected acyclic subhypergraph as follows.

Example 2. Consider the cyclic hypergraph H_1 in Example 1. Then, H_1 has three connected acyclic subhypergraphs $\{X_1, X_2\}$, $\{X_1, X_3\}$ and $\{X_2, X_3\}$, but none of them is spanning, so H_1 has no spanning connected acyclic subhypergraph.

Hence, we investigate the problem SPANNING CONNECTED ACYCLIC SUB-HYPERGRAPH.

SPANNING CONNECTED ACYCLIC SUBHYPERGRAPH
INSTANCE: A connected hypergraph $H = (V, \mathcal{E})$.
QUESTION: Does H have a spanning connected acyclic subhypergraph?

Theorem 1. SPANNING CONNECTED ACYCLIC SUBHYPERGRAPH *is NP-complete.*

Proof. It is obvious that SPANNING CONNECTED ACYCLIC SUBHYPERGRAPH is in NP, so we reduce MONOTONE 1-IN-3 3SAT to it.
First construct a set of vertices V as follows:

$$V = \{w\} \cup \{u_j, v_j, v_{j1}, v_{j2}, v_{j3} \mid 1 \leq j \leq m\}.$$

Next, for a 3-clause c_j and a variable x_i, construct hyperedges as follows.

$$E(c_j) = \{v_{j1}, v_{j2}, v_{j3}\},$$
$$E(x_i) = \{w\} \cup \{u_j, v_{jk} \mid x_i = x_{jk} \in c_j \ (k = 1, 2, 3)\}.$$

Then, let $\mathcal{E}$ be $\{E(c_j) \mid 1 \leq j \leq m\} \cup \{E(x_i) \mid 1 \leq i \leq n\}$ and H a hypergraph $(V, \mathcal{E})$. Note that H is cyclic, $|V| = 5m + 1$ and $|\mathcal{E}| = m + n$.
For example, consider the following instance of MONOTONE 1-IN-3 3SAT.

$$X = \{x_1, x_2, x_3, x_4, x_5\},$$
$$C = \left\{ \begin{array}{cccc} c_1 & c_2 & c_3 & c_4 \\ x_1 \vee x_2 \vee x_3, & x_1 \vee x_3 \vee x_4, & x_1 \vee x_4 \vee x_5, & x_3 \vee x_4 \vee x_5 \end{array} \right\}.$$

Then, H is constructed as Figure 2.
For C and H, the following statements hold.

1. $u_j \in E(x_i)$ if and only if x_i occurs in c_j ($1 \leq i \leq n, 1 \leq j \leq m$).
2. For vertices $v_{jk}, v_{jl} \in E(c_j)$ ($k, l = 1, 2, 3, \ k \neq l$), if $v_{jk} \in E(x_i)$ and $v_{jl} \in E(x_{i'})$, then a hypergraph $\{E(c_j), E(x_i), E(x_{i'})\}$ is cyclic.
3. A spanning connected subhypergraph of H always contains all of the $E(c_j)$ ($1 \leq j \leq m$) as hyperedges, because of the existence of v_j.

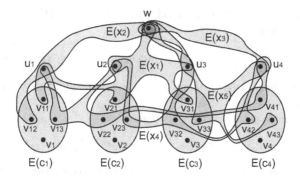

Fig. 2. A hypergraph H in the proof of Theorem 1

Suppose that C is satisfiable by $a = (a_1, \ldots, a_n)$ that makes exactly one literal of each clause in C true. Then, let $\mathcal{E}'$ be $\{E(c_1), \ldots, E(c_m)\} \cup \{E(x_i) \mid a_i = 1\}$ and H' a spanning hypergraph $(V, \mathcal{E}')$. Note that $\mathcal{E}'$ contains a hyperedge $E(c_j)$ containing vertices v_j, v_{j1}, v_{j2} and v_{j3} for each j ($1 \leq j \leq m$). Also $\mathcal{E}'$ contains a hyperedge $E(x_i)$ containing a vertex w. Since $\mathcal{E}'$ contains $E(x_i)$ such that $a_i = 1$ and by the statement 1, $E(x_i)$ contains some u_j ($1 \leq j \leq m$). By the supposition, the above assignment a makes exactly one literal of c_j true, so all of the $E(x_i)$ such that $a_i = 1$ ($1 \leq i \leq n$) contains all of the u_j ($1 \leq j \leq m$).

Furthermore, for each $E(c_j)$ ($1 \leq j \leq m$), there exists exactly one $E(x_{i_j})$ ($1 \leq i_j \leq n$) such that $a_{i_j} = 1$, and $E(c_j)$ and $E(x_{i_j})$ commonly share exactly one vertex. By the statement 2, H' is acyclic.

Since $v_j, v_{jk} \in E(c_j)$ ($k = 1, 2, 3$), there exists a path from v_j to v_{jk}. Since the existence of $E(x_{i_j})$ containing w and u_j, there exists a path from v_j or v_{jk} to w or u_j. Since these statements hold for each j ($1 \leq j \leq m$), H' is connected.

Conversely, suppose that H has a spanning connected acyclic subhypergraph $H' = (V, \mathcal{E}')$. By the statement 3, $E(c_j) \in \mathcal{E}'$ for each j ($1 \leq j \leq m$).

Since H' is spanning and connected, and by the statement 1, for each j ($1 \leq j \leq m$), there exists an i_j ($1 \leq i_j \leq n$) such that $u_j \in E(x_{i_j})$. By the statement 2 and since H' is acyclic, just $E(x_{i_j})$ is a hyperedge that shares a vertex in $E(c_j)$. Hence, $\mathcal{E}'$ contains $E(x_{i_j})$ for each j ($1 \leq j \leq m$).

Let a be $(a_1, \ldots, a_n)$ such that $a_i = 1$ if $E(x_i) \in \mathcal{E}'$; $a_i = 0$ otherwise. Since $E(x_{i_j}) \in \mathcal{E}'$ and $E(c_j) \in \mathcal{E}'$ share just a vertex v_{jk} ($k = 1, 2$ or 3) for each j, a is the assignment that makes exactly one literal of each clause in C true.   □

Theorem 1 holds even if *no hyperedge is contained by another hyperedge*. Furthermore, we can show Theorem 1 even if *the degree of every vertex is at most 2*, by replacing a hyperedge E_i with $(E_i - \{v\}) \cup \{v_i\}$ and by adding a hyperedge $\{v_1, \ldots, v_n, v\}$ for every v such that $deg(v) = n(> 2)$ and $v \in E_i$ ($1 \leq i \leq n$) in the proof of Theorem 1.

Next, we investigate the problem MAXIMUM ACYCLIC SUBHYPERGRAPH.

MAXIMUM ACYCLIC SUBHYPERGRAPH
INSTANCE: A connected hypergraph $H = (V, \mathcal{E})$ and an integer $K > 0$.

QUESTION: Does H have an acyclic subhypergraph $H' = (V', \mathcal{E}')$ such that $|\mathcal{E}'| \geq K$?

Theorem 2. MAXIMUM ACYCLIC SUBHYPERGRAPH *is NP-complete.*

Proof. It is obvious that MAXIMUM ACYCLIC SUBHYPERGRAPH is in NP, so we reduce MONOTONE 1-IN-3 3SAT [8] to it.

First construct a set of vertices V as follows:

$$V = \{u_j, v_{j1}, v_{j2}, v_{j3}, w_{j1}, w_{j2}, w_{j3} \mid 1 \leq j \leq m\} \cup \{a_{i0}, a_{i1}, b_i \mid 1 \leq i \leq n\}.$$

Next, for a 3-clause c_j and a variable x_i, construct hyperedges as follows.

$$E = \{a_{i0}, a_{i1} \mid 1 \leq i \leq n\}, \quad E(c_j) = \{v_{j1}, v_{j2}, v_{j3}\}, \quad E(x_i) = \{a_{i0}, a_{i1}\},$$
$$E_1(c_j) = \{v_{j1}, v_{j2}\}, \quad E_2(c_j) = \{v_{j2}, v_{j3}\}, \quad E_3(c_j) = \{v_{j1}, v_{j3}\},$$
$$F_1(c_j) = \{w_{j1}, w_{j2}\}, \quad F_2(c_j) = \{w_{j2}, w_{j3}\}, \quad F_3(c_j) = \{w_{j1}, w_{j3}\},$$
$$E_0(x_i) = \{a_{i0}, b_i, w_{jk} \mid x_i = x_{jk} \in c_j \ (k = 1, 2, 3)\},$$
$$E_1(x_i) = \{a_{i1}, b_i, u_j, v_{jk} \mid x_i = x_{jk} \in c_j \ (k = 1, 2, 3)\}.$$

Then, let $\mathcal{E}$ be the following set of hyperedges:

$$\mathcal{E} = \{E(c_j), E_1(c_j), E_2(c_j), E_3(c_j), F_1(c_j), F_2(c_j), F_3(c_j) \mid 1 \leq j \leq m\}$$
$$\cup \{E(x_i), E_0(x_i), E_1(x_i) \mid 1 \leq i \leq n\} \cup \{E\}.$$

Let H be a hypergraph $(V, \mathcal{E})$. Note that H is cyclic, $|V| = 7m + 3n$ and $|\mathcal{E}| = 7m + 3n + 1$. Also let K be $5m + 2n + 1$.

For example, consider the following instance of MONOTONE 1-IN-3 3SAT.

$$X = \{x_1, x_2, x_3, x_4, x_5\},$$
$$C = \left\{ \begin{array}{cccc} c_1 & c_2 & c_3 & c_4 \\ x_1 \vee x_2 \vee x_3, & x_1 \vee x_3 \vee x_4, & x_1 \vee x_4 \vee x_5, & x_3 \vee x_4 \vee x_5 \end{array} \right\}.$$

Then, H is constructed as Figure 3.

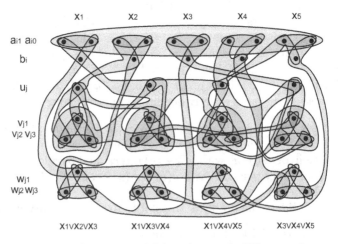

Fig. 3. A hypergraph H in the proof of Theorem 2

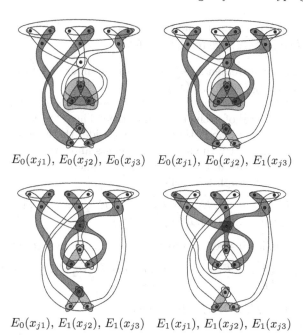

$$E_0(x_{j1}),\ E_0(x_{j2}),\ E_0(x_{j3}) \qquad E_0(x_{j1}),\ E_0(x_{j2}),\ E_1(x_{j3})$$

$$E_0(x_{j1}),\ E_1(x_{j2}),\ E_1(x_{j3}) \qquad E_1(x_{j1}),\ E_1(x_{j2}),\ E_1(x_{j3})$$

Fig. 4. The relationship between hyperedges E, $E(x_{jk})$, $E_0(x_{jk})$, $E_1(x_{jk})$, $E(c_j)$, $E_k(c_j)$ and $F_k(c_j)$ $(1 \le k \le 3)$

For C and H, let c_j be a clause $x_{j1} \vee x_{j2} \vee x_{j3} \in C$ for each j $(1 \le j \le m)$. Then, consider the relationship between hyperedges E, $E(x_{jk})$, $E_0(x_{jk})$, $E_1(x_{jk})$, $E(c_j)$, $E_k(c_j)$ and $F_k(c_j)$ $(1 \le k \le 3)$. Suppose that an acyclic subhypergraph H' of H always contains E, $E(x_{jk})$ and a hyperedge either $E_0(x_{jk})$ or $E_1(x_{jk})$ for each k $(1 \le k \le 3)$. Also see Figure 4.

1. If H' contains hyperedges $E_0(x_{j1})$, $E_0(x_{j2})$ and $E_0(x_{j3})$, then H' can contain 4 hyperedges $E(c_j)$, $E_1(c_j)$, $E_2(c_j)$ and $E_3(c_j)$. Note that H' can contain no $F_k(c_j)$ $(1 \le k \le 3)$, because $\{E_0(x_{j1}), E_0(x_{j2}), E_0(x_{j3}), F_k(c_j)\}$ is cyclic.
2. If H' contains hyperedges $E_0(x_{j1})$, $E_0(x_{j2})$ and $E_1(x_{j3})$, then H' can contain 5 hyperedges $E(c_j)$, $E_1(c_j)$, $E_2(c_j)$, $E_3(c_j)$ and $F_2(c_j)$ (or $F_3(c_j)$) as follows. Since $E_0(x_{j1})$ contains w_{j1} and $E_0(x_{j2})$ contains w_{j2}, we can add a hyperedge either $F_2(c_j) = \{w_{j2}, w_{j3}\}$ or $F_3(c_j) = \{w_{j1}, w_{j3}\}$ to H' preserving acyclicity.
3. If H' contains hyperedges $E_0(x_{j1})$, $E_1(x_{j2})$ and $E_1(x_{j3})$, then H' can contain 3 hyperedges $F_1(c_j)$, $F_2(c_j)$ and $E_1(c_j)$ (or $E_3(c_j)$). For the case selecting $E_1(c_j)$, note that H' can contain none of $E_2(c_j)$, $E_3(c_j)$ and $E(c_j)$, because $\{E_1(x_{j2}), E_1(x_{j3}), E_1(c_j), E_2(c_j)\}$, $\{E_1(x_{j2}), E_1(x_{j3}), E_1(c_j), E_3(c_j)\}$ and $\{E_1(x_{j2}), E_1(x_{j3}), E(c_j)\}$ are cyclic.
4. If H' contains hyperedges $E_1(x_{j1})$, $E_1(x_{j2})$ and $E_1(x_{j3})$, then H' can contain 2 hyperedges $F_k(c_j)$ and $F_{k'}(c_j)$ for $1 \le k < k' \le 3$. Here, H' can contain neither $E(c_j)$ nor $E_k(c_j)$ for each k $(1 \le k \le 3)$, because both $\{E_1(x_{j1}), E_1(x_{j2}), E_1(x_{j3}), E(c_j)\}$ and $\{E_1(x_{j1}), E_1(x_{j2}), E_1(x_{j3}), E_k(c_j)\}$

are cyclic. Also H' cannot contain $F_{k''}(c_j)$ such that $k'' \neq k$ and $k'' \neq k'$, because $\{F_1(c_j), F_2(c_j), F_3(c_j)\}$ is cyclic.

Note that the above statements depend on the number of 1 in a_{j1}, a_{j2} and a_{j3} for $E_{a_{j1}}(x_{j1})$, $E_{a_{j2}}(x_{j2})$ and $E_{a_{j3}}(x_{j3})$.

Suppose that C is satisfiable by the truth assignment $a = (a_1, \ldots, a_n)$ that makes exactly one literal of each clause in C true. Then, for each j $(1 \leq j \leq m)$, one of a_{j1}, a_{j2} and a_{j3} is 1 and others are 0. Consider the hyperedges $E_{a_{j1}}(x_{j1})$, $E_{a_{j2}}(x_{j2})$ and $E_{a_{j3}}(x_{j3})$. By the above statements and since one of a_{j1}, a_{j2} and a_{j3} is 1 and the others are 0, there exists exactly one k_j $(k_j = 1, 2, 3)$ such that the following subhypergraph is acyclic.

$$\left\{ \begin{array}{l} E, E(x_{j1}), E(x_{j2}), E(x_{j3}), E_{a_{j1}}(x_{j1}), E_{a_{j2}}(x_{j2}), E_{a_{j3}}(x_{j3}), \\ E(c_j), E_1(c_j), E_2(c_j), E_3(c_j), F_{k_j}(c_j) \end{array} \right\}.$$

Let $\mathcal{E}'$ be the following set of hyperedges and H' a hypergraph $(V, \mathcal{E}')$.

$$\{E\} \cup \{E(c_j), E_1(c_j), E_2(c_j), E_3(c_j), F_{k_j}(c_j) \mid 1 \leq j \leq m\}$$
$$\cup \{E_{a_i}(x_i), E(x_i) \mid 1 \leq i \leq n\}.$$

Then, it holds that $|\mathcal{E}'| = K$ and H' is acyclic.

Conversely, suppose that H has an acyclic subhypergraph $H' = (V, \mathcal{E}')$ such that $|\mathcal{E}'| \geq K$. In order to preserve acyclicity, we must select K hyperedges as:

1. at least $2n + 1$ hyperedges in $\{E\} \cup \{E_0(x_i), E_1(x_i), E(x_i) \mid 1 \leq i \leq n\}$, and
2. at least $5m$ hyperedges in $\bigcup_{j=1}^{m} (\{E(c_j) \cup \{E_k(c_j), F_k(c_j) \mid k = 1, 2, 3\})$.

For the selection 1, both $\{E\} \cup \{E_0(x_i), E_1(x_i) \mid 1 \leq i \leq n\}$ and $\{E(x_l)\} \cup \{E_0(x_i), E_1(x_i) \mid 1 \leq i \leq n\}$ $(1 \leq l \leq n)$ are cyclic, while $\{E\} \cup \{E(x_i), E_{a_i}(x_i) \mid 1 \leq i \leq n\}$ is acyclic, where a_i is either 0 or 1 $(1 \leq i \leq n)$. By the existence of H', we must select E, $E(x_i)$ and $E_{a_i}(x_i)$ for each i $(1 \leq i \leq n, a_i = 0, 1)$, of which number is $2n + 1$.

For the selection 2, consider $c_j = x_{j1} \vee x_{j2} \vee x_{j3}$ for each j $(1 \leq j \leq m)$. Then, we must select $E(c_j)$, $E_1(c_j)$, $E_2(c_j)$, $E_3(c_j)$ and $F_{k_j}(c_j)$ for some k_j $(1 \leq k_j \leq 3)$. Furthermore, we must select $E_{a_{j1}}(x_{j1})$, $E_{a_{j2}}(x_{j2})$ and $E_{a_{j3}}(x_{j3})$ such that exactly one of a_{j1}, a_{j2} and a_{j3} is 1 and the others are 0. By the existence of H', such a_{j1}, a_{j2} and a_{j3} always exist for each j $(1 \leq j \leq m)$.

Let a be a truth assignment $(a_1, \ldots, a_n)$ such that $E_{a_i}(x_i) \in \mathcal{E}'$. By the supposition of a_{j1}, a_{j2} and a_{j3} for each j $(1 \leq j \leq m)$, the assignment a makes exactly one literal of each clause in C true. □

Theorem 2 holds even if *the degree of every vertex is at most* 4. On the other hand, it remains open whether or not MAXIMUM ACYCLIC SUBHYPERGRAPH is NP-complete if no hyperedge is contained by another hyperedge.

4   Finding Maximal Acyclic Subhypergraphs

In order to avoid the intractability as shown in the previous section, in this section, we introduce a weaker concept of a *maximal* acyclic subhypergraph.

Definition 3. Let $H = (V, \mathcal{E})$ be a hypergraph and $H' = (V', \mathcal{E}')$ an acyclic subhypergraph of H. We say that H' is *maximal* if there exists no hyperedge $E \in \mathcal{E} - \mathcal{E}'$ such that $H' \cup \{E\}$ is acyclic.

In order to find a maximal acyclic subhypergraph, first we introduce the algorithm *acy* on hypergraphs designed by Tarjan and Yannakakis [15]. The algorithm $acy(H)$ for a hypergraph $H = (V, \mathcal{E})$ consists of the *restricted maximum cardinality search* (*rmcs*, for short) step and the *testing* step.

In the rmcs step, $acy(H)$ labels β- and γ-values for hyperedges and β-value for vertices, while searching for neighbors. Here, s denotes the number of vertices already searched. Initially set $\beta(E) = \gamma(E) = \beta(v) = s(E) = 0$ for each $E \in \mathcal{E}$ and $v \in V$. For the i-th iteration ($i \geq 1$) of the rmcs step, suppose that $s(E)$ is maximum. Then, $acy(H)$ selects E and sets the values $\beta(E)$, $\gamma(E)$, $s(E)$ and $\beta(v)$ for each $v \in E$ such that $\beta(v) = 0$ as follows.

$$\beta(E) = \begin{cases} i & \text{if there exists a vertex } v \in E \text{ such that } \beta(v) = 0, \\ 0 & \text{otherwise,} \end{cases}$$
$$\gamma(E) = \max\{\beta(v) \mid v \in E\}, \quad s(E) = -1, \quad \beta(v) = i.$$

Furthermore, for such a v and $F \in \mathcal{E}$ such that $v \in F$, $acy(H)$ sets $s(F)$ to $s(F) + 1$, and if $s(F) = |F|$, then $acy(H)$ sets $s(F)$ to -1. This iteration is repeated until there exists a hyperedge $E \in \mathcal{E}$ such that $s(E) \geq 0$. Here, $s(E)$ means the following formula.

$$s(E) = \begin{cases} |\{v \in E \mid \beta(v) > 0\}| & \exists v \in E \text{ such that } \beta(v) = 0, \\ -1 & \forall v \in E, \beta(v) > 0. \end{cases}$$

In the testing step, $acy(H)$ checks whether or not, for each $E \in \mathcal{E}$ and $F \in \mathcal{E}$ such that $\gamma(E) = \beta(F)$, it holds that $\{v \in E \mid \beta(v) < \gamma(E)\} \subseteq F$, that is, for each $v \in E$, if $\beta(v) < \gamma(E)$ then $v \in F$. If there exist E and F not satisfying this, then $acy(H)$ returns 'cyclic' and halts. Otherwise, $acy(H)$ returns 'acyclic.' The correctness of the testing step is due to the following theorem.

Theorem 3 (Tarjan & Yannakakis [15]). *For every hypergraph* $H = (V, \mathcal{E})$, H *is acyclic if and only if for each* $1 \leq i \leq |\mathcal{E}|$ *and each* $E \in \mathcal{E}$ *such that* $\gamma(E) = i$, *it holds that* $\{v \in E \mid \beta(v) < i\} \subseteq F$, *where* $F \in \mathcal{E}$ *and* $\beta(F) = i$.

Furthermore, the time complexity of the algorithm *acy* is represented as follows.

Theorem 4 (Tarjan & Yannakakis [15]). *For a hypergraph* $H = (V, \mathcal{E})$, *the algorithm* $acy(H)$ *determines whether or not* H *is acyclic in* $O(|V| + ||\mathcal{E}||)$ *time.*

Example 3. Consider two hypergraphs H_1 and H_2 in Figure 1 in Example 1, where H_1 is cyclic but H_2 is acyclic. Then, Figure 5 describes the results applying H_1 and H_2 to the algorithm *acy*. Note that $acy(H_1)$ describes the result selecting X_1, X_2 and X_3 in this order, and $acy(H_2)$ describes the result selecting first X_1.

Furthermore, by using the above labeling, we can construct an *elimination tree* [4,9,15] for an acyclic hypergraph. Let $H = (V, \mathcal{E})$ be an acyclic hypergraph. Then, the elimination tree T is constructed as $(\mathcal{E}, \{(E, F) \mid \gamma(E) =$

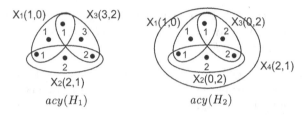

$$X_1(1,0) \qquad X_3(3,2) \qquad X_1(1,0) \qquad X_3(0,2)$$

$$X_2(2,1) \qquad X_2(0,2) \qquad X_4(2,1)$$

$$acy(H_1) \qquad acy(H_2)$$

Fig. 5. The results applying H_1 and H_2 in Example 1 to acy. Here, the label of hyperedges denotes (β-value,γ-value) and the label of vertices denotes the β-value

$\beta(F)\}$). For example, the elimination tree of the hypergraph in Figure 5 (right) is $(\{X_1, X_2, X_3, X_4\}, \{(X_1, X_4), (X_2, X_4), (X_3, X_4)\})$.

By using a simple generate-and-test method with the algorithm acy, we can find a maximal acyclic subhypergraph in $O(|E|(|V| + ||\mathcal{E}||))$ time, because it calls acy running in $O(|V| + ||\mathcal{E}||)$ time whenever we add a hyperedge to a subhypergraph. In order to find a maximal acyclic subhypergraph efficiently, we design the algorithm mas described as Figure 6, by modifying the algorithm acy. The algorithm mas is an extension of Prim's algorithm [13] to find the minimum spanning tree by searching for the neighbors in Graph Theory.

Here, for a hypergraph $(V, \mathcal{E})$ and $E \in \mathcal{E}$, the *weight* $w(E)$ of E is defined as $\sum_{v \in E} deg(v)$, and the *average weight* $aw(E)$ of E is defined as $\frac{w(E)}{|E|}$.

```
procedure mas(H) /* H = (V, E) : connected hypergraph */
1   foreach E ∈ E do (β(E), γ(E), s(E)) ← (0, 0, 0);
2   foreach v ∈ V do β(v) ← 0;
3   H' ← ∅;  i ← 0;
4   while ∃E ∈ E such that s(E) ≥ 0 do begin
5       select E ∈ {E ∈ E | s(E) is maximum} such that aw(E) is maximum;
6       i ← i + 1;  s(E) ← -1;  β(E) ← i;  γ(E) ← max{β(v) | v ∈ E};
7       foreach v ∈ E such that β(v) = 0 do begin
8           β(v) ← i;
9           foreach F ∈ E such that (s(F) ≥ 0 ∧ v ∈ F) do
10              s(F) ← s(F) + 1;
11              if s(F) = |F| then
12                  s(F) ← -1;  γ(F) ← max{β(v) | v ∈ F};  H' ← H' ∪ {F};
13      end /* foreach */
14      select F ∈ H' such that γ(E) = β(F);  /* F is unique if there exists */
15      if {v ∈ E | β(v) < γ(E)} ⊆ F then /* H' ∪ {E} is acyclic */
16          H' ← H' ∪ {E};
17      else /* H' ∪ {E} is cyclic */
18          foreach v ∈ E such that β(v) = i do β(v) ← 0;
19          i ← i - 1;  β(E) ← -1;  γ(E) ← -1;
20  end /* while */
21  return H';
```

Fig. 6. The algorithm $mas(H)$ finding a maximal acyclic subhypergraph of H

The algorithm *mas* uses the β- and γ-values of hyperedges, the β-value of vertices and the number s of vertices already searched, as similar as the algorithm *acy*. By updating s, *mas* always selects hyperedges in the neighbors.

On the other hand, the difference between *mas* and *acy* is the statements in the line 12 and the lines from 18 to 19.

By the line 12, the hyperedge F containing in E is added to H'. Note that the setting to $\gamma(F)$ in the line 12 is not essential for finding the maximal acyclic subhypergraph; It is just useful when constructing the elimination tree.

By the lines from 18 to 19, if $H' \cup \{E\}$ is cyclic, then the labeling of $v \in E$ such that $\beta(v) = i$ by the current i-th iteration is initialized to 0 and $\beta(E)$ is labeled by -1, which means that E is ignored to the line 15 in the j-th iteration $(j > i)$. Furthermore, for each $E \in H$, $\beta(E)$ is uniquely determined except the case that $\beta(E) = -1$, which is based on the line 14.

The correctness of the statement in the line 15 is due to Theorem 3. Also we can replace the statement in the line 15 with the following simple form.

> If $v \in F$ for each $v \in E$ such that $\beta(v) < \gamma(E)$ and each $F \in H'$ such that $\gamma(E) = \beta(F)$, then H' is updated to $H' \cup \{E\}$.

Theorem 5. *For every hypergraph $H = (V, \mathcal{E})$, the algorithm $mas(H)$ always finds a maximal acyclic subhypergraph of H in $O(|V| + ||\mathcal{E}||)$ time.*

Proof. Let H' be the result applying a hypergraph H to the algorithm *mas*. By Theorem 3, H' is acyclic. By the line 12, for each $E \in H'$, H' also contains each $F \in \mathcal{E}$ such that $F \subseteq E$. Hence, H' is maximal.

The running time to initialize is $O(|V|+|\mathcal{E}|)$. Let E_i $(1 \le i \le |\mathcal{E}|)$ be E in the i-th iteration of **while** loop. For the i-th iteration of **while** loop, the running time is $O(|E_i|)$, so the total running time for **while** loop is $O(\sum_{i=1}^{m} |E_i|) = O(||\mathcal{E}||)$. Hence, the total running time of the algorithm *mas* is $O(|V| + ||\mathcal{E}||)$. □

Example 4. Consider the hypergraphs H_1 and H_2, and their results applying to *mas* (described by the solid lines) in Figure 7.

For H_1, *mas* first selects X_1, for example, because $aw(X_1) = aw(X_2) = aw(X_3) = \frac{8}{3}$ and $aw(X_4) = \frac{15}{6}$. Secondly, *mas* selects X_4 and labels $\beta(v) = 2$ for each $v \in X_4 - X_1$. Then, for the second **foreach** loop, *mas* sets $\gamma(X_2)$ and $\gamma(X_3)$ to 2. Hence, *mas* returns H_1 itself.

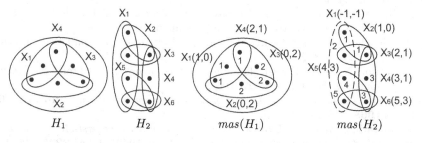

H_1 H_2 $mas(H_1)$ $mas(H_2)$

Fig. 7. Hypergraphs H_1 and H_2 in Example 4 and the results applying them to *mas*

For H_2, mas first selects X_2, for example, because $aw(X_2) = aw(X_3) = aw(X_5) = aw(X_6) = \frac{5}{2}$, $aw(X_4) = \frac{7}{3}$ and $aw(X_1) = 2$. Secondly, mas selects X_3, because $s(X_3) = s(X_1) = 1$, $s(X_i) = 0$ ($i = 4, 5, 6$) and $aw(X_3) > aw(X_1)$. Thirdly, mas selects X_1 and labels $(\beta(X_1), \gamma(X_1)) = (3, 2)$. However, there exists a vertex $v \in X_1$ such that $\beta(v) = 1 < \gamma(E) = 2 = \beta(X_2)$ and $v \notin X_2$. Hence, mas does not add X_1 to H' and labels $(\beta(X_1), \gamma(X_1)) = (-1, -1)$ and $\beta(v) = 0$ for each $v \in X_1 - (X_2 \cup X_3)$. Hence, mas returns $H_2 - \{X_1\}$.

5 Conclusion

In this paper, we have investigated the problem of finding *acyclic subhypergraphs* in a hypergraph. We have first shown that both the problem SPANNING CONNECTED ACYCLIC SUBHYPERGRAPH and MAXIMUM ACYCLIC SUBHYPERGRAPH are NP-complete. Next, we design the algorithm mas to find a *maximal* acyclic subhypergraph in linear time, based on the acyclicity test algorithm acy designed by Tarjan and Yannakakis [15].

For finding acyclic subhypergraphs, we can formulate the *maximization problems* for not only hyperedges discussed in this paper but also vertices. Then, it is a future work to investigate the complexity and approximability of these problems. Furthermore, it is a future work to apply acyclic subhypergraphs to several combinatorial problems for conjunctive queries and clauses.

References

1. C. Beeri, R. Fagin, D. Maier, M. Yannakakis: *On the desirability of acyclic database schemes*, JACM **30**, 479–513, 1983.
2. C. Berge: *Graphs and hypergraphs*, North-Holland, 1973.
3. C. Berge: *Hypergraphs*, North-Holland, 1989.
4. C. Chekuri, A. Rajaraman: *Conjunctive query containment revisited*, Theor. Comput. Sci. **239**, 211–229, 2000.
5. T. Eiter, G. Gottlob: *Identifying the minimal transversal of a hypergraph and related problems*, SIAM J. Comput. **24**, 1278–1304, 1993.
6. T. Eiter, G. Gottlob: *Hypergraph transversal computation and related problems in logic and AI*, JELIA 2002, LNCS **2424**, 549–564, 2002.
7. R. Fagin: *Degrees of acyclicity for hypergraphs and relational database schemes*, JACM **30**, 514–550, 1983.
8. M. R. Garey, D. S. Johnson: *Computers and intractability: A guide to the theory of NP-completeness* W. H. Freeman and Company, 1979.
9. G. Gottlob, N. Leone, F. Scarcello: *The complexity of acyclic conjunctive queries*, JACM **43**, 431–498, 2001.
10. G. Gottlob, N. Leone, F. Scarcello: *Hypertree decompositions and tractable queries*, J. Comput. Sys. Sci. **64**, 579–627, 2002.
11. K. Hirata: *On the hardness of learning acyclic conjunctive queries*, Proc. ALT'00, LNAI **1968**, 238–251, 2000.
12. T. Horváth, S. Wrobel: *Towards discovery of deep and wide first-order structures: A case study in the domain of mutagenicity*, Proc. DS'01, LNAI **2226**, 100–112, 2001.

13. D. Jungnickel: *Graphs, networks and algorithms*, Springer, 1999.
14. D. Maier, J. D. Ullman: *Connections in acyclic hypergraphs*, Theor. Comput. Sci. **32**, 185–199, 1984.
15. R. E. Tarjan, M. Yannakakis: *Simple linear-time algorithms to test chordality of graphs, test acyclicity of hypergraphs, and selectively reduce acyclic hypergraphs*, SIAM J. Comput. **13**, 566–579, 1984.
16. M. Yannakakis: *Algorithms for acyclic database schemes*, Proc. VLDB'81, 82–94, 1981.

An Improved Approximation Algorithm
for TSP with Distances One and Two

M. Bläser and L. Shankar Ram

Institut für Theoretische Informatik, ETH Zürich,
CH-8092 Zürich, Switzerland
{mblaeser, lshankar}@inf.ethz.ch

Abstract. The minimum traveling salesman problem with distances one and two is the following problem: Given a complete undirected graph $G = (V, E)$ with a cost function $w : E \to \{1, 2\}$, find a Hamiltonian tour of minimum cost. In this paper, we provide an approximation algorithm for this problem achieving a performance guarantee of $\frac{315}{271}$. This algorithm can be further improved obtaining a performance guarantee of $\frac{65}{56}$. This is better than the one achieved by Papadimitriou and Yannakakis [8], with a ratio $\frac{7}{6}$, more than a decade ago. We enhance their algorithm by an involved procedure and find an improved lower bound for the cost of an optimal Hamiltonian tour.

1 Introduction

The traveling salesman problem (TSP) is one of the important problems in combinatorial optimization. It is well known that, unless P = NP, there can be no efficient (polynomial) approximation algorithm for TSP that achieves some bounded approximation ratio [9]. However, for the symmetric traveling salesman problem with cost function satisfying triangle inequality (Δ-STSP), it is known that a ratio of $\frac{3}{2}$ can be achieved [3]. Improving on this ratio has been a major open problem for almost three decades.

In this paper, we study an interesting further special case of Δ-STSP, namely, when the costs on the edges are one or two. More formally, we are given an undirected complete graph $G = (V, E)$ with cost function, $w : E \to \{1, 2\}$. Min(1,2)-STSP is the problem of finding a minimum cost tour which visits every vertex in V exactly once (Hamiltonian tour). The problem is known to be NP–Hard. Papadimitriou and Yannakakis [8] show that the problem is MAXSNP-complete. Papadimitriou and Yannakakis [8], in the same paper, provide an approximation algorithm that achieves a performance guarantee of $\frac{7}{6}$. This has remained as the best performance for Min(1,2)-STSP for more than a decade.

Several other special cases of TSP are studied in the literature. Prominent amongst them is the asymmetric traveling salesman problem with triangle inequality (Δ-ATSP), which asks for a directed Hamiltonian tour in a complete digraph with costs on edges, satisfying triangle inequality. Δ-STSP is a special case of Δ-ATSP. For Δ-ATSP, the best known approximation algorithm achieves a guarantee of the order of $\log(n)$, n being the number of vertices in the given instance [4,1,6]. For Min(1,2)-ATSP, there is a $\frac{5}{4}$-factor approximation [2].

M. Liśkiewicz and R. Reischuk (Eds.): FCT 2005, LNCS 3623, pp. 504–515, 2005.
© Springer-Verlag Berlin Heidelberg 2005

In this paper, we provide an improved approximation algorithm for Min(1,2)-STSP with a guarantee of $\frac{315}{271}$. It is the first improvement over the $\frac{7}{6}$-factor approximation due to Papadimitriou and Yannakakis [8] after a decade. A still better approximation algorithm with a performance guarantee of $\frac{65}{56}$ is possible, but for lack of space we omit the details. The main contributions of our paper are: an improvement procedure achieving a better performance guarantee and a better lower bound for the cost of an optimal Hamiltonian tour.

Though the improvement is little (in terms of approximation factor), the lower bound that we obtain by our method is significantly better than the usual cycle cover one and hence there is a suggestion of better performing approximations based on our lower bound subsequently.

Our algorithm relies on the fact that a minimum cost 4-cycle cover in a graph with costs on the edges as either one or two, can be computed in $O(n^3)$ time [5], as does the $\frac{7}{6}$-algorithm of Papadimitriou & Yannakakis [8]. On the other hand, if we use a minimum cost 3-cycle cover, which can be computed in $O(n^{2.5})$ time by reduction to maximum bipartite matching, we get a $\frac{31}{26}$ factor, which is still better than the $\frac{11}{9}$ factor algorithm due to [8] obtained when they make use of a minimum cost 3-cycle cover as well.

2   Preliminaries

Let $G = (V, E)$ be a complete undirected graph, with a cost function $w : E \to \{1, 2\}$. In this section, we discuss the $\frac{7}{6}$-factor approximation algorithm of Papadimitriou and Yannakakis [8] for Min(1,2)-STSP on the graph G. We use their basic framework and devise an improved approximation algorithm in the subsequent sections.

Minimum cost 4–cycle cover. A *cycle cover* of a graph G is a collection of vertex-disjoint cycles such that every vertex of G is present in a cycle. A *4-cycle cover* is a cycle cover in which the length of every cycle is at least four. A *minimum cost 4-cycle cover* is a 4-cycle cover whose cost is the least amongst all 4-cycle covers. In [5], an algorithm for computing a so-called maximum cardinality simple 2-matching without 3-cycles is given. This algorithm can be used to compute an optimum 4-cycle cover in graphs with edge costs one and two [7]. Using this algorithm, we compute a minimum cost 4-cycle cover $\mathcal{C}$.

We normalize it as follows: firstly we may assume that there is only one cycle z with cost two edges, since we may merge two such cycles without any additional costs. Secondly, we may assume that for each cost two edge (u, v) of z there is no cost one edge (v, x) or (u, x) of G for $x \notin z$, since otherwise we may merge the cycle in which x is present with z at no extra costs. After the normalization, we call the cycle cover again $\mathcal{C}$ and let the cycles of $\mathcal{C}\backslash\{z\}$ be $\{c_1, \ldots, c_r\}$. ζ_1 is the number of cost one edges in z, ζ_2 the number of cost two edges.

Bipartite graph. We form a bipartite graph B as follows: we have the node set $\mathcal{C}\backslash\{z\}$ on the one side and V on the other side. There is an edge (c, v) in B if and only if v does not belong to the cycle c and there is a vertex $u \in c$ such that $w(u, v) = 1$.

Lemma 1 ([8]). *B has a matching of cardinality at least $r - k'$, where k' is the number of cost two edges in an optimal TSP tour that visit a cycle from $\mathcal{C}\backslash\{z\}$.*

Decomposition of functions. We compute a maximum matching M in B. From M, we compute a function $F : \mathcal{C}\backslash\{z\} \to \mathcal{C}$. We represent $F = (\mathcal{C}, A)$ as a directed graph with $(c, c') \in A$ whenever $(c, u) \in M$ and u is a vertex of c'. From the construction of B, we know that $F(c) \neq c$ for all $c \in \mathcal{C}\backslash\{z\}$.

A c-node of F is a cycle in the cycle cover $\mathcal{C}$. We will use c-nodes to mean cycles and *vertices* to mean the vertices of the graph G.

Lemma 2 ([8]). *The function F has a spanning subgraph S consisting solely of node-disjoint trees of height one, paths of length two, and isolated c-nodes, such that every c-node in the domain of F is not an isolated c-node of S. Such a spanning subgraph S can be found in polynomial time.*

Papadimitriou and Yannakakis build a cycle out of each component of S that spans the same vertices. Each of these cycles has at least one cost two edge. By breaking these edges, these cycles can then be patched together to a Hamiltonian tour without any further extra costs. Then they show that the cost of this tour is at most $\frac{7}{6}$ that of an optimal tour.

Some of the components of S allow a locally better approximation performance. In the remainder of the work, we develop an enhancement procedure that ensure that a significant number of components of S have a locally better approximation; thus yielding an overall improvement.

Definitions and notation. We fix some notation and provide some definitions before describing our approximation algorithm. A c-root of the spanning subgraph S is a c-node that is the root of a tree in S. A c-leaf is a leaf of such a tree.

Definition 1. *For an edge (c, d) in S where c, d are c-nodes, we can associate a cost one edge (u, v) in G such that $u \in c$ and $v \in d$. This edge will be denoted as $ed(c, d) = (u, v)$. Moreover, d is the parent of c and we set $par(c) = v$.*

Definition 2. *A node $v \in V$ is called* matched, *if it belongs to $ed(e)$ for some edge $e \in S$. Otherwise, it is called* unmatched.

Consider a matched vertex v belonging to a cycle c. Let $v = v_0, v_1, \ldots, v_{j-1}$ denote the vertices of c (in that order). Let $i > 0$ be the smallest index such that v_i is unmatched. Similarly, let $l > 0$ be the smallest index such that v_{j-l} is unmatched. We define $M(v) = i + l - 1$ i.e., if we divide the vertices of c into contiguous blocks of matched vertices, then $M(v)$ denotes the length of the block v belongs to. If all vertices of c are matched, then $M(v) = j$. For unmatched vertices, we set $M(v) = 0$.

We introduce the notion of good c-nodes and good c-leaves. In order to speak of left and right, we give each cycle $c_1, \ldots, c_r$ of $\mathcal{C}$ an arbitrarily chosen orientation.

Definition 3. *1. Let c be a c-node that has at least one unmatched vertex. A matched vertex v of c is called* good, *if either $M(v)$ is even or $M(v)$ is odd and the vertex to the right of v is matched. Otherwise v is called* bad. *If all vertices of c are matched and the number of vertices of c is even, then all vertices of c are good. If the number* *is odd, we arbitrarily choose one of the vertices of c. This vertex is bad, all other vertices of c are good.*

2. *A c-leaf c is called good, if the vertex par(c) is good. Otherwise c is called bad.*

For the patching procedure to work properly, the spanning subgraph S and intermediate spanning subgraphs obtained from S should have a special form.

Definition 4. *A directed graph S as above is* valid, *if it consists solely of trees with at least two c-nodes and isolated c-nodes such that*

1. *for all c-nodes c and d of a tree that are both not the c-root, $par(c) \neq par(d)$,*
2. *for each c-node c that is neither a c-leaf nor the c-root, $par(c)$ is good, and*
3. *the parent of each bad c-leaf is the c-root of a tree.*

Furthermore, for all edges e of S, $ed(e)$ has cost one in G.

The initial S is valid, if it does not contain any path of length two. As all trees in S have height one, conditions 2 and 3 in the above definition are trivially fulfilled. So, the initial S is valid in this case. In a path of length two, the c-node in the middle might have a vertex that has degree four. If it does not, we can treat it as a tree of height one by making the c-node in the middle the c-root. If it has a degree four vertex, we also treat it as a tree of height one: we pretend that one of the two c-leaves is connected to a vertex (we will call this node *virtually matched*) of the c-node in the middle that is at distance two of the vertex it was originally connected to. We will show in Section 5.3 that this does not affect the algorithm.

3   Our Approximation Algorithm

We make progress by successively improving the spanning subgraph S. The idea behind an improvement step can be described as follows: Assume for simplicity that an optimum TSP tour has cost n, i.e., consists solely of cost one edges. Consider a bad c-leaf of S. Since an optimum TSP tour has only cost one edges, there is another cost one edge leaving c (in G). If we connect c via this new edge and remove the old edge from S, then it may happen that c now is a good c-leaf. As mentioned in the patching procedure of the algorithm of [8], the extra cost of patching a good c-leaf is lesser than that of a bad c-leaf.

In a first step, we identify the cases for which one can obtain such an improvement (see Definition 5). Then we describe the improvement procedure. Finally, we obtain the approximation algorithm followed by the analysis.

Definition 5. *An unmatched vertex v in some c-node c is called* improving, *if one of the following conditions is fulfilled:*

1. *c is an interior c-node and if x is the left neighbour of v in c, then $M(x)$ is odd.*
2. *c is a c-leaf of a tree with at least three c-nodes and $par(c)$ is not good.*
3. *c is a c-leaf or c-root of a tree with two c-nodes and v is one of the two neighbours of the matched vertex of c.*
4. *c is an isolated c-node.*

Procedure 1. Improvement Procedure

Input: A valid spanning subgraph S of G.
Output: A valid spanning subgraph of G that allows no further improvements and a collection
of edges $E' \subset E$.
1: **while** at least one of the following improvement steps is possible **do**
2:      Search for a bad c-leaf c of a tree with at least four c-nodes or an isolated c-node and a
        cost one edge $(u^c v)$ in G such that $u \in c$, $v \notin c$, and v is improving. Restructure S as
        described in Lemma 3.
3:      Search for a bad c-leaf c of a tree with exactly three c-nodes and a cost one edge $(u^c v)$ in
        G such that $u \in c$, $v \notin c$, and v is improving. If v is in the other c-leaf of the tree, then v
        should be a neighbour of the matched vertex. S is restructured as described in Lemma 3.
4: **end while**
5: **while** the following improvement step is possible **do**
6:      Search for a c-node c which is the c-root of a tree with at least three c-nodes and a cost one
        edge $(u^c v)$ in G such that $u \in c$ is an unmatched vertex with its left neighbour matched,
        $v \notin c$, and $v \in d$ is improving with d as the c-root of a tree (may be the same tree as c
        is in) with at least three c-nodes. Add the edge $(u^c v)$ to E'. The two bad leaves that are
        connected to the left neighbours of u and v become good.
7: **end while**
8: return the spanning subgraph S' obtained and the edge set E'.

With this definition, we give an improvement procedure, **Procedure 1**. This will be
used as a subroutine by our main approximation algorithm. Lemma 3 shows that the
improvement procedure correctly returns a valid spanning subgraph.

Lemma 3. *If S is the initial valid spanning subgraph of G, then we can find a spanning
subgraph S' of G such that S' is a valid subgraph which does not allow any more
improvements.*

Proof. We show that at the end of every type of improvement step, the spanning subgraph obtained remains valid. For this, we need to be able to identify the current collection of rooted trees and isolated nodes after each improvement step and verify whether
the properties of a valid subgraph are indeed maintained. We distinguish several cases.
In all the cases, we assume the following: c is a c-node with $u \in c$, $v \in d$, the cost of
(u, v) is one in G and v is improving.

In the first case, c is a bad c-leaf of a tree with at least four c-nodes. If d is an
interior c-node (can be the c-root) with at least one of the neighbours of v matched,
then after the addition of edge (u, v) and removal of the matched edge incident in S at
c, d becomes the new parent of c and c will be a good c-leaf. If d is a bad c-leaf of a tree
with at least three c-nodes, then remove the matched edge incident in S at d, remove the
matched edge incident at c, and add (u, v). Now, c, d form a new tree with two c-nodes.
Arbitrarily choose one of them as the root and the other as its child. On the other hand,
if d is a c-leaf (or a c-root respectively) of a tree with exactly two c-nodes with v being
one of the two neighbours of the matched vertex of d, then by removing the matched
edge incident at c and adding (u, v), d becomes the new c-root of a tree with exactly
three c-nodes with c as a good c-leaf—the other good c-leaf being the old c-root (or the
c-leaf respectively). Finally, if d is an isolated c-node, then again we remove the edge

---

Algorithm 2. The Approximation Algorithm

Input: An undirected complete graph $G = (V, E)$ with edge costs as one or two.
Output: A Hamiltonian tour $\mathcal{H}$ in G.
1: Compute a minimum cost 4-cycle cover $\mathcal{C}$. Normalize it as described in Section 2.
2: Construct a bipartite graph B as mentioned before and find a maximum matching M in B.
3: Compute a partial function F from M and find a spanning subgraph S of F consisting of trees of height one, isolated c-nodes and paths of length two. (Lemma 2)
4: Apply **Procedure 1** on S to obtain a valid spanning subgraph S' which does not allow any further improvements and an edge set $E' \subset E$.
5: Use the patching procedure described in Lemma 4 to patch the isolated c-nodes, the trees with two c-nodes, the bad c-leaves, and finally the good c-leaves that did not become good via an edge in E'.
6: Use the patching procedure described in Lemma 4 to patch the good c-leaves that became good when inserting an edge into E'.
7: The patching procedure yields a collection of vertex disjoint paths and cycles, each containing a cost two edge. Break a cost two edge in each cycle and patch the paths arbitrarily to get a single Hamiltonian cycle $\mathcal{H}$ without incurring additional costs.

---

incident at c and add (u, v) thereby c and d form a tree with exactly two c-nodes. Thus, in all these subcases, we have maintained the properties of a valid subgraph (please refer to Definition 4).

Next consider the case when c is a bad c-leaf of a tree with exactly three c-nodes. Additionally, $u \in c$ is adjacent to the matched vertex of c. All the subcases are the same as above, but the only issue is the case when both c and d are the bad c-leaves of the same tree. We simply remove the edge incident at d in S, and add (u, v). d becomes the new c-root of this tree and c and the old c-root are now two good c-leaves.

The final case is the improvement step in the second while-loop. Since in this case we do not modify the structure of S, the validity of the spanning subgraph trivially holds. Since only bad leaves become good it cannot be the case that one of the previous improvement steps is now applicable.   □

Algorithm 2 starts with computing a 4-cycle cover $\mathcal{C}$ and a spanning subgraph S as described in Section 2. Then it invokes **Procedure 1** to get a better subgraph S' and an additional subset of edges E'. It first patches the isolated c-nodes, the trees with two c-nodes, the bad c-leaves, and finally the good c-leaves that did not become good via an edge in E'. Finally, it patches the good leaves that became good by adding an edge to E'. The patching is described in Lemma 4. This lemma also analyses the cost of the tour in terms of the following parameters:

1. t denotes the number of trees with two c-nodes.
2. p denotes the number of trees of exactly three c-nodes with two bad c-leaves.
3. b denotes the number of bad c-leaves in trees with at least four nodes.
4. g denotes the number of good c-leaves.
5. s denotes the number of isolated c-nodes.
6. f denotes the number of unmatched vertices of c-roots of all trees whose left neighbour is unmatched. These vertices will be referred to as *free* vertices.

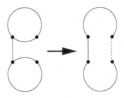

Fig. 1. Merging a tree with two c-nodes. The dashed edge may have cost two.

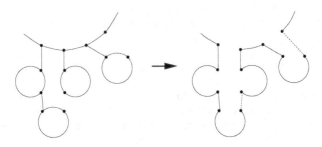

Fig. 2. With the left dashed edge, two good c-leaves are merged. With the right dashed edge, a bad c-leaf is merged.

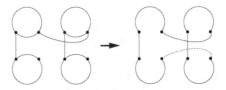

Fig. 3. Two good c-leaves joined with an edge from E' connecting two c-roots

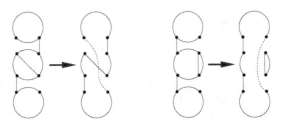

Fig. 4. Left-hand side: Two good c-leaves joined with an edge from E' that is inside a c-root. (The c-roots are the cycles in the middle.) We get one big cycle. Right-hand side: Here the patching produces two cycles. However, both of them have to contain cost two edges.

7. u denotes the number of unmatched vertices in the c-roots of trees whose left neighbour is matched.

8. m denotes the number of matched vertices in the c-roots of trees whose right neighbour is unmatched.

Lemma 4 (Total Cost). $w(\mathcal{H}) \leq n + \zeta_2 + t + s + 2p + b + \frac{1}{2}g$.

Proof. The total cost of the Hamiltonian tour $\mathcal{H}$ output by the algorithm is equal to the cost of the minimum cost 4-cycle cover and then the additional cost involved in patching the cycles.

If the component is a tree with two c-nodes, we patch the two cycles together as depicted in Figure 1. The newly formed cycle contains at most one additional cost two edge, thus we incur extra costs of one per tree yielding total costs of t.

If the component is a single c-node, we just break one edge of the cycle arbitrarily. This sums up to s.

A good c-leaf is connected, possibly via some interior c-nodes to a c-root. By construction of the improvement procedure, only two vertices of any such interior c-nodes have degree > 2, i.e., these interior nodes from a chain connecting the c-leaf to the c-root. Let u be the vertex of the c-root that this chain is attached to. If the c-leaf did not become good when adding an edge to E', the neighbour of u is also connected to such a chain ending in another good c-leaf. We patch these two good c-leaves with one cost two edge as depicted in Figure 2 (left dashed edge). This give a contribution of one per two good c-leaves

A bad c-leaf is patched as shown in Figure 2 (right dashed edge). By construction, bad c-leaves are always connected to the c-root. We get extra costs of one per bad c-leaf.

It remains to patch the good c-leaves that became good when an edge was put into E'. If such an edge connects to distinct c-roots, then we just patch it as depicted in Figure 3. But such an edge can also connect two vertices of the same c-root. The configuration shown in Figure 4 (left-hand side) can be easily patched. Patching the configuration on the right-hand side breaks the current cycle into two cycles. (Initially, the configuration on the right-hand side cannot exist, since improving vertices are always right to the bad vertices. When cycles get joined this may not be true any longer.) But both of the two cycles contain weight two edges, therefore no further costs are obtained when patching them later on to get the Hamiltonian tour. To see that this is the case, note that if we enter some cycle via some edge from E' we have to leave it via some bad leaf, since locally, the order of the vertices is preserved.

Once these patchings are done, the cycles obtained can be patched without further costs, since each cycle contains a cost two edge. The total extra cost in patching is $s + t + 2p + b + \frac{1}{2}g$. The cost of $\mathcal{C}$ is $n + \zeta_2$. □

4   Number of Isolated c-Nodes

We try to find a better bound for the number of isolated c-nodes of the spanning subgraph S' in terms of σ, the number of times an optimal tour T_{OPT} visits z. Here, T_{OPT} is an optimal tour which visits z the least number of times amongst all optimal tours.

Lemma 5. *Fix a cyclic orientation $\mathcal{O}$ of T_{OPT}. In this orientation, there is at most one edge of cost two entering cycle z, and at most one edge of cost two leaving cycle z.*

Proof. Suppose not. Then, in $\mathcal{O}$, either the cycle z is entered at least twice with edges of cost two or left at least twice with edges of cost two. Consider the former case (the latter case follows by symmetry). Let (u, v) and (u', v') be two edges of $\mathcal{O}$ of cost two each, with $u, u' \notin z$ and $v, v' \in z$. Delete these two edges and add the edges (u, u') and (v, v'). Let the new tour be denoted T'. Now T' is an optimal tour such that the number of times it visits z is lesser than that of T_{OPT}. This contradicts the choice of T_{OPT}. $\square$

Lemma 6. $\sigma \leq \zeta_1 \leq f + m + u + g - \zeta_2$.

Proof. The first inequality follows from Lemma 5 and the fact that z is *normalized* and that T_{OPT} is chosen as above. To see the second inequality, observe that if z is a c-root in the spanning subgraph S', then the vertices adjacent to cost one edges of z are either free vertices or matched by a bad c-leaf or unmatched vertices whose neighbour is matched or matched by a good c-leaf. Also, the total number of vertices in the cycle z is $\zeta_1 + \zeta_2$. If z appears as an isolated c-node, then all the vertices adjacent to cost one edges are free. $\square$

Lemma 7. $s \leq k' \leq k - \zeta_2 + \sigma$.

Proof. The first inequality follows from Lemma 1. Observe that T_{OPT} can have at most ζ_1 edges of cost one that have both endpoints in z, since otherwise, we get a cycle cover of lesser cost. The total number of edges in T_{OPT} with both endpoints in z is $\zeta_1 + \zeta_2 - \sigma$ and thus the second inequality follows. $\square$

Lemma 8. $s - f - 2b - g \leq k - 2\zeta_2$.

Proof. It follows from Lemmas 6 and 7, by noting the fact that $m + u = 2b$. $\square$

5   Performance Guarantee

5.1   Auxiliary Results

Lemma 9. *Let the parameters s, b, p, t, g, f of the spanning subgraph S' of Lemma 3 be as defined above. Then, the number of cost two edges in an optimal Hamiltonian tour k is at least $s + b + 2p - 4t - 5g - f$.*

Proof. Let T_{OPT} be an optimal Hamiltonian tour. Give an arbitrary cyclic orientation of the tour. We associate a distinct outgoing edge from each vertex u of S' depending on the c-node c that u belongs to.

If c is either a bad c-leaf or an isolated c-leaf, there is at least one edge going out of c in T_{OPT}. This becomes the associated edge of c. For every unmatched vertex $u \in c$ with c being the c-root of a tree where a neighbour of u is matched, there is one such outgoing edge.

Consider a tree with exactly three c-nodes, with two bad c-leaves. Note that if T_{OPT} leaves such a tree at least twice, then we can associate two edges with the tree. If not, i.e., if T_{OPT} leaves the tree exactly once, then we claim that there must be an internal cost two edge used by T_{OPT} (by an internal edge, we mean an edge whose both endpoints lie on the tree). This is clearly true since if every internal edge traversed by T_{OPT} is of cost one, this would mean that an additional improvement step could have been performed on the tree, a contradiction. Thus in both cases, we can associate two edges with these trees.

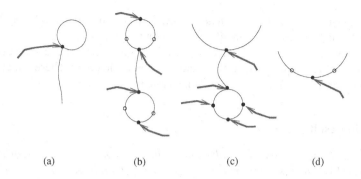

(a)   (b)   (c)   (d)

Fig. 5. [Proof of Lemma 9]. All incoming edges are possible cost one edges of an optimal tour. In (a), an edge comes into a matched vertex of a c-root. In (b), four edges enter a tree of exactly two c-nodes. Five edges enter a tree with a good c-leaf in (c), while in (d), an edge enters a free vertex.

Now, let us upper bound the number of associated edges of cost one. Since S' is not further improvable this means that cost one edges cannot come in at all vertices of a c-node, i.e., depending on the type of the c-node there are only a few possible vertices out of which cost one edges come in. For example, an associated edge may enter an already matched vertex or for a tree with exactly two c-nodes there are four possible vertices at which cost one edges can come in. For a good c-leaf c of a tree with at least three c-nodes at most five associated edges of cost one can come in and finally there may be an edge of cost one coming in at a free vertex. Please refer to Fig. 5 for an illustration of this argument. Hence, in total the number of cost two edges k in T_{OPT} is at least $s + b + u + 2p - m - 4t - 5g - f$. By noting that $u \geq m$, the Lemma follows.   □

Lemma 10 (Charging). $8t + 4s + 12p + \frac{19}{4}g + 6b \leq n - \zeta_2 - f$.

Proof. We estimate the number of vertices in G as follows:

Each tree with two nodes has at least eight vertices and each isolated node has at least four vertices, yielding total contributions of $8t$ and $4s$ respectively.

For every good leaf c, we associate at least five vertices, namely the vertices of c and the matched vertex of c. This way, we get at least $5g$ vertices. But the exception is when all the vertices (odd in number) of the root of the tree in which c is present, are matched. In such a case, we arbitrarily treat one of the good leaves as a bad c-leaf and hence we can associate only $5j - 1$ vertices with j good leaves, where $j \geq 4$. Thus, the contribution of the good leaves is at least $(5 - \frac{1}{j})g \geq \frac{19}{4}g$ vertices.

With every bad leaf c, we associate at least six vertices, namely the vertices of c, which are at least four in number, the matched vertex of the root incident to c and the vertex to the right of this vertex. The exceptional case of all the vertices of the root are matched has been handled before. Thus, the contribution of the bad leaves is at least $6b$.

For a tree with exactly three c-nodes, with two bad c-leaves we can associate twelve vertices (four per each c-node). Note that this includes paths of length two—for one bad c-leaf we charge four vertices of the c-leaf, a matched vertex of the c-root and the right neighbour of the matched vertex, while for the other bad c-leaf, we charge four vertices of the c-leaf, the *virtual matched vertex* of the c-root and the right neighbour of this virtual matched vertex. So, in all, the total contribution is $12p$.

No vertex has been counted twice. Also, at least one vertex of every cost two edge of z should not be counted, since $\mathcal{C}$ is a normalized cycle cover. There are at least ζ_2 such vertices in z. The free vertices are not charged.   □

5.2   Main Result

Theorem 1. *There exists a polynomial time approximation algorithm for* Min(1,2)-STSP *with a performance guarantee of* $\frac{315}{271}$. *Its running time is* $O(n^3)$.

Proof. Fix constants $\alpha = \frac{44}{271}$, $\beta = \frac{14}{271}$ and $\delta = \frac{7}{542}$. We shall show that Algorithm 2 has an approximation guarantee of $(1 + \alpha)$. Consider the following inequalities from Lemma 10, Lemma 9, and Lemma 8:

$$8t + 4s + 12p + \tfrac{19}{4}g + 6b \leq n - \zeta_2 - f \tag{1}$$

$$s + b + 2p - 4t - 5g \leq k + f \tag{2}$$

$$s - 2b - g \leq k - 2\zeta_2 + f \tag{3}$$

We upper bound the cost of patching, $t + s + 2p + \frac{1}{2}g + b$, (from Lemma 4), by the L.H.S of the inequality $\alpha*(1) + \beta*(2) + \delta*(3)$. To verify that this is indeed an upper bound, compare term by term on both sides of the required inequality, for example,

$$t : 8\alpha - 4\beta \geq 1; \quad p : 12\alpha + 2\beta \geq 2; \quad f : \alpha - \beta - \delta \geq 0$$

$$b : 6\alpha + \beta - 2\delta \geq 1 \text{ and } g : \tfrac{19}{4}\alpha - 5\beta - \delta \geq \tfrac{1}{2}$$

For the term s, we include all coefficients of k and ζ_2 also, since $s, \zeta_2 \leq k$. Thus, we require that $(1 - (4\alpha + \beta + \delta))s + (\beta + \delta)k + (1 - (\alpha + 2\delta))\zeta_2 \leq (1 + \alpha)k$. Equivalently, $4\alpha + \beta + 2\delta \geq 1 - 2\alpha + \beta$ It is easy to check that the given values of α, β and δ satisfy the above inequality and are optimal.

The running time of the algorithm is dominated by the time needed to compute a minimum weight 4-cycle cover.   □

5.3   Paths of Length Two

The algorithm is correct as long as there are no paths of length two in the initial S such that the c-node in the middle contains a matched vertex of degree four. Let u be such a node. Let x and y be its neighbours. Let v be the other neighbour of, say, x. So far, we treated this path also as a tree of height two with two bad c-leaves and pretended that one of them was attached to v and the other to u. All the considerations so far remain valid: whenever we would remove the edge (v, x) during patching, we remove the edge (x, u) instead and connect all cost two edges incident with u to v instead. There is only one exception: when two such paths are joined via two edges from E'. Let u', x', v', and y' the corresponding nodes of the second path. Assume that (x, x') and (y, y') are in E'. If we now patch as in Figures 3 and 4, we will get two cycles, but one of them consists solely of cost one edges. We have to break on additional edge of this cycle. This means that we introduce a total of three cost two edges (instead of two) to patch four good c-leaves, yielding an average cost of $\frac{3}{4}$ instead of $\frac{1}{2}$. However, since we have an even number of good leaves per root, we can count them with a coefficient of 5 in (1). In (2), we only have to take a factor of 3 into consideration, since the c-leaves were bad before. In (3) we get a factor of 0, since in this bound, only good leaves of z were counted. Finally, there are an additional four free nodes in the cycle consisting of cost one edges; each free node reduces the cost of patching by $\alpha - \beta - \delta = \frac{53}{542}$. But $5\alpha - 3\beta + \frac{53}{542} = \frac{409}{542} \geq \frac{3}{4}$. The details are spelled out in the forthcoming journal version of this work.

5.4 Further Improvements

The improvements of the following type are not considered in the improvement procedure: Let c be a good c-leaf of a tree with at least three c-nodes and a cost one edge (u, v) in G such that $u \in c$ is a neighbour of the matched vertex of c, $v \notin c$, and $v \in d$ is improving. Here d is a c-root with the left neighbour of v matched. When one considers these improvements also then the factor of $5g$ in Lemma 9 is reduced to $3g$. Hence, using the same arguments as in the proof of Theorem 1 we get a better approximation guarantee of $\frac{65}{56}$. Please refer to the journal version of the paper for details.

References

1. M. Bläser. A new approximation algorithm for the asymmetric TSP with triangle inequality. In *Proceedings of the 14th Annual ACM–SIAM Symposium on Discrete Algorithms (SODA)*, pages 639–647, 2003.

2. M. Bläser. A 3/4-approximation algorithm for maximum ATSP with weights zero and one. In *Proc. 7th Int. Workshop on Approximation Algorithms for Combinatorial Optimization Problems*, LNCS 3122, pages 61–71, 2004.

3. N. Christofides. Worst-case analysis of a new heuristic for the travelling salesman problem. Technical report, GSIA, Carnegie Mellon University, 1976.

4. A. Frieze, G. Galbiati, and F. Maffioli. On the worst–case performance of some algorithms for the asymmetric travelling salesman problem. *Networks*, 12:23–39, 1982.

5. D. B. Hartvigsen. *Extensions of matching theory*. PhD thesis, Carnegie–Mellon University, 1984.

6. H. Kaplan, M. Lewenstein, N. Shafrir, and M. Sviridenko. A 2/3 approximation for maximum asymmetric TSP by decomposing directed regular multi graphs. In *Proceedings of the 44th Annual IEEE Symposium on Foundations of Computer Science (FOCS)*, 2003.

7. B. Manthey, *On Approximating Restricted Cycle Covers*. arXiv:cs.CC/0504038 v1, April 11, 2005.

8. C. Papadimitriou and M. Yannakakis. The traveling salesman problem with distances one and two. *Math. of Operations Research*, 18:1–11, 1993.

9. S. Sahni and T. Gonzalez. P-complete approximation problems. *Journal of the ACM*, 23:555–565, 1976.

New Applications of Clique Separator Decomposition for the Maximum Weight Stable Set Problem

Andreas Brandstädt, Van Bang Le, and Suhail Mahfud

Institut für Informatik, Universität Rostock,
D-18051 Rostock, Germany
¶ab, le, suma◊@informatik.uni-rostock.de

Abstract. Graph decompositions such as decomposition by clique separators and modular decomposition are of crucial importance for designing efficient graph algorithms. Clique separators in graphs were used by Tarjan as a divide-and-conquer approach for solving various problems such as the Maximum Weight Stable Set (MWS) Problem, Coloring and Minimum Fill-in. The basic tool is a decomposition tree of the graph whose leaves have no clique separator (so-called *atoms*), and the problem can be solved efficiently on the graph if it is efficiently solvable on its atoms. We give new examples where the clique separator decomposition works well for the MWS problem which also improves and extends various recently published results. In particular, we describe the atom structure for some new classes of graphs whose atoms are P_5-free (the P_5 is the induced path with 5 vertices) and obtain new polynomial time results for MWS.

1 Introduction

In an undirected graph $G = (V, E)$, a *stable* (or *independent*) vertex set is a subset of mutually nonadjacent vertices. The *Maximum Weight Stable* (or *Independent*) *Set* (*MWS*) Problem asks for a stable set of maximum weight sum for a vertex weight function w on V. The *MS problem* is the MWS problem where all vertices have the same weight. Let $\alpha_w(G)$ ($\alpha(G)$) denote the maximum weight (maximum cardinality) of a stable vertex set in G.

The M(W)S problem is one of the fundamental algorithmic graph problems which frequently occurs as a subproblem in models in computer science and operations research. It is closely related to the Vertex Cover Problem and to the Maximum Clique Problem in graphs (for an extensive survey on the last one, see [10], which, at the same time, can be seen as a survey on the MWS and the Vertex Cover Problem; however, since 1999, there are many new results on this topic).

The MWS Problem is known to be NP-complete in general and remains NP-complete even on very restricted instances such as $K_{1,4}$-free graphs [48], $(K_{1,4},$diamond$)$-free graphs [26], very sparse planar graphs of maximum degree three and graphs not containing cycles below a certain length [53], in particular on triangle-free graphs [55].

M. Liśkiewicz and R. Reischuk (Eds.): FCT 2005, LNCS 3623, pp. 516–527, 2005.

On the other hand, it is known to be solvable in polynomial time on many graph classes by various techniques such as polyhedral optimization, augmenting, struction and other transformations, modular decomposition, bounded clique-width and bounded treewidth, reduction of α-redundant vertices, to mention some basic techniques; for a small selection of papers dealing with particular graph classes and such techniques for M(W)S, see[2-4,8,9,11-19,21-24,27-29,32-40,44,45,48,50-52,58]. Many of these papers deal with subclasses of P_5-free graphs, motivated by the fact that the complexity of the M(W)S problem for P_5-free graphs (and even for (P_5,C_5)-free graphs) is still unknown (for all other 5-vertex graphs H, MS is solvable in polynomial time on (P_5,H)-free graphs). For $2K_2$-free graphs, however, the following is known:

Farber in [30] has shown that a $2K_2$-free graph $G = (V, E)$ contains at most n^2 inclusion-maximal independent sets, $n = |V|$. Thus, the MWS problem on these graphs can be solved in time $\mathcal{O}(n^4)$ since Paull and Unger [54] gave a procedure that generates all maximal independent sets in a graph in $\mathcal{O}(n^2)$ time per generated set (see also [61,43]). This result has been generalized to $l \geq 2$: lK_2-free graphs have at most n^{2l-2} inclusion-maximal independent sets [1,5,31,56], and thus, MWS is solvable on lK_2-free graphs in time $\mathcal{O}(n^{2l})$.

Obviously, the MWS problem on a graph G with vertex weight function w can be reduced to the same problem on antineighborhoods of vertices in the following way:

$$\alpha_w(G) = \max\{w(v) + \alpha_w(G[\overline{N}(v)]) \mid v \in V\}$$

Now, let Π denote a graph property. A graph is *nearly Π* if for each of its vertices, the subgraph induced by the set of its nonneighbors has property Π. (Note that this notion appears in the literature in many variants, e.g., as nearly bipartite graphs [6].)

Thus, whenever MWS is solvable in time T on a class with property Π then it is solvable on nearly Π graphs in time $n \cdot T$. For example, Corneil, Perl and Stewart [27] gave a linear time algorithm for MWS on cographs along the cotree of such a graph. Thus, MWS is solvable in time $\mathcal{O}(nm)$ on nearly cographs. This simple fact, for example, immediately implies Theorem 1 of [32] (which is formulated in [32] for the Maximum Clique Problem and shown there in a more complicated way). For other examples where this approach is helpful, see [14].

A famous divide-and-conquer approach by using clique separators (also called clique cutsets) is described by Tarjan in [60] (see also [62]). For various problems on graphs such as Minimum fill-in, Coloring, Maximum Clique, and the MWS problem, it works well in a bottom-up way along a clique separator tree (which is not uniquely determined but can be constructed in polynomial time for a given graph). The leaves of such a tree, namely the subgraphs not containing clique separators are called *atoms* in [60]. Whenever MWS is solvable in time T on the atoms of a graph G, it is solvable in time $n^2 \cdot T$ on G. However, few examples are known where this approach could be applied for obtaining a polynomial time MWS algorithm on a graph class.

Modular decomposition of graphs is another powerful tool. The decomposition tree is uniquely determined and can be found in linear time [46]. The prime

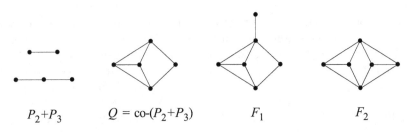

P_2+P_3 $Q = $ co-(P_2+P_3) F_1 F_2

Fig. 1. The $P_2 + P_3$, its complement co-$(P_2 + P_3)$ (called Q), and two Q extensions, called F_1 and F_2

nodes in the tree are the subgraphs having no homogeneous sets (definitions are given later). Again, various problems can be solved efficiently bottom-up along the modular decomposition tree, among them Maximum Clique, and the MWS problem, provided they can be solved efficiently on the prime nodes. In [14], it was shown that a combination of both decompositions is helpful for the MWS problem: If MWS is solvable in time T on prime atoms (i.e., prime subgraphs without clique cutset) of the graph G then it is solvable in time $n^2 \cdot T$ on G.

One of the examples where the clique separator approach works well is given by Alekseev in [3] showing that atoms of $(P_5,$co-$(P_2 + P_3))$-free graphs (the $(P_2 + P_3)$ is the graph with five vertices, say a, b, c, d, e and edges ab, cd, de) are $3K_2$-free which implies that the MWS problem is solvable in time $\mathcal{O}(n^8)$ on this graph class (see Figure 1 for the co-$(P_2 + P_3)$).

Our main results in this paper are the following ones:

(i) Atoms of (P_5,Q)-free graphs are either nearly $(P_5, \overline{P_5}, C_5)$-free or specific (i.e., a simple type of graphs defined later for which the MWS problem can be solved in the obvious way). This leads to an $\mathcal{O}(n^4 m)$ time algorithm for MWS on graphs whose atoms are (P_5,Q)-free which improves and extends Alekseev's result on these graphs [3] (and also the corresponding result of [35] on (P_5, C_5, Q)-free graphs).

(ii) Prime atoms of (P_5,F_1)-free graphs are $3K_2$-free (see Figure 1 for the F_1). By [14], this implies polynomial time for MWS on (P_5,F_1)-free graphs which extends corresponding polynomial time results on (P_5,Q)-free graphs, on $(P_5,$co-chair)-free graphs [24], and on (P_5,P)-free graphs [14,22,44] (note, however, that the time bound for (P_5,F_1)-free graphs is much worse than on the last two subclasses mentioned here).

(iii) Atoms of (P_5,F_2)-free graphs are $4K_2$-free (see again Figure 1 for the F_2). This also extends the result on (P_5,Q)-free graphs.

(iv) Finally, we show that for every fixed k, MS can be solved in polynomial time for (P_5,H_k)-free graphs (see Figure 3 for the H_k which extends F_1 and F_2).

The first three results give new examples for the power of clique separators.

For space limitations, all proofs are omitted but can be found in the full version of this paper.

2   Basic Notions

Throughout this paper, let $G = (V, E)$ be a finite undirected graph without self-loops and multiple edges and let $|V| = n$, $|E| = m$. Let $V(G) = V$ denote the vertex set of graph G. For a vertex $v \in V$, let $N(v) = \{u \mid uv \in E\}$ denote the (*open*) *neighborhood* of v in G, let $N[v] = \{v\} \cup \{u \mid uv \in E\}$ denote the (*closed*) *neighborhood* of v in G, and for a subset $U \subseteq V$ and a vertex $v \notin U$, let $N_U(v) = \{u \mid u \in U, uv \in E\}$ denote the *neighborhood* of v with respect to U. The *antineighborhood* $\overline{N}(v)$ is the set $V \setminus N[v]$ of vertices different from v which are nonadjacent to v. We also write $x \sim y$ for $xy \in E$ and $x \not\sim y$ for $xy \notin E$.

For $U \subseteq V$, let $G[U]$ denote the subgraph of G induced by U. Throughout this paper, all subgraphs are understood to be induced subgraphs. Let $\mathcal{F}$ denote a set of graphs. A graph G is $\mathcal{F}$-*free* if none of its induced subgraphs is in $\mathcal{F}$.

A vertex set $U \subseteq V$ is *stable* (or *independent*) in G if the vertices in U are pairwise nonadjacent. For a given graph with vertex weights, the Maximum Weight Stable Set (MWS) Problem asks for a stable set of maximum vertex weight.

Let co-$G = \overline{G} = (V, \overline{E})$ denote the *complement graph* of G. A vertex set $U \subseteq V$ is a *clique* in G if U is a stable set in $\overline{G}$. Let K_ℓ denote the clique with ℓ vertices, and let ℓK_1 denote the stable set with ℓ vertices. K_3 is called *triangle*. $G[U]$ is *co-connected* if $\overline{G}[U]$ is connected.

Disjoint vertex sets X, Y *form a join*, denoted by $X \text{①} Y$ (*co-join*, denoted by $X \text{⓪} Y$) if for all pairs $x \in X$, $y \in Y$, $xy \in E$ ($xy \notin E$) holds. We will also say that X has a join to Y, that there is a join between X and Y, or that X and Y are connected by join (and similarly for co-join). Subsequently, we will consider join and co-join also as operations, i.e., the co-join operation for disjoint vertex sets X and Y is the disjoint union of the subgraphs induced by X and Y (without edges between them), and the join operation for X and Y consists of the co-join operation for X and Y followed by adding all edges $xy \in E$, $x \in X$, $y \in Y$.

A vertex $z \in V$ *distinguishes* vertices $x, y \in V$ if $zx \in E$ and $zy \notin E$ or $zx \notin E$ and $zy \in E$. We also say that *a vertex z distinguishes a vertex set* $U \subseteq V$, $z \notin U$, if z has a neighbor and a non-neighbor in U.

Observation 1. *Let $v \in G[V \setminus U]$ distinguish U.*

(*i*) *If $G[U]$ is connected, then there exist two adjacent vertices $x, y \in U$ such that $v \sim x$ and $v \not\sim y$.*

(*ii*) *If $G[U]$ is co-connected, then there exist two nonadjacent vertices $x, y \in U$ such that $v \sim x$ and $v \not\sim y$.*

A vertex set $M \subseteq V$ is a *module* if no vertex from $V \setminus M$ distinguishes two vertices from M, i.e., every vertex $v \in V \setminus M$ has either a join or a co-join to M. A module is *trivial* if it is $\emptyset$, $V(G)$ or a one-elementary vertex set. A nontrivial module is also called a *homogeneous set*.

A graph G is *prime* if it contains only trivial modules. The notion of module plays a crucial role in the *modular* (or *substitution*) *decomposition* of graphs (and other discrete structures) which is of basic importance for the design of efficient algorithms - see e.g. [49] for modular decomposition of discrete structures and its algorithmic use and [46] for a linear-time algorithm constructing the modular decomposition tree of a given graph.

A *clique separator* or *clique cutset* in a connected graph G is a clique C such that $G[V \setminus C]$ is disconnected. An *atom* of G is a subgraph of G without clique cutset. See [60] for some algorithmic aspects of the clique separator decomposition.

For $k \geq 1$, let P_k denote a chordless path with k vertices and $k - 1$ edges. The $\overline{P_5}$ is also called *house*. For $k \geq 3$, let C_k denote a chordless cycle with k vertices and k edges. A *hole* is a C_k with $k \geq 5$, and an *antihole* is an $\overline{C_k}$ with $k \geq 5$. An *odd hole* (*odd antihole*, respectively) is a hole (antihole, respectively) with odd number of vertices.

The $2K_2$ is the co-C_4. More generally, the ℓK_2 consists of 2ℓ vertices, say, $x_1, \ldots, x_\ell, y_1, \ldots, y_\ell$ and edges $x_1 y_1, \ldots, x_\ell y_\ell$.

For an induced subgraph H of G, a vertex not in H is a *k-vertex* of H, if it has exactly k neighbors in H.

A graph is *chordal* if it contains no induced cycle C_k, $k \geq 4$. A graph is *weakly chordal* if it contains no hole and no antihole. See [20] for a detailed discussion of the importance and the many properties of chordal and weakly chordal graphs. Note that chordal graphs are those graphs whose atoms are cliques.

For a linear order $(v_1, \ldots, v_n)$ of the vertex set V, a well-known coloring heuristic assigns integers to the vertices from left to right such that each vertex v_i gets the smallest positive integer assigned to no neighbor v_j, $j < i$, of v_i. Chvátal defined the important notion of a *perfect order* of a graph $G = (V, E)$ as a linear order $(v_1, \ldots, v_n)$ of V such that for each $k \leq n$, the number of colors used by the preceding coloring heuristic equals the chromatic number of $G[\{v_1, \ldots, v_k\}]$.

A graph is *perfectly orderable* if it has a perfect order. See [20] for various characterizations and properties of these graphs. In particular, recognizing perfectly orderable graphs is NP-complete [47]). A graph G is *perfectly ordered* if a perfect order of G is given. Algorithmic consequences for perfectly orderable graphs rely heavily on this assumption.

3   Atoms of (P_5, Q)-Free Graphs Are Nearly $(P_5, \overline{P_5}, C_5)$-Free or Specific

In this section, we improve the following result:

Theorem 1 (Alekseev [3]). *Atoms of (P_5, Q)-free graphs are $3K_2$-free.*

Since $3K_2$-free graphs have at most n^4 maximal stable sets, the MWS problem is solvable in time $\mathcal{O}(n^8)$ on (P_5, Q)-free graphs by the clique cutset approach of Tarjan and a corresponding enumeration algorithm for all maximal stable sets in a $3K_2$-free graph. Theorem 1, however, does not give much structural insight.

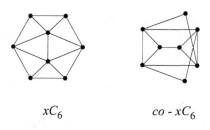

$$xC_6 \qquad\qquad co\text{-}xC_6$$

Fig. 2. A two-vertex extension xC_6 of the C_6 and its complement graph, the co-xC_6

Our main result of this section, namely Theorem 2, shows the close connection of (P_5, Q)-free graphs to known classes of perfect graphs and in particular leads to a faster MWS algorithm. Preparing this, we have to define a simple type of graphs which results from a certain extension of the $\overline{C_6}$ by two vertices (which we call $x\overline{C_6}$ or co-xC_6) and the complement of this graph (see Figure 2).

A graph is *specific* if it consists of a co-xC_6 H, a stable set consisting of 2-vertices of H having the same neighbors as one of the degree 2 vertices in H, and a clique U of universal (i.e., adjacent to all other) vertices. Note that the MWS problem for specific graphs can be solved in the obvious way.

Theorem 2. *Atoms of (P_5, Q)-free graphs are either nearly $(P_5, \overline{P_5}, C_5)$-free or specific graphs.*

The proof of Theorem 2 is based on the subsequent Lemmas 1, 2 and 3.

Lemma 1. *Atoms of (P_5, Q)-free graphs are nearly $\overline{P_5}$-free.*

Lemma 2. *Atoms of (P_5, Q)-free graphs containing an induced subgraph $x\overline{C_6}$ are specific graphs.*

Lemma 3. *Atoms of (P_5, Q)-free graphs are either nearly C_5-free or specific graphs.*

In [25], it has been observed that $(P_5, \overline{P_5}, C_5)$-free graphs are perfectly orderable, and a perfect order of such a graph can be constructed in linear time by a degree order of the vertices. Thus, also for $\overline{G}$, a perfect order can be obtained in linear time. In [42], Hoàng gave an $\mathcal{O}(nm)$ time algorithm for the Maximum Weight Clique problem on a perfectly ordered graph (i.e., with given perfect order). This means that the MWS problem on $(P_5, \overline{P_5}, C_5)$-free graphs can be solved in time $\mathcal{O}(nm)$ and consequently, it can be solved on nearly $(P_5, \overline{P_5}, C_5)$-free graphs in time $\mathcal{O}(n^2 m)$.

Now, by Theorem 2, MWS is solvable in time $\mathcal{O}(n^2 m)$ time on atoms of (P_5, Q)-free graphs. Then the clique separator approach of Tarjan implies:

Corollary 1. *The MWS problem can be solved in time $\mathcal{O}(n^4 m)$ on graphs whose atoms are (P_5, Q)-free.*

Note that this class is not restricted to (P_5, Q)-free graphs; it is only required that the atoms are (P_5, Q)-free. Thus, it contains, for example, all chordal graphs. The same remark holds for the other sections.

$(P_5, \overline{P_5}, C_5)$-free graphs are also those graphs which are Meyniel and co-Meyniel (see [20]); Meyniel graphs can be recognized in time $\mathcal{O}(m^2)$ [57]. Thus, nearly $(P_5, \overline{P_5}, C_5)$-free graphs can be recognized in time $\mathcal{O}(n^5)$ (which is even better than $\mathcal{O}(n^4 m)$).

Since $(P_5, \overline{P_5}, C_5)$-free graphs are weakly chordal, another consequence of Theorem 2 is:

Corollary 2. *Atoms of (P_5, Q)-free graphs are either nearly weakly chordal or specific.*

Note that weakly chordal graphs can be recognized in time $\mathcal{O}(m^2)$ [7,41]. Thus, recognizing whether G is nearly weakly chordal can be done in time $\mathcal{O}(nm^2)$. The time bound for MWS on weakly chordal graphs, however, is $\mathcal{O}(n^4)$ [59], and thus, worse than the one for $(P_5, \overline{P_5}, C_5)$-free graphs.

4   Minimal Cutsets in P_5-Free Graphs with ℓK_2

In this section we will collect some useful facts about P_5-free graphs that contain an induced ℓK_2. These facts will be used to prove our main results in Section 5 and represent a more detailed investigation of the background of Alekseev's Theorem 1.

Let $\ell \geq 2$ be an integer, and let G be a P_5-free graph containing an induced $H = \ell K_2$ with $E(H) = \{e_1, e_2, \ldots, e_\ell\}$. Let $S \subseteq V(G) \setminus V(H)$ be an inclusion-minimal vertex set such that, for $i \neq j$, e_i and e_j belong to distinct connected components of $G[V \setminus S]$. S is also called a *minimal cutset* for H. For $1 \leq i \leq \ell$, let H_i be the connected component of $G[V \setminus S]$ containing the edge e_i.

Observation 2.

(i) $\forall v \in S$: $N(v) \cap H_i = \emptyset$ for all $i \in \{1, 2, \ldots, \ell\}$, or $N(v) \cap H_i \neq \emptyset$ and $N(v) \cap H_j \neq \emptyset$ for at least two distinct indices i, j.

(ii) $\forall v \in S$: v distinguishes at most one H_i, $i \in \{1, 2, \ldots, \ell\}$.

By Observation 2, S can be partitioned into pairwise disjoint subsets as follows. For $L \subseteq \{1, 2, \ldots, \ell\}$, $|L| \geq 2$, let

$$S_L := \{v \in S \mid (\forall i \in L, N(v) \cap H_i \neq \emptyset) \wedge (\forall j \notin L, N(v) \cap H_j = \emptyset)\},$$

and

$$S_0 := S \setminus \left(\bigcup_{|L| \geq 2} S_L \right)$$

as well as

$$R_0 := V \setminus (S \cup H_1 \cup H_2 \cup \cdots \cup H_\ell).$$

Note that $(R_0 \cup S_0) \textcircled{0} (H_1 \cup H_2 \cup \cdots \cup H_\ell)$.

In what follows, L, M, N stand for subsets of $\{1, 2, \ldots, \ell\}$ with at least two elements. Two such subsets are called *incomparable* if each of them is not properly contained in the other. Incomparable sets L, M are *overlapping* if $L \cap M \neq \emptyset$. Note that disjoint sets are mutually incomparable.

Observation 3. *Let L and M be incomparable. Then, for all adjacent vertices $x \in S_L$, $y \in S_M$, $x \textcircled{1}(\bigcup_{i \in L \setminus M} H_i)$ and $y \textcircled{1}(\bigcup_{j \in M \setminus L} H_j)$.*

Observation 4. *Let L and M be overlapping. Then*

(i) $S_L \textcircled{1} S_M$, and
(ii) if $S_L \neq \emptyset$ and $S_M \neq \emptyset$ then $S_L \textcircled{1}(\bigcup_{j \in L \setminus M} H_j)$ and $S_M \textcircled{1}(\bigcup_{i \in M \setminus L} H_i)$.

Observation 5. *Let M be a proper subset of L. Then for all nonadjacent vertices $x \in S_M$, $y \in S_L$,*

(i) $y \textcircled{1}(\bigcup_{i \in L \setminus M} H_i)$, and
(ii) for all $j \in M$, $N(x) \cap H_j \subseteq N(y) \cap H_j$.

Observation 6. *Let $L \cap N = \emptyset$. If some vertex in S_L is nonadjacent to some vertex in S_N, then for all subsets M overlapping with L and with N, $S_M = \emptyset$.*

For each subset $L \subseteq \{1, 2, \dots, \ell\}$ with at least two elements we partition S_L into pairwise disjoint subsets as follows. Let

$$X_L := \{v \in S_L \mid \forall i \in L, v \textcircled{1} H_i\},$$

and for each $i \in L$,

$$Y_L^i := \{v \in S_L \mid v \text{ distinguishes } H_i\}.$$

By Observation 2 (ii),

$$\forall i \in L, Y_L^i \textcircled{1}\Big(\bigcup_{j \in L \setminus \{i\}} H_j \Big) \text{ and } S_L = X_L \cup \bigcup_{i \in L} Y_L^i.$$

Observation 7. *If $|L| \geq 3$ then for all distinct $i, j \in L$, $Y_L^i \textcircled{1} Y_L^j$.*

Observation 8. *If $|L| \geq 3$ and G is F_1-free or F_2-free then $X_L \textcircled{1}(S_L \setminus X_L)$.*

5 (P_5, F_1)-Free and (P_5, F_2)-Free Graphs

Theorem 3. *Prime (P_5, F_1)-free graphs without clique cutset are $3K_2$-free.*

By Theorem 3, prime (P_5, F_1)-free atoms are $3K_2$-free, hence MWS can be solved in time $O(n^5 m)$ on prime (P_5, F_1)-free atoms with n vertices and m edges. Combining with the time bound for MWS via clique separators, we obtain:

Corollary 3. *The MWS problem can be solved in time $O(n^7 m)$ for graphs whose atoms are (P_5, F_1)-free.*

Theorem 4. *(P_5, F_2)-free graphs without clique cutset are $4K_2$-free.*

By Theorem 4, (P_5, F_2)-free atoms are $4K_2$-free, hence MWS can be solved in time $O(n^7 m)$ on (P_5, F_2)-free atoms. Combining again with the clique separator time bound for MWS, we obtain:

Corollary 4. *Maximum Weight Stable Set can be solved in time $O(n^9 m)$ for graphs whose atoms are (P_5, F_2)-free graphs.*

6   Conclusion

In this paper, we give new applications of the clique separator approach, combine it in one case with modular decomposition and extend some known polynomial time results for the Maximum Weight Stable Set problem. In particular, we have shown:

(i) Atoms of (P_5,Q)-free graphs are either nearly $(P_5,\overline{P_5},C_5)$-free or specific which leads to an $\mathcal{O}(n^4 m)$ time algorithm for MWS on graphs whose atoms are (P_5,Q)-free improving a result by Alekseev [3].

(ii) Prime atoms of (P_5, F_1)-free graphs are $3K_2$-free.

(iii) Atoms of (P_5, F_2)-free graphs are $4K_2$-free.

As a consequence, the Maximum Weight Stable Set problem is polynomially solvable for graphs whose atoms are (P_5, F_1)-free $((P_5, F_2)$-free, respectively), which tremendously generalizes various polynomially solvable cases known before.

One way in trying to show that the Maximum Weight Stable Set problem can be solved in polynomial time on a large class of P_5-free graphs containing both classes of (P_5, F_1)-free graphs and of (P_5, F_2)-free graphs, is to consider the class of (P_5, H_k)-free graphs, for each fixed integer $k \geq 2$; see Figure 3.

Unfortunately, the technique used in this paper cannot be directly applied for (P_5, H_k)-free graphs. Namely, for each fixed $\ell \geq 3$, there exist prime (P_5, H_2)-free graphs that contain an induced ℓK_2 but no clique cutset. However, the unweighted case is easy:

Theorem 5. *For each fixed positive integer k, the Maximum Stable Set problem can be solved in polynomial time for (P_5, H_k)-free graphs.*

Open Problem. Let H_k^- denote the subgraph of H_k without the degree 1 vertex. Is the Maximum *Weight* Stable Set problem solvable in polynomial time for (P_5, H_k^-)-free graphs ($k \geq 3$ fixed)? If yes, the proof of Theorem 5 shows that it is also polynomially solvable for (P_5, H_k)-free graphs, for each fixed positive integer k.

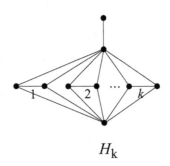

H_k

Fig. 3. The graph H_k

More generally, the following question is of interest: Suppose that MS is polynomially solvable for a certain graph class. Is MWS solvable in polynomial time on the same graph class, too?

References

1. V.E. Alekseev, On the number of maximal stable sets in graphs from hereditary classes, Combinatorial-algebraic Methods in Discrete Optimization, University of Nizhny Novgorod (1991) 5-8 (in Russian)
2. V.E. Alekseev, A polynomial algorithm for finding largest independent sets in fork-free graphs, Discrete Anal. Oper. Res. Ser. 1, 6 (1999) 3-19 (in Russian) (see also Discrete Applied Math. 135 (2004) 3-16 for the English version)
3. V.E. Alekseev, On easy and hard hereditary classes of graphs with respect to the independent set problem, Discrete Applied Math. 132 (2004) 17-26
4. C. Arbib, R. Mosca, On (P_5,diamond)-free graphs, Discrete Math. 250 (2002) 1-22
5. E. Balas, Ch.S. Yu, On graphs with polynomially solvable maximum-weight clique problem, Networks 19 (1989) 247-253
6. J. Bang-Jensen, J. Huang, G. MacGillivray, A. Yeo, Domination in convex bipartite and convex-round graphs, manuscript 2002
7. A. Berry, J.-P. Bordat, P. Heggernes, Recognizing weakly triangulated graphs by edge separability, Nordic Journal of Computing 7 (2000) 164-177
8. P. Bertolazzi, C. De Simone, A. Galluccio, A nice class for the vertex packing problem, Discrete Applied Math. 76 (1997) 3-19
9. H. Bodlaender, A. Brandstädt, D. Kratsch, M. Rao, J. Spinrad, Linear time algorithms for some NP-complete problems on (P_5,gem)-free graphs, FCT'2003, LNCS 2751, 61-72 (2003)
10. I.M. Bomze, M. Budinich, P.M. Pardalos, M. Pelillo, The maximum clique problem. In: D.-Z. Du, P.M. Pardalos, eds., Handbook of Combinatorial Optimization, Supplement Volume A, Kluwer Academic Press, Boston, MA, USA (1999)
11. A. Brandstädt, (P_5,diamond)-Free Graphs Revisited: Structure and Linear Time Optimization, Discrete Applied Math. 138 (2004) 13-27
12. A. Brandstädt, F.F. Dragan, On the linear and circular structure of (claw,net)-free graphs, Discrete Applied Math. 129 (2003) 285-303
13. A. Brandstädt, P.L. Hammer, On the stability number of claw-free P_5-free and more general graphs, Discrete Applied Math. 95 (1999) 163-167
14. A. Brandstädt, C.T. Hoàng, On clique separators, nearly chordal graphs, and the Maximum Weight Stable Set Problem, manuscript (2004); accepted for IPCO 2005, Berlin
15. A. Brandstädt, C.T. Hoàng, V.B. Le, Stability Number of Bull- and Chair-Free Graphs Revisited, Discrete Applied Math. 131 (2003) 39-50
16. A. Brandstädt, D. Kratsch, On the structure of (P_5,gem)-free graphs; Discrete Applied Math. 145 (2005) 155-166
17. A. Brandstädt, V.B. Le, H.N. de Ridder, Efficient Robust Algorithms for the Maximum Weight Stable Set Problem in Chair-Free Graph Classes, Information Processing Letters 89 (2004) 165-173
18. A. Brandstädt, H.-O. Le, V.B. Le, On α-redundant vertices in P_5-free graphs, Information Processing Letters 82 (2002) 119-122
19. A. Brandstädt, H.-O. Le, R. Mosca, Chordal co-gem-free graphs and (P_5,gem)-free graphs have bounded clique-width, Discrete Applied Math. 145 (2005) 232-241

20. A. Brandstädt, V.B. Le, and J.P. Spinrad, Graph Classes: A Survey, *SIAM Monographs on Discrete Math. Appl.*, Vol. 3, SIAM, Philadelphia (1999)

21. A. Brandstädt, H.-O. Le, J.-M. Vanherpe, Structure and Stability Number of (Chair,Co-P,Gem)-Free Graphs Revisited, *Information Processing Letters* 86 (2003) 161-167

22. A. Brandstädt and V.V. Lozin, A note on α-redundant vertices in graphs, *Discrete Applied Math.* 108 (2001) 301-308

23. A. Brandstädt, S. Mahfud, Maximum Weight Stable Set on graphs without claw and co-claw (and similar graph classes) can be solved in linear time, *Information Processing Letters* 84 (2002) 251-259

24. A. Brandstädt, R. Mosca, On the structure and stability number of P_5- and co-chair-free graphs, *Discrete Applied Math.* 132 (2004) 47-65

25. V. Chvátal, C.T. Hoàng, N.V.R. Mahadev, D. de Werra, Four classes of perfectly orderable graphs, *J. Graph Theory* 11 (1987) 481-495

26. D.G. Corneil, The complexity of generalized clique packing, *Discrete Applied Math.* 12 (1985) 233-240

27. D.G. Corneil, Y. Perl, and L.K. Stewart, Cographs: recognition, applications, and algorithms, *Congressus Numer.* 43 (1984) 249-258

28. C. De Simone, On the vertex packing problem, *Graphs and Combinatorics* 9 (1993) 19-30

29. C. De Simone, A. Sassano, Stability number of bull- and chair-free graphs, *Discrete Applied Math.* 41 (1993) 121-129

30. M. Farber, On diameters and radii of bridged graphs, *Discrete Math.* 73 (1989) 249-260

31. M. Farber, M. Hujter, Zs. Tuza, An upper bound on the number of cliques in a graph, *Networks* 23 (1993) 207-210

32. M.U. Gerber, A. Hertz, A transformation which preserves the clique number, *J. Combin. Th.* (B) 83 (2001) 320-330

33. M.U. Gerber, A. Hertz, D. Schindl, P_5-free augmenting graphs and the maximum stable set problem, *Discrete Applied Math.* 132 (2004) 109-119

34. M.U. Gerber, V.V. Lozin, Robust algorithms for the stable set problem, *Graphs and Combinatorics* 19 (2003) 347-356

35. M.U. Gerber, V.V. Lozin, On the stable set problem in special P_5-free graphs, *Discrete Applied Math.* 125 (2003) 215-224

36. V. Giakoumakis, I. Rusu, Weighted parameters in $(P_5, \overline{P_5})$-free graphs, *Discrete Applied Math.* 80 (1997) 255-261

37. M.C. Golumbic, P.L. Hammer, Stability in circular-arc graphs, *J. Algorithms* 9 (1988) 314-320

38. M. Grötschel, L. Lovász, and A. Schrijver, The ellipsoid method and its consequences in combinatorial optimization, *Combinatorica* 1 (1981) 169-197

39. P.L. Hammer, N.V.R. Mahadev, and D. de Werra, Stability in CAN-free graphs, *J. Combin. Th.* (B) 38 (1985) 23-30

40. P.L. Hammer, N.V.R. Mahadev, and D. de Werra, The struction of a graph: Application to CN-free graphs, *Combinatorica* 5 (1985) 141-147

41. R. Hayward, J.P. Spinrad, R. Sritharan, Weakly chordal graph algorithms via handles, *Proceedings* of 11th SODA'2000, 42-49 (2000)

42. C.T. Hoàng, Efficient algorithms for minimum weighted colouring of some classes of perfect graphs, *Discrete Applied Math.* 55 (1994) 133-143

43. D.S. Johnson, M. Yannakakis, C.H. Papadimitriou, On generating all maximal independent sets, *Information Processing Letters* 27 (1988) 119-123

44. V.V. Lozin, Stability in P_5- and banner-free graphs, *European J. Oper. Res.* 125 (2000) 292-297

45. V.V. Lozin, R. Mosca, Independent sets in extensions of $2K_2$-free graphs, *Discrete Applied Math.* 146 (2005) 74-80

46. R.M. McConnell, J. Spinrad, Modular decomposition and transitive orientation, *Discrete Math.* 201 (1999) 189-241

47. M. Middendorf, F. Pfeiffer, On the complexity of recognizing perfectly orderable graphs, *Discrete Math.* 80 (1990) 327-333

48. G.M. Minty, On Maximal Independent Sets of Vertices in Claw-Free Graphs, *J. Combin. Theory* (B) 28 (1980) 284-304

49. R.H. Möhring, F.J. Radermacher, Substitution decomposition for discrete structures and connections with combinatorial optimization, *Annals of Discrete Math.* 19 (1984) 257-356

50. R. Mosca, Polynomial algorithms for the maximum stable set problem on particular classes of P_5-free graphs, *Information Processing Letters* 61 (1997) 137-143

51. R. Mosca, Stable sets in certain P_6-free graphs, *Discrete Applied Math.* 92 (1999) 177-191

52. R. Mosca, Some results on maximum stable sets in certain P_5-free graphs, *Discrete Applied Math.* 132 (2003) 175-184

53. O.J. Murphy, Computing independent sets in graphs with large girth, *Discrete Applied Math.* 35 (1992) 167-170

54. M. Paull, S. Unger, Minimizing the number of states in incompletely specified sequential switching functions, *IRE Transactions on Electronic Computers* 8 (1959) 356-367

55. S. Poljak, A note on stable sets and colorings of graphs, *Commun. Math. Univ. Carolinae* 15 (1974) 307-309

56. E. Prisner, Graphs with few cliques, Graph Theory, Combinatorics and Algorithms, Kalamazoo, MI, 1992, 945-956, Wiley, New York, 1995

57. F. Roussel, I. Rusu, Holes and dominoes in Meyniel graphs, *Internat. J. Foundations of Comput. Sci.* 10 (1999) 127-146

58. N. Sbihi, Algorithme de recherche d'un stable de cardinalité maximum dans un graphe sans étoile, *Discrete Applied Math.* 29 (1980) 53-76

59. J.P. Spinrad, R. Sritharan, Algorithms for weakly chordal graphs, *Discrete Applied Math.* 19 (1995) 181-191

60. R.E. Tarjan, Decomposition by clique separators, *Discrete Math.* 55 (1985) 221-232

61. S. Tsukiyama, M. Ide, H. Ariyoshi, I. Shirakawa, A new algorithm for generating all the maximal independent sets, *SIAM J. Computing* 6 (1977) 505-517

62. S.H. Whitesides, A method for solving certain graph recognition and optimization problems, with applications to perfect graphs, in: Annals of Discrete Math. 21 (1984) Berge, C. and V. Chvátal (eds), *Topics on perfect graphs*, North-Holland, Amsterdam, 1984, 281-297

On the Expressiveness of Asynchronous Cellular Automata

Benedikt Bollig

Lehrstuhl für Informatik II, RWTH Aachen, Germany
bollig@informatik.rwth-aachen.de

Abstract. We show that a slightly extended version of asynchronous cellular automata, relative to any class of pomsets and dags without autoconcurrency, has the same expressive power as the existential fragment of monadic second-order logic. In doing so, we provide a framework that unifies many approaches to modeling distributed systems such as the models of asynchronous trace automata and communicating finite-state machines. As a byproduct, we exhibit classes of pomsets and dags for which the radius of graph acceptors can be reduced to 1.

1 Introduction

Distributed systems usually operate concurrently so that some actions do not depend on the occurrence of another. One possible single behavior of a distributed system can therefore be described naturally and in a compact manner by a partially ordered set, whose elements might depend on one another or be executed in either order, whatever the underlying partial-order relation specifies. A partially ordered set, in turn, can be represented by its covering relation or Hasse diagram, which allows to access pairs of events that may follow each other immediately. Accordingly, automata over partial orders, among them *asynchronous cellular automata* (ACAs), usually process input events along their cover relation in a transition-based manner. ACAs have been introduced originally by Zielonka in the framework of Mazurkiewicz traces [17]. They have been generalized by Droste et al. to run on pomsets without autoconcurrency and could be shown to be expressively equivalent to (existential) monadic second-order logic relative to both CROW-pomsets, which are subject to an axiom that considers concurrent read and exclusive owner write, and the more general k-pomsets [6].

However, there might be characteristics of the execution of a distributed system that cannot be captured by the cover relation of a partial order. Consider a directed acyclic graph whose edges reflect the send and receive of messages between sequential processes. Without any additional information (such as communication is both FIFO and reliable), a message exchange cannot be reconstructed from the underlying partial-order relation, which emerges from the reflexive transitive closure of the edge relation. We come across those structures within the framework of communicating systems. A communicating finite-state machine (CFM), for example, is equipped with FIFO channels for exchanging

M. Liśkiewicz and R. Reischuk (Eds.): FCT 2005, LNCS 3623, pp. 528–539, 2005.

messages. Apart from the ordering of events executed by one and the same sequential automaton, this implies an additional ordering of events that represent the send and receipt of one message. From the partial-order point of view, however, the relation of such communicating events is rather implicit and depends on semantic considerations. Modeling a behavior as a graph rather than a partial order allows for drawing additional edges to make communicating events visible so that automata can access them directly without going over the communication medium. Based on this observation, Kuske introduced the notion of directed acyclic graphs without autoconcurrency, generalizing the partial-order view. He accordingly extended ACAs and showed emptiness relative to the whole class of directed acyclic graphs without autoconcurrency to be decidable [11].

Nevertheless, ACAs are too weak to admit a general logical characterization in terms of a monadic second-order logic beyond classes such as CROW-pomsets and Mazurkiewicz traces, no matter if those structures are represented just by their partial order or by a directed acyclic graph. This is because ACAs process their actions depending solely on what happened in the past. But it is natural to provide systems with the possibility to make *communication requests*. Executing an action is then accompanied by a condition that eventually demands the occurrence of a suitable communication partner. For example, unless we deal with reliable channels anyway, sending a message might come along with the need for being received. Having in mind lossy channel systems with faulty communication [7,1], ACAs are not even able to specify the set of communication patterns where any message is received, which, however, should be easily formalizable in an appropriate logical calculus. In fact, we will show that, once ACAs are provided with the means of producing communication requests, they are exactly as expressive as the existential fragment of monadic second-order logic over graphs.

Similarly to ACAs, a *vertex-marking graph automaton*, as introduced in [13], collects states it has already read and thereupon assigns a new state to a common successor vertex. Following the idea of communication requests, an acceptance condition in terms of final states is implicitly given by the requirement that any event has to conform to the type of its state. Vertex-marking graph automata appear as a special case of Thomas' graph acceptors [14,16], which accept a given graph if it can be tiled consistently with a finite supply of patterns. As, relative to a graph class of bounded degree, any first-order definable property is *locally threshold testable* [9], graph acceptors capture precisely the projections of locally threshold testable languages. In turn, this yields a characterization of graph acceptors in terms of the existential fragment of monadic second-order logic. Hereby, the restriction to patterns of radius 1 means loss of expressivity in general. However, in characterizing ACAs (with types) logically, we exhibit many classes of graphs that are relevant for describing the behavior of distributed systems and where the use of patterns of radius 1 is sufficient. The latter property is shared by the domains of words, traces, trees, and grids [15]. The technique we apply generalizes the one used in [2] to characterize the class of CFMs.

Altogether, we combine the models of asynchronous cellular and vertex-marking graph automata towards the new notion of asynchronous cellular automata with types, which encompasses well-known automata concepts and allows a uniform embedding of many existing models of concurrency. Our main contribution is their characterization in terms of existential monadic-second order logic relative to an arbitrary class of directed acyclic graphs without autoconcurrency, covering the expressivity results of subsumed automata models.

The next section recalls the basic concepts of graphs, (existential) monadic second-order logic, and graph acceptors. Section 3 introduces asynchronous automata with types and studies its expressiveness. Finally, Section 4 provides the link between our work and established automata models.

2   Preliminaries

For this section, we fix an alphabet Σ. A *finite directed acyclic graph* (finite dag) is a pair $(V, \lhd)$ where V is its nonempty finite set of *vertices* and $\lhd \subseteq V \times V$ is the *edge relation* such that $\lhd$ is irreflexive and $\lhd^*$ is a partial-order relation. A Σ-*labeled finite dag* (Σ-dag, for short) is a triple $(V, \lhd, \lambda)$ where $(V, \lhd)$ is a finite dag and λ is a mapping $V \to \Sigma$, which we call the *vertex-labeling function*. The set of Σ-dags is denoted by $\mathbb{DAG}(\Sigma)$. We sometimes write $\leq$ for $\lhd^*$ and abbreviate $\lhd^+$ by $<$. Moreover, for $x, y \in V$, let us write $x \lessdot y$ if both $x < y$ and, for any $z \in V$, $x < z \leq y$ implies $z = y$. Then, $(V, \lessdot)$ and $\lessdot$ are called the *Hasse diagram* of $(V, \lhd)$ and, respectively, the *covering relation* of $\leq$. The degree of some $\mathcal{K} \subseteq \mathbb{DAG}(\Sigma)$ is said to be *bounded* if there is some $B \in \mathbb{N}$ such that, for any $(V, \lhd, \lambda) \in \mathcal{K}$ and any $x \in V$, $|\{y \in V \mid x \, (\lhd \cup \lhd^{-1}) \, y\}| \leq B$. As usual, we will identify isomorphic structures in the following.

Monadic Second-Order Logic. Formulas from monadic second-order (MSO) logic (over Σ) involve first-order variables $x, y, \ldots$ for vertices and second-order variables $X, Y, X_1, X_2, \ldots$ for sets of vertices. They are built up from the atomic formulas $\lambda(x) = a$ (for $a \in \Sigma$), $x \in X$, $x \lhd y$, and $x = y$ and furthermore allow the connectives $\neg$, $\vee$, $\wedge$, $\to$, $\leftrightarrow$ as well as the quantifiers $\exists$, $\forall$, which can be applied to either kind of variable. Formulas without free variables, which do not occur within the scope of a quantifier, are called *sentences*. Given $\mathcal{D} = (V, \lhd, \lambda) \in \mathbb{DAG}(\Sigma)$ and an MSO sentence φ, the validity of the satisfaction relation $\mathcal{D} \models \varphi$ is defined canonically with the understanding that first-order variables range over vertices from V and second-order variables over subsets of V. For a set $\mathcal{K} \subseteq \mathbb{DAG}(\Sigma)$ and an MSO sentence φ, the *language* of φ relative to $\mathcal{K}$, denoted by $L_\mathcal{K}(\varphi)$, is the set of Σ-dags $\mathcal{D} \in \mathcal{K}$ with $\mathcal{D} \models \varphi$. The class of subsets of $\mathbb{DAG}(\Sigma)$ that can be defined by some MSO sentence φ relative to $\mathcal{K}$ is denoted by $\mathcal{MSO}_\mathcal{K}(\Sigma)$. An important fragment of MSO logic is captured by *existential* MSO (EMSO) formulas, which are of the form $\exists X_1 \ldots \exists X_n \varphi$ where φ does not contain any set quantifier. In many cases, the restriction to EMSO formulas suffices to characterize recognizability in terms of automata, e.g., in the domains of words, trees, and Mazurkiewicz traces. Sometimes, however, we even have to restrict to EMSO formulas not to exceed recognizability, because full MSO logic is too expressive in general. The latter applies, for example, to grids and graphs [12] and MSCs [2]. The class $\mathcal{EMSO}_\mathcal{K}(\Sigma)$ is defined canonically.

Graph Acceptors. Let R be a natural. Given a Σ-dag $\mathcal{D} = (V, \lhd, \lambda)$ and vertices $x, y \in V$, the *distance* $d_{\mathcal{D}}(x, y)$ from x to y in $\mathcal{D}$ is ∞ if $(x, y) \notin (\lhd \cup \lhd^{-1})^*$ and, otherwise, the minimal natural number k such that there is a sequence $x_0, \dots, x_k \in V$ with $x_0 = x$, $x_k = y$, and $x_i (\lhd \cup \lhd^{-1}) x_{i+1}$ for each $i \in \{0, \dots, k-1\}$. Sometimes, if it is clear from the context, we omit the subscript $\mathcal{D}$ just writing $d(x, y)$. An *R-sphere* over Σ is a Σ-dag $s = (V, \lhd, \lambda, \gamma)$ together with a designated *sphere center* $\gamma \in V$ such that, for any $x \in V$, $d(x, \gamma) \leq R$. For a Σ-dag $\mathcal{D} = (V, \lhd, \lambda)$ and $x \in V$, let the *R-sphere of $\mathcal{D}$ around x*, denoted by $R\text{-Sph}(\mathcal{D}, x)$, be given by $(V', \lhd', \lambda', x)$ where $V' = \{x' \in V \mid d_{\mathcal{D}}(x', x) \leq R\}$, $\lhd' = \lhd \cap (V' \times V')$, and λ' is the restriction of λ to V'. Figure 1 (b) shows a 1-sphere over $\{a, b, c, d\}$ (with the rectangle as sphere center). It precisely deals with the 1-sphere of the graph aside (Figure 1 (a)) around the d-labeled vertex.

A graph acceptor [14,16] works on a graph as follows: it first assigns to each node one of its control states and then checks if the local neighborhood of each node (incorporating the states) corresponds to a pattern from a finite supply of spheres. More precisely, a *graph acceptor* over Σ is a structure $\mathcal{B} = (Q, R, \mathcal{S}, Occ)$ where Q is its nonempty finite set of *control states*, $R \in \mathbb{N}$ is the *radius*, $\mathcal{S}$ is a finite set of R-spheres over $\Sigma \times Q$, and Occ is a boolean combination of *conditions* of the form "sphere $s \in \mathcal{S}$ occurs at least $n \in \mathbb{N}$ times". A *run* of $\mathcal{B}$ on a Σ-dag $\mathcal{D} = (V, \lhd, \lambda)$ is a mapping $\rho : V \to Q$ such that, for any $x \in V$, the R-sphere of $(V, \lhd, (\lambda, \rho))$ around x is isomorphic to some $s \in \mathcal{S}$. We call ρ *accepting* if the tiling of $\mathcal{D}$ with spheres from $\mathcal{S}$ satisfies Occ. (In the tiling induced by ρ, sphere $s \in \mathcal{S}$ occurs $|\{x \in V \mid s = R\text{-Sph}((V, \lhd, (\lambda, \rho)), x)\}|$ times.) The *language* of $\mathcal{B}$ relative to a class $\mathcal{K} \subseteq \mathbb{DAG}(\Sigma)$, denoted by $L_{\mathcal{K}}(\mathcal{B})$, is the set of Σ-dags $\mathcal{D} \in \mathcal{K}$ on which there is an accepting run of $\mathcal{B}$. Moreover, we set $\mathcal{GA}_{\mathcal{K}}(\Sigma)$ to be $\{L \subseteq \mathbb{DAG}(\Sigma) \mid L = L_{\mathcal{K}}(\mathcal{B})$ for some graph acceptor $\mathcal{B}$ over $\Sigma\}$.

Theorem 1 ([14,16]). *For any $\mathcal{K} \subseteq \mathbb{DAG}(\Sigma)$ of bounded degree, $\mathcal{GA}_{\mathcal{K}}(\Sigma) = \mathcal{EMSO}_{\mathcal{K}}(\Sigma)$.*

An interesting class of graph languages is characterized by graph acceptors that restrict to 1-spheres [15]. So let us denote by $1\text{-}\mathcal{GA}_{\mathcal{K}}(\Sigma)$ the class $\{L \subseteq \mathbb{DAG}(\Sigma) \mid L = L_{\mathcal{K}}(\mathcal{B})$ for some graph acceptor $\mathcal{B} = (Q, R, \mathcal{S}, Occ)$ over Σ with $R = 1\}$. In general, such restricted graph acceptors are strictly weaker. However, we will identify classes of graph languages that allow to restrict to 1-spheres.

3 $\widetilde{\Sigma}$-dags and Asynchronous Cellular Automata

For the rest of this paper, we fix a nonempty finite set Ag of at least two *agents* and a *distributed alphabet* $\widetilde{\Sigma}$, which is a tuple $(\Sigma_i)_{i \in Ag}$ of (not necessarily disjoint) alphabets such that (for rather technical reasons) $\Sigma_i \nsubseteq \Sigma_j$ for any $i \neq j$. Let in the following Σ stand for $\bigcup_{i \in Ag} \Sigma_i$, the set of *actions*. Elements from Σ_i are understood to be actions that are performed by agent i. So let, for $a \in \Sigma$, $loc(a) := \{i \in Ag \mid a \in \Sigma_i\}$ denote the set of agents that are involved in a. Having this in mind, we say that actions a and b are *independent* and write $a \, I_{\widetilde{\Sigma}} \, b$ if there is no common agent that controls both of them, i.e., if $loc(a) \cap loc(b) = \emptyset$.

Fig. 1. An $(\{a\}, \{b,c\}, \{c,d\})$-dag and a 1-sphere

Otherwise, we say a and b are dependent, writing $a \, D_{\widetilde{\Sigma}} \, b$. We now introduce the model representing the behavior of a system of communicating agents. In doing so, we combine the standard models of [6] and [11]:

Definition 1. *A $\widetilde{\Sigma}$-dag is a Σ-dag $(V, \lhd, \lambda)$ such that*

- *for any $i \in Ag$, $\lambda^{-1}(\Sigma_i)$ is linearly ordered by $\leq$ and*
- *for any $(x,y), (x',y') \in \lhd$ with $\lambda(x) \, D_{\widetilde{\Sigma}} \, \lambda(x')$ and $\lambda(y) \, D_{\widetilde{\Sigma}} \, \lambda(y')$, we have $x \leq x'$ iff $y \leq y'$.*

Thus, for any $x \in V$ and $a \in \Sigma$, there is at most one vertex $y \in V$ such that both $x \lhd y$ ($y \lhd x$) and $\lambda(y) = a$. Intuitively, the second condition makes sure that communication between two processes cannot cross. When we consider communicating systems with message exchange, this corresponds to a FIFO architecture where messages (x,y) and (x',y') of equal type are received in terms of y and y' in the order they have been sent in terms of x and x', respectively.

An $(\{a\}, \{b,c\}, \{c,d\})$-dag is depicted in Figure 1 (a). We denote by $\mathbb{DAG}(\widetilde{\Sigma})$ the set of $\widetilde{\Sigma}$-dags. Note that any $\mathcal{K} \subseteq \mathbb{DAG}(\widetilde{\Sigma})$ has bounded degree. A useful subclass of $\mathbb{DAG}(\widetilde{\Sigma})$, denoted by $\mathbb{DAG}_H(\widetilde{\Sigma})$, is the set of graphs $(V, \lhd, \lambda) \in \mathbb{DAG}(\widetilde{\Sigma})$ such that $\lhd \, = \, \lessdot$, i.e., $(V, \lhd)$ is the Hasse diagram of some partially ordered set. Let $(V, \lhd, \lambda)$ be a $\widetilde{\Sigma}$-dag and let $x \in V$. For $i \in Ag$, we say that x is Σ_i-maximal if $\lambda(x) \in \Sigma_i$ and there is no $y \in \lambda^{-1}(\Sigma_i)$ such that $x < y$. Note that there is at most one Σ_i-maximal vertex. We denote by $\mathrm{R}(x) := \{a \in \Sigma \mid$ there is some $y \in V$ such that $y \lhd x$ and $\lambda(y) = a\}$ the *read domain* of x and, given $a \in \mathrm{R}(x)$, let a-pred(x) be the unique vertex y such that both $y \lhd x$ and $\lambda(y) = a$. Accordingly, let $\mathrm{W}(x) := \{a \in \Sigma \mid$ there is some $y \in V$ such that $x \lhd y$ and $\lambda(y) = a\}$ be the *write domain* of x and, for $a \in \mathrm{W}(x)$, a-succ(x) denote the unique vertex y such that both $x \lhd y$ and $\lambda(y) = a$.

Example 1 (Mazurkiewicz Traces). A *(Mazurkiewicz) trace* [5] over $\widetilde{\Sigma}$ is a $\widetilde{\Sigma}$-dag $(V, \lhd, \lambda) \in \mathbb{DAG}_H(\widetilde{\Sigma})$ such that, for any $x,y \in V$, $x \lhd y$ implies $\lambda(x) \, D_{\widetilde{\Sigma}} \, \lambda(y)$. Note that, as we consider a subclass of $\mathbb{DAG}_H(\widetilde{\Sigma})$, $\lhd$ and $\lessdot$ coincide. The set of traces over $\widetilde{\Sigma}$ is denoted by $\mathbb{TR}(\widetilde{\Sigma})$. The dag from Figure 1 (a) is clearly not a trace over $(\{a\}, \{b,c\}, \{c,d\})$: neither is it a Hasse diagram, nor are neighboring vertices consistently labeled with dependent actions.

Example 2 (Message Sequence Charts). Messages might be exchanged between the agents of a distributed system by performing send and receive actions. So set, for an agent $i \in Ag$, Γ_i to be $\{i!j \mid j \in Ag \setminus \{i\}\} \cup \{i?j \mid j \in Ag \setminus \{i\}\}$, the set of *(communication) actions* of agent i. The action $i!j$ is to be read as "i sends a message to j", while $j?i$ is the complementary action of receiving a message sent from i to j. Moreover, let Γ stand for the union of the Γ_i and set $\widetilde{\Gamma}$ to be the distributed alphabet $(\Gamma_i)_{i \in Ag}$. A *message sequence chart* (MSC) over Ag is a $\widetilde{\Gamma}$-dag $(V, \lhd, \lambda)$ such that (i) for any $i \in Ag$, $\lhd \cap (\lambda^{-1}(\Gamma_i) \times \lambda^{-1}(\Gamma_i))$ is the cover relation of some linear order, (ii) for any $(x, y) \in \lhd$ satisfying $\lambda(x) I_{\widetilde{\Gamma}} \lambda(y)$, $\lambda(x)$ is a send action and $\lambda(y)$ is its complementary receive, and (iii) for any $x \in V$, there is $y \in V$ satisfying both $\lambda(x) I_{\widetilde{\Gamma}} \lambda(y)$ and $x (\lhd \cup \lhd^{-1}) y$. We denote by $\text{MSC}(Ag)$ the set of MSCs over Ag. Note that, by the definition of a $\widetilde{\Gamma}$-dag, an MSC behaves in a FIFO manner, neglecting overtaking of messages of equal type. If we do not require a send vertex to be equipped with a corresponding receive, we obtain the class of (potentially) *lossy* MSCs over Ag, which is a superset of $\text{MSC}(Ag)$ and shall be denoted by $\text{LMSC}(Ag)$. If, in Figures 2 (a) and (b), we replace b with $1!2$ and a with $2?1$, we obtain an MSC over $\{1, 2\}$ and, respectively, a lossy MSC over $\{1, 2\}$, which is not an MSC. Note that $\text{MSC}(Ag)$ might be defined relative to $\text{LMSC}(Ag)$ by the (first-order) sentence $\forall x \bigwedge_{i \in Ag, \, j \in Ag \setminus \{i\}} (\lambda(x) = i!j \rightarrow \exists y (x \lhd y \land \lambda(y) = j?i))$.

Definition 2. *An* asynchronous cellular automaton with types (ACAT) *over* $\widetilde{\Sigma}$ *is a structure* (Q, Δ, T, F) *where*

- Q *is the nonempty finite set of* states,
- $\Delta \subseteq (Q \uplus \{-\})^{\Sigma} \times \Sigma \times Q$ *is the set of* transitions,
- $T : (\Sigma \times Q) \rightarrow 2^{\Sigma}$ *is the* type function *such that* $b \in T(a, q)$ *implies* $a \, I_{\widetilde{\Sigma}} \, b$,
- $F \subseteq (Q \uplus \{i\})^{Ag}$ *is the set of* final states.

So let $\mathcal{A} = (Q, \Delta, T, F)$ be an ACAT over $\widetilde{\Sigma}$. Note that $\overline{q} \in (Q \uplus \{-\})^{\Sigma}$ can be considered to be a subset of $\Sigma \times Q$ with the understanding that, for any $a \in \Sigma$ and $q \in Q$, $(a, q) \in \overline{q}$ iff $\overline{q}[a] = q$. In the following, we often write a transition $(\overline{q}, a, q) \in \Delta$ as $\overline{q} \rightarrow (a, q)$ with $\overline{q}$ being a subset of $\Sigma \times Q$. Let $\mathcal{D} = (V, \lhd, \lambda)$ be a Σ-dag and r be a mapping $V \rightarrow Q$. We define a mapping $r^- : V \rightarrow (Q \uplus \{-\})^{\Sigma}$ setting $r^-(x)[a]$ to be $-$ if $a \notin \text{R}(x)$ and to be $r(a\text{-pred}(x))$ if $a \in \text{R}(x)$. A *run* of $\mathcal{A}$ on $\mathcal{D}$ is a mapping $r : V \rightarrow Q$ such that, for any $x \in V$, we have $(r^-(x), \lambda(x), r(x)) \in \Delta$. It remains to constitute when r is accepting. For any $i \in Ag$ with $\lambda^{-1}(\Sigma_i) \neq \emptyset$, let f_i be given by $r(x)$ where x is the Σ_i-maximal vertex in V. For any other $i \in Ag$, set f_i to be $\imath$. The run r is *accepting* if both $(f_i)_{i \in Ag} \in F$ and, for any $x \in V$, we have $T(\lambda(x), r(x)) \subseteq \text{W}(x)$. The intuition behind the latter condition is that we require $\text{W}(x)$ to contain at least the communication requests imposed by the type function of the automaton.[1] Given $\mathcal{K} \subseteq \text{DAG}(\widetilde{\Sigma})$, we denote by $L_{\mathcal{K}}(\mathcal{A})$ the *language* of $\mathcal{A}$ relative to $\mathcal{K}$, i.e.,

---

[1] Note that we could even require $T(\lambda(x), r(x)) = \text{W}(x) \cap \{a \in \Sigma \mid a \, I_{\widetilde{\Sigma}} \, \lambda(x)\}$ without affecting the expressiveness of ACATs.

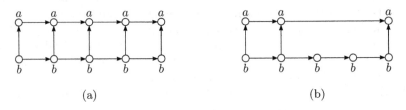

Fig. 2. The $(\{a\}, \{b\})$-dags $\mathcal{D}_5$ and $\mathcal{D}_5[2,4]$

the set of $\widetilde{\Sigma}$-dags $\mathcal{D} \in \mathcal{K}$ such that there is an accepting run of $\mathcal{A}$ on $\mathcal{D}$. An ACAT $\mathcal{A} = (Q, \Delta, T, F)$ over $\widetilde{\Sigma}$ is called an *asynchronous cellular automaton* (ACA) over $\widetilde{\Sigma}$ if $T(a, q) = \emptyset$ for any $a \in \Sigma$ and $q \in Q$. For $\mathcal{K} \subseteq \mathbb{DAG}(\widetilde{\Sigma})$, let $\mathcal{ACAT}_\mathcal{K}(\widetilde{\Sigma}) = \{L \subseteq \mathbb{DAG}(\widetilde{\Sigma}) \mid L = L_\mathcal{K}(\mathcal{A})$ for some ACAT $\mathcal{A}$ over $\widetilde{\Sigma}\}$. The class $\mathcal{ACA}_\mathcal{K}(\widetilde{\Sigma})$ is defined accordingly.

Example 3. Let L be the set of $(\{a\}, \{b\})$-dags $\mathcal{D}_n = (V_n, \lhd_n, \lambda_n)$, $n \geq 1$, where $V_n = \{x_1, \ldots, x_n, y_1, \ldots, y_n\}$, $\lhd_n$ is the union of $\{(x_i, x_{i+1}) \mid i \in \{1, \ldots, n-1\}\}$, $\{(y_i, y_{i+1}) \mid i \in \{1, \ldots, n-1\}\}$, and $\{(y_i, x_i) \mid i \in \{1, \ldots, n\}\}$, and the x_i are labeled by λ_n with a, while the y_i are labeled with b. $\mathcal{D}_5$ is depicted in Figure 2 (a). An ACAT $\mathcal{A}$ with $L_{\mathbb{DAG}((\{a\},\{b\}))}(\mathcal{A}) = L$ is (Q, Δ, T, F) where $Q = \{q_1, q_2\}$ and Δ contains the transitions $\emptyset \rightarrow (b, q_2)$, $\{(b, q_2)\} \rightarrow (b, q_2)$, $\{(b, q_2)\} \rightarrow (a, q_1)$, and $\{(a, q_1), (b, q_2)\} \rightarrow (a, q_1)$. Moreover, we set $T(a, q_1) = T(b, q_1) = T(a, q_2) = \emptyset$, $T(b, q_2) = \{a\}$, and $F = \{(q_1, q_2)\}$. An accepting run of $\mathcal{A}$ on $\mathcal{D}_n$ assigns q_1 to any a-labeled vertex and q_2 to any b-labeled one.

Lemma 1 ([6]). $\mathcal{ACA}_{\mathbb{DAG}(\widetilde{\Sigma})}(\widetilde{\Sigma}) \subset \mathcal{ACAT}_{\mathbb{DAG}(\widetilde{\Sigma})}(\widetilde{\Sigma})$

Proof. The language L from Example 3 cannot be recognized by some ACA over $(\{a\}, \{b\})$ relative to $\mathbb{DAG}((\{a\}, \{b\}))$. For suppose there is an ACA $\mathcal{A}$ over $(\{a\}, \{b\})$ with $L_{\mathbb{DAG}((\{a\},\{b\}))}(\mathcal{A}) = L$. For $n \geq 1$, suppose furthermore r to be a run of $\mathcal{A}$ on $\mathcal{D}_n$. If n has been chosen large enough, there are $1 \leq i < j < n$ such that $r(x_i) = r(x_j)$. From $\mathcal{D}_n$, we obtain the $(\{a\}, \{b\})$-dag $\mathcal{D}_n[i,j]$ by removing from V_n the vertices $x_{i+1}, \ldots, x_j$ and from $\lhd_n$ any edge touching some of these nodes and by adding instead an edge from x_i to x_{j+1}. Though $\mathcal{D}_n[i,j] \notin L$, $\mathcal{A}$ admits an accepting run on $\mathcal{D}_n[i,j]$. The reason is that, in the example, $\mathcal{A}$ has no means to impose on a b-labeled vertex an a-labeled successor. $\square$

Though ACATs are generally strictly more expressive than ACAs, many distributed systems allow for dropping types: we call a class $\mathcal{K} \subseteq \mathbb{DAG}(\widetilde{\Sigma})$ *communication closed* if, for any $(V_1, \lhd_1, \lambda_1), (V_2, \lhd_2, \lambda_2) \in \mathcal{K}$, any $a \in \Sigma$, and any $x_1 \in V_1$ and $x_2 \in V_2$ with $\lambda(x_1) = \lambda(x_2) = a$, we have $\mathrm{W}(x_1) \cap I_{\widetilde{\Sigma}}(a) = \mathrm{W}(x_2) \cap I_{\widetilde{\Sigma}}(a)$ (where $I_{\widetilde{\Sigma}}(a)$ shall contain any $b \in \Sigma$ with $a\, I_{\widetilde{\Sigma}}\, b$).

Lemma 2. *Let $\mathcal{K} \subseteq \mathbb{DAG}(\widetilde{\Sigma})$. If $\mathcal{K}$ is communication closed, then we have* $\mathcal{ACA}_\mathcal{K}(\widetilde{\Sigma}) = \mathcal{ACAT}_\mathcal{K}(\widetilde{\Sigma})$.

Proof. Let $\mathcal{A} = (Q, \Delta, T, F)$ be an ACAT over $\widetilde{\Sigma}$ and suppose $\mathcal{K} \subseteq \mathbb{DAG}(\widetilde{\Sigma})$ to be communication closed. Then, for any $a, b \in \Sigma$ with $a \ I_{\widetilde{\Sigma}} \ b$ such that $\mathcal{K}$ requires an a-labeled vertex to be directly followed by some b-labeled vertex, we can, without changing the recognized language relative to $\mathcal{K}$, remove b from any communication request $T(a, q)$. Now suppose $\mathcal{K}$ prevents a from being followed by b. Then, for every $q \in Q$ with $b \in T(a, q)$, we remove any transition $\overline{q} \rightarrow (a, q)$ and set $T(a, q) = \emptyset$. It is easily seen that we obtain an ACA, which, moreover, is equivalent to $\mathcal{A}$ relative to $\mathcal{K}$. □

The class $\mathbb{TR}(\widetilde{\Sigma})$ of Mazurkiewicz traces is trivially communication closed, as none pair of neighboring nodes can be labeled with independent actions. The class $\mathbb{MSC}(Ag)$ is communication closed, too: any sending vertex has exactly one successor vertex, labeled with the corresponding receive action, that is not controlled by the same agent. However, the class $\mathbb{LMSC}(Ag)$ of lossy MSCs can no longer do without types, as an easy adaption of the proof of Lemma 1 shows.

Theorem 2. *For any $\mathcal{K} \subseteq \mathbb{DAG}(\widetilde{\Sigma})$, $\mathcal{ACAT}_{\mathcal{K}}(\widetilde{\Sigma}) = \mathcal{EMSO}_{\mathcal{K}}(\Sigma)$. Moreover, both conversions, from automata to formulas and vice versa, are effective.*

The theorem follows from Lemma 3 and Lemma 4, which will be shown below.

Theorem 3. *For any $\mathcal{K} \subseteq \mathbb{DAG}(\widetilde{\Sigma})$, we have $\mathcal{GA}_{\mathcal{K}}(\Sigma) = 1\text{-}\mathcal{GA}_{\mathcal{K}}(\Sigma)$.*

Proof. Let $\mathcal{K} \subseteq \mathbb{DAG}(\widetilde{\Sigma})$ and let $\mathcal{B}$ be a graph acceptor over Σ. According to Theorems 1 and 2, there is an ACAT $\mathcal{A} = (Q, \Delta, T, F)$ over $\widetilde{\Sigma}$ such that $L_{\mathcal{K}}(\mathcal{A}) = L_{\mathcal{K}}(\mathcal{B})$. Without loss of generality, we shall assume that, for any $\overline{q} \in F$ and $i \in Ag$ with $\overline{q}[i] \in Q$, $\overline{q}[i]$ can be assigned at most to the Σ_i-maximal vertex of a $\widetilde{\Sigma}$-dag. A graph acceptor $\mathcal{B}'$ of radius 1 with $L_{\mathcal{K}}(\mathcal{B}') = L_{\mathcal{K}}(\mathcal{A})$ is given by $(Q, 1, \mathcal{S}, Occ)$ where, for any transition $\{(a_1, p_1), \ldots, (a_n, p_n)\} \rightarrow (b, q) \in \Delta$ and any $\{(c_1, q_1), \ldots, (c_m, q_m)\} \subseteq \Sigma \times Q$ with $c_i \neq c_j$ $(i \neq j)$ and $T(b, q) \subseteq \{c_1, \ldots, c_m\}$, $\mathcal{S}$ contains s if removing from s the edges that do not touch its center yields $\{(a_1, p_1), \ldots, (a_n, p_n)\} \rightarrow (b, q) \rightarrow \{(c_1, q_1), \ldots, (c_m, q_m)\}$ (with the expected meaning). Now let, given $q \in Q$, $\mathcal{S}_q$ contain those spheres whose sphere center is labeled with (a, q) for some a, and let, accordingly, $\mathcal{S}_a$ with $a \in \Sigma$ contain those spheres whose sphere center is labeled with (a, q) for some q. Then,

$$Occ = \bigvee_{\overline{q} \in F} \left(\bigwedge_{i \in Ag, \ \overline{q}[i] \in Q} \bigvee_{s \in \mathcal{S}_{\overline{q}[i]}} {``s \geq 1"} \wedge \bigwedge_{i \in Ag, \ \overline{q}[i] = \iota, \ a \in \Sigma_i, \ s \in \mathcal{S}_a} \neg {``s \geq 1"} \right)$$

guarantees that a run of $\mathcal{B}$ is accepting only if the corresponding run of $\mathcal{A}$ is. □

Lemma 3. *For any $\mathcal{K} \subseteq \mathbb{DAG}(\widetilde{\Sigma})$, $\mathcal{ACAT}_{\mathcal{K}}(\widetilde{\Sigma}) \subseteq \mathcal{EMSO}_{\mathcal{K}}(\Sigma)$.*

Proof. The construction of an EMSO sentence from a given ACAT follows similar instances of that problem. See, for example, [6]. Basically, an interpretation of second-order variables (actually, an interpretation that stands for a partition $(X_q)_{q \in Q}$ of the set of vertices at hand) means an assignment of states to vertices, which is then checked within the first-order fragment of the formula for being an accepting run. Hereby, the type function can be handled by the formula $\forall x \bigwedge_{a \in \Sigma, \ q \in Q}((\lambda(x) = a \wedge x \in X_q) \rightarrow \bigwedge_{b \in T(a,q)} \exists y (x \lessdot y \wedge \lambda(y) = b))$. □

Lemma 4. *For any* $\mathcal{K} \subseteq \mathbb{DAG}(\widetilde{\Sigma})$, $\mathcal{EMSO}_{\mathcal{K}}(\Sigma) \subseteq \mathcal{ACAT}_{\mathcal{K}}(\widetilde{\Sigma})$.

Proof. In the following, we generalize the technique that has been applied to CFMs in [2]. So suppose φ to be an EMSO sentence. According to Theorem 1, there is a graph acceptor $\mathcal{B} = (Q, R, \mathcal{S}, Occ)$ over Σ such that $L_{\mathcal{K}}(\mathcal{B}) = L_{\mathcal{K}}(\varphi)$. We now transform $\mathcal{B}$ into an ACAT $\mathcal{A}$ over $\widetilde{\Sigma}$ such that $L_{\mathcal{K}}(\mathcal{A}) = L_{\mathcal{K}}(\mathcal{B})$.

Let us first give an intuition of how $\mathcal{A}$, which operates rather locally, simulates the global control of $\mathcal{B}$: basically, any state of $\mathcal{A}$ makes a guess about the local environment of radius R that it is about to read and then verifies its guess by passing it through a run and checking if other components have made their guess accordingly. A guess is actually an *extended R-sphere* $\sigma = (V, \lhd, \lambda, \gamma, \alpha, i)$ where $core(\sigma) := (V, \lhd, \lambda, \gamma) \in \mathcal{S}$ is the pattern that $\mathcal{A}$ expects to see, $\alpha \in V$ is the *active vertex*, which corresponds to the vertex that $\mathcal{A}$ is about to read, and $i \in \{1, \ldots, const\}$ is the current *instance* of the pattern where $const :=$ $(2|\Sigma| + 1) \cdot (\max\{|V| \mid (V, \lhd, \lambda, \gamma) \in \mathcal{S}\})^2$. The ACAT $\mathcal{A}$, reading some vertex x and entering a state associated with a guess σ, presumes that x is to its local environment as α is to $core(\sigma)$. In other words, $\mathcal{A}$ considers x to be the analogon of α and the environment of x to look like $core(\sigma)$. To establish isomorphism between $core(\sigma)$ and the environment around x, $\mathcal{A}$ transfers σ to the immediate successor vertices x' of x except that, in σ, the active vertex α is replaced with some α' such that $\alpha \lhd \alpha'$. This is because x shall correspond to x' only if α corresponds to α'. As, in almost all cases, a tiling by $\mathcal{B}$ induces an overlapping of participating spheres, a state of $\mathcal{A}$ actually holds a set of extended R-spheres, which subsequently have to be forwarded and verified simultaneously. The state entered when reading x carries exactly one extended R-sphere whose sphere center and active vertex coincide with the understanding that the corresponding run of $\mathcal{B}$ assigns to x the control state associated with precisely that sphere center. There may even be an overlapping of isomorphic R-spheres so that a state possibly contains several instances of one and the same sphere, which then refer to distinct vertices as corresponding sphere centers. Those instances will be distinguished by means of the natural i. However, there can be at most *const* such overlappings, an order of magnitude that depends on $\mathcal{B}$ only. A second component of a state keeps track of the number of spheres used so far.

Now set $\mathcal{S}^+$ to be the set of extended R-spheres that emerge from $\mathcal{S}$. In the scope of extended spheres $\sigma, \sigma' \in \mathcal{S}^+$, we let in the following $V, \lhd, \lambda, \gamma, \alpha, i$ and $V', \lhd', \lambda', \gamma', \alpha', i'$ refer to the components of σ and σ', respectively. Given $\sigma \in \mathcal{S}^+$, and $y \in V$, $\sigma[y]$ shall denote the extended R-sphere $(V, \lhd, \lambda, \gamma, y, i) \in \mathcal{S}^+$, in which the active vertex α of σ is replaced with a new active vertex y. Finally, $\max(Occ)$ shall denote the least threshold n such that Occ does not distinguish occurrence numbers $\geq n$. Let us now turn to the construction of the ACAT $\mathcal{A} = (Q', \Delta, T, F)$, which is given as follows: a state of $\mathcal{A}$ is a pair $(\mathcal{S}, \nu)$ where ν is a mapping $\mathcal{S} \to \{0, \ldots, \max(Occ)\}$ and $\mathcal{S}$ is a subset of $\mathcal{S}^+$ such that (i) there is exactly one extended sphere $\sigma \in \mathcal{S}$ with $\gamma = \alpha$ (we set $core(\mathcal{S})$ to be $core(\sigma)$), (ii) there is $(a, q) \in \Sigma \times Q$ such that, for any $\sigma \in \mathcal{S}$, $\lambda(\alpha) = (a, q)$ (so that we can assign a well-defined unique label $\ell(\mathcal{S}) := a$ to $\mathcal{S}$), and (iii) for any $\sigma, \sigma' \in \mathcal{S}$, if $core(\sigma) = core(\sigma')$ and $i = i'$, then $\alpha = \alpha'$.

Let $(\overline{\mathcal{S}}, b, \mathcal{S}') \in \Delta$ if the following hold:

L $\ell(\mathcal{S}') = b$.

C $\mathcal{S}'(s)$ is the minimum of $\max(Occ)$ and

$$\begin{cases} \max(\{0\} \cup \{\overline{\mathcal{S}}[a](s) \mid a \in \Sigma,\ \overline{\mathcal{S}}[a] \neq -\}) & \text{if } s \neq core(\mathcal{S}') \\ \max(\{1\} \cup \{\overline{\mathcal{S}}[a](s) + 1 \mid a \in \Sigma,\ \overline{\mathcal{S}}[a] \neq -\}) & \text{if } s = core(\mathcal{S}') \end{cases}$$

For any $a \in \Sigma$ with $\overline{\mathcal{S}}[a] \neq -$, any $\sigma \in \overline{\mathcal{S}}[a]$, and any $y \in V$:

W1   If $\sigma[y] \in \mathcal{S}'$, then $\alpha \lhd y$.
W2   If $b \notin W(\alpha)$, then $d(\alpha, \gamma) = R$.
W3   If $b \in W(\alpha)$, then $\sigma[b\text{-succ}(\alpha)] \in \mathcal{S}'$.

For any $a \in \Sigma$ and any $\sigma \in \mathcal{S}'$:

R1   If $\overline{\mathcal{S}}[a] \neq -$ and $a \notin R(\alpha)$, then $d(\alpha, \gamma) = R$.
R2   If $a \in R(\alpha)$, then $\sigma[a\text{-pred}(\alpha)] \in \overline{\mathcal{S}}[a] \neq -$.

Fig. 3. Simulating a graph acceptor

The definition of Δ is given by Figure 3. For ease of notation, we hereby use $\overline{\mathcal{S}}[a]$ and $\mathcal{S}'$ to denote any component of a state, i.e., $\overline{\mathcal{S}}[a]$ and $\mathcal{S}'$ each refer to both a set of extended spheres and a mapping $\mathcal{S} \to \{0, \ldots, \max(Occ)\}$. Conditions **L** and **C** go without saying. In particular, **C** implements a simple threshold counting procedure. Now assume an extended sphere σ with active vertex α is attached to some x. Assume furthermore that $\sigma[y]$ is attached to some direct successor x' of x. As α and y have to simulate x and x' (and vice versa), they have to be joint by an edge as well. This is what condition **W1** is supposed to guarantee. Suppose now that α lacks a b-successor, while x does not. That situation is allowed only if the distance from α to γ is R, as then the scope of σ ends anyway so that, beyond x, it is no longer responsible for what will happen (**W2**). Otherwise, if b is contained in the write domain of α, then σ has to coincide with the input structure further on so that $\sigma[b\text{-succ}(\alpha)]$ is sent to the b-labeled successor of x (**W3**). The duals of **W2** and **W3**, regarding the read domain of a vertex, are guaranteed by conditions **R1** and **R2**, respectively. Note that condition **W1** lacks its dual case, as this is implicitly present.

Let us now turn to the type function and the set of final states of $\mathcal{A}$: for any $a \in \Sigma$ and $(\mathcal{S}, \nu) \in Q'$, we set $T(a, (\mathcal{S}, \nu))$ to be $\{b \in \Sigma \mid b\ I_{\overline{\Sigma}}\ a$ and there is $\sigma \in \mathcal{S}: b \in W(\alpha)\}$, i.e., if the active vertex of some extended sphere from $\mathcal{S}$ has some b-labeled successor, then so will the vertex to which $\mathcal{S}$ is attached. Finally, $\overline{q} \in (Q' \uplus \{\imath\})^{Ag}$ shall be contained in F if both, for any $i \in Ag$ and any $\sigma \in \overline{q}[i]$, $W(\alpha) \cap \Sigma_i = \emptyset$ and the union of mappings that occur in $\overline{q}$ (for each sphere, take the mapping with the maximum occurrence number) satisfies the requirements imposed by Occ. In fact, it holds $L_{\mathcal{K}}(\mathcal{A}) = L_{\mathcal{K}}(\mathcal{B})$.

Converting a formula φ into an ACAT is effective, because converting φ into a graph acceptor is: a radius R and a threshold n can be computed so that $L_{\mathcal{K}}(\varphi)$

is the finite union of $\sim_{r,n}$-equivalence classes, which do not distinguish graphs in which any sphere of radius R appears more than n times or equally often [9]. In turn, the equivalence classes of $\sim_{r,n}$ can be captured by a graph acceptor. □

4   Related Work and Applications

ACA(T)s cover many other models for concurrency, among them CFMs, lossy channel systems, and asynchronous (trace) automata. Let us discuss the former two in more detail, while the relation to the latter is studied thoroughly in [6].

Unlike the general model of an ACAT, a CFM [3] is tailored to recognizing MSCs. It comprises a collection of finite-state machines, one for each agent $i \in Ag$, which communicate with one another via unbounded FIFO channels. The machine of agent i can execute send actions $i!j$ and receive actions $i?j$, $i \neq j$, with the understanding that some message is queued into the channel from agent i to agent j and, respectively, taken out of the (distinct) channel from j to i. To increase the expressiveness, synchronization data might be sent together with the actual message, which are abstracted away in the recognized MSC.

The first logical characterization of CFMs imposed a bound on the channel capacity for any possible computation. In that case, CFMs turned out to be exactly as expressive as MSO logic [10]. Weakening this restriction and considering a fragment of MSCs that requires at least one computation to match a given channel capacity, Genest et al. showed that CFMs still capture precisely MSO logic [8]. Applying the technique that has been generalized in the present paper, CFMs could be shown to be expressively equivalent to EMSO logic over graphs if channels are assumed to be unbounded [2], which is subsumed by Theorem 2.

Not only are CFMs expressively equivalent to ACA(T)s relative to $\mathbb{MSC}(Ag)$, any CFM is precisely an ACA. For example, a transition $(q, i!j, m', q')$ (with synchronization message m') of a CFM component can be written as the collection of ACA transitions $\{(i\theta k, (m, q))\} \to (i!j, (m', q'))$ where m is a synchronization message, $k \neq i$, and $\theta \in \{!, ?\}$. In particular, a state of the CFM is paired off with a message to obtain a state of the ACA under construction. On the other hand, reading a receive vertex, the only valid ACA transition that may contribute to a run on some MSC is of the form $\{(i\theta k, q_1), (j!i, q_2)\} \to (i?j, q)$ where the exchange of a message is hidden behind the states. Altogether, we can justifiably state that CFMs are precisely ACA(T)s relative to $\mathbb{MSC}(Ag)$.

If, in a CFM, messages may get lost, we deal with a lossy channel system [1]. Paradoxically, some questions of interest become decidable in this faulty case while, wrt. reliable CFMs, corresponding problems are undecidable [4]. This fact appears less paradox if we consider a lossy channel system to be an ACA relative to $\mathbb{LMSC}(Ag)$ rather than relative to $\mathbb{MSC}(Ag)$. As $\mathbb{LMSC}(Ag)$ is not communication closed, EMSO logic is not covered by ACAs relative to $\mathbb{LMSC}(Ag)$, unless we consider typed lossy channel systems. Reachability for lossy channel systems is decidable [1] and so is emptiness for ACAs relative to $\mathbb{LMSC}(Ag)$.

Theorem 4.

(1) Emptiness for $ACAs$ relative to $\mathbb{LMSC}(Ag)$ is decidable.
(2) Emptiness for $ACAs$ relative to $\mathbb{DAG}(\widetilde{\Sigma})$ is decidable.
(3) Emptiness for $ACATs$ relative to $\mathbb{DAG}(\widetilde{\Sigma})$ is undecidable.
(4) Emptiness for $ACAs$ relative to $\mathbb{MSC}(Ag)$ is undecidable.
(5) Emptiness for $ACATs$ relative to $\mathbb{LMSC}(Ag)$ is undecidable.

Proof. Statement (1) follows from [1] and (2) follows from [11] and the fact that $\mathbb{DAG}(\widetilde{\Sigma})$ is the language of an ACA relative to $\mathbb{DAG}((\{a\})_{a \in \Sigma})$. Part (4) is due to [3] and statements (3) and (5) are reductions from (4). $\square$

References

1. P. A. Abdulla and B. Jonsson. Verifying programs with unreliable channels. In *Proceedings of LICS 1993*. IEEE Computer Society Press, 1993.
2. B. Bollig and M. Leucker. Message-Passing Automata are expressively equivalent to EMSO Logic. In *Proceedings of CONCUR 2004*, volume 3170 of *LNCS*, 2004.
3. D. Brand and P. Zafiropulo. On communicating finite-state machines. *Journal of the ACM*, 30(2), 1983.
4. G. Cécé, A. Finkel, and S. Purushothaman Iyer. Unreliable channels are easier to verify than perfect channels. *Information and Computation*, 124(1), 1996.
5. V. Diekert and G. Rozenberg, editors. *The Book of Traces*. World Scientific, Singapore, 1995.
6. M. Droste, P. Gastin, and D. Kuske. Asynchronous cellular automata for pomsets. *Theoretical Computer Science*, 247(1-2), 2000.
7. A. Finkel. Decidability of the termination problem for completely specified protocols. *Distributed Computing*, 7(3), 1994.
8. B. Genest, A. Muscholl, and D. Kuske. A Kleene theorem for a class of communicating automata with effective algorithms. In *Proceedings of DLT 2004*, volume 3340 of *LNCS*. Springer, 2004.
9. W. P. Hanf. Model-theoretic methods in the study of elementary logic. In *The Theory of Models*. North-Holland, Amsterdam, 1965.
10. J. G. Henriksen, M. Mukund, K. Narayan Kumar, and P. S. Thiagarajan. Regular collections of message sequence charts. In *Proceedings of MFCS 2000*, volume 1893 of *LNCS*. Springer, 2000.
11. D. Kuske. Emptiness is decidable for asynchronous cellular machines. In *Proceedings of CONCUR 2000*, 2000.
12. O. Matz, N. Schweikardt, and W. Thomas. The monadic quantifier alternation hierarchy over grids and graphs. *Information and Computation*, 179(2), 2002.
13. A. Potthoff, S. Seibert, and W. Thomas. Nondeterminism versus determinism of finite automata over directed acyclic graphs. *Bulletin of the Belgian Mathematical Society*, 1, 1994.
14. W. Thomas. On Logics, Tilings, and Automata. In *Proceedings of ICALP 1991*, volume 510 of *LNCS*. Springer, 1991.
15. W. Thomas. Elements of an automata theory over partial orders. In *Proceedings of POMIV 1996*, volume 29 of *DIMACS*. AMS, 1996.
16. W. Thomas. Automata theory on trees and partial orders. In *Proceedings of TAPSOFT 1997*, volume 1214 of *LNCS*. Springer, 1997.
17. Wiesław Zielonka. Notes on finite asynchronous automata. *R.A.I.R.O. — Informatique Théorique et Applications*, 21, 1987.

Tree Automata and Discrete Distributed Games

Julien Bernet and David Janin[*]

LaBRI, Université de Bordeaux I, 351, cours de la Libération,
33 405 Talence cedex France
{bernet, janin}@labri.fr

Abstract. Distributed games, as defined in [6], is a recent multiplayer
extension of discrete two player infinite games. The main motivation
for their introduction is that they provide an abstract framework for
distributed synthesis problems, in which most known decidable cases
can be encoded and solved uniformly.

In the present paper, we show that this unifying approach allows as
well a better understanding of the role played by classical results from
tree automata theory (as opposed to adhoc automata constructions) in
distributed synthesis problems. More precisely, we use alternating tree
automata composition, and simulation of an alternating automaton by
a non-deterministic one, as two central tools for giving a simple proof of
known decidable cases.

1 Introduction

Distributed games, as defined in [6], is a recent multiplayer extension of discrete
two player infinite games. The main motivation for their introduction is that
they provide an abstract framework for distributed synthesis problems, in which
most known decidable cases [1,3,4,5,10] can be encoded and solved uniformly.

In the present paper, we show that this unifying approach allows as well a
better understanding of the role played by classical results from tree automata
theory in distributed synthesis problems.

More precisely, in the above mentioned works, many decision algorithms rely
(more or less implicitly) on automata constructions that are not explicitly related
to classical automata theory.

For instance, in [3], the main construction given by the authors to solve the
pipeline synthesis problem "sounds" like the sequential composition of two tree-
automata. Similarly, one of the main construction (glue operation) defined in [6]
"sounds" like Muller and Schupp simulation of an alternating automaton by a
non deterministic one [7].

The purpose of this paper is to validate this intuition, by explicitly defining
the encountered automata (or their inputs) when they are missing in these works,
and to apply known constructions in order to reprove these synthesis results.

---

[*] This work is partially supported by the European Commission Research and Training
Network "Games and Automata for Synthesis and Validation" (RTN GAMES).

M. Liśkiewicz and R. Reischuk (Eds.): FCT 2005, LNCS 3623, pp. 540–551, 2005.

This way, it is expected that it will contribute to the foundation of a common ground into which methods and approaches can be encoded and compared one with the other.

The technical relevance of our reformulation work is illustrated by the encoding and solving of the pipeline case [3].

1.1 Other Related Works

Peterson and Reif ([8], extended in [9]) initiated the research on multiplayer games of incomplete information, considering finite games, and introducing the notion of *hierarchical games*: these games satisfy the property that one can linearly order the set of players such that $p_1 \leq p_2$ if and only if "$p2$ knows more than p_1", or equivalently "p_2 knows the state of p_1". They prove that these games are solvable, by iteratively removing the incomplete information associated with each player.

Subsequent results on distributed synthesis (such as [10], [3]) essentially used the same ideas and techniques, except in the fact that they consider infinite plays and/or branching time specifications.

The common technique is to cut out the last player from the game (i.e. the one that knows the state of all the other), modifying in the process the specification so that it reflects all moves that can be taken by this player, then do the same with the last but one, etc. ... until a "simple" 2-player game is left to solve.

Our paper rely on the same principle, making the automata constructions explicit.

1.2 Organization of the Paper

In the first section, after reviewing some of the notations used in this paper, we fix the definitions of trees, tree automata and infinite two player games. Muller and Schupp non determinization theorem is stated, and a notion of sequential composition of tree automata is also defined and analyzed.

Distributed games and distributed strategies are presented in the second section. These games are played by a team of process players versus a single environment opponent. Each process player only gets incomplete information about the position of the other processes. The existence of a winning distributed strategy in a distributed game is shown to be undecidable, even for simple winning conditions such as safety and reachability.

In the third section, we first show that using an (external) tree-automaton in order to define winning strategies in a (distributed) game is essentially equivalent to adding an additional process player (internalizing the automaton) into the game. Then, conversely, we show that when a process player has enough knowledge to deduce the positions of the other processes, then its local arena can be externalized as a tree automaton reading the strategies of the remaining processes, in which non-deterministic choices correspond to the moves of the process; this automaton can be composed with any existing external winning condition.

Under sufficient conditions, one can apply repeatedly this construction in order to reduce the number of process players and thus to solve the distributed game.

The long-term goal of this approach is to gain benefit from the high level of abstraction provided by game theory and, altogether, gain benefit from well-known constructions of automata theory (as it as been developed from Rabin's seminal result [11]), to help having a better understanding of the fundamental obstacles to the synthesis of distributed systems.

2   Trees, Automata and Games

For any alphabet A, let A^* and A^ω be the set of all finite and infinite words with letters from A. Let $A^\infty = A^* \cup A^\omega$, and $A^? = \{\epsilon\} + A$. Standard notations on words and languages of words are used. In particular, given a language $L \subseteq A^*$, we use the notations L^+ and (when the empty word $\epsilon \notin L$) L^ω that stand, respectively, for the set of words built by concatenating finitely many and infinitely many finite words of L. For any finite word $w = a_1 \dots a_n$, let $|w| = n$ be the length of w. For any infinite word w, let $\inf(w) = \{a \in A \mid w \in (A^*.a)^\omega\}$ be the set of letters that occur infinitely often in w.

For any two sets A and X, for any word $w \in A^*$, define $\pi_X(w)$ (the projection of w over X) as the word obtained by deleting any letter that is not in X from the word representation of w.

Given n numbered sets $A_1, \dots, A_n$, given $A = A_1 \times \dots \times A_n$, given any set of indices $I = \{i_1, \dots, i_k\} \subseteq \{1, \dots, n\}$ with $i_1 < \dots < i_k$, we write $A[I]$ for the set $A[I] = A_{i_1} \times \dots A_{i_k}$, for any $x = (a_1, \dots, a_n) \in A$, we write $x[I]$ for the elements $x[I] = (x_{i_1}, \dots, x_{i_k}) \in A[I]$, and, for any $P \subseteq A$, we write $P[I]$ for the set $P[I] = \{x[I] \in A[I] : x \in P\}$.

In case $I = \{i, i+1, \dots, j\}$ (where $1 \le i \le j \le n$), these notations simplify to $A[i \dots j]$, $x[i \dots j]$ and $P[i \dots j]$ respectively (and even simplify to $A[i]$, $x[i]$ and $P[i]$ when $i = j$). These notations also extend to words as follows: for any word $w = a_1.a_2 \dots \in A^\infty$, for any $I \subseteq \{1, \dots, n\}$, $w[I] = a_1[I].a_2[I].a_3[I] \dots$, and to relations: for any relation $R \subseteq A \times A$, we write $R[I]$ the relation on $A[I]$ defined by $R[I] = \{(x[I], y[I]) \in A[I] \times A[I] : (x, y) \in R\}$.

Given two finite alphabets D and Σ, a Σ-labeled D-tree (also called D, Σ-tree) is a partial function $D^* \to \Sigma$ whose domain is closed under prefix operation. In the sequel, elements of Σ are called *labels* and elements of D are called *directions*.

For any tree $t : D^* \to \Sigma$, the function $flat_t : Dom(t) \to \Sigma.(D.\Sigma)^*$ is defined by: $flat_t(\epsilon) = t(\epsilon)$ and, for any $w \in D^*$ and $d \in D$ such that $w.d \in Dom(t)$, $flat_t(w.d) = flat_t(w).d.t(w.d)$. Observe that $Dom(t)$ and $flat_t(Dom(t))$ ordered by the prefix ordering are isomorphic, and, as a consequence, $flat_t(Dom(t))$ uniquely determines tree t.

The following definition is a variation on Muller and Schupp's original definition of alternating automaton [7]. Our goal is to have a tree-transducer like automaton definition, even for alternating automaton.

Definition 1 (Alternating tree automaton). *A finite* (D, Σ)-*alternating tree automaton is a tuple:*

$$\mathcal{A} = \langle Q = Q^\forall \uplus Q^\exists, D, \Sigma, q_0, \delta = \delta^\forall \cup \delta^\exists, Acc \subseteq Q^\omega \rangle$$

where Q is a finite set of states, $q_0 \in Q^{\exists}$ is the initial state, $\delta^{\forall} : Q^{\forall} \times D \to \mathcal{P}(Q^{\exists})$ and $\delta^{\exists} : Q^{\exists} \times \Sigma \to \mathcal{P}(Q^{\forall})$ are the transition functions, and the ω-rational language Acc is the infinitary acceptance criterion.

Automaton $\mathcal{A}$ is a *non deterministic automaton* (also called non alternating) when $|\delta^{\forall}(q, d)| \leq 1$ (for any $q \in Q^{\forall}$, $d \in D$).

Definition 2 (Runs). *A run of an automaton $\mathcal{A} = \langle Q, D, \Sigma, i, \delta, Acc \rangle$ over a Σ-labeled D-tree $t : D^* \to \Sigma$ is a $Q^{\forall}$-labeled $(D \times Q^{\exists})$ tree $\rho : (D \times Q^{\exists})^* \to Q^{\forall}$ such that:*

- *$\rho(\epsilon) \in \delta^{\exists}(q_0, t(\epsilon))$,*
- *for all $w \in Dom(\rho)$, if $\rho(w) = q$, then for any direction $d \in D$ such that $a = t(w[1].d)$ is defined, and for any existential state $q_1 \in \delta^{\forall}(q, d)$, there exists a universal state $q_2 \in \delta^{\exists}(q_1, a)$ such that $\rho(w.(d, q_1)) = q_2$.*

For any infinite branch w of a run ρ of $\mathcal{A}$ over t, states$_\rho(w)$ is the sequence of (universal and existential) states encountered along w. A tree t is accepted by $\mathcal{A}$ if and only if there exists a run ρ of $\mathcal{A}$ over t such that for any infinite branch w in ρ: states$_\rho(w) \in Acc$. Denote by $L(\mathcal{A})$ the language of all trees that are accepted by $\mathcal{A}$. The size of an automaton $\mathcal{A}$ is denoted by $|\mathcal{A}|$.

Observe that these tree automata (both alternating and non alternating), if slightly unusual, have the same expressive power as their standard counterpart, as in [7]. In particular:

Theorem 1 (Simulation [7]). *Any alternating tree automaton $\mathcal{A}$ is equivalent to a non deterministic tree automaton $\mathcal{A}'$, with $|\mathcal{A}'| \leq 2^{2^{|\mathcal{A}|}}$ (with Muller acceptance condition).*

Since the runs of an automaton on trees are themselves trees, automata act as tree transducers and can be sequentially combined.

Definition 3 (Automata Composition). *Given two tree automata $\mathcal{A}_1 = \langle Q_1, D_1, \Sigma_1, q_{0,1}, \delta_1, Acc_1 \rangle$ and $\mathcal{A}_2 = \langle Q_2, D_2, \Sigma_2, q_{0,2}, \delta_2, Acc_2 \rangle$, such that automaton $\mathcal{A}_2$ is non deterministic with $D_2 = D_1 \times Q_1^{\exists}$ and $\Sigma_2 = Q_1^{\forall}$, we define the composition of $\mathcal{A}_1$ followed by $\mathcal{A}_2$ to be the automaton*

$$\mathcal{A}_2 \circ \mathcal{A}_1 = \langle \widetilde{Q}, D_1, \Sigma_1, \widetilde{q_0}, \widetilde{\delta}, \widetilde{Acc} \rangle$$

defined as follows:

- $\widetilde{Q^{\exists}} = Q_1^{\exists} \times Q_2^{\exists}$; $\widetilde{Q^{\forall}} = Q_1^{\forall} \times Q_2^{\forall}$;
- $q_0 = (q_{0,1}, q_{0,2})$,
- $(q_1', q_2') \in \widetilde{\delta^{\forall}}((q_1, q_2), d) \Leftrightarrow \begin{cases} q_1' \in \delta^{\forall}(q_1, d) \\ \{q_2'\} = \delta_2^{\forall}(q_2, (d, q_1')) \end{cases}$
- $(q_1', q_2') \in \widetilde{\delta^{\exists}}((q_1, q_2), a) \Leftrightarrow \begin{cases} q_1' \in \delta^{\exists}(q_1, a) \\ q_2' \in \delta_2^{\exists}(q_2, q_1') \end{cases}$
- $\widetilde{Acc} = \{w \in \widetilde{Q}^{\omega} \mid w[1] \in Acc_1 \wedge w[2] \in Acc_2\}$

Theorem 2. *For any tree* $t : D_1^* \to \Sigma_1$, $t \in L(\mathcal{A}_2 \circ \mathcal{A}_1)$ *if and only if there exists an accepting run* $\rho : (D_1 \times Q_1^{\exists})^* \to Q_1^{\forall}$ *of* $\mathcal{A}_1$ *over* t *such that* $\rho \in L(\mathcal{A}_2)$.

The proof, although tedious, is not complicated, and is therefore omitted here. Observe that it is crucial that $\mathcal{A}_2$ is non-alternating ; nevertheless, by applying Theorem 1, one can always assume that is is the case.

Definition 4 (Simple (or Two Player) Games). *A simple arena is a quadruple* $G = \langle P, E, T_P, T_E \rangle$, *where* P *is a finite set of Process positions,* E *is a finite set of Environment positions,* $T_P \subseteq P \times E$ *is the set of Process moves,* $T_E \subseteq E \times P$ *is the set of Environment moves. A simple game* $G = \langle P, E, T_P, T_E, e_0, \mathcal{W} \rangle$ *is built upon a simple arena* $\langle P, E, T_P, T_E \rangle$ *by equipping it with an initial position* $e_0 \in E$ *and a regular winning condition* $\mathcal{W} \subseteq (P + E)^\omega$.

As particular cases of winning condition, a *reachability condition* is a winning condition of the form $\mathcal{W} = (P + E)^*.X.(P + E)^\omega$ for some set of positions $X \subseteq P + E$ to be reached for Process to win, and a *safety condition* is a winning condition of the form $\mathcal{W} = ((P + E) - X))^\omega$ for some set of positions $X \subseteq P + E$ to be avoided for Process to win.

A *play* $w \in (P + E)^*$ in a simple game is any non-empty path in the arena beginning on e_0. A play w is winning for Process when either it is finite and ends in an Environment position, or it is infinite and belongs to $\mathcal{W}$. Otherwise, it is winning for Environment.

A *strategy* for Process is a partial function $\sigma : (E.P)^+ \to E$ such that for any $w.p \in Dom(\sigma)$, for any position $e \in \sigma(w.p)$, then $(p, e) \in T_P$, and for any successor p' of e, $w.p.e.p' \in Dom(\sigma)$. A play $w = e_0.x_1.\dots$ is *consistent* with strategy σ when, for any $i \in \mathbb{N}$, if $\sigma(e_0.\dots.x_i)$ and x_{i+1} are both defined then they are equal. A strategy σ is a *winning strategy* for Process when any maximal play (w.r.t. the prefix ordering) consistent with σ is winning for Process.

Given a strategy σ in some game G, the *strategy tree* $t_\sigma : P^* \to E$ of σ in G is defined inductively by $t_\sigma(\epsilon) = e_0$, and $t_\sigma(u.x) = \sigma(flat_{t_\sigma}(u).x)$.

Theorem 3 ([2]). *On finite two-player games with regular winning condition, either Process or Environment has a winning strategy, which can be computed effectively.*

3   Distributed Games

Definition 5 (Distributed Arena). *A distributed arena is a free asynchronous product where the possible Environment moves may have been restricted. More precisely, given two arenas* $G_1 = \langle P_1, E_1, T_{P,1}, T_{E,1} \rangle$ *and* $G_2 = \langle P_2, E_2, T_{P,2}, T_{E,2} \rangle$, *a (two-process) distributed arena built upon the arenas* G_1 *and* G_2 *is any simple arena* $G = \langle P, E, T_P, T_E \rangle$ *of the form*

- Environment positions : $E = E_1 \times E_2$,
- Processes positions : $P = (E_1 \cup P_1) \times (E_2 \cup P_2) - (E_1 \times E_2)$,

- Processes moves : T_P is the set of all pairs $(p, e) \in (P \times E)$ such that, for $i = 1$ and $i = 2$:
  - **either** $p[i] \in P_i$ and $(p[i], e[i]) \in T_{P,i}$ (Process i is active in p),
  - **or** $p[i] \in E_i$ and $p[i] = e[i]$ (Process i is inactive in p),
- and Environment moves : T_E is **some subset** of the set of all pairs $(e, p) \in (E \times P)$ such that, for $i = 1$ and $i = 2$:
  - **either** $p[i] \in P_i$ and $(e[i], p[i]) \in T_{P,i}$ (Environment activates Process i),
  - **or** $p[i] \in E_i$ and $p[i] = e[i]$ (Environment keeps Process i inactive).

When the set T_E of Environment moves is maximal, we call such an arena the free asynchronous product of arenas G_1 and G_2 and it is denoted by $G_1 \otimes G_2$. These definitions extend to n-process distributed arena.

Since a distributed arena is built upon n simple arenas, we need a definition to speak about its local components:

Definition 6 (Projection of distributed arena). Given a distributed arena $G = \langle P, E, T_P, T_E \rangle$, with $E = E_1 \times \ldots \times E_n$ and $P = ((P_1 \cup E_1) \times \ldots \times (P_n \cup E_n)) - E$, given a non empty set $I \subseteq \{1, \ldots, n\}$, define the canonical projection $G[I]$ of G on I as the arena $G[I] = \langle P', E', T_P', T_E' \rangle$ given by: $P' = P[I] - E[I]$ (possibly smaller than $P[I]$!), $E' = E[I]$, $T_P' = T_P[I] \cap (P[I] \times E[I])$, and $T_E' = T_E[I] \cap (E[I] \times P[I])$.

Remark. Observe that a n-process distributed arena G as above can always be seen as a distributed arena built upon the games $G[1], \ldots, G[n]$. Moreover, in the same way Cartesian product of sets is (up to isomorphism) associative, given an arbitrary non empty set $I \subset \{1, \ldots, n\}$, given $\overline{I} = \{1, \ldots, n\} - I$, the n-process distributed arena G can, as well, be seen as a distributed arena built upon the two (distributed) arenas $G[I]$ and $G[\overline{I}]$.

*Example 1 (**The Pipeline : Beginning**). A distributed architecture (as defined in [10], [3]) is a set of sites linked together by some communication channels. Each site can host a program, which is essentially a sequential function[1] mapping a sequence of inputs to a sequence of outputs. As a typical example, in a pipeline architecture, the sites are linearly ordered from left to right, each site taking its input from the site on its right, and writing its output to the site on its left.*

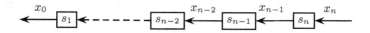

Fig. 1. A pipeline architecture

To be more precise, suppose each communication channel x_i can carry values that range over some set X_i. The site s_i receives its input from the channel x_i, and writes its outputs to the channel x_{i-1}; thus, a program for the site s_i is a

---

[1] Recall that a sequential function is a function $f : A^* \to B^*$ that is realized by a word transducer with input alphabet A and output alphabet B.

sequential function $f_i : X_i^* \to X_{i-1}^*$. The environment writes input to the system on channel x_n, and the system's output is read on channel x_0.

For any pipeline architecture $\mathbb{A}$, we can build a distributed arena $G_\mathbb{A} = \langle P, E, T_P, T_E \rangle$ where each process plays the role of a program: on its local arena, the environment's moves correspond to the possible inputs for this site, and the process moves correspond to the possible outputs:

- $P = X_1 \times \ldots \times X_n$; $E = X_0 \times \ldots \times X_{n-1}$.
- $((v_1, \ldots, v_n), (v_1', \ldots, v_n')) \in T_E$ iff $v_i' = v_{i+1}$ for each $i \in \{1, \ldots, n-1\}$ and $v_n' \in X_n$.

Observe that by restricting the Environment moves, we ensure that the environment carries correctly the values along the channels.

Definition 7 (Distributed Games). *A n-process distributed game G is a tuple*

$$G = \langle P, E, T_P, T_E, e_0, \mathcal{W} \rangle$$

where $\langle P, E, T_P, T_E \rangle$ is a n-process distributed arena, $e_0 \in E$ is the initial (Environment) position, and $\mathcal{W} \subseteq (E.P)^\omega$ is the (regular) winning infinitary condition.

A distributed game is a particular case of simple game. It follows that previous notions of plays and strategies are still defined. However, in order to avoid confusion with what may happen in the local arena a distributed game is build upon, we shall speak now of a *global play* and a *global strategy*.

The *local view* Process i has of a global play in a distributed game G is given by the map $view_i : (E.P)^*.E^? \to (E_i.P_i)^*.E_i^?$ defined in the following way:

- $view_i(\epsilon) = \epsilon$
- $view_i(x) = x[i]$
- $view_i(w.x.y) = \begin{cases} view_i(w.x) \text{ if } x[i] = y[i] \\ view_i(w.x).y[i] \text{ otherwise.} \end{cases}$

A play $w \in (E.P)^+$ is said to be active for Process i when w ends in a position $p \in P$ such that $p[i] \in P[i]$.

Definition 8 (Local and distributed Strategies). *Given a n-tuple of local strategies $(\sigma_i : (E[i].P[i])^+ \to E[i])_{i \in \{1, \ldots, n\}}$, the induced global strategy*

$$\sigma_1 \otimes \ldots \otimes \sigma_n : (E.P)^+ \to E$$

is defined as follows: for any play of the form $w.p \in (E.P)^+$, given the set $I \subseteq \{1, \ldots, n\}$ of active processes in the global Processes position p (i.e. $I = \{i \in \{1, \ldots, n\} : p[i] \in P_i\}$), define $\sigma(w.p) = e$ by:

- $e[i] = \sigma_i(view_i(w))$ *for $i \in I$*
- $e[i] = p[i]$ *for $i \in \{1, \ldots, n\} - I$*

(provided everything is well-defined, otherwise $\sigma(w.p)$ is left undefined).

A global strategy σ : $(E.P)^+ \to E$ *is a* distributed strategy *if* σ *equals the composition* $\sigma_1 \otimes \ldots \otimes \sigma_n$ *of some n local strategies.*

Note that global strategies are not always distributed. Moreover, there are distributed games in which the Processes have a winning strategy, but no winning distributed strategy.

From this, we can derive an important fact: the distributed game are not determined, in the sense that even when the environment does not have a winning strategy, the processes may not have a winning *distributed* strategy. Furthermore, using the fact that the processes do not share the same information, we are able to provide the following undecidability result:

Theorem 4. *The problem of finding a winning distributed strategy in a 3-process distributed game with safety or reachability winning condition is undecidable.*

The proof is omitted here due to space restriction. Suffice it to say that it proceeds by reduction to the Post correspondence problem, and relies heavily on the fact that there are three processes in the game. It is an open problem whether solving a 2-process distributed game is decidable or not.

4 Tree Automata and Distributed Games

We first mix games and automata, defining a winning condition by means of a tree-automaton that recognizes the set of trees of winning strategies. We illustrate this new concept by defining a pipeline game over the pipeline arena. Then, we present an algorithm to solve such a game, using the notion of leader in a distributed game.

Definition 9 (External Winning Condition). *A game with external winning condition is a tuple*

$$G = \langle P, E, T_P, T_E, e_0, \mathcal{A} \rangle$$

where $\langle P, E, T_P, T_E \rangle$ *is a simple arena,* $e_0 \in E$ *is the initial position, and* $\mathcal{A}$ *is a* (P, E)*-tree automaton. In such a game, a strategy is winning if its strategy tree belongs to* $L(\mathcal{A})$*. This definition extends to distributed games.*

In the sequel, in order to avoid confusion, a game with a winning condition defined as in section 3 is called *game with internal winning condition.*

As we are going to show, games with external winning condition are not essentially more expressive than games with internal one.

Theorem 5 (Internalization). *For any n-process game G with external winning condition, there exists a n + 1-process game G' with internal winning condition such that* $G'[1, \ldots, n] = G$*, and such that the processes have a winning strategy* σ *in G if and only if the processes have a winning strategy of the form* $\sigma \otimes \sigma'$ *in G'.*

Proof. (sketch) Let $G = \langle P, E, T_P, T_E, e_0, \mathcal{A} \rangle$ (where $\mathcal{A} = \langle Q^\forall \uplus Q^\exists, P, E, q_0, \delta = \delta^\forall \cup \delta^\exists, Acc \rangle$) be a distributed game with external winning condition. The game $G' = \langle P', E', T_P', T_E', e_0', \mathcal{W} \rangle$ is defined as follows. The positions and the winning condition are given by:

- $P' = (E \times (Q^\exists \times E)) \cup (P \times (Q^\exists \times \{\#\}))$,
- $E' = (E \times Q^\exists) \cup (E \times Q^\forall)$,
- $e_0' = (e_0, q_0)$,
- $\mathcal{W} = \{w \in (E'.P')^\omega \mid \pi_{Q^\forall \cup Q^\exists}(w) \in Acc\}$

and moves are (repeatedly) defined by: from an environment position $(e, q) \in E \times Q^\exists$ (or the initial position):

1. first, Environment (deterministically) moves to the process position $(e, (q, e)) \in E \times (Q^\exists \times E)$,
2. then, the new (automaton) process locally chooses $q' \in \delta^\exists(q, e)$, the other processes stay idle, thus the play proceeds in G', to the environment position $(e, q') \in E \times Q^\forall$,
3. then, Environment chooses $p \in T_E(e)$ and $q_1 \in \delta^\forall(q', p)$, and the play proceeds to the Process position $(p, (q_1, \#)) \in P \times Q^\exists$,
4. finally, processes 1 to n (on game G) choose some $e_1 \in T_P(p)$, the new (automaton) process stays almost idle (he simply deletes the $\#$ sign), and the play proceeds to the Environment position $(e_1, q_1) \in E \times Q^\exists$.

If ρ is an accepting run of $\mathcal{A}$ over t_σ (for some strategy σ in G), one deduce from ρ a strategy σ' such that $\sigma \otimes \sigma'$ is winning in G'. Conversely, if $\sigma \otimes \sigma'$ is a winning strategy in G', one can infer an accepting run of $\mathcal{A}$ over t_σ from σ'.

Moreover, when G is a simple game with external winning condition, the internalization procedure can be further simplified (and amounts essentially to build the product of G with the automaton), and the resulting game with internal condition is a simple game as well.

*Example 2 (**Pipeline Example Continued**).* Following the presentation from [3], the synthesis problem for distributed architectures is presented as follows: given a distributed architecture $\mathbb{A}$ and a vector of programs $(f_i)_{1 \leq i \leq n}$ (one for each site of $\mathbb{A}$), the *computation tree* of the system is a $(\prod_{1 \leq i \leq n} X_i)$-labeled X_n-tree, where each node w is labeled by the values held by the communication channels after input w to the system.

A *specification* for the system is a language of such trees specified by a tree automaton $\mathcal{A}$ (or equivalently by a MSO-formula).

The synthesis problem is then: does there exists a vector of programs such that the resulting computation tree belongs to the specification ?

Building upon the pipeline arena $G_\mathbb{A}$ as defined in example 1, we can now define a distributed game in which the processes have a winning strategy if and only if there is a solution to the synthesis problem in the pipeline architecture. Suppose the specification for $\mathbb{A}$ is given by the finite $(X_n, \prod_{0 \leq i \leq n-1} X_i)$-automaton $\mathcal{A}$, we can easily define a $(\prod_{1 \leq i \leq n} X_i, \prod_{0 \leq i \leq n-1} X_i)$-automaton $\mathcal{A}'$

that accepts a tree t' if and only if it is the widening of some tree $t \in \mathcal{L}(\mathcal{A})$ (i.e. if $t'(w) = t(w[n])$ for all $w \in Dom(t')$).

Using this automaton as an external winning condition, we get the encoding of the synthesis problem for the pipeline architecture in a distributed game.

Observe that in this game, for each $i \in \{1, \ldots, n\}$, provided that Process i knows the strategy for all the processes from 1 to $i - 1$, then he can predict the position in each of the local arenas from 1 to $i - 1$.

Using the above observation, one may ask now whether an inverse construction to internalization is possible or not. Intuitively, assuming that there is a process in an $n + 1$-distributed game that can predict, at every step, what is the global position in the game, can we *externalize* it into an external winning condition such that, the resulting n-process distributed game with external condition is equivalent, in some effective sense, to the initial game ?

The notion of leader defined below follows this intuition. In fact, it provides a local condition that is sufficient for such a global knowledge to be available to a Process player.

Definition 10 (Leader). *Given a 2-process game* $G = \langle P, E, T_P, T_E, e_0, \mathcal{A} \rangle$, *we say that Process 2 is a leader when, for any Environment position* $e \in E$, *any Processes positions* x *and* $y \in P$ *such that both* $(e, x) \in T_E$ *and* $(e, y) \in T_E$,

- *if* $x[2] = y[2]$ *then* $x[1] = y[1]$,
- *if* $x[2] \in E[2]$ *or* $y[2] \in E[2]$ *then* $x = y$.

Intuitively, Process 2 is a leader when, as soon as he knows a global Environment position then, after an Environment move (or several consecutive moves if Process 2 stays idle for some time), Process 2 can predict, from his own position, the global Processes position of the game.

This local property has the following formulation when it comes to considering plays:

Lemma 1. *Let* $G = \langle P, E, T_P, T_E, e_0 \rangle$ *be a 2-process arena with initial position* e_0. *For any strategy* σ *for the processes, the restriction of* $view_2$ *to the plays that are consistent with* σ *and active for Process 2 is one-to-one.*

Proof. Immediate from the definition.

Rephrased in a more useful way, this observation leads to the following result:

Lemma 2. *For any 2-process game* $G = \langle P, E, T_P, T_E, e_0, \mathcal{W} \rangle$ *such that Process 2 is a leader, there exists a* $(P[1], E[1])$-*automaton* $\mathcal{A}_2$ *such that for any strategies* σ *on* G, σ_1 *on* G_1, *the following propositions are equivalent:*

(1) *there exists a strategy* σ_2 *on* $G[2]$ *such that* $\sigma = \sigma_1 \otimes \sigma_2$
(2) *there is an accepting run* ρ *of* $\mathcal{A}_2$ *over* t_{σ_1} *such that* $\rho = t_{\sigma_1}$.

Proof. (sketch) We first give here a construction for $\mathcal{A}_2$ in the case both Process 1 and Process 2 are always active in the positions for Processes.

Automaton $\mathcal{A}_2 = \langle Q_2, P[1], E[1], q_{0,2}, \delta_2, Acc_2 \rangle$ can be defined as follows:

- $Q_2^{\forall} = E;\ Q_2^{\exists} = P[2] \cup \{q_{0,2}\}$,
- $\delta_2^{\forall}(q, p_1) = \{p_2 \in Q_2^{\exists} : (q, (p_1, p_2)) \in T_E\}$ $(q \in Q_2^{\forall}, p_1 \in P[1])$,
- $\delta_2^{\exists}(p_2, e_1) = \{q \in Q_2^{\forall} : q[1] = e_1 \wedge (p_2, q[2]) \in T_P[2]\}$ $(p_2 \in Q_2^{\exists}, e_1 \in E[1])$
  with $\delta_2^{\exists}(q_{0,2}, e_1) = \{e_0[2]\}$,
- $Acc_2 = Q_2^{\omega}$.

The correspondence between runs of $\mathcal{A}_2$ on strategy trees in $G[1]$ and strategy trees in G easily follows from this construction, and from the fact that Process 2 is a leader in G.

In the case Process 2 may be inactivated by Environment one can check that, since Process 2 is a leader, game G can be first normalized so that this no longer happens (details are not given due to lack of space).

In the case Process 1 may be inactivated by Environment, then the construction below can be extended, defining (quite easily though tediously) an automaton $\mathcal{A}_2$ with ϵ-transition. However, the main arguments remain the same.

Since the previous result holds for arbitrary external condition and arbitrary strategies in $G[1]$ (even if $G[1]$ is itself a distributed game), it follows:

Theorem 6 (Externalization). *For any n-process distributed game $G = \langle P, E, T_P, T_E, e_0, \mathcal{A} \rangle$ with non deterministic external winning condition $\mathcal{A}$ such that Process n is a leader, there is a $(P[1 \ldots n-1], E[1 \ldots n-1])$-automaton $\mathcal{A}_n$ such that the following propositions are equivalent:*
(1) *the processes have a distributed winning strategy on G.*
(2) *the processes have a distributed winning strategy in $\langle G[1 \ldots n-1], e_0[1 \ldots n-1], \mathcal{A} \circ \mathcal{A}_n \rangle$.*

*Example 3 (**The Pipeline: End**).* We have already mentioned that, in the n-process pipeline arena, from any initial position, Process n is a leader. It follows that Theorem 6 applies.

Moreover, observe that the resulting $(n-1)$-process game arena $G[1 \ldots n-1]$ is nothing but a $(n-1)$-process pipeline arena. This says that Theorem 6 can be applied repeatedly till the number of processes is reduced to one. Now, one can internalize the automaton, and compute a winning strategy in the resulting simple game using Theorem 3.

Transposed on our more abstract setting, this can be expressed as the following corollary of the theorem.

Corollary 1. *For any n-process ($n \geq 2$) distributed game G such that for each $i \in \{2, \ldots, n\}$ process i is a leader in $G[1 \ldots i]$, the problem of determining whether the processes have a winning strategy is decidable.*

Remark. At every step, the external condition we get from the composition is an alternating automaton that needs to be simulated by a non alternating one so that the composition can be iterated. This means that the complexity of solving the pipeline architecture synthesis problem by means of its encoding into a distributed game is a tower of exponents of depth at least the number of components in the pipeline. This (bad) complexity was expected, since this problem is non-elementary [10].

5 Concluding Remarks

We have defined a set of automata theoretic tools that can be used to solve various distributed synthesis problems, e.g. the pipeline architecture [3].

Compared to [6] we do obtain an automata theoretic interpretation of most of the operations defined there: in their approach, applying successively DIVIDE and GLUE to a game where both 0 and n are leaders amounts, in our setting, to externalize 0, to apply the simulation theorem, to externalize n, and eventually to internalize the resulting automaton.

Still, one application case presented by the authors to solve the local specification case [5] is not solved in this paper. This is left for further studies. There is a chance that tree automata theory will still provide arguments.

References

1. A. Arnold, A. Vincent, and I. Walukiewicz. Games for synthesis of controllers with partial observation. to appear in Theoretical Computer Sciences, 2002.
2. E.A. Emerson and C.S. Jutla. Tree automata, mu-calculus and determinacy. In *Proc. 32th Symp. on Foudations of Computer Sciences*, pages 368–377. IEEE, 1991.
3. O. Kupferman and M. Y. Vardi. Synthesizing distributed systems. In *Logic in Computer Sciences*, pages 389–398, 2001.
4. F. Lin and M. Wonham. Decentralized control and coordination of discrete event systems with partial observation. *IEEE Transactions on automatic control*, 33(12):1330–1337, 1990.
5. P. Madhusudan and P.S. Thiagarajan. Distributed controller synthesis for local specifications. In *28th International Colloquium on Automata, Languages and Programming (ICALP)*, volume 2076 of *LNCS*, pages 396–407, 2001.
6. S. Mohalik and I. Walukiewicz. Distributed games. In *Foundations of Software Technology and Theoretical Computer Science*, pages 338–351, 2003.
7. D.E. Muller and P.E. Schupp. Simulating alternating tree automata by nondeterministic automata. *Theoretical Computer Sciences*, 141:67–107, 1995.
8. G.L. Peterson and J.H. Reif. Multiple-person alternation. In *20th Annual IEEE Symposium on Foundations of Computer Sciences*, pages 348–363, october 1979.
9. G.L. Peterson, J.H. Reif, and S. Azhar. Decision algorithms for multiplayer noncooperative games of incomplete information. *Computers and Mathematics with Applications*, 43:179–206, january 2002.
10. Amir Pnueli and Roni Rosner. Distributed reactive systems are hard to synthesize. In *IEEE Symposium on Foundations of Computer Science*, pages 746–757, 1990.
11. M.O. Rabin. Decidability of second order theories and automata on infinite trees. *Transactions of the American Mathematical Society*, 141:1–35, 1969.

A New Linearizing Restriction in the Pattern Matching Problem*

Yo-Sub Han and Derick Wood

Department of Computer Science,
The Hong Kong University of Science and Technology
{emmous, dwood}@cs.ust.hk

Abstract. In the pattern matching problem, there can be a quadratic number of matching substrings in the size of a given text. The linearizing restriction finds, at most, a linear number of matching substrings. We first explore two well-known linearizing restriction rules, the *longest-match* rule and the *shortest-match substring search* rule, and show that both rules give the same result when a pattern is an infix-free set even though they have different semantics. Then, we introduce a new linearizing restriction, the *leftmost non-overlapping match* rule that is suitable for find-and-replace operations in text searching, and propose an efficient algorithm when the pattern is a regular language according to the new match rule.

Keywords: Automata and formal languages, design and analysis of algorithms, string pattern matching.

1 Introduction

Regular expressions are popular in many applications such as editors, programming languages and software systems in general. People often use regular expressions for searching in text editors or for UNIX command; for example, vi, emacs and grep. There are two types of questions in the pattern matching that one can ask. The first is the recognition problem: Does a string in a given text match a particular pattern? The second is the searching problem: Identify all matching substrings of a given text with respect to a particular pattern. Since a pattern is a language, regular expressions are often used to represent patterns for the pattern matching problem. If a given pattern is a single string, then we have the string matching problem [3,8]. If a given pattern is a finite language, then we have the multiple keyword matching problem [2]. If a pattern is given as a regular expression, then the first problem is the regular language membership problem and the second problem is the regular-expression matching problem.

Given a text T and a pattern L, we define a substring s of T to be a matching substring with respect to L if $s \in L$. Many researchers have investigated

---

* The authors were supported under the Research Grants Council of Hong Kong Competitive Earmarked Research Grant HKUST6197/01E.

M. Liśkiewicz and R. Reischuk (Eds.): FCT 2005, LNCS 3623, pp. 552–562, 2005.

the various regular-expression matching problems. Thompson [11] presented the first regular expression matching algorithm for his UNIX editor, ed. Aho [1] suggested an algorithm to determine whether or not T has a matching substring with respect to a given regular expression pattern E in $O(mn)$ time using $O(m)$ space, where m is the size of E and n is the size of T. Crochemore and Hancart [5] extended this result to find all end positions of matching substrings of T with the same runtime and space complexity of Aho [1]. The algorithm is a modified version of the algorithm of Aho [1] and both algorithms are based on the Thompson automata [11].

It is, in applications such as grep, sufficient to obtain the end positions of matching substrings to output lines that contain the matched substrings. However, we often need to find both the start positions and the end positions of matching substrings to replace or delete the matched strings. Myers et al. [10] solved the problem of identifying start positions and end positions of matching substrings of T with respect to E in $O(mn \log n)$ time using $O(m \log n)$ space. Recently, Han et al. [6] proposed another algorithm that runs in $O(mn^2)$ time using $O(m)$ space based on the algorithm of Crochemore and Hancart [5].

Given a regular expression pattern E and a text T, there can be at most n^2 matching substrings in T with respect to E in the worst-case. For example, $E = (a + b)^*$ and $T = abbaabaaba \cdots baba$ over the alphabet $\{a, b\}$. These matching substrings often overlap and nest with each other. To avoid this situation, researchers restrict the search to find and report only a linear subset of the matching substrings. There are two well-known *linearizing restrictions*: The *longest match* rule, which is a generalization of the leftmost longest match rule of IEEE POSIX [7] and the *shortest-match substring search* rule of Clarke and Cormack [4]. These two rules have different semantics and, therefore, identify different matching substrings in general for same E and T.

In Section 2, we define some basic notions. We revisit two linearizing restrictions in the literature and examine the relationship between them in Section 3. We observe that the two rules allow overlapping strings, which is not suitable for some applications, and we propose a new linearizing restriction, the *leftmost non-overlapping match* rule in Section 4. The new rule does not allow overlapping strings and guarantees a linear number of matching substrings. We demonstrate that the new rule is suitable for find-and-replace operations in text searching. Then, we apply the rule to the regular-expression matching problem and develop an algorithm for the problem in Section 5. The algorithm is based on the Thompson automata [11] and it is easy to implement as similar algorithms [1,5].

2   Preliminaries

Let Σ denote a finite alphabet of characters and Σ^* denote the set of all strings over Σ. A language over Σ is any subset of Σ^*. The character $\emptyset$ denotes the empty language and the character λ denotes the null string. Given two strings x and y over Σ, x is a *prefix* of y if there exists $z \in \Sigma^*$ such that $xz = y$ and x is a *suffix* of y if there exists $z \in \Sigma^*$ such that $zx = y$. Furthermore, x is said to be a *substring* or an *infix* of y if there are two strings u and v such that $uxv = y$. Given a string $x = x_1 \cdots x_n$, $|x|$ is the number of characters in x and $x(i,j) = x_i x_{i+1} \cdots x_j$ is the substring of x from position i to position j, where $i \leq j$. Given a set X of strings over Σ, X is *infix-free* if no string in X is an infix of any other string in X. Given a string x, let x^R be the reversal of x, in which case $X^R = \{x^R \mid x \in X\}$. We define a (regular) language L to be infix-free if L is an infix-free set. A regular expression E is infix-free if $L(E)$ is infix-free. We can define prefix-free and suffix-free regular expressions and languages in a similar way.

A finite-state automaton A is specified by a tuple $(Q, \Sigma, \delta, s, F)$, where Q is a finite set of states, Σ is an input alphabet, $\delta \subseteq Q \times \Sigma \times Q$ is a (finite) set of transitions, $s \in Q$ is the start state and $F \subseteq Q$ is a set of final states. Let $|Q|$ be the number of states in Q and $|\delta|$ be the number of transitions in δ. Then, the size of A is $|A| = |Q| + |\delta|$.

A string x over Σ is accepted by A if there is a labeled path from s to a final state in F that spells out x. Thus, the language $L(A)$ of a finite-state automaton A is the set of all strings spelled out by paths from s to a final state in F. We assume that A has only *useful* states; that is, each state appears on some path from the start state to some final state.

A pattern is essentially a language. Given a pattern L and a text T, we define a string x to be a *matching substring* of T with respect to L if x is a substring of T and $x \in L$. The pattern matching problem is to identify all matching substrings of T with respect to a given pattern L. If L is represented by a regular expression E, then we obtain the regular-expression matching problem. If E is prefix-free, then we obtain the prefix-free regular-expression matching problem. The size $|E|$ of a regular expression E is the total number of character appearances in E.

3   Linearizing Restrictions

In the pattern matching problem for a text T, matching substrings of T often overlap with or nest with other matching substrings. Moreover, in the worst-case, there are a quadratic number of matching substrings of T. To avoid these situations, researchers have designed methods to find a linear subset of the matching substrings while preserving specified properties for each matching string. We call such methods *linearizing restrictions*. There are two well-known linearizing restrictions in the matching problem.

3.1   Longest-Match Rule

The *leftmost longest match* rule is defined in the IEEE POSIX Standard [7] as follows:

> "*The search is performed as if all possible suffixes of the string were tested for a prefix matching the pattern; the longest suffix containing a matching prefix is chosen, and the longest possible matching prefix of the chosen suffix is identified as the matching sequence.*"

The rule reports the matching substring whose start position is leftmost and if there are several matching substrings with such a start position, then the longest string is identified. Since it is simple and easy to implement, the rule has been adopted in many tools such as `regex`, `perl` and `tcl/tk`. Note that the rule reports at most one matching string.

The *longest-match* rule is a generalization of the rule of IEEE POSIX [7] that performs a general search instead of identifying a single match string. The longest-match rule is defined as follows: Given a text T and a pattern L, we search for the longest matching prefix with respect to L from position i in T, for $1 \leq i \leq n$, where n is the size of T. Since there can be at most one longest matching prefix from each position, there are at most n matching substrings; thus, the longest-match rule guarantees a linear number of matching strings in the size of T.

Assume that we use the longest-match rule for the regular-expression matching problem. Given a regular expression E and a string w, we can find the longest prefix of w that belongs to $L(E)$ in $O(mn)$ time using $O(m)$ space based on the algorithm of Aho [1], where m is the size of E and n is the size of w. Now we search for the longest prefix from each position in T with respect to $L(E)$ and it takes $O(m|s_1|) + O(m|s_2|) + \cdots + O(m|s_n|)$ time, where $s_1, s_2, \ldots, s_n$ are suffixes of T. Since $|s_1| + |s_2| + \cdots + |s_n| = O(n^2)$, where n is the size of T, the total complexity of the regular-expression matching problem using the longest-match rule is $O(mn^2)$ time and $O(m)$ space. Note that we can improve this running time by using the algorithm of Myers [9] with additional space.

3.2   Shortest-Match Substring Search Rule

Clarke and Cormack [4] proposed a different linearizing restriction, the *shortest-match substring search*:

> "Locate the set of shortest nonnested (but possible overlapping) strings that each match the pattern."

We can rephrase the rule as follows: Given a text T and a pattern L, identify all matching substrings of T with respect to L such that each matching substring is not an infix of any other matching substrings; thus, the resulting set of matching substrings by this rule is an infix-free set. They demonstrated that the shortest-match substring search rule is appropriate for searching structured text such as SGML and XML.

Clarke and Cormack [4] showed that there are at most linear number of matching substrings in the size of T. Furthermore, they considered the case when a pattern is a regular language described by a finite-state automaton A. Let k be the maximum number of out-transitions from a state in A, m be the number of states in A and n be the size of a given text T. They proposed an $O(kmn)$ worst-case running time algorithm using $O(m)$ space. If we use the Thompson automata [11], which are often used in the regular-expression matching problem, then the running time is $O(mn)$ since k is at most 2 in the Thompson automata. Although the rule is simple and straightforward, the idea of this linearizing restriction is shown to be very useful in various cases.

3.3   Comparison of Two Linearizing Restrictions

Both the longest-match rule and the shortest-match substring search rule ensure that the number of matching substrings is linear in the size of T. However, the two rules have different semantics and, therefore, give different results for the same text and the same pattern. For example, if $T = abc$ and the pattern $L = \{a, abc\}$, then the longest-match rule outputs abc whereas the shortest-match substring search rule outputs a. Notice that both rules determine what to report for given an arbitrary text T and an arbitrary pattern L; namely, there are no restrictions on the pattern and on the text. On the other hand, Han et al. [6] showed that, if L is prefix-free, then there can be at most n matching substrings of T because of the prefix-freeness of L. From this work, we obtain:

Corollary 1. *If L is prefix-free or suffix-free, then there are at most n matching substrings of T with respect to L, where n is the size of a given text T.*

Corollary 1 demonstrates that we can apply the linearizing restriction for patterns to obtain a linear number of matching substrings. Then, one question is that whether we can compromise the semantic difference between the longest-match rule and the shortest-match substring search rule by applying the linearizing restriction on patterns.

Theorem 1. *Given a pattern L and a text T, if L is infix-free, then the longest-match rule and the shortest-match substring search rule give the same result. However, the converse does not hold.*

Proof. Assume that a set $S = \{s_1, \ldots, s_k\}$ is the set of matching substrings of T with respect to L, where k is the number of the matching substrings. Let n be the size of T. Since L is infix-free, there are at most n matching substrings; namely, $k \leq n$ [6]. By the definition of matching substrings, $s_i \in S$, for $1 \leq i \leq k$, must belong to L; it implies that S is a subset of L and, therefore, S is also infix-free. Thus, S is the output of the shortest-match substring search rule. Note that all strings in S start from different positions in T. (If any two strings s_i and s_j, for $1 \leq i \neq j \leq k$, start from the same position, then the shorter string must be a prefix of the longer string — a contradiction.) Since each string in S starts from different position, all strings in S are identified as matching substrings by the longest-match rule. Therefore, S is the output of both rules.

We demonstrate that the converse does not hold with the following counter example; $T = ab$ and $L = \{ab, c, cc\}$. Both rules output ab but L is not infix-free. □

Theorem 1 shows that we can eliminate the semantic difference between two rules by choosing an infix-free pattern. Moreover, if we know that a given pattern is an infix-free language, then an algorithm for one rule can be used for the other rule. For example, if a given pattern is an infix-free regular language, then we can use the algorithm of Clarke and Cormack [4] for the regular-expression matching problem with the longest-match rule. In additions, we can use an infix-free regular-expression matching algorithm [6] for both linearizing restriction rules; the algorithm takes $O(mn)$ time using $O(m)$ space in the worst-case.

4   Leftmost Non-overlapping Match Rule

In the pattern matching, two matching substrings of a given text T may overlap with each other. Assume that we want to find matching substrings of T and delete them from T. Then, only one of two overlapping matching substrings should be identified. For example, if T = BEFOREIGN and the pattern L = {BEFORE, FOREIGN}, then both BEFORE and FOREIGN are matching substrings with respect to L. However, if we delete BEFORE from T, then FOREIGN does not exist anymore. Similar situations can happen if we do modification or replacement for matching substrings. Therefore, if two matching substrings overlap, then only the string that starts ahead of the other string is identified. Sometimes one matching substring is nested in the other matching substring. Even in this case, we choose the string that has an earlier start position. For example, if T = AUTOPIAN and L = {TO, UTOPIA}, then UTOPIA is identified even though TO is in L and shorter than UTOPIA since UTOPIA starts ahead of TO in T. These two examples show that the previous two rules, the longest-match rule and the shortest-match substring search rule, are not suitable for such find-and-replace operations in text searching since both rules allow matching substrings to overlap. We suggest a new linearizing restriction that is suitable for find-and-replace operations by identifying only non-overlapping matching substrings.

Definition 1. *We define the leftmost non-overlapping match rule as follows:*

Given a text T, we identify the leftmost matching substring. Then, we move to the next position of the matching substring in T and repeat the identification of the leftmost matching substring in the remaining text until we cannot find it anymore. For example, if two matching strings overlap, then we choose the string whose start position is ahead of the other string's start position and discard the other string; see (a) in Fig. 1. If there are more than two matching substrings that start from the same position, then we choose the shortest string among them; see (b) in Fig. 1.

(a)   (b)

Fig. 1. The figure illustrates the leftmost non-overlapping match rule. (a) When the pattern is {BEFORE, FOREIGN}; the rule chooses BEFORE. (b) When the pattern is {EDIT, EDITOR}; the rule chooses EDIT.

Let $\mathcal{G}(L, T)$ denote the set of matching substrings of the given text T with respect to a given pattern L by the leftmost non-overlapping match rule. Let $|\mathcal{G}(L, T)|$ be the number of strings in $\mathcal{G}(L, T)$. For example, $\mathcal{G}(L = \{aa, ab, ba, bb\}, T = abcbabb) = \{(1, 2), (4, 5), (6, 7)\}$ and $|\mathcal{G}(L, T)| = 3$. Note that although the substring $T(5, 6) = ab$ is in L, it is not in $\mathcal{G}(L, T)$ since it overlaps with another matching substring $T(4,5)$. From the definition of the leftmost non-overlapping match rule, we obtain the following results.

Proposition 1. *The leftmost non-overlapping match rule ensures that the number of matching substrings of T is at most n, where n is the size of T. Namely, $|\mathcal{G}(L,T)| \le n$*

Proof. Assume that the number of matching substrings of T is greater than n. Then, by the pigeonhole principle, there must be two distinct substrings s_1 and s_2 that start from the same position in T — a contradiction. Therefore, $|\mathcal{G}(L,T)| \le n$. □

Proposition 2. *If two distinct matching pairs (u_1, v_1) and $(u_2, v_2) \in \mathcal{G}(L,T)$, then either $v_1 < u_2$ or $v_2 < u_1$.*

Proof. By the match rule of Definition 1, two strings must be non-overlapping. Then, there are only two possible cases as shown in Fig. 2. □

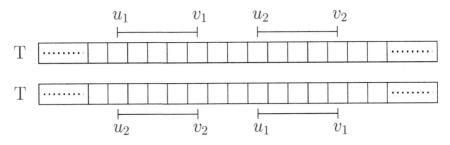

Fig. 2. Two possible cases of two non-overlapping substrings of T

Proposition 1 shows that we always have a linear number of matching substrings in the size of a given text by the leftmost non-overlapping match rule. Note that we do not require L to be a particular type of language such as a regular language or a context-free language. Similar to the longest-match rule or the shortest-match substring search rule, the leftmost non-overlapping match rule can be treated as a general principle for any text search application. Since regular expressions are often used for the matching problem, we study the regular-expression matching problem with the leftmost non-overlapping match rule in Section 5.

5   Regular-Expression Matching Problem

We consider the regular-expression matching problem using the leftmost non-overlapping match rule. Before we present an algorithm for this problem, we explain an example. Assume that we are given a regular expression $E = a(a+b)^*c$ for the text in Fig. 3. Then, $\mathcal{G}(L(E),T) = \{(1,5),(8,11),(12,14)\}$.

Note that $T(1,5), T(8,11)$ and $T(12,14)$ are not the only matching substrings of T with respect to $L(E)$. $T(3,5) = abc$ and $T(13,14) = ac$ are also

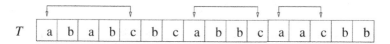

$$T \quad \boxed{a \mid b \mid a \mid b \mid c \mid b \mid c \mid a \mid b \mid b \mid c \mid a \mid a \mid c \mid b \mid b}$$

Fig. 3. The output of $\mathcal{G}(L(E), T)$, where $E = a(a+b)^*c$

in $L(E)$. Nevertheless, since both $T(3,5)$ and $T(13,14)$ overlap other matching substrings of T and they are not the leftmost matching substrings, the leftmost non-overlapping match rule does not identify them. For example, both $T(1,5)$ and $T(3,5)$ are in $L(E)$ but $T(1,5)$ is selected since $T(1,5)$ is the leftmost matching substring.

ExpressionMatching (A, T)

$Q = null(\{s\})$
if $f \in Q$ then **output** λ
for $j = 1$ **to** n
    $Q = null(goto(Q, w_j))$
    if $f \in Q$ then **output** j

Fig. 4. A regular-expression matching procedure for finding all the end positions of matching substrings of T with respect to A, where $A = (Q, \Sigma, \delta, s, f)$ is a Thompson automaton and $T = w_1 \cdots w_n$ is a text

We show that the regular-expression matching problem with the leftmost non-overlapping match rule can be solved using a double scan of T based on the algorithm of Crochemore and Hancart [5].

Theorem 2 (Crochemore and Hancart [5]). *Given a regular expression E and a text T, we can find all the end positions of matching substrings of T with respect to $L(E)$ in $O(mn)$ worst-case time with $O(m)$ space using Expression-Matching, where m is the size of E and n is the size of T.*

The algorithm ExpressionMatching (EM) in Fig. 4 is a modified version of Aho's algorithm [1] that determines whether or not a given text has a substring accepted by a given finite-state automaton. EM has two sub-functions: The function $null(Q)$ computes all states in A that can be reached from a state in the set Q of states by null transitions and the $goto(Q, w_j)$ function gives all states that can be reached from a state in Q by a transition with w_j, the current input character. For details of the algorithm, the sub-functions and the time complexity, refer to Aho [1] or Crochemore and Hancart [5].

Given a regular expression E and a text $T = w_1 \cdots w_n$, we first compute all start positions of matching substrings of T with respect to E. We prepend Σ^* to E^R; thus, allowing matching to begin at any position in T^R. We construct the Thompson automaton [11] A for $\Sigma^* E^R$ and run ExpressionMatching (A, T^R).

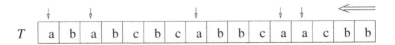

Fig. 5. The output of a single scan of T^R with respect to $\Sigma^* E^R$ using EM, where $E = a(a+b)^* c$

For example, if we run EM on the text in Fig. 3, then we obtain the following positions as indicated by "↓" in Fig. 5.

Since it takes $O(m)$ time to compute the Thompson automaton for E [11] and $O(mn)$ time to run EM, where m is $|E|$ and n is $|T|$, we can compute all start positions of matching substrings in $O(mn)$ time using $O(m)$ space. Let $P = \{q_1, \ldots, q_k\}$ be the set of the start positions of matching substrings after the single scan of T^R, where k is the number of matching start positions and $q_i < q_j$ for $i < j$. Then, we read a character from q_i position of T to find a corresponding shortest matching string with respect to E. Once we find one matching substring $T(q_i, j)$, where $q_i < j$, we move to the next start position in P that is greater than j to avoid the overlapping. A full algorithm is given in Fig. 6.

ReverseEM (A, T, P)

$Q = \{\ \}$, i = 1
for $j = q_i$ **to** n
   $Q = null(goto(Q, w_j))$
   if $f \in Q$
      output (q_i, j)
      while $(q_i < j)$
         i = i + 1
      $j = q_i$
   fi
rof

Fig. 6. A reverse-scan matching procedure for a given Thompson automaton $A = (Q, \Sigma, \delta, s, f)$ for E, a text $T = w_1 \cdots w_n$ and a set $P = \{q_1, \ldots, q_k\}$ of the start positions of matching substrings of T with respect to E

For example, if we run ReverseEM for the result in Fig. 5, where $P = \{1, 3, 8, 12, 13\}$, then the algorithm first outputs $(1, 5)$. The algorithm skips 3 in P since it makes an overlapping with the current output $(1, 5)$ and goes to 8 in P to avoid an overlapping. Fig. 7 illustrates this step.

ReverseEM is based on EM in Fig. 4 and the **while** loop in ReverseEM speeds up for finding the next matching substring by skipping inappropriate start positions and ensures that the algorithm prohibits the overlapping matching substrings. Note that the **while** loop is executed at most k times in total even

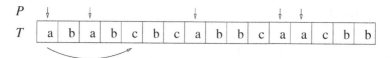

Fig. 7. An example of ReverseEM to find corresponding end positions for a given set P according to the leftmost non-overlapping match rule, where $E = a(a+b)^*c$. The algorithm skips position 3 and moves to position 8 after reporting $(1, 5)$ as a matching substring of T.

though it is inside the **for** loop. Therefore, the worst-case time complexity of ReverseEM is still $O(mn)$ using $O(m)$ space.

Theorem 3. *Given a pattern regular expression E and a text T, we can compute the set of matching substrings that conforms the leftmost non-overlapping match rule in $O(mn)$ worst-case time using $O(m)$ space, where m is the size of E and n is the size of T.*

Theorem 4. *A pair (u, v) is recognized by ReverseEM if and only if $(u, v) \in \mathcal{G}(L(E), T)$, where E is a given pattern regular expression and T is a given text.*

Proof. Assume that we have computed the set $P = \{q_1, \ldots, q_k\}$ of the start positions of matching substrings using EM in Fig. 4, where k is the number of start positions of matching substrings.

$\Longrightarrow$ If (u, v) is recognized by ReverseEM, then $T(u, v) \in L(E)$ and $u \in P$ since **output** in ReverseEM gives (q_i, j) and $q_i \in P$. It is clear that there are no matching substring $T(u, v')$, where $v' < v$, from the algorithm; namely, $T(u, v)$ is the shortest matching substring among all matching substrings that start from the same position u in T. Now assume that $T(u, v)$ overlaps with another matching substring $T(u', v')$ and $T(u, v)$ is not the leftmost matching substrings; hence, $u' < u < v'$. Then, when ReverseEM recognizes (u', v'), the value of j becomes v'. After the **output** (u', v'), ReverseEM executes the **while** loop to choose the next start position from P that is greater than the current position j. Since $u < j = v'$, u cannot be chosen as a start position because of the **while** loop. It implies that the algorithm skips the start position u and therefore (u, v) cannot be recognized by the algorithm — a contradiction; there cannot be a such matching substring $T(u', v')$ in T. Therefore, if (u, v) is recognized by ReverseEM, then $(u, v) \in \mathcal{G}(L(E), T)$.

$\Longleftarrow$ Since $(u, v) \in \mathcal{G}(L(E), T)$, $T(u, v)$ is the shortest matching substring from position u in T with respect to L and u must be in P. If u is q_1 in P, then it is clear that ReverseEM recognizes (u, v). Assume $u = q_i$, where $1 < i \leq k$. Now the only possible case that ReverseEM fails to recognize (u, v) is when u is skipped by the **while** in the algorithm; namely, $u < j$ for some j. It implies that there is an output (q', j), where $q' < u < j$ and $q' \in P$. It contradicts that $T(u, v)$ is the leftmost non-overlapping matching substring of T. Therefore, this situation is not possible and (u, v) must be recognized by ReverseEM. $\square$

6   Conclusions

We have investigated linearizing restrictions for the pattern matching problem. We have reexamined the longest-match rule that is a generalization of the rule of IEEE POSIX [7] and the shortest-match substring search rule [4] and have shown that the two rules give the same result when the given pattern is an infix-free language. Note that both rules have different semantics and give different outputs in general. Then, we have introduced a new linearizing restriction, the leftmost non-overlapping match rule, which should be useful for implementing find-and-replace operations in text searching.

Furthermore, we have proposed an $O(mn)$ worst-case running time algorithm for the regular-expression matching problem using the new linearizing rule based on the algorithm of Crochemore and Hancart [5].

Acknowledgment

We appreciate Shixiong Ma for introducing us to the idea of the linearizing restriction for the pattern matching problem.

References

1. A. Aho. Algorithms for finding patterns in strings. In J. van Leeuwen, editor, *Algorithms and Complexity*, volume A of *Handbook of Theoretical Computer Science*, 255–300. The MIT Press, Cambridge, MA, 1990.
2. A. Aho and M. Corasick. Efficient string matching: An aid to bibliographic search. *Communications of the ACM*, 18:333–340, 1975.
3. R. S. Boyer and J. S. Moore. A fast string searching algorithm. *Communications of the ACM*, 20(10):762–772, 1977.
4. C. L. A. Clarke and G. V. Cormack. On the use of regular expressions for searching text. *ACM Transactions on Programming Languages and Systems*, 19(3):413–426, 1997.
5. M. Crochemore and C. Hancart. Automata for matching patterns. In G. Rozenberg and A. Salomaa, editors, *Linear modeling: background and application*, volume 2 of *Handbook of Formal Languages*, 399–462. Springer-Verlag, 1997.
6. Y.-S. Han, Y. Wang, and D. Wood. Prefix-free regular-expression matching. In *Proceedings of CPM'05*, 298–309. Springer-Verlag, 2005. Lecture Notes in Computer Science 3537.
7. IEEE. *IEEE standard for information technology: Portable Operating System Interface (POSIX) : part 2, shell and utilities*. IEEE Computer Society Press, Sept. 1993.
8. D. Knuth, J. Morris, Jr., and V. Pratt. Fast pattern matching in strings. *SIAM Journal on Computing*, 6:323–350, 1977.
9. E. W. Myers. A four Russians algorithm for regular expression pattern matching. *Journal of the ACM*, 39(2):430–448, Apr. 1992.
10. E. W. Myers, P. Oliva, and K. S. Guimãraes. Reporting exact and approximate regular expression matches. In *Proceedings of CPM'98*, 91–103. Springer-Verlag, 1998. Lecture Notes in Computer Science 1448.
11. K. Thompson. Regular expression search algorithm. *Communications of the ACM*, 11:419–422, 1968.

Fully Incremental LCS Computation

Yusuke Ishida[1], Shunsuke Inenaga[1],
Ayumi Shinohara[2,3], and Masayuki Takeda[1,4]

[1] Department of Informatics, Kyushu University 33, Fukuoka 812-8581, Japan
{y-ishida, shunsuke.inenaga, takeda}@i.kyushu-u.ac.jp
[2] Graduate School of Information Sciences, Tohoku University,
Sendai 980-8579, Japan
ayumi@ecei.tohoku.ac.jp
[3] PRESTO, Japan Science and Technology Agency (JST)
[4] SORST, Japan Science and Technology Agency (JST)

Abstract. *Sequence comparison* is a fundamental task in pattern matching. Its applications include file comparison, spelling correction, information retrieval, and computing (dis)similarities between biological sequences. A common scheme for sequence comparison is the *longest common subsequence* (*LCS*) metric. This paper considers the *fully incremental LCS computation* problem as follows: For any strings A, B and characters a, b, compute $LCS(aA, B)$, $LCS(A, bB)$, $LCS(Aa, B)$, and $LCS(A, Bb)$, provided that $L = LCS(A, B)$ is already computed. We present an efficient algorithm that computes the four LCS values above, in $O(L)$ or $O(n)$ time depending on where a new character is added, where n is the length of A. Our algorithm is superior in both time and space complexities to the previous known methods.

1 Introduction

Pattern matching is one of the most extensively studied sub-areas of theoretical computer science [1,2], and one example of the fundamental problems on pattern matching is *sequence comparison* [3]. There are a wide range of applications for sequence comparison, including file comparison [4], spelling correction [5], information retrieval [6], and computing (dis)similarities between biological sequences [7,8]. Comparing two strings $A = a_1 a_2 \cdots a_n$ and $B = b_1 b_2 \cdots b_m$ can be done by computing an *alignment* between these strings. Standard alignment algorithms compute a dynamic programming matrix DP for the optimal alignments between the consecutive prefixes of A and B. Namely, each entry $DP[i, j]$ stores the score of the alignment between $A[1..j] = a_1 \cdots a_j$ and $B[1..i] = b_1 \cdots b_i$.

A common scheme of sequence comparison is the *longest common subsequence* (*LCS*) metric [9]. A subsequence of string A is any string obtained by removing 0 or more characters from A, and the LCS of two strings A and B (denoted by $LCS(A, B)$) is the longest subsequence that commonly appears in both A and B. In the LCS measure, matched pairs of characters are assigned score 1 and

M. Liśkiewicz and R. Reischuk (Eds.): FCT 2005, LNCS 3623, pp. 563–574, 2005.
© Springer-Verlag Berlin Heidelberg 2005

Table 1. Comparison of complexities for fully incremental LCS computation provided that $LCS(A, B)$ is already computed, where $n = |A|$ and $m = |B|$. Note that $L = LCS(A, B) \leq \min(n, m)$ always holds. The last row shows the total space requirement of each algorithm.

	Naive DP	Modified algorithm of [12]	Our algorithm
time for $LCS(aA, B)$	$O(mn)$	$O(m + n)$	$O(L)$
time for $LCS(Aa, B)$	$O(m)$	$O(m)$	$O(L)$
time for $LCS(A, bB)$	$O(mn)$	$O(m + n)$	$O(n)$
time for $LCS(A, Bb)$	$O(n)$	$O(n)$	$O(n)$
total space complexity	$O(mn)$	$O(mn)$	$O(nL + m)$

unaligned characters are assigned score 0, and the objective is to compute an optimal alignment that gives the maximum score corresponding to $LCS(A, B)$.

$LCS(A, B)$ can be obtained by computing the DP matrix in $O(mn)$ time. The DP approach is suitable for on-line incremental computation of the LCS, in such a situation where upcoming characters are appended to the tails of A and/or B. In fact, $LCS(Aa, B)$ and $LCS(A, Bb)$ can be easily computed in $O(m)$ and $O(n)$ time respectively, provided that $LCS(A, B)$ is already computed. This enables us an efficient processing of e.g. streaming data.

In recent years, the research of computing string alignments to the reversed direction (from right to left) has been a popular topic of pattern matching. Examples of motivations are to process log files backdating to the past, and to compute the alignments between not only the prefixes but also the suffixes of a biological sequence and another biological sequence [10]. However, a naive use of the DP approach is not efficient enough: Since prepending a character a to the head of A can change all the entries of the DP table, we have to recompute the whole DP table from scratch, and this obviously takes $O(mn)$ time. Significant improvement was given by Landau et al. [11] for the *edit distance* metric. For the edit distance metric, their algorithm performs in $O(m + n)$ time. Kim and Park [12] presented a simpler algorithm solving the same problem in the same complexity. Landau et al. [10] introduced the *consecutive suffix alignment problem* and showed two algorithms to solve this problem; the first one runs in $O(nL + m)$ time and space, and the second one in $O(nL)$ time and space, where $L = LCS(A, B)$, assuming that the alphabet is fixed. Note that $L \leq \min(n, m)$ always holds.

This paper treats *fully incremental LCS computation* where characters are added to any position of the heads and tails of A and B. In so doing, we pay our attention to the $O(nL + m)$ algorithm by Landau et al. in [10]. In this paper, we produce an algorithm for fast, flexible, and efficient computation of LCS. The result of this work is summarized in Table 1. It is actually possible to apply the algorithm of Kim and Park [12] to fully incremental LCS computation, which was originally designed for the edit distance metric. However, as seen in Table 1, our algorithm is superior to their algorithm in both time and space complexities.

2   Preliminaries

Let Σ be a finite *alphabet*. Throughout this paper we assume that Σ is fixed. An element of Σ^* is called a *string*. For string $A = a_1 a_2 \cdots a_n$, let $|A|$ denote its length, namely $|A| = n$. Let $A[i] = a_i$ and $A[i..j] = a_i \cdots a_j$, where $1 \leq i \leq j \leq n$. Then $A[1..j]$ is called a *prefix*, $A[i..j]$ a *substring*, and $A[i..n]$ a *suffix* of A. Sequence $A[i_1]A[i_2] \cdots A[i_\ell]$ is called a *subsequence* of A of length ℓ, where $1 \leq i_1 < i_2 \ldots < i_\ell \leq n$. Note that any substring of A is a subsequence of A. Let $B = b_1 b_2 \cdots b_m$. A subsequence occurring in both A and B is called a *common subsequence* of A and B, and the longest such subsequence is called the *longest common subsequence* (*LCS*) of A and B, which is denoted by $LCS(A, B)$.

A standard technique for computing $LCS(A, B)$ is the *dynamic programming* method, where we compute the DP matrix of size $(m + 1) \times (n + 1)$ for which $DP[i, j] = LCS(A[1..j], B[1..i])$ for $1 \leq j \leq n$ and $1 \leq i \leq m$. The recurrence of the DP matrix is the following:

$$DP[i, j] = \begin{cases} 0 & \text{if } i = 0 \text{ or } j = 0, \\ \max(DP[i-1, j], DP[i, j-1]) & \text{if } i, j > 0 \text{ and } A[j] \neq B[i], \\ DP[i-1, j-1] + 1 & \text{if } i, j > 0 \text{ and } A[j] = B[i]. \end{cases}$$

Therefore, to compute $LCS(A, B) = DP[m, n]$, we need $O(mn)$ time and space.

Pair (i, j) is said to be a *match point* between A and B, if $A[j] = B[i]$. Pair (i, j) is said to be a *partition point* of DP if $DP[i, j] = DP[i-1, j]+1$. P denotes the set of the partition points of DP. Let $(i, j) \in P$ and $DP[i, j] = v$. Then we write as $P[v, j] = i$, namely, $P[v, j]$ is the first row index i at column j of DP which bears v. See Fig. 1 for examples of match points and partition points.

3   The Landau Myers Ziv-Ukelson Algorithm

Assume that, given two strings A, B, we have already computed $L = LCS(A, B)$. In this section we recall the algorithm of [10] which, for any character a, computes $LCS(aA, B)$ in amortized $O(L)$ time. This algorithm computes only the partition points rather than the whole DP matrix, thus saving both time and space.

Let DP^{Ah} and P^{Ah} denote the DP matrix and the partition point set obtained from DP and P by adding a new character a to the head of A, respectively. Let $n = |A|$ and $m = |B|$.

Lemma 1 (Landau et al. [10]). P^{Ah} *is computed by inserting at most one new partition point at each column of P.*

See Fig. 1 for a concrete example of the above lemma.

In Lemma 2 we will show how to compute in $O(1)$ time the new partition point for each column. In so doing, we construct the *next match table* (*NM* table) as follows: $NM[i, a]$ returns $\min\{i' \mid i' > i \text{ and } B[i'] = a\}$, if such i' exists. Otherwise, it returns *null*. For fixed alphabet Σ the size of *NM* table is $O(m)$. An example of *NM* table is shown in Fig. 2.

	b	a	d	b	d	c	d
b		0	0	1	1	1	1
c		0	0	1	1	2	2
b		0	0	1	1	2	2
d		0	1	1	2	2	3

	b	a	d	b	d	c	d
b	1	1	1	1	1	1	1
c	1	1	1	1	1	2	2
b	1	1	1	2	2	2	2
d	1	1	2	2	3	3	3

Fig. 1. *DP* (left) and DP^{Ah} (right) with A = adbdcd, B = bcbd and a = b. Cells marked with a circle and rectangle are match and partition points, respectively. Grey rectangles show the new partition points inserted into DP^{Ah}.

		a	b	c	d
	0	null	1	2	4
b	1	null	3	2	4
c	2	null	3	null	4
b	3	null	null	null	4
d	4	null	null	null	null

Fig. 2. *NM* table for string B = bcbd with alphabet $\Sigma = \{\mathtt{a}, \mathtt{b}, \mathtt{c}, \mathtt{d}\}$

Lemma 2 (Landau et al. [10]). *Let $I_{j-1} = P^{Ah}[v, j-1]$ denote the row index of the new partition point in column $j-1$ of P^{Ah}. Then, the new partition point I_j at column j of DP^{Ah} is computed as follows:*

$$I_j = \begin{cases} I_{j-1} & \text{if } P^{Ah}[v, j-1] \leq P[v, j], \\ \min\{NM(P^{Ah}[v, j-1], A[j]), P^{Ah}[v+1, j-1]\} & \text{if } P^{Ah}[v, j-1] > P[v, j]. \end{cases}$$

Note that a special case occurs in Lemma 2 when v is the highest value in column $j-1$ of DP^{Ah}, and therefore partition point $P^{Ah}[v+1, j-1]$ does not exist. In this case, $P^{Ah}[v+1, j-1]$ is set to the dummy index $m+1$, so that we can proceed according to the above lemma.

The stop condition of the update procedure is as follows.

Lemma 3 (Landau et al. [10]). *If column j of DP^{Ah} is identical to colum j of DP, then all columns $j' > j$ of DP^{Ah} are also identical to columns j' of DP.*

The partition point set P is implemented by a double linked list in order that insertion of new partition points can be done in $O(1)$ time. The row indices correspond to the LCS values and the column indices correspond to the positions of string A, and each cell stores the corresponding row index of B. Fig. 3 shows an example of the update of P to P^{Ah}. It is obvious that the size of the partition point set is bounded by $O(nL)$. Since insertion of each new partition point can

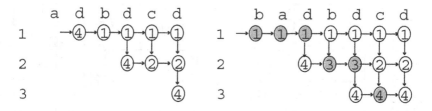

Fig. 3. Update of P with strings $A = \texttt{dbdcd}$ and $B = \texttt{bcbd}$ to P^{Ah} with new character $a = \texttt{b}$. Grey circles are the new partition points inserted to P^{Ah}.

be done in $O(1)$ time, this set can be constructed in $O(nL)$ time. Since $|A| = n$, each incrementation of a new character to the head of A takes the following time.

Theorem 1 (Landau et al. [10]). *Provided that $L = LCS(A, B)$ is already computed, for any character a, $LCS(aA, B)$ is computable in amortized $O(L)$ time.*

4   A Fully Incremental LCS Computation Algorithm

In this section we produce an efficient algorithm to solve the *fully incremental LCS computation problem*, where the problem is to compute the LCS of given two strings under the condition that characters are added to any of the heads and tails of the two strings at any time. Namely, we are to compute $LCS(aA, B)$, $LCS(Aa, B)$, $LCS(A, bB)$, or $LCS(A, Bb)$. The first one, $LCS(aA, B)$, is computable in amortized $O(L)$ time due to Theorem 1 by Landau et al. [10], as recalled in Section 3. In what follows, we will show how to compute the three others.

4.1   Computing $LCS(A, bB)$

Assume we have already computed $LCS(A, B)$. Let DP and P be the DP table and the partition point set for $LCS(A, B)$, respectively. Let DP^{Bh} and P^{Bh} denote the DP matrix and the partition point set for $LCS(A, bB)$ with character b, respectively. Let $n = |A|$ and $m = |B|$.

Where partition points are updated. This subsection is devoted to clarifying where partition points are possibly changed in the DP table when computing $LCS(A, bB)$ from $LCS(A, B)$. Fig. 4 shows an example of updating DP to DP^{Bh}.

Let $\ell = \min\{j \mid A[j] = b\}$. Namely, ℓ is the smallest column index of DP^{Bh} in which a match point exists in the first row. Then we have the following proposition.

Proposition 1. *All the entries of DP^{Bh} are identical to those of DP at the columns smaller than ℓ, except for the first row of DP^{Bh}. The scores in the first row of DP^{Bh} are 0 at columns smaller than ℓ, while the scores are 1 at the other columns.*

	a	a	a	a	b	a	c	b	a	b	c	a
b												
c	0	0	0	0	0	0	0	1	1	1	1	1
b	0	0	0	0	0	1	1	1	2	2	2	2
a	1	1	1	1	1	2	2	2	3	3	3	3
b	1	1	1	1	2	2	2	3	3	4	4	4
a	1	2	2	2	2	3	3	3	4	4	4	5
c	1	2	2	2	2	3	4	4	4	4	5	5

	a	a	a	a	b	a	c	b	a	b	c	a
b	0	0	0	0	0	0	1	1	1	1	1	1
c	0	0	0	0	0	0	1	1	2	2	2	2
b	0	0	0	0	0	0	1	1	2	3	3	3
a	0	1	1	1	1	1	1	2	2	3	4	4
b	0	1	1	1	1	2	2	2	3	4	5	5
a	0	1	2	2	2	2	3	3	3	4	5	5
c	0	1	2	2	2	2	3	4	4	4	5	6

Fig. 4. Update of DP to DP^{Bh} with $A = $ aaaabacbabca, $B = $ cbabac, and $b = $ b. Rectangles show the partition points. In DP^{Bh} on right, dashed rectangles are new partition points inserted, and circles indicate partition points deleted in updating DP to DP^{Bh}.

See Fig. 4 for concrete examples. This proposition means that we do not need to care about these entries of the DP table. In the following, we only consider the other entries than these.

Lemma 4. *For any column $j \geq \ell$, there exists row index E_j such that*

$$DP^{Bh}[i,j] = \begin{cases} DP[i,j] + 1 & \text{if } i < E_j, \\ DP[i,j] & \text{if } i \geq E_j. \end{cases}$$

Proof. Similar to the proof of Lemma 1 in [10].   □

The following lemma is derived from Lemma 4.

Lemma 5. *Column j of P^{Bh} consists of the partition points in P except for one possibly eliminated partition point from P, plus the first row index of DP^{Bh} if it has score 1 at column j. Let E_j be the smallest row index such that $\delta_{E_j} = DP^{Bh}[E_j, j] - DP[E_j, j] = 0$. Then (E_j, j) is the only partition point eliminated at column j in updating P to P^{Bh}.*

Proof. It is obvious that the first row index of DP^{Bh} becomes a partition point at each column of P^{Bh}, if it has score 1.

In what follows, we will show that (1) (E_j, j) is a partition point of DP; (2) (E_j, j) is not a partition point of DP^{Bh}.

(1) For contrary, assume (E_j, j) is not a partition point of DP. Then $DP[E_j - 1, j] = DP[E_j, j]$. Since E_j is the smallest row index such that $\delta_{E_j} = 0$, by Lemma 4 we get $\delta_{E_j - 1} = 1$ which yields $DP^{Bh}[E_j - 1, j] = DP^{Bh}[E_j, j] + 1$ but this contradicts the monotonicity of LCS. Hence (E_j, j) is a partition point of DP.

(2) For contrary, assume (E_j, j) is a partition point of DP^{Bh}. Then $DP^{Bh}[E_j - 1, j] = DP^{Bh}[E_j, j] - 1$. Since E_j is the smallest row index such that $\delta_{E_j} = 0$, by Lemma 4 we get $\delta_{E_j - 1} = 1$ which yields $DP[E_j, j] = DP[E_j - 1, j]$ which contradicts (1) above. Hence (E_j, j) is not a partition point of DP^{Bh}.

For any row $i < E_j$ of column j, we have $DP^{Bh}[i,j] = DP[i,j] + 1$ by Lemma 4. Since the first row at column j of DP^{Bh} is a new partition point of P^{Bh} with score 1, the partition point in any rows smaller than E_j are inherited from P to P^{Bh}. Similar arguments hold for the rows greater than E_j.   □

See Fig. 4 for concrete examples of Lemma 5. Each entry marked by a circle is the partition point eliminated at the column.

According to Lemma 5, at each column j of P^{Bh} at most one new partition point is inserted in the first row, and at most one partition point E_j is eliminated at a larger row. In updating P to P^{Bh}, P is processed from left column to right column. Now we show where E_j can exist at each column j.

Proposition 2. *For any column $j-1$ of DP table, let $P[v, j-1] = x$. At the next column j, we have $DP[x, j] = v$.*

Proof. Since $DP[x, j-1]$ is the partition point of score v, we know that $DP[x-1, j-1] = v-1$. There are two possible cases:

– when (x, j) is a match point.
  By the recursion of LCS computation, $DP[x, j] = DP[x-1, j-1] + 1 = v$.
– when (x, j) is not a match point.
  Since $DP[x-1, j-1] = v-1$, $DP[x-1, j]$ can assume $v-1$ or v. Thus, $DP[x, j] = \max\{DP[x-1, j], DP[x, j-1]\} = v$.   □

Lemma 6. *Let $(E_{j-1}, j-1)$ and (E_j, j) be the partition points eliminated at columns $j-1$ and j in updating P to P^{Bh}, respectively. Let $DP^{Bh}[E_{j-1}, j-1] = v$. Then we have*

$$E_{j-1} \leq E_j \leq P^{Bh}[v+1, j-1].$$

(see Fig. 5.)

Proof. Since $DP^{Bh}[E_{j-1}, j-1] = v$, $P[v, j-1] = E_{j-1}$. In what follows, we will consider three kinds of rows and show that E_j can exist in none of them. Recall that $P^{Bh}[v, j-1] < P[v, j-1] = E_{j-1}$.

– rows smaller than or equal to $P^{Bh}[v, j-1]$.
  Consider any partition point $(x, j-1)$ such that $x \leq P^{Bh}[v, j-1]$ and let $DP[x, j-1] = v'$. By Proposition 2, $DP[x, j] = v'$. On the other hand, by Lemma 4, $P[v', j-1] = P^{Bh}[v'+1, j-1] = x$. Since $DP[P^{Bh}[v'+1, j-1], j] = v'+1$ by Proposition 2, we have $DP^{Bh}[x, j] = DP[x, j]+1$ which means that, for any partition point $(x, j-1)$, we have $(x, j-1) \in P^{Bh}$, while increasing its score just by 1. Thus no partition point is eliminated in the range smaller than $P^{Bh}[v, j-1]$ at column j.
– rows greater than $P^{Bh}[v, j-1]$ and smaller than E_{j-1}.
  The scores of these rows in DP are all $v-1$, since $DP[E_{j-1}, j-1] = v$ and $(E_{j-1}, j-1)$ is a partition point. We have two cases.

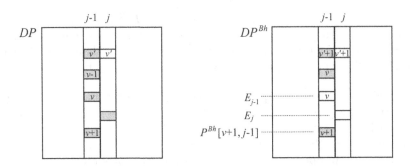

Fig. 5. The range where a partition point E_j at column j can exist, in updating P to P^{Bh}. Gray entries indicate partition points.

- when there are one or more match points in these rows at column j.
  Consider the highest such match point (of the smallest row index) and let its row index be i. Then there is a partition point (i, j) such that $P[v, j] = i$. For any row indices $P^{Bh}[v, j-1] < i' < i$, we have $DP[i', j-1] = DP[i', j] = v - 1$ and $DP^{Bh}[i', j-1] = DP^{Bh}[i', j] = v$. For any row indices $i \leq i'' < E_{j-1}$, we have $DP[i'', j-1] + 1 = DP[i'', j] = v$ and $DP^{Bh}[i'', j-1] + 1 = DP^{Bh}[i'', j] = v + 1$. Thus $P^{Bh}[v+1, j] = P[v, j] = i$.
- when there are no match points in these rows at column j.
  In this case, there are no partition points in these rows of either DP or DP^{Bh}.
- rows greater than E_j.
  Similar to the first case.

Therefore we can conclude that $E_{j-1} \leq E_j \leq P^{Bh}[v+1, j-1]$.   □

Eliminating partition points. In the last subsection we described where the partition points, which can possibly be eliminated, exist. In this section, we show how to quickly eliminate such partition points.

Lemma 7. *Let* $(E_{j-1}, j-1)$ *and* (E_j, j) *be the partition points eliminated at columns $j - 1$ and j in updating P to P^{Bh}, respectively. Let $DP^{Bh}[E_{j-1}, j-1] = v$. Then we have*

$$E_j = \begin{cases} E_{j-1} & \text{if there is no match point } (x, j) \text{ s.t. } P^{Bh}[v, j-1] < x \leq E_{j-1}, \\ P[v+1, j] & \text{otherwise.} \end{cases}$$

Proof. We begin with the first case (see Fig. 6). Since $(E_{j-1}, j-1)$ is the partition point in DP with score v, by the monotonicity of LCS we have $DP[P^{Bh}[v, j-1], j] = v-1$. Thus for any row index $P^{Bh}[v, j-1] \leq i < E_{j-1}$, $DP[i, j-1] = v-1$. By Lemma 4 $DP^{Bh}[i, j-1] = v$ for any such i. Recall $P^{Bh}[v, j] \leq P^{Bh}[v, j-1]$. Since there is no match point (x, j) such that $P^{Bh}[v, j-1] < x \leq E_{j-1}$, we get the three following properties:

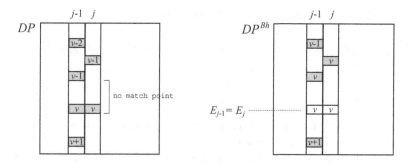

Fig. 6. $E_j = E_{j-1}$ if there is no match point between $P^{Bh}[v, j-1]$ and E_{j-1}

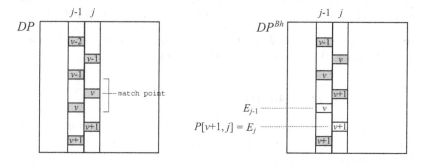

Fig. 7. $E_j = P[v+1, j]$ if there is a match point between $P^{Bh}[v, j-1]$ and E_{j-1}

- for any row index $P[v, j] \le i' < E_j$, $DP[i', j] = v - 1$,
- $P[v, j] = E_{j-1}$, and
- for any row index $P^{Bh}[v, j] \le i'' \le E_j$, $DP[i'', j] = v$,

which imply $E_j = E_{j-1}$.

Now we focus on the second case (see Fig. 7). Consider minimum row index x in range $P^{Bh}[v, j-1] < x \le E_{j-1}$, such that (x, j) is a match point. Then we know that $P[v, j] = x$. By Lemmas 4 and 6, we have $DP^{Bh}[x, j] = v + 1$. Hence $E_j = P[v+1, j]$ as it is no longer a partition point at column j of DP^{Bh}.   □

Note that a spacial case occurs in Lemma 7 when v is the highest value at column j of DP, and therefore partition point $(P[v+1, j], j)$ does not exist. In this case, $P[v, j] = P^{Bh}[v+1, j]$ as usual, but no partition point is eliminated at column j. The update of P to P^{Bh} is stopped at this point, since Lemma 3 also stands for P^{Bh}.

The initial condition to determine the first column in which a partition point is eliminated, and in which row the partition point to be eliminated exists, is given in the following lemma.

Lemma 8. *Let $\ell = \min\{j \mid A[j] = b\}$. Then we have*

$$E_\ell = \begin{cases} null & \text{if there is no partition point at column } j \text{ in } DP, \\ P[1, \ell] & \text{otherwise.} \end{cases}$$

Proof. Trivial. □

In case $E_j = null$ in Lemma 8, there occurs no partition point elimination at the greater columns than ℓ, either.

Due to the above arguments, it is possible to update each column of P in constant time using the double-linked list implementation in Section 3. Since we have to update n columns in the worst case (For instance, consider $A = $ ba^n and $B = $ b^m. Every time we add $b = $ b to the head of B, n new partition points will be added, and n old partition points will be eliminated), we conclude that:

Theorem 2. *Provided that $L = LCS(A, B)$ is already computed, for any character b, $LCS(A, bB)$ is computable in $O(n)$ time.*

4.2   Computing $LCS(Aa, B)$

Let DP^{At} and P^{At} denote the DP matrix and the partition point set which we obtain in computing $LCS(Aa, B)$ with character a, respectively.

The following proposition is obvious.

Proposition 3. *For each partition point $(P[v, j-1], j-1)$,*

$$P[v, j] = \begin{cases} NM(P[v-1, j], A[j]) & \text{if } NM(P[v-1, j], A[j]) < P[v, j-1], \\ P[v, j-1] & \text{otherwise.} \end{cases}$$

It is clear that the scores of all the existing columns of DP are inherited to DP^{At} and thus we only need to compute the partition points in the last (new) column of P^{At}, which is computable based on Proposition 3. Therefore we obtain the following result.

Theorem 3. *Provided that $L = LCS(A, B)$ is already computed, for any character a, $LCS(Aa, B)$ is computable in $O(L)$ time.*

4.3   Computing $LCS(A, Bb)$

Let DP^{Bt} and P^{Bt} denote the DP matrix and the partition point set which we obtain in computing $LCS(A, Bb)$ with character b, respectively.

It is clear that in updating DP to DP^{Bt} the scores of all rows are preserved and thus we only need to examine whether or not the last (new) row becomes a new partition point at each column. Let $P[j]$ denote the set of the partition points at column j of DP. That is, $P[j]$ is a subset of P. Then we have the following proposition and theorem.

Proposition 4. *Let partition point* $\max(P[j-1])$ *have score* v.

- *If* $\max(P[j])$ *has score* $v + 1$, *then the last row at column* j *is not in* P^{Bt}.
- *If* $\max(P[j])$ *has score* v, *then there are two sub-cases.*
  - *If the last row at column* j *of* DP^{Bt} *is a match point, then the last row at column* j *is in* P^{Bt} *with score* $v + 1$.
  - *If the last row at column* j *of* DP^{Bt} *is not a match point, then there are two further sub-cases.*
    * *If* $\max(P^{Bt}[j-1])$ *has score* v, *then the last row at column* j *is not in* P^{Bt}.
    * *If* $\max(P^{Bt}[j-1])$ *has score* $v + 1$, *then the last row at column* j *is in* P^{Bt} *with score* $v + 1$.

Theorem 4. *Provided that* $L = LCS(A, B)$ *is already computed, for any character* b, $LCS(A, Bb)$ *is computable in* $O(n)$ *time.*

Proof. By Proposition 4 we can compute each partition point in the last row of DP^{Bt} in $O(1)$ time. Since there are n column indices at the last row of DP^{Bt}, it takes $O(n)$ time in total. $\qquad\square$

4.4   Updating NM Table

The algorithms introduced in the last subsections use NM table. Recall that we construct NM table for alphabet Σ against string B. Thus, when a new character is added to the head or tail of B, NM table has to be updated accordingly. Let NM^{Bt} and NM^{Bh} denote the next match tables obtained by updating NM for $LCS(A, bB)$ and $LCS(A, Bb)$, respectively.

- computing NM^{Bh}.
  Let i be the position index of new character b added to the head of B. Then we have

$$NM^{Bh}[k, c] = \begin{cases} i & \text{if } k = i - 1 \text{ and } c = b, \\ NM[k+1, c] & \text{if } k = i - 1 \text{ and } c \neq b, \\ NM[k, c] & \text{otherwise.} \end{cases}$$

This means that we only have to update the top row $i - 1$ of NM^{Bh}. Since we have assumed that Σ is fixed, it takes $O(1)$ time.
- computing NM^{Bt}.
  Let i' be the position index of new character b appended to the tail of B. Also, let ℓ be the last occurrence of b in B. Then we have

$$NM^{Bt}[k, c] = \begin{cases} null & \text{if } k = i', \\ i' & \text{if } \ell \leq k < i' \text{ and } c = b, \\ NM[k, c] & \text{otherwise.} \end{cases}$$

Initializing row i' takes constant time as Σ is fixed. For row $\ell \leq k < i'$ at column b, in the worst case it takes linear time in the length of B. However, notice that once any entry is valued with a non-null position, its value will never change. Since the size of NM is linear in the length of B (once more recall Σ is fixed), the amortized time complexity for updating NM is $O(1)$.

In conclusion of this whole section, the following theorem stands.

Theorem 5. *Given strings A, B of length n, m respectively, and provided that $L = LCS(A, B)$ is already computed, we can compute, for any character a, b, $LCS(aA, B)$ in $O(L)$ time, $LCS(A, bB)$ in $O(n)$ time, $LCS(Aa, B)$ in $O(L)$ time, and $LCS(A, Bb)$ in $O(n)$ time. The total space complexity is $O(nL + m)$.*

References

1. Crochemore, M., Rytter, W.: Jewels of Stringology. World Scientific (2002)
2. Gusfield, D.: Algorithms on Strings, Trees, and Sequences. Cambridge University Press (1997)
3. Crochemore, M., Landau, G.M., Ziv-Ukelson, M.: A sub-quadratic sequence alignment algorithm for unrestricted cost matrices. In: Proc. 13th SIAM Symposium on Discrete Algorithms (SODA'02). (2002) 679–688
4. Hunt, J.W., Szymanski, T.G.: An algorithm for differential file comparison. Communications of the ACM **2** (1977) 417–439
5. Amir, A., Eisenberg, E., Porat, E.: Swap and mismatch edit distance. In: 12th Annual European Symposium on Algorithms (ESA'04). Volume 3221 of LNCS., Springer-Verlag (2004) 16–27
6. Wu, S., Manber, U.: Fast text searching allowing errors. Communications of the ACM **35** (1992) 83–91
7. Altschul, S.F., Madden, T.L., Schaffer, A.A., Zhang, J., Zhang, Z., Miller, W., Lipman, D.J.: Gapped BLAST and PSI-BLAST: a new generation of protein database search programs. Nucleic Acids Research **25** (1997) 3389–3402
8. Li, M., Ma, B., Kisman, D., Tromp, J.: PatternHunter II: Highly sensitive and fast homology search. Journal of Bioinformatics and Computational Biology **2** (2004) 417–439
9. Apostolico, A.: String editing and longest common subsequences. In: Handbook of Formal Languages. Volume 2., Springer-Verlag (1997) 361–398
10. Landau, G.M., Myers, E., Ziv-Ukelson, M.: Two algorithms for LCS consecutive suffix alignment. In: Proc. 15th Annual Symposium on Combinatorial Pattern Matching (CPM'04). Volume 3109 of LNCS., Springer-Verlag (2004) 173–193
11. Landau, G.M., Myers, E.W., Schmidt, J.P.: Incremental string comparison. SIAM Journal of Computing **27** (1998) 557–582
12. Kim, S.R., Park, K.: A dynamic edit distance table. In: Proc. 11th Annual Symposium on Combinatorial Pattern Matching (CPM'00). Volume 1848 of LNCS., Springer-Verlag (2000) 60–68

Author Index

Lecture Notes in Computer Science

For information about Vols. 1–3515

please contact your bookseller or Springer

.E. Christensen, M. Sonka (Eds.), Information .g in Medical Imaging. XXI, 777 pages. 2005.

564: N. Eisinger, J. Małuszyński (Eds.), Reasoning ,. IX, 319 pages. 2005.

ol. 3562: J. Mira, J.R. Álvarez (Eds.), Artificial Intelligence and Knowledge Engineering Applications: A Bioinspired Approach, Part II. XXIV, 636 pages. 2005.

Vol. 3561: J. Mira, J.R. Álvarez (Eds.), Mechanisms, Symbols, and Models Underlying Cognition, Part I. XXIV, 532 pages. 2005.

Vol. 3560: V.K. Prasanna, S. Iyengar, P.G. Spirakis, M. Welsh (Eds.), Distributed Computing in Sensor Systems. XV, 423 pages. 2005.

Vol. 3559: P. Auer, R. Meir (Eds.), Learning Theory. XI, 692 pages. 2005. (Subseries LNAI).

Vol. 3558: V. Torra, Y. Narukawa, S. Miyamoto (Eds.), Modeling Decisions for Artificial Intelligence. XII, 470 pages. 2005. (Subseries LNAI).

Vol. 3557: H. Gilbert, H. Handschuh (Eds.), Fast Software Encryption. XI, 443 pages. 2005.

Vol. 3556: H. Baumeister, M. Marchesi, M. Holcombe (Eds.), Extreme Programming and Agile Processes in Software Engineering. XIV, 332 pages. 2005.

Vol. 3555: T. Vardanega, A.J. Wellings (Eds.), Reliable Software Technology – Ada-Europe 2005. XV, 273 pages. 2005.

Vol. 3554: A. Dey, B. Kokinov, D. Leake, R. Turner (Eds.), Modeling and Using Context. XIV, 572 pages. 2005. (Subseries LNAI).

Vol. 3553: T.D. Hämäläinen, A.D. Pimentel, J. Takala, S. Vassiliadis (Eds.), Embedded Computer Systems: Architectures, Modeling, and Simulation. XV, 476 pages. 2005.

Vol. 3552: H. de Meer, N. Bhatti (Eds.), Quality of Service – IWQoS 2005. XVIII, 400 pages. 2005.

Vol. 3551: T. Härder, W. Lehner (Eds.), Data Management in a Connected World. XIX, 371 pages. 2005.

Vol. 3548: K. Julisch, C. Kruegel (Eds.), Intrusion and Malware Detection and Vulnerability Assessment. X, 241 pages. 2005.

Vol. 3547: F. Bomarius, S. Komi-Sirviö (Eds.), Product Focused Software Process Improvement. XIII, 588 pages. 2005.

Vol. 3546: T. Kanade, A. Jain, N.K. Ratha (Eds.), Audio-and Video-Based Biometric Person Authentication. XX, 1134 pages. 2005.

Vol. 3544: T. Higashino (Ed.), Principles of Distributed Systems. XII, 460 pages. 2005.

Vol. 3543: L. Kutvonen, N. Alonistioti (Eds.), Distributed Applications and Interoperable Systems. XI, 235 pages. 2005.

Vol. 3542: H.H. Hoos, D.G. Mitchell (Eds.), Theory and Applications of Satisfiability Testing. XIII, 393 pages. 2005.

Vol. 3541: N.C. Oza, R. Polikar, J. Kittler, F. Roli (Eds.), Multiple Classifier Systems. XII, 430 pages. 2005.

Vol. 3540: H. Kalviainen, J. Parkkinen, A. Kaarna (Eds.), Image Analysis. XXII, 1270 pages. 2005.

Vol. 3539: K. Morik, J.-F. Boulicaut, A. Siebes (Eds.), Local Pattern Detection. XI, 233 pages. 2005. (Subseries LNAI).

Vol. 3538: L. Ardissono, P. Brna, A. Mitrovic (Eds.), User Modeling 2005. XVI, 533 pages. 2005. (Subseries LNAI).

Vol. 3537: A. Apostolico, M. Crochemore, K. Park (Eds.), Combinatorial Pattern Matching. XI, 444 pages. 2005.

Vol. 3536: G. Ciardo, P. Darondeau (Eds.), Applications and Theory of Petri Nets 2005. XI, 470 pages. 2005.

Vol. 3535: M. Steffen, G. Zavattaro (Eds.), Formal Methods for Open Object-Based Distributed Systems. X, 323 pages. 2005.

Vol. 3534: S. Spaccapietra, E. Zimányi (Eds.), Journal on Data Semantics III. XI, 213 pages. 2005.

Vol. 3533: M. Ali, F. Esposito (Eds.), Innovations in Applied Artificial Intelligence. XX, 858 pages. 2005. (Subseries LNAI).

Vol. 3532: A. Gómez-Pérez, J. Euzenat (Eds.), The Semantic Web: Research and Applications. XV, 728 pages. 2005.

Vol. 3531: J. Ioannidis, A. Keromytis, M. Yung (Eds.), Applied Cryptography and Network Security. XI, 530 pages. 2005.

Vol. 3530: A. Prinz, R. Reed, J. Reed (Eds.), SDL 2005: Model Driven. XI, 361 pages. 2005.

Vol. 3528: P.S. Szczepaniak, J. Kacprzyk, A. Niewiadomski (Eds.), Advances in Web Intelligence. XVII, 513 pages. 2005. (Subseries LNAI).

Vol. 3527: R. Morrison, F. Oquendo (Eds.), Software Architecture. XII, 263 pages. 2005.

Vol. 3526: S. B. Cooper, B. Löwe, L. Torenvliet (Eds.), New Computational Paradigms. XVII, 574 pages. 2005.

Vol. 3525: A.E. Abdallah, C.B. Jones, J.W. Sanders (Eds.), Communicating Sequential Processes. XIV, 321 pages. 2005.

Vol. 3524: R. Barták, M. Milano (Eds.), Integration of AI and OR Techniques in Constraint Programming for Combinatorial Optimization Problems. XI, 320 pages. 2005.

Vol. 3523: J.S. Marques, N. Pérez de la Blanca, P. Pina (Eds.), Pattern Recognition and Image Analysis, Part II. XXVI, 733 pages. 2005.

Vol. 3522: J.S. Marques, N. Pérez de la Blanca, P. Pina (Eds.), Pattern Recognition and Image Analysis, Part I. XXVI, 703 pages. 2005.

Vol. 3521: N. Megiddo, Y. Xu, B. Zhu (Eds.), Algorithmic Applications in Management. XIII, 484 pages. 2005.

Vol. 3520: O. Pastor, J. Falcão e Cunha (Eds.), Advanced Information Systems Engineering. XVI, 584 pages. 2005

Vol. 3519: H. Li, P. J. Olver, G. Sommer (Eds.), Computer Algebra and Geometric Algebra with Applications. IX, 449 pages. 2005.

Vol. 3518: T.B. Ho, D. Cheung, H. Liu (Eds.), Advances in Knowledge Discovery and Data Mining. XXI, 864 pages. 2005. (Subseries LNAI).

Vol. 3517: H.S. Baird, D.P. Lopresti (Eds.), Human Interactive Proofs. IX, 143 pages. 2005.

Vol. 3516: V.S. Sunderam, G.D.v. Albada, P.M.A. Sloot, J.J. Dongarra (Eds.), Computational Science – ICCS 2005, Part III. LXIII, 1143 pages. 2005.